Translations of
MATHEMATICAL
MONOGRAPHS

Volume 239

Wavelet Theory

I. Ya. Novikov
V. Yu. Protasov
M. A. Skopina

Translated by Evgenia Sorokina

American Mathematical Society
Providence, Rhode Island

И. Я. Новиков, В. Ю. Протасов, М. А. Скопина

ТЕОРИЯ ВСПЛЕСКОВ

М.: Физматлит, 2005

This work was originally published in Russian by Физматлит under the title "Теория всплесков" © 2005. The present translation was created under license for the American Mathematical Society and is published by permission.

Translated by Evgenia Sorokina

2010 *Mathematics Subject Classification.* Primary 42C40.

For additional information and updates on this book, visit
www.ams.org/bookpages/mmono-239

Library of Congress Cataloging-in-Publication Data

Novikov, I. ÎA. (Igor′ ÎAkovlevich), 1958–

[Teoriia vspleskov. English]

Wavelet theory / I. Ya. Novikov, V. Yu. Protasov, M.A. Skopina ; translated by Evgenia Sorokina.

p. cm. — (Translations of mathematical monographs ; v. 239)

Includes bibliographical references and index.

ISBN 978-0-8218-4984-2 (alk. paper)

1. Wavelets (Mathematics) 2. Harmonic analysis. I. Protasov, V. ÎU. (Vladimir ÎUr′evich), 1970– II. Skopina, M. A. (Mariia Aleksandrovna), 1958– III. Title.

QA403.3.N6813 2010
515′.2433—dc22

2010035110

∞ The paper used in this book is acid-free and falls within the guidelines
established to ensure permanence and durability.
Visit the AMS home page at `http://www.ams.org/`

10 9 8 7 6 5 4 3 2 1 16 15 14 13 12 11

To my beloved wife and children.
I. Ya. Novikov

To my father Yurii Ivanovich Protasov who taught me to love hard sciences.
V. Yu. Protasov

My mathematical dynasty was founded by my grandfather I. A. Skopin. It was further continued by my father A. I. Skopin. I dedicate this book to them.
M. A. Skopina

Contents

Preface

Wavelet theory lies at the intersection of pure mathematics and computational mathematics, as well as audio and graphic signal processing and compression and transmission of information.

The English word *wavelet* is a translation of the French "ondelette" originally introduced by A. Grossman and J. Morlet. Under the wavelet system is usually understood dilations and shifts of a single function that form a system of representation in some sense (for example, orthogonal basis in $L_2(\mathbb{R})$). In some situations the wavelet systems consist of shifts and dilations of several functions or an entire sequence. In our monograph the notion of "wavelet" as such is not introduced; a specific meaning is given to such phrases as "wavelet function", "wavelet space", "wavelet expansion", etc. Interest in the study of wavelet systems emerged long before the appearance of the terminology and the laying of the foundation of the theory and was primarily attributable to the need of using them for signal processing. In connection with these tasks wavelet analysis was formed (in some sense as an alternative to classical Fourier analysis) in the late 1980s–early 1990s in the works of S. Mallat, Y. Meyer, P. J. Lemarie, I. Daubechies, A. Cohen, R. DeVore, W. Lawton, C. K. Chui, and others. Wavelet bases have several advantages compared to other bases which are used as tools of approximation. They have the so-called time-frequency localization; i.e., these basis functions as well as their Fourier transforms rapidly decrease at infinity. Due to this property, when we expand in such bases signals whose frequency characteristics vary over time and space (such as speech, music, and seismic signals as well as images) many coefficients of harmonics that are not essential for a space or time region turn out to be small and can be neglected, which leads to a data compression. Permissibility of this deletion is explained by another important property: wavelet expansions are unconditionally convergent series. In addition, there are efficient algorithms that allow the fast calculation of coefficients of wavelet expansions. All this attracts numerous specialists in various fields of applied and engineering mathematics to the use of wavelets. On the other hand, wavelet systems have proved useful for solving some problems of approximation theory and functional analysis. So, the wavelets provide a rare example of where the theory and its practical implementation develop in parallel.

The impact on the development of mathematical wavelet theory was made by basic works of Y. Meyer and S. Mallat that introduced the notion of multiresolution analysis, described a method of its construction based on a given (suitable) function, and found explicit formulas for finding an appropriate wavelet function whose shifts and dilations form an orthonormal basis. Due to this theory many examples of wavelet systems have been found whose basic functions are smooth and have good time-frequency localization, in particular, smooth wavelets with compact support. Just such examples were required for applications. In the fifteen-year period thousands of papers have been written on the development of wavelet theory. Major monographs include books by Y. Meyer [**5**], I. Daubechies [**2**], C. K. Chui [**6**], P. Wojtaszczyk [**8**], and E. Hernandez and G. A. Weiss [**9**]. In the Russian literature there are translations of the monographs [**2**] and [**6**], a textbook by A. P. Petukhov [**10**], chapters in monographs by B. S. Kashin and A. A. Sahakian [**1**], V. I. Berdyshev and L. V. Petrak [**11**], as well as surveys by I. Ya. Novikov and S. B. Stechkin [**43**], [**44**] and N. M. Astaf′eva [**45**]. The above list is not exhaustive; in particular, we are not citing books dedicated to narrow special issues and books in the engineering field. The book which we offer to the reader is the first monograph in the Russian literature which is devoted entirely to a systematic exposition of modern wavelet theory. It details the construction of orthogonal and biorthogonal systems of wavelets and it studies their structural and approximation properties, beginning with basic theory and ending with special issues and problems. We present some largely theoretical applications of wavelets. In the first two chapters there are the basic facts of the theory for the one-dimensional and multidimensional cases, respectively. Their presentation, however, does not duplicate any of the above-mentioned books. Many of the known fundamental assertions are represented in the most general form or have new proofs. The majority of the material presented in subsequent chapters was not given in earlier monographs. In particular, for the first time we provide a general theory of periodic multiresolution analysis for the multidimensional case with matrix dilation, the theory of nonstationary wavelets, and offer the results of the local smoothness of wavelet functions. The theory of compactly supported wavelets is offered in detail; the relevant refinement equations are investigated; the fractal properties of the wavelets and their relationship to the classical fractal curves are examined. We discuss different ways of measuring the smoothness of wavelets with compact support. In particular, in an explicit way the moduli of continuity of such wavelets are found, as well as orders of their approximation. A separate chapter is devoted to the time-frequency localization of wavelets. A new construction of modified Daubechies wavelets is given, which preserves the localization with a growth of their smoothness. Much attention is paid to convergence of wavelet expansions in various senses, as well as to the evaluation of the order of approximation by wavelets in various functional spaces.

Note, however, that this monograph cannot claim to be a complete presentation of all aspects of modern wavelet theory. For example, we did not touch upon the subject of continuous wavelet transform, applications of wavelets to differential equations, as well as many practical applications. All the facts presented in the book are provided with complete proofs except for a small number of inserts, printed in small type. Especially detailed are the proofs given in the basic chapters. Auxiliary facts are contained in the appendix also with proofs (with the exception of only those facts that can be found in widely available monographs). Historical commentary is provided for all chapters, and most chapters have exercises for the reader. This book is intended for readers who are familiar with the basics of classical and functional analysis on the level of a standard university course. Much of the material is available to engineers and, perhaps, could attract their interest. In conclusion, we would like to note that the authors are very grateful to

S. B. Stechkin—one of the first in Russia who has shown interest in wavelet theory and has in various ways contributed to the involvement of all three authors in this subject matter. The authors also express their gratitude to A. P. Petukhov, who read the manuscript fragments and made some helpful observations.

Basic notation

$\mathbb{N}$ is the set of positive integers.

$\mathbb{R}^d$ is the d-dimensional Euclidean space, $x = (x_1, \dots, x_d)$, $y = (y_1, \dots, y_d)$ are its elements (vectors), $\mathbf{0} = (0, \dots, 0) \in \mathbb{R}^d$, $(x, y) = x_1y_1 + \dots + x_dy_d$, $|x| = \sqrt{(x, x)}$.

$\mathbb{R} = \mathbb{R}^1$.

$\mathbb{Z}^d$ is the integer lattice in $\mathbb{R}^d$.

$\mathbb{Z} = \mathbb{Z}^1$.

$\mathbb{Z}^d_+ = \{x \in \mathbb{Z}^d : \ x_k \geq 0, \ k = 1, \dots, d\}$.

$\mathbb{Z}_+ = \mathbb{Z}^1_+$.

$\mathbb{T}^d = [0, 1)^d$ is the d-dimensional torus.

μ is the Lebesgue measure in $\mathbb{R}^d$.

χ_e is the characteristic function of a set $e \subset \mathbb{R}^d$; it takes the value 1 at the points $t \in e$ and 0 at all other points.

δ_{lk} is the Kronecker delta, which is equal to 1 for $l = k$; otherwise, it is equal to 0.

If $a \in \mathbb{R}$, then $[a] := \max\{n \in \mathbb{Z} : \ n \leq a\}$.

$\operatorname{span} M$ is the linear span of a set M, i.e., the set of all finite linear combinations of elements from M.

$\operatorname{supp} f$ is the support of a function f, i.e., the minimal (with respect to inclusion) closed set such that f is equal to zero almost everywhere on the complement of this set.

$\widehat{f}(k) = \int\limits_{\mathbb{T}^d} f(t)e^{-2\pi i(k,t)}\,dt$ is the k-th Fourier coefficient, $k \in \mathbb{Z}^d$, of a function $f \in L(\mathbb{T}^d)$ with respect to the trigonometric system.

$\widehat{f}(x) = \int\limits_{\mathbb{R}^d} f(t)e^{-2\pi i(x,t)}\,dt$ is the Fourier transform of a function f from $L(\mathbb{R}^d)$; $\widehat{f}$ denotes also the Fourier transform of a function f from $L_2(\mathbb{R}^d)$ or from the space of temperate distributions.

F and F^{-1} are the operators taking a function to its direct and inverse Fourier transforms, respectively.

$L \log L(\mathbb{T}^d)$ is the class of functions $f \in L(\mathbb{T}^d)$ such that

$$\int\limits_{\mathbb{T}^d} |f| \max\{0, \log|f|\} < \infty.$$

ℓ_p, $1 \le p \le \infty$, is the space of sequences of complex numbers $c = \{c_n\}_{n=1}^{\infty}$ with norm $\|c\|_p = \left(\sum_{n=1}^{\infty} |c_n|^p\right)^{1/p}$.

$\langle f, g \rangle$ is the scalar product of elements of the Hilbert space.

$\operatorname{span}\{f_n, n \in \mathbb{N}\}$ is the set of finite linear combinations of a system $\{f_n\}_{n=1}^{\infty}$ with complex coefficients.

T^* is the operator conjugate to an operator T; T^{-1} is its inverse operator.

If A is a $d \times d$ matrix, then $\|A\|$ is its Euclidean operator norm from $\mathbb{R}^d$ to $\mathbb{R}^d$, A^T is its transpose, A^* is its Hermitian conjugate matrix, $\det A$ is the determinant of A.

I_d is the unity $d \times d$ matrix.

For any trigonometric polynomial m having no positive powers (i.e., $m(\xi) = \sum_{k=0}^{N} c_k e^{-2\pi i k\xi}$), $\mathbf{m}(z) = \sum_{k=0}^{N} c_k z^k$ is the corresponding algebraic polynomial. Thus $m(\xi) = \mathbf{m}(e^{-2\pi i\xi})$.

A sequence $\dots x_{-k}, \dots, x_{-1}, x_0, x_1, \dots, x_k, \dots$ is denoted by $\{x_k\}_{k\in\mathbb{Z}}$ (sometimes we omit $k \in Z$) or (x).

$\mathcal{S}$ is the space of infinitely differentiable and rapidly decreasing functions on $\mathbb{R}$:

$$\mathcal{S} = \{f \in C^{\infty}(\mathbb{R}) \mid \forall m, k \ge 0 \ \ \|f^{(m)}(x)(1+|x|)^k\|_{\infty} < \infty\}.$$

The topology of the space $\mathcal{S}$ is defined as follows:

$$f_j \to 0 \ \Leftrightarrow \forall m, k \ge 0 \ \ \|f_j^{(m)}(x)(1+|x|)^k\|_{\infty} \to 0.$$

$\mathcal{S}'$ is the space of linear continuous functionals on the space $\mathcal{S}$ (the space of tempered distributions).

$\mathcal{D}$ is the space of infinitely differentiable, compactly supported functions on $\mathbb{R}$ with the topology defined as follows:

$$f_j \to 0 \ \Leftrightarrow \forall m \ge 0 \ \ \|f_j^{(m)}(x)\|_{\infty} \to 0, \quad \exists M > 0 : \forall j \ \operatorname{supp} f_j \subset [-M, M].$$

$\mathcal{D}'$ is the space of linear continuous functionals on this space.

$\mathbf{1}$ is the function taking the value 1 everywhere on $\mathbb{R}$.

α_f is the Hölder coefficient of a function f on a given segment $[a, b]$:

$$\alpha_f = k + \sup\Big\{\alpha \ \Big| \ |f^{(k)}(x_1) - f^{(k)}(x_2)| \le C_\alpha |x_1 - x_2|^{\alpha}, \ x_1, x_2 \in [a, b]\Big\},$$

where k is the maximal integer such that $f \in C^k[-1, 1]$. The Hölder coefficient in the space L_p is defined analogously:

$$\alpha_{f,p} = k + \sup\Big\{\alpha \ \Big| \ \|f^{(k)}(\cdot + h) - f^{(k)}(\cdot)\| \le C_\alpha h^{\alpha}, \ h > 0\Big\}.$$

Local smoothness (local Hölder coefficient) of a function f at a point x is defined as

$$\alpha_f(x) = \sup\Big\{\alpha \ \Big| \ |f^{(k)}(x + h) - f^{(k)}(x)| \le C_\alpha h^{\alpha}, \ h > 0\Big\}.$$

W_p^s is the Sobolev space.

$B_{pq}^{(s)}$ is the Besov space.

$F_{pq}^{(s)}$ is the Lizorkin-Triebel space.

$s_p(f) = \sup\left\{\alpha \,\Big|\, \int_{\mathbb{R}} |\widehat{f}(\xi)|^p(1+|\xi|^{p\alpha})\,d\xi < \infty\right\}$ is the Sobolev coefficient of smoothness.

CHAPTER 1

Wavelets on the Line

1.1. Riesz bases

In this section we introduce and discuss an important notion for wavelet theory, namely, the notion of the Riesz basis in the Hilbert space, which generalizes the notion of orthogonal basis and preserves its most essential properties. We shall primarily deal with the Riesz bases consisting of integer shifts of some function.

Let us denote by ℓ_2 the Hilbert space of sequences of complex numbers $c = \{c_n\}_{n=1}^{\infty}$ such that $\sum_{n=1}^{\infty} |c_n|^2 < \infty$ with the scalar product $\langle c^1, c^2 \rangle := \sum_{n=1}^{\infty} c_n^1 \overline{c_n^2}$.

DEFINITION 1.1.1. Let H be a Hilbert space. A system $\{f_n\}_{n=1}^{\infty} \subset H$ is called *a Riesz system with constants* A, $B > 0$ if for any $c = \{c_n\}_{n=1}^{\infty} \in \ell_2$ the series $\sum_{n=1}^{\infty} c_n f_n$ converges in H and

$$A\|c\|_{\ell_2}^2 \leq \left\| \sum_{n=1}^{\infty} c_n f_n \right\|_H^2 \leq B\|c\|_{\ell_2}^2. \tag{1.1}$$

If a Riesz system is a basis (see Appendix A.1), it is called a Riesz basis.

THEOREM 1.1.2. *Let H be a Hilbert space, and let $\{f_n\}_{n=1}^{\infty}$ be a Riesz system in H with constants A, B. Then:*

(i) *$\{f_n\}_{n=1}^{\infty}$ is a Riesz basis in the space*

$$V := \left\{ f = \sum_{n=1}^{\infty} c_n f_n, \ \sum_{n=1}^{\infty} |c_n|^2 < \infty \right\}.$$

(ii) $V = \overline{\operatorname{span}\{f_n, n \in \mathbb{N}\}}$.

(iii) *For any element $f \in V$, the following inequality holds:*

$$A\|f\|^2 \leq \sum_{n=1}^{\infty} |\langle f, f_n \rangle|^2 \leq B\|f\|^2. \tag{1.2}$$

`Proof.` By definition, any element of the space V is the sum of a series $\sum_{n=1}^{\infty} c_n f_n$; therefore to prove (i). We only have to verify the uniqueness of such a representation. Suppose that for some $f = \sum_{n=1}^{\infty} c_n f_n \in V$ there exists another representation $\sum_{n=1}^{\infty} c'_n f_n$. Let us fix $k \in \mathbb{N}$ and for given $\varepsilon > 0$ choose $N \geq k$ such that

$$\left\| \sum_{n=N+1}^{\infty} c_n f_n \right\| < \varepsilon, \quad \left\| \sum_{n=N+1}^{\infty} c'_n f_n \right\| < \varepsilon.$$

Using (1.1), we have

$$|c_k - c'_k| \le \sqrt{\sum_{n=1}^{N} |c_n - c'_n|^2} \le \frac{1}{\sqrt{A}} \left\| \sum_{n=1}^{N} (c_n - c'_n) f_n \right\|$$
$$\le \frac{1}{\sqrt{A}} \left(\left\| \sum_{n=1}^{\infty} (c_n - c'_n) f_n \right\| + \left\| \sum_{n=N+1}^{\infty} c_n f_n \right\| + \left\| \sum_{n=N+1}^{\infty} c'_n f_n \right\| \right) < \frac{2\varepsilon}{\sqrt{A}}.$$

The arbitrariness of ε implies that $c_k = c'_k$.

Let us introduce the operator $I : \ell_2 \longrightarrow H$ that takes each element $c = \{c_n\}_{n=1}^{\infty} \in \ell_2$ to the element $f = Ic = \sum\limits_{n=1}^{\infty} c_n f_n$. It is clear that I is a one-to-one map from ℓ_2 to V. Furthermore, in view of (1.1) we have

$$\|I\|^2 \le B, \quad \|I^{-1}\|^2 \le A^{-1}; \tag{1.3}$$

i.e., I is an isomorphism from ℓ_2 to V. In particular, this implies that V is closed. Taking into account the inclusions $\operatorname{span}\{f_n, n \in \mathbb{N}\} \subset V$, $V \subset \overline{\operatorname{span}\{f_n, n \in \mathbb{N}\}}$, we obtain (ii). Let $e_n = (0, \dots, 0, 1, 0, \dots)$ be the n-th basis vector in ℓ_2. Then $f_n = I(e_n)$. Since $\{e_n\}_{n=1}^{\infty}$ is an orthonormal basis in ℓ_2, using (1.3) we obtain

$$\sum_{n=1}^{\infty} |\langle f, f_n \rangle|^2 = \sum_{n=1}^{\infty} |\langle f, Ie_n \rangle|^2 = \sum_{n=1}^{\infty} |\langle I^* f, e_n \rangle|^2 = \|I^* f\|_{\ell_2}^2 \le B\|f\|^2,$$

$$\|f\|^2 = \|I^{*-1} I^* f\|^2 \le \frac{1}{A} \|I^* f\|_{\ell_2} = \frac{1}{A} \sum_{n=1}^{\infty} |\langle I^* f, e_n \rangle|^2 = \frac{1}{A} \sum_{n=1}^{\infty} |\langle f, f_n \rangle|^2,$$

which proves (iii). ◊

REMARK 1.1.3. If in the above theorem, instead of the assumption that $\{f_n\}_{n=1}^{\infty}$ is a Riesz system, we only assume that the right-hand inequality in (1.1) holds (in this case $\{f_n\}_{n=1}^{\infty}$ is said to be a *Bessel system*), then we can claim that for any $f \in H$ the following inequality holds:

$$\sum_{n=1}^{\infty} |\langle f, f_n \rangle|^2 \le B\|f\|^2.$$

To prove this assertion, it suffices to repeat the reasoning of Theorem 1.1.2 for any finite set of functions $\{f_n\}_{n=1}^{N}$ (here, we do not have to prove the closeness of the space $V_N = \left\{ f = \sum\limits_{n=1}^{N} c_n f_n \right\}$) and note that for any $f \in H$ we have

$$\sum_{n=1}^{N} |\langle f, f_n \rangle|^2 = \sum_{n=1}^{N} |\langle P_N f, f_n \rangle|^2,$$

where P_N is the orthogonal projector on V_N.

PROBLEM 1.1.4. Let $\{f_n\}_{n=1}^{\infty}$ be a basis in the Hilbert space H. Prove that if (1.2) holds for any $f \in H$, then (1.1) also holds.

We have proved that, as in the case of orthogonal bases, expansions with respect to a Riesz basis are linear combinations with coefficients from ℓ_2, while the Fourier coefficients of any element satisfy relation (1.2), which generalizes the Parseval equality. Let us show that Riesz bases have another remarkable feature of orthogonal bases: the way of numerating their elements is of no significance. Such bases are called unconditional (see Appendix A.1).

THEOREM 1.1.5. *Any Riesz basis in the Hilbert space H is an unconditional basis.*

Proof. By Theorem 1.1.2, any element $f \in H$ can be represented in the form $f = \sum_{n=1}^{\infty} c_n f_n$, $\sum_{n=1}^{\infty} |c_n|^2 < \infty$. Denote by s_n partial sums of this representation and by s'_n partial sums of the series obtained from $\sum_{n=1}^{\infty} c_n f_n$ by some rearrangement of its terms. Let I be the operator defined in Theorem 1.1.2. We set $e = I^{-1} f$, $\sigma_n = I^{-1} s_n$, $\sigma'_n = I^{-1} s'_n$. As shown above, I is an isomorphism from ℓ_2 to H; hence it follows from the relation $f = \lim_{n\to\infty} s_n$ that $e = \lim_{n\to\infty} \sigma_n = \lim_{n\to\infty} \sigma'_n$, which implies that $\lim_{n\to\infty} s'_n = f$. ◊

In view of the above theorem the way of taking partial sums in expansions with respect to Riesz systems is of no significance; hence the terms in these sums can be numerated by any countable set. In particular, in what follows, we shall assume that the summation index ranges over $\mathbb{Z}$ or $\mathbb{Z}^d$.

LEMMA 1.1.6. *Let $\varphi, \widetilde{\varphi} \in L_2(\mathbb{R})$. Then the series $\sum_{k\in\mathbb{Z}} |\widehat{\varphi}(\xi+k)\widehat{\widetilde{\varphi}}(\xi+k)|$ converges almost everywhere and its sum is summable on $(0,1)$.*

Proof. First assume that $\varphi = \widetilde{\varphi}$. Consider the series

$$\sum_{k=-\infty}^{\infty} \int_0^1 |\widehat{\varphi}(\xi+k)|^2\, d\xi = \sum_{k=-\infty}^{\infty} \int_k^{k+1} |\widehat{\varphi}(\xi)|^2\, d\xi = \int_{-\infty}^{\infty} |\widehat{\varphi}(\xi)|^2\, d\xi.$$

In view of the Plancherel theorem the right-hand side of this equality is finite; hence by the Levi theorem the function $\sum_{k\in\mathbb{Z}} |\widehat{\varphi}(\xi+k)|^2$ is finite almost everywhere and

$$\int_0^1 \sum_{k\in\mathbb{Z}} |\widehat{\varphi}(\xi+k)|^2\, d\xi = \sum_{k=-\infty}^{\infty} \int_0^1 |\widehat{\varphi}(\xi+k)|^2\, d\xi.$$

For arbitrary $\varphi, \widetilde{\varphi} \in L_2(\mathbb{R})$, we have the inequality

$$\sum_{k\in\mathbb{Z}} |\widehat{\varphi}(\xi+k)\widehat{\widetilde{\varphi}}(\xi+k)| \le \sqrt{\sum_{k\in\mathbb{Z}} |\widehat{\varphi}(\xi+k)|^2}\sqrt{\sum_{n\in\mathbb{Z}} |\widehat{\widetilde{\varphi}}(\xi+n)|^2}.$$

Using the above results and the Cauchy-Bunyakovsky inequality, we obtain the required statement. ◊

THEOREM 1.1.7. *Let $\varphi \in L_2(\mathbb{R})$. For a system of functions $\{\varphi(\cdot+n)\}_{n\in\mathbb{Z}}$ to be a Riesz system with constants A, B, it is necessary and sufficient to have*

$$A \le \sum_{k\in\mathbb{Z}} |\widehat{\varphi}(\xi+k)|^2 \le B. \tag{1.4}$$

for almost all $\xi \in \mathbb{R}$.

Proof. Sufficiency. Let $c = \{c_n\}_{n\in\mathbb{Z}} \in \ell_2$. It follows from (1.4) and the Parseval equality that

$$\int_0^1 \left|\sum_{n\in\mathbb{Z}} c_n e^{2\pi i n\xi}\right|^2 \sum_{k\in\mathbb{Z}} |\widehat{\varphi}(\xi+k)|^2 \, d\xi \;\ge\; A \sum_{n\in\mathbb{Z}} |c_n|^2, \tag{1.5}$$

$$\int_0^1 \left|\sum_{n\in\mathbb{Z}} c_n e^{2\pi i n\xi}\right|^2 \sum_{k\in\mathbb{Z}} |\widehat{\varphi}(\xi+k)|^2 \, d\xi \;\le\; B \sum_{n\in\mathbb{Z}} |c_n|^2. \tag{1.6}$$

Using Lebesgue's dominated convergence theorem to interchange integration and summation over k, we have

$$\int_0^1 \left|\sum_{n\in\mathbb{Z}} c_n e^{2\pi i n\xi}\right|^2 \sum_{k\in\mathbb{Z}} |\widehat{\varphi}(\xi+k)|^2 \, d\xi = \int_{-\infty}^{\infty} \left|\sum_{n\in\mathbb{Z}} c_n e^{2\pi i n\xi} \widehat{\varphi}(\xi)\right|^2 \, d\xi. \tag{1.7}$$

Thus, $\sum\limits_{n\in\mathbb{Z}} c_n e^{2\pi i n\cdot} \widehat{\varphi}(\cdot) \in L_2(\mathbb{R})$. By the Plancherel theorem, it follows that $\sum\limits_{n\in\mathbb{Z}} c_n \varphi(\cdot+n) \in L_2(\mathbb{R})$ and

$$\left\|\sum_{n\in\mathbb{Z}} c_n \varphi(\cdot+n)\right\|^2 = \int_{-\infty}^{\infty} \left|\sum_{n\in\mathbb{Z}} c_n e^{2\pi i n x} \widehat{\varphi}(\xi)\right|^2 \, d\xi. \tag{1.8}$$

Combining this equality with (1.7) and using (1.5), (1.6), we get

$$A \sum_{n\in\mathbb{Z}} |c_n|^2 \le \left\|\sum_{n\in\mathbb{Z}} c_n \varphi(\cdot+n)\right\|^2 \le B \sum_{n\in\mathbb{Z}} |c_n|^2,$$

which means that condition (1.1) is satisfied for the system $\{\varphi(\cdot+n)\}_{n\in\mathbb{Z}}$.

Necessity. Let the functions $\varphi(\cdot+n)$, $n \in \mathbb{Z}$, form a Riesz system. Then $\sum\limits_{n\in\mathbb{Z}} c_n \varphi(\cdot+n) \in L_2(\mathbb{R})$ for any $c = \{c_n\} \in \ell_2$; hence by the Plancherel theorem $\sum\limits_{n\in\mathbb{Z}} c_n e^{2\pi i n\cdot} \widehat{\varphi}(\cdot) \in L_2(\mathbb{R})$, and equality (1.8) holds. Since the right-hand side of (1.8) is finite, by the Levi theorem we have equality (1.7). Using (1.7), (1.8), we rewrite condition (1.1) in the following form: for any $g = \sum\limits_{n\in\mathbb{Z}} c_n e^{2\pi i n\cdot} \in L_2(0,1)$ the inequality

$$A\|g\|^2 \le \int_0^1 |g(\xi)|^2 \sigma(\xi)\, d\xi \le B\|g\|^2, \tag{1.9}$$

holds, where

$$\sigma(\xi) := \sum_{k\in\mathbb{Z}} |\widehat{\varphi}(\xi+k)|^2 .$$

The function σ is 1-periodic and by Lemma 1.1.6, it is summable on $(0,1)$. Since

$$\sup_{\|g\|_2\neq 0}\int_0^1 \frac{|g|^2}{\|g\|_2^2}\sigma = \sup_{\substack{\|h\|_1=1\\ h\geq 0}}\int_0^1 h\sigma = \sup_{\|h\|_1=1}\int_0^1 h\sigma,$$

by the Riesz theorem we have

$$\operatorname*{ess\,sup}_{[0,1]}\sigma = \|\sigma\|_\infty = \sup_{\|g\|_2\neq 0}\int_0^1 \frac{|g|^2}{\|g\|_2^2}\sigma \leq B;$$

i.e., the right-hand inequality in (1.4) holds. Let us assume that the left-hand inequality does not hold. Then there exists a set $E \subset [0,1]$ of positive measure such that $\sigma \leq A-\varepsilon$, $\varepsilon>0$, on E. Putting $g=\chi_E$, we obtain

$$\int_0^1 |g|^2\sigma = \int_E \sigma \leq (A-\varepsilon)\mu E = (A-\varepsilon)\|g\|_2^2 < A\|g\|_2^2,$$

which contradicts (1.9) ◊

REMARK 1.1.8. It is readily seen from the proof of this theorem that if instead of condition (1.4) we assume that the inequality

$$\sum_{k\in\mathbb{Z}} |\widehat{\varphi}(\xi+k)|^2 \leq B \tag{1.10}$$

holds for almost all $\xi\in\mathbb{R}$, then we can claim that the series $\sum\limits_{n\in\mathbb{Z}} c_n\varphi(\cdot+n)$ converges in $L_2(\mathbb{R})$ for any $c=\{c_n\}_{n\in\mathbb{Z}}\in\ell_2$ and

$$\left\|\sum_{n\in\mathbb{Z}} c_n\varphi(\cdot+n)\right\|^2 \leq B\|c\|_{\ell_2}^2. \tag{1.11}$$

PROBLEM 1.1.9. Let the function $\varphi\in L_2(\mathbb{R})$ be compactly supported, $\varphi_n := \varphi(\cdot+n)$, $n\in\mathbb{Z}$. Prove that there exists a constant $B>0$ such that the inequality

$$\sum_{n\in\mathbb{Z}} |\langle f,\varphi_n\rangle|^2 \leq B\|f\|^2$$

holds for any function $f\in L_2(\mathbb{R})$.

PROPOSITION 1.1.10. *Let $\varphi \in L_2(\mathbb{R})$, and let the system $\{\varphi(\cdot+n)\}_{n\in\mathbb{Z}}$ be a Riesz basis in a closed subspace V of the space $L_2(\mathbb{R})$. A function $f\in L_2(\mathbb{R})$ belongs to V if and only if there exists a 1-periodic function $m_f\in L_2(0,1)$ such that*

$$\widehat{f}(\xi) = m_f(\xi)\widehat{\varphi}(\xi) \tag{1.12}$$

for almost all $\xi\in\mathbb{R}$. In addition, the coefficients of the expansion of f with respect to the system $\{\varphi(\cdot+n)\}_{n\in\mathbb{Z}}$ coincide with the corresponding Fourier coefficients of the function m_f.

Proof. By Theorem 1.1.2, we have

$$V = \left\{f=\sum_{n\in\mathbb{Z}} c_n\varphi(\cdot+n),\ \sum_{n\in\mathbb{Z}}|c_n|^2<\infty\right\}.$$

Let $\sum_{n\in\mathbb{Z}} |c_n|^2 < \infty$ and

$$f(x) = \sum_{n\in\mathbb{Z}} c_n\varphi(x+n). \tag{1.13}$$

Since the Fourier transform and passing to the limit in L_2 commute and the series on the right-hand side of (1.13) converges in L_2, it follows that applying the Fourier transform to both sides of this equality, we get

$$\widehat{f}(\xi) = \sum_{n\in\mathbb{Z}} c_n\widehat{\varphi(\cdot+n)}(\xi) = \sum_{n\in\mathbb{Z}} c_n e^{2\pi in\xi}\widehat{\varphi}(\xi). \tag{1.14}$$

To prove the necessity, it remains to set

$$m_f(\xi) = \sum_{n\in\mathbb{Z}} c_n e^{2\pi in\xi} \tag{1.15}$$

and note that $m_f \in L_2(0,1)$ by the Parseval equality. Conversely, if (1.12) holds and $m_f \in L_2(0,1)$, then multiplying (1.15) by $\widehat{\varphi}(\xi)$, we obtain equality (1.14), which is equivalent to (1.13), as is shown above. ◊

DEFINITION 1.1.11. Let H be a Hilbert space. We say that systems $\{f_n\}, \{g_n\} \subset H$ are *biorthonormal* if

$$\langle f_n, g_k\rangle = \delta_{nk}.$$

PROPOSITION 1.1.12. *Let $\varphi, \widetilde{\varphi} \in L_2(\mathbb{R})$. Systems $\{\varphi(\cdot+n)\}_{n\in\mathbb{Z}}$, $\{\widetilde{\varphi}(\cdot+n)\}_{n\in\mathbb{Z}}$ are biorthonormal if and only if the equality*

$$\sum_{k\in\mathbb{Z}} \widehat{\varphi}(\xi+k)\overline{\widehat{\widetilde{\varphi}}(\xi+k)} = 1 \tag{1.16}$$

holds for almost all $\xi \in \mathbb{R}$, where the series on the left-hand side of (1.16) converges absolutely. In particular, the system $\{\varphi(\cdot+n)\}_{n\in\mathbb{Z}}$ is orthonormal if and only if

$$\sum_{k\in\mathbb{Z}} |\widehat{\varphi}(\xi+k)|^2 = 1 \tag{1.17}$$

for almost all $\xi \in \mathbb{R}$.

Proof. By the Plancherel theorem,

$$\int_{-\infty}^{\infty} \varphi(x)\overline{\widetilde{\varphi}(x-k)}\,dx = \int_{-\infty}^{\infty} \widehat{\varphi}(\xi)\overline{\widehat{\widetilde{\varphi}}(\xi)}e^{2\pi ik\xi}\,d\xi$$

$$= \sum_{l=-\infty}^{\infty}\int_{l}^{l+1} \widehat{\varphi}(\xi)\overline{\widehat{\widetilde{\varphi}}(\xi)}e^{2\pi ik\xi}\,d\xi = \int_0^1 \sum_{n\in\mathbb{Z}} \widehat{\varphi}(\xi+n)\overline{\widehat{\widetilde{\varphi}}(\xi+n)}e^{2\pi ik\xi}\,d\xi \tag{1.18}$$

for any $k \in \mathbb{Z}$. It is possible to change the order of summation and integration due to Lebesgue's dominated convergence theorem and Lemma 1.1.6. Again by Lemma 1.1.6 and Lebesgue's dominated convergence theorem, the 1-periodic function

$$\sigma(\xi) := \sum_{n\in\mathbb{Z}} \widehat{\varphi}(\xi+n)\overline{\widehat{\widetilde{\varphi}}(\xi+n)}$$

is summable on $(0,1)$. It remains to note that the biorthonormality of the systems $\{\varphi(\cdot - k)\}, \{\widetilde{\varphi}(\cdot - n)\}$ implies that

$$\int_{-\infty}^{\infty} \varphi(x)\overline{\widetilde{\varphi}(x-k)}\,dx = \delta_{k0}, \tag{1.19}$$

which in view of (1.18) is equivalent to the fact that all the Fourier coefficients of the function σ are equal to zero except for the zeroth coefficient. By the uniqueness theorem, this is equivalent to the fact that $\sigma(\xi) = 1$ for almost all $\xi \in \mathbb{R}$. ◊

PROBLEM 1.1.13. Prove that if functions $\varphi,\ \widetilde{\varphi} \in L_2(\mathbb{R})$ are compactly supported, then the function $\sum_{k\in\mathbb{Z}} \widehat{\varphi}(\xi+k)\overline{\widehat{\widetilde{\varphi}}(\xi+k)}$ is equivalent to a trigonometric polynomial.

THEOREM 1.1.14. *Let $\varphi \in L_2(\mathbb{R})$. If a system $\{\varphi(\cdot+n)\}_{n\in\mathbb{Z}}$ is a Riesz basis in a closed subspace V of the space $L_2(\mathbb{R})$, then there exists a function $\varphi^\sharp \in L_2(\mathbb{R})$ such that its integer shifts $\varphi^\sharp(\cdot+n)$, $n \in \mathbb{Z}$, form an orthonormal basis in V.*

Proof. Let us set

$$\widehat{\varphi^\sharp}(\xi) = \nu(\xi)\widehat{\varphi}(\xi),$$

where

$$\nu(\xi) = \left(\sum_{l\in\mathbb{Z}} |\widehat{\varphi}(\xi+l)|^2\right)^{-1/2}.$$

By Theorem 1.1.7, the function ν is essentially bounded; hence $\widehat{\varphi^\sharp} \in L_2(\mathbb{R})$ and condition (1.17) is satisfied for $\widehat{\varphi^\sharp}$. Therefore by Proposition 1.1.12 integer shifts of $\varphi^\sharp$ form an orthonormal system. Taking into account Theorem 1.1.2 and Proposition 1.1.10, we get

$$\begin{aligned} V &= \{f:\ f = \sum_{n\in\mathbb{Z}} f_n\varphi(\cdot+n),\ \sum_{n\in\mathbb{Z}} |f_n|^2 < \infty\} \\ &= \{f:\ \widehat{f} = \mu\widehat{\varphi}, \mu \in L_2(0,1)\} \\ &= \{f:\ \widehat{f} = \mu_1\widehat{\varphi^\sharp}, \mu_1 = \mu\nu^{-1} \in L_2(0,1)\} \\ &= \{f:\ f = \sum_{n\in\mathbb{Z}} f_n^\sharp\varphi^\sharp(\cdot+n),\ \sum_{n\in\mathbb{Z}} |f_n^\sharp|^2 < \infty\}, \end{aligned}$$

which implies that the system $\{\varphi^\sharp(\cdot+n)\}_{n\in\mathbb{Z}}$ is a basis in V. ◊

1.2. MRA and scaling functions

DEFINITION 1.2.1. A collection of closed spaces $V_j \subset L_2(\mathbb{R})$, $j \in \mathbb{Z}$, is called a *multiresolution analysis* in $L_2(\mathbb{R})$, for brevity referred to as *MRA*, if the following conditions (axioms) hold:

MR1. $V_j \subset V_{j+1}$ for all $j \in \mathbb{Z}$.

MR2. $\bigcup_{j\in\mathbb{Z}} V_j$ is dense in $L_2(\mathbb{R})$.

MR3. $\bigcap_{j\in\mathbb{Z}} V_j = \{0\}$.

MR4. $f \in V_j \iff f\left(2^{-j}\cdot\right) \in V_0$ for all $j \in \mathbb{Z}$.

MR5. There exists a function $\varphi \in V_0$ such that the sequence $\{\varphi(\cdot+n)\}_{n\in\mathbb{Z}}$ forms a Riesz basis in V_0.

Let us note another two properties of the spaces V_j, which immediately follow from axioms **MR4**, **MR5**. Before formulating them, we introduce the notation for shifts and dilations of a function f:

$$f_{jn} := 2^{j/2} f\left(2^j \cdot +n\right), \quad n, j \in \mathbb{Z}. \tag{1.20}$$

MR6. For any $j \in \mathbb{Z}$ the functions φ_{jn}, $n \in \mathbb{Z}$, form a Riesz basis in V_j with the same constants A, B as the functions φ_{0n}, $n \in \mathbb{Z}$.

MR7 If $f \in V_j$, then $f\left(\cdot \pm 2^{-j}\right) \in V_j$, $j \in \mathbb{Z}$.

It will be clear from the discussion below that property **MR3** also follows from **MR4**, **MR5**, though by convention we include it in the list of axioms.

The function φ in axiom **MR5** is called a *scaling* function for a given MRA $\{V_j\}_{j\in\mathbb{Z}}$.

By Theorem 1.1.14, in the space V_0 there exists a function $\varphi^\sharp$ whose integer shifts form an orthonormal basis in V_0 and, analogously to property **MR6**, the functions $\varphi^\sharp_{jn}$, $n \in \mathbb{Z}$, form an orthonormal basis in V_j for all $j \in \mathbb{Z}$.

The main reason why MRAs are introduced is as follows. Using MRAs, one can construct orthogonal (biorthogonal) bases formed by dilations and shifts of one function (two functions). Such bases will be called wavelet bases. They are studied in more detail in the next section. Now we consider the problem of existence of MRAs and methods of their construction. The main property is **MR5**, where a scaling function appears that defines the spaces V_0 and, in view of **MR6**, the whole MRA. Thus, if we have an appropriate function φ, we can define

$$V_j := \overline{\operatorname{span}\{\varphi_{jn}, n \in \mathbb{Z}\}}, \tag{1.21}$$

provided that properties **MR4**, **MR5** hold. It is clear that any function whose integer shifts form a Riesz system will not be suited since we have to satisfy **MR1**, **MR2**, **MR3**. Further we shall see that the most limiting property is the inclusion property **MR1**. Thus, to know how to construct MRAs, we have to describe the class of scaling functions. We shall do this in terms of the Fourier transform.

We say that the collection of spaces $\{V_j\}_{j\in\mathbb{Z}}$ defined by relation (1.21) *is generated* by the scaling function φ.

PROPOSITION 1.2.2. *Let $\varphi, f \in L_2(\mathbb{R})$ and let $\{V_j\}_{j\in\mathbb{Z}}$ be the collection of spaces generated by the function φ. The relation $f \perp V_0$ holds if and only if*

$$\sum_{l\in\mathbb{Z}} \widehat{f}(\xi + l)\overline{\widehat{\varphi}(\xi + l)} = 0$$

for almost all $\xi \in \mathbb{R}$.

Proof. The relation $f \perp V_0$ is equivalent to the fact that the relation

$$\int_{-\infty}^{\infty} f(x)\overline{\varphi(x+k)}\, dx = 0$$

holds for all $k \in \mathbb{Z}$. On the other hand,

$$\int_{-\infty}^{\infty} f(x)\overline{\varphi(x+k)}\, dx = \int_0^1 e^{2\pi i k\xi} \sum_{n\in\mathbb{Z}} \widehat{f}(\xi + n)\overline{\widehat{\varphi}(\xi + n)}\, d\xi.$$

Interchange of summation and integration is possible due to Lebesgue's dominated convergence theorem and Lemma 1.1.6. By the same lemma, the series on the right-hand side is a 1-periodic function and, by the uniqueness theorem, all the Fourier coefficients are equal to zero if and only if this function is equal to zero almost everywhere. ◊

PROPOSITION 1.2.3. *Let $\varphi, f \in L_2(\mathbb{R})$ and let $\{V_j\}_{j\in\mathbb{Z}}$ be the collection of spaces generated by the function φ. Suppose that (1.4) holds. The relation $f \in V_j$ holds if and only if there exists a 1-periodic function $m_f \in L_2(0,1)$ such that*

$$\widehat{f}(\xi) = 2^{-j/2} m_f(2^{-j}\xi)\widehat{\varphi}(2^{-j}\xi) \tag{1.22}$$

for almost all $\xi \in \mathbb{R}$. Moreover, the coefficients in the decomposition of f with respect to the system $\{\varphi_{jn}\}_{n\in\mathbb{Z}}$ coincide with the corresponding Fourier coefficients of the function m_f.

For $j = 0$, this statement coincides with Proposition 1.1.10; for $j \neq 0$ the proof readily follows from Proposition 1.1.10 and property **MR4**, which in turn follows from the definition of the spaces V_j.

THEOREM 1.2.4. *Suppose $\varphi \in L_2(\mathbb{R})$ and (1.4) holds. The collection of spaces $\{V_j\}_{j\in\mathbb{Z}}$ generated by the function φ satisfies property* **MR1** *if and only if there exists a 1-periodic function $m_0 \in L_2(0,1)$ such that*

$$\widehat{\varphi}(\xi) = m_0(\xi/2)\widehat{\varphi}(\xi/2) \tag{1.23}$$

for almost all $\xi \in \mathbb{R}$.

`Proof.` It follows from the definition of the spaces V_j that they satisfy properties **MR4**, **MR6**. Hence it is clear that property **MR1** is equivalent to the relation $V_0 \subset V_1$, which in turn is equivalent to the condition $\varphi \in V_1$. It remains to use Proposition 1.2.3 and put $m_0 := m_\varphi/\sqrt{2}$. ◊

Equality (1.23) is called a *refinement equation*, and the function m_0 is called a *mask*. In engineering terminology the term mask or *filter* is used for the set of the Fourier coefficients of the function m_0. In mathematical papers, the term mask is also often understood as the set of the Fourier coefficients and the function m_0 is called the symbol of the mask. However we shall not use this terminology.

Let

$$m_0(\xi) = \frac{1}{\sqrt{2}} \sum_{n\in\mathbb{Z}} h_n e^{2\pi i n\xi}.$$

Applying the Fourier inversion to both sides of equality (1.23), transform it to the form

$$\varphi = \sum_{n\in\mathbb{Z}} h_n \varphi_{1n}. \tag{1.24}$$

In engineering terminology relations (1.23) and (1.24) are referred to as refinement equations in the frequency and time domains, respectively.

COROLLARY 1.2.5. *Let φ be a scaling function that satisfies (1.4), m_0 being its mask. Then*

$$\frac{A}{B} \le |m_0(\xi)|^2 + |m_0(\xi + 1/2)|^2 \le \frac{B}{A} \tag{1.25}$$

for almost all $\xi \in \mathbb{R}$.

Proof. Combining (1.23) with (1.4) and taking into account the 1-periodicity of m_0, we get

$$\begin{aligned} A &\le \sum_{n\in\mathbb{Z}} \left| m_0\left(\frac{\xi+n}{2}\right)\widehat{\varphi}\left(\frac{\xi+n}{2}\right)\right|^2 \\ &= \left|m_0\left(\frac{\xi}{2}\right)\right|^2 \sum_{k\in\mathbb{Z}} \left|\widehat{\varphi}\left(\frac{\xi}{2}+k\right)\right|^2 + \left|m_0\left(\frac{\xi+1}{2}\right)\right|^2 \sum_{k\in\mathbb{Z}} \left|\widehat{\varphi}\left(\frac{\xi+1}{2}+k\right)\right|^2 \\ &\le B\left(\left|m_0\left(\frac{\xi}{2}\right)\right|^2 + \left|m_0\left(\frac{\xi+1}{2}\right)\right|^2\right), \end{aligned}$$

for almost all $\xi \in \mathbb{R}$, which yields the left inequality in (1.25). The right inequality is proved analogously. ◊

COROLLARY 1.2.6. *Let $\varphi, \widetilde{\varphi}$ be scaling functions (generally, for two different MRAs) with masks $m_0, \widetilde{m}_0$, respectively. If the systems $\{\varphi_{0n}\}_{n\in\mathbb{Z}}$, $\{\widetilde{\varphi}_{0n}\}_{n\in\mathbb{Z}}$ are biorthonormal, then*

$$m_0(\xi)\overline{\widetilde{m}_0(\xi)} + m_0(\xi+1/2)\,\overline{\widetilde{m}_0(\xi+1/2)} = 1 \tag{1.26}$$

for almost all $\xi \in \mathbb{R}$. In particular, if $\{\varphi_{0n}\}_{n\in\mathbb{Z}}$ is an orthonormal system, then

$$|m_0(\xi)|^2 + |m_0(\xi+1/2)|^2 = 1 \tag{1.27}$$

for almost all $\xi \in \mathbb{R}$.

The proof is analogous to that of Corollary 1.2.5, if we use equality (1.16) instead of (1.4).

PROBLEM 1.2.7. Let φ be a scaling function for some MRA and let $\varphi(x) = O((1+|x|)^{-1-\varepsilon})$, $\varepsilon > 0$. Prove that the mask m_0 is continuous at the point $1/2$ and $m_0(1/2) = 0$.

LEMMA 1.2.8. *Let $\varphi \in L_2(\mathbb{R})$ and suppose that (1.10) holds. Then*

$$\lim_{j\to-\infty} \sum_{k\in\mathbb{Z}} |\langle f, \varphi_{jk}\rangle|^2 = 0$$

for any $f \in L_2(\mathbb{R})$.

Proof. First assume that the function f is compactly supported and continuous on $\mathbb{R}$. Let $\operatorname{supp} f \subset [-R, R]$, $R > 1$. Using the Cauchy-Bunyakovsky inequality, for $j < -2\log R$, we obtain

$$\begin{aligned} \sum_{k\in\mathbb{Z}} |\langle f, \varphi_{jk}\rangle|^2 &\le 2^j \sum_{k\in\mathbb{Z}} \left(\int_{|x|\le R} |f(x)||\varphi(2^j x+k)|\,dx\right)^2 \\ &\le 2^j \|f\|_\infty^2 2R \sum_{k\in\mathbb{Z}} \int_{|x|\le R} |\varphi(2^j x+k)|^2\,dx \\ &\le 2R\|f\|_\infty^2 \int_{S(R,j)} |\varphi(y)|^2\,dy = 2R\|f\|_\infty \int_{-\infty}^{\infty} \chi_{R,j}(y)|\varphi(y)|^2\,dy, \end{aligned}$$

where

$$S(R,j) = \bigcup_{k\in\mathbb{Z}} [k - 2^j R, k + 2^j R],$$

$\chi_{R,j}$ is the characteristic function of the set $S(R,j)$. If $y \notin \mathbb{Z}$, then

$$\lim_{j\to-\infty} \chi_{R,j}(y) = 0.$$

Therefore,

$$\lim_{j\to-\infty} \int_{-\infty}^{\infty} \chi_{R,j}(y)|\varphi(y)|^2\, dy = 0.$$

The passage to the limit is possible due to Lebesgue's dominated convergence theorem.

Now let f be an arbitrary function in $L_2(\mathbb{R})$. Let us specify $\varepsilon > 0$ and find a compactly supported continuous function $\widetilde{f}$ such that

$$\|f - \widetilde{f}\| < \varepsilon. \tag{1.28}$$

It follows from (1.10), Theorems 1.1.7 and 1.1.2, and Remarks 1.1.8 and 1.1.3 that

$$\sum_{k\in\mathbb{Z}} |\langle f - \widetilde{f}, \varphi_{jk}\rangle|^2 \le B\varepsilon^2.$$

It remains to use the triangle inequality and to take into account the fact that the lemma has already been proved for $\widetilde{f}$. ◊

THEOREM 1.2.9. *Let $\varphi \in L_2(\mathbb{R})$ and suppose that* (1.4) *holds. Then the collection of spaces $\{V_j\}_{j\in\mathbb{Z}}$ generated by the function φ satisfies condition* **MR3**.

`Proof.` Let us assume that $f \in V_j$ for all $j \in \mathbb{Z}$. It follows from (1.4) and Theorems 1.1.7 and 1.1.2 that

$$\sqrt{A}\|f\| \le \left(\sum_{k\in\mathbb{Z}} |\langle f, \varphi_{jk}\rangle|^2\right)^{1/2}.$$

By Lemma 1.2.8, the right-hand side tends to zero as $j \to -\infty$. This implies that $\|f\| = 0$, which means that condition **MR3** holds. ◊

THEOREM 1.2.10. *Let* (1.4) *hold for $\varphi \in L_2(\mathbb{R})$. Condition* **MR2** *holds for the collection of spaces $\{V_j\}_{j\in\mathbb{Z}}$ generated by the function φ and satisfying* **MR1** *if and only if*

$$\bigcup_{j\in\mathbb{Z}} \operatorname{supp} \widehat{\varphi}(2^j\cdot) = \mathbb{R}. \tag{1.29}$$

In Chapter 2, this theorem will be proved in a more general situation.

To satisfy (1.29), it suffices to assume that the function $\widehat{\varphi}$ is continuous at zero and

$$\widehat{\varphi}(0) \ne 0. \tag{1.30}$$

We show that this condition can be weakened by rejecting (1.30), which is satisfied automatically.

LEMMA 1.2.11. *Let $f, \varphi \in L_2(\mathbb{R})$. Then the following equality holds for any $j \in \mathbb{Z}$:*

$$\sum_{k\in\mathbb{Z}} |\langle f, \varphi_{jk}\rangle|^2 = 2^j \int_0^1 \left|\sum_{l\in\mathbb{Z}} \widehat{f}\left(2^j(\xi+l)\right)\overline{\widehat{\varphi}\left(\xi+l\right)}\right|^2 d\xi. \tag{1.31}$$

Proof. Set

$$h(\xi) = \sum_{l\in\mathbb{Z}} \widehat{f}\left(2^j(\xi+l)\right)\overline{\widehat{\varphi}\left(\xi+l\right)}. \tag{1.32}$$

The function h is 1-periodic and summable on $(0, 1)$ (by Lemma 1.1.6). Furthermore, partial sums of the series defining this function have a summable majorant. Using the Plancherel theorem and Parseval equality, we get

$$\begin{aligned}
\sum_{k\in\mathbb{Z}} |\langle f, \varphi_{jk}\rangle|^2 &= 2^j \sum_{k\in\mathbb{Z}} \left|2^{-j}\int_{-\infty}^{\infty} \widehat{f}(\xi)e^{-2\pi i 2^{-j}k\xi}\overline{\widehat{\varphi}\left(2^{-j}\xi\right)}\,d\xi\right|^2 \\
&= 2^j \sum_{k\in\mathbb{Z}} \left|\int_{-\infty}^{\infty} \widehat{f}(2^j\xi)e^{-2\pi ik\xi}\overline{\widehat{\varphi}(\xi)}\,d\xi\right|^2 \\
&= 2^j \sum_{k\in\mathbb{Z}} \left|\sum_{l\in\mathbb{Z}}\int_0^1 \widehat{f}(2^j(\xi+l))e^{-2\pi ik\xi}\overline{\widehat{\varphi}(\xi+l)}\,d\xi\right|^2 \\
&= 2^j \sum_{k\in\mathbb{Z}} \left|\int_0^1 h(\xi)e^{-2\pi ik\xi}\,d\xi\right|^2 = 2^j \sum_{k\in\mathbb{Z}} \left|\widehat{h}(k)\right|^2 = 2^j \int_0^1 |h(\xi)|^2\,d\xi.
\end{aligned}$$

Here, the interchange of summation and integration is justified by Lemma 1.1.6 and the above-mentioned properties of the function h. ◊

LEMMA 1.2.12. *Let $f \in L_2(\mathbb{R})$, $\widehat{f} \in L_\infty(\mathbb{R})$, $\varphi \in L_2(\mathbb{R})$. Then the following equality holds for any $j \in \mathbb{Z}$:*

$$\sum_{k\in\mathbb{Z}} |\langle f, \varphi_{jk}\rangle|^2 = \int_{-\infty}^{\infty} \left|\widehat{\varphi}\left(2^{-j}\xi\right)\right|^2 \left|\widehat{f}\left(\xi\right)\right|^2 d\xi + R_j, \tag{1.33}$$

where

$$|R_j| \le \int_{-\infty}^{\infty} \sum_{l\in\mathbb{Z}, l\neq 0} \left|\widehat{f}\left(\xi\right)\right| \left|\widehat{f}\left(\xi+2^jl\right)\right| \left|\widehat{\varphi}\left(2^{-j}\xi\right)\right| \left|\widehat{\varphi}\left(2^{-j}\xi+l\right)\right| d\xi. \tag{1.34}$$

Proof. We can assume that the integral on the right-hand side of (1.34) is finite since otherwise our assertion is trivial. The integral on the right-hand side of (1.33) is also finite due to the essential boundedness of $\widehat{f}$ and the Plancherel theorem. By Lemma 1.2.11,

$$\sum_{k\in\mathbb{Z}} |\langle f, \varphi_{jk}\rangle|^2 = 2^j \int_0^1 |h(\xi)|^2\,d\xi, \tag{1.35}$$

where h is the function defined by equality (1.32). On the other hand,

$$2^j \int_0^1 |h(\xi)|^2 \, d\xi = \int_0^{2^j} \left|h(2^{-j}\xi)\right|^2 \, d\xi$$

$$= \int_0^{2^j} \sum_{k\in\mathbb{Z}} \sum_{l\in\mathbb{Z}} \overline{\widehat{f}\,(\xi + 2^j l)}\widehat{f}\,(\xi + 2^j k)\,\overline{\widehat{\varphi}\,(2^{-j}\xi + l)}\widehat{\varphi}\,(2^{-j}\xi + k)\, d\xi$$

$$= \int_{-\infty}^{\infty} \widehat{f}(\xi)\widehat{\varphi}(2^{-j}\xi) \sum_{l\in\mathbb{Z}} \overline{\widehat{f}\,(\xi + 2^j l)}\,\overline{\widehat{\varphi}\,(2^{-j}\xi + l)}\, d\xi.$$

The interchange of summation and integration is justified by Lebesgue's dominated convergence theorem, taking into account the fact that the integrands in (1.33) and (1.34) are summable. To complete the proof, it remains to extract the term with $l = 0$ in the latter sum and to take into account (1.35). ◊

LEMMA 1.2.13. *Suppose $\varphi \in L_2(\mathbb{R})$, $\widehat{\varphi}$ is continuous at zero, and* (1.10) *holds. Then*

$$\lim_{j\to+\infty} \sum_{k\in\mathbb{Z}} |\langle f, \varphi_{jk}\rangle|^2 = |\widehat{\varphi}(0)|^2 \|f\|^2 \tag{1.36}$$

for any $f \in L_2(\mathbb{R})$.

Proof. First assume that the function f is summable and thrice continuously differentiable on $\mathbb{R}$. Then its Fourier transform satisfies the estimate

$$\left|\widehat{f}(\xi)\right| \le \frac{C}{1 + |\xi|^3}, \tag{1.37}$$

which means that $\widehat{f} \in L_\infty(\mathbb{R})$. Since (1.10) implies $\widehat{\varphi} \in L_\infty(\mathbb{R})$, by Lemma 1.2.12, we have

$$\sum_{k\in\mathbb{Z}} |\langle f, \varphi_{jk}\rangle|^2 = \int_{-\infty}^{\infty} \left|\widehat{\varphi}\,(2^{-j}\xi)\right|^2 \left|\widehat{f}\,(\xi)\right|^2 \, d\xi + R_j, \tag{1.38}$$

where

$$|R_j| \le \int_{-\infty}^{\infty} \sum_{l\in\mathbb{Z}, l\ne 0} \left|\widehat{f}\,(\xi)\right| \left|\widehat{f}\,(\xi + 2^j l)\right| \left|\widehat{\varphi}\,(2^{-j}\xi)\right| \left|\widehat{\varphi}\,(2^{-j}\xi + l)\right| \, d\xi$$

$$\le \|\widehat{\varphi}\|_\infty^2 \sum_{l\in\mathbb{Z}, l\ne 0} \int_{-\infty}^{\infty} \left|\widehat{f}\,(\xi)\right| \left|\widehat{f}\,(\xi + 2^j l)\right| \, d\xi.$$

Applying (1.37) to the right-hand side and estimating the sum, we obtain

$$\sum_{l\in\mathbb{Z},l\neq 0}\int_{-\infty}^{\infty}\frac{d\xi}{(1+|\xi|^3)\left(1+|\xi+2^jl|^3\right)}$$
$$\leq \sum_{l\in\mathbb{Z},l\neq 0}\int_{|\xi+2^jl|\leq 2^{j-1}|l|}\frac{d\xi}{|\xi|^3}+\sum_{l\in\mathbb{Z},l\neq 0}\int_{|\xi+2^jl|>2^{j-1}|l|}\frac{d\xi}{(1+|\xi|^3)\left(2^{j-1}|l|\right)^3}$$
$$\leq \sum_{l\in\mathbb{Z},l\neq 0}\left(2^{j-1}|l|\right)^{-3}2^j|l|+\sum_{l\in\mathbb{Z},l\neq 0}\left(2^{j-1}|l|\right)^{-3}\int_{-\infty}^{\infty}\frac{d\xi}{1+|\xi|^3}.$$

It is not hard to verify that the right-hand side is of order $O(2^{-2j})$; hence $\lim_{j\to+\infty}R_j = 0$. Passing to the limit as $j\to+\infty$ in (1.38) and then passing to the limit under the integral sign, which is possible due to Lebesgue's dominated convergence theorem, we obtain (1.36). Now take an arbitrary $f\in L_2(\mathbb{R})$. For given $\varepsilon>0$, we find a function $\widetilde{f}\in C^3\cap L_2(\mathbb{R})\cap L(\mathbb{R})$ such that $\|f-\widetilde{f}\|<\varepsilon$. By (1.10), Theorems 1.1.7 and 1.1.2, and Remarks 1.1.8 and 1.1.3 we have

$$\sqrt{\sum_{k\in\mathbb{Z}}\left|\langle f-\widetilde{f},\varphi_{jk}\rangle\right|^2}\leq\sqrt{B}\|f-\widetilde{f}\|<\sqrt{B}\varepsilon.$$

Combining this inequality and the triangle inequality, we get

$$\left|\sqrt{\sum_{k\in\mathbb{Z}}|\langle f,\varphi_{jk}\rangle|^2}-\sqrt{\sum_{k\in\mathbb{Z}}\left|\langle\widetilde{f},\varphi_{jk}\rangle\right|^2}\right|\leq\sqrt{\sum_{k\in\mathbb{Z}}\left|\langle f-\widetilde{f},\varphi_{jk}\rangle\right|^2}<\sqrt{B}\varepsilon.$$

Taking into account that relation (1.36) is proved for $\widetilde{f}$ and ε is arbitrary, we pass to the upper and lower limits in this inequality as $j\to+\infty$ and obtain (1.36) for all f. ◊

THEOREM 1.2.14. *Let $\varphi\in L_2(\mathbb{R})$ and let $\widehat{\varphi}$ be continuous at zero, satisfy the refinement equation* (1.23) *and* (1.4). *Then the collection of spaces $\{V_j\}_{j\in\mathbb{Z}}$ generated by φ forms an MRA and φ is its scaling function.*

Proof. As already mentioned, the definition of V_j implies property **MR4**. It follows from condition (1.4) and Theorems 1.1.7 and 1.1.2 that integer shifts of the function φ form a Riesz basis in V_0, i.e., **MR5** holds. Properties **MR1** and **MR3** follow from Theorems 1.2.4 and 1.2.9, respectively. It remains to verify **MR2**. First we show that $\widehat{\varphi}(0)\neq 0$. Assume the converse. Then, by Lemma 1.2.13, we have

$$\lim_{j\to+\infty}\sum_{k\in\mathbb{Z}}|\langle f,\varphi_{jk}\rangle|^2=0 \tag{1.39}$$

for any $f\in L_2(\mathbb{R})$. Let $f\in V_0$, $\|f\|\neq 0$. In view of **MR1**, $f\in V_j$ for all $j\geq 0$. By Theorem 1.1.2, the following inequality holds for any $j\geq 0$:

$$\sum_{k\in\mathbb{Z}}|\langle f,\varphi_{jk}\rangle|^2\geq A\|f\|^2,$$

which contradicts (1.39). Assume now that **MR2** does not hold. Then there exists a function $f \in L_2(\mathbb{R})$, $\|f\| \neq 0$, such that $f \perp V_j$ for any $j \in \mathbb{Z}$. It follows that

$$\sum_{k\in\mathbb{Z}} |\langle f, \varphi_{jk}\rangle|^2 = 0$$

for all $j \in \mathbb{Z}$. Passing to the limit as $j \to +\infty$, we get (1.39), which is not true for any function $f \in L_2(\mathbb{R})$ if $\|f\| \neq 0$. ◊

PROBLEM 1.2.15. Let φ be a scaling function for some MRA with orthonormal integer shifts. Prove that the continuity of $\widehat{\varphi}$ at zero implies that $\widehat{\varphi}(0) = 1$.

1.3. Wavelet spaces

As already mentioned, the main property of MRAs is that they allow the construction of orthogonal and biorthogonal wavelet bases. First we describe such a construction for the orthogonal case. Let $\{V_j\}_{j\in\mathbb{Z}}$ be an MRA with the scaling function φ, whose integer shifts form an orthonormal system (in §1.2 it is explained why such a scaling function exists in any MRA). Denote by W_j the orthogonal complement to V_j in the space V_{j+1}. By the properties of the Hilbert space, V_{j+1} can be decomposed into the direct sum $V_{j+1} = V_j \oplus W_j$. Thus we obtain the collection $\{W_j\}_{j\in\mathbb{Z}}$ of subspaces of the space $L_2(\mathbb{R})$ such that $W_j \perp V_j$, $W_j \perp W_k$ for all $j, k \in \mathbb{Z}$, $k \neq j$, and

$$V_j = V_k \oplus W_k \oplus \cdots \oplus W_{j-1}$$

for all $j, k \in \mathbb{Z}$, $k < j$. In view of **MR2**, it follows that

$$L_2(\mathbb{R}) = V_0 \oplus W_0 \oplus W_1 \oplus \ldots,$$

and by **MR3** we have

$$V_0 = W_{-1} \oplus W_{-2} \oplus \ldots.$$

Combining these two decompositions, we get

$$L_2(\mathbb{R}) = \bigoplus_{j=-\infty}^{\infty} W_j. \tag{1.40}$$

Thus the space $L_2(\mathbb{R})$ is decomposed into the direct sum of pairwise orthogonal subspaces W_j. Furthermore, it is clear from the construction that an analog of property **MR4** holds for the spaces W_j , i.e., $f \in W_j \iff f\left(2^{-j}\cdot\right) \in W_0$. The main idea of the construction is that the scaling function φ generalizes some other function $\psi \in W_0$, which inherits many properties of φ; in particular, integer shifts of ψ form an orthonormal basis in W_0. Then, similarly to property **MR6**, the system of functions $\{\psi_{jk}\}_{k\in\mathbb{Z}}$ is an orthonormal basis in W_j. In view of (1.40), the union of these systems over all $j \in \mathbb{Z}$ is an orthonormal basis in $L_2(\mathbb{R})$ consisting of shifts and dilations of the function ψ, and any function $f \in L_2(\mathbb{R})$ can be expanded into the series

$$f = \sum_{j\in\mathbb{Z}}\sum_{n\in\mathbb{Z}} \langle f, \psi_{jn}\rangle \psi_{jn}. \tag{1.41}$$

The spaces W_j are called *wavelet spaces.*

THEOREM 1.3.1. *Let $\{V_j\}_{j\in\mathbb{Z}}$ be an MRA with a scaling function φ, and let $m_0(\xi) = \frac{1}{\sqrt{2}}\sum_{n\in\mathbb{Z}} h_n e^{2\pi i n\xi}$ be its mask. Suppose that the system $\{\varphi_{0n}\}_{n\in\mathbb{Z}}$ is orthonormal,*

$$\psi := \sum_{n\in\mathbb{Z}} (-1)^{1-n}\overline{h}_{1-n}\varphi_{1n}. \tag{1.42}$$

Then the functions ψ_{jn}, $n \in \mathbb{Z}$, $j \in \mathbb{Z}$, form an orthonormal basis in the space $L_2(\mathbb{R})$.

Actually, this assertion is a particular case of Theorem 1.3.7, which states that under the hypothesis of Theorem 1.3.1 the functions ψ_{0n}, $n \in \mathbb{Z}$, form an orthonormal system and the space $\overline{\operatorname{span}\{\psi_{0n}, n \in \mathbb{Z}\}}$ is the orthogonal complement to V_0 in the space V_1. It is clear that the functions ψ_{jn}, $n \in \mathbb{Z}$, form an orthonormal basis in the space $W_j = V_{j+1} \ominus V_j$ for any $j \in \mathbb{Z}$, and the required assertion follows from (1.40).

PROBLEM 1.3.2. Prove that (1.42) is equivalent to

$$\widehat{\psi}(2\xi) = e^{2\pi i\xi}\overline{m_0(\xi + 1/2)}\widehat{\varphi}(\xi). \tag{1.43}$$

Note that there exist orthonormal wavelet bases that cannot be generated by any MRA. Before we consider the corresponding example, let us formulate two theorems (see [**60**] for the proofs). The proof of the first theorem for a more general case will be given in Chapter 8.

THEOREM 1.3.3. *Let $\psi \in L_2(\mathbb{R})$, $\|\psi\| = 1$. The function ψ generates an orthonormal wavelet basis if and only if*

$$\sum_{m\in\mathbb{Z}} |\widehat{\psi}(2^m\cdot)|^2 = 1 \text{ a.e.},$$

$$\sum_{p=0}^{\infty} \widehat{\psi}(2^p\cdot)\overline{\widehat{\psi}(2^p(\cdot + k))} = 0 \text{ a.e. for any odd } k.$$

THEOREM 1.3.4. *Let ψ generate an orthonormal wavelet basis. A corresponding scaling function exists if and only if*

$$\sum_{p=1}^{\infty}\sum_{k\in\mathbb{Z}} |\widehat{\psi}(2^p(\cdot + k))|^2 > 0 \text{ a.e.}$$

In this case, we have

$$\sum_{p=1}^{\infty}\sum_{k\in\mathbb{Z}} |\widehat{\psi}(2^p(\cdot + k))|^2 = 1 \text{ a.e.}$$

COROLLARY 1.3.5. *If $\widehat{\psi}(\xi) \neq 0$ almost everywhere, then there exists a scaling function. In particular, this is true if ψ is compactly supported (in this case $\widehat{\psi}$ is an analytic function). Then the scaling function is also compactly supported.*

Using Theorem 1.3.3, it is easy to check that the function ψ defined by the equality

$$\widehat{\psi} = \chi_{[-4/7,-2/7]} + \chi_{[2/7,3/7]} + \chi_{[12/7,16/7]}$$

generates an orthonormal wavelet basis. It follows from Theorem 1.3.4 that the function ψ cannot be generated by any MRA. Indeed, if we assume that there exists a scaling function φ, then $|\widehat{\varphi}(\xi)|^2 - |\widehat{\varphi}(2\xi)|^2 = |\widehat{\psi}(2\xi)|^2$. This follows from (1.23), (1.27), and (1.43):

$$\begin{aligned} |\widehat{\psi}(2\xi)|^2 &= |m(\xi + 1/2)|^2|\widehat{\varphi}(\xi)|^2 \\ &= (1 - |m(\xi)|^2)|\widehat{\varphi}(\xi)|^2 = |\widehat{\varphi}(\xi)|^2 - |\widehat{\varphi}(2\xi)|^2. \end{aligned}$$

The latter equality implies that $|\widehat{\varphi}(\xi)|^2 = \sum_{j\geq 1} |\widehat{\psi}(2^j\xi)|^2$ or

$$|\widehat{\varphi}(\xi)|^2 = \chi_{[-2/7,2/7]} + \chi_{[3/7,4/7]} + \chi_{[6/7,8/7]}.$$

However the orthonormality of the system $\{\varphi(\cdot - k)\}_{k\in\mathbb{Z}}$ is equivalent to identity (1.17), which does not hold in this case.

Despite the existence of functions generating orthonormal wavelet bases without using MRA, any such function is connected with some structure similar to MRA.

THEOREM 1.3.6. *Let $\{\psi_{jk}\}_{j,k\in\mathbb{Z}}$ be an orthonormal basis in $L_2(\mathbb{R})$,*

$$W_j := \overline{\operatorname{span}\{\psi_{jk},\ k\in\mathbb{Z}\}}, \quad V_j := \bigoplus_{l<j} W_l.$$

Then the subspaces V_j, $j\in\mathbb{Z}$, possess the following properties:

(i) $V_j \subset V_{j+1}$;
(ii) $\overline{\bigcup_{j\in\mathbf{Z}} V_j} = L^2(\mathbf{R})$;
(iii) $\bigcap_{j\in\mathbf{Z}} V_j = \{0\}$;
(iv) $f\in V_j \Leftrightarrow f(2^{-j}\cdot)\in V_0$;
(v) $f\in V_0 \Leftrightarrow f(\cdot - k)\in V_0$ *for any* $k\in\mathbf{Z}$.

`Proof.` The first four properties are obvious. The fifth follows from the equality $V_0^{\perp} = \bigoplus_{j\geq 0} W_j$, where $V^{\perp}$ is the orthogonal complement to V, and from the fact that every space W_j, $j\geq 0$, is invariant under integer shifts. ◊

Let us now describe the construction of biorthogonal wavelet systems. Let $\{V_j\}_{j\in\mathbb{Z}}$ and $\{\widetilde{V}_j\}_{j\in\mathbb{Z}}$ be multiresolution analyses. To implement the same idea as in the orthogonal case, we have to define a wavelet space W_j for the first MRA $\{V_j\}_{j\in\mathbb{Z}}$ as the completion to V_j in V_{j+1} orthogonal to the space $\widetilde{V}_j$ of the second MRA $\{\widetilde{V}_j\}_{j\in\mathbb{Z}}$.

THEOREM 1.3.7. *Let $\{V_j\}_{j\in\mathbb{Z}}$, $\{\widetilde{V}_j\}_{j\in\mathbb{Z}}$ be two MRAs with scaling functions φ, $\widetilde{\varphi}$, respectively, and let*

$$m_0(\xi) = 2^{-1/2}\sum_{n\in\mathbb{Z}} h_n e^{2\pi i n\xi}, \qquad \widetilde{m}_0(\xi) = 2^{-1/2}\sum_{n\in\mathbb{Z}} \widetilde{h}_n e^{2\pi i n\xi}$$

be their masks,

$$\psi := \sum_{n\in\mathbb{Z}} (-1)^{1-n}\overline{\widetilde{h}}_{1-n}\varphi_{1n}, \tag{1.44}$$

$$\widetilde{\psi} := \sum_{n\in\mathbb{Z}} (-1)^{1-n}\overline{h}_{1-n}\widetilde{\varphi}_{1n}, \tag{1.45}$$

$$W_0 := \overline{\operatorname{span}\{\psi_{0n}, n\in\mathbb{Z}\}}.$$

Then
(i) *the system $\{\psi_{0n}\}_{n\in\mathbb{Z}}$ forms a Riesz basis in W_0,*
(ii) $W_0\subset V_1$.
Moreover, if the systems $\{\varphi_{0n}\}_{n\in\mathbb{Z}}$, $\{\widetilde{\varphi}_{0n}\}_{n\in\mathbb{Z}}$ are biorthonormal, then
(iii) $W_0 \perp \widetilde{V}_0$,
(iv) *any $f\in V_1$ can be represented in the form $f = f_1 + f_2$, $f_1\in V_0$, $f_2\in W_0$,*
(v) *$\{\psi_{0n}\}_{n\in\mathbb{Z}}$, $\{\widetilde{\psi}_{0n}\}_{n\in\mathbb{Z}}$ are biorthonormal systems.*

`Proof.` First let us note that

$$\overline{e^{2\pi i\xi}\widetilde{m}_0\left(\xi+\frac{1}{2}\right)} = \frac{1}{\sqrt{2}}\sum_{n\in\mathbb{Z}} (-1)^{1-n}\overline{\widetilde{h}}_{1-n}e^{2\pi i n\xi};$$

hence by Proposition 1.2.3,

$$\widehat{\psi}(\xi) = e^{\pi i \xi}\overline{\widetilde{m}_0\left(\frac{\xi+1}{2}\right)}\widehat{\varphi}\left(\frac{\xi}{2}\right), \tag{1.46}$$

and analogously

$$\widehat{\widetilde{\psi}}(\xi) = e^{\pi i \xi}\overline{m_0\left(\frac{\xi+1}{2}\right)}\widehat{\widetilde{\varphi}}\left(\frac{\xi}{2}\right). \tag{1.47}$$

Since $\{\varphi_{0n}\}_{n\in\mathbb{Z}}$, $\{\widetilde{\varphi}_{0n}\}_{n\in\mathbb{Z}}$ are Riesz systems, by Theorem 1.1.7, the inequalities

$$A \le \sum_{n\in\mathbb{Z}} |\widehat{\varphi}(\xi+n)|^2 \le B, \tag{1.48}$$

$$\widetilde{A} \le \sum_{n\in\mathbb{Z}} |\widehat{\widetilde{\varphi}}(\xi+n)|^2 \le \widetilde{B} \tag{1.49}$$

hold for almost all $\xi \in \mathbb{R}$. Using (1.46), we get

$$\begin{aligned}\sum_{n\in\mathbb{Z}} |\widehat{\psi}(\xi+n)|^2 &= \sum_{n\in\mathbb{Z}} \left|\widetilde{m}_0\left(\frac{\xi+n+1}{2}\right)\widehat{\varphi}\left(\frac{\xi+n}{2}\right)\right|^2 \\ &= \left|\widetilde{m}_0\left(\frac{\xi+1}{2}\right)\right|^2 \sum_{k\in\mathbb{Z}}\left|\widehat{\varphi}\left(\frac{\xi}{2}+k\right)\right|^2 + \left|\widetilde{m}_0\left(\frac{\xi}{2}\right)\right|^2 \sum_{k\in\mathbb{Z}}\left|\widehat{\varphi}\left(\frac{\xi+1}{2}+k\right)\right|^2.\end{aligned}$$

This relation, inequalities (1.48), (1.49), and (1.25) imply that

$$\frac{A\widetilde{A}}{\widetilde{B}} \le \sum_{n\in\mathbb{Z}} |\widehat{\psi}(\xi+n)|^2 \le \frac{B\widetilde{B}}{\widetilde{A}}$$

for almost all $\xi \in \mathbb{R}$. Taking into account Theorems 1.1.7 and 1.1.2, we obtain (i).

To prove (ii), it suffices to check that the basis functions ψ_{0n}, $n \in \mathbb{Z}$, belong to the space V_1, which follows from the definition of the function ψ and properties **MR6**, **MR7**.

To prove (iii), we have to check that each basis element ψ_{0n}, $n \in \mathbb{Z}$, is orthogonal to $\widetilde{V}_0$. Using (1.46) and the refinement equation for the function $\widetilde{\varphi}$, we get

$$\begin{aligned}&\sum_{k\in\mathbb{Z}} \widehat{\psi}(\xi+k)\overline{\widehat{\widetilde{\varphi}}(\xi+k)}e^{2\pi i n\xi} \\ &= \sum_{k\in\mathbb{Z}} e^{2\pi i(n+1/2)(\xi+k)}\overline{\widetilde{m}_0\left(\frac{\xi+k+1}{2}\right)\widetilde{m}_0\left(\frac{\xi+k}{2}\right)}\widehat{\varphi}\left(\frac{\xi+k}{2}\right)\overline{\widehat{\widetilde{\varphi}}\left(\frac{\xi+k}{2}\right)} \\ &= e^{2\pi i(n+1/2)\xi}\left(\overline{\widetilde{m}_0\left(\frac{\xi+1}{2}\right)\widetilde{m}_0\left(\frac{\xi}{2}\right)}\sum_{l\in\mathbb{Z}}\widehat{\varphi}\left(\frac{\xi}{2}+l\right)\overline{\widehat{\widetilde{\varphi}}\left(\frac{\xi}{2}+l\right)}\right. \\ &\qquad \left. - \overline{\widetilde{m}_0\left(\frac{\xi}{2}\right)\widetilde{m}_0\left(\frac{\xi+1}{2}\right)}\sum_{l\in\mathbb{Z}}\widehat{\varphi}\left(\frac{\xi+1}{2}+l\right)\overline{\widehat{\widetilde{\varphi}}\left(\frac{\xi+1}{2}+l\right)}\right).\end{aligned}$$

From Proposition 1.1.12 it follows that the right-hand side of this equality is equal to zero for almost all $\xi \in \mathbb{R}$. In view of Proposition 1.2.2, this implies that $\psi_{0n} \perp \widetilde{V}_0$.

Let $f \in V_1$. By Proposition 1.2.3, there exists a 1-periodic function $m_f \in L_2(0,1)$ that satisfies relation (1.22). Set

$$\alpha(\xi) := \frac{1}{\sqrt{2}}\left(m_f\left(\frac{\xi}{2}\right)\overline{\widetilde{m}_0\left(\frac{\xi}{2}\right)} + m_f\left(\frac{\xi+1}{2}\right)\overline{\widetilde{m}_0\left(\frac{\xi+1}{2}\right)}\right),$$
$$\beta(\xi) := \frac{1}{\sqrt{2}}\left(m_f\left(\frac{\xi}{2}\right)m_0\left(\frac{\xi+1}{2}\right) - m_f\left(\frac{\xi+1}{2}\right)m_0\left(\frac{\xi}{2}\right)\right)e^{-\pi i\xi}.$$

It is not hard to check that the functions α, β are 1-periodic and the equality

$$\frac{1}{\sqrt{2}}m_f\left(\frac{\xi}{2}\right) = \alpha(\xi)m_0\left(\frac{\xi}{2}\right) + \beta(\xi)\overline{\widetilde{m}_0\left(\frac{\xi+1}{2}\right)}e^{\pi i\xi} \tag{1.50}$$

holds. Let us show that $\alpha \in L_2(0,1)$. By Corollary 1.2.5, the function $\widetilde{m}_0$ is essentially bounded; hence we have

$$\left(\int_0^1 |\alpha(\xi)|^2\,d\xi\right)^{1/2} \leq C\left(\left(\int_0^1 \left|m_f\left(\frac{\xi}{2}\right)\right|^2 d\xi\right)^{1/2} + \left(\int_0^1 \left|m_f\left(\frac{\xi+1}{2}\right)\right|^2 d\xi\right)^{1/2}\right)$$
$$\leq \sqrt{2}C\left(\int_0^1 \left|m_f\left(\frac{\xi}{2}\right)\right|^2 d\xi + \int_0^1 \left|m_f\left(\frac{\xi+1}{2}\right)\right|^2 d\xi\right)^{1/2} = 2C\|m_f\|.$$

It can be checked analogously that $\beta \in L_2(0,1)$. Let

$$\alpha(\xi) = \sum_{n\in\mathbb{Z}} a_n e^{2\pi i n\xi}, \quad \beta(\xi) = \sum_{n\in\mathbb{Z}} b_n e^{2\pi i n\xi}.$$

Set

$$f_1(x) := \sum_{n\in\mathbb{Z}} a_n\varphi(x+n), \quad f_2(x) := \sum_{n\in\mathbb{Z}} b_n\psi(x+n).$$

It is clear that $f_1 \in V_0$, $f_2 \in W_0$. Using Proposition 1.1.10, the refinement equation for φ, and (1.46), we obtain

$$\widehat{f_1}(\xi) = \alpha(\xi)\widehat{\varphi}(\xi) = \alpha(\xi)m_0\left(\frac{\xi}{2}\right)\widehat{\varphi}\left(\frac{\xi}{2}\right),$$
$$\widehat{f_2}(\xi) = \beta(\xi)\widehat{\psi}(\xi) = \beta(\xi)e^{\pi i\xi}\overline{\widetilde{m}_0\left(\frac{\xi+1}{2}\right)}\widehat{\varphi}\left(\frac{\xi}{2}\right).$$

It follows from these equalities, (1.22), and (1.50) that $\widehat{f} = \widehat{f_1} + \widehat{f_2}$, which implies $f = f_1 + f_2$. This completes the proof of (iv).

In view of Proposition 1.1.12, to prove (v), it suffices to check that

$$\sum_{k\in\mathbb{Z}} \widehat{\psi}(\xi+k)\overline{\widehat{\widetilde{\psi}}(\xi+k)} = 1 \tag{1.51}$$

for almost all $\xi \in \mathbb{Z}$. Using (1.46) and (1.47), we get

$$\sum_{k\in\mathbb{Z}} \widehat{\psi}(\xi+k)\overline{\widehat{\widetilde{\psi}}(\xi+k)} = \overline{\widetilde{m}_0\left(\frac{\xi+k+1}{2}\right)} m_0\left(\frac{\xi+k+1}{2}\right)\widehat{\varphi}\left(\frac{\xi+k}{2}\right)\overline{\widehat{\widetilde{\varphi}}\left(\frac{\xi+k}{2}\right)}$$

$$= \left(\overline{\widetilde{m}_0\left(\frac{\xi+1}{2}\right)} m_0\left(\frac{\xi+1}{2}\right)\sum_{l\in\mathbb{Z}}\widehat{\varphi}\left(\frac{\xi}{2}+l\right)\overline{\widehat{\widetilde{\varphi}}\left(\frac{\xi}{2}+l\right)}\right.$$

$$\left. - \overline{\widetilde{m}_0\left(\frac{\xi}{2}\right)} m_0\left(\frac{\xi}{2}\right)\sum_{l\in\mathbb{Z}}\widehat{\varphi}\left(\frac{\xi+1}{2}+l\right)\overline{\widehat{\widetilde{\varphi}}\left(\frac{\xi+1}{2}+l\right)}\right).$$

Using (1.17) and (1.26), we obtain (1.51). ◊

Functions ψ, $\widetilde{\psi}$ are called *wavelet functions* if they satisfy properties (i)–(v) of Theorem 1.3.7. Such a pair of functions is not unique; for example, in equalities (1.46), (1.47) the factor $e^{\pi i\xi}$ can be replaced by $\pm e^{\pm\pi i\xi}$. The functions obtained in this way are also wavelet functions.

PROBLEM 1.3.8. Let $\{V_j\}_{j\in\mathbb{Z}}$, $\{\widetilde{V}_j\}_{j\in\mathbb{Z}}$ be MRAs that satisfy the conditions of Theorem 1.3.7. Describe all pairs of functions λ, $\widetilde{\lambda}$ such that $\psi, \widetilde{\psi}$ defined by the equalities

$$\widehat{\psi}(\xi) = \lambda(\xi)\overline{\widetilde{m}_0\left(\frac{\xi+1}{2}\right)}\widehat{\varphi}\left(\frac{\xi}{2}\right), \quad \widehat{\widetilde{\psi}}(\xi) = \widetilde{\lambda}(\xi)\overline{m_0\left(\frac{\xi+1}{2}\right)}\widehat{\widetilde{\varphi}}\left(\frac{\xi}{2}\right)$$

are wavelet functions.

If ψ is a wavelet function, then the collection of functions ψ_{jk}, $j, k \in \mathbb{Z}$, is called a *wavelet system*.

It is clear from Theorem 1.3.7 that if we switch the roles of the first and second MRAs and set

$$\widetilde{W}_0 := \overline{\operatorname{span}\{\widetilde{\psi}_{0n}, n\in\mathbb{Z}\}},$$

we obtain statements analogous to (i)–(iv) for the MRA $\{\widetilde{V}_j\}_{j\in\mathbb{Z}}$. Define wavelet spaces

$$W_j := \{f \in L_2(\mathbb{R}) : \ f(2^{-j}\cdot) \in W_0\},$$
$$\widetilde{W}_j := \{f \in L_2(\mathbb{R}) : \ f(2^{-j}\cdot) \in \widetilde{W}_0\}.$$

COROLLARY 1.3.9. *Under the conditions of Theorem 1.3.7 the systems $\{\psi_{jk}\}_{k\in\mathbb{Z}}$ and $\{\widetilde{\psi}_{jk}\}_{k\in\mathbb{Z}}$ are Riesz bases in the spaces W_j and $\widetilde{W}_j$, respectively; $V_{j+1} = V_j + W_j$, $\widetilde{V}_{j+1} = \widetilde{V}_j + \widetilde{W}_j$, $W_j \perp \widetilde{W}_k$ for all $j \neq k$; $V_j \perp \widetilde{W}_k$, $\widetilde{V}_j \perp W_k$ for all $k \geq j$; the systems $\{\psi_{jk}\}_{j,k\in\mathbb{Z}}$,$\{\widetilde{\psi}_{jk}\}_{j,k\in\mathbb{Z}}$ are biorthonormal.*

All the above statements immediately follow from Theorem 1.3.7 and **MR4**.

THEOREM 1.3.10. *Let $\{V_j\}_{j\in\mathbb{Z}}$, $\{\widetilde{V}_j\}_{j\in\mathbb{Z}}$ be two MRAs with scaling functions φ, $\widetilde{\varphi}$ such that the systems $\{\varphi_{0n}\}_{n\in\mathbb{Z}}$, $\{\widetilde{\varphi}_{0n}\}_{n\in\mathbb{Z}}$ are biorthonormal, and let ψ, $\widetilde{\psi}$ be the corresponding wavelet functions. Then the following decomposition holds for* *any function* $f \in L_2(\mathbb{R})$*:*

$$f = \sum_{n\in\mathbb{Z}} \langle f, \widetilde{\varphi}_{0n}\rangle \varphi_{0n} + \sum_{j=0}^{\infty}\sum_{n\in\mathbb{Z}} \langle f, \widetilde{\psi}_{jn}\rangle \psi_{jn} \tag{1.52}$$

$$= \sum_{j=-\infty}^{-1}\sum_{n\in\mathbb{Z}} \langle f, \widetilde{\psi}_{jn}\rangle \psi_{jn} + \sum_{j=0}^{+\infty}\sum_{n\in\mathbb{Z}} \langle f, \widetilde{\psi}_{jn}\rangle \psi_{jn}. \tag{1.53}$$

`Proof.` Define the operator P_j acting on $L_2(\mathbb{R})$:

$$P_j f := \sum_{n\in\mathbb{Z}} \langle f, \widetilde{\varphi}_{jn}\rangle \varphi_{jn},$$

and show that it is bounded. Since $\{\widetilde{\varphi}_{jn}\}_{n\in\mathbb{Z}}$ is a Riesz system, it follows from Theorem 1.1.2 and Remark 1.1.3 that

$$\sum_{n\in\mathbb{Z}} |\langle f, \widetilde{\varphi}_{jn}\rangle|^2 \le \widetilde{B}\|f\|^2,$$

where $\widetilde{B}$ is the upper bound of the Riesz system. Therefore, as $\{\varphi_{jn}\}_{n\in\mathbb{Z}}$ is a Riesz system, $P_j f \in V_j$ and

$$\|P_j f\|^2 \le B \sum_{n\in\mathbb{Z}} |\langle f, \widetilde{\varphi}_{jn}\rangle|^2 \le B\widetilde{B}\|f\|^2. \tag{1.54}$$

Thus,

$$\|P_j\| \le C, \tag{1.55}$$

where C does not depend on j. It is clear that

$$\langle f, g\rangle = \langle P_j f, g\rangle \tag{1.56}$$

for any $g \in \widetilde{V}_j$. It follows that P_j is a projection operator onto V_j. Analogously, the operator

$$Q_j f := \sum_{n\in\mathbb{Z}} \langle f, \widetilde{\psi}_{jn}\rangle \psi_{jn}$$

is bounded and a projector onto W_j. By Corollary 1.3.9, for any function $f \in V_{j+1}$, we have the decomposition $f = f_1 + f_2$, $f_1 \in V_j$, $f_2 \in W_j$, $f \in V_{j+1}$. Therefore,

$$P_{j+1} f = \sum_{n\in\mathbb{Z}} c_n \varphi_{jn} + \sum_{n\in\mathbb{Z}} d_n \psi_{jn}$$

for any $f \in L_2(\mathbb{R})$. Taking the scalar product of both sides by the functions $\widetilde{\varphi}_{jn}, \widetilde{\psi}_{jn}$, $n \in \mathbb{Z}$, successively and using (1.56), we find the coefficients of the decomposition: $c_n = \langle f, \widetilde{\varphi}_{jn}\rangle$, $d_n = \langle f, \widetilde{\psi}_{jn}\rangle$. Thus,

$$P_{j+1} f = P_j f + Q_j f.$$

Using this equality several times, we get

$$P_{j+1} f = P_0 f + \sum_{i=0}^{j} Q_i f, \tag{1.57}$$

$$P_0 f = P_{-j} f + \sum_{i=-j}^{-1} Q_i f. \tag{1.58}$$

According to **MR2**, for any $f \in L_2(\mathbb{R})$, there exists $g \in V_{j_0}$ such that $\|f-g\| < \varepsilon$. In view of **MR1** and (1.55), it follows that

$$\|f - P_j f\| = \|(f-g) - P_j(f-g)\| < (1+C)\varepsilon$$

for all $j \geq j_0$. Combining this with (1.57), we get

$$f = P_0 f + \sum_{i=0}^{\infty} Q_i f, \tag{1.59}$$

which proves equality (1.52). To prove (1.53), we note that the left inequality (1.54) and Lemma 1.2.8 imply that

$$\|P_j f\|^2 \leq B \sum_{n\in\mathbb{Z}} |\langle f, \widetilde{\varphi}_{jn}\rangle|^2 \underset{j\to-\infty}{\longrightarrow} 0.$$

Passing to the limit in (1.58), we get

$$P_0 f = \sum_{i=-\infty}^{-1} Q_i f.$$

Substituting this equality into (1.59), we obtain (1.53). ◊

REMARK 1.3.11. It is clear that relations (1.52), (1.53) can be replaced by

$$f = \sum_{n\in\mathbb{Z}} \langle f, \widetilde{\varphi}_{jn}\rangle \varphi_{jn} + \sum_{i=j}^{\infty}\sum_{n\in\mathbb{Z}} \langle f, \widetilde{\psi}_{in}\rangle \psi_{in} \tag{1.60}$$

$$= \sum_{i=-\infty}^{j-1}\sum_{n\in\mathbb{Z}} \langle f, \widetilde{\psi}_{in}\rangle \psi_{in} + \sum_{i=j}^{+\infty}\sum_{n\in\mathbb{Z}} \langle f, \widetilde{\psi}_{in}\rangle \psi_{in} \tag{1.61}$$

for any $j \in \mathbb{Z}$.

In Chapter 2, we show that under the conditions of Theorem 1.3.10 each of the systems $\{\psi_{jk}\}_{j,k\in\mathbb{Z}}$, $\{\widetilde{\psi}_{jk}\}_{j,k\in\mathbb{Z}}$ is a Riesz system in the space $L_2(\mathbb{R})$ if we impose some additional requirements on the functions φ, $\widetilde{\varphi}$.

In the proofs of Theorems 1.3.7 and 1.3.10 we did not use properties **MR2**, **MR3** of the second MRA $\{\widetilde{V}_j\}_{j\in\mathbb{Z}}$. Moreover, if we are not interested in the problem of convergence of wavelet expansions, restricting ourselves to the construction of biorthogonal wavelets, then, generally speaking, we need not require that the functions $\widetilde{\varphi}_{0n}$, $n \in \mathbb{Z}$, form a Riesz basis (though formally we used this condition to prove Theorem 1.3.7; in fact, we only needed the boundedness of the mask $\widetilde{m}_0$). Taking into account the aforesaid, we can extend the class of functions $\widetilde{\varphi}$ generating the MRA $\{\widetilde{V}_j\}_{j\in\mathbb{Z}}$. Moreover, if φ is continuous, then for $\widetilde{\varphi}$ we can try to take the Borel measure. Let us consider an example.

For $\widetilde{\varphi}$, take the δ-function defined on $\mathbb{R}$. Since $\widehat{\widetilde{\varphi}} \equiv 1$, the function $\widetilde{\varphi}$ is a solution of the refinement equation (1.23) with mask $\widetilde{m}_0 \equiv 1$. Suppose that the function φ is continuous, generates an MRA, and $\varphi(n) = \delta_{n0}$ for all $n \in \mathbb{Z}$. Then it is clear that condition (1.16) is satisfied; i.e., $\{\varphi_{0n}\}_{n\in\mathbb{Z}}$, $\{\widetilde{\varphi}_{0n}\}_{n\in\mathbb{Z}}$ are biorthonormal systems. Therefore the systems $\{\varphi_{jn}\}_{n\in\mathbb{Z}}$, $\{\widetilde{\varphi}_{jn}\}_{n\in\mathbb{Z}}$ are also biorthonormal for any

$j \in \mathbb{Z}$. It follows that

$$\delta_{nk} = \int_{-\infty}^{\infty} \varphi_{jn}(x)\widetilde{\varphi}_{jk}(x)\,dx$$
$$= 2^{-j/2} \int_{-\infty}^{\infty} \varphi_{jn}(2^{-j}(x-k))\delta(x)\,dx = 2^{-j/2}\varphi_{jn}(-2^{-j}k).$$

On the other hand, for any continuous function f, we have

$$\langle f, \widetilde{\varphi}_{jn}\rangle = 2^{-j/2} f(-2^{-j}k).$$

Thus the projection operator P_j onto the space V_j introduced in Theorem 1.3.10 interpolates the function f at the points $x_{jn} := 2^{-j}n$, $n \in \mathbb{Z}$; hence the MRA under consideration is called an *interpolation MRA*.

Using formulas (1.44), (1.45), we find the wavelet functions

$$\psi(x) = 2\varphi(2x+1),$$
$$\widetilde{\psi}(x) = \sqrt{2}\sum_{k\in\mathbb{Z}} (-1)^{1-k} h_{1-k}\delta(2x+k),$$

where $h_k/\sqrt{2}$ are the Fourier coefficients of the mask m_0. The projection operator onto the space W_j is not an interpolation operator, though the coefficients of the decomposition with respect to the system $\{\psi_{jn}\}_{j,n\in\mathbb{Z}}$ are expressed by the values of the function f at the points x_{jn}, $j, n \in \mathbb{Z}$. Formal calculations give

$$\langle f, \widetilde{\psi}_{jn}\rangle = 2^{j/2} \int_{-\infty}^{\infty} f(x)\widetilde{\psi}(2^j x + n)\,dx$$
$$= 2^{\frac{j+1}{2}} \int_{-\infty}^{\infty} f(x) \sum_{k\in\mathbb{Z}} (-1)^{1-k} h_{1-k}\delta(2^{j+1}x + 2n + k)\,dx$$
$$= 2^{-\frac{j+1}{2}} \sum_{k\in\mathbb{Z}} (-1)^{1-k} h_{1-k} f\left(-2^{-j-1}(2n+k)\right).$$

1.4. Haar, Battle-Lemarie, Strömberg, and Meyer systems

1. The Haar system introduced almost a century before wavelet theory is the simplest but at the same time the best model example of wavelet systems. Here we construct a Haar basis, following the general scheme described in §1.3. Set $\varphi = \varphi^H = \chi_{[0,1]}$. The Fourier transform of this function

$$\widehat{\varphi}(\xi) = e^{-\pi i\xi}\frac{\sin \pi\xi}{\pi\xi}$$

satisfies equation (1.23) with mask

$$m_0(\xi) = m^H(\xi) = \frac{1}{2} + \frac{1}{2}e^{-2\pi i\xi}.$$

Obviously, integer shifts of the function φ are orthonormal. Thus φ satisfies all the conditions of Theorem 1.2.14 and hence it is a scaling function for an MRA which is called Haar MRA. By formula (1.42) we find the corresponding wavelet function

$$\psi^H(x) = \varphi(2x+1) - \varphi(2x+2) = \begin{cases} -1, & \text{for } -1 \le x < -1/2, \\ 1, & \text{for } -1/2 < x \le 0, \\ 0, & \text{for } x = -1/2,\ x > 0,\ x < -1. \end{cases}$$

In view of Theorem 1.3.1, the functions ψ^H_{jk}, $j, k \in \mathbb{Z}$, form an orthonormal basis in $L_2(\mathbb{R})$.

A. Haar considered the orthonormal basis generated by the wavelet function $\psi(x) = \psi^H(-x)$ only in the space $L_2([0,1])$. As far as the authors know, A. Haar did not mention that all elements of the system he had constructed were binary dilations and shifts of one fixed function. Perhaps this was the reason why the development of wavelet theory was delayed for almost a century.

2. The Battle-Lemarie wavelets. The following structure is a natural generalization of the Haar system. Let $\varphi^{B,0} = \chi_{[0,1]}$ (note that it is a scaling function for the Haar system) and define the function $\varphi = \varphi^{B,N}$ recursively for any positive integer N:

$$(*) \qquad \begin{aligned} \varphi^{B,2k+1}(x) &= \int_{-\infty}^{\infty} \varphi^{B,2k}(t)\chi_{[-1,0]}(x-t)\,dt, \\ \varphi^{B,2k}(x) &= \int_{-\infty}^{\infty} \varphi^{B,2k-1}(t)\chi_{[0,1]}(x-t)\,dt. \end{aligned}$$

The functions $\varphi^{B,N}$ are B-splines. The theory of B-splines is well studied and can be found in many books. Some properties of B-splines are presented in Chapter 3 (Example 3.5.1), where we study refinement equations. Here we just note that one can easily check by induction that $\varphi^{B,N}$ is a compactly supported spline of order N; the function $\varphi^{B,2k+1}$ is supported on the segment $[-k-1, k+1]$, and the function $\varphi^{B,2k}$ on the segment $[-k, k+1]$.

Since the Fourier transform of the convolution of two functions is equal to the product of their Fourier transforms, we have

$$\widehat{\varphi^{B,N}}(\xi) = e^{-\pi i \gamma(N)\xi} \left(\frac{\sin \pi\xi}{\pi\xi}\right)^{N+1},$$

where

$$\gamma = \gamma(N) = \begin{cases} 0, & \text{if } N \text{ is odd}, \\ 1, & \text{if } N \text{ is even}. \end{cases}$$

Let us show that integer shifts of the function φ form a Riesz system. In view of Theorem 1.1.7, it suffices to check that the sum $\sum_{k\in\mathbb{Z}} |\widehat{\varphi}(\xi+k)|^2$ is bounded both above and below by some positive constants. This fact follows from the trivial inequalities

$$\begin{aligned} \sum_{k\in\mathbb{Z}} |\widehat{\varphi}(\xi+k)|^2 &\ge \min_{\xi\in[-\frac12,\frac12]} \left(\frac{\sin\pi\xi}{\pi\xi}\right)^{2(N+1)} > 0, \\ \sum_{k\in\mathbb{Z}} |\widehat{\varphi}(\xi+k)|^2 &\le \max_{\xi\in[-\frac12,\frac12]} \left(\frac{\sin\pi\xi}{\pi\xi}\right)^{2(N+1)} + \sum_{k=1}^{\infty} \frac{2}{(k-1/2)^{2(N+1)}} < \infty. \end{aligned}$$

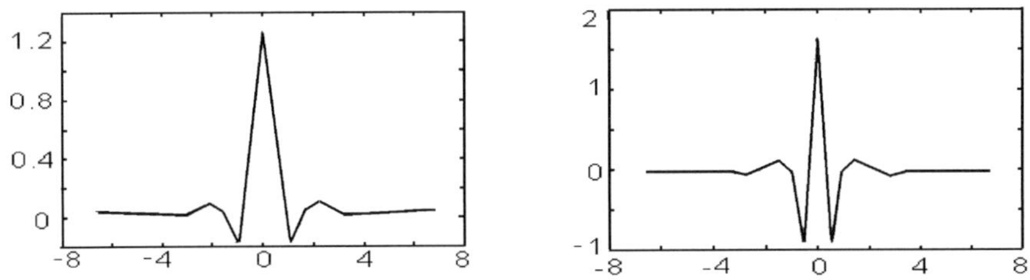

a) b)

FIGURE 1.1. Battle-Lemarie system ($N = 1$): a) scaling function $\varphi^{\#}$; b) wavelet function $\psi^{B,1}$.

It is not hard to see that the function $\widehat{\varphi}$ satisfies equation (1.23) with mask $m_0(\xi) = m^{B,N}(\xi) = e^{-\pi i \gamma \xi} \cos^{N+1} \pi\xi$. Therefore all the conditions of Theorem 1.2.14 are satisfied and $\varphi^{B,N}$ is a scaling function of the Battle-Lemarie MRA.

By Theorem 1.1.14, in this MRA there exists another scaling function $\varphi^{\sharp}$, whose integer shifts form an orthonormal system. It is clear that $\varphi^{\sharp}$ is also a spline of order N, though not compactly supported for $N > 1$. Further, by Theorem 1.3.1, one can construct the corresponding orthonormal wavelet basis ψ_{jk}, $j, k \in \mathbb{Z}$, formed by splines of order N.

Let us consider the case $N = 1$ in more detail. It is not hard to check that

$$\varphi(x) = \varphi^{B,1}(x) := (1 - |x|)\chi_{[-1,1]}(x).$$

In the time domain the refinement equation has the form

$$\varphi(x) = \frac{1}{2}\varphi(2x-1) + \varphi(2x) + \frac{1}{2}\varphi(2x+1),$$

and the equality

$$\sum_{k\in\mathbb{Z}} |\widehat{\varphi}(\xi + k)|^2 = \sum_{k\in\mathbb{Z}} \left(\frac{\sin \pi(\xi + k)}{\pi(\xi + k)} \right)^4 = \frac{1 + 2\cos^2 \pi\xi}{3}$$

holds.

The function $\varphi^{\#}$ is defined by the equation

$$\widehat{\varphi}^{\#}(\xi) = \frac{\sqrt{3}\widehat{\varphi}(\xi)}{(1 + 2\cos^2 \pi\xi)^{1/2}} = \frac{\sqrt{3}\sin^2 \pi\xi}{(\pi\xi)^2(1 + 2\cos^2 \pi\xi)^{1/2}},$$

which implies that

$$\varphi^{\#}(x) = \sqrt{3}\sum_n c_n \varphi(x + n),$$

where c_n, $n \in \mathbb{Z}$, are the Fourier coefficients of the function $[1 + 2\cos^2 \pi\xi]^{-1/2}$, and the refinement equation for $\widehat{\varphi}^{\#}$ has the form

$$\widehat{\varphi}^{\#}(2\xi) = m_0^{\#}(\xi)\widehat{\varphi}^{\#}(\xi),$$

where

$$m_0^{\#}(\xi) = \cos^2 \pi\xi \left[\frac{1+2\cos^2 \pi\xi}{1+2\cos^2 2\pi\xi} \right]^{1/2}.$$

By formula (1.46), the corresponding wavelet function $\psi = \psi^{B,1}$ is determined by the equality

$$\widehat{\psi}(\xi) = e^{i\pi\xi} \sin^2 \frac{\pi\xi}{2} \left[\frac{1+2\sin^2 \pi\xi/2}{1+2\cos^2 \pi\xi}\right]^{1/2} \widehat{\varphi}^{\#}\left(\frac{\xi}{2}\right) = \sqrt{3} m_1 \left(\frac{\xi}{2}\right) \widehat{\varphi}\left(\frac{\xi}{2}\right),$$

where

$$m_1(\xi) = e^{2i\pi\xi} \sin^2 \pi\xi \left[\frac{1+2\sin^2 \pi\xi}{(1+2\cos^2 2\pi\xi)(1+2\cos^2 \pi\xi)}\right]^{1/2}.$$

Denoting by d_n, $n \in \mathbb{Z}$, the Fourier coefficients of the function m_1, we get

$$\psi(x) = \sqrt{3} \sum_n d_n \varphi(2x+n).$$

In the case $N = 2$ we have

$$\varphi(x) = \varphi^{B,2}(x) := \begin{cases} \frac{1}{2}(x+1)^2, & \text{if } -1 \le x \le 0, \\ \frac{3}{4} - \left(x - \frac{1}{2}\right)^2, & \text{if } 0 \le x \le 1, \\ \frac{1}{2}(x-2)^2, & \text{if } 1 \le x \le 2, \\ 0, & \text{otherwise}, \end{cases}$$

the refinement equation in the time domain has the form

$$\varphi(x) = \frac{1}{4}\varphi(2x-2) + \frac{3}{4}\varphi(2x-1) + \frac{3}{4}\varphi(2x) + \frac{1}{4}\varphi(2x+1),$$

and the following equality holds:

$$\sum_{k\in\mathbb{Z}} |\widehat{\varphi}(\xi+k)|^2 = \sum_{k\in\mathbb{Z}} \left(\frac{\sin \pi(\xi+k)}{\pi(\xi+k)}\right)^6 = \frac{8}{15} + \frac{13}{30}\cos 2\pi\xi + \frac{1}{30}\cos^2 2\pi\xi.$$

One can orthogonalize the B-spline $\varphi = \varphi^{B,N}$ in a slightly different way. Note that

$$\sum_{l\in\mathbf{Z}} |\widehat{\varphi}(\xi+l)|^2 = P_N(\cos 2\pi\xi), \tag{1.62}$$

where P_N is a positive polynomial of degree N on the segment $[-1, 1]$. Indeed, it is easy to see that

$$\operatorname{supp} \varphi = \left[\frac{\gamma - N - 1}{2}, \frac{\gamma + N + 1}{2}\right]. \tag{1.63}$$

It remains to note that by the Plancherel formula we have

$$\int_0^1 e^{2\pi ik\xi} \sum_{l\in\mathbb{Z}} |\widehat{\varphi}(\xi+l)|^2 \, d\xi = \int_{\mathbb{R}} e^{2\pi ik\xi} |\widehat{\varphi}(\xi)|^2 \, d\xi = \int_{\mathbb{R}} \varphi(t)\varphi(t+k) \, dt.$$

Taking into account the equalities $\langle \varphi_{0k}, \varphi_{00}\rangle = \langle \varphi_{0,-k}, \varphi_{00}\rangle$ and (1.63), we obtain (1.62).

In view of the Riesz lemma (see Appendix A.11), we have

$$P_N(\cos 2\pi\xi) = a_N |(1 + z_1 e^{2\pi i\xi}) \cdots (1 + z_N e^{2\pi i\xi})|^2,$$

where $\{z_l\}_{l=1}^N$ can always be chosen inside the unit circle: $|z_l| < 1$. Moreover, according to Schoenberg's results [**216**], $\{z_l\}_{l=1}^N$ are positive numbers. Denote them by $s_N < s_{N-1} < \cdots < s_1$. Set

$$A_N(\xi) = \sqrt{a_N}(1 + s_1 e^{2\pi i\xi}) \cdots (1 + s_N e^{2\pi i\xi}).$$

Then the function φ^{St} defined by the equality

$$\widehat{\varphi^{St}}(\xi) = \widehat{\varphi^{St,N}}(\xi) = \frac{\widehat{\varphi^{B,N}}(\xi)}{A_N(\xi)} \tag{1.64}$$

satisfies (1.17); hence its integer shifts form an orthogonal system. The wavelet system corresponding to this function is called the *Strömberg wavelet system*. Since

$$(1 + s\, e^{2\pi i \xi})^{-1} = \sum_{l=0}^{\infty} (-s\, e^{2\pi i \xi})^l,$$

equality (1.64) means that φ^{St} can be expanded into a series with respect to the left integer shifts of the B-spline $\varphi^{B,N}$, where the coefficients of the series decrease as a geometric progression. Therefore supp $\varphi^{St} = (-\infty, \frac{\gamma+N+1}{2}]$ and $\varphi^{St}(t)$ decreases exponentially as $t \to -\infty$.

Note that similar orthonormal bases were constructed by Schauder and Chiselsky shortly after Haar's discoveries (as in the case of the Haar system, they were considered for the space $L_2([0,1])$). The main idea of this construction was multiple integration of the Haar wavelet functions in order to obtain basis elements of the required smoothness and apply the Gram-Schmidt orthogonalization procedure. However this procedure breaks the simple algebraic structure of the Haar wavelet basis, which considerably complicates the study of such systems. What is even more important, this complicates both the computation of Fourier coefficients for functions and the inverse process of reconstructing them by their Fourier coefficients.

The Battle-Lemarie and Strömberg wavelets are called spline wavelets.

3. Meyer wavelets. Let $\varepsilon \in [0, 1/6]$. Define the function $\varphi = \varphi^M$ by the equality

$$\widehat{\varphi}(\xi) = \begin{cases} 1, & \text{if } |\xi| < \frac{1}{2} - \varepsilon, \\ \cos\left(\frac{\pi}{2}\nu\left(\frac{1}{2\varepsilon}|\xi| + \frac{1}{2} - \frac{1}{4\varepsilon}\right)\right), & \text{if } \frac{1}{2} - \varepsilon \le |\xi| \le \frac{1}{2} + \varepsilon, \\ 0, & \text{if } |\xi| > \frac{1}{2} + \varepsilon, \end{cases}$$

where ν is a function defined on $\mathbb{R}$ such that $\nu(\xi) \ge 0$, $\nu(\xi) = 0$ for $\xi \le 0$, $\nu(\xi) = 1$ for $\xi \ge 1$,

$$\nu(\xi) + \nu(1 - \xi) = 1 \tag{1.65}$$

for $\xi \in [0, 1]$. Set

$$m_0(\xi) = m^M(\xi) = \sum_{l \in \mathbb{Z}} \widehat{\varphi}(2(\xi + l)). \tag{1.66}$$

Since $\widehat{\varphi}$ is such that

$$\widehat{\varphi}(2\xi) = \widehat{\varphi}(\xi)\widehat{\varphi}(2\xi)$$

for any $\xi \in \mathbb{R}$ and the supports of the functions $\widehat{\varphi}(\cdot/2)$ and $\widehat{\varphi}(\cdot + 2l)$ are disjoint for $l \ne 0$, the following identity holds:

$$m_0(\xi)\widehat{\varphi}(\xi) = \sum_{l \in \mathbb{Z}} \widehat{\varphi}(2\xi + 2l)\widehat{\varphi}(\xi) = \widehat{\varphi}(2\xi).$$

Hence the function $\widehat{\varphi}$ satisfies equation (1.23) and m_0 is its mask. Let us show that integer shifts of the function φ form an orthonormal system. In view of Proposition 1.1.12, it suffices to check that equality (1.17) holds for any $\xi \in \mathbb{R}$. If $|\xi| < \frac{1}{2} - \varepsilon$ and $|\xi| > \frac{1}{2} + \varepsilon$, this is obvious. Let $\xi = \frac{1}{2} + \delta$, $0 \le |\delta| \le \varepsilon$. Then we have

$$\begin{aligned} \sum_{k \in \mathbb{Z}} \widehat{\varphi}^2(\xi + k) &= \widehat{\varphi}^2\left(\frac{1}{2} + \delta\right) + \widehat{\varphi}^2\left(-\frac{1}{2} + \delta\right) \\ &= \cos^2\left(\frac{\pi}{2}\nu\left(\frac{\delta}{4\varepsilon} + \frac{1}{2}\right)\right) + \cos^2\left(\frac{\pi}{2}\nu\left(-\frac{\delta}{4\varepsilon} + \frac{1}{2}\right)\right) \\ &\quad + \cos^2\left(\frac{\pi}{2}\nu\left(\frac{\delta}{4\varepsilon} + \frac{1}{2}\right)\right) + \sin^2\left(\frac{\pi}{2}\left(1 - \nu\left(-\frac{\delta}{4\varepsilon} + \frac{1}{2}\right)\right)\right). \end{aligned}$$

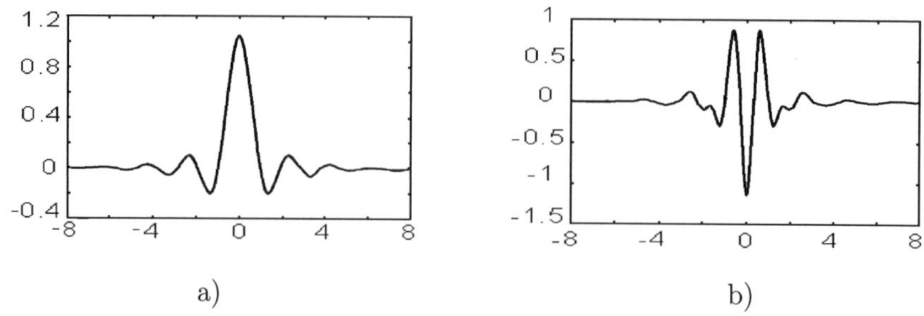

FIGURE 1.2. The Meyer system ($\varepsilon = 1/6$, $\nu(x) = -20x^7 + 70x^6 - 84x^5 + 35x^4$): a) the scaling function φ^M; b) the wavelet function ψ^M.

It remains to note that the right-hand side is identically equal to 1, since in view of (1.65), we have

$$\nu\left(\frac{1}{2} + \frac{\delta}{4\varepsilon}\right) = 1 - \nu\left(\frac{1}{2} - \frac{\delta}{4\varepsilon}\right).$$

Thus φ satisfies all the conditions of Theorem 1.2.14, which means that it is a scaling function for the Meyer MRA generated by it. By Theorem 1.3.1, the corresponding wavelet system forms an orthonormal basis in $L_2(\mathbb{R})$.

Let us consider the limiting case of the Meyer system corresponding to $\varepsilon = 0$, which is an "antipode" of a sort of the Haar system (actually, the functions φ and $\widehat{\varphi}$ switch the roles). In this case, the explicit representation of the scaling function $\varphi = \varphi^S$ is given by simple calculation:

$$\varphi(x) = \frac{\sin \pi x}{\pi x}. \tag{1.67}$$

THEOREM 1.4.1. *Let $\{V_j\}_{j\in\mathbb{Z}}$ be the MRA generated by the scaling function φ defined by equality* (1.67). *Then*

(i) *a function $f \in L_2(\mathbb{R})$ belongs to the space V_j if and only if*

$$\operatorname{supp} \widehat{f} \subset [-2^{j-1}, 2^{j-1}]; \tag{1.68}$$

(ii) *for any element f of the space V_j, the expansion*

$$f(x) = \sum_{n\in\mathbb{Z}} f(-2^{-j}n) \frac{\sin \pi(2^j x + n)}{\pi(2^j x + n)} \tag{1.69}$$

holds, where the series converges in norm in $L_2(\mathbb{R})$ and uniformly on $\mathbb{R}$ (for a continuous representative of the coset of the function f).

Proof. Relation (1.68) is equivalent to the equality

$$\widehat{f}(\xi) = \widehat{f}(\xi)\chi_{[-2^{j-1},2^{j-1}]} = m_f(2^{-j}\xi)\widehat{\varphi}(2^{-j}\xi),$$

where

$$m_f(\xi) = \sum_{n\in\mathbb{Z}} \widehat{f}(2^j(\xi + n)).$$

Combining this equality with Proposition 1.2.3, we get (i).

Let $f \in V_j$; then (1.68) holds by (i). Since $\widehat{f}$ is compactly supported, f is equivalent to some continuous function; moreover, by the Paley-Wiener theorem (see Appendix A.7), it is equivalent to an entire function of exponential type. We shall assume that f is a continuous representative of the coset. By the inversion formula of the Fourier transform, we have

$$f(x) = \int_{-\infty}^{\infty} \widehat{f}(\xi)e^{2\pi i x\xi}\, d\xi = \int_{-2^{j-1}}^{2^{j-1}} \widehat{f}(\xi)e^{2\pi i x\xi}\, d\xi \tag{1.70}$$

for any $x \in \mathbb{R}$. Using the generalized Parseval equality, we get

$$\begin{aligned}\int_{-2^{j-1}}^{2^{j-1}} \widehat{f}(\xi)e^{2\pi i x\xi}\, d\xi &= \sum_{n\in\mathbb{Z}} 2^{-j} \int_{-2^{j-1}}^{2^{j-1}} e^{2\pi i(x+2^{-j}n)\xi}\, d\xi \int_{-2^{j-1}}^{2^{j-1}} \widehat{f}(\xi)e^{-2\pi i 2^{-j}n\xi}\, d\xi \\ &= \sum_{n\in\mathbb{Z}} \frac{\sin\pi(2^j x+n)}{\pi(2^j x+n)} \int_{-2^{j-1}}^{2^{j-1}} \widehat{f}(\xi)e^{-2\pi i 2^{-j}n\xi}\, d\xi.\end{aligned}$$

To prove (1.69) for a fixed $x \in \mathbb{R}$, it remains to substitute this equality into the right-hand side of (1.70) and to note that by the inversion formula we have

$$\int_{-2^{j-1}}^{2^{j-1}} \widehat{f}(\xi)e^{-2\pi i 2^{-j}k\xi}\, d\xi = \int_{-\infty}^{\infty} \widehat{f}(\xi)e^{-2\pi i 2^{-j}k\xi}\, d\xi = f(-2^{-j}k).$$

It follows that the number $f(-2^{-j}k)$ is the k-th Fourier coefficient of the 2^j-periodic function $g(\xi) := 2^j \sum_{k\in\mathbb{Z}} \widehat{f}(\xi + 2^j k)$, which belongs to $L_2(0, 2^j)$. Hence the series $\sum_{n\in\mathbb{Z}} |f(-2^{-j}n)|^2$ converges. Furthermore, it is obvious that the series

$$\sum_{n\in\mathbb{Z}} \left(\frac{\sin\pi(2^j x+n)}{\pi(2^j x+n)}\right)^2$$

converges uniformly on $\mathbb{R}$. Hence, applying the Cauchy inequality to the series on the right-hand side of (1.69), we prove its uniform convergence.

Using the Plancherel theorem and the inversion formula and taking into account the structure of the supports for the functions $\widehat{f}, \widehat{\varphi}$, we get

$$\begin{aligned}\langle f, \varphi_{jn}\rangle &= \int_{-\infty}^{\infty} f(x)\varphi_{jn}(x)\, dx = 2^{-j/2} \int_{-\infty}^{\infty} \widehat{f}(\xi)\widehat{\varphi}(2^{-j}\xi)e^{-2\pi i 2^{-j}n\xi}\, d\xi \\ &= 2^{-j/2} \int_{-2^{j-1}}^{2^{j-1}} \widehat{f}(\xi)e^{-2\pi i 2^{-j}n\xi}\, d\xi = 2^{-j/2} \int_{-\infty}^{\infty} \widehat{f}(\xi)e^{-2\pi i 2^{-j}n\xi}\, d\xi \\ &= 2^{-j/2} f(-2^{-j}n).\end{aligned}$$

Since the functions φ_{jn}, $n \in \mathbb{Z}$, form an orthonormal basis in V_j, it follows that the series on the right-hand side of (1.69) converges in norm in $L_2(\mathbb{R})$. $\Diamond$

The MRA described in Theorem 1.4.1 is usually called the *Shannon-Kotelnikov MRA*. Let us construct the corresponding wavelet system. Note that the 1-periodic function

$$m_0(\xi) = m^S(\xi) = \begin{cases} 1, & \text{if } |\xi| \le 1/4, \\ 0, & \text{if } 1/4 \le |\xi| \le 1/2, \end{cases}$$

is the mask of the scaling function φ^S.

The coefficients $\{h_k\}_{k\in\mathbf{Z}}$ in (1.24) can be found by formula (1.69):

$$\varphi(t) = \sum_{k\in\mathbb{Z}} \varphi(-k/2)\,\varphi(2t+k).$$

Therefore,

$$h_k = h_k^S = \begin{cases} 1, & \text{if } k = 0, \\ \dfrac{2}{k\pi}(-1)^{(k-1)/2}, & \text{if } k \text{ is odd}, \\ 0, & \text{if } k \text{ is even}. \end{cases}$$

In compliance with the general scheme, we find the Fourier transform for the corresponding wavelet function $\psi = \psi^S$:

$$\begin{aligned}\widehat{\psi}(\xi) &= e^{\pi i\xi}\overline{m_0(\xi/2+1/2)}\widehat{\varphi}(\xi/2) = e^{\pi i\xi}\chi_{[-1,\,-1/2]\cup[1/2,\,1]}(\xi) \\ &= e^{\pi i\xi}(\widehat{\varphi}(\xi/2) - \widehat{\varphi}(\xi)).\end{aligned}$$

PROBLEM 1.4.2. Prove that $\psi^S(t) = 2\varphi^S(2t+1) - \varphi^S(t+1/2)$.

1.5. Uncertainty constants

The main feature of wavelet systems is that all elements of such a system are generated by dilations and shifts of one fixed function. A generating wavelet function ψ is a damped oscillation localized in both time and frequency. Dyadic dilations of the function ψ generate a partition of the frequency axis into octaves. Thus a Fourier coefficient $c_{jk} = \langle f, \psi_{jk}\rangle$ of a function f with respect to a wavelet basis provides information about the oscillation of the function f in the frequency band that depends on ψ_{jk}. Oscillations are studied in a certain local domain, where the function ψ_{jk} takes meaningful values.

In this connection an important role is played by the time-frequency localization of the wavelet function ψ. This localization is characterized by the notions of *center* and *radius*. The center $x_f^\star$ and radius Δ_f of a function $f \in L_2(\mathbb{R})$ with nonzero norm are defined by the equalities

$$x_f^\star := \frac{\int_{-\infty}^{\infty} x\,|f(x)|^2\,dx}{\|f\|^2},$$

$$\Delta_f := \frac{1}{\|f\|}\left\{\int_{-\infty}^{\infty} (x - x_f^\star)^2\,|f(x)|^2\,dx\right\}^{1/2}.$$

The center is the mathematical expectation of the random variable with density function $|f|^2/\|f\|^2$; the radius is the mean square deviation of this random variable. Roughly speaking, the function f is localized (i.e., takes meaningful values) on the segment $[x_f^\star - \Delta_f, x_f^\star + \Delta_f]$.

The center $\xi^\star_{\widehat{f}}$ and the radius $\Delta_{\widehat{f}}$ for the Fourier transform of the function f are defined analogously:

$$\xi^\star_{\widehat{f}} := \frac{\int\limits_{-\infty}^{\infty} \xi\, |\widehat{f}(\xi)|^2\, d\xi}{\|\widehat{f}\|^2},$$

$$\Delta_{\widehat{f}} := \frac{1}{\|\widehat{f}\|} \left\{ \int\limits_{-\infty}^{\infty} (\xi - \xi^\star_{\widehat{f}})^2\, |\widehat{f}(\xi)|^2\, d\xi \right\}^{1/2}.$$

The product $\Delta_f\,\Delta_{\widehat{f}}$ characterizes the time-frequency localization of f and is called the *uncertainty constant* of the function f.

PROBLEM 1.5.1. Let x^* and ξ^* be centers of f and $\widehat{f}$, respectively. Prove that the function $g(x) := e^{-2\pi i\xi^* x} f(x + x^*)$ and its Fourier transform have zero centers and equal radii $\Delta_f = \Delta_g$, $\Delta_{\widehat{f}} = \Delta_{\widehat{g}}$.

THEOREM 1.5.2. *For any function $f \in L_2(\mathbb{R})$ with nonzero norm the inequality*

$$\Delta_f \Delta_{\widehat{f}} \geq \frac{1}{4\pi} \approx 0.0795775 \tag{1.71}$$

holds. Moreover, the equality is attained if and only if

$$f(x) = ce^{2\pi i a x} g_\alpha(x - b),$$

where $g_\alpha(x) := \frac{1}{2\sqrt{\alpha}} e^{-\pi x^2/(4\alpha)}$, $c \neq 0,\ \alpha > 0,\ a, b \in \mathbb{R}$.

First we prove an auxiliary statement. Recall that a function $f(x)$ is absolutely continuous if it is differentiable almost everywhere, its derivative is locally summable, and $f(x) - f(y) = \int\limits_y^x f'(t)\, dt$ for any x, y (see Appendix A.10).

LEMMA 1.5.3. *Let f be absolutely continuous. If the functions $(1 + |x|)f(x)$ and $f'(x)$ belong to $L_2(\mathbb{R})$, then*

$$\left| \operatorname{Re} \int\limits_{-\infty}^{\infty} x f(x) \overline{f'(x)}\, dx \right|^2 \leq \left\{ \int\limits_{-\infty}^{\infty} |xf(x)|^2\, dx \right\} \left\{ \int\limits_{-\infty}^{\infty} |f'(x)|^2\, dx \right\}. \tag{1.72}$$

Moreover, the equality takes place if and only if

$$f(x) = c\, e^{-x^2/(4\alpha)},$$

where $c \in \mathbb{R}$, $\alpha > 0$.

Proof. Inequality (1.72) follows from the Cauchy-Bunyakovsky inequality. If in (1.72) we have the equality, then

$$-\operatorname{Re} x f(x) \overline{f'(x)} = |x f(x) f'(x)|, \tag{1.73}$$

$$|x f(x)| = 2\alpha |f'(x)|, \tag{1.74}$$

where α is a positive constant. It is not hard to show that the other case

$$\operatorname{Re} x f(x) \overline{f'(x)} = |x f(x) f'(x)|$$

contradicts the fact that $f \in L_2(\mathbb{R})$. Equality (1.74) implies that

$$xf(x) = 2\alpha f'(x)e^{i\theta(x)},$$

where $\theta(x)$ is a real-valued function. It follows from (1.73) that

$$-xf(x)\overline{f'(x)} \geq 0,$$

which implies that

$$-2\alpha|f'(x)|^2 e^{i\theta(x)} \geq 0;$$

hence $e^{i\theta(x)} = -1$. Finally, we have

$$xf(x) = -2\alpha f'(x),$$

so f' is continuous and

$$f(x) = c\,e^{-x^2/4\alpha}$$

for some constant $c \neq 0$. ◊

`Proof of Theorem 1.5.2.` First of all note that without loss of generality we can assume that the functions f and $\widehat{f}$ have zero centers (see Problem 1.5.1) and $\|f\| = 1$; i.e., the inequality to be proved has the form

$$\int\limits_{\mathbb{R}} x^2|f(x)|^2\,dx \int\limits_{\mathbb{R}} \xi^2|\widehat{f}(\xi)|^2\,d\xi \geq \frac{1}{16\pi^2}. \tag{1.75}$$

Moreover, we can assume that $\Delta_f\Delta_{\widehat{f}} < \infty$ (otherwise the statement is trivial), which implies that each integral in (1.75) is finite. Let us show that the function f is absolutely continuous. Consider the distributional derivative $f' \in \mathcal{S}'$ (see Appendix A.3), for which the equality

$$\widehat{f'}(\xi) = 2\pi i\xi\widehat{f}(\xi)$$

holds (in the sense of distributions). The function on the right-hand side of this equality belongs to $L_2(\mathbb{R})$, hence $\widehat{f'} \in L_2(\mathbb{R})$, and the equality holds almost everywhere on $\mathbb{R}$. Therefore $f' \in L_2(\mathbb{R})$. Hence $f' \in L[a,b]$ for each segment $[a,b]$. Thus the function f' has a primitive f_0 on [a,b] that coincides with f. Thus the function f_0 is absolutely continuous; hence its derivative $f_0' = f'$ coincides with its distributional derivative. So the distributional derivative of the function $f_0 - f$ is equal to zero, and $f_0 - f$ is identically constant on the segment $[a,b]$; hence f is absolutely continuous on this segment. Thus f is absolutely continuous on any segment and hence on the whole $\mathbb{R}$; i.e., f coincides with the primitive of the function f'. We see that the function f is absolutely continuous and its derivative belongs to the space $L_2(\mathbb{R})$. Therefore the function $g(x) := |f(x)|^2$ is also absolutely continuous and its derivative is summable on $\mathbb{R}$. The equality $g(x) - g(0) = \int\limits_0^x g'(t)\,dt$ and summability of g' imply that the right-hand side has a limit as $x \to \pm\infty$, while the summability of g implies that this limit is equal to zero.

It is clear that

$$(\Delta_f \Delta_{\widehat{f}})^2 = \left(\int_{-\infty}^{\infty} x^2 |f(x)|^2 dx\right)\left(\int_{-\infty}^{\infty} \xi^2 |\widehat{f}(\xi)|^2 d\xi\right)$$
$$= \frac{1}{4\pi^2}\left(\int_{-\infty}^{\infty} x^2 |f(x)|^2 dx\right)\left(\int_{-\infty}^{\infty} |\widehat{f'}(\xi)|^2 d\xi\right)$$
$$= \frac{1}{4\pi^2}\left(\int_{-\infty}^{\infty} x^2 |f(x)|^2 dx\right)\|f'\|^2.$$

By Lemma 1.5.3, integrating by parts, we get

$$\left(\int_{-\infty}^{\infty} x^2 |f(x)|^2 dx\right)\|f'\|^2 \geq \left|\operatorname{Re}\int_{-\infty}^{\infty} x f(x)\overline{f'}(x)dx\right|^2$$
$$= \left|\frac{1}{2}\int_{-\infty}^{\infty} x \frac{d}{dx}|f(x)|^2 dx\right|^2 = \left(\frac{1}{2}\int_{-\infty}^{\infty} |f(x)|^2 dx\right)^2 = \frac{1}{4}. \qquad \diamond$$

Note that $\Delta_{\psi_{jk}} = 2^{-j}\Delta_{\psi}$, $\Delta_{\widehat{\psi_{jk}}} = 2^j \Delta_{\widehat{\psi}}$, $j,k \in \mathbf{Z}$. Thus the uncertainty constant is the same for all elements of a wavelet basis.

Let us analyze the examples of wavelet bases considered in §1.4 from the point of view of their time-frequency localization. It is not hard to verify that for the Haar system (see p. 23) $\Delta_{\psi^H} = \sqrt{5/6}$, $\Delta_{\widehat{\psi^H}} = \infty$, which means that the localization is very good in time but very bad in frequency. For the Shannon-Kotelnikov system (see p. 29) the situation is directly opposite: $\Delta_{\psi^S} = \infty$, $\Delta_{\widehat{\psi^S}} = \sqrt{\frac{7}{12}}$. Historically, the first example of orthonormal bases in $L_2(\mathbb{R})$, whose elements have uniformly bounded uncertainty constants, was Meyer's wavelet system (see p. 27) with the sufficiently smooth Fourier transform. It is clear that

$$\Delta_{\psi^M}\Delta_{\widehat{\psi^M}} < \infty$$

for $\widehat{\psi^M} \in C^2(\mathbb{R})$. Meyer's wavelets are very well localized in frequency (their Fourier transform has compact support), and in the case when $\widehat{\psi^M} \in C^\infty(\mathbb{R})$ they are well localized in time (they decrease at $\pm\infty$ more rapidly than any power though not exponentially). On the contrary, Battle-Lemarie's and Strömberg's wavelets (see p. 24) decrease exponentially while their Fourier transforms have polynomial rate of decrease corresponding to finite smoothness.

One can obtain more precise lower estimates of localization characteristics for wavelets (using mutual orthogonality of ψ_{jk}).

THEOREM 1.5.4 ([**189**]). *If ψ is a wavelet function generating an orthonormal wavelet basis, then*

$$\left\{\int_{\mathbb{R}} \xi^2 |\widehat{\psi}(\xi)|^2 \, d\xi \int_{\mathbb{R}} t^2 |\psi(t)|^2 \, dt\right\}^{1/2} \geq \frac{\sqrt{2}}{4(\sqrt{2}-1)\pi} \approx 0.271694.$$

THEOREM 1.5.5 ([**189**]). *If ψ is a wavelet function that has the Fourier transform with zero center and generates an orthonormal wavelet basis, then*

$$\Delta_\psi \Delta_{\widehat{\psi}} \geq \frac{3}{4\pi}.$$

For any normalized function φ we have

$$t^*_\varphi := \int_{\mathbb{R}} t|\varphi(t)|^2\,dt,$$

$$\Delta_\varphi := \left(\int_{\mathbb{R}} (t - t^*_\varphi)^2|\varphi(t)|^2\,dt\right)^{1/2},$$

$$\xi^*_{\widehat{\varphi}} := \int_{\mathbb{R}} \xi|\widehat{\varphi}(\xi)|^2\,d\xi,$$

$$\Delta_{\widehat{\varphi}} := \left(\int_{\mathbb{R}} (\xi - \xi^*_{\widehat{\varphi}})^2|\widehat{\varphi}(\xi)|^2\,d\xi\right)^{1/2}.$$

In what follows, we shall need the following characteristics of the 1-periodic function $g(\xi) := \sum\limits_{l\in\mathbb{Z}} c_l e^{2\pi i l\xi}$. For simplicity, we shall denote these characteristics by the symbols used in the nonperiodic case:

$$t^*_g := \sum_{l\in\mathbb{Z}} l|c_l|^2 \Big/ \sum_{l\in\mathbb{Z}} |c_l|^2,$$

$$\Delta_{ti,g} := \left(\sum_{l\in\mathbb{Z}} (l - t^*_g)^2|c_l|^2 \Big/ \sum_{l\in\mathbb{Z}} |c_l|^2\right)^{1/2},$$

$$\xi^*_g := \int_{-\frac{1}{2}}^{\frac{1}{2}} \xi|g(\xi)|^2\,d\xi \Big/ \int_{-\frac{1}{2}}^{\frac{1}{2}} |g(\xi)|^2\,d\xi,$$

$$\Delta_{fr,g} := \left(\int_{-\frac{1}{2}}^{\frac{1}{2}} (\xi - \xi^*_g)^2|g(\xi)|^2 d\xi \Big/ \int_{-\frac{1}{2}}^{\frac{1}{2}} |g(\xi)|^2 d\xi\right)^{1/2}.$$

For the mask

$$m_0(\xi) = \frac{1}{\sqrt{2}}\sum_{l\in\mathbb{Z}} h_l e^{2\pi i l\xi}$$

of a refinable function with orthonormal integer shifts (see (1.23) and (1.24)) we have

$$\sum_{l\in\mathbb{Z}} |h_l|^2 = 1. \tag{1.76}$$

Hence,

$$t^*_{m_0} = \sum_{l\in\mathbb{Z}} l|h_l|^2,$$

$$\Delta_{ti,m_0} = \left(\sum_{l\in\mathbb{Z}}(l - t^*_{m_0})^2|h_l|^2\right)^{1/2},$$

$$\xi^*_{m_0} = 2\int_{-\frac{1}{2}}^{\frac{1}{2}} \xi|m_0(\xi)|^2\, d\xi,$$

$$\Delta_{fr,m_0} = \left(2\int_{-\frac{1}{2}}^{\frac{1}{2}} (\xi - \xi^*_{m_0})^2|m_0(\xi)|^2 d\xi\right)^{1/2}.$$

If m_0 is a differentiable function and

$$m_0'(\xi) = i\sqrt{2}\pi\sum_{l\in\mathbb{Z}} l\, h_l\, e^{2\pi il\xi}, \tag{1.77}$$

then it is clear that

$$t^*_{m_0} = \frac{1}{i\pi}\int_{-\frac{1}{2}}^{\frac{1}{2}} m_0'(\xi)\overline{m_0(\xi)}\, d\xi,$$

$$\Delta_{ti,m_0} = \frac{1}{\sqrt{2}\pi}\left(\int_{-\frac{1}{2}}^{\frac{1}{2}} |m_0'(\xi)|^2\, d\xi - (t^*_{m_0})^2\right)^{1/2}.$$

Let us now find the connection between the radii of a scaling function and auxiliary characteristics of its mask.

LEMMA 1.5.6. *Let φ be a refinable function with mask m_0 and orthonormal integer shifts. If $\xi^*_{m_0} = \xi^*_{\widehat{\varphi}} = 0$, then*

$$\Delta_{fr,m_0} \le \frac{\Delta_{\widehat{\varphi}}}{2}.$$

Proof. It is clear that

$$\Delta^2_{\widehat{\varphi}} = \int_{\mathbb{R}} \xi^2|\widehat{\varphi}(\xi)|^2\, d\xi = \int_{-\frac{1}{2}}^{\frac{1}{2}} \sum_{l\in\mathbb{Z}}(\xi + l)^2|\widehat{\varphi}(\xi + l)|^2\, d\xi.$$

Let

$$P(\xi) := \sum_{l\in\mathbb{Z}}(\xi + l)^2|\widehat{\varphi}(\xi + l)|^2.$$

As $\widehat{\varphi}(\xi) = m_0(\xi/2)\widehat{\varphi}(\xi/2)$ (see (1.23)), we have

$$P(2\xi) = 4\left(|m_0(\xi)|^2P(\xi) + |m_0(\xi + 1/2)|^2P(\xi + 1/2)\right). \tag{1.78}$$

Hence,

$$\int_{-\frac{1}{2}}^{\frac{1}{2}} P(\xi)\,d\xi = \int_{-\frac{1}{2}}^{\frac{1}{2}} P(2\xi)\,d\xi$$

$$= 4\int_{-\frac{1}{2}}^{\frac{1}{2}} \left(|m_0(\xi)|^2 P(\xi) + |m_0(\xi+1/2)|^2 P(\xi+1/2)\right) d\xi \tag{1.79}$$

$$= 8\int_{-\frac{1}{2}}^{\frac{1}{2}} \xi^2 |m_0(\xi)|^2 \frac{P(\xi)}{\xi^2}\,d\xi \geq 8\left(\inf_{\xi\in[-\frac{1}{2},\frac{1}{2}]} \frac{P(\xi)}{\xi^2}\right)\int_{-\frac{1}{2}}^{\frac{1}{2}} \xi^2 |m_0(\xi)|\,d\xi.$$

Since for any $l \in \mathbb{Z}$, $|\xi + l| \geq |\xi|$ for $\xi \in [-\frac{1}{2}, \frac{1}{2}]$, then by (1.17)

$$\inf_{\xi\in[-\frac{1}{2},\frac{1}{2}]} \frac{P(\xi)}{\xi^2} \geq 1.$$

Finally, we get

$$\Delta^2_{\widehat{\varphi}} \geq 4\Delta^2_{fr,m_0}. \qquad \Diamond$$

REMARK 1.5.7. If φ is a real-valued refinable function, then $\{h_k\}_{k\in\mathbb{Z}}$ are real numbers; hence $m_0(\xi) = \overline{m_0(-\xi)}$ and $|m_0(\xi)|$ is an even function as well as $|\widehat{\varphi}(\xi)|$. Obviously, in this case $\xi^*_{fr,m_0} = \xi^*_{\widehat{\varphi}} = 0$.

REMARK 1.5.8. Let φ be a refinable function with mask m_0 and orthonormal integer shifts. If $\xi^*_{fr,m_0} = \xi^*_{\widehat{\varphi}} = 0$, then it follows from (1.79) that

$$\Delta_{\widehat{\varphi}} \leq 2\sqrt{2}\left(\sup_{\xi\in[-\frac{1}{2},\frac{1}{2}]} \frac{P(\xi)}{\xi^2}\right)^{1/2} \Delta_{fr,m_0}.$$

LEMMA 1.5.9. *Suppose a refinable function φ with orthonormal integer shifts is defined by the equality*

$$\widehat{\varphi}(\xi) = \prod_{l=1}^{\infty} m_0(2^{-l}\xi) \tag{1.80}$$

and the infinite product admits termwise differentiation in the sense of almost everywhere convergence. Then

$$\left(\int_{\mathbb{R}} t^2|\varphi(t)|^2\,dt\right)^{1/2} \leq \frac{2\pi}{\sqrt{2}-1}\left(\int_{-\frac{1}{2}}^{\frac{1}{2}} |m_0'(\xi)|^2\,d\xi\right)^{1/2}.$$

Proof. Since $\widehat{\cdot\varphi(\cdot)}(\xi) = -2\pi i\widehat{\varphi}'(\xi)$, then in view of Plancherel's identity we have

$$\int_{\mathbb{R}} t^2|\varphi(t)|^2\,dt = 4\pi^2\int_{\mathbb{R}} |\widehat{\varphi}'(\xi)|^2 d\xi.$$

It follows from (1.80) that

$$\widehat{\varphi}'(\xi) = \left(\sum_{l\in\mathbb{N}} 2^{-l}\frac{m_0'(2^{-l}\xi)}{m_0(2^{-l}\xi)}\right)\prod_{p=1}^{\infty} m_0(2^{-p}\xi).$$

By (1.27), $|m_0(\xi)| \le 1$; therefore,

$$|\widehat{\varphi}'(\xi)| \le \sum_{l\in\mathbb{N}} 2^{-l}|m_0'(2^{-l}\xi)|\cdot|\prod_{p=l+1}^{\infty} m_0(2^{-p}\xi)|$$

and

$$\begin{aligned}\left(\int_{\mathbb{R}} |\widehat{\varphi}'(\xi)|^2\,d\xi\right)^{1/2} &\le \sum_{l\in\mathbb{N}} 2^{-l}\left(\int_{\mathbb{R}} |m_0'(2^{-l}\xi)|^2\cdot|\prod_{p=l+1}^{\infty} m_0(2^{-p}\xi)|^2 d\xi\right)^{1/2}\\ &= \sum_{l\in\mathbb{N}} 2^{-l/2}\left(\int_{\mathbb{R}} |m_0'(\xi)|^2\cdot|\prod_{p=1}^{\infty} m_0(2^{-p}\xi)|^2 d\xi\right)^{1/2}\\ &= \frac{1}{\sqrt{2}-1}\left(\int_{\mathbb{R}} |m_0'(\xi)|^2\cdot|\prod_{p=1}^{\infty} m_0(2^{-p}\xi)|^2 d\xi\right)^{1/2}.\end{aligned}$$

Taking into account (1.17), we get

$$\int_{\mathbb{R}} |m_0'(\xi)|^2|\prod_{p=1}^{\infty} m_0(2^{-p}\xi)|^2 d\xi = \int_{-\frac{1}{2}}^{\frac{1}{2}} |m_0'(\xi)|^2 \sum_{l\in\mathbb{Z}} |\widehat{\varphi}(\xi+l)|^2\,d\xi = \int_{-\frac{1}{2}}^{\frac{1}{2}} |m_0'(\xi)|^2 d\xi. \quad \diamond$$

In the following lemmas both the Fourier transform of a refinable function and its mask are supposed to be differentiable.

LEMMA 1.5.10. *Suppose a refinable function φ with orthonormal integer shifts is defined by the equality*

$$\widehat{\varphi}(\xi) = \prod_{l=1}^{\infty} m_0(2^{-l}\xi)$$

and the series $\sum_{l\in\mathbb{Z}} |\widehat{\varphi}(\xi+l)|^2$ admits termwise differentiation in the sense of almost everywhere convergence. Then

$$\pi\left(2\int_{-\frac{1}{2}}^{\frac{1}{2}} |m_0'(\xi)|^2\,d\xi\right)^{1/2} \le \left(\int_{\mathbb{R}} t^2|\varphi(t)|^2\,dt\right)^{1/2} \le 2\pi\left(\int_{-\frac{1}{2}}^{\frac{1}{2}} |m_0'(\xi)|^2\,d\xi\right)^{1/2}.$$

Proof. Using (1.23) and (1.17), we get

$$\begin{aligned}\int_{\mathbb{R}} |\widehat{\varphi}'(\xi)|^2 d\xi &= \int_{\mathbb{R}} \left|\frac{1}{2}m_0'(\xi/2)\widehat{\varphi}(\xi/2) + \frac{1}{2}m_0(\xi/2)\widehat{\varphi}'(\xi/2)\right|^2 d\xi \\ &= \frac{1}{4}\int_{\mathbb{R}} |m_0'(\xi/2)\widehat{\varphi}(\xi/2)|^2 d\xi + \frac{1}{4}\int_{\mathbb{R}} |m_0(\xi/2)\widehat{\varphi}'(\xi/2)|^2 d\xi \\ &+ \frac{1}{2}\int_{\mathbb{R}} \mathrm{Re}\left(m_0(\xi/2)\overline{m_0'(\xi/2)\widehat{\varphi}(\xi/2)}\widehat{\varphi}'(\xi/2)\right) d\xi \\ &= \frac{1}{2}\int_{\mathbb{R}} |m_0'(\xi)\widehat{\varphi}(\xi)|^2 d\xi + \frac{1}{2}\int_{\mathbb{R}} |m_0(\xi)\widehat{\varphi}'(\xi)|^2 d\xi \\ &+ \int_{\mathbb{R}} \mathrm{Re}\left(m_0(\xi)\overline{m_0'(\xi)\widehat{\varphi}(\xi)}\widehat{\varphi}'(\xi)\right) d\xi \\ &= \frac{1}{2}\int_{-\frac{1}{2}}^{\frac{1}{2}} |m_0'(\xi)|^2 d\xi + \frac{1}{2}\int_{\mathbb{R}} |m_0(\xi)\widehat{\varphi}'(\xi)|^2 d\xi \\ &+ \int_{-\frac{1}{2}}^{\frac{1}{2}} \mathrm{Re}\left(m_0(\xi)\overline{m_0'(\xi)}\right)\left(\sum_{l\in\mathbb{Z}} \mathrm{Re}\left(\widehat{\varphi}'(\xi+l)\overline{\widehat{\varphi}(\xi+l)}\right)\right) d\xi.\end{aligned}$$

In view of the assumption of the lemma and (1.17), we have

$$\sum_{l\in\mathbb{Z}} \left(|\widehat{\varphi}(\xi+l)|^2\right)' = 2\sum_{l\in\mathbb{Z}} \mathrm{Re}\left(\widehat{\varphi}'(\xi+l)\overline{\widehat{\varphi}(\xi+l)}\right) = 0.$$

Hence,

$$\int_{\mathbb{R}} |\widehat{\varphi}'(\xi)|^2 d\xi = \frac{1}{2}\int_{-\frac{1}{2}}^{\frac{1}{2}} |m'(\xi)|^2 d\xi + \frac{1}{2}\int_{\mathbb{R}} |m(\xi)\widehat{\varphi}'(\xi)|^2 d\xi.$$

Since $|m(\xi)| \leq 1$ (see (1.27)), the latter equality implies the statement of the lemma. ◊

Under some additional assumptions on the refinable mask one can obtain another lower estimate for $\left(\int_{\mathbb{R}} t^2|\varphi(t)|^2\, dt\right)^{1/2}$.

LEMMA 1.5.11. *Suppose a refinable function φ with orthonormal shifts is defined by the equality*

$$\widehat{\varphi}(\xi) = \prod_{l=1}^{\infty} m_0(2^{-l}\xi)$$

and the modulus of its mask is an even function decreasing on $[0, \frac{1}{2}]$. *Then*

$$\left(\int_{\mathbb{R}} t^2|\varphi(t)|^2\,dt\right)^{1/2} \geq \sqrt{2}\pi\left|\widehat{\varphi}\left(\frac{1}{2}\right)\right|\left(\int_{-\frac{1}{2}}^{\frac{1}{2}} \left(|m_0(\xi)|'\right)^2\,d\xi\right)^{1/2}.$$

Proof. It is easy to see that $||\widehat{\varphi}(\xi)|'| \leq |\widehat{\varphi}'(\xi)|$. Therefore, using (1.23), we get

$$\begin{aligned}
\int_{\mathbb{R}} |\widehat{\varphi}'(\xi)|^2 d\xi &\geq \int_{-1}^{1} |\widehat{\varphi}'(\xi)|^2 d\xi \geq \int_{-1}^{1} \left(|\widehat{\varphi}|'(\xi)\right)^2 d\xi \\
&= \int_{-1}^{1} \left(\frac{1}{2}|m_0(\xi/2)|'|\widehat{\varphi}(\xi/2)| + \frac{1}{2}|m_0(\xi/2)||\widehat{\varphi}(\xi/2)|'\right)^2 d\xi \\
&= \frac{1}{2}\int_{-\frac{1}{2}}^{\frac{1}{2}} \left(|m_0(\xi)|'|\widehat{\varphi}(\xi)| + |m_0(\xi)||\widehat{\varphi}(\xi)|'\right)^2 d\xi.
\end{aligned}$$

By assumption, $|m_0(\xi)|$ is an even function and $|m_0(\xi)|' \leq 0$ for $\xi \in [0, \frac{1}{2}]$. Then $|\widehat{\varphi}(\xi)|' \leq 0$ for $\xi \in [0, \frac{1}{2}]$ and

$$\begin{aligned}
&\int_{-\frac{1}{2}}^{\frac{1}{2}} \left(|m_0(\xi)|' \cdot |\widehat{\varphi}(\xi)| + |m_0(\xi)| \cdot |\widehat{\varphi}(\xi)|'\right)^2 d\xi \\
&= 2\int_{0}^{\frac{1}{2}} \left(|m_0(\xi)|' \cdot |\widehat{\varphi}(\xi)| + |m_0(\xi)| \cdot |\widehat{\varphi}(\xi)|'\right)^2 d\xi \\
&\geq 2\int_{0}^{\frac{1}{2}} \left(|m_0(\xi)|' \cdot |\widehat{\varphi}(\xi)|\right)^2 d\xi \geq |\widehat{\varphi}(1/2)|^2 \int_{-\frac{1}{2}}^{\frac{1}{2}} \left(|m_0(\xi)|'\right)^2 d\xi.
\end{aligned}$$

◊

LEMMA 1.5.12. *Let* φ *be a refinable function with orthonormal integer shifts. Then*

$$\int_{-\frac{1}{2}}^{\frac{1}{2}} m_0'(\xi)\overline{m_0(\xi)}\,d\xi = \int_{\mathbb{R}} |m_0(\xi + 1/2)|^2\widehat{\varphi}'(\xi)\overline{\widehat{\varphi}(\xi)}\,d\xi.$$

Proof. It follows from (1.23) and (1.17) that

$$\begin{aligned}
&\int_{\mathbb{R}} \widehat{\varphi}'(\xi)\overline{\widehat{\varphi}(\xi)}\,d\xi \\
&= \int_{\mathbb{R}} \frac{1}{2}\left[m_0'(\xi/2)\widehat{\varphi}(\xi/2) + m_0(\xi/2)\widehat{\varphi}'(\xi/2)\right]\overline{m_0(\xi/2)\widehat{\varphi}(\xi/2)}\,d\xi \\
&= \int_{\mathbb{R}} m_0'(\xi)\overline{m_0(\xi)}|\widehat{\varphi}(\xi)|^2 d\xi + \int_{\mathbb{R}} |m_0(\xi)|^2\widehat{\varphi}'(\xi)\overline{\widehat{\varphi}(\xi)}d\xi
\end{aligned}$$

$$= \int_{\frac{1}{2}}^{\frac{1}{2}} m'(\xi)\overline{m_0(\xi)}\, d\xi + \int_{\mathbb{R}} |m_0(\xi)|^2 \widehat{\varphi}'(\xi)\overline{\widehat{\varphi}(\xi)} d\xi.$$

The statement of the lemma follows from this equality and (1.27). ◊

1.6. Computational algorithms

To use wavelet expansions in practice, we need to compute coefficients of these expansions. One of the advantages of MRA structure is that coefficients of wavelet expansions can be computed recursively. Suppose that we want to expand a function $f \in L_2(\mathbb{R})$ with respect to orthonormal wavelet systems ψ_{jn}, $\widetilde{\psi}_{jn}$ generated by scaling functions φ, $\widetilde{\varphi}$, respectively. By Theorem 1.3.10,

$$f = \sum_{n\in\mathbb{Z}} \langle f, \widetilde{\varphi}_{0n}\rangle \varphi_{0n} + \sum_{j=0}^{\infty} \sum_{n\in\mathbb{Z}} \langle f, \widetilde{\psi}_{jn}\rangle \psi_{jn}.$$

In applications, one can compute only a finite number of coefficients, so we choose some positive integer j_0 and replace the function f by the element

$$\widetilde{f} = \sum_{n\in\mathbb{Z}} \langle f, \widetilde{\varphi}_{0n}\rangle \varphi_{0n} + \sum_{j=0}^{j_0} \sum_{n\in\mathbb{Z}} \langle f, \widetilde{\psi}_{jn}\rangle \psi_{jn} \tag{1.81}$$

which is sufficiently close to f in $L_2(\mathbb{R})$. Note that in view of Corollary 1.3.9, we have

$$\langle f, \widetilde{\varphi}_{0n}\rangle = \langle \widetilde{f}, \widetilde{\varphi}_{0n}\rangle, \quad \langle f, \widetilde{\psi}_{jn}\rangle = \langle \widetilde{f}, \widetilde{\psi}_{jn}\rangle \tag{1.82}$$

for all $n \in \mathbb{Z}$, $j = 0, \ldots, j_0$. Set

$$c_{jn} = \langle f, \widetilde{\varphi}_{jn}\rangle, \quad d_{jn} = \langle f, \widetilde{\psi}_{jn}\rangle.$$

Our aim is to find the coefficients of expansion (1.81), i.e., to compute c_{0n}, d_{jn} for all $n \in \mathbb{Z}$, $j = 0, \ldots, j_0$. To do this, we find $c_{j_0 n}$, $n \in \mathbb{Z}$, by the formulas

$$c_{j_0 n} = \langle \widetilde{f}, \widetilde{\varphi}_{j_0 n}\rangle = 2^{j_0/2} \int_{-\infty}^{\infty} f(t)\overline{\widetilde{\varphi}(2^{j_0}t + n)}\, dt 2^{-j_0/2} = \int_{-\infty}^{\infty} f(t-2^{-j_0}n) 2^{j_0} \overline{\widetilde{\varphi}(2^{j_0}t)}\, dt.$$

If the function $\widetilde{\varphi}$ is well localized (e.g., compactly supported), then for sufficiently large j_0 the function $2^{j_0}\widetilde{\varphi}(2^{j_0}t)$ is close to the δ function, and the corresponding integral on the right-hand side is close to $f(-2^{-j_0}n)$; i.e., the following approximate equality holds:

$$c_{j_0 n} \approx 2^{-j_0/2} f(-2^{-j_0}n).$$

We shall consider these quantities as initial data. Moreover, in applications we do not usually have complete information of the function considered. As a rule, it is defined by a set of values at nodes of some net, which is usually taken to be uniform. The points $2^{-j_0}n$, $n \in \mathbb{Z}$, are well suited as nodes of such a net. Thus the function f can be identified with the sequence $\{c_{j_0 n}\}_{n\in\mathbb{Z}}$. Suppose the functions φ, $\widetilde{\varphi}$ satisfy the refinement equations

$$\varphi = \sum_{k\in\mathbb{Z}} h_k \varphi_{1k}, \quad \widetilde{\varphi} = \sum_{k\in\mathbb{Z}} \widetilde{h}_k \widetilde{\varphi}_{1k}.$$

By Theorem 1.3.7, the corresponding wavelet functions are determined by the equalities

$$\psi = \sum_{k\in\mathbb{Z}} (-1)^{1-k}\overline{\widetilde{h}_{1-k}}\varphi_{1k}, \quad \widetilde{\psi} = \sum_{k\in\mathbb{Z}} (-1)^{1-k}\overline{h_{1-k}}\widetilde{\varphi}_{1k}.$$

By the dilation by 2^j and shift by n, transform these equalities to the form

$$\varphi_{jn} = \sum_{k\in\mathbb{Z}} h_{k-2n}\varphi_{j+1,k}, \quad \psi_{jn} = \sum_{k\in\mathbb{Z}} (-1)^{1-k}\overline{\widetilde{h}_{1-k+2n}}\varphi_{j+1,k}, \tag{1.83}$$

$$\widetilde{\varphi}_{jn} = \sum_{k\in\mathbb{Z}} \widetilde{h}_{k-2n}\widetilde{\varphi}_{j+1,k}, \quad \widetilde{\psi}_{jn} = \sum_{k\in\mathbb{Z}} (-1)^{1-k}\overline{h_{1-k+2n}}\widetilde{\varphi}_{j+1,k}. \tag{1.84}$$

Note that in the case of compactly supported φ, $\widetilde{\varphi}$ only finitely many coefficients $h_k, \widetilde{h}_k$ are nonzero; i.e., the sums in (1.83), (1.84) are finite. Multiplying scalarly both left and right sides of (1.84) by f, we get

$$c_{jn} = \sum_{k\in\mathbb{Z}} \widetilde{h}_{k-2n}c_{j+1,k}, \quad d_{jn} = \sum_{k\in\mathbb{Z}} (-1)^{1-k}\overline{h_{1-k+2n}}c_{j+1,k}.$$

If $c_{j+1,k}$, $k \in \mathbb{Z}$, are known, then using these formulas, we can easily find the coefficients $c_{j,k}$, $d_{j,k}$, $k \in \mathbb{Z}$. In the case of compactly supported φ, $\widetilde{\varphi}$, we only have to compute finite sums with a fixed number of summands for all j. Thus we have described one step of the recursive algorithm. Starting with the initial data $c_{j_0,k}$, $k \in \mathbb{Z}$, and making j_0 steps, we find all the coefficients. This algorithm is called a cascade algorithm.

Let us now consider the inverse problem: to reconstruct a function f by the coefficients of its wavelet expansion. As before, we identify the function f with the sequence $\{c_{j_0n}\}_{n\in\mathbb{Z}}$. Thus we have to find the elements of this sequence for the initial data $c_{0,k}$, $d_{j,k}$, $k \in \mathbb{Z}$, $j = 0, \dots, j_0$. Let $j < j_0$. Using the above notation for $\widetilde{f}$, we rewrite (1.60) in the form

$$\widetilde{f} = \sum_{k\in\mathbb{Z}} c_{jk}\varphi_{jk} + \sum_{i=j}^{j_0}\sum_{k\in\mathbb{Z}} d_{ik}\psi_{ik}.$$

Taking the scalar product of both sides of this equality and $\widetilde{\varphi}_{j+1,n}$, in view of Corollary 1.3.9 and (1.83), we get

$$c_{j+1,n} = \sum_{k\in\mathbb{Z}} c_{jk}\langle\varphi_{jk}, \widetilde{\varphi}_{j+1,n}\rangle + \sum_{k\in\mathbb{Z}} d_{jk}\langle\psi_{jk}, \widetilde{\varphi}_{j+1,n}\rangle.$$

Using (1.83), in view of the orthogonality of the systems $\{\varphi_{j+1,k}\}_{k\in\mathbb{Z}}$, $\{\widetilde{\varphi}_{j+1,n}\}_{k\in\mathbb{Z}}$, we obtain

$$\begin{aligned} c_{j+1,n} &= \sum_{k\in\mathbb{Z}} c_{jk}\left\langle \sum_{l\in\mathbb{Z}} h_{l-2k}\varphi_{j+1,l}, \widetilde{\varphi}_{j+1,n}\right\rangle \\ &\quad + \sum_{k\in\mathbb{Z}} d_{jk}\left\langle \sum_{l\in\mathbb{Z}} (-1)^{1-l}\overline{\widetilde{h}_{1-l+2k}}\varphi_{j+1,l}, \widetilde{\varphi}_{j+1,n}\right\rangle \\ &= \sum_{k\in\mathbb{Z}} c_{jk}h_{n-2k} + (-1)^{1-n}\sum_{k\in\mathbb{Z}} d_{jk}\overline{\widetilde{h}_{1-n+2k}}. \end{aligned}$$

By these formulas, given coefficients $c_{j,k}$, $d_{j,k}$, $k \in \mathbb{Z}$, we can compute $c_{j+1,n}$, $n \in \mathbb{Z}$. So we have described one step of the recursive algorithm. Starting from $j = 0$ and making j_0 steps, we find the sequence $\{c_{j_0n}\}_{n\in\mathbb{Z}}$. As before, in the case of compactly supported functions φ, $\widetilde{\varphi}$, we have to compute only finite sums with a fixed number of summands.

1.7. Convergence of wavelet expansions

Theorem 1.3.10 proves that any function $f \in L_2(\mathbb{R})$ can be expanded with respect to a system of biorthogonal wavelets $\psi_{jk}, \widetilde{\psi}_{jk}$, and partial sums of this expansion can be represented in the form

$$\sum_{n\in\mathbb{Z}} \langle f, \widetilde{\varphi}_{jn}\rangle \varphi_{jn} = P_j f,$$

where φ, $\widetilde{\varphi}$ are scaling functions generating the corresponding MRAs, P_j is the projection operator onto the space V_j. Thus the sequence $\{P_j(f)\}_{j\in\mathbb{Z}}$ converges to f in L_2-norm as $j \to \infty$. In this section, we study some other types of convergence of such a sequence. We shall impose additional conditions on the functions $\varphi, \widetilde{\varphi}$ to provide for their belonging to the spaces $L_p(\mathbb{R})$, $1 \le p \le \infty$. Thus we extend the class of functions f, for which the scalar products $\langle f, \widetilde{\varphi}_{jn}\rangle$ make sense and hence the operator $P_j f$ is defined.

LEMMA 1.7.1. *Let η be an even, bounded, and summable function decreasing on $[0, \infty)$. Then there exists a constant C depending only on η and such that*

$$\sum_{k\in\mathbb{Z}} \eta(x+k)\eta(y+k) \le C\eta\left(\frac{x-y}{8}\right) \tag{1.85}$$

for all $x, y \in \mathbb{R}$.

Proof. First of all we note that both left- and right-hand sides of inequality (1.85) are invariant under the operation $(x, y) \longrightarrow (x+t, y+t)$; hence we can assume that $|x| \le 1/2$. Let $|y| \le 2$. Then we have

$$\eta\left(\frac{x-y}{8}\right) \ge \eta\left(\frac{5}{16}\right) =: C_1.$$

If $C_1 = 0$, then $\eta(t) = 0$ for all $t \ge 5/16$; hence

$$\sum_{k\in\mathbb{Z}} \eta(x+k)\eta(y+k) = \eta(x)\eta(y) \le C_2\eta\left(\frac{x-y}{8}\right).$$

If $C_1 > 0$, then

$$\sum_{k\in\mathbb{Z}} \eta(x+k) \le 2\left(\eta(0) + \int_0^\infty \eta(x)\,dx\right) =: C_3. \tag{1.86}$$

Suppose now that $|y| \ge 2$. In this case $|x-y| \le 2|y|$, and (1.86) implies that

$$\sum_{|k|\le|y|/2} \eta(x+k)\eta(y+k) \le \eta\left(\frac{y}{2}\right) \sum_{k\in\mathbb{Z}} \eta(x+k) \le C_3\eta\left(\frac{x-y}{4}\right),$$

$$\begin{aligned}\sum_{|k|\ge|y|/2} \eta(x+k)\eta(y+k) &\le \sum_{|k|\ge|y|/2} \eta\left(\frac{|k|}{2}\right)\eta(y+k)\\ &\le C_3\eta\left(\frac{y}{4}\right) \le C_3\eta\left(\frac{x-y}{8}\right).\end{aligned}$$

◊

COROLLARY 1.7.2. *Suppose* η *satisfies the conditions of Lemma* 1.7.1 *and the inequality*

$$|\varphi(x)|, |\widetilde{\varphi}(x)| \le \eta(x) \tag{1.87}$$

holds for functions $\varphi, \widetilde{\varphi}$ *for almost all* $x \in \mathbb{R}$. *Then* P_j *is a continuous operator from* $L_p(\mathbb{R})$ *to* $L_p(\mathbb{R})$, $1 \le p \le \infty$, *and*

$$\|P_j\| \le A, \tag{1.88}$$

where A *is a constant that depends only on* η, *and*

$$P_j f(x) = \int_{-\infty}^{\infty} f(y) \sum_{k\in\mathbb{Z}} \varphi_{jk}(x)\overline{\widetilde{\varphi}_{jk}(y)}\, dy \tag{1.89}$$

for any function $f \in L_p(\mathbb{R})$, $1 \le p \le \infty$, *and for almost all* $x \in \mathbb{R}$. *If* $|\varphi(x)| \le \eta(x)$ *for any* $x \in \mathbb{R}$, *then equality* (1.89) *holds for all* $x \in \mathbb{R}$.

Proof. Let $|\varphi(2^j x + k)| \le \eta(2^j x + k)$ for all $k \in \mathbb{Z}$; then, by Lemma 1.7.1, for almost all y partial sums of the series

$$\sum_{k\in\mathbb{Z}} |\varphi(2^j x + k)\overline{\widetilde{\varphi}(2^j y + k)}|$$

the partial sums of the series are majorized by the function $C\eta(2^{j-3}(x-y))$, which (as a function of y) belongs to any $L_q(\mathbb{R})$, $1 \le q \le \infty$. Thus, by Lebesgue's dominated convergence theorem, equality (1.89) holds and we have

$$\begin{aligned}\|P_j f\|_p &\le C \left\| 2^j \int_{-\infty}^{\infty} f(y)\eta(2^{j-3}(\cdot - y))\, dy \right\|_p \\ &= C \left\| 2^j \int_{-\infty}^{\infty} f(\cdot - y)\eta(2^{j-3} y)\, dy \right\|_p \le 8C\|f\|_p \int_{-\infty}^{\infty} \eta(y)\, dy. \qquad \diamond\end{aligned}$$

Note that in all the examples of orthogonal wavelet systems presented in §1.4, except for the Shannon-Kotelnikov system, the generating functions φ satisfy the conditions of Corollary 1.7.2.

In what follows, we shall consider functions $\varphi, \widetilde{\varphi}$ that satisfy the conditions of Theorem 1.3.10. They will be called scaling functions generating a biorthogonal wavelet system (according to the statement of this theorem).

LEMMA 1.7.3. *Let* $\varphi, \widetilde{\varphi}$ *be scaling functions generating a biorthogonal wavelet system and satisfying the conditions of Corollary* 1.7.2,

$$K_j(x, y) := \sum_{k\in\mathbb{Z}} \varphi_{jk}(x)\overline{\widetilde{\varphi}_{jk}(y)}.$$

Then for all $j \in \mathbb{Z}$ *and almost all* $x \in \mathbb{R}$, *the equality*

$$\int_{-\infty}^{\infty} K_j(x, y)\, dy = 1 \tag{1.90}$$

holds. If the function $\varphi(x)$ *is continuous, then* (1.90) *holds for all* $x \in \mathbb{R}$.

Proof. It is clear that

$$\int_{-\infty}^{\infty} K_j(x,y)\,dy = \int_{-\infty}^{\infty} K_0(2^j x, y)\,dy,$$

so it suffices to prove (1.90) for $j = 0$. Suppose $\int_{-\infty}^{\infty} K_0(x,y)\,dy \neq 1$ on a set of positive measure. Then there exists $\delta > 0$ such that

$$\mu\left\{x \in \mathbb{R} : \left|\int_{-\infty}^{\infty} K_0(x,y)\,dy - 1\right| > \delta\right\} > 0,$$

and by 1-periodicity of the function $\int_{-\infty}^{\infty} K_0(\cdot,y)\,dy$, it follows that

$$\mu\left\{x \in \mathbb{R} : \left|\int_{-\infty}^{\infty} K_0(x,y)\,dy - 1\right| > \delta\right\} = \infty. \tag{1.91}$$

Let us show that the sequence $\int_{-\infty}^{\infty} K_j(\cdot,y)\,dy - 1$ converges to zero in norm on $[-1,1]$ as $j \to +\infty$. Using Lemma 1.7.1, we get

$$\begin{aligned}\left|\int_{|y|\geq 2} K_j(x,y)\,dy\right| &\leq C2^j \int_{|y|\geq 2} \eta(2^{j-3}(x-y))\,dy \\ &\leq C2^j \int_{|y|\geq 2} \eta(2^{j-4}y)\,dy = 16C \int_{|y|\geq 2^{j-3}} \eta(y)\,dy,\end{aligned}$$

for almost all $x \in [-1,1]$, which implies the convergence to zero of the left-hand side almost everywhere on $[-1,1]$ and thus the convergence in norm. Let $f_1 = \chi_{[-2,2]}$. Since $f_1 \in L_2(\mathbb{R})$, by Theorem 1.3.10, the sequence $P_j f_1$ converges to f_1 in the space $L_2(\mathbb{R})$ as $j \to +\infty$. Therefore it converges in $L_2[-1,1]$ and converges in measure on $[-1,1]$. It remains to note that the equality

$$\int_{|y|<2} K_j(\cdot,y)\,dy - 1 = P_j f_1 - f_1 \tag{1.92}$$

holds on $[-1,1]$. Thus there exists a sequence of positive integers j_k such that

$$\mu\left\{x \in [-1,1] : \left|\int_{-\infty}^{\infty} K_{j_k}(x,y)\,dy - 1\right| > \delta\right\} < \frac{1}{k^2}.$$

This implies that

$$\mu\left\{x \in \left[-2^{j_k}, 2^{j_k}\right] : \left|\int_{-\infty}^{\infty} K_0(x,y)\,dy - 1\right| > \delta\right\} < \frac{1}{k^2},$$

which contradicts (1.91).

Now let the function φ be continuous. Then, obviously, $|\varphi(x)| \le \eta(x)$ for any $x \in \mathbb{R}$. By Corollary 1.7.2, the equality

$$\int_{-\infty}^{\infty} K_0(x,y)\,dy = \sum_{k\in\mathbb{Z}} \varphi(x+k) \int_{-\infty}^{\infty} \widetilde{\varphi}\,dy$$

holds for all $x \in \mathbb{R}$. In view of (1.86), the series on the right-hand side converges uniformly on $\mathbb{R}$; hence its sum is continuous. ◊

THEOREM 1.7.4. *Let $\varphi, \widetilde{\varphi}$ be scaling functions generating a biorthogonal wavelet system and satisfying the conditions of Corollary* 1.7.2, *$f \in C(\mathbb{R}) \cap L_\infty(\mathbb{R})$. Then we have*

$$P_j f(x) \underset{j\to+\infty}{\longrightarrow} f(x) \tag{1.93}$$

for almost all $x \in \mathbb{R}$. If f is uniformly continuous on $\mathbb{R}$, then the convergence in (1.93) *is uniform. Furthermore, if the function $\varphi(x)$ is continuous, then* (1.93) *holds for all $x \in \mathbb{R}$.*

`Proof.` Suppose that equalities (1.89) and (1.90) hold at some point x; then we have

$$|P_j f(x) - f(x)| \le \int_{-\infty}^{\infty} |f(x) - f(y)||K_j(x,y)|\,dy. \tag{1.94}$$

Given $\varepsilon > 0$, choose $\delta > 0$ such that $|f(x) - f(y)| < \varepsilon$ for all $y \in (x-\delta, x+\delta)$. Decompose the integral on the right-hand side of (1.94) as follows:

$$\int_{-\infty}^{\infty} = \int_{|x-y|<\delta} + \int_{|x-y|>\delta} =: I_1 + I_2.$$

In view of (1.88),

$$|I_1| \le A\varepsilon,$$

for all $j \in \mathbb{Z}$. Using Lemma 1.7.1, we get

$$|I_2| \le C\|f\|_\infty 2^{j+1} \int_{|x-y|>\delta} \eta\left(2^{j-3}(x-y)\right)\,dy = C\|f\|_\infty 2^4 \int_{|y|>2^{j-3}\delta} \eta\,(y)\,dy \underset{j\to+\infty}{\longrightarrow} 0.$$

The arbitrariness of ε implies (1.93). If f is uniformly continuous on $\mathbb{R}$, all the above estimates are independent of the point x. It remains to note that in view of

Lemma 1.7.3 and Corollary 1.7.2, equalities (1.89) and (1.90) hold for almost all $x \in \mathbb{R}$. If φ is continuous, these equalities hold for all $x \in \mathbb{R}$. $\lozenge$

THEOREM 1.7.5. *Let φ, $\widetilde{\varphi}$ be scaling functions that generate a biorthogonal wavelet system and satisfy the conditions of Corollary* 1.7.2, $f \in L_p(\mathbb{R})$, $1 \le p < \infty$. *Then*

$$P_j f \xrightarrow[j\to+\infty]{L_p} f. \tag{1.95}$$

Proof. First assume that the function f is continuous and compactly supported, $\operatorname{supp} f \subset [-R, R]$. Theorem 1.7.4 implies that

$$\int_{-2R}^{2R} |f(x) - P_j f(x)|^p \, dx \underset{j\to+\infty}{\longrightarrow} 0. \tag{1.96}$$

Using Lemmas 1.7.1 and 1.7.3 and Corollary 1.7.2 and taking into account the monotonicity of η, we get

$$\begin{aligned}
\int_{|x|>2R} |f(x) - P_j f(x)|^p \, dx &= \int_{|x|>2R} dx \left| \int_{-R}^{R} f(y) K_j(x,y)\, dy \right|^p \\
&\le \left(2^j C \|f\|_\infty\right)^p \int_{|x|>2R} dx \left| \int_{-R}^{R} \eta\left(2^{j-3}(x-y)\right) dy \right|^p \\
&\le \left(2^{j+1} R C \|f\|_\infty\right)^p \int_{|x|>2R} \eta^p\left(2^{j-4}x\right) dx \\
&\le 16\left(2RC\|f\|_\infty\right)^p \left(2^j \eta\left(2^{j-3}R\right)\right)^{p-1} \int_{|x|>2^{j-3}R} \eta(x)\, dx \\
&\le 2R\left(16C\|f\|_\infty\right)^p \left(\int_0^{2^{j-3}R} \eta(t)\, dt \right)^{p-1} \int_{|x|>2^{j-3}R} \eta(x)\, dx \underset{j\to+\infty}{\longrightarrow} 0.
\end{aligned}$$

To prove the theorem in the general case, it remains to approximate an arbitrary function $f \in L_p(\mathbb{R})$ by some compactly supported continuous function $\tilde{f} \in L_p(\mathbb{R})$ and to take into account (1.88). $\lozenge$

THEOREM 1.7.6. *Let φ, $\widetilde{\varphi}$ be scaling functions that generate a biorthogonal wavelet system and satisfy the conditions of Corollary* 1.7.2, $f \in L_p(\mathbb{R})$, $1 \le p < \infty$.

Then

$$P_j f(x) \underset{j\to+\infty}{\longrightarrow} f(x)$$

for almost all $x \in \mathbb{R}$.

Proof. Suppose x is a Lebesgue point of the function f, i.e., the relation

$$\lim_{h\to 0}\frac{1}{h}\int_{-h/2}^{h/2} |f(x+y)-f(x)|\,dy = 0$$

holds and equalities (1.89), (1.90) are satisfied at this point; then we have

$$|P_j f(x) - f(x)| \le \int_{-\infty}^{\infty} |f(x)-f(y)||K_j(x,y)|\,dy. \tag{1.97}$$

Given $\varepsilon > 0$, we choose δ such that

$$\frac{1}{h}\int_{-h/2}^{h/2} |f(x+y)-f(x)|\,dy < \varepsilon \tag{1.98}$$

for all $h \in (0,\delta)$. Let $2^{-j} < \delta$, $2^{k_0} \le \delta < 2^{k_0+1}$, $k_0 \in \mathbb{Z}$. Represent the integral on the right-hand side of (1.97) in the form

$$\int_{-\infty}^{\infty} = \int_{|x-y|<2^{-j}} + \sum_{k=-j}^{k_0} \int_{2^k<|x-y|<2^{k+1}} + \int_{|x-y|>2^{k_0+1}} =: I_0 + \sum_{k=-j}^{k_0} I_k + J.$$

By Lemma 1.7.1 and (1.98), taking into account the monotonicity of η, we get

$$\begin{aligned}\left|\sum_{k=-j}^{k_0} I_k\right| &\le C2^j \sum_{k=-j}^{k_0} \int_{2^k<|x-y|<2^{k+1}} |f(y)-f(x)|\eta(2^{j-3}(x-y))\,dy \\ &= C2^j \sum_{k=-j}^{k_0} \int_{2^k<|y|<2^{k+1}} |f(x+y)-f(x)|\eta(2^{j-3}y)\,dy \\ &\le C\sum_{k=-j}^{k_0} 2^{-k-2}\int_{-2^{k+1}}^{2^{k+1}} |f(x+y)-f(x)|\,dy 2^{j+k+2}\eta(2^{j+k-3}) \\ &\le 32C\varepsilon \sum_{k=-j}^{k_0} 2^{j+k-3}\eta(2^{j+k-3}) \le 64C\varepsilon\int_0^{\infty}\eta(t)\,dt,\end{aligned}$$

$$|I_0| \le C2^j \int_{|x-y|<2^{-j}} |f(y)-f(x)|\eta(2^{j-3}(x-y))\,dy \le 2C\eta(0)\varepsilon.$$

Thus the sum of all I_k is small for any j. To estimate J, let us use Lemma 1.7.1 again:

$$|J| \le C\left(2^j \int_{|y|>2^{k_0+1}} |f(x+y)|\eta(2^{j-3}y)\,dy + |f(x)|\int_{|y|>2^{j+k_0-2}} \eta(y)\,dy\right).$$

The second integral tends to zero as j increases since η is summable. Since η is monotonic, we have

$$2^j \int\limits_{|y|>2^{k_0+1}} |f(x+y)|\eta(2^{j-3}y)\,dy \le 2^j\eta(2^{j+k_0-2})\|f\|_1,$$

for $p=1$. If $p>1$, using the Hölder inequality, we obtain

$$2^j \int\limits_{|y|>2^{k_0+1}} |f(x+y)|\eta(2^{j-3}y)\,dy \le 2^j\|f\|_p \left(\int\limits_{|y|>2^{k_0+1}} \eta^q(2^{j-3}y)\,dy\right)^{1/q}$$
$$\le 2^{3/q}\|f\|_p \left(2^j\eta(2^{j+k_0-2})\right)^{1/p} \left(\int\limits_{|y|>2^{j+k_0-2}} \eta(y)\,dy\right)^{1/q}.$$

The right-hand sides of both inequalities tend to zero as j increases since

$$2^j\eta(2^j+k_0-2) \le 2^{3-k_0} \int\limits_{2^{j+k_0-3}}^{2^{j+k_0-2}} \eta(t)\,dt \underset{j\to+\infty}{\longrightarrow} 0.$$

It remains to note that, by Lemma 1.7.3 and Corollary 1.7.2, equalities (1.89) and (1.90) are satisfied for almost all $x\in\mathbb{R}$ and almost all $x\in\mathbb{R}$ are Lebesgue points of the function f. $\Diamond$

Theorems 1.7.4, 1.7.5, and 1.7.6 prove the convergence but provide no information about its rate. For smooth functions, good tools of approximation provide the rate of convergence proportional to the order of smoothness. We show that wavelet expansions will possess this property if we require not only the corresponding smoothness of a wavelet function ψ but the sufficiently fast decrease of both functions ψ, $\widetilde{\psi}$ as well. The following statement reveals the essence of the matter.

THEOREM 1.7.7. *Suppose $m\in\mathbb{N}$, $\psi\in C^m(\mathbb{R})\cap L_\infty(\mathbb{R})$, and the inequality*

$$|\widetilde{\psi}(x)| \le \frac{M}{(1+|x|)^{m+1+\alpha}} \tag{1.99}$$

holds for almost all $x\in\mathbb{R}$, where $\alpha>0$, M is an absolute constant, $\psi,\widetilde{\psi}\in L_2(\mathbb{R})$, and the systems of functions $\{\psi_{jk}\}_{j,k\in\mathbb{Z}}$, $\{\widetilde{\psi}_{jk}\}_{j,k\in\mathbb{Z}}$ are biorthogonal. Then

$$\int\limits_{-\infty}^{\infty} x^l\widetilde{\psi}(x)\,dx = 0, \qquad l=0,\dots,m. \tag{1.100}$$

Proof. The proof is by induction on l. First we prove the statement of the theorem for $l=0$. Fix $a\in\mathbb{R}$ such that $\psi(a)\neq 0$. Then due to the biorthogonality of the systems $\{\psi_{jk}\}_{j,k\in\mathbb{Z}}$, $\{\widetilde{\psi}_{jk}\}_{j,k\in\mathbb{Z}}$, we have

$$\int\limits_{-\infty}^{\infty} \widetilde{\psi}(x)\,dx = 2^j\int\limits_{-\infty}^{\infty} \widetilde{\psi}(2^jx+k)\,dx = \frac{2^j}{\psi(a)}\int\limits_{-\infty}^{\infty} (\psi(a)-\psi(x))\widetilde{\psi}(2^jx+k)\,dx \tag{1.101}$$

for any $j, k \in \mathbb{Z}$, $j \neq 0$. Given $\varepsilon > 0$, we choose $\delta > 0$ such that $|\psi(a) - \psi(x)| < \varepsilon$ for $|x - a| < \delta$. Since the function ψ is summable on $\mathbb{R}$, there exists sufficiently large N such that

$$\int\limits_{|x|\geq N} |\widetilde{\psi}(x)|\, dx < \varepsilon. \tag{1.102}$$

Choosing $j, k \in \mathbb{Z}$ to satisfy the inequality $|2^j x + k| > N$, we get

$$\left| 2^j \int\limits_{|x-a|<\delta} (\psi(a) - \psi(x))\widetilde{\psi}(2^j x + k)\, dx \right|$$
$$\leq \varepsilon 2^j \int\limits_{-\infty}^{\infty} |\widetilde{\psi}(2^j x + k)|\, dx = \varepsilon \int\limits_{-\infty}^{\infty} |\widetilde{\psi}(x)|\, dx,$$
$$\left| 2^j \int\limits_{|x-a|>\delta} (\psi(a) - \psi(x))\widetilde{\psi}(2^j x + k)\, dx \right|$$
$$\leq \|\psi\|_\infty 2^{j+1} \int\limits_{|x-a|>\delta} |\widetilde{\psi}(2^j x + k)|\, dx \leq 2\|\psi\|_\infty \int\limits_{|x|>N} |\widetilde{\psi}(x)|\, dx$$

for $|x - a| > \delta$. Combining these relations with (1.101) and (1.102), in view of the arbitrariness of ε we obtain (1.100) for $l = 0$.

Assume now that (1.100) holds for $l = 0, \ldots, n-1$, $n \leq m$. Fix $b \in \mathbb{R}$ such that $\psi^{(n)}(b) \neq 0$. By the inductive assumption, taking into account the biorthogonality of the systems $\{\psi_{jk}\}_{j,k\in\mathbb{Z}}$, $\{\widetilde{\psi}_{jk}\}_{j,k\in\mathbb{Z}}$, we get

$$\int\limits_{-\infty}^{\infty} \left(\psi(x) - \sum_{r=0}^{n} \frac{1}{r!}\psi^{(r)}(b)(x-b)^r \right) \widetilde{\psi}(2^j x + k)\, dx$$
$$= -\frac{\psi^{(n)}(b)}{n!} 2^{-j(n+1)} \int\limits_{-\infty}^{\infty} x^n \widetilde{\psi}(x)\, dx \tag{1.103}$$

for any $j, k \in \mathbb{Z}$, $j \neq 0$. Given $\varepsilon > 0$, apply the Taylor formula with the remainder in Peano's form, choosing $\delta \in (0, 1)$ such that

$$\left| \psi(x) - \sum_{r=0}^{n} \frac{1}{r!}\psi^{(r)}(b)(x-b)^r \right| \leq \varepsilon |x - b|^n$$

for $|x - b| < \delta$. In view of (1.99) we can select N such that

$$\int\limits_{|x|\geq N} (1 + |x|)^n |\widetilde{\psi}(x)|\, dx < \delta^n \varepsilon.$$

Furthermore, if we choose $j, k \in \mathbb{Z}$, $j \neq 0$, such that $|2^j x + k| > N$ for $|x - b| > \delta$ and $|2^j b + k| \leq 1$, we obtain

$$2^{j(n+1)} \left| \int\limits_{|x-b|<\delta} \left(\psi(x) - \sum_{r=0}^{n} \frac{1}{r!} \psi^{(r)}(b)(x-b)^r \right) \widetilde{\psi}(2^j x + k)\, dx \right|$$

$$\leq 2^{j(n+1)} \varepsilon \int\limits_{|x-b|<\delta} |x-b|^n |\widetilde{\psi}(2^j x + k)|\, dx \leq \varepsilon \int\limits_{-\infty}^{\infty} (|y|+1)^n |\widetilde{\psi}(y)|\, dy$$

$$\leq M\varepsilon \int\limits_{-\infty}^{\infty} \frac{dy}{(|y|+1)^{m-n+1+\alpha}} \leq C_1 \varepsilon,$$

$$2^{j(n+1)} \left| \int\limits_{|x-b|>\delta} \left(\psi(x) - \sum_{r=0}^{n} \frac{1}{r!} \psi^{(r)}(b)(x-b)^r \right) \widetilde{\psi}(2^j x + k)\, dx \right|$$

$$\leq 2^{j(n+1)} \int\limits_{|x-b|>\delta} \left(|\psi(x)| + \sum_{r=0}^{n} |\psi^{(r)}(b)| \delta^{-r} |x-b|^r \right) |\widetilde{\psi}(2^j x + k)|\, dx$$

$$\leq \delta^{-n} 2^{j(n+1)} \left(\|\psi\|_\infty + \sum_{r=0}^{n} |\psi^{(r)}(b)| \right) \int\limits_{|x-b|>\delta} |x-b|^n |\widetilde{\psi}(2^j x + k)|\, dx$$

$$\leq \delta^{-n} \left(\|\psi\|_\infty + \sum_{r=0}^{n} |\psi^{(r)}(b)| \right) \int\limits_{|y|>N} (1+|y|)^n |\widetilde{\psi}(y)|\, dy$$

$$\leq \left(\|\psi\|_\infty + \sum_{r=0}^{n} |\psi^{(r)}(b)| \right) \varepsilon = C_2 \varepsilon.$$

Combining these relations with (1.103) and taking into account the arbitrariness of ε, we prove (1.100) for $l = n$. $\Diamond$

THEOREM 1.7.8. *Suppose $\varphi, \widetilde{\varphi}$ are scaling functions that generate a biorthogonal wavelet system ψ, $\widetilde{\psi}$ such that $\psi \in C^m(\mathbb{R})$, $m \in \mathbb{N}$, and*

$$|\psi(x)|, |\widetilde{\psi}(x)| \leq \frac{M}{(1+|x|)^{m+1+\alpha}}, \quad \alpha > 0, \tag{1.104}$$

for almost all $x \in \mathbb{R}$. Then the inequality

$$\|P_j f - f\|_\infty \leq C \frac{\|f^{(m)}\|_\infty}{2^{jm}} \tag{1.105}$$

holds for any $j \in \mathbb{Z}$ and any function $f \in C^m(\mathbb{R}) \cap L_2(\mathbb{R})$, where C is a constant depending only on m, α, and M.

Proof. From Remark 1.3.11 it follows that

$$f - P_j f = \sum_{i=j}^{\infty} Q_i f, \tag{1.106}$$

where

$$Q_i f := \sum_{n\in\mathbb{Z}} \langle f, \widetilde{\psi}_{in}\rangle \psi_{in}. \tag{1.107}$$

By Corollary 1.7.2,

$$Q_i f(x) := \int_{-\infty}^{\infty} f(y) \sum_{n\in\mathbb{Z}} \psi_{in}(x)\overline{\widetilde{\psi}_{in}(y)}\, dy$$

for all $x \in \mathbb{R}$. Fix $x \in \mathbb{R}$. Applying the Taylor formula with the remainder term in Lagrange's form and Theorem 1.7.7, we get

$$\begin{aligned}\langle f, \widetilde{\psi}_{in}\rangle &= \int_{-\infty}^{\infty} \left(\sum_{l=0}^{m-1} \frac{f^{(l)}(x)}{l!}(y-x)^l + \frac{f^{(m)}(\xi)}{m!}(y-x)^m\right) \overline{\widetilde{\psi}_{in}(y)}\, dy \\ &= \frac{1}{m!}\int_{-\infty}^{\infty} f^{(m)}(\xi)(y-x)^m \overline{\widetilde{\psi}_{in}(y)}\, dy. \qquad (1.108)\end{aligned}$$

We can assume that $f^{(m)} \in L_\infty(\mathbb{R})$; otherwise the statement is trivial. Lemma 1.7.1 implies that the series

$$\sum_{n\in\mathbb{Z}} |\psi_{in}(x)\overline{\widetilde{\psi}_{in}(y)}(y-x)^m|$$

and all its partial sums have a summable majorant. Substituting (1.108) into (1.107) and interchanging summation and integration (which is possible due to Lebesgue's dominated convergence theorem), we get

$$Q_i f(x) = \frac{1}{m!}\int_{-\infty}^{\infty} f^{(m)}(\xi)(y-x)^m \sum_{n\in\mathbb{Z}} \psi_{in}(x)\overline{\widetilde{\psi}_{in}(y)}\, dy$$

for almost all $x \in \mathbb{R}$. Combining this with Lemma 1.7.1 and (1.104), we have

$$\begin{aligned}|Q_i f(x)| &\le \frac{C_1}{m!} 2^i \|f^{(m)}\|_\infty \int_{-\infty}^{\infty} |x-y|^m \frac{dy}{(1+2^{i-3}|x-y|)^{m+1+\alpha}} \\ &\le \frac{C_1}{m!} 2^{3(m+1)} \frac{\|f^{(m)}\|_\infty}{2^{im}} \int_{-\infty}^{\infty} \frac{dy}{(1+|y|)^{1+\alpha}} = C_2 \frac{\|f^{(m)}\|_\infty}{2^{im}},\end{aligned}$$

which implies that

$$\sum_{i=j}^{\infty} |Q_j f(x)| \le C_2 \sum_{i=j}^{\infty} \frac{\|f^{(m)}\|_\infty}{2^{im}} = C\frac{\|f^{(m)}\|_\infty}{2^{jm}}. \tag{1.109}$$

The series on the right-hand side of (1.106) converges in $L_2(\mathbb{R})$; thus it converges in measure on an arbitrary segment and by the Riesz theorem it contains a subsequence that converges almost everywhere on this segment. Taking into account the uniformity of estimate (1.109), we get (1.105). ◊

REMARK 1.7.9. It is clear that instead of the smoothness of ψ we can impose the condition of vanishing moments (1.100).

In §4.2 (Theorem 4.2.6), we shall prove stronger versions of Theorems 1.7.7 and 1.7.8 for compactly supported wavelets.

1.8. Wavelet frames

In §1.3, it was shown that if integer shifts of a scaling function φ are orthogonal, this generates a wavelet system $\{\psi_{jk}\}_{j,k\in\mathbb{Z}}$ that forms an orthonormal basis in $L_2(\mathbb{R})$. Now we shall study wavelet systems generated by some φ, for which the requirement of the orthogonality of integer shifts can be weakened. This is interesting for the following reasons.

Integer shifts of the function

$$\varphi(x) = \begin{cases} \frac{1}{2}, & |x| \le 1, \\ 0, & |x| \ge 1, \end{cases}$$

are not orthogonal; moreover, they do not even form a Riesz basis. However it is easy to see that this function generates a Haar MRA and thereby an orthonormal wavelet basis.

Let m_0 be the mask of a scaling function φ. The equality

$$|m_0|^2 + |m_0(\cdot + 1/2)|^2 = 1, \quad \text{a.e.} \tag{1.110}$$

is a necessary condition for the system $\{\varphi_{jk}\}$ to be orthonormal (see Corollary 1.2.6), but it is not sufficient. In practice, it may be rather complicated to satisfy the sufficient conditions (see Chapter 2). The simple method for orthogonalization described by Theorem 1.1.14 does not usually work in applications since it does not preserve compactness of the support of a function φ. On the other hand, application researchers do not often need wavelet systems that form bases, since redundant systems can be used instead of bases if any function can be expanded with respect to such a system. Moreover, redundancy may even prove useful in engineering investigations.

We shall study wavelet systems generated by a function φ that satisfies condition (1.110) instead of the orthonormality condition, or even only the condition

$$|m_0|^2 + |m_0(\cdot + 1/2)|^2 \le 1, \quad \text{a.e.} \tag{1.111}$$

We need a notion of frame which is in a sense a generalization of the Riesz basis and has some of its advantages.

DEFINITION 1.8.1. Let H be a Hilbert space. A system $\{f_n\}_{n=1}^{\infty} \subset H$ is called a *frame* if there exist constants $A, B > 0$ such that the inequality

$$A\|f\|^2 \le \sum_{n=1}^{\infty} |\langle f, f_n\rangle|^2 \le B\|f\|^2 \tag{1.112}$$

holds for any element $f \in H$.

The constants A and B are called the upper and lower frame bounds, respectively. If the bounds of a frame coincide, this frame is said to be *tight*. Sometimes tight frames with unit bounds ($A = B = 1$) are called orthogonal-like systems or Parseval frames, which is justified by the following statement.

PROPOSITION 1.8.2. *A system $\{f_n\}_{n=1}^{\infty} \subset H$ is a tight frame with unit bounds if and only if the following expansion holds for any element $f \in H$:*

$$f = \sum_{n=1}^{\infty} \langle f, f_n \rangle f_n, \tag{1.113}$$

where the series converges in H.

Proof. Sufficiency is obvious. Let us prove necessity. Let $f \in H$. Then by (1.112) the equality

$$\sum_{n=1}^{\infty} |\langle f, f_n \rangle|^2 = \|f\|^2 \tag{1.114}$$

holds. First of all we note that the series on the right-hand side of (1.113) converges:

$$\left\| \sum_{n=N}^{M} \langle f, f_n \rangle f_n \right\|^2 = \left| \sum_{n=N}^{M} \langle f, f_n \rangle \langle f_n, g_{NM} \rangle \right|^2 \le \sum_{n=N}^{M} |\langle f, f_n \rangle|^2 \sum_{n=1}^{\infty} |\langle f_n, g_{NM} \rangle|^2,$$

where $\|g_{NM}\| \le 1$. It follows from the definition of frame that the right-hand side tends to zero as $N, M \to \infty$. Let us check equality (1.113). We can assume that $\|f\| = 1$. Set

$$s_N = \sum_{n=1}^{N} \langle f, f_n \rangle f_n, \qquad \sigma_N = \sum_{n=1}^{N} |\langle f, f_n \rangle|^2.$$

Then it is obvious that

$$\|f - s_N\|^2 = 1 - 2\sigma_N + \|s_N\|^2.$$

Since (1.114) implies that $\lim_{N \to \infty} \sigma_N = 1$, it remains to check that $\|s_N\| \to 1$. Given $\varepsilon > 0$, we choose N such that $\sigma_N \ge 1 - \varepsilon$. Then we have

$$\|s_N\| = \sup_{\|g\| \le 1} |\langle s_N, g \rangle| \ge |\langle s_N, f \rangle| = \left| \sum_{n=1}^{N} \langle f, f_n \rangle \langle f_n, f \rangle \right| = \sigma_N \ge 1 - \varepsilon.$$

On the other hand,

$$\|s_N\|^2 = |\langle s_N, g_N \rangle|^2 = \left| \sum_{n=1}^{N} \langle f, f_n \rangle \langle f_n, g_N \rangle \right|^2$$

where $\|g_N\| \le 1$. Applying the Cauchy-Bunyakovsky equality and (1.114) to the right-hand side, we obtain $\|s_N\| \le 1$. ◊

COROLLARY 1.8.3. *If a tight frame with unit bounds forms a basis, this basis is orthonormal.*

To prove this fact, it suffices to set $f = f_k$ in (1.113) and to take into account the uniqueness of this expansion.

Thus tight frames with unit bounds can be considered as a generalization of orthonormal bases for redundant systems (i.e., systems that are not minimal). Now we show that the notion of frame generalizes the notion of the Riesz basis to redundant systems.

THEOREM 1.8.4. *A system $\{f_n\}_{n=1}^{\infty}$ is a frame with bounds A, B in the Hilbert space H if and only if it is an orthogonal projection of some Riesz basis with constants A, B in a Hilbert space H^* that contains H. In particular, if elements f_n, $n = 1, 2, \dots$, form a basis in the space H, then it is a Riesz basis.*

Proof. Sufficiency. Suppose H^* is a Hilbert space, H is its subspace, and $\{F_n\}_{n=1}^{\infty}$ is a Riesz basis in H^* with constants A, B. Denote by P the operator of orthogonal projection of H^* on H and set $f_n = PF_n$, $n = 1, 2, \dots$. By Theorem 1.1.2, the system $\{F_n\}_{n=1}^{\infty}$ forms a frame with bounds A, B. It remains to note that if $f \in H$, then $\|f\|_{H^*} = \|f\|_H$ and $\langle f, F_n\rangle = \langle f, f_n\rangle$.

Necessity. Let the system $\{f_n\}_{n=1}^{\infty}$ satisfy (1.112). First we show that the series $\sum\limits_{n=1}^{\infty} c_n f_n$ converges in H for any $c = \{c_n\}_{n=1}^{\infty} \in \ell_2$. Indeed,

$$\left\|\sum_{n=N}^{M} c_n f_n\right\| = \left|\left\langle \sum_{n=N}^{M} c_n f_n, g_{NM}\right\rangle\right| = \left|\sum_{n=N}^{M} c_n \langle f_n, g_{NM}\rangle\right|,$$

where $\|g_{NM}\| \le 1$. It follows from the Cauchy inequality and (1.112) that

$$\begin{aligned}\left|\sum_{n=N}^{M} c_n \langle f_n, g_{NM}\rangle\right|^2 &\le \sum_{n=N}^{M} |c_n|^2 \sum_{n=1}^{\infty} |\langle f_n, g_{NM}\rangle|^2 \\ &\le B \sum_{n=N}^{M} |c_n|^2 \|g_{NM}\|^2 \underset{N,M\to\infty}{\longrightarrow} 0,\end{aligned}$$

which implies that $\sum\limits_{n=1}^{\infty} c_n f_n \in H$ and the operator

$$T: \ c \to \sum_{n=1}^{\infty} c_n f_n$$

is bounded. Let h be the kernel of the operator T, i.e., the set of points $c \in \ell_2$ such that $\sum\limits_{n=1}^{\infty} c_n f_n = 0$. Since the operator T is continuous, h is a closed subspace of the space ℓ_2. Denote by h' its orthogonal complement. Let us define the operator I that maps H to ℓ_2 and takes every $f \in H$ to the sequence $\{\langle f, f_n\rangle\}_{n=1}^{\infty}$. It follows from the left inequality (1.112) that I is a one-to-one map of H to $I(H) =: h_1$; therefore there exists the inverse operator I^{-1} taking h_1 onto H. Moreover, (1.112) implies that the operators I, I^{-1} are bounded, where

$$\|I\|^2 \le B, \quad \|I^{-1}\|^2 \le 1/A, \tag{1.115}$$

i.e., I is an isomorphism from H to h_1. Let us show that $h_1 = h'$. If $c \in h$, $c' \in h_1$, then there exists $f \in H$ such that $c'_n = \langle f, f_n\rangle$ for all $n = 1, 2, \dots$, and consequently,

$$\langle c, c'\rangle = \sum_{n=1}^{\infty} c_n \overline{\langle f, f_n\rangle} = \left\langle \sum_{n=1}^{\infty} c_n f_n, f\right\rangle = 0.$$

Therefore $h_1 \subset h'$. If $h_1 \ne h'$, then there exists a nonzero element $c \in h' \ominus h_1$, and hence $\sum\limits_{n=1}^{\infty} c_n \langle f, f_n\rangle = 0$ for any f from H, in particular, for $f = f_0 := \sum\limits_{n=1}^{\infty} c_n f_n$, and thus $f_0 = 0$, which contradicts the assumption $c \in h'$, $c \neq 0$. Denote by T_1 the restriction of the operator T to h'. For any $c \in h'$ and $g \in H$, we have the equality

$$\langle T_1 c, g\rangle = \left\langle \sum_{n=1}^{\infty} c_n f_n, g \right\rangle = \sum_{n=1}^{\infty} c_n \langle f_n, g\rangle = \langle c, Ig\rangle,$$

which implies that $T_1 = I^*$. Therefore for any $c \in h'$, $I^* c = T_1 c = \sum_{n=1}^{\infty} c_n f_n$; using (1.115), we get

$$A \sum_{n=1}^{\infty} |c_n|^2 \leq \left\| \sum_{n=1}^{\infty} c_n f_n \right\|^2 \leq B \sum_{n=1}^{\infty} |c_n|^2, \quad \text{for} \quad c \in h'. \tag{1.116}$$

Let $\{\alpha_n\}_{n\in\Lambda}$ be an orthogonal basis in h such that $\|\alpha_n\| = \sqrt{A}$ for all $n \in \Lambda$. The set Λ may be empty, finite, or countable. Assume first that Λ is countable. Then, for any $c \in \ell_2$, the following equality holds:

$$\left\| \sum_{n=1}^{\infty} c_n \alpha_n \right\|^2 = A \sum_{n=1}^{\infty} |c_n|^2. \tag{1.117}$$

Set $\tilde{h} := \{\sum_{n=1}^{\infty} c_n \alpha_n, \quad c \in h\}$. It is clear that the orthogonal complement to the space $\tilde{h}$ in h consists of elements of the form $\sum_{n=1}^{\infty} c_n \alpha_n$, where $c \in h'$. Denote by β_n the orthogonal projection of α_n onto $\tilde{h}$. For any $c \in h$ we have

$$\sum_{n=1}^{\infty} c_n(\alpha_n - \beta_n) = \sum_{n=1}^{\infty} c_n \alpha_n - \sum_{n=1}^{\infty} c_n \beta_n \in \tilde{h},$$

which implies that

$$\sum_{n=1}^{\infty} c_n(\alpha_n - \beta_n) = 0 \quad \text{for} \quad c \in h. \tag{1.118}$$

Similarly, we get

$$\sum_{n=1}^{\infty} c_n \beta_n = 0 \quad \text{for} \quad c \in h'. \tag{1.119}$$

Combining (1.117) and (1.118), we ascertain that the equality

$$\left\| \sum_{n=1}^{\infty} c_n \beta_n \right\|^2 = A \sum_{n=1}^{\infty} |c_n|^2 \quad \text{for} \quad c \in h \tag{1.120}$$

holds for all $c \in h$.

Define the space H^* as the direct sum of the spaces H and h. The space H^* consists of pairs $(f, c) =: f \oplus c$, where $f \in H$, $c \in h$, and the scalar product of elements $F = f \oplus c$, $F' = f' \oplus c'$ is given by the formula

$$\langle F, F'\rangle_{H^*} = \langle f, f'\rangle_H + \langle c, c'\rangle_h.$$

Let us show that the elements $F_n := f_n \oplus \beta_n$ form a Riesz basis in H^*. Any $c \in \ell_2$ can be represented in the form $c = c^0 + c'$, where $c^0 \in h$, $c' \in h'$. Using (1.119) and the definition of the space h, we get

$$\left\| \sum_{n=1}^{\infty} c_n F_n \right\|^2 = \left\| \sum_{n=1}^{\infty} c_n f_n \right\|^2 + \left\| \sum_{n=1}^{\infty} c_n \beta_n \right\|^2 = \left\| \sum_{n=1}^{\infty} c'_n f_n \right\|^2 + \left\| \sum_{n=1}^{\infty} c_n^0 \beta_n \right\|^2.$$

Using (1.116) and (1.120), we obtain

$$A\sum_{n=1}^{\infty}|c_n|^2 \le \left\|\sum_{n=1}^{\infty} c_n F_n\right\|^2 = B\sum_{n=1}^{\infty}|c_n|^2, \quad \text{for } c \in \ell_2.$$

If Λ is finite, we set

$$\beta_n = \frac{1}{\sqrt{A}}\sum_{k\in\Lambda}\overline{(\alpha_k)_n}\alpha_k.$$

It is not hard to check that relations (1.119), (1.120) hold. It remains to note that the elements f_n, $n = 1, 2, \ldots$, form a basis in H if the space h is empty; therefore H^* coincides with H and $F_n = f_n$, $n = 1, 2, \ldots$. ◊

COROLLARY 1.8.5. *If a system $\{f_n\}_{n=1}^{\infty}$ forms a frame with bounds A, B in a Hilbert space H, then in H there exists a dual frame $\{g_n\}_{n=1}^{\infty}$ with bounds $1/B, 1/A$ such that any element $f \in H$ can be represented in the form*

$$f = \sum_{n=1}^{\infty}\langle f, g_n\rangle f_n, \tag{1.121}$$

where the series converges in norm in H.

Proof. Let $\{F_n\}_{n=1}^{\infty}$ be the Riesz basis in H^* constructed in Theorem 1.8.4 and let P be the orthogonal projection of H^* on H. Thus $f_n = PF_n$, $n = 1, 2, \ldots$. For any $F \in H^*$, we have the expansion

$$F = \sum_{n=1}^{\infty}\langle F, G_n\rangle F_n,$$

where $\{G_n\}_{n=1}^{\infty}$ is the dual system (see Appendix A.1, Theorem A.1.3). In particular, if $f \in H$, we have

$$f = \sum_{n=1}^{\infty}\langle f, G_n\rangle F_n = \sum_{n=1}^{\infty}\langle f, PG_n\rangle F_n = \sum_{n=1}^{\infty}\langle f, PG_n\rangle f_n.$$

Let $g_n = PG_n$, $n = 1, 2, \ldots$, and verify that $\{g_n\}_{n=1}^{\infty}$ is the required frame. In view of Theorem 1.8.4 it suffices to show that the system $\{G_n\}_{n=1}^{\infty}$ is a Riesz basis in H^*. Let $c \in \ell_2$. Using Theorem 1.1.2 and the fact that the systems $\{F_n\}_{n=1}^{\infty}$, $\{G_n\}_{n=1}^{\infty}$ are biorthonormal, we get

$$\begin{aligned}\left\|\sum_{n=N}^{M} c_n G_n\right\|^2 &= \sup_{\|F\|\le 1}\left|\left\langle\sum_{n=N}^{M} c_n G_n, F\right\rangle\right| \le \sup_{\|c'\|^2\le 1/A}\left|\left\langle\sum_{n=N}^{M} c_n G_n, \sum_{k=1}^{\infty} c'_k F_k\right\rangle\right| \\ &= \sup_{\|c'\|^2\le 1/A}\left|\sum_{n=N}^{M} c_n c'_n\right| \le \frac{1}{\sqrt{A}}\sqrt{\sum_{n=N}^{M}|c_n|^2}.\end{aligned}$$

It follows that the series $\sum\limits_{n=1}^{\infty} c_n G_n =: G$ converges in norm. By Theorem 1.1.2, the system $\{F_n\}_{n=1}^{\infty}$ is a frame with bounds A, B. Therefore,

$$\frac{1}{B}\sum_{n=1}^{\infty}|\langle G, F_n\rangle|^2 \le \|G\|^2 \le \frac{1}{A}\sum_{n=1}^{\infty}|\langle G, F_n\rangle|^2.$$

It remains to take into account that $c_n = \langle G, F_n\rangle$. ◊

COROLLARY 1.8.6. *Let $\{f_n\}_{n=1}^{\infty}$, $\{\tilde{f}_n\}_{n=1}^{\infty}$ be biorthonormal systems in a Hilbert space H. If one of these systems forms a frame in H, then both systems are Riesz bases in H.*

Proof. Let the system $\{f_n\}_{n=1}^{\infty}$ satisfy (1.112). By Corollary 1.8.5, expansion (1.121) holds for any $f \in H$. The coefficients of this expansion can be found by multiplying scalarly (1.121) by $\tilde{f}_n$, $n = 1, 2, \ldots$. It follows that the expansion with respect to the system $\{f_n\}_{n=1}^{\infty}$ is unique and the equality

$$\langle f, g_n \rangle = \langle f, \tilde{f}_n \rangle, \quad n = 1, 2, \ldots, \tag{1.122}$$

holds for all $f \in H$. Thus $\{f_n\}_{n=1}^{\infty}$ is a basis in H. By Theorem 1.8.4, it is a Riesz basis. Since $\{g_n\}_{n=1}^{\infty}$ is a frame, it follows from (1.122) that the system $\{\tilde{f}_n\}_{n=1}^{\infty}$ is also a frame and, as we have already proved, it is a Riesz basis. ◊

We now turn to the study of wavelet frames. Let φ be a function in $L_2(\mathbb{R})$ that satisfies the refinement equation

$$\widehat{\varphi}(\xi) = m_0(\xi/2)\widehat{\varphi}(\xi/2) \text{ a.e.}, \tag{1.123}$$

where m_0 is a mask. Let $m_1, \ldots, m_r$ be 1-periodic functions in $L_2(0,1)$ (we shall also call them masks) such that for the matrix

$$M(\xi) = \begin{pmatrix} m_0(\xi) & m_1(\xi) & \ldots & m_r(\xi) \\ m_0(\xi+1/2) & m_1(\xi+1/2) & \ldots & m_r(\xi+1/2) \end{pmatrix}$$

the following equality holds for almost all $\xi \in \mathbb{R}$:

$$M(\xi)M^*(\xi) = I_2. \tag{1.124}$$

Introduce wavelet functions $\psi^{(1)}, \ldots, \psi^{(r)}$, defining their Fourier transform by the formulas

$$\widehat{\psi^{(k)}}(\xi) = m_k(\xi/2)\widehat{\varphi}(\xi/2), \quad k = 1, \ldots, r. \tag{1.125}$$

This construction generalizes the construction of wavelet functions in the orthonormal case, where $m_1(\xi) = e^{\pi i \xi}\overline{m_0(\xi+1/2)}$, $r = 1$ (see §1.3).

LEMMA 1.8.7. *Let $\varphi \in L_2(\mathbb{R})$ and let $\widehat{\varphi}$ be continuous at zero and satisfy (1.123), where the mask m_0 satisfies (1.111). Then*

$$\sum_{k\in\mathbb{Z}} |\widehat{\varphi}(\xi+k)|^2 \le |\widehat{\varphi}(0)|^2 \tag{1.126}$$

for almost all $\xi \in \mathbb{R}$.

Proof. First note that (1.123) and (1.111) imply that $|\widehat{\varphi}(\xi)| \le |\widehat{\varphi}(0)|$ for almost all $\xi \in \mathbb{R}$. Hence for any $l \in \mathbb{Z}_+$ we get

$$\begin{aligned} \sum_{k=-2^l}^{2^l-1} |\widehat{\varphi}(\xi+k)|^2 &= \sum_{k=-2^l}^{2^l-1} \prod_{n=1}^{l+1} |m_0(2^{-n}(\xi+k))|^2 \left|\widehat{\varphi}(2^{-n}(\xi+k))\right|^2 \\ &\le |\widehat{\varphi}(0)|^2 \sum_{k=-2^l}^{2^l-1} \prod_{n=1}^{l+1} |m_0(2^{-n}(\xi+k))|^2 \\ &= |\widehat{\varphi}(0)|^2 \sum_{k=0}^{2^l-1} \left(\prod_{n=1}^{l+1} |m_0(2^{-n}(\xi+k))|^2 + \prod_{n=1}^{l+1} |m_0(2^{-n}(\xi+k-2^l))|^2 \right). \end{aligned}$$

It follows from (1.111) and 1-periodicity of the function m_0 that the right-hand side does not exceed

$$|\widehat{\varphi}(0)|^2 \sum_{k=0}^{2^l-1} \prod_{n=1}^{l} |m_0\left(2^{-n}(\xi+k)\right)|^2 =: \mu(l).$$

Using (1.111) and 1-periodicity of the function m_0 again, we obtain

$$\begin{aligned}\mu(l) &= |\widehat{\varphi}(0)|^2 \sum_{k=0}^{2^{l-1}-1} \left(\prod_{n=1}^{l} |m_0\left(2^{-n}(\xi+k)\right)|^2 + \prod_{n=1}^{l} |m_0\left(2^{-n}(\xi+k+2^{l-1})\right)|^2 \right) \\ &\le \mu(l-1).\end{aligned}$$

Applying this estimate l times, we get $\mu(l) \le |\widehat{\varphi}(0)|^2$. ◊

COROLLARY 1.8.8. *Let a function φ satisfy the conditions of Lemma* 1.8.7, $f \in L_2(\mathbb{R})$. *Then*

$$\sum_{k\in\mathbb{Z}} |\langle f, \varphi_{jk}\rangle|^2 \le |\widehat{\varphi}(0)|^2 \|f\|^2 \tag{1.127}$$

for all $j \in \mathbb{Z}$ and

$$\lim_{j\to-\infty} \sum_{k\in\mathbb{Z}} |\langle f, \varphi_{jk}\rangle|^2 = 0. \tag{1.128}$$

$$\lim_{j\to+\infty} \sum_{k\in\mathbb{Z}} |\langle f, \varphi_{jk}\rangle|^2 = |\widehat{\varphi}(0)|^2 \|f\|^2. \tag{1.129}$$

Proof. In view of Theorems 1.1.7 and 1.1.2 and Remarks 1.1.8 and 1.1.3, relation (1.126) implies (1.127), while relations (1.128) and (1.129) follow from Lemmas 1.2.8 and 1.2.13, respectively. ◊

THEOREM 1.8.9. *Let $\varphi \in L_2(\mathbb{R})$ and let $\widehat{\varphi}$ be continuous at zero and satisfy* (1.123). *Suppose that wavelet functions $\psi^{(1)}, \dots, \psi^{(r)}$ are defined by relation* (1.125) *and their masks $m_0, m_1, \dots, m_r$ satisfy* (1.124). *Then the system of functions $\psi_{jk}^{(\nu)}$, $j, k \in \mathbb{Z}$, $\nu = 1, \dots, r$, forms a tight frame in $L_2(\mathbb{R})$ with bounds $A = B = |\widehat{\varphi}(0)|^2$.*

Proof. First of all we note that (1.124) implies that $|m_\nu(\xi)| \le 1$ for almost all $\xi \in \mathbb{R}$ and $\nu = 1, \dots, r$; hence $\widehat{\psi^{(\nu)}} \in L_2(\mathbb{R})$ and by the Plancherel theorem $\psi^{(\nu)} \in L_2(\mathbb{R})$. Let us show that (1.111) holds for m_0. Suppose that equality (1.124) holds for some $\xi \in \mathbb{R}$. Then the $(r+1)$-dimensional vectors

$$a := (m_0(\xi), \dots, m_r(\xi)), \qquad b := (m_0(\xi+1/2), \dots, m_r(\xi+1/2))$$

are orthogonal and have unit lengths. Extend the set of these vectors to an orthonormal basis of the space $\mathbb{R}^{r+1}$ and make M a square matrix by filling all the consequent rows with the remaining basis vectors. The matrix thus constructed is unitary; therefore its columns are unit vectors, which implies that $|m_0(\xi)|^2 + |m_0(\xi+1/2)|^2 \le 1$. Thus the function φ satisfies the conditions of Lemma 1.8.7; hence (1.126) holds and (1.127) follows from Corollary 1.8.8.

Let us prove that the following equality holds for any $j, j' \in \mathbb{Z}$, $j < j'$:

$$\sum_{k\in\mathbb{Z}} |\langle f, \varphi_{jk}\rangle|^2 + \sum_{\nu=1}^{r} \sum_{i=j}^{j'-1} \sum_{k\in\mathbb{Z}} |\langle f, \psi_{ik}^{(\nu)}\rangle|^2 = \sum_{k\in\mathbb{Z}} |\langle f, \varphi_{j',k}\rangle|^2. \tag{1.130}$$

Clearly, it suffices to verify (1.130) for the case $j' = j+1$. Lemma 1.2.11, relations (1.123), (1.125), and 1-periodicity of the functions $m_0, m_1, \ldots, m_r$ imply that

$$\sum_{k\in\mathbb{Z}} |\langle f, \varphi_{jk}\rangle|^2 + \sum_{\nu=1}^{r} \sum_{i=j}^{j'-1} \sum_{k\in\mathbb{Z}} |\langle f, \psi_{ik}^{(\nu)}\rangle|^2$$

$$= 2^j \sum_{\nu=0}^{r} \int_0^1 \left| \sum_{k\in\mathbb{Z}} \widehat{f}\left(2^j(\xi+k)\right) \overline{m_\nu\left(\frac{\xi+k}{2}\right)\widehat{\varphi}\left(\frac{\xi+k}{2}\right)} \right|^2 d\xi$$

$$= 2^j \sum_{\nu=0}^{r} \int_0^1 \left| \sigma(\xi)\overline{m_\nu\left(\frac{\xi}{2}\right)} + \sigma(\xi+1)\overline{m_\nu\left(\frac{\xi+1}{2}\right)} \right|^2 d\xi, \tag{1.131}$$

where

$$\sigma(\xi) = \sum_{k\in\mathbb{Z}} \widehat{f}\left(2^{j+1}\left(\frac{\xi}{2}+k\right)\right) \overline{\widehat{\varphi}\left(\frac{\xi}{2}+k\right)}.$$

The right-hand side of (1.131) can be rewritten in the form

$$2^j \int_0^1 |\sigma(\xi)|^2 \sum_{\nu=0}^{r} \left| m_\nu\left(\frac{\xi}{2}\right)\right|^2 d\xi + 2^j \int_0^1 |\sigma(\xi+1)|^2 \sum_{\nu=0}^{r} \left| m_\nu\left(\frac{\xi+1}{2}\right)\right|^2 d\xi$$

$$+ 2^j \int_0^1 \overline{\sigma(\xi+1)}\sigma(\xi) \sum_{\nu=0}^{r} m_\nu\left(\frac{\xi+1}{2}\right)\overline{m_\nu\left(\frac{\xi}{2}\right)} d\xi$$

$$+ 2^j \int_0^1 \sigma(\xi+1)\overline{\sigma(\xi)} \sum_{\nu=0}^{r} \overline{m_\nu\left(\frac{\xi+1}{2}\right)} m_\nu\left(\frac{\xi}{2}\right) d\xi.$$

Taking into account (1.124), we get

$$\sum_{k\in\mathbb{Z}} |\langle f, \varphi_{jk}\rangle|^2 + \sum_{\nu=1}^{r} \sum_{i=j}^{j'-1} \sum_{k\in\mathbb{Z}} |\langle f, \psi_{ik}^{(\nu)}\rangle|^2$$

$$= 2^j \int_0^1 \left(|\sigma(\xi)|^2 + |\sigma(\xi+1)|^2\right) d\xi = 2^j \int_0^2 |\sigma(\xi)|^2 \, d\xi$$

$$= 2^{j+1} \int_0^1 |\sigma(2\xi)|^2 \, d\xi = 2^{j+1} \int_0^1 \left| \sum_{k\in\mathbb{Z}} \widehat{f}\left(2^{j+1}(\xi+k)\right) \overline{\widehat{\varphi}(\xi+k)} \right|^2 d\xi.$$

To prove (1.130) for $j' = j+1$, it remains to use Lemma 1.2.11 once more.

Passing to the limit in (1.130) first as $j \to -\infty$, then as $j' \to +\infty$, by Corollary 1.8.8, we derive

$$\sum_{\nu=1}^{r} \sum_{i,k\in\mathbb{Z}} |\langle f, \psi_{ik}^{(\nu)}\rangle|^2 = |\widehat{\varphi}(0)|^2 \|f\|^2,$$

which was to be proved. ◊

COROLLARY 1.8.10. *If a function $\varphi \in L_2(\mathbb{R})$ satisfies the conditions of Theorem 1.8.9 and its mask m_0 satisfies (1.111), then there exist $\psi^{(1)}, \psi^{(2)} \in L_2(\mathbb{R})$ such that the functions $\psi_{jk}^{(\nu)}$, $j,k \in \mathbb{Z}$, $\nu = 1,2$, form a tight frame in $L_2(\mathbb{R})$.*

Proof. Let

$$m_1(\xi) = \begin{cases} \frac{e^{2\pi i\xi}\overline{m_0(\xi+1/2)}}{B(\xi)}, & |m_0(\xi)|^2 + |m_0(\xi+1/2)|^2 \neq 0, \\ \frac{\sqrt{2}}{2}e^{2\pi i\xi}, & |m_0(\xi)|^2 + |m_0(\xi+1/2)|^2 = 0, \end{cases}$$

$$m_2(\xi) = \begin{cases} \frac{A(\xi)m_0(\xi)}{B(\xi)}, & |m_0(\xi)|^2 + |m_0(\xi+1/2)|^2 \neq 0, \\ \frac{\sqrt{2}}{2}, & |m_0(\xi)|^2 + |m_0(\xi+1/2)|^2 = 0, \end{cases}$$

where $A = \sqrt{1 - |m_0(\xi)|^2 - |m_0(\xi+1/2)|^2}$, $B = \sqrt{|m_0(\xi)|^2 + |m_0(\xi+1/2)|^2}$. It is not hard to verify that the corresponding matrix M satisfies (1.124); hence the required statement follows from Theorem 1.8.9. ◊

The following assertion is an important particular case of Theorem 1.8.9. In fact, it strengthens Theorem 1.3.1.

THEOREM 1.8.11. *Let $\varphi \in L_2(\mathbb{R})$ and let $\widehat{\varphi}$ be continuous and different from zero at zero and satisfy (1.123). Suppose its mask $m_0(\xi) = \frac{1}{\sqrt{2}} \sum_{n\in\mathbb{Z}} h_n e^{2\pi i n\xi}$ satisfies (1.110) and*

$$\psi := \sum_{n\in\mathbb{Z}} (-1)^{1-n} \overline{h_{1-n}} \varphi_{1n}. \tag{1.132}$$

Then the functions ψ_{jk}, $j,k \in \mathbb{Z}$, form a tight frame in $L_2(\mathbb{R})$ with bounds $A = B = |\widehat{\varphi}(0)|^2$.

Moreover, if the system of functions φ_{0k}, $k \in \mathbb{Z}$, is a Riesz system with constants a, b, then $a \le |\widehat{\varphi}(0)|^2 \le b$.

Proof. First of all we note that by Proposition 1.2.3, equality (1.132) can be rewritten in the form

$$\widehat{\psi}(\xi) = m_1\left(\frac{\xi}{2}\right)\widehat{\varphi}\left(\frac{\xi}{2}\right),$$

where

$$m_1(\xi) = e^{2\pi i\xi}\overline{m_0\left(\xi + \frac{1}{2}\right)}.$$

It is easy to verify that the matrix

$$M(\xi) = \begin{pmatrix} m_0(\xi) & m_1(\xi) \\ m_0(\xi+1/2) & m_1(\xi+1/2) \end{pmatrix}$$

is unitary, i.e., it satisfies (1.124), and the first statement follows from Theorem 1.8.9.

Now let φ_{0k}, $k \in \mathbb{Z}$, be a Riesz system and let $\{V_j\}$ be the MRA generated by it. It follows from Theorems 1.1.2 and 1.1.7 that

$$a\|f\|^2 \le \sum_{k\in\mathbb{Z}} |\langle f, \varphi_{0k}\rangle|^2 \le b\|f\|^2$$

for any $f \in V_0$. Then the inequality

$$a\|f\|^2 \leq \sum_{k\in\mathbb{Z}} |\langle f, \varphi_{jk}\rangle|^2 \leq b\|f\|^2$$

holds for any $j \in \mathbb{Z}$, $f \in V_j$. By Corollary 1.8.8, it follows that the second statement of the theorem holds. $\Diamond$

CHAPTER 2

Multivariate Wavelets

2.1. Separable MRA

In this chapter, we generalize the notion of wavelet system to the multidimensional case. To do this, one can choose various approaches. The simplest and quite natural way is to define a multidimensional wavelet basis as the tensor product of one-dimensional bases. It is in this way that many one-dimensional orthonormal bases (for example, the trigonometric one) are extended to the multidimensional case.

The scheme of the construction is as follows. Let $\{f_n^{(\nu)}\}_n$, $\nu = 1, \dots, d$, be orthonormal bases in the space $L_2(\mathbb{R})$. Define functions f_N on $\mathbb{R}^d$, $N = (N_1, \dots, N_d)$, as the tensor products of the functions $f_{N_1}, \dots, f_{N_d}$, i.e., for $x \in \mathbb{R}^d$, we set

$$f_N(x) = f_{N_1}(x_1) \cdot \dots \cdot f_{N_d}(x_d).$$

In the sequel we shall use the following notation for the tensor products:

$$f_N = f_{N_1} \otimes \dots \otimes f_{N_d} = \bigotimes_{\nu=1}^{d} f_{N_\nu}.$$

It is clear that the functions f_N are pairwise orthogonal in the space $L_2(\mathbb{R}^d)$. It is not hard to check that the system $\{f_N\}_N$ is complete in $L_2(\mathbb{R}^d)$. Thus using the wavelet function ψ such that the system $\{\psi_{jk}\}_{jk}$ is an orthonormal basis in $L_2(\mathbb{R})$, we can form the d-dimensional wavelet system

$$\Psi_{JK} := \psi_{J_1K_1} \otimes \dots \otimes \psi_{J_dK_d}, \tag{2.1}$$

where $J = (J_1, \dots, J_d)$, $K = (K_1, \dots, K_d)$, which constitutes an orthonormal basis in $L_2(\mathbb{R}^d)$. Similarly, using a pair of wavelet functions ψ, $\tilde{\psi}$ such that the systems $\{\psi_{jk}\}_{jk}$, $\{\tilde{\psi}_{jk}\}_{jk}$ are biorthogonal, one can construct d-dimensional biorthogonal wavelet systems Ψ_{JK}, $\tilde{\Psi}_{JK}$. Moreover, we can use various one-dimensional wavelet functions as the components of the tensor product in (2.1); i.e., we can set

$$\Psi_{JK} := \psi^{(1)}_{J_1K_1} \otimes \dots \otimes \psi^{(d)}_{J_dK_d}. \tag{2.2}$$

The wavelet systems described above are of considerable interest and can be useful in various situations. Nevertheless, they do not possess all the advantages of one-dimensional wavelet bases, in particular, the localization property, which is of great value for applications. Let us illustrate this by the example of the Haar system. In the one-dimensional case, basis functions ψ_{jk} with large indices j have small supports. The support of the multidimensional basis function Ψ_{JK} may be large along one or several directions for arbitrarily large $|J|$. Therefore, if we replace the multidimensional index J by a one-dimensional index in some natural way, for instance, over cubes or balls, the support diameter of functions with large indices will not necessarily be small.

To avoid these drawbacks, we use a different approach. The main idea is that instead of the tensor product of available wavelet bases we use the tensor product of the MRAs generating these bases.

Let $\{V_j^{(\nu)}\}_{j\in\mathbb{Z}}$, $\nu = 1,\dots,d$, be a one-dimensional MRA. We define subspaces V_j, $j \in \mathbb{Z}$, of the space $L_2(\mathbb{R}^d)$ by

$$V_j := \bigotimes_{\nu=1}^{d} V_j^{(\nu)} = \overline{\operatorname{span}\{f = f_1 \otimes \cdots \otimes f_d,\ f_\nu \in V_j^{(\nu)}\}}. \tag{2.3}$$

Let $\varphi^{(\nu)}$ be a scaling function of the ν-th MRA $\{V_j^{(\nu)}\}_{j\in\mathbb{Z}}$ and let

$$\varphi = \varphi^{(1)} \otimes \cdots \otimes \varphi^{(d)}.$$

By the definition of the scaling function, the integer shifts of the function φ form a Riesz system. Theorem 1.1.7 implies that for almost all ξ, the following inequality holds:

$$A^{(\nu)} \le \sum_{k\in\mathbb{Z}} |\widehat{\varphi}^{(\nu)}(\xi + k)|^2 \le B^{(\nu)}, \quad A^{(\nu)}, B^{(\nu)} > 0.$$

This inequality combined with the following trivial relation $\widehat{\varphi} = \widehat{\varphi}^{(1)} \otimes \cdots \otimes \widehat{\varphi}^{(d)}$ yields that for almost all $\xi \in \mathbb{R}^d$ we have

$$A \le \sum_{k\in\mathbb{Z}^d} |\widehat{\varphi}(\xi + k)|^2 \le B, \tag{2.4}$$

where

$$A := \prod_{\nu=1}^{d} A^{(\nu)}, \quad B := \prod_{\nu=1}^{d} B^{(\nu)}.$$

Reasoning as in the proof of Theorem 1.1.7, we obtain

PROPOSITION 2.1.1. *Let $\varphi \in L_2(\mathbb{R}^d)$. The system $\{\varphi(\cdot + n)\}_{n\in\mathbb{Z}^d}$ is a Riesz system with constants A, B if and only if inequality* (2.4) *holds for almost all $\xi \in \mathbb{R}^d$.*

Since for each ν, the system $\{\varphi^{(\nu)}(\cdot + n)\}_{n\in\mathbb{Z}}$ is a basis of the space $V_0^{(\nu)}$ (see axiom **MR5** in Definition 1.2.1), it is clear that

$$V_0 = \overline{\operatorname{span}\{\varphi(\cdot + n), n \in \mathbb{Z}^d\}},$$

and by Theorem 1.1.2, the functions $\varphi(\cdot + n)$, $n \in \mathbb{Z}^d$, form a Riesz basis in the space V_0. It is easy to verify that if the integer shifts of all the functions $\varphi^{(\nu)}$, $\nu = 1,\dots,d$, are pairwise orthogonal, then $\{\varphi(\cdot + n)\}_{n\in\mathbb{Z}^d}$ is an orthogonal system. From the definition of the spaces V_j it is clear that the conditions $f \in V_0$ and $f(2^j\cdot) \in V_j$ are equivalent. It follows easily from the inclusion property for each MRA $\{V_j^{(\nu)}\}_j$ (see axiom **MR1** in Definition 1.2.1) that $\varphi(2^j \cdot + n) \in V_{j+1}$ for any $n \in \mathbb{Z}^d$, which implies the inclusion $V_j \subset V_{j+1}$.

Observe that any element F of the space $L_2(\mathbb{R}^d)$ can be approximated, to an arbitrary degree of accuracy, by a finite linear combination of the elements of the form

$$f = f_1 \otimes \cdots \otimes f_d, \ f_\nu \in L_2(\mathbb{R}), \ \nu = 1,\dots,d$$

(to check this, it suffices to expand the function F with respect to a basis that is the tensor product of one-dimensional bases, for instance, one of the orthonormal wavelet bases described above). In turn, it follows from the completeness property for each MRA $\{V_j^{(\nu)}\}_j$ (see axiom **MR2** in Definition 1.2.1) that any element $f \in L_2(\mathbb{R})$ can be approximated by a finite linear combination of the functions $\varphi^{(\nu)}(2^j \cdot +n)$, $n \in \mathbb{Z}$, for sufficiently large $j \in \mathbb{Z}$. Hence, the element F can be approximated by an element of the space V_j for some j; i.e., the collection of all the spaces V_j, $j \in \mathbb{Z}$, is dense in $L_2(\mathbb{R}^d)$.

Let us show that the intersection of all V_j does not contain a nonzero element. Assume that there exists a function F that belongs to all the spaces V_j, $j \in \mathbb{Z}$, and $\|F\| \neq 0$. Let $\psi_{jk}^{(\nu)}$, $\nu = 1, \dots, d$, be orthonormal wavelet bases in the space $L_2(\mathbb{R})$ generated by the corresponding MRAs $\{V_j^{(\nu)}\}_j$, and let Ψ_{JK} be the functions defined by equality (2.2). If $j < \min\{J_1, \dots, J_d\}$, then $\Psi_{JK} \perp V_j$; consequently, $\Psi_{JK} \perp F$ for all $J, K \in \mathbb{Z}^d$. As we already noted, the system $\{\Psi_{JK}\}_{J,K \in \mathbb{Z}^d}$ is an orthonormal basis in $L_2(\mathbb{R}^d)$, which contradicts the assumption $\|F\| \neq 0$.

Let us now summarize our observations in the following theorem.

THEOREM 2.1.2. *Let $\{V_j^{(\nu)}\}_{j \in \mathbb{Z}}$, $\nu = 1, \dots, d$, be MRAs in $L_2(\mathbb{R})$. Then the subspaces V_j of the space $L_2(\mathbb{R}^d)$ given by* (2.3) *possess the following properties:*

MR1. $V_j \subset V_{j+1}$ *for all* $j \in \mathbb{Z}$.

MR2. $\bigcup_{j \in \mathbb{Z}} V_j$ *is dense in* $L_2(\mathbb{R}^d)$.

MR3. $\bigcap_{j \in \mathbb{Z}} V_j = \{0\}$.

MR4. $f \in V_j \iff f\left(2^{-j} \cdot\right) \in V_0$ *for all* $j \in \mathbb{Z}$.

MR5. *There exists a function $\varphi \in V_0$ such that the sequence $\{\varphi(\cdot + n)\}_{n \in \mathbb{Z}^d}$ forms a Riesz basis in V_0.*

The collections of the spaces V_j, $j \in \mathbb{Z}$, satisfying the conditions of Theorem 2.1.2 will be called a *separable MRA* in the space $L_2(\mathbb{R}^d)$.

Now, following the standard scheme, we define the wavelet spaces by the formulas

$$W_j = V_{j+1} \ominus V_j.$$

Since

$$\begin{aligned} V_{j+1} &= \bigotimes_{\nu=1}^{d} V_{j+1}^{(\nu)} = \bigotimes_{\nu=1}^{d} \left(V_j^{(\nu)} \oplus W_j^{(\nu)}\right) \\ &= \bigoplus_{\substack{e \subset \{1,\dots,d\} \\ e \neq \emptyset}} \left(\bigotimes_{\nu \in e} W_j^{(\nu)}\right) \left(\bigotimes_{\mu \notin e} V_j^{(\mu)}\right) \oplus \left(\bigotimes_{\nu=1}^{d} V_j^{(\nu)}\right), \end{aligned}$$

we can see that the space W_j is factorized into a direct sum of $2^d - 1$ subspaces denoted as $W_{j,e}$, $e \subset \{1, \dots, d\}, e \neq \emptyset$. Let $\psi^{(\nu)}$ be a wavelet function and assume that its integer shifts form a Riesz basis in $W_j^{(\nu)}$. It is clear that the integer shifts of the function

$$\psi_e := \left(\bigotimes_{\nu \in e} \psi^{(\nu)}\right) \left(\bigotimes_{\mu \notin e} \varphi^{(\mu)}\right), \quad e \subset \{1, \dots, d\}, \ e \neq \emptyset,$$

form a Riesz basis in $W_{0,e}$; moreover, the multidimensional bases are orthogonal, whenever the one-dimensional bases are. Thus, in contrast to the one-dimensional case, in each wavelet space we have found several wavelet functions, whose integer shifts form its basis.

2.2. Matrix dilation

In the previous section we considered separable MRAs. Scaling functions generating these MRAs have rather special form; namely, they are tensor products of one-dimensional functions. In this situation, it seems natural to extend the class of generating scaling functions. This can be done if we consider properties **MR1**–**MR5** of Theorem 2.1.2 as the axioms to define MRAs. Such a definition of MRA will not essentially differ from the one-dimensional definition. However the existence of different directions in the multidimensional case allows more generalizations of one-dimensional theories.

In our case, a possibility for generalization of this type is provided by axiom **MR4**, where the dilation 2^j appears. Multiplication of each component of the argument by the same factor means multiplication of the vector argument by a diagonal matrix with equal diagonal elements, i.e., the same dilation along each coordinate axis. It is natural to consider other matrices as coefficients of dilation. Moreover, it is not necessary to assume the matrix to be dilating. However, to provide the completeness property (axiom **MR2**), repeated multiplication by the matrix must result in dilation along all directions. We consider the class of square integer matrices M with eigenvalues greater than one in modulus. For such matrices the relation

$$\lim_{n\to+\infty} \|M^{-n}\| = 0 \tag{2.5}$$

holds, which provides the above-mentioned condition. Moreover, we have

$$\sum_{n=1}^{\infty} \|M^{-n}\|^{\delta} < \infty \tag{2.6}$$

for any $\delta > 0$ (see Appendix A.5).

Before turning to the general definition of multidimensional MRA, we present some notions and facts connected with the matrix dilation of functions.

If A is a nonsingular integer $d \times d$ matrix, we say that vectors $k, n \in \mathbb{Z}^d$ *are congruent modulo* A and write $k \equiv n \pmod{A}$ if $k - n = Al$, $l \in \mathbb{Z}^d$. The integer lattice $\mathbb{Z}^d$ is partitioned into cosets with respect to the congruence introduced above. Any set containing only one representative of each coset is called *a set of digits of the matrix* A. When it does not matter which set of digits is chosen, we shall assume that it is chosen arbitrarily and denote it by $D(A)$.

PROPOSITION 2.2.1. *Let A be a nonsingular integer $d \times d$ matrix. Then the number of cosets with respect to the congruence modulo A is equal to $|\det A|$, the set $H(A) := \mathbb{Z}^d \cap A[0,1)^d$ being a set of digits.*

`Proof.` Let us show that the number of integer points in the set $A[0,1)^d$ is equal to its volume. Denote by a and b the volume of the set $A[0,1)^d$ and the number of its integer points, respectively. Consider a d-dimensional cube with side N and find the number of parallelepipeds of the form $A([0,1)^d + k)$, $k \in \mathbb{Z}^d$, inside the cube $[0,N]^d$ in two ways. Let S denote the sum of all such parallelepipeds. Obviously, for large N, the volume of S is $N^d + o(N^d)$. The number of integer points in S is $N^d + o(N^d)$. The volume of a parallelepiped is a and the number of integer points in it is b. Expressing the number of parallelepipeds in S in two ways and equating these expressions, we get

$$\frac{N^d + o(N^d)}{a} = \frac{N^d + o(N^d)}{b}.$$

Dividing both sides by N^d and passing to the limit as N tends to infinity, we get $a = b$. Let us show that the set $H(A)$ does not contain any congruent elements. Suppose integer points $s_1, s_2 \in A[0,1)^d$ are congruent modulo A. Then $s_1 - s_2 = Al$, where $l \in \mathbb{Z}^d$. Multiplying both sides of this equality by A^{-1}, we deduce that the difference between two elements from $[0,1)^d$ is equal to an integer vector, which is impossible unless $l = \mathbf{0}$; thus $s_1 = s_2$. Show that the set $H(A)$ contains representatives of all cosets of the matrix A. Let an integer vector $s \notin A[0,1)^d$ represent a coset. The vector $A^{-1}s$ belongs to a cube of the form $[0,1)^d + k$, $k \in \mathbb{Z}^d$ (since such cubes cover the whole space $\mathbb{R}^d$). Set $s' = A(A^{-1}s - k)$. It is clear that s' and s are congruent modulo A and $s' \in A[0,1)^d$; i.e., in $A[0,1)^d$ we have found a representative of the coset that contains s. Thus $H(A)$ contains exactly one representative of each coset; therefore the number of cosets is a. It remains to note that $a = |\det A|$. ◊

LEMMA 2.2.2. *Let A be a nonsingular integer $d \times d$ matrix and let $D(A)$ be its set of digits. Then the equality $p = Ak + s$ gives a one-to-one mapping of the set of all $p \in \mathbb{Z}^d$ to the set of pairs (k, s), $k \in \mathbb{Z}^d$, $s \in D(A)$.*

`Proof.` First let $D(A) = H(A)$. Any $p \in \mathbb{Z}^d$ can be represented in the form $p = Ak + Ar$, where the vector k is the integral part of the vector $A^{-1}p$ and r is its fractional part. Setting $s = Ar$, we find the required representation, since $s \in A[0,1)^d$ and $s \in \mathbb{Z}^d$ (the difference $p - Ak$ is integer). Expressing the digit $s \in H(A)$ by the digit $s' \in D(A)$ from the same coset, we obtain the required representation for an arbitrary $D(A)$. The uniqueness of the representation for any $D(A)$ is obvious. ◊

LEMMA 2.2.3. *Let A be a nonsingular integer $d \times d$ matrix, $|\det A| > 1$. Then the set $\{r + A^j p\}$ for all possible $r \in D(A^j)$ and $p \in D(A)$ is a set of digits of the matrix A^{j+1}.*

`Proof.` The number of all possible pairs $(r,\ p)$ for $r \in D(A^j)$ and $p \in D(A)$ is $|\det A|^{j+1}$; i.e., it suffices to prove that for different pairs we cannot obtain any vectors congruent modulo A^{j+1}. Let $r,\ r_1 \in D(A^j)$ and $p,\ p_1 \in D(A)$. Suppose $r + A^j p$ and $r_1 + A^j p_1$ are congruent modulo A^{j+1}; i.e., $(r - r_1) + A^j(p - p_1) = A^{j+1}n$ for some $n \in \mathbb{Z}^d$. Multiplying both sides of this equality by A^{-j} from the left, we get $A^{-j}(r - r_1) = -(p - p_1) + An \in \mathbb{Z}^d$; i.e., $r \equiv r_1 \pmod{A^j}$. However both r and r_1 belong to $D(A^j)$, which contains only one representative of each coset; hence $r = r_1$. The equality $(p - p_1) = An$ implies that $p = p_1$. Thus $r + A^j p$ and $r_1 + A^j p_1$ can be congruent modulo A^{j+1} if and only if $r = r_1$ and $p = p_1$. ◊

DEFINITION 2.2.4. Let A be a nonsingular integer $d \times d$ matrix. We say that *a set K is congruent to a set L modulo A (modulo $\mathbb{Z}^d$ for $A = I_d$)* if the set K can be partitioned into a finite number of mutually disjoint measurable subsets K_n, $n = 1, \ldots, N$, in such a way that there exist integer vectors $l_1, \ldots, l_N$ such that $L = \bigcup_{n=1}^{N} (K_n + Al_n)$, where $(K_n + Al_n) \cap (K_{n_1} + Al_{n_1}) = \emptyset$ for $n \neq n_1$.

It is obvious that if a set K is congruent to a set L, then L is congruent to K. Congruence modulo $\mathbb{Z}^d$ means that by shifting parts of the set K by integer vectors we can "compose" the set L. Congruence modulo A means that by shifting parts of the set K by vectors of the form Al, where l is an integer vector, we can "compose" the set L. Obviously, if two sets are congruent, they have the same measure.

LEMMA 2.2.5. *Let A be a nonsingular integer $d \times d$ matrix. Then the set*

$$K := \bigcup_{r \in D(A)} (A^{-1}[0,1)^d + A^{-1}r)$$

is congruent to $[0,1)^d$ modulo $\mathbb{Z}^d$, and for any function $f \in L(\mathbb{T}^d)$, the following equality holds:

$$\int_{[0,1)^d} f = \sum_{r \in D(A)} \int_{A^{-1}[0,1)^d + A^{-1}r} f.$$

Proof. Set

$$K_n := ([0,1)^d + n) \cap \bigcup_{r \in D(A)} (A^{-1}[0,1)^d + A^{-1}r).$$

It is obvious that $K = \bigcup_{n \in \mathbb{Z}^d} K_n$, where $K_n \cap K_{n_1} = \emptyset$ for $n \neq n_1$. As K is bounded, the number of nonempty sets K_n is finite. Let us show that $K_n \cap (K_{n_1} - l) = \emptyset$ for any $l \in \mathbb{Z}^d$. Let $u \in K_n \subset K$. Then $u = A^{-1}v + A^{-1}r$, where $v \in [0,1)^d$, $r \in D(A)$. Suppose there exists a vector $u_1 \in K_{n_1}$, $n \neq n_1$, such that $u = u_1 - l$, $l \in \mathbb{Z}^d$, or (equivalently) $u - u_1 \in \mathbb{Z}^d$. Let $u_1 = A^{-1}v_1 + A^{-1}r_1$. Then the difference $u - u_1$ has the form

$$A^{-1}(v - v_1) + A^{-1}(r - r_1) = l, \quad l \in \mathbb{Z}^d.$$

Multiplying this equality by A from the left, we obtain $v - v_1 + r - r_1 = Al$, $l \in \mathbb{Z}^d$. Since the vectors r, r_1 and Al are integer and the vectors v and v_1 belong to $[0,1)^d$, this equality is satisfied only for $v = v_1$. However, since both vectors r, r_1 belong to $D(A)$, they cannot be congruent modulo A. Hence $r = r_1$, i.e., $u = u_1$, which contradicts the assumption $n \neq n_1$.

It remains to verify that $[0,1)^d = \bigcup\limits_n (K_n - n)$. Note that by definition the set $K_n - n$ belongs to $[0,1)^d$ for any n. Moreover, $(K_n - n) \cap (K_{n_1} - n_1) = \emptyset$.

Let us verify that for any u from $[0,1)^d$ there exist n and $w \in K_n$ such that $u = w - n$. Multiplying the vector u by the matrix A from the left, we represent Au in the form $Au = p + v$, where $p \in \mathbb{Z}^d$, $v \in [0,1)^d$. The vector p is congruent to one of the digits; i.e., there exist $r \in D(A)$ and $l \in \mathbb{Z}^d$ such that $p = r + Al$. Expressing u from this relation, we get $u = A^{-1}r + l + A^{-1}v$; therefore $u - l \in K$. Consequently, $u - l \in K_n$, where $n = -l$. Setting $w := u - l = u + n$, $w \in K_n$, we obtain $u = w - n$.

The second statement of the lemma immediately follows from 1-periodicity of the function f with respect to each variable. ◊

LEMMA 2.2.6. *Let A be a nonsingular integer $d \times d$ matrix. Then*

$$\sum_{s \in D(A^*)} e^{2\pi i (A^{-1}r,\, s)} = \begin{cases} |\det A|, & \text{if } r \equiv \mathbf{0} \pmod{A}, \\ 0, & \text{if } r \not\equiv \mathbf{0} \pmod{A}. \end{cases} \tag{2.7}$$

Proof. Set $m := |\det A| = |\det A^*|$. The cosets with respect to A^* form a group under addition, which contains m elements. If $r \equiv \mathbf{0} \pmod{A}$, i.e., $r = Al$, $l \in \mathbb{Z}^d$, then (2.7) is obvious. If $r \not\equiv \mathbf{0} \pmod{A}$, we take any vector $a \in \mathbb{Z}^d$

such that $(A^{-1}r, a) \notin \mathbb{Z}$, in particular, $a \not\equiv \mathbf{0} \pmod{A^*}$. Consider the vectors $a,\ 2a,\ 3a, \dots$. Let m_1 be the least integer such that $m_1 a \equiv 0 \pmod{A^*}$. Then $1 < m_1 \le m$ (if all elements ka, $k = 1, \dots, m$, are not congruent to zero, they are not mutually congruent, while the group contains only m elements). Thus the vectors $a,\ 2a, \dots, m_1 a$ form a subgroup of m_1 elements; hence by Lagrange's theorem m is divisible by m_1. Let $m = m_1 n$. Set $a_1 = 0$ and let $a_2, \dots, a_n$ be representatives of elements from the factor group. Then $a_k + ja$, $k = 1, \dots, n$, $j = 1, \dots, m_1$, range over all elements of the group and we get

$$\begin{aligned}\sum_{k=1}^{m} e^{2\pi i(A^{-1}r, s_k)} &= \sum_{k=1}^{n}\sum_{j=1}^{m_1} e^{2\pi i(A^{-1}r, a_k + ja)} \\ &= \sum_{k=1}^{n} e^{2\pi i(A^{-1}r, a_k)} \sum_{j=1}^{m_1} e^{2\pi i(A^{-1}r, ja)} = C\frac{1 - e^{2\pi i m_1 (A^{-1}r, a)}}{1 - e^{2\pi i(A^{-1}r, a)}}.\end{aligned}$$

The equality $m_1 a = A^* l$, $l \in \mathbb{Z}^d$, together with the assumption $(A^{-1}r, a) \notin \mathbb{Z}$ implies that the latter fraction is equal to zero. ◊

COROLLARY 2.2.7. *Under the conditions of Lemma 2.2.6 the matrix*

$$\left\{\frac{1}{\sqrt{|\det A|}} e^{2\pi i(A^{-1}n, r)}\right\}_{n \in D(A),\, r \in D(A^*)}$$

is unitary.

`Proof.` The scalar product of the columns with numbers n_1, n_2 is

$$\sum_{r \in D(A^*)} e^{2\pi i(A^{-1}(n_1 - n_2), r)}.$$

By Lemma 2.2.6, this sum is equal to $|\det A|$ for $n_1 = n_2$; otherwise it is equal to zero, because $n_1 \not\equiv n_2 \pmod{A}$. ◊

PROBLEM 2.2.8. Let A satisfy the conditions of Lemma 2.2.6, $f \in C(\mathbb{T}^d)$. Prove that $f(A^{-1}n) = 0$ for all $n \in \mathbb{Z}^d$ not congruent to zero modulo A if and only if all sums of the Fourier coefficients of the function f with numbers from the same coset of the matrix A^* are equal.

2.3. Nonseparable MRAs

Here and in the sequel we shall assume that M is a fixed integer $d \times d$ matrix such that all its eigenvalues are greater than one in modulus, $m = |\det M|$. Obviously, for this matrix $m \in \mathbb{N}$, $m > 1$ and, as mentioned above, relations (2.5), (2.6) hold. In particular, they imply that

$$\lim_{n \to \infty} |M^n x| = \infty \tag{2.8}$$

for any $x \in \mathbb{R}^d$, $x \neq \mathbf{0}$, i.e., the operator associated with this matrix realizes dilation in all directions when applied repeatedly.

DEFINITION 2.3.1. A collection of closed spaces $V_j \subset L_2(\mathbb{R}^d)$, $j \in \mathbb{Z}$, is called a *multiresolution analysis (MRA) in* $L_2(\mathbb{R}^d)$ *with matrix dilation* M if the following conditions (axioms) hold:

MR1. $V_j \subset V_{j+1}$ for all $j \in \mathbb{Z}$.

MR2. $\bigcup_{j\in\mathbb{Z}} V_j$ is dense in $L_2(\mathbb{R}^d)$.

MR3. $\bigcap_{j\in\mathbb{Z}} V_j = \{0\}$.

MR4. $f \in V_j \iff f\left(M^{-j}\cdot\right) \in V_0$ for all $j \in \mathbb{Z}$.

MR5. There exists a function $\varphi \in V_0$ such that the sequence $\{\varphi(\cdot + n)\}_{n\in\mathbb{Z}^d}$ forms a Riesz basis in V_0.

As in the one-dimensional case, we say that φ is a *scaling* function. If a scaling function of an MRA is not the tensor product of one-variable functions, this MRA is called *nonseparable*.

We introduce the notation

$$f_{jn} := m^{j/2} f\left(M^j \cdot + n\right), \quad n \in \mathbb{Z}^d,\ j \in \mathbb{Z}. \tag{2.9}$$

Then axioms **MR4**, **MR5** imply two obvious properties:

MR6. For any $j \in \mathbb{Z}$, the functions φ_{jn}, $n \in \mathbb{Z}^d$, form a Riesz basis in V_j with the same constants A, B as the functions φ_{0n}, $n \in \mathbb{Z}^d$.

MR7 If $f \in V_j$, then $f\left(\cdot + M^{-j}n\right) \in V_j$, for all $n \in \mathbb{Z}^d$.

Let φ be a scaling function for an MRA. Using properties **MR1**, **MR6** and Theorem 1.1.2, we get the *refinement equation*

$$\varphi = \sum_{n\in\mathbb{Z}^d} h_n \varphi_{1n}, \quad \sum_{n\in\mathbb{Z}^d} |h_n|^2 < \infty. \tag{2.10}$$

Applying the Fourier transform to both sides of this equality, we get

$$\widehat{\varphi}(\xi) = m^{1/2} \sum_{n\in\mathbb{Z}^d} h_n \int_{\mathbb{R}^d} \varphi(Mx + n) e^{-2\pi i(x,\xi)}\, dx.$$

After the change of variables in the integral the equality takes the form

$$\widehat{\varphi}(\xi) = m_0(M^{*-1}\xi)\widehat{\varphi}(M^{*-1}\xi), \tag{2.11}$$

where

$$m_0(\xi) = m^{-1/2} \sum_{n\in\mathbb{Z}^d} h_n e^{2\pi i(n,\xi)}.$$

As in the one-dimensional case, the function m_0 is called a *mask*. It follows from (2.10) and the Parseval equality that $m_0 \in L_2(\mathbb{T}^d)$.

Let us discuss the problem of constructing an MRA. Following the same reasoning as in the one-dimensional case, we shall seek an appropriate generating function φ among the functions that satisfy (2.4). Set

$$V_j := \overline{\operatorname{span}\{\varphi_{jn}, n \in \mathbb{Z}^d\}}. \tag{2.12}$$

Axiom **MR5** follows from Proposition 2.1.1 and Theorem 1.1.2. By the definition of the spaces V_j, **MR4** also holds, and **MR1** is equivalent to the fact that φ is a solution of the refinement equation (2.10). Hence it is necessary that a required function φ satisfies (2.10) or, equivalently, its Fourier transform satisfies (2.11).

THEOREM 2.3.2. *Suppose $\varphi \in L_2(\mathbb{R}^d)$ and (2.4) holds. Then condition* **MR3** *holds for the collection of the spaces V_j, $j \in \mathbb{Z}$, defined by equality* (2.12).

Proof. Let us show that

$$\lim_{j\to-\infty}\sum_{k\in\mathbb{Z}^d}|\langle f,\varphi_{jk}\rangle|^2=0 \tag{2.13}$$

for any function $f\in L_2(\mathbb{R}^d)$. First suppose that f is continuous on $\mathbb{R}^d$ and compactly supported, with the support in the ball $B_R:=\{|x|<R\}$, $R>1$. Using the Cauchy-Bunyakovsky inequality, we get

$$\begin{aligned}\sum_{k\in\mathbb{Z}^d}|\langle f,\varphi_{jk}\rangle|^2 &\le m^j\sum_{k\in\mathbb{Z}^d}\left(\int_{B_R}|f(x)||\varphi(M^jx+k)|\,dx\right)^2\\ &\le m^j(2R)^d\|f\|_\infty\sum_{k\in\mathbb{Z}^d}\int_{B_R}|\varphi(M^jx+k)|^2\,dx\\ &= (2R)^d\|f\|_\infty\sum_{k\in\mathbb{Z}^d}\int_{M^jB_R+k}|\varphi(y)|^2\,dy. \end{aligned}\tag{2.14}$$

It follows from (2.5) that the diameter of the set M^jB_R tends to zero as $j\to-\infty$; hence the sets M^jB_R+k, $k\in\mathbb{Z}^d$, are mutually disjoint for negative j sufficiently large in modulus. So we set

$$S(R,j)=\bigcup_{k\in\mathbb{Z}^d}(M^jB_R+k)$$

and rewrite the right-hand side of (2.14) in the form

$$(2R)^d\|f\|_\infty\int_{S(R,j)}|\varphi(y)|^2\,dy=(2R)^d\|f\|_\infty\int_{\mathbb{R}^d}\chi_{R,j}(y)|\varphi(y)|^2\,dy$$

where $\chi_{R,j}$ is the characteristic function of the set $S(R,j)$. If $y\notin\mathbb{Z}^d$, then

$$\lim_{j\to-\infty}\chi_{R,j}(y)=0.$$

Consequently,

$$\lim_{j\to-\infty}\int_{\mathbb{R}^d}\chi_{R,j}(y)|\varphi(y)|^2\,dy=0.$$

Here we can pass to the limit due to Lebesgue's dominated convergence theorem.

Let f be an arbitrary function from $L_2(\mathbb{R}^d)$. Given $\varepsilon>0$, we find a compactly supported continuous function $\widetilde{f}$ such that

$$\|f-\widetilde{f}\|<\varepsilon. \tag{2.15}$$

Inequality (2.4), Proposition 2.1.1, Theorem 1.1.2, and Remark 1.1.3 imply that

$$\sum_{k\in\mathbb{Z}^d}|\langle f-\widetilde{f},\varphi_{jk}\rangle|^2\le B\varepsilon.$$

To prove (2.13), it remains to take into account that it holds for $\widetilde{f}$ and to use the triangle inequality.

Suppose that there exists a function $f \in L_2(\mathbb{R}^d)$ such that $f \in V_j$ for all $j \in \mathbb{Z}$. It follows from (2.4), Proposition 2.1.1, and Theorem 1.1.2 that

$$A\|f\| \le \left(\sum_{k\in\mathbb{Z}^d} |\langle f, \varphi_{jk}\rangle|^2 \right)^{1/2}.$$

In view of (2.13), the right-hand side tends to zero as $j \to -\infty$. It follows that $\|f\| = 0$, which means that condition **MR3** holds. ◊

The proof of Theorem 2.3.2 is practically the same as that of the corresponding one-dimensional assertion. In the same way, repeating the reasoning for the one-dimensional case, one can prove the multidimensional analog of Theorem 1.2.14. Instead of doing this, we shall prove another criterion for axiom **MR2** to be satisfied.

LEMMA 2.3.3. *Suppose φ, $f \in L_2(\mathbb{R}^d)$ and (2.4) holds. Let $\{V_j\}_{j\in\mathbb{Z}}$ be the collection of spaces generated by the function φ. The relation $f \in V_j$ holds if and only if there exists a function $m_f \in L_2(\mathbb{T}^d)$ such that*

$$\widehat{f}(\xi) = m_f(M^{*-j}\xi)\widehat{\varphi}(M^{*-j}\xi)$$

for almost all $\xi \in \mathbb{R}^d$.

To prove this lemma, we can repeat almost word for word the proofs of Propositions 1.1.10 and 1.2.3.

THEOREM 2.3.4. *Let (2.4) be true for $\varphi \in L_2(\mathbb{R}^d)$.* **MR2** *holds for the collection of spaces $\{V_j\}_{j\in\mathbb{Z}}$ generated by the function φ and satisfying* **MR1** *if and only if*

$$\bigcup_{j\in\mathbb{Z}} \operatorname{supp} \widehat{\varphi}(M^{*j}\cdot) = \mathbb{R}^d. \tag{2.16}$$

`Proof.` First of all we note that the spaces V_j generated by the functions φ satisfy axioms **MR4** and **MR5**; hence they satisfy axiom **MR7** as well. Therefore each space V_j is invariant under the shifts $t = M^{-j}n$, $n \in \mathbb{Z}^d$. Let us show that the space $\overline{\bigcup_{j\in\mathbb{Z}} V_j}$ is invariant under any shift $t \in \mathbb{R}^d$. Any $t \in \mathbb{R}^d$ can be approximated by vectors of the form $M^{-j}n$, $n \in \mathbb{Z}^d$, $j \in \mathbb{N}$, for arbitrarily large j. Indeed, it follows from (2.5) that for any $\varepsilon > 0$, there exists $j_0 \in \mathbb{N}$ such that $|M^{-j}x| < \varepsilon$ for all $j \ge j_0$ and all $x \in [0,1)^d$. If $j \ge j_0$, then choosing $n \in \mathbb{Z}^d$ in such a way that $M^j t - n \in [0,1)^d$, we obtain $|t - M^{-j}n| \le \varepsilon$. If $f \in \bigcup_{j\in\mathbb{Z}} V_j$, then in view of axiom **MR1**, $f \in V_j$ for all $j \ge j_1$ and the continuity of the function $\|f(\cdot + t)\|$ implies that $f(\cdot + t) \in \overline{\bigcup_{j\in\mathbb{Z}} V_j}$, $t \in \mathbb{R}^d$. If $g \in \overline{\bigcup_{j\in\mathbb{Z}} V_j}$, then approximating g by the functions $f \in \bigcup_{j\in\mathbb{Z}} V_j$, using the continuity of shifts and the invariance of norm under shifts in $L_2(\mathbb{R}^d)$, we deduce that $g(\cdot + t) \in \overline{\bigcup_{j\in\mathbb{Z}} V_j}$.

Put $\widehat{X} := \{\widehat{f} : f \in X\}$ for $X \subset L_2(\mathbb{R}^d)$. By the Wiener theorem for L_2 (see Appendix A.8), the closed subset X of the space $L_2(\mathbb{R}^d)$ is invariant under shifts if and only if $\widehat{X} = L_2(\Omega)$ for some set $\Omega \subset \mathbb{R}^d$. Let $X := \overline{\bigcup_{j\in\mathbb{Z}} V_j}$. Then $\widehat{X} = L_2(\Omega)$. Therefore $X = L_2(\mathbb{R}^d)$ if and only if $\Omega = \mathbb{R}^d$. Set

$$\varphi_j := \varphi(M^j\cdot), \quad \Omega_0 := \bigcup_{j\in\mathbb{Z}} \operatorname{supp} \widehat{\varphi}_j.$$

Let us prove that $\Omega = \Omega_0$. Since $\varphi_j \in V_j$, $j \in \mathbb{Z}$, we have $\operatorname{supp} \widehat{\varphi}_j \subset \Omega$; hence $\Omega_0 \subset \Omega$. Suppose now that $\Omega \setminus \Omega_0$ contains a set of positive measure Ω_1. By Lemma 2.3.3, the Fourier transform of any element from V_j is equal to zero almost everywhere on Ω_1. Consequently, the same is true for any element from $\bigcup_{j\in\mathbb{Z}} V_j$. Passing to the limit, we deduce that the Fourier transform of any element from X is equal to zero almost everywhere on Ω_1, which contradicts the fact that $\widehat{X} \supset L_2(\Omega_1)$. It remains to note that $\operatorname{supp} \widehat{\varphi}_j = \operatorname{supp} \widehat{\varphi}(M^{*-1})$. ◊

REMARK 2.3.5. Analyzing the proof of Theorem 2.3.4 and the auxiliary facts, it is easy to see that the theorem remains true without the assumption that φ satisfies (2.4) if we consider the spaces

$$V_j := \overline{\operatorname{span}\{\varphi(M^j \cdot +n), n \in \mathbb{Z}^d\}}$$

as the spaces generated by this function.

PROBLEM 2.3.6. Let φ be a scaling function for an MRA in $L_2(\mathbb{R}^d)$. Prove that if $\widehat{\varphi}$ is continuous at zero, then $\widehat{\varphi}(\mathbf{0}) \neq 0$.

PROBLEM 2.3.7. Suppose φ is a scaling function for an MRA in $L_2(\mathbb{R}^d)$ and m_0 is its mask, $\varphi(x) = O((1+|x|)^{-d-\varepsilon})$, $\varepsilon > 0$. Prove that the equality $\widehat{\varphi}(n) = 0$ holds for all $n \in \mathbb{Z}^d$, $n \neq \mathbf{0}$, if and only if $m_0(M^{*-1}k) = 0$ for all $k \in \mathbb{Z}^d$, $k \not\equiv \mathbf{0}$ (mod M^*).

PROBLEM 2.3.8. Suppose a function $\varphi \in L(\mathbb{R}^d)$ satisfies the refinement equation (1.23) and the series $\sum\limits_{l\in\mathbb{Z}^d} |\widehat{\varphi}(\xi+l)|^2$ converges uniformly. Prove that the function $\sum\limits_{k\in\mathbb{Z}^d} \varphi(x+k)$ is equivalent to a constant.

EXAMPLE 2.3.9. Let M be a $d \times d$ diagonal matrix with 2's on the diagonal. Define the function B_N, $N \geq d$, by its Fourier transform

$$\widehat{B}_N(\xi) = 2^d \prod_{k=1}^{N} \frac{1 - e^{2\pi i(a_k,\xi)}}{2\pi i(a_k,\xi)},$$

where $a_k \in \mathbb{R}^d$, $0 < |a_k| < 2$, $k = 1, \dots, N$, and the range of the matrix $A := (a_1, \dots, a_N)$ is d. The equality

$$\widehat{B}_N(\xi) = 2^d \prod_{k=1}^{N} \frac{(1 - e^{\pi i(a_k,\xi)})(1 + e^{\pi i(a_k,\xi)})}{2\pi i(a_k,\xi)} = \widehat{B}_N(\xi/2) \prod_{k=1}^{N} \frac{1 + e^{\pi i(a_k,\xi)}}{2}$$

implies that the function B_N satisfies the refinement equation with mask

$$m_0(\xi) = \prod_{k=1}^{N} \frac{1 + e^{2\pi i(a_k,\xi)}}{2}.$$

By Proposition 2.1.1, in order to show that integer shifts of the function B_N form a Riesz basis, it suffices to verify that the sum $\sum_{n\in\mathbb{Z}^d} \left|\widehat{B}_N(\xi+n)\right|^2$ is bounded almost everywhere and bounded away from zero. It is clear that for $|a_k| < 2$, there exists a constant $C = C(A) > 0$ such that

$$\left|\widehat{B}_N(\xi)\right|^2 = \left|2^d \prod_{k=1}^{N} \frac{\sin \pi(a_k,\xi)}{\pi i(a_k,\xi)}\right|^2 \geq C$$

for all $\xi \in [-1/2, 1/2]^d$; hence,

$$\sum_{n\in\mathbb{Z}^d} \left|\widehat{B}_N(\xi+n)\right|^2 \geq C > 0.$$

There are at least d linearly independent vectors in the set $a_1, \dots, a_N$ (we assume that these are $a_1, \dots, a_d$); therefore,

$$\sum_{n\in\mathbb{Z}^d} \left|\widehat{B}_N(\xi+n)\right|^2 \leq \left(\frac{2}{\pi}\right)^{2d} \sum_{n\in\mathbb{Z}^d} \prod_{k=1}^{d} \left|\frac{\sin \pi((a_k,\xi)+(a_k,n))}{(a_k,\xi)+(a_k,n)}\right|^2.$$

It is not hard to verify that the series on the right-hand side converges uniformly on $[-1/2, 1/2]^d$, so we get

$$\sum_{n\in\mathbb{Z}^d} \left|\widehat{B}_N(\xi+n)\right|^2 \leq C'.$$

Thus B_N generates a nonseparable MRA.

The functions B_N are multivariate generalizations of B-splines. They are called *box splines*. Clearly, B_d is the characteristic function of the set $A[-1/2, 1/2]^d$; i.e., B_d is a spline of order zero. It can be shown that for an arbitrary N, the function B_N is a spline of order $N-d$. $\Diamond$

2.4. Construction of refinable functions

A function $\varphi \in L_2(\mathbb{R}^d)$ is called a *refinable* function if it satisfies the refinement equation (2.10) . Let us recall that any scaling function is refinable and the function φ satisfies (2.10) if and only if its Fourier transform $\widehat{\varphi}$ satisfies (2.11). Any refinable function defines its mask. The converse is also true. Taking an appropriate function $m_0 \in L_2(\mathbb{T}^d)$ as a mask, we can construct a function $\widehat{\varphi}$ that satisfies equation (2.11) with this mask. Indeed, the right-hand side of (2.11) depends on the function $\widehat{\varphi}$, to which we again can apply (2.11). Continuing this process, for any positive integer n, we get

$$\widehat{\varphi}(\xi) = \widehat{\varphi}(M^{*-n}\xi) \prod_{j=1}^{n} m_0(M^{*-j}\xi).$$

If the function $\widehat{\varphi}$ is continuous at zero and $\widehat{\varphi}(\mathbf{0}) = 1$, then passing to the limit in this equality as $n \to \infty$ and taking into account (2.5), we get

$$\widehat{\varphi}(\xi) = \prod_{j=1}^{\infty} m_0(M^{*-j}\xi). \tag{2.17}$$

For the infinite product to converge at zero, it is necessary to have

$$m_0(\mathbf{0}) = 1. \tag{2.18}$$

Certainly, the convergence at zero is not sufficient. It makes sense to consider only functions m_0 that provide the convergence of the infinite product at least almost everywhere. The following assertion shows that the class of such masks is sufficiently wide.

PROPOSITION 2.4.1. *Suppose a function* $m_0(\xi) = \sum_{k\in\mathbb{Z}^d} c_k e^{2\pi i(k,\xi)}$ *satisfies* (2.18) *and*

$$c_k = O(|k|^{-d-1-\varepsilon}), \quad \varepsilon > 0. \tag{2.19}$$

Then the infinite product (2.17) *converges*[1] *absolutely and uniformly on any compact set.*

Proof. From (2.18) and (2.19) we get

$$|m_0(\xi) - 1| \le 2 \sum_{k \in \mathbb{Z}^d} |c_k \sin \pi(k, \xi)| \le C \sum_{k \in \mathbb{Z}^d} |k|^{-d-\varepsilon} |\xi| =: C_1 |\xi|. \tag{2.20}$$

This relation and (2.6) imply the uniform convergence of the series

$$\sum_{j=1}^{\infty} |m_0(M^{*-j}\xi) - 1|$$

on any compact set. It follows that the infinite product (2.17) converges uniformly on any compact set.

In particular, condition (2.19) holds for trigonometric polynomials, which are of most interest for applications, since computational algorithms for finding coefficients are finite, as in the one-dimensional case (see §1.6).

COROLLARY 2.4.2. *Under the conditions of Proposition* 2.4.1, *the function* $\widehat{\varphi}$ *defined by equality* (2.17) *is continuous on* $\mathbb{R}^d$.

The proof follows from the uniform convergence of partial products to $\widehat{\varphi}$ on compact sets and the continuity of the function m_0, which takes place due to the uniform convergence of its Fourier series.

PROBLEM 2.4.3. Suppose φ is a refinable function, its Fourier transform is continuous, and the series $\sum_{l \in \mathbb{Z}} |\widehat{\varphi}(\xi + l)|^2$ converges uniformly. Prove that if $\widehat{\varphi}(\mathbf{0}) \neq 0$, then $\widehat{\varphi}(n) = 0$ for all $n \in \mathbb{Z}^d$, $n \neq \mathbf{0}$.

It appears that a refinement equation with polynomial mask always has a unique (up to a constant factor) solution in the space of distributions S'; moreover, this solution is compactly supported. In particular, if there exists a solution in $L^2(\mathbb{R}^d)$, it is unique (up to normalization) and compactly supported. To prove this, let us introduce a *transition operator* on the space of distributions with compact support:

$$[Tf](x) = \sum_{k \in \mathbb{Z}^d} c_k f(Mx + k). \tag{2.21}$$

This is a linear continuous operator in the space S'. The refinable function is its eigenvector corresponding to the eigenvalue 1.

THEOREM 2.4.4. *A refinement equation with polynomial mask that satisfies* (2.18) *always has a unique (up to a constant factor) compactly supported solution* $\varphi \in S'$. *This solution is given by formula* (2.17). *Moreover, for any compactly supported distribution* $f \in S'$, *the sequence* $f_n = T^n f$ *converges to the function* $c \cdot \varphi$ *in the space* S', *where the normalizing factor* c *is* $\widehat{f}(0)$.

Proof. First we show that the infinite product $\prod_{k=1}^{\infty} m_0\left(M^{*-k}\xi\right)$ converges in the space S'. It follows that the function φ defined as the inverse Fourier transform of this product belongs to S'. This will also imply (if we use the inverse Fourier

[1] Unlike in the traditional terminology, the infinite product will be considered convergent even in the case when it is equal to zero.

transform) convergence of $T^n f \to \varphi$ in the space $\mathcal{S}'$. The convergence takes place for any compactly supported function $f \in S'$ such that $\widehat{f}(0) = 1$. In particular, it follows that the solution φ is unique. Applying Lemma A.5.1 (see Appendix A.5) to the operator $B = {M^*}^{-1}$, we deduce that the operator T preserves the space $\{f \in \mathcal{S}' : \operatorname{supp} f \subset K\}$. Hence if we choose f such that $\operatorname{supp} f \subset K$, $\widehat{f}(0) = 1$, then the limit $\lim\limits_{n\to\infty} T^n f$ has support on the compact set K, which completes the proof. Thus it remains to prove the convergence of

(2.22)

$$\widehat{T^n f}(\xi) = m_0\left({M^*}^{-1}\xi\right)\cdots m_0\left({M^*}^{-n}\xi\right)\widehat{f}\left({M^*}^{-n}\xi\right) \to \widehat{f}(0)\prod_{k=1}^{\infty} m_0\left({M^*}^{-k}\xi\right)$$

in the space $\mathcal{S}'$ for any compactly supported distribution $f \in \mathcal{S}'$. We have already proved convergence on each compact set in $\mathbb{R}^d$, which however is not sufficient for the convergence in $\mathcal{S}'$.

Let us estimate the rate of growth of the function $\widehat{T^n f}(\xi)$ as $\xi \to \infty$. Let $\rho = \rho({M^*}^{-1})$ be the spectral radius of the operator ${M^*}^{-1}$ (the maximal absolute value of its eigenvalues; see Appendix A.5). By assumption, $\rho < 1$. Fix an arbitrary $q \in (\rho, 1)$ and find a constant C_q such that $\|{M^*}^{-k}\| \le C_q q^k$, $k \in \mathbb{N}$. As is known, there exists a constant C such that $|1 - m_0(\xi)| \le C|\xi|$ for all $\xi \in \mathbb{R}^d$. Now take an arbitrary $\xi \in \mathbb{R}^d$, $|\xi| > \frac{1}{C_q C}$. Let p be the minimal integer such that $|{M^*}^{-p}\xi| \le \frac{1}{C_q C}$. For each $n \ge p+1$, we have

$$(2.23)\quad \left|\widehat{f}\left({M^*}^{-p}\xi\right)\prod_{k=1}^{n} m_0\left({M^*}^{-k}\xi\right)\right| \le C_1 \left|\prod_{k=1}^{p} m_0\left({M^*}^{-k}\xi\right)\right| \cdot \left|\prod_{j=1}^{n-p}\left(1+q^j\right)\right|$$

$$\le C_1 C_2 \|m_0\|_\infty^p < C_1 C_2 \|m_0\|_\infty^{\log_q \frac{q}{C_q C\|\xi\|}} = C_1 C_2 \left(\frac{C_q C|\xi|}{q}\right)^{\log_{1/q}\|m_0\|_\infty},$$

where

$$C_1 = \max\left\{|\widehat{f}(\eta)|,\ |\eta| \le \frac{1}{C_q C}\right\}, \quad C_2 = \left|\prod_{j=1}^{\infty}\left(1+q^j\right)\right|.$$

To estimate the number p, we used the relation $|{M^*}^{-p+1}\xi| > \frac{1}{C_q C}$, whence it follows that

$$q^{p-1} > \frac{1}{C_q C|\xi|},$$

which implies that

$$p < \log_q\left(\frac{q}{C_q C|\xi|}\right).$$

Thus for sufficiently large n, the rate of growth of the function $\widehat{T^n f}(\xi)$ cannot exceed that of a polynomial of degree $\log_{1/q}\|m_0\|_\infty$ as $\xi \to \infty$. Let us now take an arbitrary test function $h \in \mathcal{S}$. Since it decreases at infinity more rapidly than any power of $|\xi|$, it follows that for any $\varepsilon > 0$, there exists a sufficiently large $R = R(\varepsilon)$ such that $\left|\int_{|\xi|>R} \widehat{T^n f}\, h\, d\xi\right| < \varepsilon$ for all $n \ge p+1$. On the other hand, in the ball $\{\xi \in \mathbb{R}^d : |\xi| \le R\}$ the product (2.22) uniformly converges to $\widehat{f}(0)\widehat{\varphi}(\xi)$; hence $(\widehat{T^n f}, h) \to (\widehat{f}(0)\widehat{\varphi}, h)$ as $n \to \infty$, which completes the proof. ◊

Thus, if a function $\varphi \in \mathcal{S}'$, whose Fourier transform is continuous at zero and for which $\widehat{\varphi}(0) = 1$, satisfies a refinement equation with polynomial mask, then it is compactly supported. Moreover, for any function $f \in \mathcal{S}'$, whose Fourier transform is continuous at zero, the sequence $f_n = T^n f$ converges to $\widehat{f}(0) \cdot \varphi$ in the space of distributions.

COROLLARY 2.4.5. *If a refinement equation with polynomial mask has a summable solution φ normalized by the condition $\int_{-\infty}^{\infty} \varphi \, dx = 1$, then this solution is compactly supported and defined by formula* (2.17). *Moreover, for any summable function f, the sequence $f_n = T^n f$ converges to the function $\widehat{f}(0) \cdot \varphi$ in the space $\mathcal{S}'$.*

Let us now study the infinite products (2.17) for masks m_0 with some specific properties.

LEMMA 2.4.6. *Let m_0 be a continuous* 1*-periodic function that satisfies condition* (2.18),

$$F(\xi) := \prod_{j=1}^{\infty} m_0(M^{*-j}\xi).$$

If the function F is continuous and

$$\sum_{l \in \mathbb{Z}^d} |F(\xi + l)| \neq 0 \tag{2.24}$$

for all $\xi \in \mathbb{R}^d$, then there exists a set K congruent to the cube $[-\frac{1}{2}, \frac{1}{2}]^d$ modulo $\mathbb{Z}^d$, containing a neighborhood of zero and such that

$$\inf_{k \in \mathbb{N}} \inf_{\xi \in K} |m_0(M^{*-k}\xi)| \neq 0. \tag{2.25}$$

Proof. Set

$$C_\xi = \sum_{l \in \mathbb{Z}^d} |F(\xi + l)|$$

if the series on the right-hand side converges. Otherwise, set $C_\xi = 1$. It follows from (2.24) that for any $\xi \in [-1/2, 1/2]^d$, there exists $l_\xi \in \mathbb{N}$ such that

$$\sum_{\|l\|_\infty \leq l_\xi} |F(\xi + l)| \geq \frac{1}{2} C_\xi.$$

The function $|F|$ is continuous, so the finite sum $\sum_{\|l\|_\infty \leq l_\xi} |F(\cdot + l)|$ is also continuous. Hence for any ξ, there exists a neighborhood $N_\xi = \{\theta \in \mathbb{R}^d : |\xi - \theta| < R_\xi\}$ such that

$$\sum_{\|l\|_\infty \leq l_\xi} |F(\theta + l)| \geq \frac{1}{4} C_\xi$$

for all $\theta \in N_\xi$. Since $[-1/2, 1/2]^d$ is a compact set and $\{N_\xi\}_{\xi \in \mathbb{R}^d}$ is its open covering, one can select in it a finite subcovering $\{N_\xi\}_{\xi \in \Omega}$. Denote by l_0 the maximal element among l_ξ, $\xi \in \Omega$, and by $4C$ the minimal element among C_ξ, $\xi \in \Omega$. Clearly, for any $\xi \in [-1/2, 1/2]^d$, there exists $l \in \mathbb{Z}^d$, $\|l\|_\infty \leq l_0$, such that

$$|F(\xi + l)| \geq \frac{C}{(2l_0 + 1)^d} =: C_1.$$

Let us renumber the vectors $l \in \mathbb{Z}^d$, $\|l\|_\infty \le l_0$, in such a way that $l(0) = \mathbf{0}$. Thus we obtain the sequence $l(n)$, $n = 0, \ldots, (2l_0+1)^d =: L$. Define sets S_n as follows:

$$S_0 = \{\xi \in [-1/2, 1/2]^d : F(\xi) \ge C_1\},$$
$$S_n = \{\xi \in [-1/2, 1/2]^d \setminus \bigcup_{k=0}^{n-1} S_k : |F(\xi + l(n))| \ge C_1\}.$$

It is clear that the sets S_n, $n = 0, \ldots, L$, are measurable and form a partition of $[-1/2, 1/2]^d$. The function F is continuous and $F(0) = 1$; hence S_0 contains a neighborhood of zero. Set

$$K = \bigcup_{n=0}^{L} (S_n + l(n)). \tag{2.26}$$

It is not hard to see that the set K is congruent to the cube $[-1/2, 1/2]^d$. By construction, $F(\xi) \ge C_1$ on K and K contains a neighborhood of zero.

Let us show that K satisfies (2.25). Let $K \subset [-R, R]^d$. The function m_0 is continuous; hence relations (2.18), (2.5) imply that

$$|m_0(M^{*-k}\xi)| > \frac{1}{2}$$

for all $\xi \in [-R, R]^d$ for $k > k_0$. Thus it suffices to show that

$$\inf_{\xi \in K} |m_0(M^{*-k}\xi)| \ne 0$$

for a finite number of k, $1 \le k \le k_0$. For $\xi \in K$, we have

$$F(\xi) = \left(\prod_{k=1}^{k_0} m_0(M^{*-k}\xi)\right) F(M^{*-k_0}\xi).$$

The left-hand side of this equality is bounded away from zero; the second factor on the right-hand side is bounded since the function F is continuous and thus is bounded on the compact set $\overline{K}$. It follows that

$$\prod_{k=1}^{k_0} |m_0(M^{*-k}\xi)| \ge C_2 > 0$$

for any $\xi \in K$. Since the function m_0 is continuous, each factor in this product is bounded away from zero, i.e.,

$$|m_0(M^{*-k}\xi)| \ge C_3 > 0$$

for all $k = 1, \ldots, k_0$ and any $\xi \in K$, which completes the proof. ◊

LEMMA 2.4.7. *Suppose m_0 satisfies the conditions of Proposition 2.4.1 and there exists a set K congruent to the cube $[-1/2, 1/2]^d$ modulo $\mathbb{Z}^d$ containing a neighborhood of zero and such that*

$$C := \inf_{k \in \mathbb{N}} \inf_{\xi \in K} |m_0(M^{*-k}\xi)| \ne 0. \tag{2.27}$$

Let

$$F_n(\xi) = \prod_{j=1}^{n} |m_0(M^{*-j}\xi)| \chi_K(M^{*-n}\xi), \quad F(\xi) = \prod_{j=1}^{\infty} |m_0(M^{*-j}\xi)|.$$

If $F \in L(\mathbb{R}^d)$, then F_n converges to F in $L(\mathbb{R}^d)$, and if $F \in L_2(\mathbb{R}^d)$, then F_n converges to F in $L_2(\mathbb{R}^d)$.

Proof. First of all we note that by Proposition 2.4.1 the infinite product defining the function F converges at every point. Using the inequality $e^{-2x} \le 1-x$, which is true for $0 \le x \le 1/2$, and relations (2.27), (2.20), for sufficiently large N (which we fix) and any $\xi \in K$, we get

$$|F(\xi)| \ge C^N \prod_{j=N+1}^{\infty} \Big(1 - |1 - m_0(M^{*-j}\xi)|\Big)$$

$$\ge C^N \exp\left(-2C_1 \max_{\xi\in K} |\xi| \sum_{k=N+1}^{\infty} \|M^{-k}\|\right) =: C_2 > 0.$$

This can be reformulated as follows:

$$\chi_K(\xi) \le |F(\xi)|/C_2 \tag{2.28}$$

for all $\xi \in \mathbb{R}^d$. The compact set K contains a neighborhood of zero; hence (2.5) implies that $\chi_K(M^{*-n}\xi) \to 1$ as $n \to +\infty$ for any $\xi \in \mathbb{R}^d$. Combining this with Proposition 2.4.1, we conclude that $F_n \to F$ converges pointwise on $\mathbb{R}^d$. From inequality (2.28) we get

$$|F_n(\xi)| \le \frac{1}{C_2} \prod_{j=1}^{n} |m_0(M^{*-j}\xi)||F(M^{*-n}\xi)| = \frac{1}{C_2}|F(\xi)|$$

for all $\xi \in \mathbb{R}^d$. Thus, if $F \in L(\mathbb{R}^d)$, the functions F_n have a summable majorant. By Lebesgue's dominated convergence theorem, it follows that F_n converges to F in $L(\mathbb{R}^d)$. If $F \in L_2(\mathbb{R}^d)$, then we get

$$|F_n(\xi) - F(\xi)|^2 \le |F_n(\xi)|^2 + 2|F_n(\xi)|\,|F(\xi)| + |F(\xi)|^2$$

$$\le \left(\frac{1}{C_2^2} + \frac{2}{C_2} + 1\right)|F(\xi)|^2 =: C_4|F(\xi)|^2.$$

The function $|F_n - F|^2$ converges pointwise to zero and has the summable majorant $C_4|F(\xi)|^2$; therefore $F_n \to F$ in $L_2(\mathbb{R}^d)$. ◊

2.5. Conditions of biorthogonality

In this section, we consider refinable functions $\varphi, \widetilde{\varphi}$ with masks

$$m_0(\xi) := m^{-1/2} \sum_{k\in\mathbb{Z}^d} h_k e^{2\pi i(k,\xi)}, \qquad \widetilde{m_0}(\xi) := m^{-1/2} \sum_{k\in\mathbb{Z}^d} \widetilde{h}_k e^{2\pi i(k,\xi)}, \tag{2.29}$$

respectively. We want to know when integer shifts of the functions $\varphi, \widetilde{\varphi}$ form a biorthonormal system in $L_2(\mathbb{R}^d)$ (in particular, when integer shifts of one function φ form an orthonormal system).

Repeating Proposition 1.1.12 almost word for word, we obtain the following assertion.

PROPOSITION 2.5.1. *Let φ, $\widetilde{\varphi} \in L_2(\mathbb{R}^d)$. The systems $\{\varphi(\cdot + n)\}_{n\in\mathbb{Z}^d}$ and $\{\widetilde{\varphi}(\cdot + n)\}_{n\in\mathbb{Z}^d}$ are biorthonormal if and only if the equality*

$$\sum_{k\in\mathbb{Z}^d} \widehat{\varphi}(\xi + k)\overline{\widehat{\widetilde{\varphi}}(\xi + k)} = 1 \tag{2.30}$$

holds for almost all $\xi \in \mathbb{R}^d$, *where the series on the left-hand side of* (2.30) *converges absolutely.*

PROPOSITION 2.5.2. *Let* $\varphi, \widetilde{\varphi}$ *be refinable functions with masks* $m_0, \widetilde{m_0}$, *respectively. If integer shifts of* φ, $\widetilde{\varphi}$ *form a biorthonormal system, then*

$$\sum_{s\in D(M^*)} m_0(\xi + {M^*}^{-1}s)\overline{\widetilde{m_0}(\xi + {M^*}^{-1}s)} = 1 \tag{2.31}$$

for almost all $\xi \in \mathbb{R}^d$ *and any set of digits* $D(M^*)$ *of the matrix* M^*.

Proof. Using formula (2.30) and the refinement equations for the functions $\varphi, \widetilde{\varphi}$, we get

$$\begin{aligned} 1 &= \sum_{p\in\mathbb{Z}^d} \widehat{\varphi}(M^*\xi + p)\overline{\widehat{\widetilde{\varphi}}(M^*\xi + p)} \\ &= \sum_{p\in\mathbb{Z}^d} m_0(\xi + {M^*}^{-1}p)\widehat{\varphi}(\xi + {M^*}^{-1}p)\overline{\widetilde{m_0}(\xi + {M^*}^{-1}p)\widehat{\widetilde{\varphi}}(\xi + {M^*}^{-1}p)}. \end{aligned} \tag{2.32}$$

Using Lemma 2.2.2, we make the change of the index of summation $p = M^*k + s$, $k \in \mathbb{Z}^d$, $s \in D(M^*)$, on the right-hand side of (2.32), reducing it to the form

$$\sum_{k\in\mathbb{Z}^d}\sum_{s\in D(M^*)} m_0(\xi + \theta_{ks})\overline{\widetilde{m_0}(\xi + \theta_{ks})}\widehat{\varphi}(\xi + \theta_{ks})\overline{\widehat{\widetilde{\varphi}}(\xi + \theta_{ks})},$$

where $\theta_{ks} = {M^*}^{-1}s + k$. Changing the order of summation, using 1-periodicity of the functions $m_0, \widetilde{m_0}$ in each variable and (2.30), we get

$$\begin{aligned} \sum_{s\in D(M^*)} m_0(\xi + \theta_{\mathbf{0}s})\overline{\widetilde{m_0}(\xi + \theta_{\mathbf{0}s})} & \sum_{k\in\mathbb{Z}^d} \widehat{\varphi}(\xi + \theta_{ks})\overline{\widehat{\widetilde{\varphi}}(\xi + \theta_{ks})} \\ &= \sum_{s\in D(M^*)} m_0(\xi + {M^*}^{-1}s)\overline{\widetilde{m_0}(\xi + {M^*}^{-1}s)}. \end{aligned} \qquad \diamond$$

Condition (2.31) can be easily verified for a given pair of masks, but it is not convenient to use this condition for finding an appropriate pair m_0, $\widetilde{m}_0$. One can use the equivalent form of (2.31) in terms of the Fourier coefficients of the functions $m_0, \widetilde{m}_0$.

PROPOSITION 2.5.3. *Let there be given* 1*-periodic functions*

$$m_0(\xi) = m^{-1/2}\sum_{k\in\mathbb{Z}^d} h_k e^{2\pi i(k,\xi)}, \qquad \widetilde{m}_0(\xi) = m^{-1/2}\sum_{k\in\mathbb{Z}^d} \widetilde{h}_k e^{2\pi i(k,\xi)}$$

with absolutely convergent Fourier series. Then condition (2.31) *is equivalent to the fact that the equality*

$$\sum_{k\in\mathbb{Z}^d} h_k\overline{\widetilde{h}_{k-Mn}} = \delta_{n0} \tag{2.33}$$

holds for any $n \in \mathbb{Z}^d$.

Proof. We have the equality

$$\sum_{s\in D(M^*)} m_0(\xi + {M^*}^{-1}s)\overline{\widetilde{m_0}(\xi + {M^*}^{-1}s)}$$
$$= m^{-1} \sum_{s\in D(M^*)} \sum_{k\in\mathbb{Z}^d} \sum_{l\in\mathbb{Z}^d} h_k \overline{\widetilde{h}_l} e^{2\pi i(k-l,\xi)} e^{2\pi i(k-l,{M^*}^{-1}s)}$$
$$= m^{-1} \sum_{n\in\mathbb{Z}^d} \sum_{k\in\mathbb{Z}^d} e^{2\pi i(n,\xi)} h_k \overline{\widetilde{h}_{k-n}} \sum_{s\in D(M^*)} e^{2\pi i(M^{-1}n,s)},$$

where the change of the order of summation is justified by absolute convergence over the $2d$-dimensional index (n,k) of the series on the right-hand side, while the sum over s is finite. From Lemma 2.2.6 it follows that

$$\sum_{s\in D(M^*)} m_0(\xi + {M^*}^{-1}s)\overline{\widetilde{m_0}(\xi + {M^*}^{-1}s)} = \sum_{n\in\mathbb{Z}^d} e^{2\pi i(Mn,\xi)} \sum_{k\in\mathbb{Z}^d} h_k \overline{\widetilde{h}_{k-Mn}} \tag{2.34}$$

for all $\xi \in \mathbb{R}^d$. Thus condition (2.31) is equivalent to the fact that

$$1 = \sum_{n\in\mathbb{Z}^d} e^{2\pi i(n,\omega)} \left(\sum_{k\in\mathbb{Z}^d} h_k \overline{\widetilde{h}_{k-Mn}} \right), \qquad \omega = M^*\xi,$$

for all $\omega \in \mathbb{R}^d$. Thus we have obtained the Fourier expansion of unity. Taking the Fourier coefficients of this series, we get the required equalities. ◊

We now show how to choose a polynomial mask $\widetilde{m}_0$ satisfying conditions (2.31), (2.18), given a polynomial mask m_0.

Example 2.5.4. Let $d = 2$, $M = \begin{pmatrix} 1 & 1 \\ 1 & -1 \end{pmatrix}$, $m_0(\xi) = \sum_{k\in\mathbb{Z}^2} c_k e^{2\pi i(k,\xi)}$, $c_{00} = 1/2$, $c_{10} = 1/4$, $c_{11} = 1/4$, and let the rest of c_k be equal to zero. We want to find $\widetilde{m}_0(\xi) = \sum_{k\in\mathbb{Z}^2} x_k e^{2\pi i(k,\xi)}$ to satisfy conditions (2.33), (2.18). First of all we need to choose a suitable spectrum Ω of the function $\widetilde{m}_0$. It is clear that for a finite spectrum, condition (2.33) is a finite set of equalities. Adding to them the equality

$$\sum_{k\in\mathbb{Z}^2} x_k = 1,$$

which makes $\widetilde{m}_0$ satisfy condition (2.18), we obtain a linear system in variables x_k, $k \in \Omega$. Certainly, this system is not compatible for any choice of spectrum. For a suitable spectrum, take that of the function m_0. It is not hard to see that for the vectors congruent to zero modulo M, the sums of coordinates are even and the right-hand sides of the equalities in (2.33) are nonzero only for the vectors $Mn = (0,0)^T$, $(1,1)^T$, $(-1,-1)^T$ (which correspond to $n = (0,0)^T, (1,0)^T, (-1,0)^T$). Thus the system takes the form

$$\begin{cases} c_{00}x_{00} + c_{10}x_{10} + c_{11}x_{11} = \frac{1}{2}, \\ c_{11}x_{00} = 0, \\ c_{00}x_{11} = 0, \\ x_{00} + x_{10} + x_{11} = 1. \end{cases}$$

Substituting the values of the coefficients for c_k in the above system, we see that it has no solutions. Extend the spectrum of the function $\widetilde{m}_0$ by adding the unknown coefficients x_{21}, x_{22}. We get the system

$$\begin{cases} c_{00}x_{00} & +c_{10}x_{10} & +c_{11}x_{11} & & & = \frac{1}{2}, \\ c_{11}x_{00} & & & & & = 0, \\ & & c_{00}x_{11} & +c_{10}x_{21} & +c_{11}x_{22} & = 0, \\ & & & & c_{00}x_{22} & = 0, \\ x_{00} & +x_{10} & +x_{11} & +x_{21} & +x_{22} & = 1, \end{cases}$$

which has a unique solution

$$x_{00} = 0, \qquad x_{10} = 3/2, \qquad x_{11} = 1/2, \qquad x_{21} = -1, \qquad x_{22} = 0. \qquad \Diamond$$

Proposition 2.5.2 generalizes the one-dimensional result formulated in Corollary 1.2.6. Thus we have obtained a necessary condition for integer shifts of a refinable function to be biorthonormal, which generally is not sufficient. This is demonstrated by the following example.

EXAMPLE 2.5.5. Let $\varphi = \chi_{[0,3]}$. Obviously, this is a refinable function for $M = 2$ and for its mask $m_0(\xi) = \frac{1}{2}\left(1 + e^{6\pi i\xi}\right)$. The following identity holds:

$$|m_0(\xi)|^2 + |m_0(\xi + 1/2)|^2 = 1.$$

However, it is clear that all integer shifts of the function φ are not orthogonal to each other. Moreover, these integer shifts do not form a Riesz system. Indeed, direct calculation gives

$$\sum_{l\in\mathbb{Z}} |\widehat{\varphi}(\xi + l)|^2 = 3 + 4\cos 2\pi\xi + 2\cos 4\pi\xi.$$

It is readily seen that the right-hand side of this equality vanishes for $\xi = 1/3$. $\Diamond$

Unfortunately, no sufficient conditions that would be simply formulated and easily checked in practice have been found so far. The sufficient conditions described below in various terms do not seem very good in this sense. Still, they can often be verified in some specific cases.

THEOREM 2.5.6 (Cohen's criterion). *Let $\varphi, \widetilde{\varphi}$ be refinable functions with masks $m_0, \widetilde{m_0}$, respectively, and let these masks satisfy the conditions of Proposition* 2.4.1 *and* (2.31). *A sufficient condition for the biorthonormality of integer shifts of the functions φ, $\widetilde{\varphi}$ is the existence of a set K congruent to the cube $[-\frac{1}{2}, \frac{1}{2}]^d$ modulo $\mathbb{Z}^d$, containing a neighborhood of zero and such that*

$$\inf_{k\in\mathbb{N}} \inf_{\xi\in K} |m_0(M^{*^{-k}}\xi)| \neq 0, \quad \inf_{k\in\mathbb{N}} \inf_{\xi\in K} |\widetilde{m_0}(M^{*^{-k}}\xi)| \neq 0. \tag{2.35}$$

If, in addition,

$$\varphi(x), \widetilde{\varphi}(x) = O\left(|x|^{-(d+\varepsilon)}\right), \quad \varepsilon > 0, \quad |x| \to \infty, \tag{2.36}$$

then this condition is necessary and sufficient.

REMARK 2.5.7. In practice, to verify that condition (2.35) holds, it suffices to find the minimal value of a finite number of functions on $\overline{K}$. Indeed, the compactness of $\overline{K}$, continuity of m_0 and $\widetilde{m_0}$, equalities $m_0(\mathbf{0}) = \widetilde{m}_0(\mathbf{0}) = 1$, and condition (2.5) imply that $|m_0(M^{*^{-k}}\xi)\widetilde{m_0}(M^{*^{-k}}\xi)| > 1/2$ for all $\xi \in [-R, R]^d$ for $k > k_0$. Therefore it remains to show that $2k_0$ functions

$$m_0(M^{*^{-1}}\xi),\ \widetilde{m_0}(M^{*^{-1}}\xi), \ldots, m_0(M^{*^{-k_0}}\xi),\ \widetilde{m_0}(M^{*^{-k_0}}\xi)$$

have no zeros on $\overline{K}$. The set $\overline{K}$ is called Cohen's compact set.

Proof of Theorem 2.5.6. Sufficiency. Applying Lemma 2.4.7 to the function

$$m_0^*(\xi) := m_0(\xi)\overline{\widetilde{m}_0(\xi)},$$

we deduce that the sequence

$$\mu_k(\xi) := \prod_{j=1}^{k} m_0(M^{*^{-j}}\xi)\overline{\widetilde{m_0}(M^{*^{-j}}\xi)}\ \chi_K(M^{*^{-k}}\xi), \quad k \in \mathbb{N}, \tag{2.37}$$

converges to $\widehat{\varphi} \cdot \overline{\widehat{\widetilde{\varphi}}}$ in $L(\mathbb{R}^d)$. Since the set K is congruent to the cube $[-\frac{1}{2}, \frac{1}{2}]^d$ modulo $\mathbb{Z}^d$, we get

$$\int_K f = \int_{[-1/2,1/2]^d} f = \int_{[0,1]^d} f$$

for any function f 1-periodic in each variable. In particular, using (2.37), for any $n \in \mathbb{Z}^d$, we obtain

$$\begin{aligned}
\int_{\mathbb{R}^d} \mu_k(\xi)e^{2\pi i(n,\xi)}d\xi &= m^k \int_{\mathbb{R}^d} \mu_k(M^{*^k}\theta)e^{2\pi i(n,M^{*^k}\theta)}d\theta \\
&= m^k \int_K \prod_{j=1}^{k} m_0(M^{*^{k-j}}\theta)\overline{\widetilde{m_0}(M^{*^{k-j}}\theta)}e^{-2\pi i(n,M^{*^k}\theta)}d\theta \\
&= m^k \int_{[0,1]^d} \prod_{l=1}^{k-1} m_0(M^{*^l}\theta)\overline{\widetilde{m_0}(M^{*^l}\theta)}e^{2\pi i(n,M^{*^k}\theta)}m_0(\theta)\overline{\widetilde{m_0}(\theta)}d\theta.
\end{aligned}$$

Under the integral sign on the right-hand side we have a function 1-periodic in each variable; hence using Lemma 2.2.5 for $A = M^*$, we represent this integral in the form

$$m^k \sum_{r\in D(M^*)} \int_{Q_1+M^{*^{-1}}r} \prod_{l=1}^{k-1} m_0(M^{*^l}\theta)\overline{\widetilde{m_0}(M^{*^l}\theta)}e^{2\pi i(n,M^{*^k}\theta)}m_0(\theta)\overline{\widetilde{m_0}(\theta)}d\theta,$$

where $Q_1 = M^{*^{-1}}[0,1)^d$. Making the change of variables $\theta = \xi + M^{*^{-1}}r$ in each summand, putting the sum under the integral sign, and using (2.31), we get

$$m^k \int_{Q_1} \prod_{l=1}^{k-1} m_0(M^{*^l}\xi)\overline{\widetilde{m_0}(M^{*^l}\xi)}e^{2\pi i(n,M^{*^k}\xi)}d\xi.$$

The change of variables $\theta = M^*\xi$ in the integral and the change of index $l-1 \to l$ in the product give

$$\begin{aligned}
m^{k-1} \int_{[0,1)^d} \prod_{l=0}^{k-2} m_0(M^{*^l}\theta)\overline{\widetilde{m_0}(M^{*^l}\theta)}e^{2\pi i(n,M^{*^{k-1}}\theta)}d\theta \\
= \int_{\mathbb{R}^d} \mu_{k-1}(\theta)e^{2\pi i(n,\theta)}d\theta.
\end{aligned}$$

Thus for any $k,\ n \in \mathbb{N}$, we have

$$\int_{\mathbb{R}^d} \mu_k(\theta)e^{2\pi i(n,\theta)}d\theta = \int_{\mathbb{R}^d} \mu_{k-1}(\theta)e^{2\pi i(n,\theta)}d\theta.$$

Constructing an analogous chain of equalities for $k = 1$, we obtain

$$\int_{\mathbb{R}^d} \mu_1(\xi) e^{2\pi i(n,\xi)} d\xi = \cdots = \int_{[0,1)^d} e^{2\pi i(n,\theta)} d\theta = \delta_{n0} \quad \forall n \in \mathbb{N};$$

therefore,

$$\int_{\mathbb{R}^d} \widehat{\varphi}(\xi)\, \overline{\widehat{\widetilde{\varphi}}(\xi)} e^{2\pi i(n,\xi)} d\xi = \lim_{k\to\infty} \int_{\mathbb{R}^d} \mu_k(\xi) e^{2\pi i(n,\xi)} d\xi = \delta_{n0},$$

which by the Plancherel theorem yields the equality

$$\int_{\mathbb{R}^d} \varphi(x)\, \overline{\widetilde{\varphi}}(x-n)\, dx = \delta_{n0}.$$

Necessity. Assume that (2.36) holds and prove that the function

$$\sum_{l\in\mathbb{Z}^d} \widehat{\varphi}(\xi + l)\overline{\widehat{\widetilde{\varphi}}(\xi + l)} =: \Phi(\xi) \tag{2.38}$$

is continuous on $\mathbb{R}^d$. Set

$$G(t) = \sum_{k\in\mathbb{Z}^d} \varphi(t+k) e^{-2\pi i(t+k,\xi)}, \quad \widetilde{G}(t) = \sum_{k\in\mathbb{Z}^d} \widetilde{\varphi}(t+k) e^{-2\pi i(t+k,\xi)}.$$

By the Poisson summation formula we have

$$\widehat{G}(l) = \widehat{\varphi}(\xi + l), \quad \widehat{\widetilde{G}}(l) = \widehat{\widetilde{\varphi}}(\xi + l)$$

for any $l \in \mathbb{Z}^d$. Using the generalized Parseval equality, we get

$$\begin{aligned}
\Phi(\xi) &= \sum_{l\in\mathbb{Z}^d} \widehat{G}(l)\overline{\widehat{\widetilde{G}}(l)} = \int_{[0,1]^d} G(t)\overline{\widetilde{G}(t)}\, dt \\
&= \int_{[0,1]^d} \sum_{k\in\mathbb{Z}^d} \varphi(t+k) e^{-2\pi i(t+k,\xi)} \overline{\widetilde{G}(t)}\, dt = \int_{\mathbb{R}^d} \varphi(t) e^{-2\pi i(t,\xi)} \overline{\widetilde{G}(t)}\, dt \\
&= \int_{\mathbb{R}^d} \varphi(t) \sum_{k\in\mathbb{Z}^d} \overline{\widetilde{\varphi}(t+k)} e^{-2\pi i(k,\xi)}\, dt.
\end{aligned}$$

Since in view of condition (2.36),

$$\sum_{k\in\mathbb{Z}^d} \int_{\mathbb{R}^d} |\varphi(t)\widetilde{\varphi}(t+k)|\, dt = \int_{\mathbb{R}^d} |\varphi(t)| \sum_{k\in\mathbb{Z}^d} |\widetilde{\varphi}(t+k)|\, dt < \infty,$$

interchanging summation and integration, represent $\Phi(\xi)$ as the sum of absolutely convergent trigonometric series, which is continuous.

The continuity of the function Φ and Proposition 2.5.1 imply that the equality $\Phi(\xi) = 1$ holds for all $\xi \in \mathbb{R}^d$, so F satisfies all the conditions of Lemma 2.4.6, which proves our statement. ◊

COROLLARY 2.5.8. *Suppose $\varphi, \widetilde{\varphi}$ are refinable functions with masks $m_0, \widetilde{m_0}$, respectively, satisfying* (2.31) *and the conditions of Proposition* 2.4.1. *If m_0 and $\widetilde{m}_0$ do not vanish on the set $\bigcup_{k=1}^{\infty} M^{*-k}[-1/2, 1/2]^d$, then integer shifts of the functions φ, $\widetilde{\varphi}$ form a biorthonormal system.*

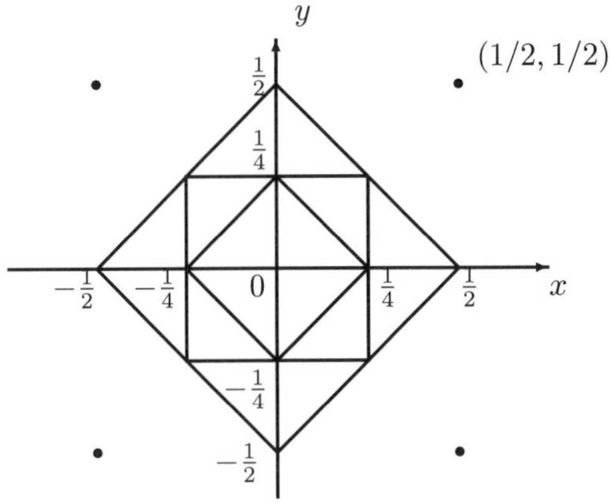

FIGURE 2.1. Sets $M^{*-k}[-1/2,1/2]^2$ for $k=1,2,3$.

The assumptions of this corollary are a special case of the assumptions of Theorem 2.5.6, where the cube $[-1/2,1/2]^d$ is taken as a set K.

In view of (2.5), there exists k_0 such that the relation $M^{*-k}[-1/2,1/2]^d \subset M^{*-1}[-1/2,1/2]^d$ holds for any $k > k_0$; hence under the conditions of Corollary 2.5.8 it suffices to check that the functions m_0 and $\widetilde{m}_0$ have no zeros on the sets $M^{*-k}[-1/2,1/2]^d$, $k=1,\dots,k_0$. If $M^{*-k}[-1/2,1/2]^d \subset M^{*-k+1}[-1/2,1/2]^d$, then it is sufficient that m_0 and $\widetilde{m}_0$ have no zeros on $M^{*-1}[-1/2,1/2]^d$.

EXAMPLE 2.5.9. Let $d=2$, $M = \begin{pmatrix} 1 & 1 \\ 1 & -1 \end{pmatrix}$,

$$\begin{aligned} m_0(\xi) &= \frac{1}{2} + \frac{1}{4}\cos 2\pi\xi_1 + \frac{1}{4}\cos 2\pi\xi_2, \\ \widetilde{m}_0(\xi) &= 1 + \frac{1}{4}(\cos 2\pi\xi_1 + \cos 2\pi\xi_2) \\ &\quad - \frac{1}{8}(\cos^2 2\pi\xi_1 + \cos^2 2\pi\xi_2) - \frac{1}{4}\cos 2\pi\xi_1 \cos 2\pi\xi_2. \end{aligned}$$

The coefficients of the polynomials m_0 and $\widetilde{m}_0$ in the matrix form are the following:

$$\begin{pmatrix} 0 & 1/8 & 0 \\ 1/8 & 1/2 & 1/8 \\ 0 & 1/8 & 0 \end{pmatrix}, \quad \begin{pmatrix} 0 & 0 & -1/32 & 0 & 0 \\ 0 & -1/16 & 1/8 & -1/16 & 0 \\ -1/32 & 1/8 & 7/8 & 1/8 & -1/32 \\ 0 & -1/16 & 1/8 & -1/16 & 0 \\ 0 & 0 & -1/32 & 0 & 0 \end{pmatrix}.$$

It is not hard to verify that (2.18) and (2.33) (and thus (2.31)) hold. Let us show that these masks satisfy the sufficient condition of biorthogonality of Corollary 2.5.8. Indeed, the polynomials m_0 and $\widetilde{m}_0$ vanish only at the point $(1/2,1/2)^T$ on $\mathbb{T}^d$, since the equalities $m_0(\xi)=0$, $\widetilde{m}_0(\xi)=0$ hold only if all cosines in them are equal to -1. Hence no zero of the functions $m_0, \widetilde{m}_0$ belongs to the set $M^{-1}[-1/2,1/2]^d$, and it is easily seen that the inclusion $M^{-k}[-1/2,1/2]^d \subset M^{-1}[-1/2,1/2]^d$ takes place for all $k \geq 2$ (see Figure 2.1). ◊

Consider two refinable functions $\varphi, \widetilde{\varphi} \in L(\mathbb{R}^d)$ and set

$$\varphi^*(x) := \int_{\mathbb{R}^d} \varphi(t)\overline{\widetilde{\varphi}(x-t)}dt, \quad x \in \mathbb{R}^d, \tag{2.39}$$

$$h_k^* := \sum_{n\in\mathbb{Z}^d} h_n \overline{\widetilde{h}_{n-k}}, \quad k \in \mathbb{Z}^d. \tag{2.40}$$

The Plancherel theorem implies the equality

$$\varphi^*(x) = \int_{\mathbb{R}^d} \widehat{\varphi}(-\xi)\overline{\widehat{\widetilde{\varphi}}(\xi)}e^{-2\pi i(\xi,x)}\,d\xi,$$

from which it is easy to see that the function φ^* is continuous on $\mathbb{R}^d$. Moreover, by the condition $\varphi, \widetilde{\varphi} \in L_2(\mathbb{R}^d)$, the function φ^* is bounded, and $\varphi^* \in L(\mathbb{R}^d)$ by the condition $\varphi, \widetilde{\varphi} \in L(\mathbb{R}^d)$. Hence $\varphi^* \in L_2(\mathbb{R}^d)$. Let us find the Fourier transform of this function. Since the function under the integral sign in

$$\int_{\mathbb{R}^d}\int_{\mathbb{R}^d} \varphi(t)\overline{\widetilde{\varphi}(x-t)}e^{-2\pi i(\xi,x)}\,dt\,dx$$

is summable on $\mathbb{R}^d \times \mathbb{R}^d$, changing the order of integration, we get

$$\widehat{\varphi}^*(\xi) = \widehat{\varphi}(\xi)\overline{\widehat{\widetilde{\varphi}}(-\xi)}.$$

Applying equality (2.11) to both factors on the right-hand side, we obtain

$$\begin{aligned}\widehat{\varphi}^*(\xi) &= m_0({M^*}^{-1}\xi)\overline{\widetilde{m}_0(-{M^*}^{-1}\xi)}\widehat{\varphi}({M^*}^{-1}\xi)\overline{\widehat{\widetilde{\varphi}}(-{M^*}^{-1}\xi)} \\ &= m_0({M^*}^{-1}\xi)\overline{\widetilde{m}_0(-{M^*}^{-1}\xi)}\widehat{\varphi}^*({M^*}^{-1}\xi).\end{aligned}$$

Thus φ^* is a refinable function with mask

$$m_0(\xi)\overline{\widetilde{m}_0(-\xi)} =: m_0^*(\xi).$$

It is easy to verify that

$$m_0^*(\xi) = m^{-1}\sum_{k\in\mathbb{Z}^d} h_k^* e^{2\pi i(k,\xi)}.$$

LEMMA 2.5.10. *Suppose $\varphi, \widetilde{\varphi} \in L(\mathbb{R}^d)$ are refinable functions with masks m_0, $\widetilde{m_0}$, respectively. The Fourier series of these masks are absolutely convergent, and the function φ^* is given by equality* (2.39). *Then for any $\xi \in \mathbb{R}^d$, we have the following equality:*

$$\varphi^*(\xi) = \sum_{n\in\mathbb{Z}^d} h_n^*\varphi^*(M\xi + n). \tag{2.41}$$

Proof. The absolute convergence of the Fourier series of the functions $m_0, \widetilde{m_0}$ and relation (2.40) imply that

$$\sum_{n\in\mathbb{Z}^d} |h_n^*| < \infty. \tag{2.42}$$

Since the function φ^* is bounded, we can state that the series on the right-hand side of (2.41) converges uniformly. Hence in view of the continuity of φ^*, its sum is continuous. On the other hand, this series converges to the function φ^* in $L_2(\mathbb{R}^d)$

since relation (2.41) is a refinement equation for φ^*, which implies that (2.41) holds at each point. $\Diamond$

We introduce the operator W_{φ^*}, which takes ℓ_1 to ℓ_1:

$$(W_{\varphi^*}c)_k := \sum_{n\in\mathbb{Z}^d} h_n^* c_{n+Mk}, \quad c = \{c_n\}_{n\in\mathbb{Z}^d} \in \ell_1. \tag{2.43}$$

THEOREM 2.5.11 (Lawton's criterion). *Let $\varphi, \widetilde{\varphi} \in L(\mathbb{R}^d)$ be refinable functions with masks $m_0, \widetilde{m_0}$, respectively. Suppose the Fourier series of these masks converge absolutely and satisfy (2.18) and the function φ^* is defined by equality (2.39). A necessary condition for integer shifts of the functions φ, $\widetilde{\varphi}$ to be biorthonormal is that the sequence $\delta := \{\delta_{k0}\}_{k\in\mathbb{Z}^d}$ be the only eigenvector of the operator W_{φ^*} corresponding to eigenvalue 1 in the space ℓ_1. If, in addition,*

$$\varphi(x), \widetilde{\varphi}(x) = O\left(|x|^{-d-\varepsilon}\right), \quad \varepsilon > 0, \quad |x| \to \infty, \tag{2.44}$$

then the above necessary condition is sufficient.

Proof. Sufficiency. Let δ be the only eigenvector of the operator W_{φ^*} corresponding to the eigenvalue 1 in the space ℓ_1. By Lemma 2.5.10,

$$\varphi^*(k) = \sum_{n\in\mathbb{Z}^d} h_n^* \varphi^*(Mk+n)$$

for any $k \in \mathbb{Z}^d$; i.e., the sequence $\{\varphi^*(k)\}_{k\in\mathbb{Z}^d}$ is an eigenvector of the operator W_{φ^*} corresponding to the eigenvalue 1. Let us show that

$$\sum_{n\in\mathbb{Z}^d} |\varphi^*(n)| = \sum_{n\in\mathbb{Z}^d} \left| \int_{\mathbb{R}^d} \varphi(x)\overline{\widetilde{\varphi}}(x+n)\,dx \right| < \infty. \tag{2.45}$$

To do this, it suffices to verify the convergence of the series

$$\sum_{k\in\mathbb{Z}^d}\sum_{n\in\mathbb{Z}^d} \int_{\mathbb{T}^d} |\varphi(x+k)\widetilde{\varphi}(x+k+n)|\,dx. \tag{2.46}$$

From (2.44) it follows that there exist $C > 0$ and $R > 1$ such that

$$|\varphi(x)|, |\widetilde{\varphi}(x)| \le \frac{C}{|x|^{d+\varepsilon}}$$

for $|x| \ge R$. Therefore,

$$\sum_{|k|\ge R+1}\sum_{n\in\mathbb{Z}^d} \int_{\mathbb{T}^d} |\varphi(x+k)\widetilde{\varphi}(x+k+n)|\,dx \le \sum_{|k|\ge R+1} \frac{2^{d+\varepsilon}C}{|k|^{d+\varepsilon}}\|\widetilde{\varphi}\|_1 < \infty.$$

To prove the convergence of series (2.46), it remains to note that for any fixed $k \in \mathbb{Z}^d$,

$$\sum_{|n+k|\ge R+1} \int_{\mathbb{T}^d} |\varphi(x+k)\widetilde{\varphi}(x+k+n)|\,dx \le \sum_{|n+k|\ge R+1} \frac{2^{d+\varepsilon}C}{|n+k|^{d+\varepsilon}}\|\varphi\|_1 < \infty.$$

Relation (2.45) means that $\{\varphi^*(k)\}_{k\in\mathbb{Z}^d} \in \ell_1$. Consequently, $\varphi^*(k) = \delta_{k0}$ for any $k \in \mathbb{Z}^d$, which means that integer shifts of the functions φ, $\widetilde{\varphi}$ are biorthonormal.

Necessity. Suppose $\varphi^*(k) = \delta_{k0}$ for any $k \in \mathbb{Z}^d$ and $c = \{c_k\}_{k\in\mathbb{Z}^d} \in \ell_1$ is an eigenvector of the operator W_{φ^*} corresponding to the eigenvalue 1. Let us show that the equality

$$\sum_{k\in\mathbb{Z}^d} c_{-k}\varphi^*(r + M^{-N}k) = c_r \tag{2.47}$$

holds for any positive integer N and any $r \in \mathbb{Z}^d$. The proof is by induction. First take $N = 1$. Apply (2.41) to the left-hand side of (2.47), which gives

$$\sum_{k\in\mathbb{Z}^d} c_{-k} \sum_{n\in\mathbb{Z}^d} h_n^*\varphi^*(Mr + k + n) = \sum_{k\in\mathbb{Z}^d} c_{-k}h^*_{-k-Mr} = (W_{\varphi^*}c)_r = c_r.$$

Now take $N+1$ in place of N in (2.47). Applying (2.41), we get

$$\sum_{k\in\mathbb{Z}^d} c_{-k}\varphi^*(r + M^{-N-1}k) = \sum_{k\in\mathbb{Z}^d} c_{-k} \sum_{n\in\mathbb{Z}^d} h_n^*\varphi^*(Mr + M^{-N}k + n).$$

Changing the order of summation on the right-hand side (which is possible since the double series converges absolutely in view of relation (2.42) and boundedness of φ^*) and using the induction hypothesis, we obtain

$$\sum_{k\in\mathbb{Z}^d} c_{-k}\varphi^*(r + M^{-N-1}k) = \sum_{n\in\mathbb{Z}^d} h_n^* c_{Mr+n} = (W_{\varphi^*}c)_r = c_r.$$

It follows from the absolute convergence of the series $\sum\limits_{k\in\mathbb{Z}^d} c_k$ and boundedness of φ^* that

$$\lim_{N\to+\infty} \sum_{k\in\mathbb{Z}^d} c_{-k}\varphi^*(r + M^{-N-1}k) = \sum_{k\in\mathbb{Z}^d} c_{-k} \lim_{N\to+\infty} \varphi^*(r + M^{-N-1}k).$$

Passing to the limit in (2.47) and taking into account (2.5), we get

$$c_r = \varphi^*(r) \sum_{k\in\mathbb{Z}^d} c_{-k} = \delta_{r0} \sum_{k\in\mathbb{Z}^d} c_k,$$

which obviously implies that $c_r = \delta_{r0}$ for any $r \in \mathbb{Z}^d$. It remains to note that Propositions 2.5.2 and 2.5.3 and equality (2.40) imply that δ is an eigenvector of the operator W_{φ^*} corresponding to the eigenvalue 1. ◊

In practice, Theorem 2.5.11 suits better to check that orthogonality or biorthogonality does not hold (though even by this theorem it is usually rather hard to do). As an example, we check the absence of orthogonality for the mask considered in Example 2.5.5.

EXAMPLE 2.5.12. Let $d = 1$, $M = 2$, $m_0(\xi) = \frac{1}{2}\left(1 + e^{6\pi i\xi}\right)$. Let us show that this mask does not generate a refinable function φ, whose integer shifts form an orthonormal system. The mask m^* of the function φ^* has three nonzero elements $h_0^* = 1$, $h_3^* = 1/2$, $h_{-3}^* = 1/2$; hence the eigenvectors of the operator W_{φ^*} corresponding to eigenvalue 1 can be found from the system

$$c_k = c_{2k} + \frac{1}{2}c_{2k+3} + \frac{1}{2}c_{2k-3}, \quad k \in \mathbb{Z}.$$

It is clear that δ is a solution of this system. Consider the vector c with only four nonzero components $c_1 = c_{-1} = 1$, $c_2 = c_{-2} = 1/2$. These components appear only in the equations

$$\begin{cases} c_1 & = c_2 & +\frac{1}{2}c_{-1} & +\frac{1}{2}c_5, \\ c_2 & = c_4 & +\frac{1}{2}c_1 & +\frac{1}{2}c_7, \\ c_{-1} & = c_{-2} & +\frac{1}{2}c_1 & +\frac{1}{2}c_{-5}, \\ c_{-2} & = c_{-4} & +\frac{1}{2}c_{-1} & +\frac{1}{2}c_{-7}, \end{cases}$$

which turn into equalities. Thus we have found another eigenvector of the operator W_{φ^*} corresponding to the eigenvalue 1 in ℓ_1. ◊

In the above criteria of orthogonality and biorthogonality both refinable functions were assumed to belong to the space $L_2(\mathbb{R}^d)$. In practice, these functions are usually constructed by masks, which makes it rather difficult to verify whether they belong to $L_2(\mathbb{R}^d)$ or not. Thus the question arises of how to determine whether a solution of this equation belongs to L_2 if we know the coefficients of a refinement equation (or, equivalently, the Fourier coefficients of the mask). In the case of compactly supported scaling functions, i.e., for equations with polynomial masks, there are two algorithms which allow us to reduce our problem to the computation of eigenvalues for some finite-dimensional operators. Here we describe both algorithms. As for the proofs, they can be found in the corresponding articles. In both algorithms, we first have to find a so-called invariant set Ω corresponding to a given mask m_0. We say that a finite set of integer points $\Omega \subset \mathbb{Z}^d$ is invariant for a given polynomial mask $m_0(\xi) = \frac{1}{\sqrt{m}} \sum\limits_{k\in\mathbb{Z}^d} h_k e^{2\pi i(k,\xi)}$ if

$$M^{-1}\Big(k + \operatorname{Conv}\Omega\Big) \subset \operatorname{Conv}\Omega$$

for any $k \in \mathbb{Z}^d$ such that $h_k \neq 0$ (recall that by Conv we denote the convex hull of a set). In the case $d = 1$, for Ω, one can take the set of all integer points of the segment $\left[\frac{N_1}{M-1}, \frac{N_2}{M-1}\right]$. In the case $M = 2$, $N_1 = 0, N_2 = N$, we have $\Omega = \{0, 1, \ldots, N\}$. In the multidimensional case, it is more difficult to find Ω. For instance, we can always take the set of integer points contained in the support of the refinable function. However, here the problem is that this support is initially unknown. Therefore we shall proceed in another way; namely, we construct a convex bounded set $D \subset \mathbb{R}^d$ such that $M^{-1}\Big(D + k\Big) \subset D$ and set $\Omega = D \cap \mathbb{Z}^d$. For instance, we can take D to be a ball of sufficiently large radius in some norm such that M^{-1} is a dilation operator. Such a norm can be found since all eigenfunctions of this operator are less than 1 in modulus (Appendix A.5, Lemma A.5.1). In fact, we constructed such a set D to prove Theorem 2.4.4. If M has a basis of eigenvectors, then for D, we can take a sufficiently large parallelepiped spanned by these vectors. For diagonal matrixes M, a cube is suitable. One of the algorithms of constructing D is described in [**86**]. However, in practice it is easier to guess a suitable set D.

Suppose we have an invariant set Ω. Let $s_0, \ldots, s_{m-1}$ be all the points in the image MI_d of a half-open cube

$$I_d = \big\{(x_1, \ldots, x_d)^T \in \mathbb{R}^d,\ x_i \in [0, 1),\ i = 1, \ldots, d\big\}.$$

Recall that the number of these points is exactly $m = |\det M|$. Let us now construct m $N \times N$ matrices $T_0, \ldots, T_{m-1}$, where N is the number of elements in Ω. We determine the elements of these matrices by the formula

$$(2.48) \qquad \big(T_k\big)_{ij} = \sqrt{m}\, h_{Mi-j+s_k}, \quad i, j \in \Omega, \qquad k = 0, \ldots, m-1.$$

The corresponding operators T_k act in $\mathbb{R}^N$. If integer shifts of the refinable function form a Riesz basis, then all these operators have a common eigenspace

$$W = \big\{(x_1, \ldots, x_N)^T \in \mathbb{R}^N,\ \sum x_i = 0\big\}$$

(for the proof of this fact, see, e.g., [**85**]). Let $\tilde{T}_k = T_k|_W$. Finally, introduce the operator

$$\mathcal{A}_2 X = \frac{1}{m}\Big(\tilde{T}_0 X \tilde{T}_0^* + \cdots + \tilde{T}_{m-1} X \tilde{T}_{m-1}^*\Big)$$

on the space X of self-adjoint operators in $\mathbb{R}^{N-1}$. This operator acts in $\mathbb{R}^{\frac{N(N-1)}{2}}$.

THEOREM 2.5.13. *Let us have a refinement equation with a polynomial mask. If all the eigenvalues of the operator $\mathcal{A}_2$ are smaller than 1 in modulus, then the solution belongs to $L_2(\mathbb{R}^d)$. If the mask satisfies (2.27), then the converse is also true: all the eigenvalues of $\mathcal{A}_2$ are smaller than 1 in modulus, whenever $\varphi \in L_2(\mathbb{R}^d)$.*

Comments, an explanation, and the idea of the proof of this theorem can be found in Appendix A.6.

Another way to check if a refinable function belongs to the space $L_2(\mathbb{R}^d)$ is to use the operator W_φ defined by equality (2.43). Let us take the invariant set Ω corresponding to the mask $|m_0(\xi)|^2 = \sum_{k\in\mathbb{Z}^d} a_k e^{2\pi i(k,\xi)}$, and let L be the number of elements in this set. Consider the restriction of the operator W_φ to the L-dimensional space $\ell_2(\Omega)$. The matrix of this operator is given by the equality

$$\left(W_\varphi\right)_{ij} = m \cdot a_{Mi-j}, \qquad i,j \in \Omega. \tag{2.49}$$

THEOREM 2.5.14. *Let us have a refinement equation with a polynomial mask. If operator (2.49) has a simple eigenvalue 1 and all other eigenvalues are smaller than 1 in modulus, then the solution belongs to $L_2(\mathbb{R}^d)$. If the mask satisfies (2.27), then the converse is also true.*

For the proof, see [**87**].

Thus both criteria reduce the question of whether a refinable function belongs to the space L_2 to checking whether the spectral radius ρ of some finite-dimensional operator is strictly less than 1. In most cases, it is more convenient to use Theorem 2.5.14, since the size of the matrix W_φ, as a rule, is less than the size of $\mathcal{A}_2$. However, sometimes both criteria prove useful. For example, in the one-dimensional case the value of $-\frac{1}{2}\log_2 \rho(\mathcal{A}_2)$ is equal to the Hoĺder coefficient of the function φ in the space L_2 (see, for example, the review [**116**]), and the value $-\frac{1}{2}\log_2 \rho(W_\varphi)$ is equal to the Sobolev coefficient of smoothness (see [**173**]).

2.6. Construction of wavelet functions

In this section, we shall study biorthogonal wavelet systems corresponding to a pair of scaling functions with biorthogonal integer shifts generating a pair of MRAs. As shown in §2.1, the fundamental difference between the one-dimensional and multidimensional cases is the existence of several wavelet functions instead of one. It is an essentially more complicated problem to find wavelet functions in a nonseparable MRA.

DEFINITION 2.6.1. Suppose we have two MRAs $\{V_j\}_{j\in\mathbb{Z}}$, $\{\widetilde{V}_j\}_{j\in\mathbb{Z}}$. We say that functions $\psi^{(\nu)} \in V_1, \widetilde{\psi}^{(\nu)} \in \widetilde{V}_1$, $\nu = 1,\dots,m-1$, form a set of wavelet functions if $\psi^{(\nu)} \perp \widetilde{V}_0$, $\widetilde{\psi}^{(\nu)} \perp V_0$ for all $\nu = 1,\dots,m-1$, and

$$\langle \psi^{(\nu)}(\cdot + k), \widetilde{\psi}^{(\kappa)}(\cdot + l)\rangle = \delta_{\nu\kappa}\delta_{kl}$$

for all $k, l \in \mathbb{Z}^d$ and all $\nu, \kappa = 1,\dots,m-1$.

THEOREM 2.6.2. *Suppose MRAs $\{V_j\}_{j\in\mathbb{Z}}$, $\{\widetilde{V}_j\}_{j\in\mathbb{Z}}$ are generated by scaling functions $\varphi, \widetilde{\varphi}$ with masks $m_0, \widetilde{m}_0$, respectively, and the systems $\{\varphi(\cdot + n)\}_{n\in\mathbb{Z}^d}$, $\{\widetilde{\varphi}(\cdot + n)\}_{n\in\mathbb{Z}^d}$ are orthonormal. If for the matrices*

$$\mathcal{M} := \{m_\nu(\xi + M^{*-1}s_k)\}_{\nu,k=0}^{m-1}, \qquad \widetilde{\mathcal{M}} := \{\widetilde{m}_\nu(\xi + M^{*-1}s_k)\}_{\nu,k=0}^{m-1},$$

where $m_\nu, \widetilde{m}_\nu \in L_2(\mathbb{T}^d)$, $\nu = 1, \dots, m-1$, $\{s_0, \dots, s_{m-1}\}$ *is an arbitrary collection of digits of the matrix* M^*, *the equality*

$$\mathcal{M}\widetilde{\mathcal{M}}^* = I_m \tag{2.50}$$

holds for almost all $\xi \in \mathbb{R}^d$, *then the functions* $\psi^{(\nu)}, \widetilde{\psi}^{(\nu)}$, $\nu = 1, \dots, m-1$, *defined by the equalities*

$$\widehat{\psi}^{(\nu)}(\xi) = m_\nu({M^*}^{-1}\xi)\widehat{\varphi}({M^*}^{-1}\xi), \qquad \widehat{\widetilde{\psi}^{(\nu)}}(\xi) = \widetilde{m}_\nu({M^*}^{-1}\xi)\widehat{\widetilde{\varphi}}({M^*}^{-1}\xi) \tag{2.51}$$

form a set of wavelet functions.

Proof. First of all we note that Lemma 2.3.3 implies the inclusions $\psi^{(\nu)} \in V_1, \widetilde{\psi}^{(\nu)} \in \widetilde{V}_1$, $\nu = 1, \dots, m-1$.

For the sake of convenience, let us rename the functions $\varphi, \widetilde{\varphi}$ by $\psi^{(0)}, \widetilde{\psi}^{(0)}$, respectively. We need to verify the equality

$$\langle \psi^{(\nu)}(\cdot + k), \widetilde{\psi}^{(\kappa)}(\cdot + l)\rangle = \delta_{\nu\kappa}\delta_{kl} \tag{2.52}$$

for all $k, l \in \mathbb{Z}^d$ and all $\nu, \kappa = 0, \dots, m-1$ (by assumption, it holds for $\nu = \kappa = 0$). Let $\nu \neq 0$. If $\nu = \kappa$, then in view of Proposition 2.5.1, (2.52) is equivalent to the equality

$$\sum_{p\in\mathbb{Z}^d} \widehat{\psi}^{(\nu)}(M^*\xi + p)\overline{\widehat{\widetilde{\psi}^{(\kappa)}}(M^*\xi + p)} = \delta_{\nu\kappa} \tag{2.53}$$

for almost all ξ. Reasoning as in the proof of Proposition 1.1.12, it is not hard to show that relations (2.52) and (2.53) are equivalent for $\nu \neq \kappa$ as well.

Using (2.51) and the refinement equations for the functions $\varphi, \widetilde{\varphi}$, we get

$$\begin{aligned} &\sum_{p\in\mathbb{Z}^d} \widehat{\psi}^{(\nu)}(M^*\xi + p)\overline{\widehat{\widetilde{\psi}^{(\kappa)}}(M^*\xi + p)} \\ &\quad = \sum_{p\in\mathbb{Z}^d} m_\nu(\xi + {M^*}^{-1}p)\widehat{\varphi}(\xi + {M^*}^{-1}p)\overline{\widetilde{m}_\kappa(\xi + {M^*}^{-1}p)\widehat{\widetilde{\varphi}}(\xi + {M^*}^{-1}p)}. \end{aligned} \tag{2.54}$$

Using Lemma 2.2.2, set the index of summation $p = M^*k + s$, $k \in \mathbb{Z}^d$, $s \in D(M^*)$, on the right-hand side of (2.54), which reduces it to the form

$$\sum_{k\in\mathbb{Z}^d}\sum_{s\in D(M^*)} m_\nu(\xi + \theta_{ks})\overline{\widetilde{m}_\kappa(\xi + \theta_{ks})}\widehat{\varphi}(\xi + \theta_{ks})\overline{\widehat{\widetilde{\varphi}}(\xi + \theta_{ks})},$$

where $\theta_{ks} = {M^*}^{-1}s + k$. Changing the order of summation, using 1-periodicity of the functions $m_\nu, \widetilde{m}_\kappa$ in each variable and relation (2.30), we get

$$\begin{aligned} &\sum_{s\in D(M^*)} m_\nu(\xi + \theta_{\mathbf{0}s})\overline{\widetilde{m}_\kappa(\xi + \theta_{\mathbf{0}s})} \sum_{k\in\mathbb{Z}^d} \widehat{\varphi}(\xi + \theta_{ks})\overline{\widehat{\widetilde{\varphi}}(\xi + \theta_{ks})} \\ &\qquad = \sum_{s\in D(M^*)} m_\nu(\xi + {M^*}^{-1}s)\overline{\widetilde{m}_\kappa(\xi + {M^*}^{-1}s)}. \end{aligned}$$

It remains to note that (2.50) implies that

$$\sum_{s\in D(M^*)} m_\nu(\xi + {M^*}^{-1}s)\overline{\widetilde{m}_\kappa(\xi + {M^*}^{-1}s)} = \delta_{\nu\kappa}$$

for almost all $\xi \in \mathbb{R}^d$. ◊

Theorem 2.6.2 gives no answer to the question of existence of a set of wavelet functions. It only reduces this question to the problem of finding periodic functions $m_\nu, \widetilde{m}_\nu$, $\nu = 1, \dots, m-1$, that satisfy equalities (2.50). It is rather complicated to find such functions. The following assertion simplifies the problem to some extent.

LEMMA 2.6.3. *Suppose* $m_\nu, \widetilde{m}_\kappa \in L_2(\mathbb{T}^d)$,

$$m_\nu(\xi) = \frac{1}{\sqrt{m}} \sum_{k\in\mathbb{Z}^d} h_k e^{2\pi i(k,\xi)}, \qquad \widetilde{m}_\kappa(\xi) = \frac{1}{\sqrt{m}} \sum_{k\in\mathbb{Z}^d} \widetilde{h}_k e^{2\pi i(k,\xi)},$$

$D(M) = \{s_0, \dots, s_{m-1}\}$, $D(M^*)$ *are sets of digits of matrices* M, M^*, *respectively, and*

$$\mu_{\nu k}(\xi) := \sum_{l\in\mathbb{Z}^d} h_{Ml+s_k} e^{2\pi i(l,\xi)}, \qquad \widetilde{\mu}_{\kappa k}(\xi) := \sum_{l\in\mathbb{Z}^d} \widetilde{h}_{Ml+s_k} e^{2\pi i(l,\xi)}.$$

Then

$$\sum_{s\in D(M^*)} m_\nu(\xi + {M^*}^{-1}s)\overline{\widetilde{m}_\kappa(\xi + {M^*}^{-1}s)} = \sum_{k=0}^{m-1} \mu_{\nu k}(M^*\xi)\overline{\widetilde{\mu}_{\kappa k}(M^*\xi)}. \tag{2.55}$$

Proof. In view of Lemma 2.2.2, for any $s \in D(M^*)$, we have

$$m_\nu(\xi + {M^*}^{-1}s) = \frac{1}{\sqrt{m}} \sum_{k=0}^{m-1} \sum_{l\in\mathbb{Z}^d} h_{Ml+s_k} e^{2\pi i(Ml+s_k,\xi+{M^*}^{-1}s)}$$

$$= \frac{1}{\sqrt{m}} \sum_{k=0}^{m-1} e^{2\pi i(M^{-1}s_k,s)} e^{2\pi i(s_k,\xi)} \mu_{\nu k}(M^*\xi). \tag{2.56}$$

The functions $\widetilde{m}_\nu(\xi + {M^*}^{-1}s)$ can be represented in a similar form. The required relation follows from these equalities and Corollary 2.2.7. ◊

It follows from the above lemma that if the periodic functions $m_\nu, \widetilde{m}_\nu$, $\nu = 1, \dots, m-1$, from $L_2(\mathbb{T}^d)$ satisfy (2.50), then there exist functions $\mu_{\nu k}, \widetilde{\mu}_{\nu k} \in L_2(\mathbb{T}^d)$ such that the matrices

$$\{\mu_{\nu k}(\xi)\}_{\nu,k=0}^{m-1}, \quad \{\overline{\widetilde{\mu}_{k\nu}(\xi)}\}_{\nu,k=0}^{m-1} \tag{2.57}$$

are mutually inverse for almost all $\xi \in \mathbb{R}^d$. These matrices are called *polyphase*. The proof of Lemma 2.6.3 shows that the converse is also true; i.e., if for the functions $\mu_{\nu k}, \widetilde{\mu}_{\nu k} \in L_2(\mathbb{T}^d)$, matrices (2.57) are mutually inverse and for $\nu = 0, \dots, m-1$

$$m_\nu(\xi) := \frac{1}{\sqrt{m}} \sum_{k=0}^{m-1} e^{2\pi i(s_k,\xi)} \mu_{\nu k}(M^*\xi),$$

$$\widetilde{m}_\nu(\xi) := \frac{1}{\sqrt{m}} \sum_{k=0}^{m-1} e^{2\pi i(s_k,\xi)} \widetilde{\mu}_{\nu k}(M^*\xi),$$

then (2.50) holds. Thus, in order to construct a set of wavelet functions, we have to extend two matrices with elements from $L_2(\mathbb{T}^d)$ to mutually inverse matrices, given their first row $\{\mu_{0k}\}_{k=0}^{m-1}$ and first column $\{\overline{\widetilde{\mu}_{0k}}\}_{k=0}^{m-1}$, respectively. Note that for $m = 2$, this problem has a trivial solution

$$\mu_{10} = -\overline{\widetilde{\mu}_{01}}, \qquad \mu_{11} = \overline{\widetilde{\mu}_{00}}, \qquad \widetilde{\mu}_{10} = -\overline{\mu_{01}}, \qquad \widetilde{\mu}_{11} = \overline{\mu_{00}}.$$

Further we shall discuss a possibility of such an extension for an arbitrary m.

First consider the situation (corresponding to the construction of an orthogonal wavelet system) where we have functions $\mu_{0k} = \widetilde{\mu}_{0k}$, $k = 1, \ldots, m$, from $L_2(\mathbb{T}^d)$ such that

$$\sum_{k=0}^{m-1} |\mu_{0k}(\xi)|^2 = 1 \quad \text{a.e.} \tag{2.58}$$

We need to extend this row to a unitary matrix. This can be done by Householder's transform. Set

$$\mu_{k0} = \overline{\mu_{0k}} \frac{1-\mu_{00}}{1-\overline{\mu_{00}}}, \qquad \mu_{kl} = \delta_{kl} - \frac{\mu_{0l}\overline{\mu_{0k}}}{1-\overline{\mu_{00}}}$$

for $k, l = 1, \ldots, m-1$. Direct calculation shows that this matrix is unitary. Let us prove that the functions constructed above belong to the space $L_2(\mathbb{T}^d)$. Indeed, from (2.58) it follows that

$$|\mu_{0k}| \le \sqrt{1 - |\mu_{00}|^2}$$

for all $k = 1, \ldots, m-1$, which implies that $|\mu_{kl}| \le 2 + |\mu_{00}|$.

Let us now consider the biorthogonal case. The scaling functions φ, $\widetilde{\varphi}$ determine the set of functions $\mu_{0k}, \widetilde{\mu}_{0k} \in L_2(\mathbb{T}^d)$, $k = 0, \ldots, m-1$. Since integer shifts of the functions φ, $\widetilde{\varphi}$ are biorthonormal, by Proposition 2.5.2 and Lemma 2.6.3, we get

$$\sum_{k=0}^{m-1} \mu_{0k}(\xi)\overline{\widetilde{\mu}_{0k}(\xi)} = 1 \quad \text{a.e.} \tag{2.59}$$

Moreover, since integer shifts of the function φ form a Riesz system, using Proposition 2.1.1 and reasoning as in the proof of Corollary 1.2.5, we see that the sum

$$\sum_{s \in D(M^*)} |m_0(\xi + {M^*}^{-1}s)|^2$$

is bounded away from zero almost everywhere. By Lemma 2.6.3, we have

$$\sum_{k=0}^{m-1} |\mu_{0k}(\xi)|^2 \ge C > 0 \quad \text{a.e.}$$

It follows that the functions

$$\mu'_{0k}(\xi) := \frac{\mu_{0k}(\xi)}{\left(\sum\limits_{l=0}^{m-1} |\mu_{0l}(\xi)|^2\right)^{1/2}}$$

are in $L_2(\mathbb{T}^d)$. The row $\mu'_{00}, \ldots, \mu'_{0(m-1)}$ can be extended to a unitary matrix with elements from $L_2(\mathbb{T}^d)$ (for instance, by Householder's transform described above). Denote by $\overline{\widetilde{Q}}_1, \ldots, \overline{\widetilde{Q}}_{m-1}$ the rows of this matrix from the second to the last one and set

$$Q_0 = (\mu_{00}, \ldots, \mu_{0(m-1)}), \quad \widetilde{Q}_0 = (\overline{\widetilde{\mu}}_{00}, \ldots, \overline{\widetilde{\mu}}_{0(m-1)}).$$

Since the matrix is unitary, it follows from relation (2.59) that

$$\overline{\widetilde{Q}}_n \widetilde{Q}_k^T = \delta_{nk}, \quad Q_0 \widetilde{Q}_l^T = \delta_{0l} \quad \text{a.e.}$$

for all $n, k = 1, \ldots, m-1$, $l = 0, \ldots, m-1$. Setting

$$Q_k = \overline{\widetilde{Q}}_k - \overline{\widetilde{Q}}_k \widetilde{Q}_0^T Q_0, \quad k = 1, \ldots, m-1,$$

we obtain

$$Q_k \widetilde{Q}_l^T = \delta_{kl} \quad \text{a.e.}$$

for all $k, l = 0, \dots, m-1$. Thus, given the first rows $\{\mu_{0k}\}_{k=0}^{m-1}$, $\{\overline{\widetilde{\mu}_{0k}}\}_{k=0}^{m-1}$, we have found the matrices $Q, \widetilde{Q}$ such that

$$Q\widetilde{Q}^T = I_m \quad \text{a.e.}$$

From the above reasoning and Theorem 2.6.2 we get the following assertion.

THEOREM 2.6.4. *Let MRAs $\{V_j\}_{j\in\mathbb{Z}}$, $\{\widetilde{V}_j\}_{j\in\mathbb{Z}}$ be generated, respectively, by scaling functions $\varphi, \widetilde{\varphi}$ with masks $m_0, \widetilde{m}_0$. If the systems $\{\varphi(\cdot + n)\}_{n\in\mathbb{Z}^d}$ and $\{\widetilde{\varphi}(\cdot + n)\}_{n\in\mathbb{Z}^d}$ are orthonormal, there exist $m_\nu, \widetilde{m}_\nu \in L_2(\mathbb{T}^d)$, $\nu = 1, \dots, m-1$, such that the functions $\psi^{(\nu)}, \widetilde{\psi}^{(\nu)}$, $\nu = 1, \dots, m-1$, given by equalities* (2.51) *form a set of wavelet functions.*

The functions $m_\nu, \widetilde{m}_\nu$, $\nu = 1, \dots, m-1$, are called masks of the corresponding wavelet functions.

If masks of scaling functions are smooth, the masks of wavelet functions constructed by the above algorithm are not necessarily smooth (or even continuous). Let us show that in the case $d < 2m-1$ smoothness can be preserved. Suppose that the functions $\mu_{0k}, \widetilde{\mu}_{0k}$ have continuous partial derivatives of order r, $r \geq 1$. Obviously, the functions μ'_{0k} are smooth. The smooth mapping $(\mu'_{00}, \dots, \mu'_{0(m-1)})$ takes the d-dimensional torus $\mathbb{T}^d$ to a unit sphere S in the complex m-dimensional space $\mathbf{C}^{\mathbf{m}}$. Let us cover $\mathbb{T}^d$ with finitely many sets, whose diameters do not exceed certain $\varepsilon > 0$. Then the diameters of the images of these sets do not exceed $C\varepsilon$, where C is a constant that depends only on the functions μ'_{0k}. Since the dimension of $\mathbb{T}^d$ is less than the dimension of S, it is clear that for sufficiently small ε, the union of the images of these sets does not cover S. Hence there exists a point $\alpha \in S$ which together with some of its neighborhood does not belong to the image of the torus. Suppose U is a unitary numerical matrix and U^T takes the vector α to the vector $(1, 0, \dots, 0)^T$. Set

$$(\mu''_{00}, \dots, \mu''_{0(m-1)}) = (\mu'_{00}, \dots, \mu'_{0(m-1)})U,$$
$$(\widetilde{\mu}''_{00}, \dots, \widetilde{\mu}''_{0(m-1)}) = (\widetilde{\mu}_{00}, \dots, \widetilde{\mu}_{0(m-1)})U.$$

Both new rows preserve smoothness of order r; the row $\mu''_{00}, \dots, \mu''_{0(m-1)}$ can be extended to a unitary matrix by Householder's transform, which also preserves smoothness since all values of the function μ''_{00} are bounded away from unity. The consequent steps of the construction are based on linear transformations. Thus we can construct mutually inverse matrices $\{\mu''_{\nu k}(\xi)\}_{\nu,k=0}^{m-1}$, $\{\overline{\widetilde{\mu}''_{k\nu}(\xi)}\}_{\nu,k=0}^{m-1}$, consisting of functions with continuous derivatives of order r. To construct the required matrices, it remains to set

$$\begin{aligned}(\mu_{\nu 0}, \dots, \mu_{\nu(m-1)}) &= (\mu''_{\nu 0}, \dots, \mu''_{\nu(m-1)})\overline{U^T},\\ (\widetilde{\mu}_{\nu 0}, \dots, \widetilde{\mu}_{\nu(m-1)}) &= (\widetilde{\mu}''_{\nu 0}, \dots, \widetilde{\mu}''_{\nu(m-1)})\overline{U^T}\end{aligned}$$

for $\nu = 1, \dots, m-1$.

Of special interest for engineering applications are sets of wavelet functions with polynomial masks. Such wavelet functions, as well as the generating scaling functions, are compactly supported. Indeed, if two scaling functions $\varphi, \widetilde{\varphi}$ with biorthogonal integer shifts are compactly supported, it is obvious that their masks are trigonometric polynomials. Suppose we have found a corresponding set of wavelet functions with polynomial masks. Applying the Fourier transform to equalities (2.51), represent $\psi^{(\nu)}$ and $\widetilde{\psi}^{(\nu)}$ in the form of finite linear combinations of the functions $\widetilde{\varphi}_{1k}$ and φ_{1k}, respectively, which implies that they are compactly supported. Let us now discuss the problem of the existence of compactly supported wavelet functions.

Let $\varphi, \widetilde{\varphi}$ be scaling functions with polynomial masks $m_0, \widetilde{m}_0$. Then the corresponding functions $\mu_{0k}, \widetilde{\mu}_{0k}$ are trigonometric polynomials. Our aim is to find matrices $Q, \widetilde{Q}$, whose elements are trigonometric polynomials such that $Q\widetilde{Q}^T = I_m$, given their first rows $\{\mu_{0k}\}_{k=0}^{m-1}$, $\{\overline{\widetilde{\mu}}_{0k}\}_{k=0}^{m-1}$. This can be done in the following way. Set

$$Q_0 = (\mu_{00}, \dots, \mu_{0(m-1)}), \quad \widetilde{Q}_0 = (\overline{\widetilde{\mu}}_{00}, \dots, \overline{\widetilde{\mu}}_{0(m-1)}).$$

Suppose we have extended Q_0 to a *unimodular* matrix, i.e., a quadratic matrix whose elements are trigonometric polynomials and the absolute value of the determinant is identically equal to 1 (this is essential and the most complicated part of the construction). Denote by $Q'_1, \dots, Q'_{m-1}$ the rows of the matrix from the second to the last one. It is clear that the inverse matrix exists and its elements are also trigonometric polynomials. Denote by $\widetilde{Q}_1^T, \dots, \widetilde{Q}_{m-1}^T$ the columns of the inverse matrix from the second to the last one, and set

$$Q_k = Q'_k - Q'_k \widetilde{Q}_0^T Q_0, \quad k = 1, \dots, m-1.$$

It is not hard to see that

$$Q_k \widetilde{Q}_l^T = \delta_{kl}$$

for all $k, l = 0, \dots, m-1$; i.e., $Q, \widetilde{Q}$ are the required matrices.

Let us now return to the first step of the construction and reformulate the problem in algebraic terms. Setting $z = (z_1, \dots, z_d)$, $z_k = e^{2\pi i \xi_k}$, we shall consider the Laurent polynomials instead of trigonometric polynomials, using the same notation. We have a unimodular row of Laurent polynomials (i.e., a row Q_0 for which there exists a row $\widetilde{Q}_0$ such that $Q_0\widetilde{Q}_0^T = 1$ for all $z \neq \mathbf{0}$). We need to extend the row Q_0 to a unimodular matrix (i.e., a quadratic matrix whose elements are Laurent polynomials and the determinant is equal to 1 in modulus for all $z \neq \mathbf{0}$). Problems of this sort have been of algebraic interest for a long time. In particular, the following problem (so-called Serre's problem) has been studied: is it always possible to construct a unimodular matrix such that its elements are multivariate algebraic polynomials, given its first unimodular row? This problem was solved in 1976 by Quillen and Suslin independently (see [**98**]). Moreover, in 1977 Suslin [**99**] extended the result to a wider class of rings, in particular, to the ring of the Laurent polynomials. Combining Suslin's result with the process of orthogonalization described above, we obtain the following assertion.

THEOREM 2.6.5. *Let MRAs* $\{V_j\}_{j\in\mathbb{Z}}$, $\{\widetilde{V}_j\}_{j\in\mathbb{Z}}$ *be generated, respectively, by scaling functions* $\varphi, \widetilde{\varphi}$ *with polynomial masks* $m_0, \widetilde{m}_0$. *If the systems* $\{\varphi(\cdot+n)\}_{n\in\mathbb{Z}^d}$, $\{\widetilde{\varphi}(\cdot+n)\}_{n\in\mathbb{Z}^d}$ *are biorthonormal, then there exist trigonometric polynomials* $m_\nu, \widetilde{m}_\nu \in L_2(\mathbb{T}^d)$, $\nu = 1, \dots, m-1$, *such that the functions* $\psi^{(\nu)}, \widetilde{\psi}^{(\nu)}$, $\nu = 1, \dots, m-1$, *given by equalities* (2.51) *form a set of wavelet functions.*

Suslin's result, on which the proof of Theorem 2.6.5 is based, is of theoretical character. Park and Woodburn [**100**] presented a constructive proof of this result, which formally gives an algorithm for constructing a unimodular matrix from a given row. This algorithm is based on the construction of Gröbner bases, which is rather complicated and thus cannot be used in practice except for some rare cases of polynomials of sufficiently small degree.

In conclusion, we present two results concerning the possibility of extending matrices.

THEOREM 2.6.6. *If* $m > 2$, *then there exist no functions* $f_{\nu k}$, $\nu = 1, \dots, m-1$, $k = 0, \dots, m-1$, *continuous on the unit sphere* S *in the complex* m*-dimensional space* $\mathbf{C}^m$ *and such that the matrix* $\{f_{\nu k}\}_{\nu,k=0}^{m-1}$, *where* $f_{0k}(z_0, \dots, z_{m-1}) = z_k$, *is invertible for all* $z = (z_0, \dots, z_{m-1}) \in S$.

This theorem follows from the results presented in [**92**, p. 24]; see also [**91**].

Packer and Rieffel [**101**] proved that for the 5×5 matrix $M = \begin{pmatrix} 0 & 3 \\ I_4 & 0 \end{pmatrix}$ there exists a smooth mask m_0 such that the row of its polyphase functions $(\mu_{01}, \mu_{02}, \mu_{03})$ satisfies equality (2.58) but cannot be extended to a unitary matrix with continuous elements.

The question of the possibility of extending any suitable row to a unitary matrix, whose elements are Laurent polynomials, still remains unsolved.

In spite of considerable difficulties, which may occur in the implementation of the method of extending a row to a unimodular matrix, the process of constructing biorthogonal compactly supported wavelets described above is useful since in some particular cases one can sometimes guess a required unimodular matrix (naturally, this is more likely than to guess two mutually inverse matrices).

EXAMPLE 2.6.7. Let $M = \begin{pmatrix} 2 & 0 \\ 0 & 2 \end{pmatrix}$. As a scaling function, we take a box spline B_3 (see Example 2.3.9) defined by the equality

$$\widehat{B}_3(\xi) = 2^2 \prod_{k=1}^{3} \frac{1 - e^{2\pi i(a_k,\xi)}}{2\pi i(a_k, \xi)},$$

where $a_1 = (1,0), a_2 = (0,1), a_3 = (1,1)$. Its mask is the polynomial

$$m_0(\xi) = \frac{1}{2} \prod_{k=1}^{3} \left(1 + e^{2\pi i(a_k,\xi)}\right).$$

As a set of digits, consider the vectors

$$s_0 = (0,0), \qquad s_1 = (0,1), \qquad s_2 = (1,0), \qquad s_3 = (1,1)$$

and represent the mask in the form

$$m_0(\xi) = \frac{1}{2}\left(e^{2\pi i(s_0,\xi)}\left(1 + e^{4\pi i(\xi_1+\xi_2)}\right) + e^{2\pi i(s_1,\xi)}\left(1 + e^{4\pi i\xi_1}\right) + e^{2\pi i(s_2,\xi)}\left(1 + e^{4\pi i\xi_2}\right) + 2e^{2\pi i(s_3,\xi)}\right).$$

Setting $z_1 = e^{2\pi i\xi_1}, z_2 = e^{2\pi i\xi_2}$, we find the polyphase functions

$$\mu_{00}(\xi) = \frac{1+z_1z_2}{2}, \quad \mu_{01}(\xi) = \frac{1+z_1}{2}, \quad \mu_{02}(\xi) = \frac{1+z_2}{2}, \quad \mu_{03}(\xi) = 1.$$

This row can easily be extended to the diagonal matrix

$$\begin{pmatrix} \frac{1+z_1z_2}{2} & \frac{1+z_1}{2} & \frac{1+z_2}{2} & 1 \\ 0 & 0 & 1 & 0 \\ 0 & 1 & 0 & 0 \\ 1 & 0 & 0 & 0 \end{pmatrix}$$

with the determinant equal to 1. $\diamond$

Let us give a general scheme of constructing a unimodular matrix, given a unimodular row for $d = 1$. This scheme is based on the Euclidean algorithm. Let there be given a row of Laurent polynomials Q_0, for which there exists a row $\widetilde{Q}_0$ such that $Q_0\widetilde{Q}_0^T = 1$. Multiplying these rows, respectively, by z^N, z^M, where N, M are suitable integers, we obtain rows of the algebraic polynomials $P_0 = (p_1, \ldots, p_m)$, $\widetilde{P}_0 = (\widetilde{p}_1, \ldots, \widetilde{p}_m)$, where $P_0\widetilde{P}_0^T = z^{N+M}$. Find a nonzero polynomial of the minimal degree from $p_1, \ldots, p_m$. Let us assume that this polynomial is p_1. If $p_2, \ldots, p_m$

are identically equal to zero, then $p_1\widetilde{p}_1 = z^{N+M}$; therefore p_1 and $\widetilde{p}_1$ are monomials. In this case the problem is trivial. Otherwise, there exists a nonzero polynomial p_k whose degree is not smaller than the degree of p_1. Dividing p_k by p_1 with remainder, we get $p_k = qp_1 + r$, $\deg r < \deg p_1$. Replace p_k by r in the row P_0 and denote the new row by P_0'. It is not hard to see that $P_0' = P_0E$, where E is an elementary matrix, i.e., a quadratic matrix with each diagonal element equal to 1 and only one other nonzero element (in this case it is equal to $-q$). For this new row P_0', there again exists a row $\widetilde{P}_0' := \widetilde{P}_0(E^{-1})^T$ such that $P_0'(\widetilde{P}_0')^T = z^{N+M}$. Continuing the Euclidean algorithm

$$p_1 = n_1 r + r_1, \quad r = n_2 r_1 + r_2, \dots, \quad r_L = n_{L+2} r_{L+1},$$

we obtain a row of polynomials, where one of the coordinates is identically equal to zero. Take the row of the minimal degree, choose a nonzero element in it, and repeat our reasoning. We shall continue this process until we obtain a row of polynomials P_0'' with only one nonzero component p_n. Here $P_0'' = P_0X$, where X is the product of elementary matrices, and there exists a row $\widetilde{P}_0''$ such that $P_0''(\widetilde{P}_0'')^T = z^{N+M}$. Combining this with the equality $P_0''(\widetilde{P}_0'')^T = p_n\widetilde{p}_n$, we obtain that p_n is a monomial; therefore the row P_0'' can be extended to a unimodular matrix P. The matrix $Q := ZPX^{-1}$ is a solution of the initial problem, where Z is a diagonal matrix with diagonal elements $z^{-N}, z^N, 1, \dots, 1$. ◊

The above construction does not work for $d > 1$ since in the multidimensional case we do not have the Euclidean algorithm for polynomials.

Theorem 2.6.5 does not solve the problem of the existence of orthogonal compactly supported wavelet systems in the case $\mu_{0k} = \widetilde{\mu}_{0k}$, $k = 0, \dots, m-1$. Let us show how to construct such a system for $d = 1$.

Let there be given a row of Laurent polynomials Q_0 such that

$$Q_0 Q_0^* = 1 \tag{2.60}$$

for all $z \neq \mathbf{0}$. We need to extend this row to a unitary matrix. Represent Q_0 in the form

$$Q_0 = \sum_{n=N}^{M} a_n z^n, \quad a_n \in \mathbb{R}^m, \quad n = N, \dots, M, \quad |a_N| \neq 0.$$

Computing the leading coefficient of the polynomial $Q_0Q_0^*$, we obtain from (2.60)

$$(a_N, a_M) = 0. \tag{2.61}$$

Let U be a numeric unitary matrix such that

$$a_N U = (|a_N|, 0, \dots, 0).$$

From (2.61) we get $(a_N U, a_M U) = 0$; hence the first component of the vector $a_M U$ is equal to zero. Let Z be a diagonal matrix with the diagonal elements $1, z^{-1}, \dots, z^{-1}$. Then

$$Q_0 U Z = \sum_{n=N}^{M-1} a_n' z^n =: Q_0',$$

where the matrix UZ is unitary, so the new row Q_0' also satisfies (2.60). Repeat the same reasoning for the row Q_0' and continue the process until we obtain the row

$$Q_0'' = a_N'' z^N = Q_0 Y,$$

where Y is a unitary matrix and Q_0'' satisfies relation (2.60), which is equivalent to the equality $|a_N''| = 1$. Let U' be a numerical unitary matrix such that $a_N''U' = (1, 0, \dots, 0)$ and let Z' be a diagonal matrix with the diagonal elements $z^{-N}, \dots, z^{-N}$. Then we have

$$Q_0 Y U' Z' = Q_0'' U' Z' = (1, 0, \dots, 0);$$

hence $Q := (YU'Z')^{-1}$ is the required unitary matrix. ◊

2.7. Wavelet bases

In this section, we show that for a wide class of MRAs satisfying the conditions of Theorem 2.6.4 the sets of wavelet functions $\psi^{(\nu)}, \widetilde{\psi}^{(\nu)}$, $\nu = 1, \dots, m-1$, constructed in this theorem generate biorthogonal bases $\{\psi_{jk}^{(\nu)}\}, \{\widetilde{\psi}_{jk}^{(\nu)}\}$ of the space $L_2(\mathbb{R}^d)$.

THEOREM 2.7.1. *Let there be given MRAs $\{V_j\}_{j\in\mathbb{Z}}$, $\{\widetilde{V}_j\}_{j\in\mathbb{Z}}$ in $L_2(\mathbb{R}^d)$, and let $\psi^{(\nu)}, \widetilde{\psi}^{(\nu)}$, $\nu = 1, \dots, m-1$, be the corresponding sets of wavelet functions. Then the systems $\{\psi_{jk}^{(\nu)}\}_{j,k,\nu}, \{\widetilde{\psi}_{jk}^{(\nu)}\}_{j,k,\nu}$ are biorthonormal.*

Proof. It follows from the definitions of MRA and wavelet sets that for all $j \in \mathbb{Z}$,

$$\psi_{jk}^{(\nu)} \in V_{j+1}, \qquad \widetilde{\psi}_{jk}^{(\nu)} \in \widetilde{V}_{j+1}$$

for all $k \in \mathbb{Z}^d$, $\nu = 1, \dots, m-1$;

$$\langle f, \widetilde{\psi}_{jk}^{(\nu)} \rangle = 0, \qquad \langle \widetilde{f}, \psi_{jk}^{(\nu)} \rangle = 0$$

for all $f \in V_i$, $\widetilde{f} \in \widetilde{V}_i$ for $i \le j$, $k \in \mathbb{Z}^d$, $\nu = 1, \dots, m-1$;

$$\langle \psi_{jk}^{(\nu)}, \widetilde{\psi}_{jl}^{(\kappa)} \rangle = \delta_{kl}\delta_{\nu\kappa}$$

for all $k, l \in \mathbb{Z}^d$, $\nu, \kappa = 1, \dots, m-1$. It follows that the systems $\{\psi_{jk}\}, \{\widetilde{\psi}_{jk}^{(\nu)}\}$ are biorthonormal. ◊

To find out if the systems $\{\psi_{jk}^{(\nu)}\}, \{\widetilde{\psi}_{jk}^{(\nu)}\}$ form bases, we shall need some auxiliary results.

LEMMA 2.7.2. *Let $f, \varphi, \widetilde{\varphi} \in L_2(\mathbb{R}^d)$. Then for any $j \in \mathbb{Z}$, the following equality holds:*

$$\sum_{k\in\mathbb{Z}^d} \langle f, \varphi_{jk} \rangle \overline{\langle f, \widetilde{\varphi}_{jk} \rangle} = m^j \int_{\mathbb{T}^d} h(\xi)\overline{\widetilde{h}(\xi)}\, d\xi,$$

where

$$h(\xi) = \sum_{l\in\mathbb{Z}^d} \widehat{f}\left(M^{*j}(\xi + l)\right) \overline{\widehat{\varphi}\,(\xi + l)}, \quad \widetilde{h}(\xi) = \sum_{l\in\mathbb{Z}^d} \widehat{f}\left(M^{*j}(\xi + l)\right) \overline{\widehat{\widetilde{\varphi}}\,(\xi + l)}.$$

Proof. First of all let us note that the functions $h, \widetilde{h}$ are 1-periodic in each variable. Next, the Bunyakovsky inequality and the argument analogous to the proof of Lemma 1.1.6 imply that these functions are summable on $\mathbb{T}^d$ and partial sums of the series defining $h, \widetilde{h}$ have a summable majorant. Using the Plancherel theorem and Parseval equality, we have

$$\sum_{k\in\mathbb{Z}^d} \langle f, \varphi_{jk}\rangle \overline{\langle f, \widetilde{\varphi}_{jk}\rangle}$$

$$= \sum_{k\in\mathbb{Z}^d} \int_{\mathbb{R}^d} \widehat{f}(\xi) e^{-2\pi i (M^{*-j}k,\xi)} \overline{\widehat{\varphi}\left(M^{*-j}\xi\right)}\, d\xi \overline{\int_{\mathbb{R}^d} \widehat{f}(\omega) e^{-2\pi i (M^{*-j}k,\omega)} \overline{\widehat{\widetilde{\varphi}}\left(M^{*-j}\omega\right)}\, d\omega}$$

$$= m^j \sum_{k\in\mathbb{Z}^d} \int_{\mathbb{R}^d} \widehat{f}(M^{*j}\xi) e^{-2\pi i (k,\xi)} \overline{\widehat{\varphi}(\xi)}\, d\xi \overline{\int_{\mathbb{R}^d} \widehat{f}(M^{*j}\omega) e^{-2\pi i (k,\omega)} \overline{\widehat{\widetilde{\varphi}}(\omega)}\, d\omega}$$

$$= m^j \sum_{k\in\mathbb{Z}^d} \sum_{l\in\mathbb{Z}^d} \int_{\mathbb{T}^d} \widehat{f}(M^{*j}(\xi+l)) e^{-2\pi i (k,\xi)} \overline{\widehat{\varphi}(\xi+l)}\, d\xi$$

$$\times \overline{\sum_{m\in\mathbb{Z}^d} \int_{\mathbb{T}^d} \widehat{f}(M^{*j}(\omega+m)) e^{-2\pi i (k,\omega)} \overline{\widehat{\varphi}(\omega+m)}\, d\omega}$$

$$= m^j \sum_{k\in\mathbb{Z}^d} \int_{\mathbb{T}^d} h(\xi) e^{-2\pi i k\xi}\, d\xi \overline{\int_{\mathbb{T}^d} \widetilde{h}(\omega) e^{-2\pi i k\omega}\, d\omega} = m^j \sum_{k\in\mathbb{Z}^d} \widehat{h}(k) \overline{\widehat{\widetilde{h}}(k)}$$

$$= m^j \int_{\mathbb{T}^d} h(\xi) \overline{\widetilde{h}(\xi)}\, d\xi.$$

Here, the interchange of summation and integration is justified by Lebesgue's dominated convergence theorem and the above-mentioned properties of $h, \widetilde{h}$. ◊

LEMMA 2.7.3. *Let the Fourier transform of a function $\psi \in L^2(\mathbb{R}^d)$ satisfy the conditions*

$$|\widehat{\psi}(\xi)| \le C_0|\xi|, \tag{2.62}$$

$$|\widehat{\psi}(\xi)| \le C_1(1+|\xi|)^{-d/2-\varepsilon}, \quad \varepsilon > 0, \tag{2.63}$$

for almost all $\xi \in \mathbb{R}^d$. Then there exists $C > 0$ such that

$$\sum_{j,k} |\langle f, \psi_{jk}\rangle|^2 \le C\|f\|^2 \tag{2.64}$$

for any function f in $L^2(\mathbb{R}^d)$.

Proof. Let $\delta \in (0,1)$. Using Lemma 2.7.2 and the Cauchy inequality, we get

$$\sum_{k} |\langle f, \psi_{jk}\rangle|^2 = \int_{M^{*j}[0,1)^d} \left| \sum_{l\in\mathbb{Z}^d} \widehat{f}(\xi + M^{*j}l) \overline{\widehat{\psi}(M^{*-j}\xi + l)} \right|^2 d\xi$$

$$\le \int_{M^{*j}[0,1)^d} \left(\sum_{l\in\mathbb{Z}^d} |\widehat{f}(\xi + M^{*j}l)|^2 |\widehat{\psi}(M^{*-j}\xi + l)|^{2\delta} \right)$$

$$\times \left(\sum_{k\in\mathbb{Z}^d} |\widehat{\psi}(M^{*-j}\xi + k)|^{2(1-\delta)} \right) d\xi$$

$$= \int_{\mathbb{R}^d} |\widehat{f}(\xi)|^2 |\widehat{\psi}(M^{*-j}\xi)|^{2\delta} \sum_{k\in\mathbb{Z}^d} |\widehat{\psi}(M^{*-j}\xi + k)|^{2(1-\delta)} d\xi.$$

From condition (2.63) it follows that the function $\sum_k |\widehat{\psi}(\xi + k)|^{2(1-\delta)}$ is essentially bounded on $\mathbb{R}^d$ for $\delta < 2\varepsilon/(d + 2\varepsilon)$ (since $2(1-\delta)(d/2+\varepsilon) > d$ and the series $\sum_{k\neq \mathbf{0}} 1/|k|^n$ converges for $n > d$). On the other hand, for such δ, we have

$$\sum_{j,k} |\langle f, \psi_{jk} \rangle|^2 \le C_2 \int_{\mathbb{R}^d} |\widehat{f}(\xi)|^2 \sum_{j\in\mathbb{Z}} |\widehat{\psi}(M^{*-j}\xi)|^{2\delta} d\xi$$

$$\le C_2 \operatorname*{ess\,sup}_{\xi\in\mathbb{R}^d} \sum_{j\in\mathbb{Z}} |\widehat{\psi}(M^{*-j}\xi)|^{2\delta} \|\widehat{f}\|^2.$$

Set $\alpha(\xi) = \sum_{j\in\mathbb{Z}} |\widehat{\psi}(M^{*-j}\xi)|^{2\delta}$. Since $\alpha(\xi) = \alpha(M^{*n}\xi)$ for any $n \in \mathbb{Z}$, the essential supremum over $\mathbb{R}^d$ can be replaced by the essential supremum over the set

$$\Omega := M^{*k_0}[-1,1]^d \backslash (-1,1)^d,$$

where k_0 is the minimal positive integer for which the inclusion $[-1,1]^d \subset M^{*k_0}[-1,1]^d$ holds. Indeed, if $\xi \in (-1,1)^d$, $\xi \neq \mathbf{0}$, then the vector $M^{*k_0}\xi$ is contained either in Ω or in $(-1,1)^d$. Consider the vectors $M^{*2k_0}\xi$, $M^{*3k_0}\xi$, etc. For some number n, the vector $M^{*nk_0}\xi$ will belong to Ω. Therefore,

$$\alpha(\xi) = \alpha(M^{*nk_0}\xi) < \operatorname*{ess\,sup}_{\xi\in\Omega} \alpha(\xi)$$

for almost all $\xi \in (-1,1)^d$. If $\xi \in \mathbb{R}^d \backslash M^{*k_0}[-1,1]^d$, then there exists a positive integer j_0 such that $M^{*-j_0}\xi \in (-1,1)^d$; i.e., we have the case considered above. Thus,

$$\sum_{j,k} |\langle f, \psi_{jk}\rangle|^2 \le C_2 \operatorname*{ess\,sup}_{\xi\in\Omega} \sum_{j\in\mathbb{Z}} |\widehat{\psi}(M^{*-j}\xi)|^{2\delta} \|\widehat{f}\|^2.$$

Using (2.62) and (2.6), for almost all $\xi \in \Omega$, we have

$$\sum_{j=0}^{\infty} |\widehat{\psi}(M^{*-j}\xi)|^{2\delta} \le C_0^{2\delta} \sum_{j=0}^{\infty} |M^{*-j}\xi|^{2\delta} \le C_0^{2\delta} \sup_{\Omega} |\xi|^{2\delta} \sum_{j=0}^{\infty} \|M^{*-j}\|^{2\delta} := C_3.$$

Now employing (2.63), we estimate the second part of the sum for $\xi \in \Omega$:

$$\sum_{j=-\infty}^{-1} |\widehat{\psi}(M^{*-j}\xi)|^{2\delta} \le C_1^{2\delta} \sum_{j=1}^{\infty} (1 + |M^{*-j}\xi|)^{2\delta(-d/2-\varepsilon)}.$$

Since $2\delta(-d/2 - \varepsilon) = -\delta(d + 2\varepsilon) \le -\delta$, it follows that

$$\sum_{j=-\infty}^{-1} |\widehat{\psi}(M^{*-j}\xi)|^{2\delta} \le C_1^{2\delta} \sum_{j=1}^{\infty} (1 + |M^{*j}\xi|)^{-\delta} \le C_1^{2\delta} \sum_{j=1}^{\infty} |M^{*j}\xi|^{-\delta}.$$

Note that $1/|M^{*j}\xi| \le \|M^{*-j}\|/|\xi|$; hence continuing this chain of inequalities and using (2.6), we obtain

$$\sum_{j=-\infty}^{-1} |\widehat{\psi}(M^{*-j}\xi)|^{2\delta} \le C_1^{2\delta} \sup_{\Omega} \frac{1}{|\xi|^{\delta}} \sum_{j=1}^{\infty} \|M^{*-j}\|^{\delta} := C_4.$$

Combining these estimates, we get (2.64). $\Diamond$

Further we shall discuss the situation where integer shifts of generating scaling functions $\varphi, \widetilde{\varphi}$ are biorthonormal. Recall that in this case wavelet functions are defined by equalities (2.51) and their masks $m_\nu, \widetilde{m}_\nu$, $\nu = 1, \dots, m-1$, are such that for the matrices

$$\mathcal{M} := \{m_\nu(\xi + {M^*}^{-1} s_k)\}_{\nu,k=0}^{m-1}, \qquad \widetilde{\mathcal{M}} := \{\widetilde{m}_\nu(\xi + {M^*}^{-1} s_k)\}_{\nu,k=0}^{m-1},$$

where $\{s_0, \dots, s_{m-1}\} = D(M^*)$, equality (2.50) holds almost everywhere.

LEMMA 2.7.4. *Suppose MRAs $\{V_j\}_{j\in\mathbb{Z}}$, $\{\widetilde{V}_j\}_{j\in\mathbb{Z}}$ are generated, respectively, by scaling functions $\varphi, \widetilde{\varphi}$ with masks $m_0, \widetilde{m}_0$ satisfying (2.31) and functions $\psi^{(\nu)}, \widetilde{\psi}^{(\nu)}$, $\nu = 1, \dots, m-1$, are defined by equalities (2.51) with masks $m_\nu, \widetilde{m}_\nu$, $\nu = 1, \dots, m-1$, satisfying (2.50). Assume that $\widehat{\varphi}, \widehat{\widetilde{\varphi}}$ are continuous at zero and $\widehat{\varphi}(0) = \widehat{\widetilde{\varphi}}(0) = 1$. If there exist constants B, $\widetilde{B}$ such that*

$$\sum_{j,k,\nu} |\langle f, \psi_{jk}^{(\nu)}\rangle|^2 \le B\|f\|^2, \tag{2.65}$$

$$\sum_{j,k,\nu} |\langle f, \widetilde{\psi}_{jk}^{(\nu)}\rangle|^2 \le \widetilde{B}\|f\|^2 \tag{2.66}$$

for any function $f \in L^2(\mathbb{R}^d)$, then

$$\sum_{j,k,\nu} |\langle f, \psi_{jk}^{(\nu)}\rangle|^2 \ge \frac{1}{\widetilde{B}}\|f\|^2 \tag{2.67}$$

for any function $f \in L^2(\mathbb{R}^d)$.

Proof. First show that for any function f from $L^2(\mathbb{R}^d)$ and any $j, j' \in \mathbb{Z}$, $j < j'$, the following equality holds:

(2.68)

$$\sum_{k\in\mathbb{Z}^d} \langle f, \varphi_{jk}\rangle \overline{\langle f, \widetilde{\varphi}_{jk}\rangle} + \sum_{i=j}^{j'} \sum_{\nu=1}^{m-1} \sum_{k\in\mathbb{Z}^d} \langle f, \psi_{ik}^{(\nu)}\rangle \overline{\langle f, \widetilde{\psi}_{ik}^{(\nu)}\rangle} = \sum_{k\in\mathbb{Z}^d} \langle f, \varphi_{j',k}\rangle \overline{\langle f, \widetilde{\varphi}_{j',k}\rangle}.$$

Clearly, it suffices to check (2.68) for $j' = j+1$. Using Lemmas 2.2.2 and 2.7.2 and the refinement equations for $\varphi, \widetilde{\varphi}$ and taking into account 1-periodicity of the functions $m_0, m_1, \dots, m_{m-1}$, we get

$$\sum_{k\in\mathbb{Z}^d} \langle f, \varphi_{jk}\rangle \overline{\langle f, \widetilde{\varphi}_{jk}\rangle} + \sum_{\nu=1}^{m-1} \sum_{k\in\mathbb{Z}^d} \langle f, \psi_{jk}^{(\nu)}\rangle \overline{\langle f, \widetilde{\psi}_{jk}^{(\nu)}\rangle}$$

$$= m^j \sum_{\nu=0}^{m-1} \int_{\mathbb{T}^d} \sum_{k\in\mathbb{Z}^d} \widehat{f}\left(M^{*j}(\xi+k)\right) \overline{\widetilde{m}_\nu\left({M^*}^{-1}(\xi+k)\right) \widehat{\varphi}\left({M^*}^{-1}(\xi+k)\right)}$$

$$\times \overline{\overline{\sum_{l\in\mathbb{Z}^d} \widehat{f}\left(M^{*j}(\xi+l)\right) \widetilde{m}_\nu\left({M^*}^{-1}(\xi+l)\right) \widehat{\widetilde{\varphi}}\left({M^*}^{-1}(\xi+l)\right)}}\, d\xi$$

$$= m^j \sum_{\nu=0}^{m-1} \int_{\mathbb{T}^d} \left(\sum_{s\in D(M^*)} \sigma(\xi+s) \overline{m_\nu\left({M^*}^{-1}(\xi+s)\right)} \right)$$

$$\times \overline{\left(\sum_{r\in D(M^*)} \widetilde{\sigma}(\xi+r) \overline{\widetilde{m}_\nu\left({M^*}^{-1}(\xi+r)\right)} \right)}\, d\xi, \tag{2.69}$$

where

$$\begin{aligned}\sigma(\xi) &= \sum_{k\in\mathbb{Z}^d} \widehat{f}\left(M^{*(j+1)}\left(M^{*-1}\xi+k\right)\right)\overline{\widehat{\varphi}\left(M^{*-1}\xi+k\right)},\\ \widetilde{\sigma}(\xi) &= \sum_{k\in\mathbb{Z}^d} \widehat{f}\left(M^{*(j+1)}\left(M^{*-1}\xi+k\right)\right)\overline{\widehat{\widetilde{\varphi}}\left(M^{*-1}\xi+k\right)}.\end{aligned}$$

By (2.50) the right-hand side of (2.69) can be represented as

$$m^j \int_{\mathbb{T}^d} \sum_{s\in D(M^*)} \sigma(\xi+s)\overline{\widetilde{\sigma}(\xi+s)}\,d\xi.$$

Using Lemma 2.2.5, we see that

$$\begin{aligned}&\sum_{k\in\mathbb{Z}^d} \langle f,\varphi_{jk}\rangle\overline{\langle f,\widetilde{\varphi}_{jk}\rangle} + \sum_{\nu=1}^{m-1}\sum_{k\in\mathbb{Z}^d} \langle f,\psi_{jk}^{(\nu)}\rangle\overline{\langle f,\widetilde{\psi}_{jk}^{(\nu)}\rangle}\\ &\quad= m^{j+1}\sum_{s\in D(M^*)}\int_{M^{*-1}(\mathbb{T}^d+s)} \sigma(M^*\xi)\overline{\widetilde{\sigma}(M^*\xi)}\,d\xi\\ &\quad= m^{j+1}\int_{\mathbb{T}^d}\sum_{k\in\mathbb{Z}^d} \widehat{f}\left(M^{*(j+1)}(\xi+k)\right)\overline{\widehat{\varphi}(\xi+k)}\\ &\qquad\times\overline{\sum_{l\in\mathbb{Z}^d} \widehat{f}\left(M^{*(j+1)}(\xi+l)\right)\overline{\widehat{\widetilde{\varphi}}(\xi+l)}}\,d\xi.\end{aligned}$$

To prove (2.68) for $j' = j+1$, it remains to use Lemma 2.7.2 once more.

It follows from the Cauchy inequality and relation (2.13) that

$$\lim_{j\to-\infty}\sum_{k\in\mathbb{Z}^d} \langle f,\varphi_{jk}\rangle\overline{\langle f,\widetilde{\varphi}_{jk}\rangle} = 0.$$

Passing to the limit as $j\to-\infty$ in (2.68), for any $j'\in\mathbb{Z}$, we get the equality

$$\sum_{i=-\infty}^{j'}\sum_{\nu=1}^{m-1}\sum_{k\in\mathbb{Z}^d} \langle f,\psi_{ik}^{(\nu)}\rangle\overline{\langle f,\widetilde{\psi}_{ik}^{(\nu)}\rangle} = \sum_{k\in\mathbb{Z}^d} \langle f,\varphi_{j',k}\rangle\overline{\langle f,\widetilde{\varphi}_{j',k}\rangle}.$$

Similarly to the proof of Lemma 1.2.13, with using (2.65), (2.66), and Lemma 2.7.2, it is not difficult to show that

$$\lim_{j\to+\infty}\sum_{k\in\mathbb{Z}^d} \langle f,\varphi_{jk}\rangle\overline{\langle f,\widetilde{\varphi}_{jk}\rangle} = \|f\|^2.$$

Thus, if $\|f\|\neq 0$, given $\varepsilon>0$, there exists j' such that

$$\sum_{i=-\infty}^{j'}\sum_{\nu=1}^{m-1}\sum_{k\in\mathbb{Z}^d} \langle f,\psi_{ik}^{(\nu)}\rangle\overline{\langle f,\widetilde{\psi}_{ik}^{(\nu)}\rangle} \ge \|f\|^2(1-\varepsilon).$$

It follows from (2.66) and the Cauchy inequality that

$$\begin{aligned}\sqrt{\sum_{i=-\infty}^{\infty}\sum_{\nu=1}^{m-1}\sum_{k\in\mathbb{Z}^d} |\langle f,\psi_{ik}^{(\nu)}\rangle|^2} &\ge \frac{1}{\sqrt{\widetilde{B}}\|f\|}\left|\sum_{i=-\infty}^{j'}\sum_{\nu=1}^{m-1}\sum_{k\in\mathbb{Z}^d} \langle f,\psi_{ik}^{(\nu)}\rangle\overline{\langle f,\widetilde{\psi}_{ik}^{(\nu)}\rangle}\right|\\ &\ge \frac{\|f\|}{\sqrt{\widetilde{B}}}\sqrt{1-\varepsilon}.\end{aligned}$$

Passing to the limit as $\varepsilon \to 0$, we obtain (2.67). ◊

THEOREM 2.7.5. *Let MRAs $\{V_j\}_{j\in\mathbb{Z}}$, $\{\widetilde{V}_j\}_{j\in\mathbb{Z}}$ be generated, respectively, by scaling functions $\varphi, \widetilde{\varphi}$ with masks $m_0, \widetilde{m}_0$ satisfying* (2.31). *Suppose that for some constants $C > 0, \varepsilon > 0$ and almost all $\xi \in \mathbb{R}^d$ the inequality*

$$|\widehat{\varphi}(\xi)|, |\widehat{\widetilde{\varphi}}(\xi)| \le C(1 + |\xi|)^{-d/2-\varepsilon} \tag{2.70}$$

holds; the functions $\widehat{\varphi}, \widehat{\widetilde{\varphi}}$ are continuously differentiable in a neighborhood of zero. Suppose the functions $\psi^{(\nu)}, \widetilde{\psi}^{(\nu)}$, $\nu = 1, \dots, m-1$, are given by equalities (2.51), *their masks $m_\nu, \widetilde{m}_\nu$, $\nu = 1, \dots, m-1$, are bounded, continuously differentiable in a neighborhood of zero, satisfy* (2.50), *and*

$$m_\nu(\mathbf{0}) = \widetilde{m}_\nu(\mathbf{0}) = 0, \quad \nu = 1, \dots, m-1. \tag{2.71}$$

Then each of the systems $\{\psi_{jk}^{(\nu)}\}, \{\widetilde{\psi}_{jk}^{(\nu)}\}$ forms a frame in $L_2(\mathbb{R}^d)$. Moreover, if the systems $\{\varphi(\cdot + n)\}_{n\in\mathbb{Z}^d}$, $\{\widetilde{\varphi}(\cdot + n)\}_{n\in\mathbb{Z}^d}$ are biorthonormal, then $\psi^{(\nu)}, \widetilde{\psi}^{(\nu)}$, $\nu = 1, \dots, m-1$, form a set of wavelet functions and each of the systems $\{\psi_{jk}^{(\nu)}\}, \{\widetilde{\psi}_{jk}^{(\nu)}\}$ forms a Riesz basis in the space $L_2(\mathbb{R}^d)$.

Proof. Let us show that the functions ψ^ν, $\nu = 1, \dots, m-1$, satisfy the conditions of Lemma 2.7.3. Indeed, (2.63) follows from (2.70), (2.51) and boundedness of the masks. In addition, the function $\widehat{\varphi}$ and the masks $m_1, \dots, m_{m-1}$ are bounded and have bounded partial derivatives in a neighborhood of zero. Hence (2.62) follows from (2.71) and (2.51). Lemma 2.7.3 gives a frame upper estimate for the system $\{\psi_{jk}^{(\nu)}\}$ for each $\nu = 1, \dots, m-1$, which implies relations (2.65) and (2.66). By Lemma 2.7.4, these relations yield (2.67). Thus the system $\{\psi_{jk}^{(\nu)}\}$ is a frame in $L^2(\mathbb{R}^d)$.

If $\varphi, \widetilde{\varphi}$ have biorthonormal integer shifts, then by Theorem 2.7.1 the systems $\{\psi_{jk}^{(\nu)}\}, \{\widetilde{\psi}_{jk}^{(\nu)}\}$ are orthonormal. Moreover, in view of Corollary 1.8.6, a sufficient condition for these systems to be Riesz bases is that one of them be a frame. ◊

Generally speaking, condition (2.70) in Theorem 2.7.5 is not necessary. If scaling functions $\varphi, \widetilde{\varphi}$ and wavelet functions $\psi^{(\nu)}, \widetilde{\psi}^{(\nu)}$, $\nu = 1, \dots, m-1$, are compactly supported, then assumption (2.70) in Theorem 2.7.5 is not necessary. This fact follows from the following results obtained by B. Han:

Theorem A ([**107**, Theorem 2.3]). *Let a compactly supported function $\psi \in L^2(\mathbb{R}^d)$ satisfy the condition*

$$\int_{\mathbb{R}^d} \psi(x)\, dx = 0$$

and let it belong to the Sobolev space W_2^ε for some $\varepsilon > 0$. Then there exists $C > 0$ such that

$$\sum_{j,k} |\langle f, \psi_{jk}\rangle|^2 \le C\|f\|^2$$

for any function f from $L^2(\mathbb{R}^d)$.

Theorem B ([**107**, Theorem 2.2]). *Let $\varphi \in L^2(\mathbb{R}^d)$ be a refinable function with a polynomial mask. Then $\varphi \in W_2^\varepsilon$ for some $\varepsilon > 0$.*

In view of the above results, a slight modification of the proof of Theorem 2.7.5 gives the following statement.

THEOREM 2.7.6. *Let MRAs* $\{V_j\}_{j\in\mathbb{Z}}$, $\{\widetilde{V}_j\}_{j\in\mathbb{Z}}$ *be generated by scaling functions* $\varphi, \widetilde{\varphi}$ *with polynomial masks* $m_0, \widetilde{m}_0$ *satisfying* (2.31). *Suppose functions* $\psi^{(\nu)}, \widetilde{\psi}^{(\nu)}$, $\nu = 1, \dots, m-1$, *are given by equalities* (2.51) *with polynomial masks* $m_\nu, \widetilde{m}_\nu$, $\nu = 1, \dots, m-1$, *satisfying* (2.50) *and such that*

$$m_\nu(\mathbf{0}) = \widetilde{m}_\nu(\mathbf{0}) = 0, \quad \nu = 1, \dots, m-1. \tag{2.72}$$

Then each of the systems $\{\psi^{(\nu)}_{jk}\}, \{\widetilde{\psi}^{(\nu)}_{jk}\}$ *forms a frame in* $L_2(\mathbb{R}^d)$. *If, in addition, the systems* $\{\varphi(\cdot + n)\}_{n\in\mathbb{Z}^d}$, $\{\widetilde{\varphi}(\cdot + n)\}_{n\in\mathbb{Z}^d}$ *are biorthonormal, then* $\psi^{(\nu)}, \widetilde{\psi}^{(\nu)}$, $\nu = 1, \dots, m-1$, *form a set of wavelet functions and each of the systems* $\{\psi^{(\nu)}_{jk}\}, \{\widetilde{\psi}^{(\nu)}_{jk}\}$ *forms a Riesz basis in the space* $L_2(\mathbb{R}^d)$.

In Theorem 2.7.5, we imposed certain conditions on masks $m_1, \dots, m_{m-1}$, $\widetilde{m}_1, \dots, \widetilde{m}_{m-1}$. Now discuss the question of how to provide these conditions.

LEMMA 2.7.7. *Suppose functions* $m_0, \dots, m_{m-1}, \widetilde{m}_0, \dots, \widetilde{m}_{m-1} \in L_2(\mathbb{T}^d)$ *satisfy condition* (2.50), $m_1, \dots, m_{m-1}, \widetilde{m}_0$ *are continuous in a neighborhood of the set* $M^{*-1}(D(M^*))$, *and*

$$\widetilde{m}_0(\mathbf{0}) = 1. \tag{2.73}$$

The equalities

$$m_\nu(\mathbf{0}) = 0, \quad \nu = 1, \dots, m-1, \tag{2.74}$$

hold if and only if

$$\widetilde{m}_0(M^{*-1}s) = 0, \quad s \in D(M^*), \ s \neq \mathbf{0}. \tag{2.75}$$

Proof. By Lemma 2.6.3, equalities (2.50) hold if and only if the polyphase matrices $\{\mu_{\nu k}(\xi)\}_{\nu,k=0}^{m-1}$, $\{\widetilde{\mu}_{k\nu}(\xi)\}_{\nu,k=0}^{m-1}$ are mutually inverse, which gives

$$\sum_{k=0}^{m-1} \overline{\widetilde{\mu}_{0k}(\xi)} \mu_{\nu k}(\xi) = 0 \ \text{ a.e.}, \quad \nu = 1, \dots, m-1. \tag{2.76}$$

From (2.56) and Corollary 2.2.7 it follows that

$$\mu_{\nu k}(M^*\xi) = \frac{1}{\sqrt{m}} \sum_{s\in D(M^*)} e^{-2\pi i (M^{-1}s_k, s - M^*\xi)} m_\nu(\xi + M^{*-1}s), \quad k = 0, \dots, m-1.$$

Hence the functions $\mu_{\nu k}$ (and, analogously, $\widetilde{\mu}_{\nu k}$), $\nu, k = 0, \dots, m-1$, are continuous in a neighborhood of zero. Therefore (2.76) holds for $\xi = \mathbf{0}$; i.e., the vectors $(\mu_{\nu 0}(\mathbf{0}), \dots, \mu_{\nu,m-1}(\mathbf{0}))$, $\nu = 1, \dots, m-1$, are orthogonal to the vector $(\widetilde{\mu}_{00}(\mathbf{0}), \dots, \widetilde{\mu}_{0,m-1}(\mathbf{0}))$. On the other hand, it follows from (2.56) that equalities (2.73) and (2.74) can be rewritten, respectively, in the form

$$\sum_{k=0}^{m-1} \overline{\widetilde{\mu}_{0k}(\mathbf{0})} = \sqrt{m}, \qquad \sum_{k=0}^{m-1} \mu_{\nu k}(\mathbf{0}) = 0, \quad \nu = 1, \dots, m-1.$$

Thus the vectors $(\mu_{\nu 0}(\mathbf{0}), \dots, \mu_{\nu 0}(\mathbf{0}))$, $\nu = 1, \dots, m-1$, are orthogonal to the vector $(1, \dots, 1)$. Taking into account the linear independence of these vectors, we see that equalities (2.74) hold if and only if

$$\widetilde{\mu}_{0k}(\mathbf{0}) = \frac{1}{\sqrt{m}}, \quad k = 0, \dots, m-1. \tag{2.77}$$

Using (2.56) again, we get

$$\frac{1}{\sqrt{m}}\sum_{k=0}^{m-1} e^{2\pi i(M^{-1}s_k,s)}\widetilde{\mu}_{0k}(\mathbf{0}) = \widetilde{m}_0(M^{*-1}s), \quad s \in D(M^*).$$

Consider these equalities as a system of linear equations with the unknowns $\widetilde{\mu}_{0k}(\mathbf{0})$, $k = 1, \dots, m-1$. By Corollary 2.2.7, this system has a unique solution, and from Lemma 2.2.6 it follows that equalities (2.73) and (2.75) are equivalent to (2.77). ◊

It is now clear that if masks of wavelet functions are continuously differentiable in a neighborhood of the set $M^{*-1}(D(M^*))$ and are bounded, then condition (2.71) is provided by the required properties of masks of scaling functions. It is not hard to see that the functions $\mu_{\nu k}, \widetilde{\mu}_{\nu k}$, $\nu = 1, \dots, m-1$, constructed as described in §2.6 inherit the properties of the functions $\mu_{0k}, \widetilde{\mu}_{0k}$; i.e., they are bounded and continuously differentiable in a neighborhood of zero whenever (2.77) holds. As was proved above, the latter relation is equivalent to equalities (2.73) and (2.75). Summarizing the aforesaid, we obtain the following assertion.

THEOREM 2.7.8. *Let MRAs $\{V_j\}_{j\in\mathbb{Z}}$, $\{\widetilde{V}_j\}_{j\in\mathbb{Z}}$ be generated, respectively, by scaling functions $\varphi, \widetilde{\varphi}$ with masks $m_0, \widetilde{m}_0$ continuously differentiable in a neighborhood of the set $M^{*-1}(D(M^*))$, bounded, and satisfying conditions* (2.73), (2.75). *Suppose the functions $\widehat{\varphi}, \widehat{\widetilde{\varphi}}$ are continuously differentiable in a neighborhood of zero and satisfy condition* (2.70)*; systems $\{\varphi(\cdot+n)\}_{n\in\mathbb{Z}^d}$, $\{\widetilde{\varphi}(\cdot+n)\}_{n\in\mathbb{Z}^d}$ are biorthonormal. Then there exists a set of wavelet functions $\psi^{(\nu)}, \widetilde{\psi}^{(\nu)}$, $\nu = 1, \dots, m-1$, such that each of the systems $\{\psi_{jk}^{(\nu)}\}$, $\{\widetilde{\psi}_{jk}^{(\nu)}\}$ forms a Riesz basis in the space $L_2(\mathbb{R}^d)$.*

PROBLEM 2.7.9. Let $\psi^{(\nu)}, \widetilde{\psi}^{(\nu)}$, $\nu = 1, \dots, m-1$, be a set of wavelet functions, and let the functions $\widehat{\psi}^{(\nu)}$ be continuous at zero. Prove that for the system $\{\psi_{jk}^{(\nu)}\}$ to be a Riesz basis in the space $L_2(\mathbb{R}^d)$ it is necessary to have $\widehat{\psi}^{(\nu)}(\mathbf{0}) = 0$, $\nu = 1, \dots, m-1$.

In the orthogonal case, it is much easier to make out if a wavelet system forms a basis.

THEOREM 2.7.10. *Suppose an MRA $\{V_j\}_{j\in\mathbb{Z}}$ is generated by a scaling function φ with mask m_0 and the system $\{\varphi(\cdot+n)\}_{n\in\mathbb{Z}^d}$ is orthonormal. Let $\{s_0, \dots, s_{m-1}\} = D(M^*)$. If $\psi^{(\nu)}$, $\nu = 1, \dots, m-1$, is a set of wavelet functions given by equalities* (2.51) *with masks m_ν, $\nu = 1, \dots, m-1$, such that the matrix*

$$\mathcal{M} := \{m_\nu(\xi + M^{*-1}s_k)\}_{\nu,k=0}^{m-1}$$

is unitary for almost all $\xi \in \mathbb{R}^d$, then the system $\{\psi_{jk}^{(\nu)}\}$ is an orthonormal basis in the space $L_2(\mathbb{R}^d)$.

Proof. By Theorem 2.7.1, the system $\{\psi_{jk}^{(\nu)}\}$ is orthonormal. It follows from Bessel's inequality that relation (2.65) holds with $B = 1$. By Lemma 2.7.4, this relation implies (2.67) with $\widetilde{B} = 1$. Thus for a function $f \in L_2(\mathbb{R}^d)$, the Parseval equality holds; i.e., the system $\{\psi_{jk}^{(\nu)}\}$ is complete in the space $L^2(\mathbb{R}^d)$. ◊

2.8. Haar MRAs

In the classical Haar MRA the scaling function is the characteristic function of the segment $[0, 1]$. Let us see what kind of sets in $\mathbb{R}^d$ have characteristic functions that generate orthogonal wavelet bases as scaling functions. It is clear that integer shifts of such sets must be mutually disjoint (otherwise, integer shifts of their characteristic functions will not form orthogonal systems). Certainly, this condition is not sufficient for the characteristic function of a set to generate an MRA.

In this section, equality of sets from $\mathbb{R}^d$ will be understood up to a null set. Accordingly, we shall say that two sets are disjoint if their intersection is a null set.

THEOREM 2.8.1. *Let E be a measurable set in $\mathbb{R}^d$ such that its integer shifts are mutually disjoint. The function χ_E is scaling for an MRA if and only if*

(i) *there exists a set of digits D_E of a matrix M such that* $ME = \bigcup_{n\in D_E} (E+n)$,

(ii) $\bigcup_{l\in\mathbb{Z}^d} (E+l) = \mathbb{R}^d$.

Proof. Let χ_E be the scaling function for an MRA. It follows from axiom **MR1** that $\chi_{ME} \in V_{-1} \subset V_0$; hence by axiom **MR5** we have the expansion

$$\chi_{ME} = \sum_{n\in\mathbb{Z}^d} c_n \chi_E(\cdot + n) = \sum_{n\in\mathbb{Z}^d} c_{-n}\chi_{E+n}.$$

Taking into account that the sets $E+n$ are disjoint for different n, we conclude that the coefficients c_n can take only two values: 1 and 0. Moreover, since $\mu(ME) = m\mu E$, it follows that the number of nonzero coefficients c_n is m and

$$\chi_{ME} = \sum_{k=1}^{m} \chi_{E+n_k}. \tag{2.78}$$

Let us show that the integer vectors n_k, $k = 1, \dots, m$, belong to different cosets. Suppose that $n_{k_1} \equiv n_{k_2} \pmod M$ for $k_1 \neq k_2$. Then $n_{k_1} = n_{k_2} + Ml$, where $l \in \mathbb{Z}^d$, $l \neq \mathbf{0}$; hence,

$$M(E+l) = ME + Ml = \bigcup_{k=1}^{m} (E + n_k + Ml) \supset E + n_{k_1}.$$

On the other hand, from (2.78) it follows that

$$ME = \bigcup_{k=1}^{m} (E + n_k); \tag{2.79}$$

therefore $ME \cap M(E+l) \neq \emptyset$. Since the map M is one-to-one, we get the relation $E \cap (E+l) \neq \emptyset$, which contradicts the assumptions of the theorem. Thus $\{n_1, \dots, n_m\}$ is the required set of digits D_E, which completes the proof of (i).

To prove (ii), we set $S = \bigcup_{l\in\mathbb{Z}^d} (E+l)$, $\Omega = \mathbb{R}^d \setminus S$. Suppose that (ii) is not true, i.e., $\mu\Omega > 0$. From (2.79) it follows that

$$MS = M\left(\bigcup_{l\in\mathbb{Z}^d} (E+l)\right) = \bigcup_{l\in\mathbb{Z}^d} (ME + Ml) = \bigcup_{l\in\mathbb{Z}^d} \bigcup_{k=1}^{m} (E + n_k + Ml). \tag{2.80}$$

Taking into account Lemma 2.2.2, we obtain the equality $MS = S$, which implies that $M\Omega = \Omega$, since M is a one-to-one mapping. It follows that

$$\int_\Omega \chi_E(M^j x + n)\,dx = \frac{1}{m^j}\int_\Omega \chi_E(x+n)\,dx = \frac{1}{m^j}\int_\Omega \chi_{E-n}(x)\,dx = 0$$

for all $j \in \mathbb{Z}$ and $n \in \mathbb{Z}^d$. Let Ω_0 be a bounded subset of positive measure in Ω. Let $f = \chi_{\Omega_0}$. Clearly, $\|f\|_2 > 0$ and $f \perp V_j$ for any $j \in \mathbb{Z}$, which contradicts axiom **MR2**.

Now let E satisfy conditions (i), (ii). Consider the function

$$\Phi := \sum_{n \in \mathbb{Z}^d} \chi_E(\cdot + n). \tag{2.81}$$

This function is 1-periodic in each variable. Since the sets $E + n$, $n \in \mathbb{Z}^d$, are mutually disjoint, (ii) implies that $\Phi(x) = 1$ for almost all $x \in \mathbb{T}^d$. Therefore we get

$$\|\chi_E\|_2^2 = \|\chi_E\|_1 = \int_{\mathbb{R}^d} \chi_E(x)\, dx = \int_{\mathbb{T}^d} \Phi(x)\, dx = 1.$$

Thus $\chi_E \in L_2(\mathbb{R}^d)$ and integer shifts of the function χ_E form an orthonormal system. Moreover, we have proved the relation $\chi_E \in L(\mathbb{R}^d)$, which implies the continuity of the function $\widehat{\chi}_E$ and the equality

$$\widehat{\chi}_E(\mathbf{0}) = \int_{\mathbb{R}^d} \chi_E(x)\, dx = \|\chi_E\|_1 = 1,$$

i.e.,

$$\lim_{x \to \mathbf{0}} \widehat{\chi}_E(x) = \widehat{\chi}_E(\mathbf{0}) = 1. \tag{2.82}$$

Condition (i) and the fact that the sets $E + n$, $n \in \mathbb{Z}^d$, are mutually disjoint imply the equality

$$\chi_E(x) = \sum_{n \in D_E} \chi_E(Mx + n), \tag{2.83}$$

which is a refinement equation for the function χ_E; hence axiom **MR1** holds for the spaces V_j, $j \in \mathbb{Z}$, generated by this function. Then by Theorems 2.3.2 and 2.3.4, in view of (2.82), axioms **MR2**, **MR3** also hold. ◊

By Theorem 2.8.1, it is clear that to construct an analog of the classical Haar system, we need a set with mutually disjoint integer shifts that satisfies conditions (i), (ii). For simple sets, it is not hard to check whether their integer shifts are mutually disjoint. However, we shall further present a wide class of sets that satisfy (i), (ii), but for which it is very complicated to understand if disjointness holds. In this regard, the following assertion will be useful.

PROPOSITION 2.8.2. *Let E be a measurable set satisfying* (ii) *of Theorem* 2.8.1 *in $\mathbb{R}^d$. Then the following conditions are equivalent:*

OR1. *The system $\{\chi_E(\cdot + n)\}_{n \in \mathbb{Z}^d}$ is orthonormal.*

OR2. *Integer shifts of the set E are mutually disjoint.*

OR3. $\mu E = 1$.

`Proof.` Let Φ be the function given by relation (2.81). It is easy to see that under condition (ii) of Theorem 2.8.1 the equality $\Phi(x) = 1$ holds for almost all $x \in \mathbb{T}^d$ if and only if the sets $E + n$, $n \in \mathbb{Z}^d$, are mutually disjoint. Otherwise, we have $\Phi(x) \geq 1$ for almost all $x \in \mathbb{R}^d$ and there exists a set of positive measure, where $\Phi(x) \geq 2$. Hence the equivalence of conditions **OR2** and **OR3** follows from the equality

$$\mu E = \int_{\mathbb{R}^d} \chi_E(x)\,dx = \int_{\mathbb{T}^d} \Phi(x)\,dx.$$

It remains to note that condition **OR2** together with **OR3** are equivalent to condition **OR1**. $\diamondsuit$

An MRA generated by a function χ_E that satisfies condition **OR1** is called the *Haar MRA*.

THEOREM 2.8.3. *Let E be a measurable set in $\mathbb{R}^d$ that satisfies conditions* (i), (ii) *of Theorem* 2.8.1, *and let $\mu E = 1$. Then the function χ_E generates a Haar MRA.*

The proof follows from Theorem 2.8.1 and Proposition 2.8.2.

If it is not easy to check conditions **OR2**, **OR3**, we can use the results from §2.5 to check **OR1**. Note that to prove equality (2.83), we do not need mutual orthogonality of all sets $E+n$, $n \in \mathbb{Z}$. It suffices to have orthogonal pairs $E+n$, $E+k$ for $n, k \in D_E$, $n \neq k$, which is always the case if condition (i) holds, since

$$\mu ME = m\mu E = \sum_{n \in D_E} \mu E = \sum_{n \in D_E} \mu(E+n).$$

Thus we see that condition (i) plays the key role in the construction of the Haar MRA. This condition implies that the function χ_E satisfies the refinement equation (2.83). Its mask is the trigonometric polynomial

$$m_0(\xi) = \frac{1}{m} \sum_{n \in D_E} e^{2\pi i(n,\xi)}. \tag{2.84}$$

Let us show that for m_0 the necessary condition of orthonormality (2.31) is satisfied, i.e., the identity

$$\sum_{s \in D_E} |m_0(\xi + M^{*-1}s)|^2 = 1 \tag{2.85}$$

holds. Indeed,

$$\sum_{s \in D_E} |m_0(\xi + M^{*-1}s)|^2 = \frac{1}{m^2} \sum_{n \in D_E} \sum_{k \in D_E} \sum_{s \in D_E} e^{2\pi i(n-k,\xi+M^{*-1}s)}.$$

By Corollary 2.2.7, the inner sum on the right-hand side is nonzero only for $k \equiv n \pmod{M}$, which in this case is equivalent to $k = n$. Therefore the sum over n, k has exactly m nonzero summands, each being equal to m. In §2.5, it was shown that equality (2.85) is not sufficient for integer shifts of the function χ_E to be orthogonal. To provide **OR1**, we need some sufficient condition of orthogonality. If **OR1** is satisfied, then, as already shown, it follows from (i) that χ_E is a scaling function; hence the spaces V_j, $j \in \mathbb{Z}$, generated by it satisfy **MR1**. Theorems 2.3.2 and 2.3.4 and the equality

$$\widehat{\chi}_E(0) = \|\chi_E\|_2^2 = 1$$

imply that axioms **MR2**, **MR3** hold. $\diamondsuit$

Summarizing the aforesaid and using Theorem 2.5.6, we obtain the following assertion.

THEOREM 2.8.4. *Let E be a compact set in $\mathbb{R}^d$ satisfying condition* (i) *of Theorem* 2.8.1 *and let m_0 be a trigonometric polynomial defined by equality* (2.84). *If there exists a set K congruent to the cube $[-1/2, 1/2]^d$ modulo $\mathbb{Z}^d$, containing a neighborhood of zero and such that*

$$\inf_{k\in\mathbb{N}} \inf_{\xi\in K} |m_0(M^{*^{-k}}\xi)| \neq 0, \tag{2.86}$$

then χ_E generates a Haar MRA.

Now we construct a set of wavelet functions in the Haar MRA. Let us extend the row $\alpha_{00}, \dots, \alpha_{0,m-1}$, where $\alpha_{0k} = 1/\sqrt{m}$, $k = 0, \dots, m-1$, to the orthogonal matrix $A := \{\alpha_{\nu k}\}_{\nu,k}$. Let $D_E = \{s_0, \dots, s_{m-1}\}$. For $\nu = 1, \dots, m-1$, we set

$$m_\nu(\xi) = \frac{1}{\sqrt{m}} \sum_{k=0}^{m-1} \alpha_{\nu k} e^{2\pi i (s_k, \xi)}$$

and define functions $\psi^{(\nu)}$ by the equalities

$$\widehat{\psi}^{(\nu)}(\xi) = m_\nu(M^{*^{-1}}\xi)\widehat{\chi}_E(M^{*^{-1}}\xi).$$

Using the Fourier transform, we get the explicit representations for these functions:

$$\psi^{(\nu)}(x) = \sqrt{m} \sum_{k=0}^{m-1} \alpha_{\nu k} \chi_E(Mx + s_k).$$

To verify that the functions $\psi^{(\nu)}$, $\nu = 1, \dots, m-1$, form a set of wavelet functions, let us check that the matrix

$$\mathcal{M} := \{m_\nu(\xi + M^{*^{-1}} s_k)\}_{\nu,k=0}^{m-1}$$

is unitary. Using the orthogonality of the matrix A and Corollary 2.2.7, we get

$$\sum_{l=0}^{m-1} m_\nu(\xi + M^{*^{-1}} s_l)\overline{m_\kappa(\xi + M^{*^{-1}} s_l)}$$
$$= \frac{1}{m} \sum_{k=0}^{m-1} \sum_{n=0}^{m-1} \alpha_{\nu k}\alpha_{\kappa n} \sum_{l=0}^{m-1} e^{2\pi i (s_k - s_n, \xi + M^{*^{-1}} s_l)} = \sum_{k=0}^{m-1} \alpha_{\nu k}\alpha_{\kappa k} = \delta_{\nu\kappa},$$

which means that the matrix $\mathcal{M}$ is unitary. By Theorem 2.7.10, the system $\{\psi_{jk}^{(\nu)}\}$ is an orthonormal basis in the space $L_2(\mathbb{R}^d)$. These bases are a generalization of the classical Haar bases.

Let us now give a constructive description of the sets whose characteristic functions generate Haar MRAs. Let D be an arbitrary set of digits of a matrix M. Set

$$E = E_D := \{x \in \mathbb{R}^d : \ x = \sum_{k=1}^{\infty} M^{-k} s_k, \ s_k \in D\}. \tag{2.87}$$

PROPOSITION 2.8.5. *The set E_D is compact in $\mathbb{R}^d$.*

Proof. Let $\{x_n\}$ be a Cauchy sequence in E. Then

$$x_n = \sum_{k=1}^{\infty} M^{-k} s_k^{(n)}, \quad s_k^{(n)} \in D.$$

Since D is a finite set, we can choose a subsequence $\{x_{n_k}\}$ of elements with the same first digit s_1^*. Then we find another subsequence of elements with the same second digit s_2^*, etc. Select a diagonal subsequence $\{x_{n_k^*}\}$ and set

$$x^* = \sum_{k=1}^{\infty} M^{-k} s_k^*.$$

By construction, it is clear that $x_{n_k^*} \longrightarrow x^* \in E$. ◊

PROPOSITION 2.8.6. *The set E_D satisfies conditions* (i), (ii) *of Theorem* 2.8.1.

Proof. Obviously (i) is satisfied for $D_E = D$. To prove (ii), set $S = \bigcup_{l\in\mathbb{Z}^d}(E+l)$, $\Omega = \mathbb{R}^d \backslash S$. Using (i) and repeating the chain of equalities (2.80), we get the equality $MS = S$. It follows from the definition of the set S that $S+n = S$ for any $n \in \mathbb{Z}^d$. It is clear that the set Ω has analogous properties; i.e.,

$$\Omega + n = \Omega \quad \text{for all} \quad n \in \mathbb{Z}^d, \tag{2.88}$$

$$M^j\Omega = \Omega \quad \text{for all} \quad j \in \mathbb{Z}. \tag{2.89}$$

Note that equalities (2.88) and (2.89) are pointwise equalities (not up to a null set as they have been understood so far). Let us show that the set S is closed. Let $x_n \to x$, $\{x_n\} \subset S$. Then $x_n = y_n + k_n$, where $y_n \in E$, $k_n \in \mathbb{Z}^d$. The sequence $\{x_n\}$ is bounded due to its convergence, and the sequence $\{y_n\}$ is bounded because the set E is compact. Hence the sequence $\{k_n\}$ is also bounded; i.e., there are only finitely many different k_n. Select a subsequence $\{x_{n_k}\}$ such that $x_{n_k} = y_{n_k} + k^*$, $k^* \in \mathbb{Z}^d$, for all $k = 1, 2, \dots$. The convergence of the sequence $\{x_{n_k}\}$ implies the convergence of the sequence $\{y_{n_k}\}$ and, as E is a compact set, $y_{n_k} \to y \in E$; hence $x_{n_k} \to y + k^* \in S$. Thus the set S is closed and the set Ω is open. Suppose that $\Omega \neq \emptyset$. Then Ω contains a ball Δ. It follows from (2.89) that $M^j\Delta \subset \Omega$ for arbitrarily large $j \in \mathbb{Z}$. Taking into account (2.5), we obtain that Ω contains a set $[0,1]^d + n$, $n \in \mathbb{Z}^d$. In view of (2.88), it follows that $\Omega = \mathbb{R}^d$, which contradicts the fact that the set S is not empty. ◊

Combining Propositions 2.8.2 and 2.8.6 with Theorem 2.8.1, we see that the characteristic function of the set E_D generates a Haar MRA if one of the equivalent conditions **OR1**, **OR2**, **OR3** holds. These conditions do not hold for each set of digits D. Let $d = 1$, $M = 2$, $D = \{0, 3\}$. Then it is clear that the set E_D is the segment $[0, 3]$, but here **OR3** does not take place. Moreover, in §2.5, it was shown that integer shifts of the characteristic function of this segment do not even form a Riesz system. By Theorems 2.8.3 and 2.8.4, to prove that the characteristic function of the set E_D generates a Haar MRA, it suffices either to verify condition **OR3** or to find a set K that is congruent to the cube $[-1/2, 1/2]^d$, contains a neighborhood of zero, and such that the functions $m_0(M^{*-k}\xi)$, $k = 1, 2, \dots$, do not vanish on $\overline{K}$.

It can be shown that if the characteristic function of a set E generates a Haar MRA, then E can be represented in the form (2.87).

EXAMPLE 2.8.7. Suppose $d = 2$, $M = \begin{pmatrix} -1 & 4 \\ -1 & 1 \end{pmatrix}$, the set D consists of the digits $r_0 = (0, 0)$, $r_1 = (1, 0)$, $r_2 = (-1, 0)$. Let us show that the set E_D is

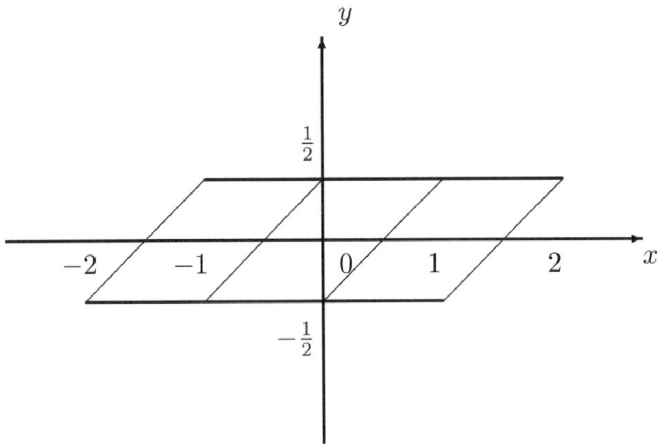

FIGURE 2.2. The set P is the small central parallelogram; the set MP is the large central parallelogram (Example 2.8.7).

the parallelogram with vertices $(0,1/2)$, $(1,1/2)$, $(0,-1/2)$, $(-1,-1/2)$, which we denote by P. We have the equality (see Figure 2.2)

$$MP = P \cup (P + r_1) \cup (P + r_2), \tag{2.90}$$

which implies that for any $N \in \mathbb{N}$, an arbitrary $x \in P$ can be represented in the form

$$x = \sum_{k=1}^{N} M^{-k} s_k + M^{-N} y,$$

where $s_k \in D$, $k = 1, \dots, N$, $y \in P$. Passing to the limit in this equality as $N \to \infty$ and taking into account (2.5), we deduce that $x \in E_D$. Thus $P \subset E_D$.

Now let $x \in E_D$. Since $M^{-1} r_k \in P$, $k = 0,1,2$, it follows from (2.90) by induction that $\sum\limits_{k=1}^{N} M^{-k} s_k \in P$ for any $N \in \mathbb{N}$ and any $s_1, \dots, s_N \in D$. Since P is closed, we have $x \in P$, i.e., $E_D \subset P$. It is clear that for E_D, conditions (i), (ii) of Theorem 2.8.1 and condition OR3 of Proposition 2.8.2 are satisfied. By Theorem 2.8.3, the function χ_{E_D} generates a Haar MRA. $\lozenge$

EXAMPLE 2.8.8. Let $d = 1$, $M = 3$. If we compose D from the digits $0, 1, 2$, then the set E_D coincides with the segment $[0,1]$. It is not hard to check that for any three sequential digits, the set E_D is an interval. Take $D = \{-4, 0, 1\}$. In this case, E_D is not an interval. The graph of its characteristic function (see Figure 2.3) looks like a "torn segment".

The equality

$$|m_0(\xi)|^2 = \frac{1}{9}(3 + 2\cos 2\pi\xi + 2\cos 8\pi\xi + 2\cos 6\pi\xi)$$

shows that the functions $m_0(3^{-k}\xi)$, $k = 1, 2, \dots$, do not have zeros on the set $K = [-1/2, 1/2]$. Hence, by Theorem 2.8.4, the function χ_{E_D} generates a Haar MRA. $\lozenge$

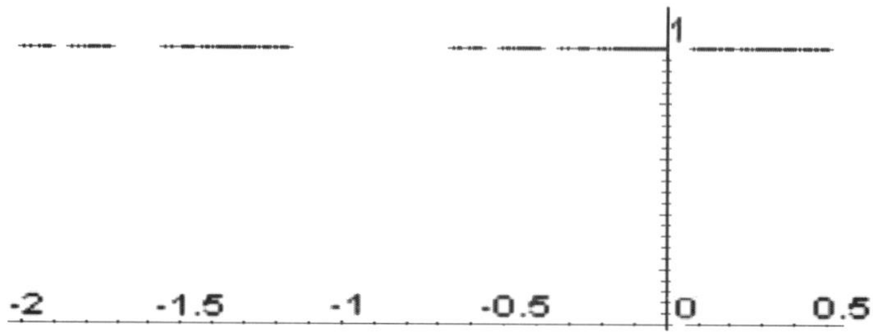

FIGURE 2.3. Graph of the function χ_{E_D} (Example 2.8.8).

EXAMPLE 2.8.9. Suppose $d = 2$, $M = \begin{pmatrix} 1 & -1 \\ 1 & 1 \end{pmatrix}$, the set D consists of the digits $r_0 = (0,0)$, $r_1 = (1,0)$. In this case E_D is a fractal set represented in Figure 2.4, and the mask of the function χ_{E_D} has the form

$$m_0(\xi) = \frac{1}{2}(1 + e^{2\pi i \xi_1}).$$

Zeros of the function

$$|m_0(M^{*-1}\xi)|^2 = \frac{1}{2}(1 + \cos \pi(\xi_1 - \xi_2))$$

are located on the straight lines $\xi_1 - \xi_2 = 2k + 1$, $k \in \mathbb{Z}$; hence we cannot take the square $[-1/2, 1/2]^2$ as a set K of Theorem 2.8.4. Instead, we can take a parallelogram congruent to this square, with vertices $(0, 1/2)$, $(1, 1/2)$, $(0, -1/2)$,

FIGURE 2.4. Set E_D (double-headed dragon) (Example 2.8.9).

$(-1, -1/2)$. It can easily be seen that the functions $m_0(M^{*-k}\xi)$ do not vanish on this parallelogram for $k = 2, 3, \ldots$. Hence by Theorem 2.8.4, the function χ_{E_D} generates a Haar MRA. $\Diamond$

PROBLEM 2.8.10. Suppose $M = \begin{pmatrix} 2 & 1 \\ 0 & 2 \end{pmatrix}$, the set D consists of digits $(0, 0)$, $(1, 0), (0, 1), (1, 1)$. Verify that the function χ_{E_D} generates a Haar MRA.

CHAPTER 3

Compactly Supported Refinable Functions

As is known, the construction of one-dimensional orthogonal compactly supported wavelets can be reduced to finding solutions of the refinement equation

$$\varphi(x) = \sqrt{2} \sum_{k \in \mathbb{Z}} h_k \varphi(2x + k), \tag{3.1}$$

where only finitely many coefficients h_k are nonzero.

Without loss of generality we can assume that h_{-N} and h_0 are the first and the last nonzero coefficients, respectively. Denoting $c_k = \sqrt{2} h_{-k}$, we reduce this equation to the form

$$\varphi(x) = \sum_{k=0}^{N} c_k \varphi(2x - k), \tag{3.2}$$

where $c_0, \dots, c_N$ are arbitrary complex coefficients and $c_0 c_N \neq 0$, $N \geq 1$.

3.1. Existence, uniqueness, and weak convergence

If equation (3.2) has a summable compactly supported solution, then integrating both parts of this equation, we get

$$\int_{\mathbb{R}} \varphi(x)\, dx = \Big(\frac{1}{2} \sum_k c_k\Big) \int_{\mathbb{R}} \varphi(x)\, dx;$$

hence either $\sum_k c_k = 2$ or $\int_{\mathbb{R}} \varphi(x)\, dx = 0$. In the latter case the primitive $\varphi_1(x) = \int_{-\infty}^{x} \varphi(t)\, dt$ also has compact support and satisfies the equality

$$\varphi_1(x) = \sum_{k=0}^{N} \frac{c_k}{2} \varphi_1(2x - k).$$

Integrating both parts, we again have two possibilities: either $\sum_k c_k = 4$ or $\int_{\mathbb{R}} \varphi_1(x)\, dx = 0$. In the latter case we go to the next primitive $\varphi_2(x) = \int_{-\infty}^{x} \varphi_1(t)\, dt$, etc. Thus we obtain a sequence (finite or infinite) $\varphi, \varphi_1, \varphi_2, \dots$, where each function φ_k has compact support and $\varphi_k^{(k)} = \varphi$. We have $\widehat{\varphi}(\xi) = (2\pi i \xi)^k \widehat{\varphi}_k(\xi)$. Both functions $\widehat{\varphi}, \widehat{\varphi}_k$ are entire; therefore the latter equality means that the function $\widehat{\varphi}(\xi)$ has zero of order no less than k at the point $\xi = 0$. Hence the sequence $\varphi, \varphi_1, \varphi_2, \dots$ will come to an end at some φ_n. This means that

$$\widehat{\varphi}_n(0) = \int_{\mathbb{R}} \varphi_n(x)\, dx \neq 0;$$

consequently $\sum_{k=1}^{N} c_k = 2^{n+1}$. Thus we have the following statement:

PROPOSITION 3.1.1. *If equation (3.2) has a nonzero compactly supported summable solution, then* $\sum_{k=0}^{N} c_k = 2^{n+1}$ *for some integer* $n \geq 0$. *In the case* $n = 0$ *we have* $\int_{\mathbb{R}} \varphi(x)\,dx \neq 0$. *If* $n \geq 1$, *then our solution has the n-th primitive* φ_n *with compact support, which solves another refinement equation:*

$$\varphi_n(x) = \sum_{k=0}^{N} \frac{1}{2^n} c_k \varphi_n(2x - k), \tag{3.3}$$

where $\int_{\mathbb{R}} \varphi_n(x)\,dx \neq 0$.

This fact allows us to confine our consideration to the case $\sum c_k = 2$, $\int_{\mathbb{R}} \varphi(x)\,dx \neq 0$, since the general case can be reduced to it by passing to the primitive finitely many times.

Here and in the sequel we assume $\sum c_k = 2$. *Under this condition, for any solution of the refinement equation,* $\int_{\mathbb{R}} \varphi(x)\,dx \neq 0$. *Further we shall assume that all solutions are normalized by the condition* $\int_{\mathbb{R}} \varphi(x)\,dx = 1$.

REMARK 3.1.2. Proposition 3.1.1 holds in a more general situation, not only for summable solutions but for any generalized solutions with compact support. The proof is slightly different. It suffices to use two facts:

1) The Fourier transform of a compactly supported distribution is an entire function.

2) If a distribution $\varphi \in \mathcal{S}'$ with compact support is such that $\langle \varphi, \mathbf{1} \rangle = 0$, then it has a primitive with compact support.

Recall that by the symbol $\mathbf{1}$ we denote the function that is identically equal to 1 on $\mathbb{R}$. Though $\mathbf{1} \notin \mathcal{S}$, for a compactly supported function $\varphi \in \mathcal{S}'$ the expression $\langle \varphi, \mathbf{1} \rangle$ is well-defined and coincides with $\widehat{\varphi}(0)$, while for regular (ordinary) functions φ, it coincides with the integral $\int_{\mathbb{R}} \varphi(x)\,dx$. In different situations we shall use any of these three designations.

Consider again the transition operator (2.21) on the space S', which is now written as

$$[Tf](x) = \sum_{k=0}^{N} c_k f(2x - k). \tag{3.4}$$

Using Theorem 2.4.4 for the one-dimensional operator $Mx = 2x$, we deduce that if $\sum c_k = 2$, refinement equation (3.2) has a unique (up to a constant factor) generalized solution φ with compact support. This solution is represented by the Fourier transform using formula (2.17), which takes the form

$$\widehat{\varphi}(\xi) = \prod_{k=1}^{\infty} m_0\left(\frac{\xi}{2^k}\right). \tag{3.5}$$

For any initial function f_0 with compact support, the sequence $f_n = T^n f_0$ converges in S' to the function $\widehat{f_0}(0)\varphi$. The solution φ is concentrated on the segment $[0, N]$

(i.e., $\operatorname{supp}\varphi \subseteq [0, N]$). Indeed, the operator T preserves the space of functions concentrated on this segment. Consequently, choosing the initial function f_0 such that $\operatorname{supp} f_0 \subset [0, N]$, we see that all functions f_n as well as the limit of the sequence are also concentrated on this segment.

Formula (3.5) defines a unique (up to a constant factor) solution of equation (3.2) in the class of distributions with the Fourier transform that is continuous at zero. In particular, this solution is unique in the class of distributions with compact support and (if $\varphi \in L_1(\mathbb{R})$) in the class of summable functions. If there exists a summable solution of equation (3.2), it has compact support and thus coincides with φ (up to a constant factor). In what follows, we shall call φ a *canonical solution* and assume (unless otherwise specified) that any refinable function is a canonical solution of the corresponding equation. The canonical solution is normalized by the condition $\widehat{\varphi}(0) = 1$ and is concentrated on the segment $[0, N]$.

Thus a refinable function is uniquely defined by coefficients of the equation or, equivalently, by a mask of the refinement equation.

3.2. Strang-Fix conditions

A distribution $f \in \mathcal{S}'$ with compact support satisfies *the Strang-Fix condition of order* L ($L \geq 0$) if

$$\widehat{f}^{(r)}(n) = 0, \quad r = 0, \dots, L, \quad n \in \mathbb{Z} \setminus \{0\};$$

in other words, the Fourier transform $\widehat{f}$ (which is an entire function) has zeros of order no less than L at all integer nonzero points.

For a given compactly supported function f such that $\widehat{f}(0) \neq 0$, the Strang-Fix condition is equivalent to the fact that the integer shifts of this function generate all polynomials of degree at most L; i.e., $\mathcal{P}_L \subset V_f$, where

$$V_f = \left\{ \sum_{k \in \mathbb{Z}} \lambda_k f(x-k), \ \lambda_k \in \mathbb{C} \right\}$$

is the space of functions generated by integer shifts of the function f and $\mathcal{P}_L$ is the space of algebraic polynomials of degree at most L. Note that since f has compact support, the functional series $\sum_k \lambda_k f(\cdot - k)$ converges on each compact set in $\mathbb{R}$ for any choice of sequence $\{\lambda_k\}$. If f is a regular (summable on any compact set) function, then the sum of the series is also a regular function. If $f \in \mathcal{S}'$, the sum of the series is a distribution from $\mathcal{D}'$ (from the space of linear continuous functionals on the space $\mathcal{D}$ of test functions with compact support). In addition, if the sequence $\{\lambda_k\}$ does not increase more rapidly than a polynomial, the sum of the series is a function from $\mathcal{S}'$.

THEOREM 3.2.1. *For given $L \geq 0$ and a given function $f \in \mathcal{S}'$ with compact support such that $\widehat{f}(0) = 1$, the following properties are equivalent:*

(a) $\widehat{f}^{(r)}(n) = 0$, $r = 0, \dots, L$, $n \in \mathbb{Z} \setminus \{0\}$ *(the Strang-Fix condition of order L).*

(b) $\sum_{k \in \mathbb{Z}} (x-k)^r f(x-k) \equiv C$ *for each $r = 0, \dots, L$. In particular, $\sum_{k \in \mathbb{Z}} f(x-k) \equiv 1$ (partition of unity).*

(c) *For every $r = 0, \dots, L$ the function $F_r(x) = \sum_{k \in \mathbb{Z}} k^r f(x-k)$ is an algebraic polynomial.*

(d) *For every* $r = 0, \dots, L$ *the function* $F_r(x)$ *is a polynomial of degree* r *with the leading coefficient* 1.

(e) $\mathcal{P}_L \subset V_f$.

`Proof.` (a) $\Leftrightarrow$ (b). By the Poisson summation formula, for an arbitrary function $g \in \mathcal{S}'$, we have

$$\sum_{k\in\mathbb{Z}} g(x-k) = \sum_{n\in\mathbb{Z}} \widehat{g}(n) e^{2\pi i n x}.$$

Therefore $\sum_{k\in\mathbb{Z}} g(x-k) \equiv C$ if and only if

$$\sum_{n\in\mathbb{Z}} \widehat{g}(n) e^{2\pi i n x} \equiv C\,,$$

which means that all coefficients of the trigonometric series, perhaps except the first one, are equal to zero, i.e., $\widehat{g}(n) = 0$, $n \neq 0$. It remains to set $g(x) = x^r f(x)$ and to note that $\widehat{g} = (-2\pi i)^{-r} \widehat{f}^{(r)}$. For $L = 0$, we have

$$\sum_{k\in\mathbb{Z}} f(x-k) = \sum_{n\in\mathbb{Z}} \widehat{f}(n) e^{-2i\pi n x} = \widehat{f}(0) = 1.$$

(b) $\Rightarrow$ (d). We have

$$\sum_{k\in\mathbb{Z}} k^r f(x-k) = \sum_{k\in\mathbb{Z}} \Big(x - (x-k)\Big)^r f(x-k)$$

$$= \sum_{k\in\mathbb{Z}} \sum_{j=0}^{r} (-1)^j \binom{r}{j} x^{r-j} (x-k)^j f(x-k)$$

$$= \sum_{j=0}^{r} (-1)^j \binom{r}{j} x^{r-j} \sum_{k\in\mathbb{Z}} (x-k)^j f(x-k), \tag{3.6}$$

where $\binom{r}{j}$ are binomial coefficients. Since the sum

$$\sum_{k\in\mathbb{Z}} (x-k)^j f(x-k)$$

is identically constant for each $j = 0, \dots, r$, sum (3.6) is a polynomial of degree r, and the coefficient of x^r is equal to

$$\binom{r}{0} \sum_{k\in\mathbb{Z}} f(x-k) = 1.$$

(d) $\Rightarrow$ (c). This is obvious.

(d) $\Rightarrow$ (e). This is obvious.

(e) $\Rightarrow$ (c). This property is the most complicated to prove. If a polynomial $p(x)$ belongs to the space V_f, the coefficients of its expansion $p(x) = \sum \lambda_k f(x-k)$ may increase more rapidly than any polynomial, i.e., the series does not converge in the space $\mathcal{S}'$, and the termwise Fourier transform cannot be applied. We prove this statement in a different way, using difference equations.

First note that we can confine our consideration to the case $f \in C(\mathbb{R})$; i.e., assume that f is a compactly supported continuous function. Otherwise, instead of f, we take its convolution with some compactly supported function $h \in \mathcal{D}$ such that $\widehat{h}(0) \neq 0$. If $\mathcal{P}_L \subset V_f$, then $\mathcal{P}_L \subset V_{f*h}$. Indeed, if $x^r \in V_f$, then $x^r = \sum_{k\in\mathbb{Z}} a_k f(x-k)$ and we have

$$\begin{aligned}
\sum_{k\in\mathbb{Z}} a_k [f*h](x-k) &= \sum_{k\in\mathbb{Z}} a_k [f(\cdot - k) * h](x) \\
&= \left(\sum_{k\in\mathbb{Z}} a_k f(s-k)\right) * h = [s^r * h(s)](x) = \int\limits_{\mathbb{R}} h(s)(x-s)^r ds \\
&= x^r \int\limits_{\mathbb{R}} h(s)ds + \sum_{j=1}^{r} x^{r-j}(-1)^j \binom{r}{r-j} \int\limits_{\mathbb{R}} h(s)s^j ds.
\end{aligned}$$

The latter expression is a polynomial of degree r, since the coefficient of x^r is equal to $\int\limits_{\mathbb{R}} h(s)ds = \widehat{h}(0) \neq 0$. Thus the space V_{f*h} contains $\mathcal{P}_L$. If property (e) is satisfied for a function f, it is satisfied for the convolution $f*h$ as well. Furthermore, if property (c) holds for every such convolution $f*h$, it is satisfied for the function f. Indeed, if for each $h \in \mathcal{D}$ such that $\int\limits_{\mathbb{R}} h \neq 0$ the series $\sum_k k^r [f*h](x-k)$ is a polynomial of x, then it is a polynomial for every $h \in \mathcal{S}'$ (due to the completeness of the space $\mathcal{D}$ in $\mathcal{S}'$), in particular, for $h = \delta$. Hence property (c) is satisfied for $f * \delta = f$. Thus it suffices to prove the statement (e) $\Rightarrow$ (c) for smooth functions f. Assume also that $\operatorname{supp} f \subset [0, N]$, $\int_{\mathbb{R}} f = 1$.

For each $r = 0, \dots, L$, let the polynomial x^r be expanded into the series

$$x^r = \sum_{k\in\mathbb{Z}} a_k f(x-k). \tag{3.7}$$

As f is continuous, this equality is satisfied for each point $x \in \mathbb{R}$. Choose $y \in [N-1, N]$ such that

$$f(y) + f(y-1) + \cdots + f(y-N+1) \neq 0.$$

Such y exists, since

$$\int\limits_{N-1}^{N} (f(y) + f(y-1) + \cdots + f(y-N+1))\, dy = \int\limits_{\mathbb{R}} f(x)\, dx = 1.$$

Set $f(y-j) = v_j$, $j \in \mathbb{Z}$. It is clear that $v_j = 0$ for $j \leq -1$ and $j \geq N$. Consider equality (3.7) for $x = y+n$, $k = j+n$. For any n, we have

$$\sum_{j=0}^{N-1} v_j a_{j+n} = (y+n)^r, \quad n \in \mathbb{Z}.$$

This is a difference equation (for the sequence $\{a_k\}$) with a polynomial right-hand side. The solution is represented as the sum

$$a_k = a_k^{\text{hom}} + a_k^{\text{par}}$$

of the general solution to the homogeneous equation a_k^{hom} and a particular solution a_k^{par}. We have

$$\begin{aligned}
a_k^{\text{hom}} &= \sum_{q=1}^{d} \lambda_q^k p_q(k), \\
a_k^{\text{par}} &= p_0(k) k^{\mu},
\end{aligned} \tag{3.8}$$

where $\lambda_1, \ldots, \lambda_d$ are roots of the characteristic polynomial

$$\chi_y(\lambda) = \sum_{j=0}^{N-1} v_j \lambda^j,$$

$p_1, \ldots, p_d$ are arbitrary polynomials, whose degrees do not exceed the multiplicities of the corresponding roots, p_0 is some fixed polynomial of degree r, μ is the multiplicity of the root $\lambda = 1$ (see Appendix A.18). Thus,

$$a_k = p_0(k) k^\mu + \sum_{q=1}^{d} \lambda_q^k p_q(k). \tag{3.9}$$

In our case, $\mu = 0$ since

$$\chi_y(1) = \sum_{j=0}^{N-1} v_j = \sum_{j=0}^{N-1} f(y-j) \neq 0.$$

Note that for each $y \in [N-1, N]$, the coefficients $\{a_k\}$ are expressed by formula (3.9); however the roots λ_q of the polynomial $\chi_y(\lambda)$ may be different for different y. Using the continuity of the function f, we show that in our case these roots are equal for all y. For any two different values of y, expressions (3.9) are identically equal on the set of integers; hence they have the same quantities λ_k and the corresponding coefficients of the polynomials p_q at nonzero λ_k (see Appendix A.18). Thus expressions (3.9) are identically equal for all y. In particular, the polynomials $p_0(k)k^\mu$ coincide for all y. As $\mu = 0$ for one of the values of y, we have $\mu = 0$ for all y. Since for each y the sequence $a_k = p_0(k)$ is a particular solution of the difference equation $\sum_{j=0}^{N-1} v_j a_{j+n} = (y+n)^r$, it satisfies (3.7). Thus for each $r = 0, \ldots, L$ there exists a polynomial p_0 of degree r such that

$$\sum_{k \in \mathbb{Z}} p_0(k) f(x-k) = x^r.$$

Property (c) immediately follows.

(c) $\Rightarrow$ (a). Let

$$\sum_{k \in \mathbb{Z}} k^r f(x-k) = \alpha_0 + \alpha_1 x + \ldots + \alpha_d x^d.$$

Applying the Fourier transform, we get

$$\left(\sum_k k^r e^{-2\pi i k \xi} \right) \widehat{f}(\xi) = \sum_{q=0}^{d} \alpha_q (2\pi i)^{-q} \delta^{(q)}(\xi).$$

Taking into account that

$$\sum_k k^r e^{-2\pi i k \xi} = \sum_k (-2\pi i)^{-r} \delta^{(r)}(\xi - k),$$

we transform the left-hand side of the latter relation to the form

$$\begin{aligned} \left(\sum_k k^r e^{-2\pi i k \xi} \right) \widehat{f}(\xi) &= \left(\sum_k (-2\pi i)^{-r} \delta^{(r)}(\xi - k) \right) \widehat{f}(\xi) \\ &= \sum_k (-2\pi i)^{-r} \widehat{f}(\xi) \delta^{(r)}(\xi - k). \end{aligned}$$

It follows that

$$(-2\pi i)^{-r}\sum_k \widehat{f}(\xi)\delta^{(r)}(\xi-k)=\sum_{q=0}^{d}\alpha_q(2\pi i)^{-q}\delta^{(q)}(\xi).$$

Hence the function on the left-hand side of this equality has support at the point $\xi=0$ and for each $k\neq 0$ we have

$$\widehat{f}(\xi)\delta^{(r)}(\xi-k)=0,\quad r=0,\dots,L;$$

therefore $\widehat{f}^{(r)}(k)=0,\ r=0,\dots,L$. This completes the proof. ◊

It is not hard to find explicitly the constants arising in (b) of Theorem 3.2.1 and the polynomials from item (c). In what follows we shall use the notation

$$\langle f,g\rangle=\int_{\mathbb{R}} f(x)g(x)\,dx \quad\text{and}\quad f*g\,(x)=\int_{\mathbb{R}} f(x-t)g(t)\,dt.$$

PROPOSITION 3.2.2. *If a distribution f with compact support and such that $\langle f,\mathbf{1}\rangle=1$ satisfies the Strang-Fix condition of order $L\geq 0$, then for an arbitrary polynomial $p\in\mathcal{P}_L$ we have*

$$\sum_{k\in\mathbb{Z}} p(x-k)f(x-k)=\langle p,f\rangle, \tag{3.10}$$

$$\sum_{k\in\mathbb{Z}} p(k)f(x-k)=p*f\,(x). \tag{3.11}$$

`Proof.` According to item (b) of Theorem 3.2.1, we have

$$\sum_k p(x-k)f(x-k)\equiv C;$$

hence

$$C=\int_0^1 C\,dx=\int_0^1\sum_k p(x-k)f(x-k)\,dx=\sum_k\int_0^1 p(x-k)f(x-k)\,dx=\langle p,f\rangle,$$

which proves (3.10). Set $p_x(\cdot)=p(x-\cdot)$. It is clear that for any x, the polynomial p_x has the same degree as p. Using (3.10), we get

$$\sum_k p_x(t-k)f(t-k)=\int p_x(t)f(t)\,dt=p*f\,(x).$$

This equality holds for any t, in particular, for $t=x$. It remains to note that $p_x(x-k)=p(k)$, which implies (3.11). ◊

Thus for each $r\leq L$, the polynomial

$$(t^r)*f\,(x)=\sum_{j=0}^{r}(-1)^j\binom{r}{j}\langle f,x^j\rangle x^{r-j}=x^r-r\cdot\langle f,x\rangle x^{r-1}+\dots+(-1)^r\langle f,x^r\rangle$$

can be expanded with respect to integer shifts of the function f as $\sum_k k^r f(x-k)$. Using this expression, we obtain the expansion for an arbitrary polynomial $P\in\mathcal{P}_L$:

$$P(x)=\sum_{k\in\mathbb{Z}}\left[S_f^{-1}P\right](k)f(x-k),\qquad \deg P\leq L, \tag{3.12}$$

where S_f^{-1} is a linear operator on the space $\mathcal{P}$ (the space of algebraic polynomials). This operator is the inverse of the convolution operator S_f with the function f. It is easily seen that all the subspaces $\mathcal{P}_L,\ L \ge 0$, are invariant under S_f; hence they are invariant under the inverse operator S_f^{-1} if the latter exists. In the general case, the convolution operator is not necessarily invertible. It is invertible on the space of polynomials if and only if $\langle f, \mathbf{1}\rangle \ne 0$. In our case this condition is satisfied since $\langle f, \mathbf{1}\rangle = 1$.

Let us return to refinable functions. We shall show that any smooth solution of a refinement equation satisfies the Strang-Fix condition (of order no less than the order of smoothness of the function). Here, smoothness of functions can be evaluated in both spaces $C(\mathbb{R})$ (the Hölder exponent) and $L_p(\mathbb{R})$ for any $p \in [1, +\infty]$, or according to Sobolev (coefficient s_p) for any p as well. The result remains the same, namely, smoothness of a refinable function cannot exceed the order of the Strang-Fix condition satisfied by this function.

THEOREM 3.2.3. *Suppose a compactly supported function $\varphi(x)$ is an L_1-solution of refinement equation* (3.2), *where $\varphi \in W_1^L(\mathbb{R})$ or $s_p(\varphi) > L$ for some $L \ge 0,\ p \ge 1$; then φ satisfies the Strang-Fix condition of order L.*

Proof. We assume that φ is a canonical solution given by formula (3.5). Applying the Fourier transform to both parts of the refinement equation, we obtain equation (1.23). After k iterations we get

$$\widehat{\varphi}(\xi) = \widehat{\varphi}\left(\frac{\xi}{2^k}\right)\prod_{l=1}^{k} m_0\left(\frac{\xi}{2^l}\right). \tag{3.13}$$

Since $\widehat{\varphi}(0) = 1$, the function $\widehat{\varphi}(\xi)$ does not vanish in a sufficiently small circle $\Omega_\varepsilon = \{\xi \in \mathbb{C},\ |\xi| < \varepsilon\}$. Choose some $\delta \in [0, \varepsilon)$ and integer $n \ne 0,\ k \ge 1$. Substituting $2^k n + \delta$ for ξ in equality (3.13), we obtain

$$\widehat{\varphi}(2^k n + \delta) = \widehat{\varphi}\left(n + \frac{\delta}{2^k}\right)\prod_{l=1}^{k} m_0\left(2^{k-l}n + \frac{\delta}{2^l}\right) = \widehat{\varphi}\left(n + \frac{\delta}{2^k}\right)\prod_{l=1}^{k} m_0\left(\frac{\delta}{2^l}\right).$$

It follows from equality (3.13) that there exists a constant $C_\varepsilon > 0$ such that for all $k \ge 1$ we have

$$\inf_{z \in \Omega_\varepsilon}\left|\prod_{l=1}^{k} m_0\left(\frac{z}{2^l}\right)\right| > C_\varepsilon.$$

Consequently,

$$|\widehat{\varphi}(2^k n + \delta)| > C_\varepsilon |\widehat{\varphi}(n + 2^{-k}\delta)|. \tag{3.14}$$

Suppose now that $\varphi \in W_1^L(\mathbb{R})$; then

$$\xi^L \widehat{\varphi}(\xi) \to 0 \qquad \text{as } \xi \to \infty$$

along the real axis. Combining this with (3.14), we deduce that for any $\delta \in [0, \varepsilon)$

$$(2^k n + \delta)^L \widehat{\varphi}(n + \frac{\delta}{2^k}) \to 0 \qquad \text{as } k \to \infty.$$

Therefore the function $\widehat{\varphi}(\xi)$ has a zero of order at least $L + 1$ at the point $\xi = n$, which proves the theorem in the case $\varphi \in W_1^L$.

In the case $s_p > L$ we have

$$\int_{\mathbb{R}} |\widehat{\varphi}(\xi)|^p (1 + |\xi|^{pL}) d\xi < \infty.$$

On the other hand,

$$\int |\widehat{\varphi}(\xi)|^p (1 + |\xi|^{pL}) d\xi \geq \sum_{k\in\mathbb{N}} \int_{\varepsilon/2}^{\varepsilon} |\widehat{\varphi} 2^k n + \delta|^p (1 + |2^k n + \delta|^{pL}) d\delta$$

$$> \sum_{k\in\mathbb{N}} C_\varepsilon \int_{\varepsilon/2}^{\varepsilon} |\widehat{\varphi}(n + \delta 2^{-k})|^p (1 + |2^k n + \delta|^{pL}) d\delta.$$

This series converges; hence,

$$\int_{\varepsilon/2}^{\varepsilon} |\widehat{\varphi}(n + \delta 2^{-k})|^p (1 + |(2^k n + \delta)|^{pL}) d\delta \to 0 \quad \text{as } k \to \infty.$$

By the mean value theorem, we find $\eta_k \in (\varepsilon/2, \varepsilon)$ such that

$$|\widehat{\varphi}(n + 2^{-k}\eta_k)|^p (1 + |(2^k n + \eta_k)|^{pL}) \to 0 \quad \text{as } k \to \infty.$$

Therefore the function $\widehat{\varphi}(\xi)$ has a zero of order at least $L+1$ at the point $\xi = n$. ◊

COROLLARY 3.2.4. *If a refinable function with compact support either belongs to $W_1^L(\mathbb{R})$ or its Sobolev exponent is greater than L, then it has properties* (a)–(e) *of Theorem* 3.2.1 *($L \geq 0$).*

Thus any *smooth refinable function satisfies the Strang-Fix condition and its integer shifts generate all polynomials of the corresponding degree.* In particular, for $L = 0$, we get

COROLLARY 3.2.5. *Any summable refinable function with compact support has the property of partition of unity:*

$$\sum_{k\in\mathbb{Z}} \varphi(x - k) \equiv 1.$$

Actually, by proving Theorem 3.2.3, we established a stronger assertion:

COROLLARY 3.2.6. *If $\widehat{\varphi}(\xi) = o(\xi^{-L})$ for $\xi \to \infty$, then φ satisfies the Strang-Fix condition of order L.*

It is natural that the Strang-Fix conditions for refinable functions should be characterized in terms of coefficients of the equation $\{c_k\}$ or in terms of the mask m_0. Let us consider a *binary tree* $\mathcal{T}$ constructed in the following way: the number $\frac{1}{2}$ corresponds to the root of the tree, and the numbers $\frac{1}{4}$ and $\frac{3}{4}$ correspond to the vertices next to the root of the first level. Further, the tree is constructed by induction: if a number γ corresponds to a vertex of the n-th level, then its neighbors at the $(n+1)$-st level are the numbers $\gamma/2$ and $\frac{1}{2} + \gamma/2$. Thus at the n-th level the tree has the numbers

$$2^{-n-1}(2k+1), \quad k = 0, \dots, 2^n - 1$$

(in what follows, we shall identify vertices with the corresponding numbers).

Let $\mathcal{A}$ be a set of vertices of the tree. Some of the elements of $\mathcal{A}$ may coincide, in which case they are counted according to their multiplicity. We say that $\mathcal{A}$ is a *blocking set of multiplicity* r ($r \geq 1$) if any infinite path $\alpha_0 \to \alpha_1 \to \dots$ along the tree going out of the root (here, a vertex α_k belongs to the k-th level, and all paths go in the increasing order of levels; i.e., there is no way back) has r elements that belong to the set $\mathcal{A}$ as well, counting according to their multiplicities. For example, the root of the tree $\{\frac{1}{2}\}$ is a blocking set of multiplicity 1.

PROBLEM 3.2.7. Prove that any blocking set is finite.

PROBLEM 3.2.8. Prove that any blocking set has a pair of symmetric points $\{\alpha,\, \frac{1}{2} + \alpha\}$ except for the case when it is trivial, i.e., is contained in the tree root.

PROBLEM 3.2.9. Prove that a blocking set of multiplicity $r \geq 1$ contains at least r elements (taking into account multiplicity). For which sets is the minimum attained? What is the minimal number of elements a blocking set of multiplicity r can contain if all its elements are different?

THEOREM 3.2.10. *A refinable function satisfies the Strang-Fix conditions of order $L \geq 0$ if and only if there exists a blocking set of order $L+1$ that consists of the roots of the mask $m_0(\xi)$.*

Proof. Let φ satisfy the Strang-Fix conditions of order L. Choose any infinite path $\alpha_0 \to \alpha_1 \to \cdots$ from the tree root. Since the function $\widehat{\varphi}$ is entire, there exists α_s such that $\widehat{\varphi}(\alpha_s) \neq 0$. Iterating the equality $\widehat{\varphi}(\xi) = m_0(\xi/2)\widehat{\varphi}(\xi/2)$ $s+1$ times, we get

$$\widehat{\varphi}(\xi) = \widehat{\varphi}(2^{-s-1}\xi) \prod_{k=1}^{s+1} m_0(2^{-k}\xi). \tag{3.15}$$

Setting $\xi = 2^{s+1}\alpha_s$ and taking into account that $2\alpha_{k+1} \equiv \alpha_k \pmod 1$, we obtain

$$\widehat{\varphi}(2^{s+1}\alpha_s) = \widehat{\varphi}(\alpha_s) \prod_{k=0}^{s} m_0(\alpha_k).$$

As $2^{s+1}\alpha_s = r$ for some $r \in \mathbb{Z}\setminus\{0\}$, the function $\widehat{\varphi}(\xi)$ has a zero of order $\geq L+1$ at the point $\xi = 2^{s+1}\alpha_s$. Since $\widehat{\varphi}(\alpha_s) \neq 0$, the set of points α_k, $k = 0, \dots, s$, contains at least $L+1$ zeros of the polynomial m_0 (taking into account their multiplicities).

Conversely, let each infinite path starting from the root contain at least $L+1$ zeros of the polynomial m_0. Prove that for any integer $r \neq 0$, the point r is a zero of the function $\widehat{\varphi}$ of multiplicity $\geq L+1$. Let $r = 2^l q$ for some odd q. Substituting $\xi = r$ into equality (3.5) and taking into account that $m_0(2^{-k}r) = 1$, we get for all $k = 1, \dots, l$:

$$\widehat{\varphi}(r) = \prod_{k=1}^{\infty} m_0(2^{-k}q) = \prod_{k=1}^{\infty} m_0(\alpha_k),$$

where $\alpha_{k-1} \equiv 2^{-k}q \pmod 1$. Since the infinite path $\alpha_0 \to \alpha_1 \to \cdots$ contains at least $L+1$ zeros of the polynomial m_0, $\widehat{\varphi}$ has a zero of multiplicity $\geq L+1$ at the point r. ◊

COROLLARY 3.2.11. *Let a compactly supported function $\varphi(x)$ be an L_1-solution of the refinement equation (3.2). Let either $\varphi \in W_1^L(\mathbb{R})$ or $s_p(\varphi) > L$ for some $L \geq 0$, $p \geq 1$; then there exists a blocking set of order $L+1$ that consists of the roots of the mask $m_0(\xi)$.*

Thus the existence of a blocking set of order $L+1$ is a necessary condition for a refinable function to be smooth. Generally, this condition is not sufficient (see Examples 3.5.5 and 7.3.7–7.3.9).

Since a blocking set of order r contains at least r elements (Problem 3.2.9), we obtain

COROLLARY 3.2.12. *If a refinable function satisfies the Strang-Fix condition of order L, then* $\deg m_0 \geq L+1$.

Using Theorem 3.2.3, we get

COROLLARY 3.2.13. *If $\varphi \in W_1^L(\mathbb{R})$, then* $\deg m_0 \geq L+1$.

Therefore a solution of the refinement equation (3.1) of order N can satisfy the Strang-Fix conditions of order at most $N-1$, and its smoothness (in the space $L_1(\mathbb{R})$) does not exceed N. For each N, there exists a unique refinable function, for which these values are attained (cardinal B-spline; see Example 3.5.1).

Any refinable function with compact support has finite smoothness.

Using Corollary 3.2.6, we obtain

COROLLARY 3.2.14. *If $\widehat{\varphi}(\xi) = o(\xi^{-L})$ as $\xi \to \infty$, then* $\deg m_0 \geq L+1$. *The mask $m_0(\xi) = \left(\frac{e^{-2\pi i \xi}+1}{2}\right)^n$ corresponding to the B-spline is a unique mask of a given degree $n \geq 0$, for which $\widehat{\varphi}(\xi) = o(\xi^{1-n})$.*

Next, we show the relation between smoothness of refinable functions and their approximation qualities. To do this, we have to prove an auxiliary result concerning arbitrary functions with compact support. It appears that the property of a function to generate the space of polynomials (i.e., the Strang-Fix condition of a corresponding order) is equivalent to its property of approximating smooth functions sufficiently well.

3.3. Approximation by shifts of refinable functions

Consider an arbitrary distribution $f \in S'$ with compact support. If f satisfies the Strang-Fix condition of order $L \geq 0$, then its shifts generate the space of polynomials of degree L. In general, the function f does not have to be continuous; moreover, it is even not necessarily regular. The order L for any function is bounded, as the order of any zero of an entire function $\widehat{f}$ is bounded. Let L be the maximal order of the Strang-Fix condition satisfied by f. By Theorem 3.2.1, integer shifts of f do not generate the polynomial x^{L+1}. Therefore they cannot generate all smooth functions. Moreover, the restriction of the space V_f to any segment $[a,b]$ is finite-dimensional; hence this space can neither contain any class of functions (C^l or W_p^l) nor approximate it with any degree of accuracy. Hence we consider a different problem: how to approximate functions from some class by dilations of integer shifts of the function f.

DEFINITION 3.3.1. For a given function $f \in S'$ with compact support and given $h \in (0,1)$, we set $V_f(h) = \left\{\sum_{k\in\mathbb{Z}} \lambda_k f\left(h^{-1}\cdot - k\right)\right\}$. If f is a regular bounded function, then its *order of approximation* in the metric of $C(\mathbb{R})$ is defined as the supremum of the values $\gamma > 0$ such that

$$\inf_{g\in V_f(h)} \|\psi - g\|_{C[-M,M]} \leq C(M,\psi)\, h^{\gamma}$$

for any $\psi \in C^l(\mathbb{R})$ $(l = [\gamma])$, $M > 0$, $h \in (0,1)$, where $C(M,\psi)$ is a positive constant. The order of approximation in the metric of L_p is defined analogously.

In other words, the order of approximation is greater than or equal to γ if the distance from any sufficiently smooth compactly supported function to the space $V_f(h)$ (in a corresponding metric) is equal to $O(h^\gamma)$ as $h \to 0$. The order of approximation in the space C will be denoted by w_f, and in the space L_p by $w_{f,p}$. For any fixed smooth function $g(x)$, its order of approximation $w_f(g)$ by the spaces $V_f(h)$ is defined analogously.

THEOREM 3.3.2. *Suppose f is a bounded function with compact support and $L \geq 0$ is an integer. If the order of approximation of f is greater than L, then f satisfies the Strang-Fix condition of order L. Conversely, if f satisfies the Strang-Fix condition of order L, its order of approximation is greater than or equal to $L+1$. Moreover, for any function $\psi \in C^L(\mathbb{R})$ that does not belong to $C^{L+1}(\mathbb{R})$, we have*

$$w_f(\psi) = \alpha_\psi, \tag{3.16}$$

where α_ψ is the Hölder exponent of the function ψ. The same result holds for approximation in the metric of L_p if we replace the space C^L by W_p^L and use the corresponding smoothness exponent. Here, we assume that $f \in L_p(\mathbb{R})$.

Before we prove this theorem, let us present its corollaries:

COROLLARY 3.3.3. *The order of approximation of any compactly supported function f is $L+1$, where L is the maximal number, for which $\mathcal{P}_L \subset V_f$ (if V_f does not contain $\mathcal{P}_0$ (the space of identical constants), then we set $L = -1$).*

COROLLARY 3.3.4. *The order of approximation of any compactly supported function is an integer.*

We prove Theorem 3.3.2 for the space C; for L_p the proof is analogous.

Sufficiency. For any function $g \in C^L \setminus C^{L+1}$ and some small $h > 0$, we need to find a linear combination

$$\sum_k a_k f(h^{-1} \cdot - k)$$

approximating that function on the segment $[-M, M]$ up to $C_M h^\alpha$, where $\alpha = \alpha_g$ is the Hölder exponent of g. If the Strang-Fix condition of order L is satisfied, we have (formula (3.12))

$$P(x) = \sum_k \left[S_f^{-1} P\right](k) f(x-k)$$

for any polynomial P of degree $\leq L$. Let A_k be the shift operator defined by the relation $A_k g(\cdot) = g(\cdot + k)$. Then $a_k = S_f^{-1}P(k) = A_k S_f^{-1} P(0)$. It is easy to show that the shift operator commutes with the convolution operator; hence,

$$A_k S_f^{-1} P(0) = S_f^{-1} A_k P(0).$$

Thus we have

$$a_k = S_f^{-1} A_k P(0) = [S_f^{-1} P(\cdot + k)](0) = \langle \delta\,,\, S_f^{-1} P(\cdot + k)\rangle,$$

where δ is the delta function. Further, we have

$$\langle \delta\,,\, S_f^{-1} P(\cdot + k)\rangle = \langle (S_f^{-1})^* \delta\,,\, P(\cdot + k)\rangle.$$

Hence $a_k = \big\langle (S_f^{-1})^*\delta\,,\, P(\cdot + k)\big\rangle$. Thus the coefficients of the expansion of a polynomial with respect to integer shifts of the function f are the values of the linear functional $(S_f^{-1})^*\delta$ for integer shifts of the given polynomial. Using the Hahn-Banach theorem, we extend this functional to the space $C[-M, M]$ and obtain a formula for the decomposition of an arbitrary continuous function, analogous to (3.12). This formula is exact for polynomials of degree at most L. For an arbitrary segment $[a, b]$, the functional $(S_f^{-1})^*\delta$ is defined on the space $\mathcal{P}_L[a, b]$ even if $0 \notin [a, b]$. To each continuous function $\psi \in C(\mathbb{R})$, assign the series

$$\psi(x) \;\sim\; \sum_k \langle \lambda(\cdot - k), \psi(\cdot)\rangle f(x - k). \tag{3.17}$$

This expansion is exact for polynomials of order at most L and preserves compactness of support; i.e., if ψ has compact support, then the corresponding series (3.17) has compact support as well. Let us show that the expansion dilated by the coefficient $h > 0$,

$$\psi(x) \;\sim\; \psi_{\lambda,h}(x) = \sum_k h^{-1}\Big\langle \lambda(h^{-1}\cdot - k), \psi\Big\rangle f(h^{-1}x - k), \tag{3.18}$$

approximates ψ on any segment up to $O(h^\alpha)$, where

$$\alpha \;=\; \min\{\alpha_\psi\,,\, L + 1\}.$$

Choose N so large that

$$\operatorname{supp} f \subset [-N, N], \quad \operatorname{supp} \lambda \subset [-N, N].$$

Suppose $\psi \in C^L(\mathbb{R})$ (otherwise we replace L by the maximal integer l, for which $\psi \in C^l$). Consider an arbitrary point x_0 and the segment $[x_0 - 2Nh,\, x_0 + 2Nh]$. It is obvious that the value of the function $\psi_{\lambda,h}$ at the point x_0 is completely defined by values of the function ψ on this segment. The function ψ in turn can be uniformly approximated on this segment by a polynomial of order at most L up to $O(h^\alpha)$. In particular, for the Taylor polynomial

$$p_{x_0,L}(x) = \psi(x_0) + \cdots + \frac{\psi^{(L)}(x_0)}{L!}(x - x_0)^L,$$

we have

$$\|\psi - p_{x_0,L}\|_{C[x_0 - 2Nh, x_0 + 2Nh]} \le C_\varepsilon h^{\alpha - \varepsilon}$$

for any $\varepsilon > 0$ (see [**19**]). Then for all $k \in [\frac{x_0}{h} - N, \frac{x_0}{h} + N]$ (for such k we have $\operatorname{supp} \lambda(h^{-1}\cdot - k) \subset [x_0 - 2Nh, x_0 + 2Nh]$), we obtain

$$\begin{aligned}
&\big|h^{-1}\langle \lambda(h^{-1}\cdot - k),\ \psi\rangle - h^{-1}\langle \lambda(h^{-1}\cdot - k),\ p_{x_0,L}\rangle\big| \\
&\quad = \big|h^{-1}\langle \lambda(h^{-1}\cdot - k),\ \psi - p_{x_0,L}\rangle\big| \\
&\quad \le \big\|h^{-1}\lambda(h^{-1}\cdot - k)\big\|_{C^*[x_0 - 2Nh, x_0 + 2Nh]}\, \|\psi - p_{x_0,L}\|_{C[x_0 - 2Nh, x_0 + 2Nh]} \\
&\quad \le \|\lambda\| \cdot C_\varepsilon h^{\alpha - \varepsilon},
\end{aligned}$$

where $\|\lambda\| = \|\lambda\|_\infty$ is total variation of the charge λ. Then we have

$$\begin{aligned}
&|p_{x_0,L}(x_0) - \psi_{\lambda,h}(x_0)| \\
&\quad \le \sum_{k \in [\frac{x_0}{h} - N, \frac{x_0}{h} + N]} \big|h^{-1}\langle \lambda(h^{-1}\cdot - k),\ p_{x_0,L}\rangle - h^{-1}\langle \lambda(h^{-1}\cdot - k),\ \psi\rangle\big| \cdot \|f\| \\
&\quad \le 2N\|f\| \cdot \|\lambda\| C_\varepsilon h^{\alpha - \varepsilon}.
\end{aligned}$$

As $\psi(x_0) = p_{x_0,L}(x_0)$, we finally get

$$|\psi(x_0) - \psi_{\lambda,h}(x_0)| \le 2N\|f\| \cdot \|\lambda\| C_\varepsilon h^{\alpha-\varepsilon}$$

uniformly with respect to all points x_0 on any compact set. Sufficiency is thus proved.

Necessity. Let us show that an arbitrary compactly supported distribution, whose order of approximation is greater than L, satisfies the Strang-Fix condition of the same order. For the approximated function, take a polynomial x^r of degree at most L. Then we have

$$\inf_{g\in V_f(h)} \|x^r - g\|_{C[-M,M]} = o(h^L), \qquad h \to 0$$

(as $w_f > L$). On the other hand,

$$\begin{aligned}\inf_{g\in V_f(h)} \|x^r - g\|_{C[-M,M]} &= \inf_{g\in V_f} \|(hx)^r - g\|_{C[-M/h,M/h]} \\ &= h^r \inf_{g\in V_f} \|x^r - g\|_{C[-M/h,M/h]} \ge h^r \inf_{g\in V_f} \|x^r - g\|_{C[-M,M]}.\end{aligned}$$

Thus we have

$$h^r \inf_{g\in V_f} \|x^r - g\|_{V[-M,M]} = o(h^L), \quad h \to 0.$$

Consequently,

$$\inf_{g\in V_f} \|x^r - g\|_{C[-M,M]} = o(h^{L-r});$$

hence,

$$\inf_{g\in V_f} \|x^r - g\|_{C[-M,M]} = o(1), \quad h \to 0,$$

for all $r = 0, \dots, L$. Therefore the space V_f approximates all polynomials of degree at most L, but as this space is finite-dimensional on any segment $[-M, M]$, it contains all these polynomials. Hence $\mathcal{P}_L \subset V_f$ and by Theorem 3.2.1, the function f satisfies the Strang-Fix condition of order L. $\Diamond$

REMARK 3.3.5. It is not hard to construct a functional λ for decomposition (3.18). According to Riesz's theorem about the general form of elements from $C^*[a, b]$ (see Appendix A.2), this functional is defined by some Borel charge λ. Therefore,

$$(S_f^{-1})^* \delta P = \int_a^b P d\lambda = \langle \lambda, P \rangle$$

for any $P \in \mathcal{P}_L$. Note that for λ, we can take any Borel functional that satisfies the relations

$$[S_f^{-1} p](0) = \langle \lambda, p \rangle$$

for $p = 1, x, \dots, x^L$, or else,

$$p(0) = \langle \lambda,\ f * p \rangle, \quad p = 1,\ x, \dots,\ x^L.$$

It follows that

$$\langle \lambda,\ f*[x^r] \rangle = \sum_{j=0}^{r} (-1)^{r-j} \binom{r}{j} \langle \lambda, x^j \rangle \cdot \langle f,\ x^{r-j} \rangle = \begin{cases} 1, & \text{for } r = 0, \\ 0, & \text{for } r = 1, \dots, L. \end{cases} \tag{3.19}$$

Thus λ is a Borel charge such that

$$\langle \lambda,\ x^j \rangle = \alpha_j, \quad j = 0, \dots, L, \tag{3.20}$$

where the constants $\alpha_0, \dots, \alpha_L$ are determined by the linear system (3.19). In particular, $\alpha_0 = 1$; hence $\int_a^b d\lambda = 1$. One can take a discrete charge:

$$(\lambda, g) = \sum_{s=0}^{L} \beta_s g(x_s),$$

where $x_0, \dots, x_L$ are different points of the segment $[a, b]$ and the coefficients are found from the system (3.20). For each set of points $x_0, \dots, x_L$, this system has a unique solution because its determinant is the corresponding Vandermonde determinant (see Lemma A.18.1 and Remark A.18.2 in Appendix A.18). For λ, we can take an arbitrary absolutely continuous function that satisfies the relation

$$\int_a^b x^j \lambda'(x) dx = \alpha_j, \qquad j = 0, \dots, L.$$

Such a function is certainly not unique.

REMARK 3.3.6. For an arbitrary function g, whose L-th derivative is bounded on the whole real axis, there is an analog of Theorem 3.3.2 for uniform approximation on the whole of $\mathbb{R}$ with no restriction on segments $[-M, M]$. If f satisfies the Strang-Fix condition of order L, then

$$\inf_{f \in V_f(h)} \|g - f\|_\infty = O(h^L).$$

The proof if similar to that of Theorem 3.3.2.

REMARK 3.3.7. Approximation properties of a compactly supported function f entirely depend on the Strang-Fix conditions. Consequently, any function that can be locally approximated by polynomials of order at most L can be approximated by the functions $\{f(h^{-1} \cdot -k)\}_{k \in \mathbb{Z}}$ with the same accuracy (provided f satisfies the Strang-Fix condition of order L). Hence for the function g, the value $w_f(g)$ is characterized by its order of approximation by algebraic polynomials. This fact can be expressed, for example, by means of finite differences or some other means of the approximation theory (see Appendix A.15 and [**19**]).

REMARK 3.3.8. A compactly supported function f has approximation properties whenever it satisfies the Strang-Fix condition. No other requirements are needed. In particular, f need not be smooth. It can be an arbitrary bounded Borel function with compact support. However boundedness is essential. The reader can easily find corresponding examples.

REMARK 3.3.9. As decomposition (3.18) locally depends on the function g, one can prove the following local analog of Theorem 3.3.2:

If a bounded function f with compact support satisfies the Strang-Fix condition of order L, then an arbitrary function g continuous at the point x can be uniformly approximated by the system of functions $\{f(h^{-1} \cdot -k)\}$ in a neighborhood of the point x. The order of approximation is equal to $\min\{L+1, \alpha_g(x)\}$.

The necessary conditions of Theorem 3.3.2 can also be formulated locally. Note that in the proof of the theorem we used only the condition that the system $\{f(h^{-1} \cdot -k)\}$ approximates the power function x^r on some segment $[-M, M]$, which can be arbitrarily small.

Suppose a bounded function f has compact support and $[a, b]$ is an arbitrary segment. If for each function $g \in C^L[a, b]$ $(L \geq 0)$ there exists a sequence $h_j \to +0$ such that

$$\inf_{v \in V_f(h_j)} \|v - g_{|[a,b]}\|_{C[a,b]} = o(h_j^L), \quad j \to \infty,$$

then f satisfies the Strang-Fix condition of order L.

REMARK 3.3.10 (**Approximation in the space L_p**). Let us show how to prove Theorem 3.3.2 for the case of metric L_p. The proof of necessity remains the same. Let us prove sufficiency. Suppose a function $f \in L_p$ satisfies the Strang-Fix condition of order $L \geq 0$. Take an arbitrary charge $\lambda \in L_q$ with compact support, where $1/q + 1/p = 1$, that satisfies (3.20), and for an arbitrary function $g \in W_p^L$, construct its decomposition $g_{\lambda,h}$ by formula (3.18).

Restrict the function g to a segment $[-M, M]$ and fix small $\varepsilon > 0$. Take a segment $[-N, N]$ that contains the supports of both function f and charge λ. For an arbitrary $x \in [-M, M]$, approximate the function g on the segment $[x - 2hN, x + 2hN]$ by a suitable polynomial $P(x)$ of degree at most L, so that

$$\|g - P\|_{L_p[x-2hN, x+2hN]} \leq C_1 h^{\alpha-\varepsilon} \|g\|_{L_p[x-2hN, x+2hN]}$$

(this is possible due to Jackson's inequality; see Appendix A.15, Theorem A.15.2). Next, using the fact that decomposition (3.18) is exact for polynomials of degree at most L and arguing as in the proof of Theorem 3.3.2, we obtain

$$\|g - g_{\lambda,h}\|_{L_p[-M,M]} \leq C_2 h^{\alpha-\varepsilon} \|g\|_{L_p[-M,M]}.$$

It remains to divide the real axis into segments of length $2M$ and to sum this inequality (or, more correctly, the p-th power of this inequality) over all segments.

Combining Theorems 3.2.3 and 3.3.2, we obtain the following theorem.

THEOREM 3.3.11. *If a refinable function belongs to C^L (W_p^L), its order of approximation in the corresponding space is at least $L + 1$.*

3.4. Linear independence, stability, and orthogonality of integer shifts

We shall distinguish between linear dependence of integer shifts of a function f over the space of all sequences ℓ (there exists a sequence such that $\sum_k \lambda_k f(\cdot - k) = 0$) and over the space of bounded sequences ℓ_∞ (when there exists a bounded sequence λ with the above-mentioned property). If it is not specifically mentioned, linear dependence and independence will be understood in the general sense, i.e., over the space of all sequences ℓ. As a rule, linear independence of integer shifts over ℓ_∞ is sufficient to study wavelets. Nevertheless, we shall formulate all results in the general situation.

DEFINITION 3.4.1. We say that a sequence of complex numbers $\Theta = \{\theta_0 + k, k \in \mathbb{Z}\}$ that consists of zeros of an entire function g is a periodic zero of this function. A periodic zero consisting of real elements is called a real periodic zero.

It is clear that a periodic zero is real if it contains at least one real number. If a function g can be represented as $g(\xi) = (e^{2\pi i \xi} - a)h(\xi)$, where h is an entire function, then $\frac{1}{2\pi i} \operatorname{Ln} a$ is a periodic zero for g. Conversely, if $\Theta = \{\theta_0 + k, k \in \mathbb{Z}\}$ is a periodic zero of a function g, then g is divisible by $e^{2\pi i \xi} - e^{2\pi i \theta_0}$, i.e., can be represented in the form $\left(e^{2\pi i \xi} - e^{2\pi i \theta_0}\right) h(\xi)$, where h is an entire function. By definition, multiplicity of a periodic zero is the maximal power of r such that g is divisible by $\left(e^{2\pi i\xi} - e^{2\pi i\theta_0}\right)^r$. For all elements $\theta \in \Theta$, the value of the exponent $e^{2\pi i\theta}$ is the same. It will be denoted by $e^{2\pi i\Theta}$.

Let a distribution $f \in \mathcal{S}'$ have compact support. If its Fourier transform has a periodic zero, integer shifts of the function f are linearly dependent:

$$\sum_{k\in\mathbb{Z}} \lambda^k f(x-k) = 0, \tag{3.21}$$

where $\lambda = e^{2\pi i\Theta}$, $\Theta = \{\theta_0 + k,\ k \in \mathbb{Z}\}$ is a periodic zero. To prove this statement, it suffices to apply the Poisson summation formula to the function $g(x) = e^{-2\pi i\theta_0 x} f(x)$. We have

$$\begin{aligned} e^{-2\pi i\theta_0 x}\left[\sum_k \lambda^k f(x-k)\right] &= \sum_k e^{-2\pi i\theta_0(x-k)} f(x-k) = \sum_k g(x-k) \\ &= \sum_n \widehat{g}(n) e^{-2\pi i n x} = \sum_n \widehat{f}(\theta_0 + n) e^{-2\pi i n x} = 0. \end{aligned}$$

Analogously, one can prove that if a periodic zero Θ has multiplicity r, then

$$\sum_{k\in\mathbb{Z}} k^n \lambda^k f(x-k) = 0, \quad \lambda = e^{2\pi i\Theta}, \tag{3.22}$$

for each $n = 0, \ldots, r-1$. The sequence $\{\lambda^k\}$ is bounded if the periodic zero Θ is real. Hence if the function $\widehat{f}$ has a real periodic zero, then integer shifts of f are linearly dependent over ℓ_∞.

An entire function can have infinitely many periodic zeros. However, if this function is the Fourier transform of a distribution with compact support, the number of its periodic zeros is finite and cannot exceed the length of support. This follows from formula (3.21). Moreover, from formula (3.22) we conclude that the sum of multiplicities of all periodic zeros cannot exceed the length N of support.

Consider the operator of convolution with a function f on the space of sequences $\mathcal{S}_f : \ell \to V_f$. Here, we shall omit the index if it is clear which function f is considered. $\mathcal{S}^p$ is the restriction of this operator to the space ℓ_p. Denote by $\mathcal{N}$ (respectively, $\mathcal{N}_p$) the kernel of this operator. Thus, if $\widehat{f}$ has a periodic zero, the operator $\mathcal{S}$ is degenerate ($\mathcal{N} \neq \{0\}$). If there exists a real periodic zero, then $\mathcal{N}_\infty \neq \{0\}$. As will be shown below (Theorem 3.4.2), the kernel $\mathcal{N}$ is finite-dimensional, its dimension does not exceed the length of support of the function f.

THEOREM 3.4.2. *Integer shifts of a distribution f with compact support are linearly dependent if and only if its Fourier transform $\widehat{f}$ has a periodic zero. They are linearly independent over ℓ_∞ if and only if there exists a real periodic zero. The kernel of the operator $\mathcal{S}$ has the form*

$$\mathcal{N} = \operatorname{span}\left\{(\lambda_1^k), \ldots, (k^{r_1-1}\lambda_1^k), \ldots, (\lambda_q^k), \ldots, (k^{r_q-1}\lambda_q^k)\right\}, \tag{3.23}$$

where $\lambda_n = e^{2\pi i\Theta_n}$, $\Theta_1, \ldots, \Theta_q$ are periodic zeros of the function $\widehat{f}$, $r_1, \ldots, r_q$ are their multiplicities.

COROLLARY 3.4.3. *In the notation of Theorem 3.4.2 we have*

$$\dim \mathcal{N} = r_1 + \cdots + r_q;$$

i.e., the dimension of the kernel of the operator $\mathcal{S}$ is equal to the sum of multiplicities of all periodic zeros of the function $\widehat{f}$. This sum does not exceed the length of support of the function f.

COROLLARY 3.4.4. *For the kernel of the operator $\mathcal{S}^\infty$, we have*

$$\mathcal{N}_\infty = \operatorname{span}\Big\{(\lambda_1^k), \ldots, (\lambda_j^k)\Big\},$$

where $\lambda_n = e^{2\pi i \Theta_n}$, $\Theta_1, \ldots, \Theta_j$ are real periodic zeros of the function $\widehat{f}$. Thus the dimension of the kernel $\dim \mathcal{N}_\infty$ is equal to the number of different real periodic zeros.

COROLLARY 3.4.5. *For each $p \in [1, +\infty)$, the operator S^p is nondegenerate; i.e.; $\mathcal{N}_p = \{0\}$.*

To prove Corollary 3.4.4, note that a linear combination of different sequences of the form $(k^r \lambda^k)_{k\in\mathbb{Z}}$ is bounded only if all λ are equal to 1 in modulus and all powers of r are equal to zero (see Appendix A.18, Proposition A.18.4). Using representation (3.23), we deduce that an element of the kernel $\mathcal{N}$ belongs to ℓ_∞ if all powers of r_n are equal to 0 and all λ_n are equal to 1 in modulus, i.e., correspond to the real periodic zeros. This completes the proof.

Now use the fact that no finite linear combination of sequences of the form $(\lambda^k)_{k\in\mathbb{Z}}$, where all λ are equal to 1 in modulus, can tend to zero as $k \to \infty$ (see Appendix A.18, Corollary A.18.5). Hence no such sequence can belong to ℓ_p for $p < \infty$, which proves Corollary 3.4.5.

`Proof of Theorem 3.4.2.` It follows from (3.22) that each linear combination of sequences

$$(k^l \lambda_n^k), \quad l = 0, \ldots, r_n - 1, \quad n = 1, \ldots, q,$$

belongs to the kernel $\mathcal{N}$. It remains to prove the converse, namely, that each element of the kernel is such a linear combination. Without loss of generality one can assume that $\operatorname{supp} f \subset [0, N]$. First we regularize the function f (replace f by its convolution with an appropriate smooth function) and thus reduce the problem to the case of a continuous function f. Further, as in the proof of Theorem 3.2.1, we choose an arbitrary $y \in [N-1, N]$ and conclude that any sequence $\{a_k\}$ such that

$$\sum_{j=0}^{N-1} f(y-j) a_{j+n} = 0, \quad n \in \mathbb{Z}, \tag{3.24}$$

is given by formula (3.8), i.e.,

$$a_k = \sum_{n=1}^{q} \lambda_n^k p_n(k)$$

where λ_n are roots of the characteristic polynomial

$$\chi_y(\lambda) = \sum_{j=0}^{N-1} f(y-j)\lambda^j,$$

and the degrees of polynomials p_n do not exceed the multiplicities of these roots. If the sequence $\{a_k\}$ belongs to the kernel $\mathcal{N}$, then it satisfies (3.24) for all $y \in [0, 1]$.

Hence each λ_n is a common root of all polynomials χ_y, $y \in [0,1]$, and its multiplicity is at least $\deg p_n$. Then for each λ_n, we have

$$\sum_{j=0}^{N-1} k^l \lambda^j f(y-j) = 0, \qquad y \in [0,1];$$

therefore,

$$\sum_k k^l \lambda_n^k f(x-k) \equiv 0, \qquad l = 0, \dots, \deg p_n.$$

According to (3.22), this means that $\frac{1}{2\pi i} \mathrm{Ln}\, \lambda_n$ is a periodic zero of the function $\widehat{f}$ of multiplicity at least $\deg p_n$. ◊

REMARK 3.4.6. In fact, we have proved a stronger statement: if a function $\sum_k a_k f(x-k)$ is identically equal to zero on some ray $x \geq A$, then either $a_k = 0$ for all sufficiently large k or the function $\widehat{f}$ has a periodic zero. If in addition $\{a_k\} \in \ell_\infty$, then $\widehat{f}$ has a real periodic zero. Thus we have established the following proposition.

PROPOSITION 3.4.7. *If for some sequence $\{a_k\}$ the function $\sum_k a_k f(x-k)$ has compact support, then either this sequence is finite or integer shifts of f are linearly dependent.*

Let us apply these results to refinable functions. Now that we have Theorem 3.4.2, it seems natural to expect that independence of integer shifts is determined only by location of zeros of the polynomial $m_0(\xi)$ (mask). This is indeed the case. However, the set of zeros of the mask may have a complicated structure, and to describe it, we shall need some new definitions.

Consider an arbitrary trigonometric polynomial $p(\xi)$. If for some $\alpha \in \mathbb{T}$ we have $p(\frac{\alpha}{2}) = p(\frac{1}{2} + \frac{\alpha}{2}) = 0$, then the pair $\{\frac{\alpha}{2}, \frac{1}{2} + \frac{\alpha}{2}\}$ is called *a pair of symmetric roots of the polynomial* $p(\xi)$. For definiteness, we shall assume that for an arbitrary $\alpha \in \mathbb{T}$, to the element $\alpha/2 \in \mathbb{T}$ there corresponds a number from the half-interval $[0, \frac{1}{2})$ and to the element $\frac{1}{2} + \frac{\alpha}{2}$ there corresponds a number from $[\frac{1}{2}, 1)$. Analogously, we can define a pair of symmetric roots for the corresponding algebraic polynomial $\mathbf{p}(z)$: it is a pair of numbers $\{\pm\sqrt{z}\}$ such that $\mathbf{p}(\sqrt{z}) = \mathbf{p}(-\sqrt{z}) = 0$.

A set of different nonzero complex numbers $\mathbf{b} = \{b_1, \dots, b_n\}$ is called *cyclic* if $b_{j+1} = b_j^2$, $j = 1, \dots, n$ (here we set $b_{n+1} = b_1$). It is clear that any cyclic set belongs to the unit circle $\{z \in \mathbb{C},\ |z| = 1\}$. The simplest cyclic set is $\{1\}$; next is the set of two elements $\{e^{2\pi i/3}, e^{4\pi i/3}\}$; etc. For each n, there exist finitely many cyclic sets consisting of n elements. The smallest cyclic set $\{1\}$ is called *trivial*; all other cyclic sets are called nontrivial. We shall consider only irreducible cyclic sets, i.e., cyclic sets consisting of different elements. Note that if any two cyclic sets do not coincide, they are disjoint. A complex number $b \neq 0$ is called cyclic if it is an element of some cyclic set, i.e., $b^{2^n} = b$. The least of such numbers n is called the order of a cyclic number. A cyclic set $\mathbf{b}$ is a *cycle* of a polynomial $\mathbf{p}$ if $\mathbf{p}(-\mathbf{b}) = 0$; i.e., $\mathbf{p}(-b_j) = 0$ for all $b_j \in \mathbf{b}$. If $\mathbf{b} = \{1\}$, the cycle is called *trivial*; all other cycles are called nontrivial. We shall need the following

LEMMA 3.4.8. *Suppose A, B are two sets of nonzero complex numbers, B is nonempty, and for any $z \in B$, both numbers $\pm\sqrt{z}$ belong to $A \cup B$. Then either A contains a pair of symmetric numbers $\{a, -a\}$ or B contains a cyclic set $\mathbf{b}$ such that $-\mathbf{b} \subset A$.*

Proof. Suppose A does not contain any symmetric pairs. Then for an arbitrary $z_1 \in B$, at least one of the numbers $\pm\sqrt{z_1}$ belongs to B. Denote this number by z_2. At least one of the numbers $\pm\sqrt{z_2}$ belongs to B; denote it by z_3, etc. Thus we obtain a sequence $\{z_k\} \subset B$ and, as B is finite, $z_j = z_{j+n}$ for some $j, n \in \mathbb{N}$. Then $z_j^{2^n} = z_j$; hence $z_1^{2^n} = z_1$. Thus the set $\mathbf{b} = \{z_1, \dots, z_n\} \subset B$ is cyclic and without loss of generality we can assume that it is irreducible. If $-\mathbf{b} \subset A$, then the proof is complete. Let $-z_j \notin A$ for some $z_j \in \mathbf{b}$. Then $-z_j \in B$, since $-z_j \in \{\pm\sqrt{z_{j-1}}\} \subset A \cup B$. Arguing as above, we show that the number $-z_j$ belongs to some irreducible cyclic set $\mathbf{b}' \subset B$. Note that $z_{j-1} = (-z_j)^2 \in \mathbf{b}'$. Thus the sets $\mathbf{b}'$ and $\mathbf{b}$ intersect; hence $\mathbf{b}' = \mathbf{b}$, which is impossible, as $-z_j \notin \mathbf{b}$. Therefore $-\mathbf{b} \subset A$, which completes the proof. ◊

PROPOSITION 3.4.9. *The Fourier transform of a refinable function has a periodic zero if and only if the mask* $\mathbf{m}_0$ *has either a nontrivial cycle or a pair of symmetric zeros. This Fourier transform has a real periodic zero if and only if* $\mathbf{m}_0$ *has either a nontrivial cycle or a pair of symmetric zeros on the unit circle.*

Proof. If $\mathbf{m}_0$ has a pair of symmetric zeros, which we denote by $e^{-2\pi i\alpha/2}$, $-e^{-2\pi i\alpha/2}$ $(\alpha \in \mathbb{C})$, then $m_0\left(\frac{\alpha}{2} + \frac{k}{2}\right) = 0$ for any integer k; therefore,

$$\widehat{\varphi}(\alpha + k) = \widehat{\varphi}\Big(\frac{\alpha}{2} + \frac{k}{2}\Big) m_0\Big(\frac{\alpha}{2} + \frac{k}{2}\Big) = 0.$$

It follows that $\{\alpha+k\}$ is a periodic zero. It is real in the case $\alpha \in \mathbb{R}$. If $\mathbf{m}_0$ has a cycle $\mathbf{b} = \{z_1, \dots, z_n\}$, then $\widehat{\varphi}$ has n periodic zeros $\frac{-1}{2\pi i}\mathrm{Ln} z_r = \{\alpha_r + k\}$, $r = 1, \dots, n$. Indeed, fix some k and choose the least integer s for which $e^{-2\pi i 2^{-s}(\alpha_r + k)} \notin \mathbf{b}$. Clearly, $s \ge 1$. Then we have

$$e^{-2\pi i 2^{-s+1}(\alpha_r + k)} \in \mathbf{b};$$

therefore,

$$-e^{-2\pi i 2^{-s}(\alpha_r + k)} \in \mathbf{b}.$$

Consequently, $m_0\big(2^{-s}(\alpha_r + k)\big) = 0$. Using decomposition (3.5), we get $\widehat{\varphi}(\alpha_r + k) = 0$, which was to be proved.

Conversely, let $\widehat{\varphi}$ have a periodic zero. Then the sets

$$A = \{z \in \mathbb{C},\ \mathbf{m}_0(z) = 0\},$$
$$B = \{z \in \mathbb{C},\ \tfrac{-1}{2\pi i}\,\mathrm{Ln}\, z \text{ is a periodic zero of the function } \widehat{\varphi}\}$$

satisfy the conditions of Lemma 3.4.8. Indeed, both sets are finite (the function $\widehat{\varphi}$ can have only finitely many periodic zeros), and it follows from the equality $\widehat{\varphi}(\xi) = m_0(\xi/2)\widehat{\varphi}(\xi/2)$ that $\{\pm\sqrt{z}\} \subset A \cup B$ for each $z \in B$. Using Lemma 3.4.8, we deduce that $\mathbf{m}_0$ has either a pair of symmetric zeros or a cycle that is not trivial, as $\mathbf{m}_0(1) \ne 0$. ◊

Thus we have proved the following

THEOREM 3.4.10. *Integer shifts of a refinable function are linearly independent if and only if the mask has neither cycles nor symmetric zeros. They are linearly independent over* ℓ_∞ *if and only if the mask has neither cycles nor symmetric zeros on the unit circle.*

Let us now find out when integer shifts of a refinable function have the property of the Riesz basis. First we introduce the following generalization of the notion of the Riesz basis to the spaces $L_p(\mathbb{R})$:

DEFINITION 3.4.11. For any given $p \in [1, +\infty]$, a function f is called *ℓ_p-stable*, or, equivalently, it has the property of the Riesz basis in the space $L_p(\mathbb{R})$, if its integer shifts form a Riesz basis (in the metric of L_p) of their linear span. This means that $f \in L_p(\mathbb{R})$ and there exist positive constants A_p, B_p such that for each sequence $a = \{a_k\} \in \ell_p$

$$A_p\|a\|_p \le \Big\|\sum_{k\in\mathbb{Z}} a_k f(x-k)\Big\|_p \le B_p\|a\|_p. \tag{3.25}$$

For $p = 2$, we have the classical definition of a Riesz system in $L_2(\mathbb{R})$ (Definition 1.1.1). The property of ℓ_p-stability means that the operator $\mathcal{S}^p(a) = a * f$ is *boundedly invertible* in the metric of L_p. Recall that linear independence of integer shifts means the usual (set-theoretic) invertibility. Consequently, ℓ_p-stability implies linear independence, but the converse is not true. In general, a function may be stable for some p, being nonstable for other p. However, it appears that for functions with compact support, stability for all $p \in [1, +\infty]$ is equivalent to linear independence over ℓ_∞.

THEOREM 3.4.12. *For any compactly supported function f the following properties are equivalent:*

1) *the function f is ℓ_p-stable for some $p \in [1, +\infty]$;*
2) *the function f is ℓ_p-stable for all $p \in [1, +\infty]$ such that $f \in L_p(\mathbb{R})$;*
3) *integer shifts of f are linearly independent over ℓ_∞;*
4) *the Fourier transform of the function f has no real periodic zeros.*

REMARK 3.4.13. For $p = 2$, we obtain the statement of Theorem 1.1.7 (though only for compactly supported functions φ). Indeed, if φ is compactly supported, then its Fourier transform $\widehat{\varphi}$ is an entire function and the periodic function $\Phi(\xi) = \sum_k |\varphi(\xi + k)|^2$ is continuous (in fact, it is a trigonometric polynomial; see Problem 1.1.13). Therefore condition 4) of Theorem 3.4.12 means that the function $\Phi(\xi)$ is bounded below by a positive constant (boundedness above follows from the fact that the function is periodic and continuous).

Proof. If the shifts $f(\cdot - k)$ are linearly dependent over ℓ_∞, then the function f is nonstable for $p = \infty$. Let us prove that it is nonstable for the rest of the values of p as well. By Theorem 3.4.2, there exists a real number α such that $\sum_k e^{2\pi i k\alpha} f(x-k) = 0$. Without loss of generality we can assume that $\operatorname{supp} f \subset [0, N]$. Choose an arbitrary $M \ge 2N$. The function $\sum_{k=0}^M e^{2\pi i k\alpha} f(x-k)$ vanishes outside of the segments $[0, N-1]$ and $[M+1, M+N]$. Then

$$\begin{aligned}\Big\|\sum_{k=0}^{M} e^{2\pi i k\alpha} f(x-k)\Big\|_p &\le \Big\|\sum_{k=0}^{N-1} e^{2\pi i k\alpha} f(x-k)\Big\|_p + \Big\|\sum_{k=M-N+1}^{M} e^{2\pi i k\alpha} f(x-k)\Big\|_p \\ &\le N\|f\|_p + N\|f\|_p = 2N\|f\|_p.\end{aligned}$$

Thus for the sequence

$$\{a_k\} = (0, \ldots, 1, e^{2\pi i\alpha}, \ldots, e^{2\pi i M\alpha}, 0, \ldots),$$

we get

$$\Big\|\sum_k a_k f(x-k)\Big\|_p / \|a\|_p \ \to\ 0, \quad M \to \infty.$$

Therefore the function f is not p-stable for any p.

Conversely, assume that $\widehat{f}$ has no real periodic zeros; i.e., integer shifts of the function f are linearly independent over ℓ_∞.

First we prove that the function is p-stable for $p = \infty$. Suppose f is nonstable. Then there exists a set of bounded sequences $a^k \in \ell_\infty$, $\|a^k\|_\infty = 1$, for which $\|a^k * f\|_\infty \to 0$, $k \to \infty$. Each of the sequences can be shifted by an integer, so we can assume that $|a_0^k| \geq 1/2$ for each k. Next, by the theorem about *-weak compactness of the unit ball (see Appendix A.2), we can assume that the sequence a^k converges weakly to a limit $a \in \ell_\infty$. In the spaces ℓ_p for $p > 1$ this means the coordinatewise convergence (see Appendix A.2). In our case the limit is nonzero since $|a_0| \geq 1/2$. It is clear that $a^k * f$ converges to $a * f$ on each compact set. On the other hand, $a^k * f \to 0$. Hence $a * f = 0$; i.e., shifts of the function f are linearly dependent over ℓ_∞.

Consider now the case $\mathbf{p} < \infty$. We shall need a discrete analog of Theorem 3.4.2, which we formulate in the form of a lemma.

LEMMA 3.4.14. *Let $f_0, \ldots, f_{N-1}$ be given complex numbers. Consider the operator $\Phi : \ell_p \to \ell_p$ defined by the formula*

$$\Phi[a]_l = \sum_{k \in \mathbb{Z}} a_k f_{l-k}$$

(we set $f_k = 0$ for $k \notin \{0, \ldots, N-1\}$).

If $\inf_{\|a\|_p = 1} \Phi[a] = 0$, then the polynomial $F(\xi) = \sum_k f_k e^{2\pi i k \xi}$ has a real root.

Proof. First we show that the operator Φ is bounded in ℓ_p. As $\Phi = \sum_{r=0}^{N-1} \Phi_r$, where $\Phi_r[a] = f_r a(\cdot - r)$, we have

$$\|\Phi\|_p \leq \sum_{r=0}^{N-1} \|\Phi_r\|_p = \sum_{r=0}^{N-1} |f_r|.$$

Let us find the inverse operator Φ^{-1}. Applying the Fourier transform $a \to \widehat{a}(\xi) = \sum a_k e^{-2\pi i k \xi}$, we get

$$\widehat{\Phi[a]}(\xi) = \sum_l e^{-2\pi i l} \sum_k a_k f_{l-k} = \widehat{a}(\xi) F(-\xi).$$

If a polynomial F has no real roots, we define $1/F(-\xi) = \sum_k \beta_k e^{2\pi i k \xi}$ (the function $1/F$ is analytical on the real axis; hence its Fourier coefficients β_k decrease exponentially; in particular, $\sum_k |\beta_k| < \infty$). Applying the inverse Fourier transform, we obtain

$$\left(\Phi^{-1}[a]\right)_l = \sum_{k \in \mathbb{Z}} a_k \beta_{l-k}.$$

Using the inequality $\sum_k |\beta_k| < \infty$ and arguing as above for the operator Φ, we prove that the inverse operator is also bounded in ℓ_p. Therefore,

$$\inf_{\|a\|_p = 1} \|\Phi[a]\| = 1/\|\Phi^{-1}\| > 0,$$

which was to be proved.

Continue the proof of Theorem 3.4.12 by reducing it to the discrete case. Suppose that

$$\left\| \inf_{\|a\|_p = 1} \sum_k a_k f(\cdot - k) \right\| = 0.$$

We choose an integer j and consider the operator $F_j : \mathbb{R}^{2N+1} \to L_1[j, j+N]$ defined by the equality

$$F_j[(b_0, \dots, b_{2N})] = \left(\sum_{k=0}^{2N} b_k f(x - j + N - k) \right) |_{[j,j+N]}.$$

The kernel of this operator is generated by the vectors

$$\left(e^{2\pi i k \theta} k^l \right)_{k=s}^{s+2N}, \quad s \in \mathbb{Z},\ l = 0, \dots, r,$$

where $\{\theta + k,\ k \in \mathbb{Z}\}$ is a periodic zero of the function $\widehat{f}(\xi)$ and r is its multiplicity. The kernel is the same for all operators F_j; we denote it by K. Since this operator is finite-dimensional, we have

$$F_j(b) \geq C_0 \operatorname{dist}(b, K), \quad b \in \mathbb{R}^{2N+1},$$

where $\operatorname{dist}(b, K) = \inf_{x \in K} \|b - x\|_p$ and the constant C_0 is the same for all j. Then we get

$$\Big\| \sum_k a_k f(x-k) \Big\|_p^p = \frac{1}{N} \sum_{l \in \mathbb{Z}} \Big\| \Big(\sum_k a_k f(x-k) \Big) |_{[l,l+N]} \Big\|_p^p$$
$$= \frac{1}{N} \sum_l \Big\| F_l(a_{-l-N}, \dots, a_{-l+N}) \Big\|_p^p \geq \frac{C_0}{N} \sum_l \operatorname{dist}\Big((a_{-l-N}, \dots, a_{-l+N})^T, K \Big)^p.$$

Thus,

$$\inf_{\|a\|_p = 1} \sum_j \operatorname{dist}\Big((a_{-j-N}, \dots, a_{-j+N})^T, K \Big)^p = 0. \tag{3.26}$$

Consider a trigonometric polynomial

$$q(\xi) = \sum_{k=-N}^{N} q_k e^{-2\pi i k \xi},$$

which has roots (of the same multiplicities) at all real periodic zeros of the function $\widehat{f}(\xi)$ and no other real roots. Consider the operator

$$(Qa)_j = \sum_{k=-N}^{N} a_{-j+k} q_k$$

in the space ℓ_p. It is clear that in the case when $(a_{-j-N}, \dots, a_{j+N})^T \in K$ we have $(Qa)_j = 0$. Hence for any vector $(a_{-j-N}, \dots, a_{j+N})^T \in \mathbb{R}_p^{2N+1}$, we get

$$|(Qa)_j| \leq C_1 \operatorname{dist}\left((a_{-j-N}, \dots, a_{j+N})^T, K \right).$$

Using (3.26), we obtain

$$\inf_{\|a\|_p} \|Qa\|_p = \inf_{\|a\|_p = 1} \sum_j \big| (Qa)_j \big|^p = 0.$$

Apply Lemma 3.4.14 to the operator Q. We see that the polynomial $q(\xi)$ has at least one real root. This means that the function $\widehat{f}(\xi)$ has a real periodic zero, which completes the proof. $\diamond$

Thus we have proved the following assertion.

COROLLARY 3.4.15. *Integer shifts of a refinable function are stable if and only if the mask* $\mathbf{m}_0$ *has neither nontrivial cycles nor symmetric zeros on the unit circle.*

We turn now to the criterion of the Strang-Fix conditions for refinable functions (Theorem 3.2.10). Recall that a nontrivial blocking set always contains a pair of symmetric roots (Problem 3.2.8). Hence, if a refinable function is stable and the mask $\mathbf{m}_0$ has a blocking set of order r, this set is trivial. This means that the mask $\mathbf{m}_0$ has a root $z = -1$ of multiplicity $\geq r$; i.e., it can be represented in the form

$$\mathbf{m}_0(z) = \left(\frac{z+1}{2}\right)^r q(z), \tag{3.27}$$

where $q(z)$ is an algebraic polynomial, $q(1) = 1$. Using Theorem 3.2.10, we get

THEOREM 3.4.16. *A stable refinable function satisfies the Strang-Fix conditions of order* L *if and only if the mask* $\mathbf{m}_0$ *has a root* $z = -1$ *of multiplicity* $\geq L+1$, *i.e., can be represented in the form*

$$\mathbf{m}_0(z) = \left(\frac{z+1}{2}\right)^{L+1} q(z). \tag{3.28}$$

In particular (Theorem 3.3.2), for a stable function, condition (3.28) is a criterion of the fact that its integer shifts generate the space of polynomials $\mathcal{P}_L$ and its order of approximation is $\geq L+1$. Moreover (Corollary 3.2.11), this condition is necessary (though not sufficient) for a stable refinable function to be smooth (to belong to the space $W_1^L(\mathbb{R})$). For $L = 0$, we have the following corollary.

COROLLARY 3.4.17. *If a refinement equation has a stable summable solution, then* $\sum_k c_{2k} = \sum_k c_{2k+1} = 1$.

This condition is called the *sum rule*, or *zero sum rule*. It is equivalent to the condition $\mathbf{m}_0(-1) = 0$. Analogously, the *sum rule of order* L,

$$\sum_k (-1)^k k^s c_k = 0, \qquad s = 0, \dots, L,$$

is equivalent to factorability of mask (3.28).

Orthogonality of integer shifts. According to Proposition 1.1.12, a compactly supported function $f \in L_2(\mathbb{R})$ has orthogonal integer shifts if and only if

$$\sum_{q \in \mathbb{Z}} |\widehat{f}(\xi + k)|^2 \equiv C\,; \tag{3.29}$$

integer shifts are orthogonal if and only if this constant is equal to 1. For a refinable function φ, Corollary 1.2.6 gives necessary conditions of orthogonality in terms of the mask of the equation: if integer shifts of a refinable function $\varphi \in L_2(\mathbb{R})$ are orthonormal, the mask satisfies condition (1.27):

$$|m_0(\xi)|^2 \;+\; |m_0(\xi + 1/2)|^2 \;\equiv\; 1.$$

This condition is not sufficient (see Example 2.5.5). It appears that for compactly supported refinable functions, this condition combined with the condition of stability (i.e., the mask has neither cycles nor symmetric roots on the unit circle) is sufficient for orthonormality.

THEOREM 3.4.18. *Integer shifts of a compactly supported refinable function $\varphi \in L^2$ are orthonormal if and only if $\widehat{\varphi}(0) = 1$, the mask has no nontrivial cycles, and it satisfies* (1.27).

Proof. *Necessity* follows from Theorem 1.1.14 and Corollary 3.4.15 (note that under condition (1.27) the mask can have no symmetric zeros).

Prove *sufficiency*. Suppose property (1.27) holds and the refinable function φ is stable. Set

$$\Phi(\xi) = \sum_{k\in\mathbb{Z}} \bigl|\widehat{\varphi}(\xi + k)\bigr|^2.$$

This function is periodic with period 1. Expand it into the Fourier series. Note that $|\widehat{\varphi}|^2 \in L_1(\mathbb{R})$ as $\widehat{\varphi} \in L_2(\mathbb{R})$. Then the function $f = \varphi(\cdot) * \varphi(-\cdot)$, which is the inverse Fourier transform of the function $|\widehat{\varphi}|^2$, is continuous. Next, since φ is supported on the segment $[0, N]$, f is concentrated on the segment $[-N, N]$. Applying the Poisson summation formula

$$\Phi(\xi) = \sum_{k\in\mathbb{Z}} \widehat{f}(\xi + k) = \sum_{k\in\mathbb{Z}} f(-k)e^{-2\pi i k\xi} = \sum_{k=-N}^{N} f(-k)e^{-2\pi i k\xi},$$

we obtain a trigonometric polynomial. Note that $\Phi(0) = 1$ since $\widehat{\varphi}(0) = 1$ and $\widehat{\varphi}(k) = 0$, $k \in \mathbb{Z} \setminus \{0\}$ (the Strang-Fix condition of order zero, Theorem 3.2.3). Show that $\sup_{\xi\in\mathbb{T}} \Phi(\xi) = \inf_{\xi\in\mathbb{T}} \Phi(\xi) = 1$, which will imply that $\Phi \equiv 1$. Prove the equality $\sup_{\xi\in\mathbb{T}} \Phi(\xi) = 1$; the second equality can be proved analogously. Assuming the converse, we deduce that $\|\Phi\|_\infty > 1$ and Φ is not an identical constant. Hence the set

$$\Lambda = \{\xi \in \mathbb{T},\ \Phi(\xi) = \|\Phi\|_\infty\}$$

is finite. Then the sets

$$A = \{z \in \mathbb{C},\ \mathbf{m}_0(z) = 0\}, \qquad B = \{e^{-2\pi i\xi},\ \xi \in \Lambda\}$$

satisfy the conditions of Lemma 3.4.8. Indeed, if $\xi \in \Lambda$, then

$$\Phi(\xi) = \|\Phi\|_\infty = M;$$

therefore,

$$\begin{aligned} M &= \Phi(\xi/2)|m_0(\xi/2)|^2 + \Phi(\xi/2 + 1/2)|m_0(\xi/2 + 1/2)|^2 \\ &\le M\Bigl(|m_0(\xi/2)|^2 + |m_0(\xi/2 + 1/2)|^2\Bigr) = M. \end{aligned}$$

The equality is attained if either $\Phi(\xi/2) = \Phi(\xi/2 + 1/2) = M$ (in this case $\pm e^{-2\pi i\xi/2} \in B$) or at one of the points $\frac{\xi}{2}$, $\frac{1}{2} + \frac{\xi}{2}$ we have $\Phi = M$ and at the other one $m_0 = 0$. In this case, $\pm e^{-2\pi i\xi/2} \in A \cup B$. Using Lemma 3.4.8, we deduce that the polynomial $\mathbf{m}_0$ either has a symmetric root (which is impossible in view of (1.27)) or a cycle, which is nontrivial as $0 \notin \Lambda$. This contradicts our assumption. Thus Φ is an identical constant, which completes the proof. ◊

REMARK 3.4.19. Example 2.5.5 shows that condition (1.27) is not sufficient for orthonormality. It does not even provide stability of a refinable function. In that example, a so-called "contracted" mask is considered. Choose an arbitrary mask m_0 such that $m_0\bigl(\frac{1}{2}\bigr) = 0$ and contract it p times, where $p \neq 1$ is an arbitrary odd number. We obtain the mask $m_p(\xi) = m_0(p\xi)$ and the refinable function

$\varphi_p(x) = \varphi(x/p)$. This function is not stable. Indeed, since $m_0(\frac{1}{2}) = 0$, for all nonzero integers k, we have

$$\widehat{\varphi}_p(k/p) = \widehat{\varphi}(k) = 0.$$

Therefore the function $\widehat{\varphi}_p$ has at least p real periodic zeros $\{\frac{1}{p}+k\}, \dots, \{\frac{p-1}{p}+k\}$; hence φ_p is nonstable. Note that if $\varphi \in L_2(\mathbb{R})$, this function satisfies the Strang-Fix condition of order zero (Theorem 3.2.3), and if (1.27) holds, the mask has no real symmetric zeros; hence $m_0(\frac{1}{2}) = 0$. Thus a contraction by an odd integer ratio "spoils" any mask; i.e., it acquires cycles (Corollary 3.4.15). In Example 2.5.5, we take $m_0(\xi) = (1+e^{-2\pi i\xi})/2$ and $p = 3$.

3.5. Examples and applications

EXAMPLE 3.5.1 (B-splines). The best known refinable function is B-spline $B_n(x)$, which was considered long before the theory of refinement equations was developed. B-splines were considered in §1.4 to construct the Battle-Lemarie wavelets. To represent B-splines in the form of refinable functions, we shall slightly change their definition. Namely, we shall consider the B-spline $B_n(x)$ defined by formulas (∗) in §1.4 (page 24) with the argument shifted in such a way that $\operatorname{supp} B_n = [0, n+1]$. Therefore,

$$B_{2k+1}(x) = \varphi^{B,2k+1}(x-k-1), \qquad B_{2k}(x) = \varphi^{B,2k}(x-k).$$

It is not hard to show that

$$B_n(x) = \chi_{[0,1]} * \cdots * \chi_{[0,1]}$$

(the convolution of $n+1$ characteristic functions of the segment $[0,1]$). For example, $B_0(x) = \chi_{[0,1]}$. The function $B_n(x)$ is a spline of order n with integer nodes. It is concentrated on the segment $[0,\, n+1]$; on each of the segments $[k, k+1]$, $k = 0, \dots, n$, it is a polynomial of degree n. The B-spline $B_n(x)$ is a refinable function, the mask of the corresponding equation $\mathbf{m}_0(z) = \left(\frac{z+1}{2}\right)^{n+1}$, the coefficients of this equation $c_k = 2^{-n}\binom{n+1}{k}$, $k = 0, \dots, n+1$. This is the only equation of order $n+1$ that satisfies the Strang-Fix conditions of order n (Corollary 3.2.12). Consequently, of all refinable functions with support of a given length, the B-spline has the maximal order of approximation $n+1$ and maximal smoothness ($B_n \in W_1^n(\mathbb{R})$, Corollary 3.2.13). B-splines play such an important role in the theory of refinement equations not only due to their optimal properties. It follows from Theorem 3.4.16 that if a refinement equation has a stable W_1^L-solution, then its mask contains a factor $\left(\frac{z+1}{2}\right)^{L+1}$, i.e., the mask of the B-spline B_L. It follows that a stable smooth refinable function can be represented in the form of the convolution of some refinable function with a B-spline of the corresponding order (in Chapter 6, we present a more detailed study of this problem).

For any $n \geq 0$, the function B_n is stable (Corollary 3.2.5), but its integer shifts are orthogonal only for $n = 0$ (Theorem 3.4.18). Note that B-splines are not unique refinement splines; for each $n \geq 1$, there exist a finite number of refinement spline functions of order n (they are classified in [**46**, Section 8] and [**110**]). However all such functions are combinations of integer shifts of a B-spline and all of them (except for B-splines) are nonstable (see [**110**]). Therefore of all refinement splines, only B-splines can be used to construct wavelets.

In the sequel we shall use the following assertion (which the reader can easily prove):

LEMMA 3.5.2. *For any positive integer N on the segment $[0, N]$, there exists a unique, up to a constant factor, spline of order $N-1$ with integer nodes that belongs to $W_1^{N-1}(\mathbb{R})$. This is the cardinal B-spline of order $N-1$.*

EXAMPLE 3.5.3. The equation

$$\varphi(x) = \omega\, \varphi(2x) + (1-\omega)\, \varphi(2x-1) + (1-\omega)\, \varphi(2x-2) + \omega\, \varphi(2x-3),$$

where $\omega \in (0,1)$, corresponds to the so-called generalized *Chaikin algorithm* for the approximation of functions by their values on a uniform lattice ([**111, 112**]). Its solution is concentrated on the segment $[0,3]$; it is continuous for any $\omega \in (0,1)$ and differentiable on the whole segment only for $\omega = \frac{1}{4}$ (it is easy to show that in this case $\varphi(x) = B_2(x)$ is a quadratic spline). For all other

ω, this function has varying local smoothness. On a set of full measure, local smoothness is the same; it is greater than 1 (hence $\varphi'(x) = 0$ almost everywhere).

EXAMPLE 3.5.4 (De Rham's curves). Let there be given a set of points $\{p_j\}$ on the plane, $p_j = (x_j, y_j)$. Construct sequences $\{p_j^1\}$, $\{p_j^2\}, \dots$ following the recurrent rule:

$$\left\{\begin{array}{lllll} p_{2r}^{n+1} & = & \omega p_{r-1}^n & + & (1-\omega)p_r^n, \\ p_{2r+1}^{n+1} & = & (1-\omega)p_{r-1}^n & + & \omega p_r^n, \end{array}\right. \tag{3.30}$$

where $p_j^0 = p_j$. A limit curve $\gamma(t) = (x(t), y(t))$ such that

$$\sup_{k\in\mathbb{Z}} \|\gamma(2^{-n}k) - p_k^n\| \to 0 \quad \text{as } n \to \infty$$

is called de Rham's curve corresponding to the parameter $\omega \in (0, 1/2)$ and the initial sequence of points $\{p_j\}$. Geometrically, de Rham's curve is obtained as the limit of the broken line with given vertices $\{p_j\}$, by successive smoothing its angles (at every step, each side of the broken line is divided into three parts in the ratio $\omega : (1-2\omega) : \omega$). In parametric representation of the curve $\gamma(t)$, both coordinate functions $x(t), y(t)$ are linear combinations of integer shifts of Chaikin's refinable function $\varphi(t)$ of the previous example. However, the geometric smoothness of the curve γ (i.e., its smoothness in the natural parametrization) exceeds the smoothness of φ. Thus for $\omega \le \frac{1}{3}$, de Rham's curve is continuously differentiable and its second derivative vanishes almost everywhere on the curve. For $\omega = \frac{1}{4}$, de Rham's curve is a quadratic spline that consists of pieces of parabolas joined smoothly at the midpoints of the sides of the initial broken line; for $\omega > \frac{1}{3}$, the curve is differentiable at all points except for those corresponding to dyadic rational t. We shall return to de Rham's curves in Chapter 5, Example 5.5.3.

EXAMPLE 3.5.5 (Dubuc's interpolation scheme). In 1986 S. Dubuc first suggested a subdivision scheme for interpolation of functions. This scheme is as follows:

$$\left\{\begin{array}{lll} f_{2r}^{n+1} & = & f_r^n, \\ f_{2r+1}^{n+1} & = & -\frac{1}{16}\left(f_{r-1}^n + f_{r+2}^n\right) + \frac{9}{16}\left(f_r^n + f_{r+1}^n\right). \end{array}\right. \tag{3.31}$$

For any initial sequence $\{f_k^0\}$, this scheme converges; i.e., there exists a continuous function f such that

$$\sup_{k\in\mathbb{Z}} \|f(2^{-n}k) - f_k^n\|_\infty \to 0\,, \quad n \to \infty$$

([**113**]). This scheme is exact for polynomials of degree at most 3; i.e., if $f_k^0 = f^0(k)$, where $f^0(x) \in \mathcal{P}_3$, then $f(x) \equiv f^0(x)$. If for an initial sequence f^0 we take the δ function (for $k = 0$ $\delta_k = 1$; otherwise $\delta_k = 0$), then the limit function $f = \varphi$ solves the symmetric refinement equation

$$\varphi(x) = -\frac{1}{16}\varphi(2x-3) + \frac{9}{16}\varphi(2x-1) + \varphi(2x) + \frac{9}{16}\varphi(2x+1) - \frac{1}{16}\varphi(2x+3). \tag{3.32}$$

The function φ is concentrated on the segment $[-3, 3]$. The mask of the equation has a root of multiplicity 4 at the point $z = -1$; hence (Theorem 3.4.16) the function φ satisfies the Strang-Fix conditions of order 3. In particular, its integer shifts generate the space of polynomials $\mathcal{P}_3$ (this fact accounts for the precision of the algorithm for functions from this space). The order of approximation of φ is equal to 4 (Theorem 3.3.2). Therefore, if we take values of some smooth function g on a uniform lattice with step h (setting $f_j^0 = g(a + hj)$) and then apply Dubuc's scheme to this sequence, the limit function f will uniformly approximate g up to $O(h^4)$; more precisely, $\|g - f\|_\infty \le C\|g^{(IV)}\|_\infty h^4$.

In fact, the coefficients of Dubuc's scheme are nothing more than coefficients of the Lagrange interpolation polynomial of degree 3 at the points $-1, 0, 1, 2$. Actually, at each iteration of this algorithm, given four adjacent points

$$2^{-n}(k-1), \quad 2^{-n}k, \quad 2^{-n}(k+1), \quad 2^{-n}(k+2),$$

we construct a Lagrange polynomial; for the value of the next function, we take its value at the point $2^{-n}k + 2^{-n-1}$.

This scheme was studied by S. Dubuc in [**113**], and later, in [**114**], jointly with G. Deslauriers, it was generalized to an interpolation algorithm of arbitrary order. This scheme is optimal of all subdivision schemes reproducing polynomials of a given degree; i.e., it has the least number of nonzero coefficients. The mask of equation (3.32) can be reduced to the form

$$\mathbf{m}_0(z) = z^{-3}\left(\frac{1+z}{2}\right)^4\left(\frac{-1+4z-z^2}{2}\right). \tag{3.33}$$

Thus the mask has zero $z = -1$ of the fourth order and no other zeros on the unit circle. Therefore (Theorem 3.4.10) the refinable function φ is stable. In addition, $\varphi \in C^1(\mathbb{R})$ (Example 7.3.14) and it "almost" belongs to C^2 in the sense that it belongs to $C^{2-\varepsilon}$ for any $\varepsilon > 0$. Indeed, the modulus of continuity of the derivative of the function φ is equal to $Ct\left|\ln t\right|$.

CHAPTER 4

Wavelets with Compact Support

4.1. Construction of orthogonal wavelets

This chapter deals with one-dimensional real orthonormal wavelets with compact support, which are of most interest for applications since they allow simple numerical implementation for algorithms of decomposition and syntheses of signals.

In this chapter, we shall use the terms *orthogonal wavelet function* and *orthogonal scaling function*. They are abridged forms of the following expressions: a wavelet function whose dilations and shifts form an orthonormal basis in $L^2(\mathbb{R})$; a scaling function whose integer shifts form an orthonormal basis in the null space of the MRA generated by this function.

As noted in Corollary 1.3.5, any orthogonal wavelet function with compact support has a corresponding orthogonal scaling function with compact support. The converse is obvious. If an orthogonal scaling function has compact support, then in view of (1.42), the corresponding wavelet is a finite linear combination of functions with compact support, i.e., has compact support as well.

As is known, if an orthogonal wavelet function has compact support, then there exists a corresponding multiresolution analysis (Corollary1.3.5). Hence the construction of an MRA can be reduced to the construction of a generating orthogonal scaling function φ with compact support. The space V_0 generated by its integer shifts and all spaces V_j obtained by binary dilations of V_0 must possess properties **MR1**–**MR5** of Definition 1.2.1. Suppose a function $\varphi \in L_2(\mathbb{R})$ is such that

$$\langle \varphi(\cdot), \varphi(\cdot - k)\rangle = \delta_{0k}.$$

Set

$$V_0 = \overline{\operatorname{span}\{\varphi(\cdot - k),\ k \in \mathbb{Z}\}}, \qquad V_j = \{f(2^j\cdot),\ f \in V_0\}.$$

Recall the conditions for φ, under which these spaces form an MRA. Properties **MR4**, **MR5** are already satisfied. Property **MR1** is equivalent to the inclusion $V_0 \subset V_1$, since by property **MR4**, this inclusion can be extended to all integer j. In turn, this property means that $\varphi \in V_1$. As the functions $\sqrt{2}\varphi(2\cdot -k)$ form an orthonormal basis of the space V_1, the function φ can be decomposed in this basis. In other words,

$$\varphi(x) = \sum_{k\in\mathbb{Z}} \sqrt{2}h_k\varphi(2x - k).$$

Since $h_k = \langle \sqrt{2}\varphi(2\cdot -k), \varphi(\cdot)\rangle$ and the function φ has compact support, it follows that only finitely many coefficients h_k are nonzero. Thus the sum on the right-hand side of the equality contains finitely many terms.

REMARK 4.1.1. Linear independence of integer shifts of the function φ is not as obvious as it may seem. The stability of φ implies only independence over ℓ_∞, but not over ℓ. In fact, a stable function with compact support can satisfy a refinement equation with infinitely many terms. For example, the stable function

$$\varphi(x) = \chi_{[0,1)}(x) + \frac{1}{2}\chi_{[0,1)}(x-1)$$

satisfies the refinement equation

$$\varphi(x) = \sum_{k=0}^{+\infty} 2^{-k}\varphi(2x-k).$$

Hence in the proof we use not only the stability but the orthogonality of integer shifts as well.

Thus $\varphi(x) = \sum_{k=n_1}^{n_2} \sqrt{2}h_k\varphi(2x-k)$. Without loss of generality, shifting the argument by n_1 if needed, we can assume that $n_1 = 0$. Now setting $c_k = \sqrt{2}h_k$, we get

$$\varphi(x) = \sum_{k=0}^{N} c_k\varphi(2x-k).$$

Therefore φ satisfies some refinement equation with finitely many terms. Further, condition **MR2** means that for an arbitrary function $f \in L_2(\mathbb{R})$, we have

$$\operatorname{dist}(f, V_j) \to 0, \quad j \to +\infty.$$

Using Theorem 3.3.2, we deduce that the function φ satisfies the Strang-Fix condition of order 0. In particular, $\int_{\mathbb{R}} \varphi(x)\,dx \neq 0$ and we have $\sum_k c_k = 2$ (Proposition 3.1.1). Thus, if the function φ with compact support generates an MRA, then (maybe after an integer shift of the argument) it is an L_2-solution of refinement equation (3.1), where $\sum_k c_k = 2$. As above, the function φ is normalized by the condition $\widehat{\varphi}(0) = \int_{\mathbb{R}} \varphi(x)\,dx = 1$. The conditions of orthogonality of integer shifts of φ are given by Theorem 3.4.18: the mask m_0 of the equation has no cycles and $|m_0(\xi)|^2 + |m_0(\xi+1/2)|^2 \equiv 1$. In particular, the latter condition implies that $m(1/2) = 0$; hence the Strang-Fix condition of order 0 is satisfied and we have

$$\operatorname{dist}(f, V_j) \to 0, \quad j \to +\infty,$$

which provides condition **MR2**. It remains to check the last condition, namely, the emptiness of the intersection of the sample sets V_j (**MR3**). It is also satisfied. Indeed, if a function f belongs to all V_j, then (property **MR4**) we have

$$f(2^{-j}\cdot) \in V_0 \Leftrightarrow \widehat{f}(2^j\xi) = \widehat{\varphi}(\xi)p_j(\xi), \quad j \in \mathbb{Z},$$

where p_j is a 1-periodic function. This equality for all integer j means $f \equiv 0$ (see Theorem 1.2.9 for details).

THEOREM 4.1.2. *If a compactly supported $\varphi \in L_2(\mathbb{R})$ with orthogonal integer shifts generates an MRA, it satisfies the condition $\widehat{\varphi}(0) = 1$ and solves the refinement equation* (3.1), *whose mask has no cycles and which satisfies the conditions*

$$m_0(0) = 1, \qquad |m_0(\xi)|^2 + |m_0(\xi+1/2)|^2 \equiv 1. \tag{4.1}$$

Conversely, if the mask of a refinement equation satisfies the above conditions, then its solution normalized by the condition $\widehat{\varphi}(0) = 1$ belongs to $L_2(\mathbb{R})$, has orthonormal integer shifts, and generates an MRA.

Thus we have proved all the statements of the theorem except for the sufficiency of conditions (4.1) for the solution to belong to $L_2(\mathbb{R})$. We shall prove a stronger statement, which holds even for nonpolynomial masks:

LEMMA 4.1.3. *If the mask of the refinement equation*

$$\varphi(x) = \sum_{k\in\mathbb{Z}} c_k \varphi(2x-k)$$

satisfies the conditions

$$c_k = O(|k|^{-2-\varepsilon}),\ \varepsilon > 0, \quad m_0(0) = 1, \quad |m_0(\xi)|^2 + |m_0(\xi+1/2)|^2 \equiv 1,$$

then its solution φ belongs to $L_2(\mathbb{R})$ and $\|\varphi\|_2 \le |\widehat{\varphi}(0)|$.

Proof. We shall assume that $\widehat{\varphi}(0) = 1$. Then (Proposition 2.4.1)

$$|\widehat{\varphi}(\xi)|^2 = \prod_{k=1}^{\infty} |m_0(2^{-k}\xi)|^2.$$

As $\|m_0^2\|_\infty = 1$, for any n, we have

$$|\widehat{\varphi}(\xi)|^2 \le \prod_{k=1}^{n} |m_0(2^{-k}\xi)|^2$$

for all ξ; hence,

$$\int_{-2^{n-1}}^{2^{n-1}} |\widehat{\varphi}(\xi)|^2\, d\xi \le \int_{-2^{n-1}}^{2^{n-1}} \prod_{k=1}^{n} |m_0(2^{-k}\xi)|^2 d\xi = 2^n \int_{-\frac{1}{2}}^{\frac{1}{2}} \prod_{k=0}^{n-1} |m_0(2^k s)|^2\, ds. \tag{4.2}$$

Let $|m_0(s)|^2 = \sum_{k\in\mathbb{Z}} a_k e^{-2\pi i k s}$. It follows from the conditions for the mask that

$$a_{2k} = \frac{1}{2}\delta_{0k}.$$

Thus,

$$|m_0(s)|^2 = a_0 + \sum_{k\in 2\mathbb{Z}+1} a_k e^{-2\pi i k s},$$

where $a_0 = \frac{1}{2}$. Then we have

$$\prod_{k=0}^{n-1} |m_0(2^k s)|^2 = \sum_{k_0+2k_1+\ldots+2^{n-1}k_{n-1}=r} a_{k_0}\cdots a_{k_{n-1}} e^{-2\pi i r s},$$

where for each j, either $k_j = 0$ or k_j is an odd number. Consequently, if $r = 0$, then $k_0 = \cdots = k_{n-1} = 0$. Therefore,

$$\prod_{k=0}^{n-1} |m_0(2^k s)|^2 = 2^{-n} + \sum_{r\in\mathbb{Z}\setminus\{0\}} b_r e^{-2\pi i s r}.$$

Since

$$\int_{-\frac{1}{2}}^{\frac{1}{2}} b_r e^{-2\pi i s r}\, ds = 0$$

for all $r \neq 0$, it follows that

$$\int_{-\frac{1}{2}}^{\frac{1}{2}} \prod_{k=0}^{n-1} |m_0(2^k s)|^2 \, ds = 2^{-n}.$$

Substituting this in (4.2), we get

$$\int_{-2^{n-1}}^{2^{n-1}} |\widehat{\varphi}(\xi)|^2 \, d\xi \leq 1.$$

As $n \to +\infty$, we obtain

$$\|\widehat{\varphi}\|_2 \leq 1;$$

hence $\|\varphi(\xi)\|_2 \leq 1$, which was to be proved. ◊

Thus, if a mask satisfies the conditions of Theorem 4.1.2, it generates an L_2-refinable function with orthonormal integer shifts. Now the wavelet function ψ is constructed by the formula

$$\widehat{\psi}(\xi) = \nu(\xi) e^{-\pi i \xi} \overline{m_0\Big(\frac{\xi}{2} + \frac{1}{2}\Big)} \widehat{\varphi}\Big(\frac{\xi}{2}\Big), \tag{4.3}$$

where ν is an arbitrary 1-periodic function such that $|\nu(\xi)| = 1$ almost everywhere. This formula gives all wavelet systems corresponding to a given MRA.

LEMMA 4.1.4. *Let an MRA be generated by a compactly supported function $\varphi \in L_2(\mathbb{R})$ with orthogonal integer shifts. A function $\psi \in L_2(\mathbb{R})$ is a wavelet function corresponding to that MRA if and only if it satisfies* (4.3).

Proof. Let us characterize the space W_0 in terms of the Fourier transform. The inclusion $f \in W_0$ is equivalent to both relations $f \in V_1$ and $f \perp V_0$. The first of them means that

$$f(x) = \sum_{k \in \mathbb{Z}} a_k \varphi(2x - k)$$

for some sequence $\{a_k\} \in \ell_2$. After the Fourier transform we get

$$\widehat{f}(\xi) = \frac{1}{2} \widehat{a}(\xi/2) \widehat{\varphi}(\xi/2), \tag{4.4}$$

where $\widehat{a}(\xi) = \sum_{k \in \mathbb{Z}} a_k e^{-2\pi k \xi}$. Property $f \perp V_0$ means that $\langle f, \varphi(\cdot - k) \rangle$ for all $k \in \mathbb{Z}$. Using the Parseval equality, we obtain

$$\int_{\mathbb{R}} e^{2\pi i k \xi} \widehat{f}(\xi) \overline{\widehat{\varphi}(\xi)} \, dx = 0, \qquad k \in \mathbb{Z}.$$

Finally, after periodization we get

$$\sum_{k \in \mathbb{Z}} \widehat{f}(\xi + k) \overline{\widehat{\varphi}(\xi + k)} \equiv 0,$$

which holds for almost all $\xi \in \mathbb{R}$. Expressing $\widehat{f}$ from equality (4.4) and $\widehat{\varphi}$ from the equality $\widehat{\varphi}(\xi) = m_0(\xi/2)\widehat{\varphi}(\xi/2)$ and grouping the sums with even and odd summands, respectively, we obtain

$$\frac{1}{2} \widehat{a}\Big(\frac{\xi}{2}\Big) \overline{m_0\Big(\frac{\xi}{2}\Big)} + \frac{1}{2} \widehat{a}\Big(\frac{\xi}{2} + \frac{1}{2}\Big) \overline{m_0\Big(\frac{\xi}{2} + \frac{1}{2}\Big)} = 0.$$

As $m_0\left(\frac{\xi}{2}\right)$ and $m_0\left(\frac{\xi}{2}+\frac{1}{2}\right)$ do not vanish simultaneously, there exists a 1-periodic function $\lambda(\xi)$ such that

$$\frac{1}{2}\widehat{a}(\xi) = \lambda(\xi)\overline{m_0\left(\xi+\frac{1}{2}\right)}$$

almost everywhere and in addition,

$$\lambda(\xi) + \lambda\left(\xi+\frac{1}{2}\right) = 0$$

almost everywhere. The latter property is equivalent to the relation

$$\lambda(\xi) = e^{-2\pi i\xi}\nu(\xi)$$

for almost all ξ, where ν is a 1-periodic function. Substituting this in (4.4), we get

$$\widehat{f}(\xi) = e^{-\pi i\xi}\nu(\xi)\overline{m_0\left(\frac{\xi}{2}+\frac{1}{2}\right)}\varphi(\xi/2). \tag{4.5}$$

Thus we have found the general form of functions from the space W_0. Now we must characterize the functions ψ that belong to this space and have orthonormal integer shifts (i.e., wavelet functions). Thus $\sum_{k\in\mathbb{Z}}|\widehat{\psi}(\xi+k)|^2 = 1$ almost everywhere (Proposition 1.1.12). Using (1.27), we get

$$\begin{aligned}
\sum_k |\widehat{\psi}(\xi+k)|^2 &= \sum_k |\nu(\xi+k)|^2 \left|m_0\left(\frac{\xi}{2}+\frac{1}{2}+\frac{k}{2}\right)\right|^2 \left|\varphi\left(\frac{\xi}{2}+\frac{k}{2}\right)\right|^2 \\
&= |\nu(\xi)|^2 \sum_l \left(\left|m_0\left(\frac{\xi}{2}+\frac{1}{2}+l\right)\right|^2 \left|\varphi\left(\frac{\xi}{2}+l\right)\right|^2\right. \\
&\qquad\qquad \left. + \left|m_0\left(\frac{\xi}{2}+l+1\right)\right|^2 \left|\varphi\left(\frac{\xi}{2}+l+\frac{1}{2}\right)\right|^2\right) \\
&= |\nu(\xi)|^2 \left(\left|m_0\left(\frac{\xi}{2}+\frac{1}{2}\right)\right|^2 \sum_l \left|\varphi\left(\frac{\xi}{2}+l\right)\right|^2 + \left|m_0\left(\frac{\xi}{2}\right)\right|^2 \sum_l \left|\varphi\left(\frac{\xi}{2}+l+\frac{1}{2}\right)\right|^2\right) \\
&= |\nu(\xi)|^2 \left(\left|m_0\left(\frac{\xi}{2}+\frac{1}{2}\right)\right|^2 + \left|m_0\left(\frac{\xi}{2}\right)\right|^2\right) = |\nu(\xi)|^2.
\end{aligned}$$

Hence $|\nu(\xi)|^2 = 1$ almost everywhere. ◊

Thus a wavelet function constructed above for a given MRA is not unique since we can choose a function ν, following our own way. However, we shall obtain wavelets with compact support only for $\nu(\xi) = e^{-2\pi i\xi n}$ for some integer n. In order to show this, choose $\nu(\xi) \equiv 1$ and get

$$\widehat{\psi}(\xi) = e^{-\pi i\xi}\overline{m_0\left(\frac{\xi}{2}+\frac{1}{2}\right)}\widehat{\varphi}\left(\frac{\xi}{2}\right), \tag{4.6}$$

or, after the inverse Fourier transform,

$$\psi(x) = \sum_{k=0}^{N}(-1)^k\bar{c}_k\varphi(2x+k-1). \tag{4.7}$$

As $\operatorname{supp}\varphi = [0,N]$, we have $\operatorname{supp}\psi = [-\frac{N-1}{2}, \frac{N+1}{2}]$. Thus the wavelet function ψ, just like its scaling function, is concentrated on a segment of length N. If we now return to the general formula (4.3), i.e., multiply the Fourier transform of our wavelet function by some function ν, we obtain a new wavelet function

$$\psi_\nu(x) = \sum_k \hat{\nu}(k)\psi(x-k),$$

where $\hat{\nu}(k)$ are the Fourier coefficients of the function ν : $\nu(\xi) = \sum_k \hat{\nu}(k)e^{-2\pi ik\xi}$. If ψ_ν has compact support, then for sufficiently large M, we have

$$\hat{\nu}(j) = \langle \psi_\nu, \psi(\cdot - j)\rangle = 0$$

for all j such that $|j| > M$. Therefore the sequence $\hat{\nu}(k)$ is finite; i.e., ν is a trigonometric polynomial. But as $|\nu(\xi)| = 1$ almost everywhere, the trigonometric polynomial $|\nu(\xi)|^2$ is identically equal to 1; hence $\nu(\xi) = e^{-2\pi i\xi n}$.

Consequently, $\psi_\nu(\cdot) = \psi(\cdot - n)$. Thus all wavelets with compact support generated by a given scaling function coincide up to an integer shift. So, in what follows, we shall confine our consideration to the wavelets constructed by formula (4.7), i.e., corresponding to the case $\nu \equiv 1$.

Now the problem of the construction of wavelets with compact support is reduced to the construction of the corresponding scaling function, i.e., the construction of a mask that satisfies the conditions of Theorem 4.1.2.

We have $m_0(\frac{1}{2}) = 0$. Moreover, if we are interested in smooth solutions, we must assume that at the point $\frac{1}{2}$ the mask has a zero of high order—no less than the smoothness exponent (Theorem 3.4.16). Therefore we factorize the mask in the form

$$m_0(\xi) = \left(\frac{1+e^{-2\pi i\xi}}{2}\right)^n q(\xi),$$

where $n \geq 1$ and $q(\frac{1}{2}) \neq 0$. Then we have

$$|m_0(\xi)|^2 = \left(\cos^2 \pi\xi\right)^n |q(\xi)|^2 .$$

Let us try to find the trigonometric polynomial $|q(\xi)^2|$ under the addition assumption that it has real coefficients. In this case $|q(\xi)|^2$ is an algebraic polynomial with real coefficients in $\cos 2\pi\xi$ (see Appendix A.11), or in $(1 - \cos 2\pi\xi)/2 = \sin^2 \pi\xi$. Thus,

$$|m_0(\xi)|^2 = \left(\cos^2 \pi\xi\right)^n P\left(\sin^2 \pi\xi\right),$$

where $P(y)$ is an algebraic polynomial nonnegative on the segment $[0, 1]$ and such that $P(0) = 1$. Now the condition

$$|m_0(\xi)|^2 + \left|m_0\left(\xi + \frac{1}{2}\right)\right|^2 \equiv 1$$

becomes an equation for the polynomial P:

$$(1-y)^n P(y) + y^n P(1-y) = 1, \tag{4.8}$$

which must be satisfied identically for all $y \in \mathbb{R}$. This is a linear equation on the polynomial P. Therefore, every solution is a sum $P = P_0 + P_n$, where P_0 is a solution of the corresponding homogeneous equation $(1-y)^n P_0(y) + y^n P_0(1-y) = 0$, and P_n is some particular solution of (4.8). Let us start with the homogeneous equation. Since $P_0(y)$ is divisible by y^n, one can write $P_0(y) = y^n R_0(y)$, where R_0 is a polynomial. Thus, $(1-y)^n y^n R_0(y) + y^n (1-y)^n P_0(1-y) = 0$, and hence $R_0(y) + R_0(1-y) = 0$. Consequently, $R_0 = R\left(\frac{1}{2} - y\right)$, where R is an arbitrary odd polynomial. Thus, the general solution of the homogeneous equation is $P_0(y) = y^n R\left(\frac{1}{2} - y\right)$. In particular, the minimal degree of the solution is $n+1$. Now let us find a particular solution P_n. Let us first observe that such a solution P of degree $\le n-1$ exists. Indeed, the Diophantine equation $(1-y)^n P_1(y) + y^n P_2(1-y) = 1$ possesses a solution (P_1, P_2) such that $\deg P_1$, $\deg P_2 \le n-1$, which can be found, for instance, by the Euclidean algorithm. Moreover, this solution is unique; otherwise, the difference of two solutions would satisfy the homogeneous equation $(1-y)^n \tilde{P}_1(y) + y^n \tilde{P}_2(1-y) = 0$. This is impossible, because the degree of $\tilde{P}_1$ is smaller than n and hence it is not divisible by y^n. On the other hand, if $\big(P_1(y), P_2(y)\big)$ is a solution, then $\big(P_1(1-y), P_2(1-y)\big)$ also is. Hence, from the uniqueness it follows that $P_2(y) = P_1(1-y)$. Writing $P_n(y) = P_1(y)$, we see that equation (4.8) has a unique solution of degree less than or equal to $n-1$. To find this solution, we write $P_n(y) = (1-y)^{-n}\big(1 - y^n P_n(1-y)\big)$ and decompose the function on the right-hand side in its Taylor series at zero up to the power $n-1$. Since P_n is a polynomial of degree $\le n-1$, it coincides with this expansion. We have $(1-y)^{-n} = \sum_{k=0}^{n-1} \binom{n-1+k}{k} y^k$; we finally have

$$P_n(y) = \sum_{k=0}^{n-1} \binom{n-1+k}{k} y^k. \tag{4.9}$$

It is easy to see that this solution is positive on $[0, 1]$.

Thus the general solution of equation (4.8) has the form

$$P(y) = \sum_{k=0}^{n-1} \binom{n-1+k}{k} y^k + y^n R\left(\frac{1}{2} - y\right), \tag{4.10}$$

where R is an odd polynomial. Hence the mask $m_0(\xi)$ satisfies the conditions of Theorem 4.1.2 if and only if

$$|m_0(\xi)|^2 = \left(\cos^2 \pi\xi\right)^n P\left(\sin^2 \pi\xi\right), \tag{4.11}$$

where the polynomial P is given by formula (4.10) if for R we take an arbitrary odd polynomial such that the polynomial $P(y)$ is nonnegative on the segment $[0,1]$ and the polynomial $P\left(\sin^2 \pi\xi\right)$ has no cycles.

To construct a wavelet function with compact support, we take an arbitrary positive integer n and an appropriate polynomial R in formula (4.10). Thus we obtain a function $|m_0(\xi)|^2$. Strictly speaking, what we have found is not the mask but its squared modulus. To find the mask, we have to take the square root of the latter function, which is possible due to the Riesz lemma (see Appendix A.11, Theorem A.11.3).

REMARK 4.1.5. In practice, the Riesz lemma is realized by using roots of some auxiliary algebraic polynomial. For this reason, to take the square root even of "good" polynomials, some numerical technique is usually used. Therefore in spite of the fact that we have the explicit formula for the squared mask, the coefficients of this mask will be found only approximately (see [**2**, §6.4]).

In general, the polynomial m_0, which exists by the Riesz lemma, is not unique. Indeed, let us represent a trigonometric polynomial h in the form

$$h(\xi) = e^{-2\pi i N\xi} G(e^{2\pi i\xi}),$$

where G is an algebraic polynomial of degree $2N$. If the polynomial G has roots that do not belong to the unit circle, then there exist several (finitely many) essentially different polynomials m (which differ from each other not only by constant factors) such that $|m_0|^2 = h$.

Let

$$B(\cos(2\pi\xi)) := |m_0(\xi)|^2, \quad m_0(0) = 1, \tag{4.12}$$

where $B(t) = \left(\frac{1+t}{2}\right)^N M(t)$, $M(-1) \neq 0$. All solutions of m_0 can be described in the following way. Let $\{t_k\}$ be the roots of the polynomial M. As $m_0(0) = 1$, we have $B(1) = 1$ and

$$M(\cos 2\pi\xi) = \prod_k \frac{\cos 2\pi\xi - t_k}{1 - t_k}.$$

Partition the roots into three groups: $\{t_k\}_{k\in K_1} \subset (-1,1)$; $\{t_k\}_{k\in K_2} \subset \mathbb{R}\backslash[-1,1]$; $\{t_k\}_{k\in K_3} \subset \mathbb{C}\backslash\mathbb{R}$. As the polynomial $B(\cos(2\pi\xi))$ is nonnegative, the roots $\{t_k\}_{k\in K_1}$ have even multiplicity. Hence,

$$\prod_{k\in K_1} \frac{\cos 2\pi\xi - t_k}{1 - t_k} = \prod_{k\in K_1'} \left(\frac{\cos 2\pi\xi - t_k}{1 - t_k}\right)^2$$

for some $K_1' \subset K_1$. The trigonometric polynomial m_0 has the form

$$m_0(\xi) = \left(\frac{1 + e^{-2\pi i\xi}}{2}\right)^N \prod_{k\in K_1'} \frac{\cos 2\pi\xi - t_k}{1 - t_k} \prod_{k\in K_2\cup K_3} S_k^{\varepsilon_k}(\xi), \tag{4.13}$$

where $\varepsilon_k = \pm 1$,

$$S_k^1(\xi) = \frac{e^{-2\pi i\xi} - z_k}{1 - z_k}, \quad S_k^{-1}(\xi) = \frac{1 - z_k e^{-2\pi i\xi}}{1 - z_k}.$$

Here, z_k is a solution of the equation $\frac{1}{2}\left(z_k + \frac{1}{z_k}\right) = t_k$, where $|z_k| > 1$.

Since the polynomial M has real coefficients, its complex roots $\{t_k\}_{k\in K_3}$ are symmetric with respect to the real axis. To a complex conjugate pair of roots $t_k, \overline{t_k}$ there corresponds the quadruple z_k, z_k^{-1}, $\bar{z}_k$, $\bar{z}_k^{-1}$. Formula (4.13) shows that the polynomial m_0 is determined by the choice of two roots from each quadruple ($\varepsilon_k = 1$ means the choice z_k; $\varepsilon_k = -1$ means the choice $1/z_k$). For $k \in K_2$, the root z_k is a real number.

The Riesz lemma is based on the following relations:

$$\begin{aligned}
\frac{\cos(2\pi\xi) - t_k}{1 - t_k} &= \frac{e^{2\pi i\xi} + e^{-2\pi i\xi} - 2t_k}{2 - z_k - 1/z_k} = \frac{e^{2\pi i\xi} + e^{-2\pi i\xi} - z_k - \frac{1}{z_k}}{(1 - z_k)(1 - 1/z_k)} \\
&= \frac{\left(e^{-2\pi i\xi} - z_k\right)}{(1 - z_k)} \frac{\left(1 - e^{2\pi i\xi}\frac{1}{z_k}\right)}{\left(1 - \frac{1}{z_k}\right)} = e^{2\pi i\xi} \frac{\left(e^{-2\pi i\xi} - z_k\right)}{(1 - z_k)} \frac{\left(e^{-2\pi i\xi} - \frac{1}{z_k}\right)}{\left(1 - \frac{1}{z_k}\right)} \\
&= e^{2\pi i\xi} S_k^1(\xi) S_k^{-1}(\xi).
\end{aligned}$$

For the polynomial m_0 to satisfy equation (4.12), we cannot choose z_k and $\bar{z}_k^{-1}$ simultaneously, since

$$\overline{S_k^1(\xi)} = \overline{\left(\frac{e^{-2\pi i\xi} - z_k}{1 - z_k}\right)} = \frac{e^{2\pi i\xi}\left(e^{-2\pi i\xi} - \frac{1}{\bar{z}_k}\right)}{1 - \frac{1}{\bar{z}_k}}.$$

For m_0 to have real coefficients, it is necessary to take both z_k and $\overline{z}_k$ for $k \in K_3$. Thus for real-valued wavelets, a pair of conjugate roots of the quadruple z_k, z_k^{-1}, $\bar{z}_k$, $\bar{z}_k^{-1}$ should be taken either inside the unit circle or outside of it. For complex-valued wavelets, one of the two opposite pairs can be taken. A special computer program has been developed [**194**] to compute various scaling functions and corresponding wavelets for the classical case $R = 0$ and for various sets of roots. That program can be used to find wavelets with various extremal properties: maximal regularity, minimal uncertainty constant, minimal or maximal asymmetry, etc.

Thus equation (4.11) does not define the mask uniquely. From the standpoint of wavelet construction, it is not essential which square root of the polynomial $|m_0|^2$ is taken for a mask. Each of them satisfies the conditions of Theorem 4.1.2 and generates its own wavelet system.

Moreover, for given n, the choice of polynomial R is not unique. The simplest case corresponds to $R \equiv 0$. Here, the polynomial P is determined by formula (4.9), and for the corresponding mask m_n we have

$$|m_n(\xi)|^2 = \left(\cos^2 \pi\xi\right)^n P_n\left(\sin^2 \pi\xi\right),$$

where P_n is defined in (4.9). This mask has no cycles (since the polynomial $P_n(y)$ is strictly positive for all $y \geq 0$, the mask m_n has no roots different from $\xi = \frac{1}{2}$ on the fundamental period). Hence this mask meets the requirements of Theorem 4.1.2 and the scaling function φ_n generates a wavelet function ψ_n by formula (4.7). Such wavelets are called the *Daubechies wavelets*. For any n, there are several corresponding *Daubechies masks* m_n and Daubechies wavelet functions ψ_n (Remark 4.1.5).

The degree of the polynomial m_n is equal to $N = 2n-1$; hence the Daubechies wavelet functions ψ_n have support of length $2n - 1$, which is the least length of support for a wavelet function for given n. The wavelet function ψ_1 is the Haar function. For all other n, the Daubechies wavelets not only belong to $L_2(\mathbb{R})$ but are continuous. This fact follows from the more general theorem:

THEOREM 4.1.6. *For each positive integer n, the smoothness of the wavelet function ψ_n is at least $n/5$.*

The proof is given in Chapter 7 (§7.5). Note that this statement holds regardless of the choice of mask m_n obtained by taking the square root of the polynomial. Daubechies' theorem provides the construction of a wavelet function with compact support of any smoothness. However, the length of support grows linearly with respect to the smoothness. This is inevitable: since the smoothness of a scaling function with support of length N is at most $N - 1$, the smoothness of the wavelet function cannot exceed this number. Moreover, according to Corollary 3.2.13, smoothness of the wavelets corresponding to a given n, including the Daubechies wavelets, does not exceed $n - 1$. Thus for each n, smoothness of the Daubechies wavelets lies between $n/5$ and $n - 1$. So $1/5$ can be changed only by a greater constant. For small n, the smoothness of ψ_n is estimated more precisely; it exceeds $n/5$ considerably. The values of smoothness of the wavelets for $n = 2, 3, 4, 5$ are given in the table in (7.14).

For a different choice of polynomial R, we obtain wavelets with greater lengths of support. Sometimes it is possible to increase smoothness by choosing a particular polynomial (see §4.3).

4.2. Wavelets generated by a compactly supported scaling function

Let us return to Theorem 1.1.14. Suppose that integer shifts of a scaling function $\varphi \in L_2(\mathbb{R})$ with compact support are not orthogonal but only stable (form a Riesz basis). In other words, property **MR5** holds in its usual form. As is known (Theorem 1.1.14), in this case there exists a scaling function $\varphi^\sharp \in V_0$ whose integer shifts form an orthonormal basis of V_0. In the case of wavelets with compact support, we can describe all such functions, in particular, functions with compact support.

PROPOSITION 4.2.1. *Suppose a scaling function $\varphi \in L_2(\mathbb{R})$, $\|\varphi\|_2 = 1$ has compact support and is stable but its integer shifts are not orthogonal. If the function $\varphi^\sharp$ is defined by the formula*

$$\widehat{\varphi^\sharp}(\xi) = \widehat{\varphi(\xi)}\nu(\xi)\,[\Phi(\xi)]^{-\frac{1}{2}}, \tag{4.14}$$

where $\Phi(\xi) = \sum_{k\in\mathbb{Z}} |\widehat{\varphi}(\xi+k)|^2$ and $\nu(\xi)$ is an arbitrary measurable 1-periodic function such that $|\nu(\xi)| = 1$ almost everywhere, then integer shifts of $\varphi^\sharp$ form an orthonormal basis of V_0. Conversely, if integer shifts of a function $\varphi^\sharp$ form an orthonormal basis of V_0, then $\varphi^\sharp$ is defined by formula (4.14) for an appropriate ν, $|\nu(\xi)| = 1$ almost everywhere. This function satisfies the refinement equation with mask

$$m_0^\sharp(\xi) = m_0(\xi)\left[\frac{\Phi(\xi)}{\Phi(2\xi)}\right]^{1/2}\frac{\nu(2\xi)}{\nu(\xi)}.$$

If the mask m_0 has no symmetric roots (i.e., its integer shifts φ are linearly independent over ℓ), then the support of the function $\varphi^\sharp$ is not compact.

Proof. Let us show that $\widehat{\varphi^\sharp}$ belongs to V_0 and generates this space. Recall that the function Φ is a trigonometric polynomial $\Phi(\xi) = \sum_{k=-N}^{N} f(-k)e^{-2\pi i k\xi}$, where $f = \varphi(\cdot)*\varphi(-\cdot)$. Here, we take into account that the function φ has compact support. We assume that $\operatorname{supp}\varphi \subset [0,N]$ and this polynomial does not vanish on the real axis if the function φ is stable. Therefore the function

$$g(\xi) = \nu(\xi)\,[\Phi(\xi)]^{-\frac{1}{2}}$$

is uniformly bounded almost everywhere on the fundamental period. If $\{b_k\} \in \ell_2$ are its Fourier coefficients, then

$$\varphi^\sharp(x) = \sum_{k\in\mathbb{Z}} b_k\varphi(x-k). \tag{4.15}$$

Hence $\varphi^\sharp \in V_0$. On the other hand, the inverse formula, which gives the expression of φ in terms of integer shifts of the function $\varphi^\sharp$, has the same form with the only difference that for the coefficients b_k, we take the Fourier coefficients of the function $g(\xi) = \nu(\xi)\,[\Phi(\xi)]^{\frac{1}{2}}$. Thus the function $\varphi^\sharp$ generates φ; hence it generates the whole space V_0. The function $\varphi^\sharp$ satisfies equality (3.29), so it has orthonormal integer shifts. As integer shifts of φ are independent, the function $\varphi^\sharp$ represented in the form of sum (4.15) can have compact support only if the sequence $\{b_k\}$ is finite (Proposition 3.4.7), i.e., the function $g(\xi)$ is a trigonometric polynomial. This is possible only if $\Phi \equiv C$; otherwise the polynomial $|g(\xi)|\Phi(\xi)$ would be identically equal to 1. $\diamond$

REMARK 4.2.2. To determine the function $\varphi^\sharp$, one can choose various functions $\nu(\xi)$. If we set $\nu \equiv 1$, the coefficients b_k in (4.15) decrease exponentially: $|b_k| \le C_\varepsilon e^{(-\beta+\varepsilon)|k|}$, $\beta > 0$, since the function $[\Phi(\xi)]^{-1/2}$ is holomorphic in some region $|\mathrm{Im}\,\xi| < \beta$. Therefore the function $\varphi^\sharp$ decreases exponentially at infinity though its support is not compact. The same result will be obtained if we take ν such that the function $[\Phi(\xi)]^{1/2}/\nu(\xi)$ is a trigonometric polynomial. Such a choice is possible due to Riesz's lemma (Theorem A.11.3), since $\Phi(\xi)$ is a positive trigonometric polynomial. Thus $\widehat{\varphi}^\sharp(\xi) = \widehat{\varphi}(\xi)/g(\xi)$. The choice of polynomial g is not unique (Remark 4.1.5). Nevertheless, for any g, the function $1/g$ is holomorphic in some region $|\mathrm{Im}\,\xi| < \beta$; hence the coefficients $\{b_k\}$ and the function $\varphi^\sharp$ increase exponentially at infinity.

REMARK 4.2.3. The condition of linear independence for integer shifts of the function φ in Proposition 4.2.1 is essential. It follows from the stability of φ that its Fourier transform $\widehat{\varphi}$ has no real periodic zeros. However, it may have complex periodic zeros (this is equivalent to the fact that the mask $\mathbf{m}_0$ has symmetric roots that do not belong to the unit circle). In this case, for an appropriate choice of ν, the function $\varphi^\sharp$ can have compact support. As an example, consider the scaling function φ_n generating the Daubecies wavelets of order n and suppose that $a \in \mathbb{C}$, $|a| \ne 1$, is a root of the mask $\mathbf{m}_n$. Set $\varphi(x) = \varphi_n(x-1) - a\varphi_n$. This function has compact support and satisfies the refinement equation with mask $\mathbf{m}_0(z) = \frac{z^2-a}{z-a}\mathbf{m}_n(z)$. Since $\Phi(\xi) = |e^{-i\xi} - a|^2$, for an appropriate function ν, we have $\varphi^\sharp = \varphi_n$; i.e., $\varphi^\sharp$ has compact support. However, the case when integer shifts of φ are linearly dependent is reduced to the case when the scaling function has a smaller support and integer shifts are linearly independent (Chapter 6, §6.2). Hence all functions $\varphi^\sharp$ with compact support that can be obtained in this case coincide with those constructed in Theorem 4.1.2. In what follows, we shall assume that the mask $\mathbf{m}_0$ has no symmetric roots and the support of $\varphi^\sharp$ is not compact.

Thus, any stable scaling function generates an MRA; hence it generates a wavelet system. To construct a wavelet function, one must take the function $\varphi^\sharp$ and apply formula (4.7), thus obtaining a wavelet function ψ. The reasoning remains the same. In the above proofs we did not take into account the orthonormality of the system $\{\varphi(\cdot - k)\}$, using only properties of the Riesz basis. The only difference is that the function $\varphi^\sharp$ does not have compact support; hence the corresponding wavelet function ψ does not have it either. Now employing Corollary 3.4.15, we deduce that each L_2-refinable function, whose mask has neither cycles nor symmetric zeros on the unit circle, generates an MRA. In addition, for different refinement equations, sample spaces V_0 are also different. We can say that the number of multiresolution analyses generated by functions with compact support is equal to the number of stable refinable functions. However, we can obtain wavelets with compact support only for a rather restricted class of masks that satisfy the conditions of Theorem 4.1.2.

The condition of stability for a refinable function cannot be omitted. If integer shifts of a function φ are linearly dependent over ℓ_∞, then linear shifts of any function $f \in V_0$ are linearly dependent over ℓ_∞; hence they do not form an orthonormal system.

Thus, from the point of view of wavelets, only stable refinable functions are of interest.

THEOREM 4.2.4. *A function $\varphi \in L_2(\mathbb{R})$, $\|\varphi\|_2 = 1$, with compact support generates an MRA if and only if it satisfies a refinement equation with mask that has neither cycles nor symmetric zeros on the unit circle. The corresponding wavelet function has compact support if and only if* $|m(\xi)|^2 + |m(\xi + \frac{1}{2})|^2 \equiv 1$.

REMARK 4.2.5. Though any scaling function whose integer shifts are nonorthogonal generates a wavelet function supported on the whole real axis, this function still has "good" localization. The function $\varphi^{\#}$ decreases exponentially at infinity; therefore the corresponding wavelet function ψ decreases exponentially as well.

Using results of Chapter 3 (Theorems 3.3.2 and 3.4.16), we obtain an assertion which accounts for the approximation properties of wavelets generated by a scaling function with compact support.

Recall the following notation for a compactly supported function φ:

$$V_\varphi = \left\{ \sum_{k\in\mathbb{Z}} \lambda_k \varphi(x-k), \ \lambda_k \in \mathbb{C} \right\}$$

denotes the space of functions generated by integer shifts of φ. If a function φ generates an MRA $\{V_j\}$, then $V_0 = V_\varphi \cap L_2(\mathbb{R})$. Analogous formulas hold for the spaces V_j for any $j \in \mathbb{Z}$.

THEOREM 4.2.6. *Let there be given an MRA $\{V_j\}$ with a compactly supported scaling function φ (not necessarily orthogonal). Then for any $r \geq 0$, the following conditions are equivalent:*

a) *The space V_φ contains all algebraic polynomials of degree $\leq r$.*

b) *The order of approximation of the spaces V_j in the metrics of C and L_p is at least $r+1$; i.e., for any smooth compactly supported function f, we have* $\operatorname{dist}(f, V_j) \leq C(f)2^{-j(r+1)}$.

c) *The function φ satisfies the Strang-Fix condition of order r.*

d) *The point $\xi = \frac{1}{2}$ is a zero of the polynomial m_0 of order at least $r+1$. In other words, m_0 admits the factorization*

$$m_0(\xi) = \left(\frac{1 + e^{-2\pi i \xi}}{2} \right)^{r+1} q(\xi),$$

where q is a trigonometric polynomial.

REMARK 4.2.7. In Theorems 1.7.7 and 1.7.8 we show that if a wavelet function ψ is smooth (or, more generally, condition (1.100) holds), then the scaling function has the corresponding order of approximation in L_2 (relation (1.105)). In the case of compact support we obtain both sufficient and necessary conditions (Theorem 4.2.6).

REMARK 4.2.8. Condition a) of Theorem 4.2.6 can be formulated in terms of the spaces $\{V_j\}$, without using V_φ, in the following way: each space V_j contains all algebraic polynomials of degree $\leq r$ locally; i.e., for each compact set $K \subset \mathbb{R}$, the restrictions of functions from V_j to this compact set contain the restrictions of all polynomials of order $\leq r$ to the same set.

Thus the factorization of a mask determines the approximation properties of wavelets. This fact is of extreme importance since it shows the rate of convergence of the expansion of an arbitrary smooth function with respect to a wavelet system,

to this function. Consider a sufficiently smooth function f, for example, $f \in C^r$, bounded on $\mathbb{R}$ together with its first r derivatives. We have

$$f = \sum_{j,k \in \mathbb{Z}} c_{jk} \psi_{jk},$$

where $c_{jk} = \langle f, \psi_{jk} \rangle$. The series $\sum\limits_{k \in \mathbb{Z}, i \le j} c_{ik} \psi_{ik}$ is an orthogonal projection of the function f to the space V_j. Therefore,

$$\operatorname{dist}_{L_2}(f, V_j) = \left\| f - \sum_{k \in \mathbb{Z}, i \le j} c_{ik} \psi_{ik} \right\|_2,$$

which, by the Parseval equality, is equal to

$$\left(\sum_{k \in \mathbb{Z}, i \ge j+1} |c_{ik}|^2 \right)^{1/2}.$$

Hence the order of approximation of the spaces V_j in the metric of $L_2(\mathbb{R})$ is less than r if and only if

$$\left\| f - \sum_{k \in \mathbb{Z}, i \le j} c_{ik} \psi_{ik} \right\|_2 \le C(f) 2^{-j(r+1)},$$

or, equivalently, the "tail of the series" c_{jk} has sufficiently small norm in ℓ_2 :

$$\sum_{k \in \mathbb{Z}, i \ge j+1} |c_{ik}|^2 \le C^2(f) 2^{-2j(r+1)}.$$

Thus the order of factorization of the mask, i.e., the order of zero at the point $\frac{1}{2}$, contains all the information about the rate of convergence for expansions of smooth functions with respect to a wavelet system and the rate of decrease of the coefficients for such expansions. The wavelet function itself need not be smooth, only the order of factorization of the mask is important. Nevertheless, for scaling functions, as well as for wavelets, smoothness is *sufficient* for "good" factorization. Using Theorem 3.3.11, we readily obtain the following statement.

COROLLARY 4.2.9. *Suppose a wavelet function is generated by a compactly supported scaling function φ. If $\psi \in W_1^r$ or $\varphi \in W_1^r$, then properties* a) *and* b) *of Theorem* 4.2.6 *are satisfied.*

Let n be the maximal number for which the conditions of Theorem 4.2.6 hold. Note that an expansion with respect to wavelets cannot converge more rapidly than $o(2^{-j(r+1)})$ for all sufficiently smooth functions (Theorem 4.2.6). A higher rate or more rapid decrease of coefficients cannot be achieved even if the expanded function is more smooth. Thus we can say that the algorithm of expansion with respect to wavelets is *"saturable"*.

The aforesaid concerns wavelets with compact support as well. Recall that they can be obtained from scaling functions with orthogonal integer shifts. The corresponding mask is constructed by formula (4.11) for some positive integer n under the assumption that $P(0) \neq 0$.

COROLLARY 4.2.10. *For a compactly supported wavelet function (in particular, for the Daubechies wavelets ψ_n) the order of approximation by the spaces V_j, as well as the order of convergence for smooth functions, is equal to n.*

For compactly supported wavelets, the condition of mask factorization is equivalent to the condition of moments (1.100). Since in this case the mask satisfies the equation $|m_0(\xi)|^2 + |m_0(\xi + \frac{1}{2})|^2 \equiv 1$, the condition

$$m_0\left(\frac{1}{2}\right) = m_0'\left(\frac{1}{2}\right) = \cdots = m_0^{(n)}\left(\frac{1}{2}\right) = 0$$

is equivalent to the relation

$$m_0'(0) = \cdots = m_0^{(n)}(0) = 0.$$

This relation in turn means that (in view of formula (3.5))

$$\widehat{\varphi}'(0) = \cdots = \widehat{\varphi}^{(n)}(0) = 0;$$

in other words,

$$\int_{\mathbb{R}} x\varphi(x)\,dx = \int_{\mathbb{R}} x^n\varphi(x)\,dx = 0. \tag{4.16}$$

The length of support for a wavelet function is greater than or equal to $2n - 1$ (the equality takes place only for Daubechies wavelets). Hence the order of approximation for wavelets is at most $(N+1)/2$. In this sense, the Daubechies wavelets are optimal.

As for the upper estimate for smoothness of wavelets, we can again use the necessary condition of smoothness.

COROLLARY 4.2.11. *Smoothness of wavelet functions cannot exceed the length of support N of the scaling function. Smoothness of compactly supported wavelet functions does not exceed $n \leq (N+1)/2$.*

4.3. Time-frequency localization

In this section, we present results concerning the time-frequency localization of orthonormal wavelets with compact support. In the first subsection, it is proved that the time localization of the classical Daubechies wavelets becomes worse as their smoothness increases. In the second subsection, we give a modification of Daubechies' construction to obtain orthonormal compactly supported wavelets whose localization is independent of their smoothness.

In §4.1, we showed that the equation (4.8) has a unique solution P_n of degree at most $n-1$, and this polynomial is given by formula (4.9). In this section it will be more convenient to use another form of the polynomial P_n:

$$P_n(y) = \sum_{k=0}^{n-1} \binom{2n-1}{k} y^k (1-y)^{n-k-1}.$$

It is an easy exercise for the reader to check that this polynomial is a solution of (4.8). Now it follows from the uniqueness that this polynomial coincides with polynomial (4.9).

4.3.1. Classical Daubechies wavelets. Suppose $\varphi^{D,N}$ is an orthogonal scaling function for the Daubechies wavelet functions $\{\psi_{j,k}^{D,N}\}_{j\in\mathbb{Z},\, k\in\mathbb{Z}}$ and d_N is the corresponding mask; i.e.,

$$\widehat{\varphi^{D,N}}(\xi) := \prod_{l=1}^{\infty} d_N(\xi 2^{-l}),$$

$$|d_N(\xi)|^2 := D_N(\xi) := \left(\cos^2 \pi\xi\right)^N P_N\left(\sin^2 \pi\xi\right), \tag{4.17}$$

where the polynomial P_N is defined in (4.9) (see also p. 151).

To compute the uncertainty constants for the Daubechies wavelet functions, we need to investigate the roots of the polynomials that define the squared modulus of masks (see Appendix A.11). It is much easier to study uncertainty constants of autocorrelation functions of wavelet functions.

Autocorrelation function Φ of a function φ is defined by the equality

$$\Phi(t) := \int_{-\infty}^{\infty} \varphi(s)\overline{\varphi(s-t)}\, ds.$$

PROBLEM 4.3.1. Prove that $\widehat{\Phi}(\xi) = |\widehat{\varphi}(\xi)|^2$.

To prove the results of this section, we need the following assertion.

LEMMA 4.3.2. *For any positive integer N, the functions $D_N(\xi)$ satisfy the inequalities*

$$|\cos^2(\pi\xi)|^N \le D_N(\xi) \le |Q(\xi)|^N, \tag{4.18}$$

where $Q(\xi)$ is a 1-periodic function that is equal to 1 for $|\xi| \le 1/4$ and $\sin^2(2\pi\xi)$ for $1/4 \le |\xi| \le 1/2$. The function $Q(\xi)$ can be written as

$$Q(\xi) = \cos^2(\pi\xi)\widetilde{Q}(\xi),$$

where the function $\widetilde{Q}$ is bounded, satisfies the Lipschitz condition at zero, and satisfies the inequality

$$\sigma_2 := \sup_{\xi} |\widetilde{Q}(\xi)\widetilde{Q}(2\xi)| < 2^4 = 16. \tag{4.19}$$

Proof. The left inequality in (4.18) is obvious since $P_N\left(\sin^2 \pi\xi\right) \ge 1$. The upper estimate for $|\xi| \le 1/4$ follows from (4.1). To prove the upper estimate for $|\xi| \in [1/4, 1/2]$, note that for $t \in [1/2, 1]$, the following inequality holds:

$$\begin{aligned} P_N(t) &= \sum_{k=0}^{N-1} \binom{N-1+k}{k} (2t)^k 2^{-k} \le (2t)^{N-1} \sum_{k=0}^{N-1} \binom{N-1+k}{k} 2^{-k} \\ &= (2t)^{N-1} P_N(1/2) = (4t)^{N-1} \le (4t)^N, \end{aligned}$$

since by (4.8), $P_N(1/2) = 2^{N-1}$. Hence for $1/4 \le |\xi| \le 1/2$,

$$\begin{aligned} D_N(\xi) &= \cos^{2N}(\pi\xi) P_N(\sin^2(\pi\xi)) \le (4\cos^2(\pi\xi)\sin^2(\pi\xi))^N \\ &= (\sin^2(2\pi\xi))^N = (Q(\xi))^N. \end{aligned}$$

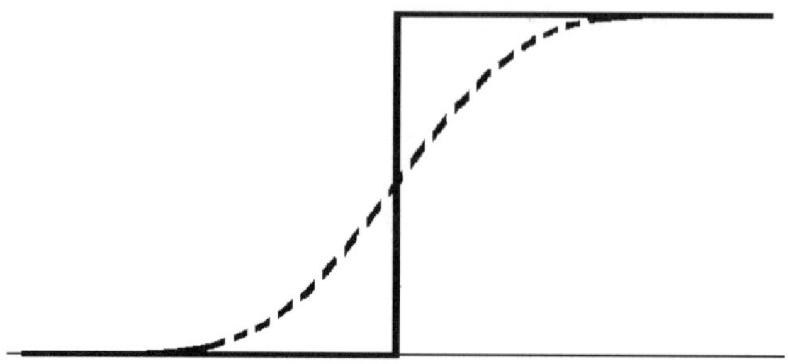

FIGURE 4.1

To prove (4.19), note that $\sup_\xi \widetilde{Q}(\xi) = 4$ and $\widetilde{Q}(\xi) = 4$ if and only if $\xi = (2n + 1),\ n \in \mathbb{N}$. Therefore,

$$\sup_\xi \left(\widetilde{Q}(\xi)\widetilde{Q}(2\xi)\right) < \sup_\xi \left(\widetilde{Q}(\xi)\right)^2 = 16. \qquad \Diamond$$

THEOREM 4.3.3. *Let $\Phi^{D,N}$ be an autocorrelation function of a scaling function $\varphi^{D,N}$. Then*

$$\lim_{N\to\infty} \widehat{\Phi^{D,N}}(\xi) = \chi_{[-1/2,1/2]}(\xi) \quad \textit{for any } \xi \in \mathbb{R}\backslash\{\pm 1/2\}, \tag{4.20}$$

$$\lim_{N\to\infty} \|\widehat{\Phi^{D,N}} - \chi_{[-1/2,1/2]}\|_p = 0, \quad 1 \le p < \infty, \tag{4.21}$$

$$\lim_{N\to\infty} \Delta_{\widehat{\Phi^{D,N}}} = \frac{1}{2\sqrt{3}}, \tag{4.22}$$

$$\lim_{N\to\infty} \Delta_{\Phi^{D,N}} = \infty. \tag{4.23}$$

Thus,

$$\lim_{N\to\infty} \Delta_{\widehat{\Phi^{D,N}}} \Delta_{\Phi^{D,N}} = \infty. \tag{4.24}$$

Proof. We have

$$|d_N(\xi)|^2 = D_N(\xi) = \mathcal{D}_N(\cos 2\pi\xi) \tag{4.25}$$

where

$$\begin{aligned} \mathcal{D}_N(t) &:= \left(\frac{1+t}{2}\right)^N \sum_{l=0}^{N-1} \binom{N-1+l}{l} \left(\frac{1-t}{2}\right)^l \\ &= \sum_{l=N}^{2N-1} \binom{2N-1}{l} \left(\frac{1+t}{2}\right)^l \left(\frac{1-t}{2}\right)^{2N-1-l}. \end{aligned} \tag{4.26}$$

Formula (4.26) shows that $\mathcal{D}_N(t)$ is Bernstein's polynomial on $[-1,1]$ approximating the function $\chi_{[0,1]}$. The dashed curve in Figure 4.1 is the graph of the function $\mathcal{D}_7$. By the properties of Bernstein's polynomials, we have

$$\lim_{N\to\infty} \mathcal{D}_N(t) = \chi_{[0,1]}(t) \quad \text{for any } \xi \in [-1,1]\backslash\{0\}. \tag{4.27}$$

Note that

$$\chi_{[-1/2,1/2]}(\xi) = \prod_{l=1}^{\infty} \chi_{[0,1]}(\cos 2^{-l+1}\pi\xi).$$

Fix ξ and choose $L \in \mathbb{N}$ so that $|\xi| < 2^{L-2}$. Then we have

$$\prod_{l=L+1}^{\infty} \chi_{[0,1]}(\cos \pi\xi 2^{-l+1}) = 1,$$

$$\prod_{l=1}^{L} \chi_{[0,1]}(\cos \pi\xi 2^{-l+1}) = \chi_{[0,1]}(\cos \pi\xi)\chi_{[-1,1]}(\xi),$$

and

$$\begin{aligned}
&\left|\widehat{\Phi^{D,N}}(\xi) - \chi_{[-1/2,1/2]}(\xi)\right| \\
&\quad \le \left|\left(\prod_{l=1}^{L} D_N(\xi 2^{-l}) - \prod_{l=1}^{L} \chi_{[0,1]}(\cos \pi\xi 2^{1-l})\right) \cdot \widehat{\Phi^{D,N}}(\xi 2^{-L})\right| \\
&\qquad + \left|\left(1 - \prod_{l=1}^{\infty} D_N(\xi 2^{-L-l})\right) \cdot \chi_{[0,1]}(\cos \pi\xi)\chi_{[-1,1]}(\xi)\right|.
\end{aligned}$$

Set

$$I_1(\xi) := \left|\left(\prod_{l=1}^{L} D_N(\xi 2^{-l}) - \prod_{l=1}^{L} \chi_{[0,1]}(\cos \pi\xi 2^{1-l})\right) \cdot \widehat{\Phi^{D,N}}(\xi 2^{-L})\right|,$$

$$I_2(\xi) := \left|\left(1 - \prod_{l=1}^{\infty} D_N(\xi 2^{-L-l})\right) \cdot \chi_{[0,1]}(\cos \pi\xi)\chi_{[-1,1]}(\xi)\right|.$$

Obviously, $0 \le \widehat{\Phi^{D,N}}(\xi) \le 1$. So it follows from (4.27) that

$$I_1(\xi) \to 0 \quad \text{as } N \to \infty. \tag{4.28}$$

To estimate I_2, we use the inequalities:

$$1 - e^{-t} \le t \quad \text{for } t \ge 0 \tag{4.29}$$

and

$$-\ln(1-t) \le \frac{t}{1-t} \quad \text{for } 0 < t < 1. \tag{4.30}$$

Then we get

$$I_2(\xi) \le \sum_{l=1}^{\infty} \frac{1 - D_N(\xi 2^{-l-L})}{D_N(\xi 2^{-l-L})}. \tag{4.31}$$

Since $|\xi| < 2^{L-2}$, it follows that $\left|\xi 2^{-L-l}\right| \le \dfrac{1}{4}$ for any $l \in \mathbb{N}$. Therefore,

$$D_N(\xi 2^{-l-L}) \ge D_N(1/4) = \tfrac{1}{2}, \quad N \in \mathbb{N}. \tag{4.32}$$

It follows from (4.1) and (4.18) that for $|\xi| \le 1/4$,

$$0 \le D_N(0) - D_N(\xi) = 1 - D_N(\xi) = D_N(\xi \pm 1/2) \le \sin^{2N}(2\pi\xi).$$

The latter inequality together with (4.31) and (4.32) implies that

$$I_2 \leq 2\sum_{l=1}^{\infty}(1 - D_N(\xi 2^{-l-L})) \leq \sum_{l=1}^{\infty}\left(\pi\xi 2^{2-l-L}\right)^{2N} \to 0 \quad \text{as } N \to \infty. \tag{4.33}$$

Relation (4.20) follows from (4.28) and (4.33).

Let us prove (4.21). By Lebesgue's dominated convergence theorem, it follows from (4.20) that

$$\lim_{N\to\infty} \int_{-1}^{1} \widehat{\Phi^{D,N}}(\xi)\, d\xi = \int_{-1}^{1} \chi_{[-1/2,1/2]}(\xi)\, d\xi = 1. \tag{4.34}$$

Since

$$\int_{\mathbb{R}} \widehat{\Phi^{D,N}}(\xi)\, d\xi = \Phi^{D,N}(0) = \|\varphi^{D,N}\|_2^2 = 1,$$

it follows that

$$\lim_{N\to\infty} \int_{\{|\xi|\geq 1\}} \widehat{\Phi^{D,N}}(\xi)\, d\xi = 0.$$

Since $0 \leq \widehat{\Phi^{D,N}}(\xi) \leq 1$, we have $(\widehat{\Phi^{D,N}}(\xi))^p \leq (\widehat{\Phi^{D,N}}(\xi))$ and

$$\lim_{N\to\infty} \int_{\{|\xi|\geq 1\}} |\widehat{\Phi^{D,N}}(\xi)|^p\, d\xi = 0 \tag{4.35}$$

for any $p \in [1, \infty)$. Finally, we get

$$\begin{aligned}
\|\widehat{\Phi^{D,N}} - \chi_{[-1/2,1/2]}\|_p &\leq \left(\int_{-1}^{1} \left|\widehat{\Phi^{D,N}}(\xi) - \chi_{[-1/2,1/2]}(\xi)\right|^p d\xi\right)^{1/p} \\
&\quad + \left(\int_{\{|\xi|\geq 1\}} \left|\widehat{\Phi^{D,N}}(\xi) - \chi_{[-1/2,1/2]}(\xi)\right|^p d\xi\right)^{1/p} \\
&= \left(\int_{-1}^{1} \left|\widehat{\Phi^{D,N}}(\xi) - \chi_{[-1/2,1/2]}(\xi)\right|^p d\xi\right)^{1/p} \\
&\quad + \left(\int_{\{|\xi|\geq 1\}} \left|\widehat{\Phi^{D,N}}(\xi)\right|^p d\xi\right)^{1/p}.
\end{aligned}$$

By (4.34) and (4.35), the latter relation tends to 0 as $N \to \infty$.

To prove (4.22), remark that by (4.1) we have

$$|\widehat{\Phi^{D,N}}(\xi)| \leq \prod_{l=1}^{2L} |d_N(\xi 2^{-l})|^2. \tag{4.36}$$

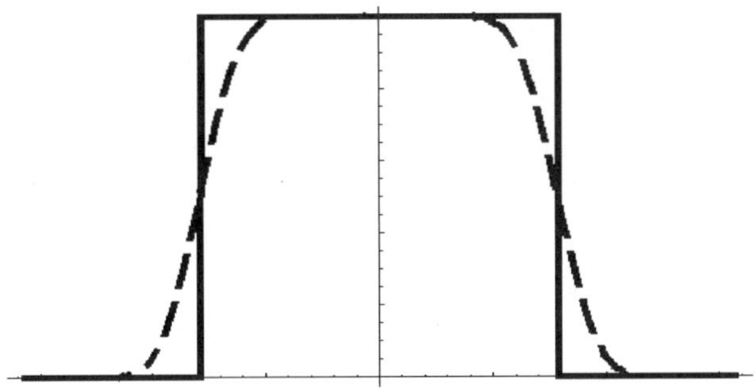

FIGURE 4.2

Taking into account Lemma 4.3.2, we get

$$\begin{aligned}\prod_{l=1}^{2L}|d_N(\xi 2^{-l})|^2 &= \prod_{l=1}^{L} D_N(\xi 2^{-2l})D_N(\xi 2^{1-2l})\\ &\le \prod_{l=1}^{L}(\widetilde{Q}(\xi 2^{-2l})\widetilde{Q}(\xi 2^{1-2l}))^N \left|\cos\pi\xi 2^{1-2l}\cos\pi\xi 2^{-2l}\right|^{2N} \qquad (4.37)\\ &= (\sigma_2)^{L\,N}\left|\frac{\sin\pi\xi}{2^{2L}\sin\pi\xi 2^{-2L}}\right|^{2N} \le \left(\frac{\sigma_2}{16}\right)^{L\,N}|\sin\pi\xi 2^{-2L}|^{-2N}.\end{aligned}$$

It is clear that $|\sin\pi\xi 2^{-2L}| \ge \sin(\pi/8)$ for ξ such that $|\xi| \in [2^{2L-3}, 2^{2L-1}]$, $L \in \mathbb{N}$. Hence from (4.37) we get

$$\prod_{l=1}^{2L}|d_N(\xi 2^{-l})|^2 \le \left(\frac{\sigma_2}{16}\right)^{L\,N}|\sin\pi/8|^{-2N}. \qquad (4.38)$$

This estimate obviously gives

$$\lim_{N\to\infty}\int_{\{|\xi|\ge 1/2\}} \xi^2\,|\widehat{\Phi^{D,N}}(\xi)|^2\,d\xi = 0. \qquad (4.39)$$

Using (4.39), (4.20), and Lebesgue's dominated convergence theorem, we obtain (4.22).

To prove (4.23), note that for $p = 2$, it follows from (4.21) and the Parseval equality that $L_2(\mathbb{R})$ $\Phi^{D,N}$ converges to $\frac{\sin\pi x}{\pi x}$ in $L_2(R)$-norm as $N \to \infty$. Since the center of $\Phi^{D,N}$ is 0, then by Fatou's lemma we obtain

$$\liminf_{N\to\infty}\Delta^2_{\Phi^{D,N}} = \liminf_{N\to\infty}\int_{\mathbb{R}} x^2\left(\Phi^{D,N}(x)\right)^2\,dx \ge \int_{\mathbb{R}} x^2\left(\frac{\sin\pi x}{\pi x}\right)^2\,dx = \infty,$$

which completes the proof of (4.23). $\Diamond$

The Daubechies wavelets can be considered as the compactification of the Kotelnikov-Shannon wavelets (see p. 29). The scaling function of the latter wavelets is an ideal filter for low frequencies $\widehat{\varphi^S}(\xi) := \chi_{[-\frac{1}{2},\frac{1}{2}]}(\xi)$. The dashed curve in Figure 4.2 is the graph of the function $\mathcal{D}_7(\cos(2\pi\xi))$.

Theorem 4.3.3 shows that the Daubechies wavelets lose their time localization as their smoothness increases.

4.3.2. Modified Daubechies wavelets. Let us describe the construction of compactly supported wavelet functions whose scaling functions tend to the Meyer scaling function in $L_2(\mathbb{R})$-norm, as the smoothness increases (see p. 27).

Let $a \in (0,1)$ and let $f_a(t)$ be an infinitely differentiable, nondecreasing function on $[-1,1]$ that is equal to 0 for $t \in [-1,-a]$ and which satisfies the identity

$$f_a(t) + f_a(-t) = 1, \quad t \in [-1,1]. \tag{4.40}$$

Denote by

$$b_l^N(t) := \binom{N}{l}\left(\frac{1+t}{2}\right)^l \left(\frac{1-t}{2}\right)^{N-l}, \quad l = 0,1,\ldots,N,$$

Bernstein's monomials on the segment $[-1,1]$, $t_{N,l} := \frac{2l-N}{N}$.

Consider the trigonometric polynomials with real coefficients $h_N^a(l)$

$$m_N^a(\xi) = 2^{-1/2}\sum_{l=0}^{2N-1} h_N^a(l)e^{i2\pi l\xi}, \quad N \in \mathbb{N}, \tag{4.41}$$

which satisfy the equation

$$|m_N^a(\xi)|^2 = B_{2N-1}^a(\cos 2\pi\xi), \quad m_N^a(0) = 1, \tag{4.42}$$

where

$$B_K^a(t) := \sum_{l=0}^{K} f_a(t_{K,l})b_l^K(t). \tag{4.43}$$

The polynomials B_K^a are Bernstein's polynomials approximating the function f_a. The polynomials m_N^a exist by Theorem A.11.3.

It follows from (4.40) that

$$|m_N^a(\xi)|^2 + |m_N^a(\xi+1/2)|^2 = 1. \tag{4.44}$$

Indeed,

$$\begin{aligned}B_N^a(-\cos 2\pi\xi) &= \sum_{l=0}^{2N-1} f_a(t_{2N-1,l})\binom{2N-1}{l}(\sin\pi\xi)^{2l}(\cos\pi\xi)^{2(2N-1-l)}\\ &= \sum_{l=0}^{2N-1} f_a(t_{2N-1,2N-1-l})\binom{2N-1}{l}(\cos\pi\xi)^{2l}(\sin\pi\xi)^{2(2N-1-l)}.\end{aligned}$$

Therefore,

$$\begin{aligned}&|m_N^a(\xi)|^2 + |m_N^a(\xi+1/2)|^2 = B_{2N-1}^a(\cos 2\pi\xi) + B_{2N-1}^a(-\cos 2\pi\xi)\\ &= \sum_{l=0}^{2N-1}(f_a(t_{2N-1,l}) + f_a(t_{2N-1,2N-1-l}))\\ &\qquad\times\binom{2N-1}{l}(\cos\pi\xi)^{2l}(\sin\pi\xi)^{2(2N-1-l)}\\ &= \left(\sin^2(\pi\xi)+\cos^2(\pi\xi)\right)^{2N-1} = 1.\end{aligned}$$

For convenience, set $h_N^a(l) = 0$ for $l < 0$ or $l > 2N-1$. In terms of coefficients $\{h_N^a(l)\}_{l=0}^N$, identity (4.44) is equivalent to the equalities

$$\sum_{l\in\mathbb{N}} h_N^a(2l) = \sum_{l\in\mathbb{N}} h_N^a(2l+1) = 2^{-1/2}, \tag{4.45}$$

$$\sum_{l\in\mathbb{N}} h_N^a(l-2p)h_N^a(l-2q) = \delta_{p,q}, \quad p,q\in\mathbb{N}. \tag{4.46}$$

By Proposition 2.4.1, the infinite products

$$G_N^a(\xi) = \prod_{l=1}^{\infty} m_N^a(\xi 2^{-l}), \quad N\in\mathbb{N},$$

converge absolutely for any $\xi\in\mathbb{R}$, and the convergence is uniform on any compact set.

Define functions $\varphi^{a,N}$ and $\psi^{a,N}$ by their Fourier transforms:

$$\widehat{\varphi^{a,N}}(\xi) = \prod_{l=1}^{\infty} m_N^a(\xi 2^{-l}), \tag{4.47}$$

$$\widehat{\psi^{a,N}}(\xi) = e^{i\pi\xi}\cdot\overline{m_N^a(\xi/2+1/2)}\cdot\prod_{l=2}^{\infty} m_N^a(\xi 2^{-l}), \quad \xi\in\mathbb{R}.$$

THEOREM 4.3.4. *For any $a\in(0,1)$ and any positive integer N, the functions $\varphi^{a,N}$, $\psi^{a,N}$ belong to $L_2(\mathbb{R})$ and have compact support:*

$$\operatorname{supp}\varphi^{a,N} = [-2N+1,0], \quad \operatorname{supp}\psi^{a,N} = [-N,N-1]. \tag{4.48}$$

They satisfy the relations

$$\varphi^{a,N}(t) = \sqrt{2}\sum_{l=0}^{2N-1} h_N^a(l)\varphi^{a,N}(2t+l), \tag{4.49}$$

$$\psi^{a,N}(t) = \sqrt{2}\sum_{l=0}^{2N-1} h_N^a(l)\varphi^{a,N}(2t+1-l).$$

The wavelet function $\psi^{a,N}$ generates an orthonormal basis $\{\psi_{jk}^{a,N}\}_{j,k\in\mathbb{Z}}$ in $L_2(\mathbb{R})$.

`Proof.` Equality (4.49) follows from Theorem 2.4.4. For the first equality in (4.48), see p. 117. The second equality is readily obtained from (1.42). Note that for any $N\in\mathbb{N}$, the trigonometric polynomials m_N^a have no zeros on $(-\frac{1}{2},\frac{1}{2})$, which follows from (4.42) and the fact that $B_{2N-1}(t) > 0$ for $t\in(-1,1)$. Moreover, they satisfy (4.44). Hence by Corollary 2.5.8, $\varphi^{a,N}$ is an orthonormal scaling function. ◊

The smoothness of the wavelets obtained above increases linearly as the parameter N increases. To prove this fact, we need an auxiliary estimate.

For simplicity, we replace the segment $[-1,1]$ by $[0,1]$ making the change of variable $t\to 2t-1$:

$$\widetilde{b}_l^N(t) := b_l^N(2t-1) := \binom{N}{l} t^l(1-t)^{N-l}, \quad l=0,1,\ldots,N,$$

which are Bernstein's monomials on the segment $[0,1]$;

$$\widetilde{t}_{N,l} := \frac{l}{N}; \quad \widetilde{f}_a(t) = f_a(2t-1);$$

$$\widetilde{B}^a_K(t) := \sum_{l=0}^{K} \widetilde{f}_a(\widetilde{t}_{K,l})\widetilde{b}^K_l(t).$$

LEMMA 4.3.5. *For any positive integers K, k, m, $m \le k < K$, and any $t \in [0,1]$,*

$$t^{-m}\sum_{l=k}^{K} \widetilde{b}^K_l(t) = \sum_{l=k}^{K} \binom{K}{l} t^{l-m}(1-t)^{K-l} \le \frac{\binom{K}{m}}{\binom{k}{m}}. \tag{4.50}$$

Proof.

$$\begin{aligned}\sum_{l=k}^{K} \binom{K}{l} t^{l-m}(1-t)^{K-l} &= \sum_{l=k}^{K} \frac{\binom{K}{l}}{\binom{K-m}{K-l}} \binom{K-m}{K-l} t^{l-m}(1-t)^{K-l}\\ &\le \sup_{l=k,\dots,K} \frac{\binom{K}{l}}{\binom{K-m}{K-l}} = \sup_{l=k,\dots,K} \frac{\binom{K}{m}}{\binom{l}{m}} = \frac{\binom{K}{m}}{\binom{k}{m}}.\end{aligned}$$ ◊

THEOREM 4.3.6. *If $a \in (0,1/2)$, then there exists a constant $\mu > 0$ such that*

$$\varphi^{a,N},\ \psi^{a,N} \in C^{\mu N} \quad \textit{for any } N \in \mathbb{N},$$

where

$$C^\alpha := \{f : \int_{\mathbb{R}} \hat{f}(\xi)(1+|\xi|)^\alpha d\xi < \infty\}, \quad \alpha > 0.$$

Proof. Let

$$l_a(K) := \left[\frac{K(1-a)}{2}\right] + 1,$$

where $[t]$ is the integral part of t. Fix $m < l_a(K)$. Let

$$L^a_N(\xi) := \left(\frac{1+\cos 2\pi\xi}{2}\right)^{-m} B^a_{2N-1}(\cos 2\pi\xi).$$

Using the well-known formula

$$\prod_{l=1}^{\infty} \cos(2^{-l}\xi) = \frac{\sin \xi}{\xi},$$

we get

$$\begin{aligned}\left|\widehat{\varphi^a_N}(\xi)\right|^2 &= \prod_{l=1}^{\infty} B^a_{2N-1}(\cos 2^{1-l}\pi\xi)\\ &= \prod_{l=1}^{\infty}\left(\frac{1+\cos 2^{1-l}\pi\xi}{2}\right)^m \prod_{l=1}^{\infty} L^a_N(2^{-l}\xi) = \left(\frac{\sin \pi\xi}{\pi\xi}\right)^{2m} \prod_{l=1}^{\infty} L^a_N(2^{-l}\xi).\end{aligned} \tag{4.51}$$

By Proposition 2.4.1, the infinite product $\prod_{l=1}^{\infty} L^a_N(2^{-l}\xi)$ converges; hence there exists a constant C such that for $|\xi| \le 1/2$,

$$\prod_{l=1}^{\infty} L^a_N(2^{-l}\xi) \le C.$$

Consider now $|\xi| > 1/2$. Define $l_0 \in \mathbb{N}$ so that

$$2^{-l_0}|\xi| < 1 \le 2^{1-l_0}|\xi|.$$

Then

$$\begin{aligned}\prod_{l=1}^{\infty} L_N^a(2^{-l}\xi) &= \prod_{l=1}^{l_0} L_N^a(2^{-l}\xi) \prod_{l=1}^{\infty} L_N^a(2^{-l_0-l}\xi) \\ &\le C\|L_N^a\|_\infty^{l_0} \le 2C|\xi|^{\log_2(\|L_N^a\|_\infty)}.\end{aligned} \tag{4.52}$$

It follows from (4.51) and (4.52) that

$$\left|\widehat{\varphi_N^a}(\xi)\right|^2 \le 2C|\xi|^{-2m+\log_2(\|L_N^a\|_\infty)}. \tag{4.53}$$

To complete the proof, it remains to estimate $\|L_N^a\|_\infty$. It is clear that $f_a(t_{K,l}) = 0$ for $l < l_a(K)$. Therefore,

$$B_K^a(t) \le \sum_{l=l_a(K)}^{K} b_l^K(t). \tag{4.54}$$

By (4.50),

$$\left(\frac{1+t}{2}\right)^{-m} \sum_{l=l_a(K)}^{K} b_l^K(t) = \left(\frac{1+t}{2}\right)^{-m} \sum_{l=l_a(K)}^{K} \widetilde{b}_l^K\left(\frac{1+t}{2}\right) \le \frac{\binom{K}{m}}{\binom{l_a(K)}{m}}. \tag{4.55}$$

By the Stirling formula, there exists a constant C_1 such that

$$\left(\frac{\binom{K}{m}}{\binom{l_a(K)}{m}}\right)^{1/K} \le C_1 \frac{(\widetilde{a}-\gamma)^{\widetilde{a}-\gamma}}{(1-\gamma)^{1-\gamma}\,\widetilde{a}^{\widetilde{a}}}, \tag{4.56}$$

where $\widetilde{a} := \frac{1-a}{2}$, $\gamma := \frac{m}{K}$.

It is easy to see that for any $\widetilde{a} \in \left(\frac{1}{4}, \frac{1}{2}\right)$, there exists $\gamma \in (0, \widetilde{a})$ such that

$$2\gamma - \log_2\left(\frac{(\widetilde{a}-\gamma)^{\widetilde{a}-\gamma}}{(1-\gamma)^{1-\gamma}\,\widetilde{a}^{\widetilde{a}}}\right) > 0,$$

which means that the statement of Theorem 4.3.6 follows from (4.53). ◊

To prove the further results for modified wavelets, we need to obtain an estimate similar to that presented in Lemma 4.3.2.

LEMMA 4.3.7. *For any positive integers K, l, $l \le K$, the following equality holds:*

$$\sum_{s=l}^{K} \binom{K}{s} (1-t)^{s-l} t^{K-s} = \sum_{s=0}^{K-l} \binom{l-1+s}{s} t^s, \quad t \in \mathbb{R}. \tag{4.57}$$

Proof. Since

$$\widetilde{b}_s^K(1-t) = \widetilde{b}_{K-s}^K(t), \tag{4.58}$$

we have

$$\sum_{s=l}^{K}\binom{K}{s}(1-t)^{s-l}t^{K-s} = (1-t)^{-l}\sum_{s=l}^{K}\widetilde{b}_s^K(1-t)$$
$$= (1-t)^{-l}\sum_{s=0}^{K-l}\widetilde{b}_s^K(t) = \sum_{s=0}^{K-l}\binom{l-1+s}{s}t^s.$$

The last equality in this relation is based on Theorem 4.4.5. ◊

Let $b(a)\in(0,1)$ be a root of the equation

$$2^{a+\frac{2}{K}}(1-b)^{\frac{1-a}{2}}(1+b)^{\frac{1+a}{2}} = 1.$$

Denote by $\xi(a)\in[1/4,1/2]$ the root of the equation

$$\cos 2\pi\xi(a) = -b(a). \tag{4.59}$$

LEMMA 4.3.8. *For $K\ge\frac{1}{a}$, we have*

$$(B_K^a(\cos 2\pi\xi))^{1/K} \le \begin{cases} 1, & |\xi|\le\xi(a), \\ 2^{1+a+\frac{2}{K}}\left|\cos(\pi\xi)\right|^{1-a}\left|\sin(\pi\xi)\right|^{1+a}, & |\xi|\in(\xi(a),1/2]. \end{cases} \tag{4.60}$$

Proof. Using (4.57), we get

$$\begin{aligned} &B_K^a(\cos 2\pi\xi) \\ &\le \cos^{2l_a(K)}(\pi\xi)\sum_{l=l_a(K)}^{K}\binom{K}{l}\left(\frac{1+\cos 2\pi\xi}{2}\right)^{l-l_a(K)}\left(\frac{1-\cos 2\pi\xi}{2}\right)^{K-l} \\ &= \cos^{2l_a(K)}(\pi\xi)\sum_{s=0}^{K-l_a(K)}\binom{l_a(K)-1+s}{s}\sin^{2s}(\pi\xi). \end{aligned} \tag{4.61}$$

It follows from the condition $K\ge\frac{1}{a}$ that $l_a(K)-1\le K-l_a(K)$. Hence (see (4.61)),

$$\begin{aligned} B_K^a(\cos 2\pi\xi) &\le \cos^{4l_a(K)-2K-2}(\pi\xi)\cos^{2(K-l_a(K)+1)}(\pi\xi) \\ &\quad\times\sum_{s=0}^{K-l_a(K)}\binom{K-l_a(K)+s}{s}\sin^{2s}(\pi\xi) \\ &= \cos^{4l_a(K)-2K-2}(\pi\xi)D_{K-l_a(K)+1}(\xi), \end{aligned}$$

where the polynomials $D_K(\xi)$ are defined in (4.17).

By Lemma 4.3.2 for $|\xi|\in[1/4,1/2]$, we have

$$\begin{aligned} B_K^a(\cos 2\pi\xi) &\le \cos^{2(2l_a(K)-K-1)}(\pi\xi)\sin^{2(K-l_a(K)+1)}(2\pi\xi) \\ &= 2^{2(K-l_a(K)+1)}\cos^{2l_a(K)}(\pi\xi)\sin^{2(K-l_a(K)+1)}(\pi\xi) \end{aligned}$$

or

$$(B_K^a(\cos 2\pi\xi))^{1/K} \le 2^{2(1-\frac{l_a(K)}{K}+\frac{1}{K})}\cos^{2\frac{l_a(K)}{K}}(\pi\xi)\sin^{2(1-\frac{l_a(K)}{K}+\frac{1}{K})}(\pi\xi).$$

Note that $\frac{l_a(K)}{K}\ge\frac{1-a}{2}$ and $1-\frac{l_a(K)}{K}+\frac{1}{K}\ge 1-\frac{1-a}{2}=\frac{1+a}{2}$. Therefore,

$$B_K^a(\cos 2\pi\xi)^{1/K} \le 2^{1+a+\frac{2}{K}}|\cos(\pi\xi)|^{1-a}|\sin(\pi\xi)|^{1+a}. \tag{4.62}$$

Recall that (4.62) holds for $|\xi|\in[1/4,1/2]$. Obviously, for any ξ, we have

$$0\le B_K^a(\cos 2\pi\xi)\le 1. \tag{4.63}$$

Hence it is necessary to find out for which ξ the right-hand side of (4.62) is less than 1.

Let $t := -\cos\xi$. Simple computation shows that the expression

$$2^{1+a+\frac{2}{K}}\left(\frac{1-t}{2}\right)^{\frac{1-a}{2}}\left(\frac{1+t}{2}\right)^{\frac{1+a}{2}} = 2^{a+\frac{2}{K}}(1-t)^{\frac{1-a}{2}}(1+t)^{\frac{1+a}{2}}$$

is less than 1 for $t > b(a)$. Thus (4.60) follows from (4.62) and (4.63). $\diamond$

For further estimates, we need the inequality $\xi(a) < 1/3$, which is equivalent to the inequality $b(a) < \frac{1}{2}$. The latter inequality means that a and K must satisfy the condition

$$a < \frac{4(1-\frac{1}{K})\ln 2 - \ln 12}{\ln 12}. \tag{4.64}$$

In the sequel we assume that

$$a < \frac{\ln 4 - \ln 3}{\ln 4 + \ln 3} \approx 0.115772. \tag{4.65}$$

Set

$$R_K(\xi) := 2^{1+a+\frac{2}{K}}\left|\cos(\pi\xi)\right|^{1-a}\left|\sin(\pi\xi)\right|^{1+a}.$$

From (4.60) it follows that

$$(B_K^a(\cos 2\pi\xi)B_K^a(\cos 4\pi\xi))^{\frac{1}{K}} \leq \begin{cases} 1, & |\xi| \in \left[0, \frac{\xi(a)}{2}\right], \\ R_K^a(2\xi), & |\xi| \in \left(\frac{\xi(a)}{2}, \frac{1}{4}\right], \\ R_K^a(1-2\xi), & |\xi| \in \left(\frac{1}{4}, \xi(a)\right], \\ R_K^a(\xi)R_K^a(1-2\xi), & |\xi| \in \left(\xi(a), \frac{1-\xi(a)}{2}\right], \\ R_K^a(\xi), & |\xi| \in \left(\frac{1-\xi(a)}{2}, \frac{1}{2}\right]. \end{cases}$$

Bulky though not complicated computation shows that under condition (4.65) for sufficiently large K, we get

$$(B_K^a(\cos 2\pi\xi)B_K^a(\cos 4\pi\xi))^{1/K} \leq \alpha 2^{2\beta}|\cos\pi\xi|^{\beta}|\cos 2\pi\xi|^{\beta}, \tag{4.66}$$

where $\alpha \in (0,1)$ is a constant depending only on a, $\beta := 1-a$.

REMARK 4.3.9. For α, one can take

$$\alpha := \max\{2^{3a-1}, 3^{\frac{a-1}{2}}, 2^{\frac{5a-1}{2}}, \left(\frac{3}{4}\right)^{\frac{1-a}{2}}, 2^{\frac{5a-3}{2}}(\sqrt{2}+1)^{\frac{1+a}{2}}\}.$$

Let us study the asymptotic behavior of modified wavelets. Define a phase $w_{a,N}(\xi)$ as an infinitely differentiable real-valued function that satisfies the conditions

$$w_{a,N}(0) = 0, \quad m_N^a(\xi) = e^{-iw_{a,N}(\xi)}|m_N^a(\xi)|.$$

Then

$$\widehat{\varphi^{a,N}}(\xi) = e^{-iW_N^a(\xi)}\prod_{l=1}^{\infty}|m_N^a(\xi 2^{-l})|,$$

where $W_N^a(\xi) := \sum_{l=1}^{\infty} w_{a,N}(\xi 2^{-l})$. Consider the Meyer scaling functions (their Fourier transforms differ from those for scaling functions considered on page 27 only by a phase factor):

$$\widehat{\varphi^{a,M,N}}(\xi) = e^{-iW_N^a(\xi)}(f_a(\cos\pi\xi))^{1/2}\chi_{[-1,1]}(\xi), \quad N \in \mathbb{N}.$$

LEMMA 4.3.10. *For $a \in (0, \frac{1}{2})$ and any $\xi \in \mathbb{R}$,*

$$\lim_{N\to\infty} \widehat{\varphi^{a,N}}(\xi) = \widehat{\varphi^{a,M,N}}(\xi).$$

Proof. By the properties of Bernstein's polynomials, we have

$$\lim_{K\to\infty} B_K^a(\cos 2\pi\xi) = f_a(\cos 2\pi\xi), \tag{4.67}$$

so

$$\lim_{K\to\infty} |m_K^a(\xi)| = (f_a(\cos 2\pi\xi))^{1/2}. \tag{4.68}$$

Note that for $a < \frac{1}{2}$,

$$\widehat{\varphi^{a,M,N}}(\xi) = \prod_{l=1}^{\infty} e^{-iw_{a,N}(\xi 2^{-l})} \left(f_a(\cos\pi\xi 2^{-l+1})\right)^{1/2}.$$

Fix ξ and choose $L \in \mathbb{N}$ in such a way that $|\xi| < 2^{L-2}$. Then

$$\prod_{l=L+1}^{\infty} f_a(\cos\pi\xi 2^{-l+1}) = 1,$$

$$\prod_{l=1}^{L} f_a(\cos\pi\xi 2^{-l+1}) = f_a(\cos\pi\xi)\chi_{[-1,1]}(\xi),$$

and

$$\begin{aligned}
&\left|\widehat{\varphi^{a,N}}(\xi) - \widehat{\varphi^{a,M,N}}(\xi)\right| \\
&\quad \le \left|\left(\prod_{l=1}^{L} |m_N^a(\xi 2^{-l})| - \prod_{l=1}^{L} \left(f_a(\cos\pi\xi 2^{1-l})\right)^{1/2}\right) \cdot \left|\widehat{\varphi^{a,N}}(\xi 2^{-L})\right|\right| \\
&\quad\quad + \left|\left(1 - \prod_{l=1}^{\infty} |m_N^a(\xi 2^{-L-l})|\right) \cdot (f_a(\cos(\pi\xi)))^{1/2}\,\chi_{[-1,1]}(\xi)\right|.
\end{aligned}$$

Set

$$I_1(\xi) := \left|\left(\prod_{l=1}^{L} |m_N^a(\xi 2^{-l})| - \prod_{l=1}^{L} \left(f_a(\cos\pi\xi 2^{1-l})\right)^{1/2}\right) \cdot \left|\widehat{\varphi^{a,N}}(\xi 2^{-L-1})\right|\right|,$$

$$I_2(\xi) = \left(1 - \prod_{l=1}^{\infty} |m_N^a(\xi 2^{-L-l})|\right) \cdot (f_a(\cos(\pi\xi)))^{1/2}\,\chi_{[-1,1]}(\xi).$$

Obviously, $0 \le |\widehat{\varphi^{a,N}}(\xi)| \le 1$. Hence it follows from (4.68) that $I_1(\xi) \to 0$ as $N \to \infty$.

To estimate I_2, use estimates (4.29) and (4.30). Then we have

$$I_2(\xi) \le \sum_{l=1}^{\infty} \frac{1 - |m_N^a(\xi 2^{-l-L})|}{|m_N^a(\xi 2^{-l-L})|}. \tag{4.69}$$

Since $|\xi| < 2^{L-2}$, it follows that $\left|\frac{\xi}{2^{L+l}}\right| \le \frac{1}{4}$ for any $l \in \mathbb{N}$. Hence,

$$
\begin{aligned}
|m_N^a(\xi 2^{-l-L})| &= \left(B_{2N-1}^a(\cos(\pi\xi 2^{1-l-L}))\right)^{1/2} \\
&\ge \left(B_{2N-1}^a(0)\right)^{1/2} = \tfrac{1}{\sqrt{2}}, \quad N \in \mathbb{N}.
\end{aligned} \tag{4.70}
$$

Using (4.44) and (4.62), we get

$$
\begin{aligned}
1 - \left|m_N^a(\xi 2^{-l-L})\right| &\le 1 - \left|m_N^a(\xi 2^{-l-L})\right|^2 \\
&\le 2^{(2N-1)(1+a)+2} \sin^{(2N-1)(1-a)}(\pi\xi 2^{-l-L}).
\end{aligned} \tag{4.71}
$$

The latter inequality together with (4.69) and (4.70) imply that $I_2(\xi) \to 0$ as $N \to \infty$. ◊

THEOREM 4.3.11. *Let $a \in (0, \frac{1}{2})$. Then*

$$
\lim_{N\to\infty} \|\varphi^{a,N} - \varphi^{a,M,N}\|_2 = 0. \tag{4.72}
$$

Proof. By Lebesgue's dominated convergence theorem and Lemma 4.3.10 we have

$$
\lim_{N\to\infty} \int_{-1}^{1} \left|\widehat{\varphi^{a,N}}(\xi)\right|^2 d\xi = \int_{-1}^{1} f_a(\cos(\pi\xi))\, d\xi = \int_{0}^{1} (f_a(\cos(\pi\xi)) + f_a(-\cos(\pi\xi)))\, d\xi = 1.
$$

Since

$$
\int_{\mathbb{R}} \left|\widehat{\varphi^{a,N}}(\xi)\right|^2 d\xi = \|\varphi^{a,N}\|_2 = 1,
$$

it follows that

$$
\lim_{N\to\infty} \int_{\{|\xi|\ge 1\}} \left|\widehat{\varphi^{a,N}}(\xi)\right|^2 d\xi = 0. \tag{4.73}
$$

Finally, we get

$$
\begin{aligned}
&\|\widehat{\varphi^{a,N}} - \widehat{\varphi^{a,M,N}}\|_2 = \||\widehat{\varphi^{a,N}}| - |\widehat{\varphi^{a,M,N}}|\|_2 \\
&\le \left(\int_{-1}^{1} \left||\widehat{\varphi^{a,N}}(\xi)| - |\widehat{\varphi^{a,M,N}}(\xi)|\right|^2 d\xi\right)^{1/2} \\
&\quad + \left(\int_{|\xi|\ge 1} \left||\widehat{\varphi^{a,N}}(\xi)| - |\widehat{\varphi^{a,M,N}}(\xi)|\right|^2 d\xi\right)^{1/2} \\
&= \left(\int_{-1}^{1} \left||\widehat{\varphi^{a,N}}(\xi)| - |\widehat{\varphi^{a,M,N}}(\xi)|\right|^2 d\xi\right)^{1/2} + \left(\int_{|\xi|\ge 1} \left|\widehat{\varphi^{a,N}}(\xi)\right|^2 d\xi\right)^{1/2}.
\end{aligned}
$$

By Lemma 4.3.10 and (4.73), the latter expression tends to 0 as $N \to \infty$. ◊

Let us now prove that the uncertainty constants of the modified Daubechies wavelets are uniformly bounded with respect to the parameter N determining their smoothness.

Define the Meyer scaling function (see §1.4) φ^M by using the function f_a:

$$\widehat{\varphi^M}(\xi) := \sqrt{f_a(\cos(\pi\xi))}\chi_{[-1,1]}(\xi). \tag{4.74}$$

Denote the autocorrelation functions for the Meyer scaling functions and modified Daubechies wavelets in the following way:

$$\Phi^M(t) := \int_{\mathbb{R}} \varphi^M(s)\varphi^M(s-t)\,ds,$$

$$\Phi^{a,N}(t) := \int_{\mathbb{R}} \varphi^{a,N}(s)\varphi^{a,N}(s-t)\,ds.$$

THEOREM 4.3.12. *Suppose* $a < \frac{\ln 4 - \ln 3}{\ln 4 + \ln 3}$ *and a function* f_a *satisfies the conditions*

$$(f_a)''(t) \geq 0 \quad \text{for } t \leq 0, \qquad (f_a)''(t) \leq 0 \quad \text{for } t > 0, \tag{4.75}$$

$$(f_a)'(t) \leq C\sqrt{f_a(t)} \quad \text{for some constant } C. \tag{4.76}$$

Then

$$\lim_{N\to\infty} \|\widehat{\Phi}^{a,N} - \widehat{\Phi}^M\|_p = 0, \quad 1 \leq p < \infty, \tag{4.77}$$

$$\lim_{N\to\infty} \Delta_{\widehat{\Phi}^{a,N}} = \Delta_{\widehat{\Phi}^M}, \tag{4.78}$$

$$\lim_{N\to\infty} \Delta_{\Phi^{a,N}} \leq \frac{8\pi^2\sqrt{2}\,C}{(\sqrt{2}-1)\|\Phi^M\|_2}; \tag{4.79}$$

consequently,

$$\lim_{N\to\infty} \Delta_{\widehat{\Phi}^{a,N}}\Delta_{\Phi^{a,N}} \leq \frac{8\pi^2\sqrt{2}\,C\Delta_{\widehat{\Phi}^M}}{(\sqrt{2}-1)\|\Phi^M\|_2} < \infty. \tag{4.80}$$

REMARK 4.3.13. By (4.40), the second condition

$$(f_a)''(t) \leq 0 \quad \text{for } t > 0$$

in (4.75) follows from the first one and vice versa.

The proof of Theorem 4.3.12 is divided into 6 lemmas.

LEMMA 4.3.14. *Under the assumptions of Theorem* 4.3.12 *we have*

$$\lim_{N\to\infty} \widehat{\Phi}^{a,N}(\xi) = f_a(\cos(\pi\xi))\chi_{[-1,1]}(\xi).$$

`Proof.` Since

$$\widehat{\Phi}^{a,N}(\xi) = |\widehat{\varphi}^{a,N}(\xi)|^2 = \prod_{l=1}^{\infty} B^a_{2N-1}(\cos(\pi\xi 2^{1-l}))$$

(see (4.42) and (4.47)), the statement follows from Lemma 4.3.10. ◊

LEMMA 4.3.15. *For any* $p \in [1,\infty)$,

$$\lim_{N\to\infty} \|\widehat{\Phi}^{a,N} - \widehat{\Phi}^M\|_p = 0.$$

Proof. By Lebesgue's dominated convergence theorem, it follows from Lemma 4.3.14 that

$$\lim_{N\to\infty}\int_{-1}^{1}\widehat{\Phi}^{a,N}(\xi)\,d\xi=\int_{-1}^{1}f_a(\cos(\pi\xi))\,d\xi=\int_{0}^{1}(f_a(\cos(\pi\xi))+f_a(-\cos(\pi\xi)))\,d\xi=1.$$

Since

$$\int_{\mathbb{R}}\widehat{\Phi}^{a,N}(\xi)\,d\xi=\Phi^{a,N}(0)=\|\varphi^{a,N}\|_2^2=1,$$

it follows that

$$\lim_{N\to\infty}\int_{\{|\xi|\geq 1\}}\widehat{\Phi}^{a,N}(\xi)\,d\xi=0.$$

As $0\leq\widehat{\Phi}^{a,N}(\xi)\leq 1$, we have $(\widehat{\Phi}^{a,N}(\xi))^p\leq\widehat{\Phi}^{a,N}(\xi)$ and

$$\lim_{N\to\infty}\int_{\{|\xi|\geq 1\}}|\widehat{\Phi}^{a,N}(\xi)|^p\,d\xi=0 \tag{4.81}$$

for any $p\in[1,\infty)$. Finally, we get

$$\begin{aligned}\|\widehat{\Phi^{a,N}}-\widehat{\Phi^M}\|_p&\leq\left(\int_{-1}^{1}\left|\widehat{\Phi^{a,N}}(\xi)-\widehat{\Phi^M}(\xi)\right|^p d\xi\right)^{1/p}\\&\quad+\left(\int_{\{|\xi|\geq 1\}}\left|\widehat{\Phi^{a,N}}(\xi)-\widehat{\Phi^M}(\xi)\right|^p d\xi\right)^{1/p}\\&=\left(\int_{-1}^{1}\left|\widehat{\Phi^{a,N}}(\xi)-\widehat{\Phi^M}(\xi)\right|^p d\xi\right)^{1/p}+\left(\int_{\{|\xi|\geq 1\}}\left|\widehat{\Phi^{a,N}}(\xi)\right|^p d\xi\right)^{1/p}.\end{aligned}$$

By Lemma 4.3.14 and (4.81), the latter expression tends to 0 as $N\to\infty$. ◊

LEMMA 4.3.16.

$$\lim_{N\to\infty}\Delta_{\widehat{\Phi}^{a,N}}=\Delta_{\widehat{\Phi}^M}. \tag{4.82}$$

Proof. By (4.44), we have

$$|\Phi^{a,N}(\xi)|\leq\prod_{l=1}^{2L}|m_N^a(\xi 2^{-l})|^2. \tag{4.83}$$

Taking into account (4.66), we get

$$\begin{aligned}\prod_{l=1}^{2L}|m_N^a(\xi 2^{-l})|^2 &= \prod_{l=1}^{L} B_{2N-1}(\cos\pi\xi 2^{-2l+2})B_{2N-1}(\cos\pi\xi 2^{1-2l})\\ &\le \prod_{l=1}^{L}(\alpha 2^{2\beta})^{2N-1}\left|\cos\pi\xi 2^{1-2l}\cos\pi\xi 2^{-2l}\right|^{\beta(2N-1)}\\ &= (\alpha 2^{2\beta})^{L(2N-1)}\left|\frac{\sin\pi\xi}{2^{2L}\sin\pi\xi 2^{-2L}}\right|^{\beta(2N-1)}\\ &\le \alpha^{L(2N-1)}|\sin\pi\xi 2^{-2L}|^{-\beta(2N-1)}.\end{aligned}\tag{4.84}$$

It is clear that $|\sin\pi\xi 2^{-2L}|\ge\sin(\pi/8)$ for ξ such that $|\xi|\in[2^{2L-3},2^{2L-1}]$, $L\in\mathbb{N}$. Hence continuing inequality (4.84), we obtain

$$\prod_{l=1}^{2L}|m_N^a(\xi 2^{-l})|^2\le\alpha^{L(2N-1)}(\sin(\pi/8))^{-\beta(2N-1)}=(\alpha^L(\sin(\pi/8))^{-\beta})^{2N-1}.\tag{4.85}$$

This estimate implies that there exists $L_0\in\mathbb{N}$ such that

$$\lim_{N\to\infty}\int\limits_{\{|\xi|\ge 2^{L_0}\}}\xi^2\,|\widehat{\Phi}^{a,N}(\xi)|^2\,d\xi=\lim_{N\to\infty}\int\limits_{\{|\xi|\ge 2^{L_0}\}}|\widehat{\Phi}^{a,N}(\xi)|^2\,d\xi=0.\tag{4.86}$$

Using (4.86), Lemma 4.3.14, and Lebesgue's dominated convergence theorem, we get (4.82). $\Diamond$

To prove (4.79), we use the following lemma, which is proved similarly to Lemma 1.5.9.

LEMMA 4.3.17. *Let*

$$\widehat{\Phi}(\xi):=\prod_{l=1}^{\infty}f(\cos(\pi\xi 2^{1-l})),$$

where $f(t)$ is a nonnegative differentiable function such that the infinite product admits termwise differentiation in the sense of almost evrywhere convergence. Suppose that

$$\sum_{m\in\mathbb{Z}}\widehat{\Phi}(\xi+m)=1.\tag{4.87}$$

Then

$$\left(\int\limits_{\mathbb{R}}|t\Phi(t)|^2\,dt\right)^{1/2}\le\frac{4\pi^2}{\sqrt{2}-1}\|\sin(2\pi\xi)f'(\cos\pi\xi)\|_{L_2([-\frac{1}{2},\frac{1}{2}])}.\tag{4.88}$$

Proof. First of all let us note that

$$\int\limits_{\mathbb{R}}|t\Phi(t)|^2\,dt=4\pi^2\int\limits_{\mathbb{R}}|\left(\widehat{\Phi}\right)'(\xi)|^2\,d\xi$$

$$=4\pi^2\int\limits_{\mathbb{R}}\left(\sum_{m=1}^{\infty}\pi 2^{1-m}|\sin(\pi\xi 2^{1-m})f'(\cos(\pi\xi 2^{1-m}))|\prod_{l\in\mathbb{N}:l\ne m}|f(\cos(\pi\xi 2^{1-l}))|\right)^2\,d\xi.$$

It follows from (4.87) (see the proof of (1.27)) that $f(t)+f(-t)=1$ for $t\in[-1,1]$ and as f is nonnegative, $0\le f(t)\le 1$. Hence using the triangle inequality in $L_2(\mathbb{R})$, we obtain

$$\left(\int_{\mathbb{R}}|t\Phi(t)|^2\,dt\right)^{1/2}\le 2\pi^2\sum_{m=1}^{\infty}\Bigg(\int_{\mathbb{R}}2^{2-2m}|\sin(\pi\xi 2^{1-m})f'(\cos(\pi\xi 2^{1-m}))|^2 \times\prod_{l\in\mathbb{N}:l\neq m}|f(\cos(\xi 2^{-l}))|^2\,d\xi\Bigg)^{1/2}$$

$$\le 2\pi^2\sum_{m=1}^{\infty}\Bigg(\int_{\mathbb{R}}2^{2-2m}|\sin(\pi\xi 2^{1-m})f'(\cos(\pi\xi 2^{1-m}))|^2 \times\prod_{l=m+1}^{\infty}|f(\pi\xi 2^{1-l})|^2\,d\xi\Bigg)^{1/2}$$

$$\le 2\pi^2\sum_{m=1}^{\infty}\left(\int_{\mathbb{R}}2^{2-m}|\sin(\pi\xi 2^{1-m})f'(\cos(\pi\xi 2^{1-m}))\widehat{\Phi}(\xi 2^{-m})|^2 d(\xi 2^{-m})\right)^{1/2}$$

$$=2\pi^2\sum_{m=1}^{\infty}2^{1-m/2}\left(\int_{\mathbb{R}}\left|\sin(2\pi\xi)f'(\cos 2\pi\xi)\widehat{\Phi}(\xi)\right|^2\,d\xi\right)^{1/2}$$

$$\le\frac{4\pi^2}{\sqrt{2}-1}\left(\int_{\mathbb{R}}\left|\sin(2\pi\xi)f'(\cos 2\pi\xi)\widehat{\Phi}(\xi)\right|^2\,d\xi\right)^{1/2}.$$

Using the periodicity of the first cofactor in the latter integral and identity (4.87), we get (4.88). ◊

LEMMA 4.3.18.

$$|(B_N^a)'(\cos 2\pi\xi)\sin(2\pi\xi)|\le 2\sqrt{2}C. \tag{4.89}$$

Proof. Let us use the formula for the derivative of Bernstein's polynomials:

$$(B_N^a(t))'=\sum_{l=0}^{N-1}\frac{N}{2}\partial(f_a(t_{Nl}))b_l^{N-1}(t), \tag{4.90}$$

where $\partial(f_a(t_{Nl})):=f_a(t_{N,l+1})-f_a(t_{Nl})$. Taking into account the properties of the function f_a (see p. 162), we obtain

$$(B_N^a(t))'=\sum_{l=l_N(a)-1}^{N-l_N(a)}\frac{N}{2}\partial(f_a(t_{Nl}))b_l^{N-1}(t), \tag{4.91}$$

where $l_N(a):=[\frac{N(1-a)}{2}]+1$. Here, $[t]$ is the integral part of t. By (4.75), we have

$$\partial(f_a(t_{Nl}))\le\tfrac{2}{N}(f_a)'(t_{N,l+1})\quad\text{for } l<\tfrac{N}{2}, \tag{4.92}$$

$$\partial(f_a(t_{Nl}))\le\tfrac{2}{N}(f_a)'(t_{N,l})\quad\text{for } l\ge\tfrac{N}{2}. \tag{4.93}$$

Simple calculation gives

$$b_l^{N-1}(\cos 2\pi\xi)\sin^2(2\pi\xi) \le 4b_{l+1}^N(\cos 2\pi\xi),$$
$$b_l^{N-1}(\cos 2\pi\xi)\sin^2(2\pi\xi) \le 4b_l^N(\cos 2\pi\xi).$$

Hence inequalities (4.92)–(4.93), condition (4.76), and the Cauchy inequality imply that

(4.94)

$$\begin{aligned}
&((B_N^a)'(\cos 2\pi\xi))^2\sin^2 2\pi\xi\\
&\le C^2\left(\sum_{l=l_N(a)-1}^{[\frac{N}{2}]-1}\sqrt{f_a(t_{N,l+1})b_l^{N-1}(\cos 2\pi\xi)}\right.\\
&\qquad\left.+\sum_{l=[\frac{N}{2}]}^{N-l_N(a)}\sqrt{f_a(t_{N,l})b_l^{N-1}(\cos 2\pi\xi)}\right)^2\sin^2 2\pi\xi\\
&\le C^2\left(\sum_{l=l_N(a)-1}^{[\frac{N}{2}]-1}f_a(t_{N,l+1})b_l^{N-1}(\cos 2\pi\xi)\right.\\
&\qquad\left.+\sum_{l=[\frac{N}{2}]}^{N-l_N(a)}f_a(t_{N,l})b_l^{N-1}(\cos 2\pi\xi)\right)\sin^2 2\pi\xi\sum_{l=l_N(a)-1}^{N-l_N(a)}b_l^{N-1}(\cos 2\pi\xi)\\
&\le 4C^2\left(\sum_{l=l_N(a)-1}^{[\frac{N}{2}]-1}f_a(t_{N,l+1})b_{l+1}^N(\cos 2\pi\xi)+\sum_{l=[\frac{N}{2}]}^{N-l_N(a)}f_a(t_{N,l})b_l^N(\cos 2\pi\xi)\right)\\
&\le 8C^2\sum_{l=l_N(a)-1}^{N-l_N(a)}f_a(t_{N,l})b_l^N(\cos 2\pi\xi)=8C^2B_N^a(\cos 2\pi\xi).
\end{aligned}$$

As $0\le B_N^a(\cos 2\pi\xi)\le 1$, (4.94) implies (4.89). ◊

LEMMA 4.3.19.

$$\left(\int_{\mathbb{R}}|t\Phi^{a,N}(t)|^2\,dt\right)^{1/2}\le\frac{8\pi^2\sqrt{2}C}{\sqrt{2}-1}. \tag{4.95}$$

Proof. Inequality (4.95) follows from (4.88) and (4.89). ◊

The simplest example of modified wavelets can be obtained if for f_a we take the piecewise linear function

$$f(t):=\begin{cases}0, & t\in[-1,-\frac{1}{2}],\\ t+\frac{1}{2}, & t\in(-\frac{1}{2},\frac{1}{2}],\\ 1, & t\in(\frac{1}{2},1].\end{cases}$$

Let $tfu(m):=\Delta_{ti,m}\Delta_{fr,m}$ be the uncertainty constant for the mask m. Denote by s_L, $L\in\mathbb{N}$, the modified masks with the minimal uncertainty constant (it is convenient to take the mask of degree $4L-1$); $d_K^{\min}$ are the Daubechies masks with the minimal uncertainty constant (whose degree is $2K-1$). Here, the minimization is performed over masks with the same modulus (see Remark 4.1.5). The following table gives the uncertainty constants of these masks for equal degrees:

L	$tfu(s_L)$	K	$tfu\left(d_K^{\min}\right)$
1	0.592	2	0.592
2	0.657	4	0.635
3	0.620	6	0.694
4	0.659	8	0.748
5	0.643	10	0.765
6	0.640	12	0.800
7	0.661	14	0.830
8	0.671	16	0.853
9	0.671	18	0.873
10	0.679	20	0.895
11	0.680	22	0.915

4.4. Asymptotics of zeros of Bernstein's polynomials

In §4.3.2, we constructed the modified Daubechies wavelets and proved that they preserve localization as their smoothness increases. To study these wavelets and to find the coefficients of a refinement equation, we need to find complex zeros of special polynomials (see Remark 4.1.5).

In this section, we study the asymptotics of zeros of the following Bernstein polynomials:

(4.96)

$$B_L(x) = \sum_{l=0}^{L-1} \binom{4L}{l} x^l (1-x)^{4L-l} + \sum_{l=L}^{3L-1} \left(\frac{3}{2} - \frac{l}{2L}\right) \binom{4L}{l} x^l (1-x)^{4L-l}, \quad L \in \mathbb{N},$$

which approximate the piecewise linear function

$$f(x) := \begin{cases} 1, & x \in [0, \frac{1}{4}], \\ \frac{3}{2} - 2x, & x \in (\frac{1}{4}, \frac{3}{4}], \\ 0, & x \in (\frac{3}{4}, 1], \end{cases}$$

on the segment $[0, 1]$.

Figure 4.3 shows the function f and Bernstein's polynomial B_7.

To simplify the notation, the segment $[-1, 1]$ corresponding to the definition of masks is replaced by $[0, 1]$.

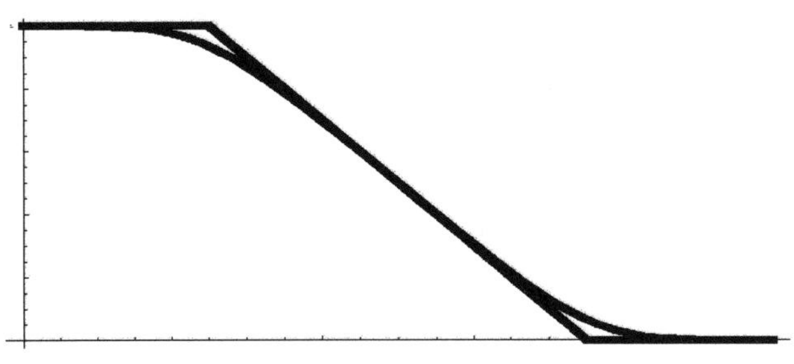

FIGURE 4.3

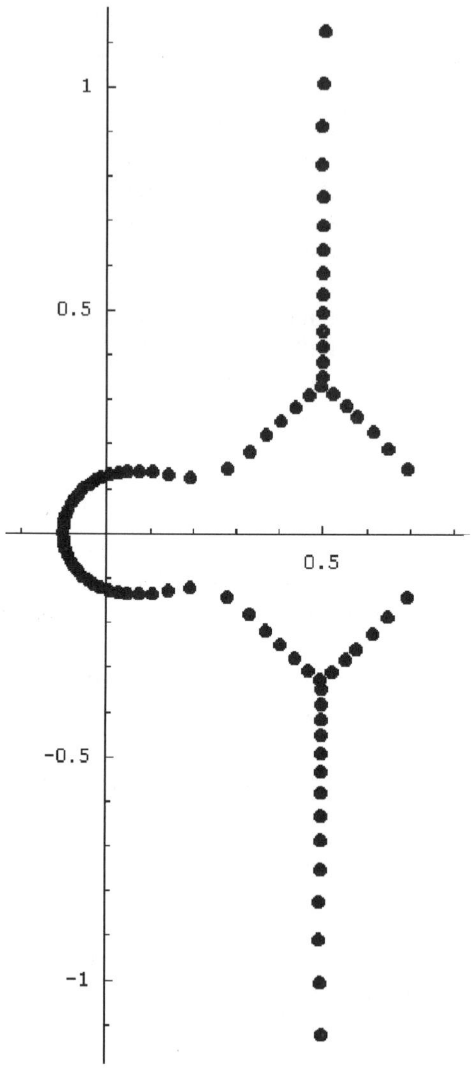

FIGURE 4.4. Zeros of B_{37}.

Using the methods described in §4.3.2, one can prove that the modified Daubechies wavelets constructed using the function f instead of f_a preserve the localization as their smoothness increases. Figure 4.4 shows the roots of B_{37}. The aim of §4.4 is to find a curve where the roots are located and to study their behavior as $L \to \infty$. It appears that the limit curve consists of parts of two lemniscates.

The first lemniscate is defined by the equation

$$\mathcal{L}_1 := \left\{ x : \left| \frac{x}{1/4} \right|^{1/4} \left| \frac{1-x}{3/4} \right|^{3/4} = 1 \right\}$$

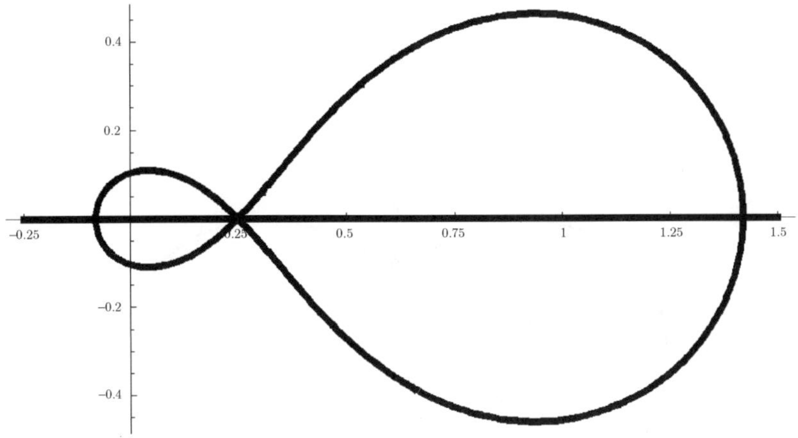

FIGURE 4.5. Lemniscate $\mathcal{L}_1$.

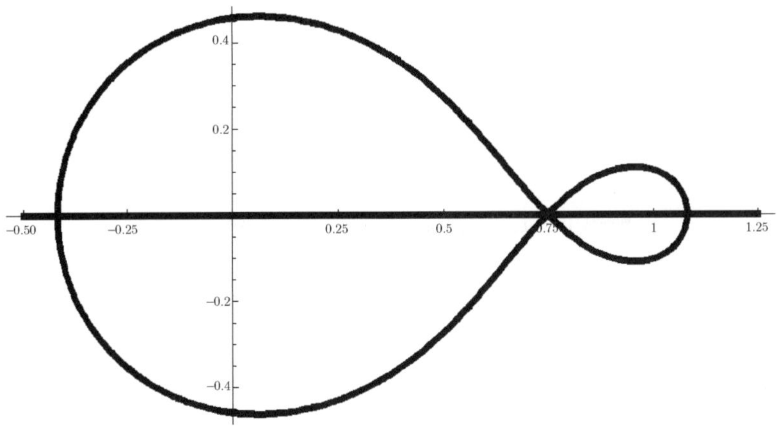

FIGURE 4.6. Lemniscate $\mathcal{L}_2$.

(Figure 4.5); the second one is defined by the equation

$$\mathcal{L}_2 := \left\{ x : \left| \frac{x}{3/4} \right|^{3/4} \left| \frac{1-x}{1/4} \right|^{1/4} = 1 \right\}$$

(Figure 4.6).

Figure 4.7 shows the zeros of B_{37} together with $\mathcal{L}_1$ and $\mathcal{L}_2$. The small circle in Figure 4.7 has center at $\frac{36}{35}$ and radius $\frac{6}{35}$. It is proved that the polynomials B_L have no roots different from 1 inside this circle. In §4.4, we prove some exact results concerning the roots of B_L contained in the left-hand loop of the lemniscate $\mathcal{L}_2$ and having the real part less than 1/4. Similar results are obtained concerning other domains inside the loops of the lemniscates $\mathcal{L}_1$ and $\mathcal{L}_2$. Numerical experiments show that the vertical straight line $\operatorname{Re} x = 1/2$ is the limit curve for the zeros of B_L located outside the lemniscates.

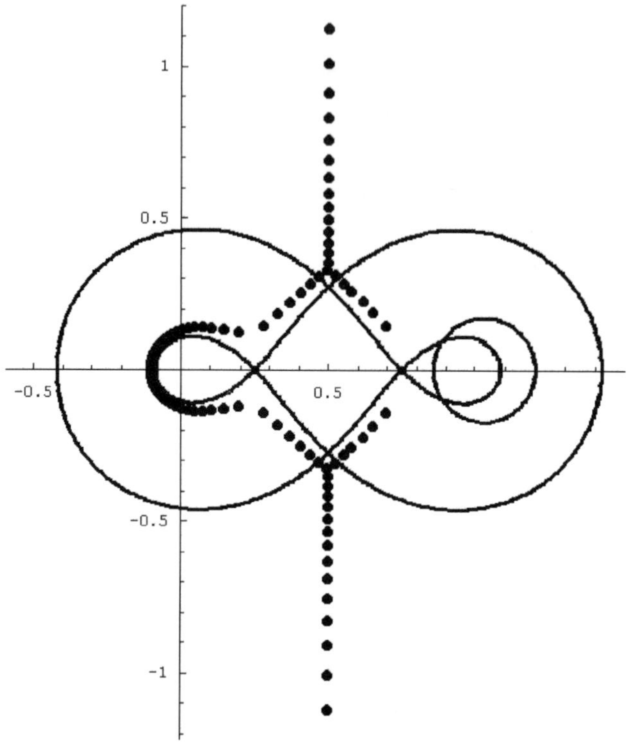

FIGURE 4.7. Roots of B_{37} and lemniscates.

THEOREM 4.4.1. *All the roots of the polynomials B_L, $L \in \mathbb{N}$, different from* 1, *lie outside the circle* $\left|x - \frac{36}{35}\right| = \frac{6}{35}$.

Proof. It is clear that for $x \neq 1$,

$$B_L(x) = (1-x)^{4L}\left[\sum_{l=0}^{L-1}\binom{4L}{l}\left(\frac{x}{1-x}\right)^l + \sum_{l=L}^{3L-1}\left(\frac{3}{2}-\frac{l}{2L}\right)\binom{4L}{l}\left(\frac{x}{1-x}\right)^l\right].$$

Denote $z := \frac{x}{1-x}$ and apply Theorem A.12.1 to the polynomial

$$\sum_{l=0}^{L-1}\binom{4L}{l}z^l + \sum_{l=L}^{3L-1}\left(\frac{3}{2}-\frac{l}{2L}\right)\binom{4L}{l}z^l.$$

According to this theorem, the zeros of the polynomial are located in the ring:

$$\min\left\{\min_{0\le l\le L-1}\frac{\binom{4L}{l}}{\binom{4L}{l+1}}, \min_{L\le l\le 3L-2}\frac{\left(\frac{3}{2}-\frac{l}{2L}\right)\binom{4L}{l}}{\left(\frac{3}{2}-\frac{l}{2L}\right)\binom{4L}{l+1}}\right\} \le |z|$$

$$\le \max\left\{\max_{0\le l\le L-1}\frac{\binom{4L}{l}}{\binom{4L}{l+1}}, \max_{L\le l\le 3L-2}\frac{\left(\frac{3}{2}-\frac{l}{2L}\right)\binom{4L}{l}}{\left(\frac{3}{2}-\frac{l}{2L}\right)\binom{4L}{l+1}}\right\}.$$

Direct calculation shows that

$$\frac{1}{4L} \le |z| \le 6 - \frac{14}{L+2} < 6.$$

In terms of the variable x we have

$$\left\{x : \left|\frac{x}{1-x}\right| < 6\right\} = \left\{x : \left|x - \frac{36}{35}\right| > \frac{6}{35}\right\}. \qquad \Diamond$$

Let α_1 and β_1 be domains inside the left and right loops of the lemniscate $\mathcal{L}_1$; let α_2 and β_2 be the same for the lemniscate $\mathcal{L}_2$:

$$\begin{aligned} \alpha_1 &:= \left\{x : \left|\tfrac{x}{1/4}\right|^{1/4} \left|\tfrac{1-x}{3/4}\right|^{3/4} < 1;\ \operatorname{Re} x < 1/4\right\}, \\ \beta_1 &:= \left\{x : \left|\tfrac{x}{1/4}\right|^{1/4} \left|\tfrac{1-x}{3/4}\right|^{3/4} < 1;\ \operatorname{Re} x > 1/4\right\}, \end{aligned} \tag{4.97}$$

$$\begin{aligned} \alpha_2 &:= \left\{x : \left|\tfrac{x}{3/4}\right|^{3/4} \left|\tfrac{1-x}{1/4}\right|^{1/4} < 1;\ \operatorname{Re} x < 3/4\right\}, \\ \beta_2 &:= \left\{x : \left|\tfrac{x}{3/4}\right|^{3/4} \left|\tfrac{1-x}{1/4}\right|^{1/4} < 1;\ \operatorname{Re} x > 3/4\right\}. \end{aligned} \tag{4.98}$$

THEOREM 4.4.2. *The zeros of the polynomials $B_L(x)$, $L \in \mathbb{N}$, located in the set $\{x : |x| \le 1/4\}$ satisfy the inequality*

$$\left|\frac{x}{1/4}\right|^{1/4} \left|\frac{1-x}{3/4}\right|^{3/4} > a(L), \tag{4.99}$$

where

$$a(L) = \left(\frac{1 - \left(\frac{5}{27}\right)^L \frac{9}{128L}}{\frac{\sqrt{3}}{4\sqrt{2\pi L}}\Phi(L)}\right)^{3/(12L+4)},$$

$$\Phi(L) := \frac{\Gamma^*(4L)}{\Gamma^*(L)\Gamma^*(3L)}, \qquad \Gamma^*(r) := \sqrt{\frac{r}{2\pi}} e^r r^{-r} \Gamma(r),$$

Γ is the gamma function.

Proof. Let

$$\begin{aligned} b_{N,l}(x) &:= \binom{N}{l} x^l (1-x)^{N-l}, \quad l = 0, 1, \ldots, N, \\ B_{L,1}(x) &:= \sum_{l=0}^{3L-1} \left(\frac{3}{2} - \frac{l}{2L}\right) b_{4L,l}(x), \\ R_{L,1}(x) &:= \sum_{l=3L+1}^{4L} \left(\frac{l}{2L} - \frac{3}{2}\right) b_{4L,l}(x), \\ B_{L,2}(x) &:= \sum_{l=0}^{L-1} \left(\frac{1}{2} - \frac{l}{2L}\right) b_{4L,l}(x). \end{aligned} \tag{4.100}$$

It is known that

$$\sum_{l=0}^{N} b_{N,l}(x) = 1, \quad \sum_{l=0}^{N} \frac{l}{N} b_{N,l}(x) = x.$$

Therefore,

$$B_L(x) = B_{L,1}(x) - B_{L,2}(x) = \frac{3}{2} - 2x + R_{L,1}(x) - B_{L,2}(x). \tag{4.101}$$

Applying formulas (4.113) and (4.116) to $B_{L,2}$, we obtain

$$B_{L,2}(x) = \frac{1}{2} - 2x + (6L)\binom{4L-1}{L-1} I_{4L+1,L-1}(x)\, x^{L+1}(1-x)^{3L+1}, \tag{4.102}$$

where

$$I_{N,k}(x) := \int_0^1 s(1-xs)^{-N}(1-s)^k ds.$$

(4.101) and (4.102) imply

$$B_L(x) = 1 + R_{L,1}(x) - (6L)\binom{4L-1}{L-1} I_{4L+1,L-1}(x)\, x^{L+1}(1-x)^{3L+1}. \tag{4.103}$$

It is obvious that for $|x| \le 1/4$ and $s \in [0,1]$,

$$\left|(1-xs)^{-1}\right| \le \left(1 - \frac{s}{4}\right)^{-1}. \tag{4.104}$$

If $B_L(x) = 0$ and $|x| \le 1/4$, then in view of (4.103) and (4.104),

$$\left|(1 + R_{L,1}(x))\, x^{-L-1}(1-x)^{-3L-1}\right| < (6L)\binom{4L-1}{L-1} I_{4L+1,L-1}(1/4). \tag{4.105}$$

It follows from (4.102) that

$$(6L)\binom{4L-1}{L-1} I_{4L+1,L-1}(1/4) = B_{L,2}(1/4)(1/4)^{-L-1}(3/4)^{-3L-1}. \tag{4.106}$$

Now, (4.105) and (4.106) imply that

$$\left|\frac{x}{1/4}\right|^{L+1} \left|\frac{1-x}{3/4}\right|^{3L+1} > \frac{|1 + R_{L,1}(x)|}{B_{L,2}(1/4)}.$$

Taking into account Corollary 4.4.9 and (4.122), we get

$$\left|\frac{x}{1/4}\right|^{L+1} \left|\frac{1-x}{3/4}\right|^{3L+1} > \frac{1 - \left(\frac{5}{27}\right)^L \frac{9}{128L}}{\frac{\sqrt{3}}{4\sqrt{2\pi L}}\Phi(L)}.$$

Raising both parts of the latter inequality to the power $\frac{3}{4(3L+1)}$, we obtain (4.99). ◊

Let $\mathcal{L}_{11}$ denote the lemniscate

$$\mathcal{L}_{11}(L) := \left\{x : \left|\frac{x}{1/4}\right|^{1/4} \left|\frac{1-x}{3/4}\right|^{3/4} = a(L)\right\}.$$

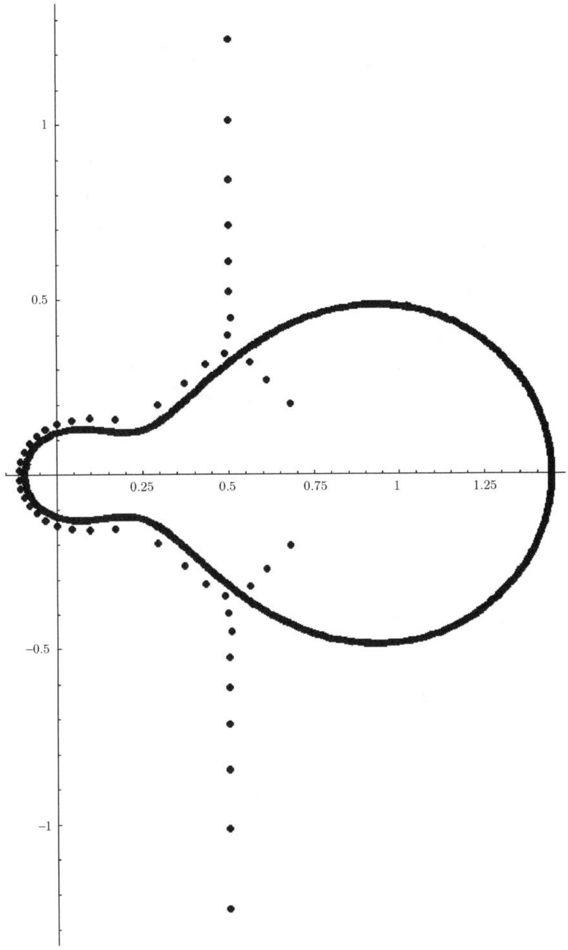

FIGURE 4.8

Figure 4.8 shows the roots of the polynomials B_{20} and the lemniscate $\mathcal{L}_{11}(20)$. Fix an arbitrary $\delta > 0$ and consider the domain

$$\alpha_{2,1} := \alpha_2 \bigcap \left\{ |x - \frac{1}{4}| > \delta \right\} \bigcap \left\{ \operatorname{Re} x < \frac{1}{4} \right\}.$$

By Corollary 4.4.15,

$$I_{4L+1,L-1}(x) = \frac{1}{L^2(1-4x)^2} + O(L^{-3}). \tag{4.107}$$

If $B_L(x) = 0$, then in view of (4.103) we have

$$x^{-L-1}(1-x)^{-3L-1} = \frac{(6L)\binom{4L-1}{L-1} I_{4L+1,L-1}(x)}{1 + R_{L,1}(x)}. \tag{4.108}$$

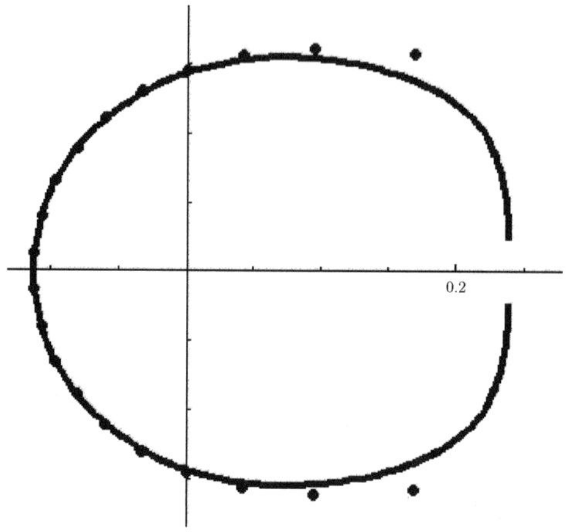

FIGURE 4.9

According to the Stirling formula,

$$\binom{4L-1}{L-1} \sim \frac{4^{4L}}{2\sqrt{6\pi L}\,3^{3L}}.$$

Hence (4.108), (4.107), and Corollary 4.4.9 imply that

$$x^{-L-1}(1-x)^{-3L-1} = \frac{4^{4L}}{\sqrt{6\pi L}\,L\,3^{3L-1}} \cdot \frac{1}{(1-4x)^2} \cdot \left(1 + O\left(L^{-1}\right)\right).$$

Raising both parts of this equality to the power $\left(-\frac{1}{4L+4}\right)$, we get

(4.109)

$$\left|\frac{x}{1/4}\right|^{1/4} \left|\frac{1-x}{3/4}\right|^{3/4} = \left(\frac{2^{17}\pi L^3}{3^7}\right)^{\frac{1}{8(L+1)}} \cdot \left(|1-x| \cdot |1-4x|\right)^{\frac{1}{2(L+1)}} \cdot \left(1 + O\left(L^{-2}\right)\right).$$

THEOREM 4.4.3. *Let $\delta > 0$, $L \in \mathbb{N}$. Then the distance from any root of the polynomial $B_L(x)$ contained in $\alpha_{2,1}$ to the curve*

(4.110)

$$\mathcal{L}_{1,L} := \left\{x : \left|\frac{x}{1/4}\right|^{1/4} \left|\frac{1-x}{3/4}\right|^{3/4} = \left(\frac{2^{17}\pi L^3}{3^7}\right)^{\frac{1}{8(L+1)}} \left(|1-x| \cdot |1-4x|\right)^{\frac{1}{2(L+1)}}\right\}$$

does not exceed $c(\delta)L^{-2}$.

Figure 4.9 shows the roots of the polynomial B_{20} and the curve $\mathcal{L}_{1,20}$ inside the domain $\alpha_{2,1}$.

Proof. Let X be a root of B_L located in $\alpha_{2,1}$, and let x be the point on $\mathcal{L}_{1,L}$ nearest to X, $\varepsilon := x - X$. Since $|1+\varepsilon|^a = 1 + O(a|\varepsilon|)$ for any $a \in (0,1)$, it follows

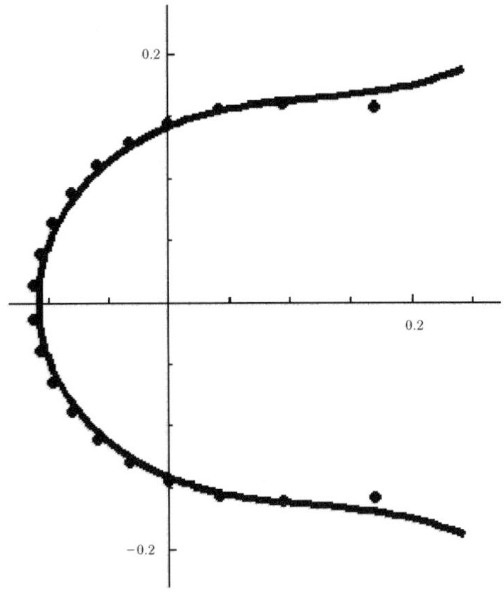

FIGURE 4.10

that

$$|1-X|^{\frac{2}{(L+1)}} = |1-x|^{\frac{2}{(L+1)}} \cdot \left|1+\frac{\varepsilon}{1-x}\right|^{\frac{2}{(L+1)}} = |1-x|^{\frac{2}{(L+1)}}\left(1+O\left(\frac{|\varepsilon|}{L}\right)\right),$$

$$|1-4X|^{\frac{2}{(L+1)}} = |1-4x|^{\frac{2}{(L+1)}}\left(1+O\left(\frac{|\varepsilon|}{L}\right)\right),$$

$$\left|X(1-X)^3\right| = \left|x(1-x)^3\right| \cdot \left|1+\frac{2+x}{x(1-x)}\varepsilon + O(\varepsilon^2)\right|.$$

Let $E := \frac{2+x}{x(1-x)}$. Since $x \in \mathcal{L}_{1,L}$, it follows that

$$\frac{\left|\frac{X}{1/4}\right| \cdot \left|\frac{1-X}{3/4}\right|^3}{\left(\frac{2^{17}\pi L^3}{3^7}\right)^{\frac{1}{2(L+1)}} \left(|1-X|\cdot|1-4X|\right)^{\frac{1}{(L+1)}}} = \frac{\left|1+E\varepsilon+O(\varepsilon^2)\right|}{1+O\left(\frac{|\varepsilon|}{L}\right)}$$
$$= \left|1+E\varepsilon+O(|\varepsilon|)\right| = 1+O\left(L^{-2}\right).$$

The latter equality follows from (4.109) and implies that $\varepsilon = O(L^{-2})$. ◊

COROLLARY 4.4.4. *Let $\delta > 0$. The distances from the zeros of $B_L(x)$, $L \in \mathbb{N}$, in $\alpha_{2,1}$ to the curve*

$$\mathcal{L}'_{1,L} := \left\{ x : \left|\frac{x}{1/4}\right|^{1/4} \left|\frac{1-x}{3/4}\right|^{3/4} = 1 + \frac{\log\left(\frac{2^{17}\pi L^3}{3^7}\right)}{(8L+1)} \right\}$$

do not exceed $c'(\delta)L^{-1}$.

Figure 4.10 shows the zeros of B_{20} and $\mathcal{L}'_{1,20}$ in $\alpha_{2,1}$.

`Proof.` This statement follows from the relation $t = 1 + \log t + O\left((t-1)^{-2}\right)$. ◊

Let us prove representations of some special Bernstein polynomials in the form of Taylor polynomials which have been used above.

Let $T_{g,k}(x)$ be the Taylor polynomial of degree k for the function g at the point 0:

$$T_{g,k}(x) := \sum_{l=0}^{k} \frac{g^{(l)}(0)}{l!} x^l.$$

Let

$$A_{N,k}(x) := \sum_{l=0}^{k} b_{N,l}(x).$$

THEOREM 4.4.5. *For any $N \in \mathbb{N}$ and $k \in \{0, 1, 2, \ldots, N\}$,*

$$A_{N,k}(x) = (1-x)^{N-k} T_{h_{N-k},k}(x) = (1-x)^{N-k} \sum_{l=0}^{k} \binom{N-k-1+l}{l} t^k, \tag{4.111}$$

where

$$h_l(x) := (1-x)^{-l}.$$

Proof. It is clear that

$$h_l^{(m)}(0) = l \cdot (l+1) \cdots (l+m-1), \quad m \in \mathbb{N}, \tag{4.112}$$

where for $m = 0$ the expression on the right is set equal to 1 by definition. Thus,

$$T_{h_{N-k},k}(x) = \sum_{l=0}^{k} \binom{N-k-1+l}{l} x^l.$$

The proof of formula (4.111) is by induction. For $k = 0$, it is trivial. Let us assume that formula (4.111) holds for k. For $k+1$, we have

$$\begin{aligned} A_{N,k+1}(x) &= A_{N,k}(x) + \tbinom{N}{k+1} x^{k+1} (1-x)^{N-k-1} \\ &= (1-x)^{N-k} T_{h_{N-k},k}(x) + \tbinom{N}{k+1} x^{k+1} (1-x)^{N-k-1} \\ &= (1-x)^{N-k-1} \left(T_{h_{N-k},k}(x) - x T_{h_{N-k},k}(x) + \tbinom{N}{k+1} x^{k+1} \right). \end{aligned}$$

Taking into account (4.112), we get (4.111) for $k+1$. ◊

COROLLARY 4.4.6. *For any $N \in \mathbb{N}$ and $k \in \{0, 1, 2, \ldots, N\}$, we have*

$$A_{N,k}(x) = 1 - (N-k) \binom{N}{k} \int_0^1 (1-xs)^{-N-1} (1-s)^k ds \; x^{k+1} (1-x)^{N-k}. \tag{4.113}$$

Proof. Taylor's formula with integral remainder gives

$$T_{h_{N-k},k}(x) = (1-x)^{k-N} - \frac{(N-k) \cdots N}{k!} \int_0^x (1-s)^{-N-1} (x-s)^k ds. \tag{4.114}$$

The change of variables $s = xs$ gives (4.113). ◊

Set

$$C_{N,k}(x) := \sum_{l=1}^{k} \frac{l}{N} b_{N,l}(x).$$

COROLLARY 4.4.7. *For any $N \in \mathbb{N}$ and $k \in \{0, 1, 2, \dots, N\}$,*

$$C_{N,k}(x) = x(1-x)^{N-k} T_{h_{N-k},k-1}(x) \tag{4.115}$$

$$= x - (N-k)\binom{N-1}{k-1} \int_0^1 (1-xs)^{-N}(1-s)^{k-1} ds \; x^{k+1}(1-x)^{N-k}. \tag{4.116}$$

Proof. As $\frac{l}{N}\binom{N}{l} = \binom{N-1}{l-1}$, using (4.111), we obtain (4.115):

$$C_{N,k}(x) = x \sum_{l=1}^{k} \binom{N-1}{l-1} x^{l-1}(1-x)^{(N-1)-(l-1)}$$
$$= xA_{N-1,k-1}(x) = x(1-x)^{N-k} T_{h_{N-k},k-1}(x).$$

As in the proof of (4.114), we get

$$T_{h_{N-k},k-1}(x) = (1-x)^{k-N} - \frac{(N-k)\cdots(N-1)}{(k-1)!} \int_0^x (1-s)^{-N}(x-s)^{k-1} ds$$
$$= (1-x)^{k-N} - (N-k)\binom{N-1}{k-1} \int_0^1 (1-xs)^{-N}(1-s)^{k-1} ds \; x^k,$$

which implies (4.116). ◊

Let us now prove some auxiliary estimates.

LEMMA 4.4.8. *For $x \in \alpha_2$,*

$$|R_{L,1}(x)| \le \frac{q(x)^{4L}\tau(x)}{2L(1-\tau(x))^2}, \tag{4.117}$$

where α_2, $R_{L,1}$ are defined by (4.98), (4.100),

$$q(x) := \left|\frac{x}{3/4}\right|^{3/4} \left|\frac{1-x}{1/4}\right|^{1/4}, \quad \tau(x) := \frac{|x|}{3|1-x|}.$$

Proof. If $x \in \alpha_2$, then

$$q(x) = \left|\frac{x}{3/4}\right|^{3/4} \left|\frac{1-x}{1/4}\right|^{1/4} = \frac{|1-x|}{1/4}\left(\frac{|x|}{3|1-x|}\right)^{3/4} < 1,$$

which implies that

$$\tau(x) = \frac{|x|}{3|1-x|} = \left(q(x)\frac{1/4}{|1-x|}\right)^{4/3} < 1,$$

because $|1-x| > 1/4$ for $x \in \alpha_2$.

Since for any $\lambda \in (0,1)$, $\binom{N}{l} \le \lambda^{-l}(1-\lambda)^{l-N}$, it follows that for $x \in \alpha_2$ and $l > 3L$, we have

$$\left|\binom{4L}{l} x^l (1-x)^{4L-l}\right| \le \left|\frac{x}{3/4}\right|^l \left|\frac{1-x}{1/4}\right|^{4L-l}$$
$$= \left(\frac{|1-x|}{1/4}\right)^{4L} \left(\frac{|x|}{3|1-x|}\right)^{3L} \left(\frac{|x|}{3|1-x|}\right)^{l-3L} = q(x)^{4L}\tau(x)^{l-3L}. \tag{4.118}$$

Therefore,

$$\begin{aligned} R_{L,1}(x) &:= \sum_{l=3L+1}^{4L} \left(\frac{l}{2L} - \frac{3}{2}\right) b_{4L,l}(x) \\ &\le q(x)^{4L} \sum_{l=3L+1}^{4L} \left(\frac{l}{2L} - \frac{3}{2}\right) \tau(x)^{l-3L} = \frac{q(x)^{4L}}{2L} \sum_{l=1}^{L} l\tau(x)^l \\ &= \frac{q(x)^{4L}\tau(x)\left(1-(L+1)\tau(x)^L + L\tau(x)^{L+1}\right)}{2L(1-\tau(x))^2} < \frac{q(x)^{4L}\tau(x)}{2L(1-\tau(x))^2}. \end{aligned}$$ ◊

COROLLARY 4.4.9. *For* $|x| \le 1/4$,

$$|R_{L,1}(x)| < \frac{9}{128L} \cdot \left(\frac{5}{27}\right)^L.$$

Proof. If $|x| \le 1/4$, then $3/4 \le |1-x| \le 5/4$. Hence,

$$q(x) = \left|\frac{x}{3/4}\right|^{3/4} \left|\frac{1-x}{1/4}\right|^{1/4} \le \frac{1}{3^{3/4}} 5^{1/4} = \left(\frac{5}{27}\right)^{1/4},$$

$$\tau(x) = \frac{|x|}{3|1-x|} \le \frac{1/4}{9/4} = \frac{1}{9}.$$ ◊

The following estimates for $B_L(1/4)$, $L \in \mathbb{N}$, are based on the representation of some Bernstein polynomials using the incomplete beta function:

$$\mathrm{Inc}_{a,b}(x) := \frac{\Gamma(a+b)}{\Gamma(a)\Gamma(b)} \int_0^x s^{a-1}(1-s)^{b-1} ds.$$

THEOREM 4.4.10. *For any* $N \in \mathbb{N}$ *and* $k \in \{0, 1, 2, \ldots, N\}$,

$$A_{N,k}(x) = 1 - \mathrm{Inc}_{k+1,N-k}(x). \tag{4.119}$$

Proof. It is known that

$$(b_{N,k}(x))' = N(b_{N-1,k-1}(x) - b_{N-1,k}(x)),$$

where $b_{N,l}(x)$ is set identically equal to zero for $l < 0$ or $l > N$. Therefore,

$$(A_{N,k}(x))' = -Nb_{N-1,k}(x).$$

Since $A_{N,k}(0) = 1$, it follows that

$$A_{N,k}(x) = 1 - N \int_0^x b_{N-1,k}(s)\, ds,$$

which implies (4.119). ◊

COROLLARY 4.4.11.

$$B_{L,2}(x) = \frac{1}{2} - 2x + 2x\, \mathrm{Inc}_{L,3L}(x) - \frac{1}{2} \mathrm{Inc}_{L+1,3L}(x). \tag{4.120}$$

Proof. It is easy to see that

$$B_{L,2}(x) = \frac{1}{2} A_{4L,L}(x) - 2C_{4L,L}(x) = \frac{1}{2} A_{4L,L}(x) - 2xA_{4L-1,L-1}(x).$$

Relation (4.120) follows from (4.119). ◊

COROLLARY 4.4.12.

$$B_{L,1}(x) = \frac{3}{2} - 2x + 2x\,\mathrm{Inc}_{3L,L}(x) - \frac{3}{2}\,\mathrm{Inc}_{3L+1,L}(x). \tag{4.121}$$

`Proof.` It is clear that

$$B_{L,1}(x) = \frac{3}{2}A_{4L,3L-1}(x) - 2C_{4L,3L-1}(x) = \frac{1}{2}A_{4L,3L}(x) - 2xA_{4L-1,3L-1}(x).$$

Now, (4.121) follows from (4.119). ◊

LEMMA 4.4.13.

$$B_{L,2}(1/4) = \frac{\sqrt{3}}{4\sqrt{2\pi L}}\Phi(L) < \frac{\sqrt{3}}{4\sqrt{2\pi L}}, \tag{4.122}$$

where

$$\Phi(L) := \frac{\Gamma^*(4L)}{\Gamma^*(L)\Gamma^*(3L)}, \qquad \Gamma^*(r) := \sqrt{\frac{r}{2\pi}}e^r r^{-r}\Gamma(r).$$

`Proof.` In view of (4.120), we have

$$\begin{aligned} B_{L,2}(1/4) &= \frac{1}{2}\left(\mathrm{Inc}_{L,3L}(1/4) - \mathrm{Inc}_{L+1,3L}(1/4)\right) \\ &= \frac{(4L-1)!}{(2L-1)!(3L-1)!}\int_0^{1/4}(1-4s)s^{L-1}(1-s)^{3L-1}ds. \end{aligned} \tag{4.123}$$

Make the change of variables

$$-\frac{\zeta^2}{2} = \ln\left[\left(\frac{s}{1/4}\right)^{1/4}\left(\frac{1-s}{3/4}\right)^{3/4}\right],$$

where ζ is negative. Then we get

$$B_{L,2}(1/4) = \sqrt{\frac{3L}{2\pi}}\Phi(L)\int_0^{\infty}\exp(-2L\zeta^2)\zeta\,d\zeta = \frac{\sqrt{3}}{4\sqrt{2\pi L}}\Phi(L).$$

It is known that $\Gamma^*(r) \searrow 1$ as $r \to \infty$ and $\Phi(L) \nearrow 1$ as $L \to \infty$. ◊

LEMMA 4.4.14. *Let* $k, N \in \mathbb{N}$; $\delta > 0$; $k < N-2$; $\mathrm{Re}\,x < \frac{k+1}{N-1}$, $\left|x - \frac{k+1}{N-1}\right| > \delta$. *Then*

$$\mathrm{In}_{N,k}(x) := \int_0^1 (1-xs)^{-N}(1-s)^k ds = \frac{1}{k+1-(N-1)x} + O(k^{-3}). \tag{4.124}$$

`Proof.` Set $a := \frac{N-1}{k+1}$. Make the change of variables

$$w := w(s) := \frac{1-s}{(1-xs)^a}$$

in the integral $\mathrm{In}_{N,k}$. Then

$$\frac{dw}{ds} = \frac{(a-(a-1)s)x-1}{(1-xs)^{a+1}}$$

and

$$\mathrm{In}_{N,k}(x) = \int_0^1 \frac{w^k}{1-(a-(a-1)s)x} dw.$$

Prove that $|w(s)| < 1$ for sufficiently small $s \in [0,1]$. Let $x := \rho e^{i\varphi}$. Then

$$|w(s)| = \frac{1-s}{(1-2\rho s\cos\varphi + \rho^2 s^2)^{a/2}}.$$

Therefore the inequality $|w(s)| < 1$ is equivalent to

$$\frac{\left(1-2\rho s\cos\varphi+\rho^2 s^2\right)^{a/2}-1}{s} = \frac{f(s)-f(0)}{s} > -1,$$

where

$$f(s) := \left(1-2\rho s\cos\varphi+\rho^2 s^2\right)^{a/2}.$$

The latter inequality holds for sufficiently small $s \in [0,1]$ because $f'(0) = -a\rho\cos\varphi > -1$.

Let a constant $\varepsilon \in (0,1)$ be chosen so that

$$\varepsilon < \frac{\delta^2}{(k+1)^2(a-1)|x|}$$

and $|\tau| < 1$, where $\tau := w(\varepsilon)$. Then

$$\mathrm{In}_{N,k}(x) = \int_0^\tau \frac{w^k}{1-(a-(a-1)s)x} dw + \int_\tau^1 \frac{w^k}{1-(a-(a-1)s)x} dw.$$

Set

$$I_1 := \int_0^\tau \frac{w^k}{1-(a-(a-1)s)x} dw,$$

$$I_2 := \int_\tau^1 \frac{w^k}{1-(a-(a-1)s)x} dw.$$

Since $s \in [0,1]$, it follows that $(a-(a-1)s) \in [1,a]$. It is clear that the segment $[x, ax]$ does not intersect the circle $\{z : |1-z| \le \delta\}$; hence $|1-(a-(a-1)s)x| > \delta$ for any $s \in [0,1]$ and

$$|I_1| < \frac{|\tau|^{k+1}}{(k+1)\delta}.$$

For I_2, we have

$$I_2 = \frac{1}{1-ax}\int_\tau^1 w^k\,dw - \int_\tau^1 \frac{w^k(a-1)xs}{(1-ax)(1-(a-(a-1)s)x)} dw$$

$$= \frac{1-\tau^{k+1}}{(1-ax)(k+1)} + \int_\tau^1 \frac{w^k(a-1)xs}{(1-ax)(1-(a-(a-1)s)x)} dw.$$

Therefore,

$$\left|\mathrm{In}_{N,k}(x) - \frac{1}{(1-ax)(k+1)}\right| < \frac{2|\tau|^{k+1}}{\delta(k+1)} + \frac{(a-1)|x|\varepsilon}{(k+1)\delta^2}$$
$$< \frac{2|\tau|^{k+1}}{\delta(k+1)} + \frac{1}{(k+1)^3}. \qquad \diamond$$

COROLLARY 4.4.15.

$$\begin{aligned} I_{N,k}(x) &:= \mathrm{In}_{N,k}(x) - \mathrm{In}_{N,k+1}(x) \\ &= \frac{1}{(k+1-(N-1)x)(k+2-(N-1)x)} + O(k^{-3}) \\ &= \frac{1}{(k+1)^2(1-\frac{N-1}{k+1}x)^2} + O(k^{-3}). \end{aligned}$$

CHAPTER 5

Fractal Properties of Wavelets

5.1. Refinable functions and fractal curves

One of the main methods of studying local and global smoothness of compactly supported wavelets is to represent a corresponding refinable function in the form of a fractal curve for an appropriate pair of finite-dimensional affine operators.

Let A_n be an n-dimensional affine space and let $\tilde{A}_n$ be the corresponding linear space. Denote by $\mathcal{A}(A_n)$ the algebra of affine operators on the space A_n and by $\mathcal{L}(\tilde{A}_n)$ the algebra of linear operators on $\tilde{A}_n$. For an arbitrary $B \in \mathcal{A}(A_n)$, we denote by $\tilde{B} \in L(\tilde{A}_n)$ the linear part (the differential) of the operator B.

A nonempty compact set $K \subset A_n$ is called a *fractal* corresponding to a given collection of affine operators $\{B_0, B_1\}$ if $B_0K \cup B_1K = K$ (see Appendix A.17). For the sake of simplicity we consider collections of two operators, though the theory remains the same in the case of an arbitrary finite collection. According to Hutchinson's classical theorem (Theorem A.17.2 in Appendix A.17), if both B_0, B_1 are contraction operators (more precisely, their linear parts $\tilde{B}_0, \tilde{B}_1$ are contractions in some norm), then there exists a unique fractal.

The existence of a norm in the space $\mathbb{R}^n$ where both operators are contraction operators can be verified in terms of the joint spectral radius of operators.

DEFINITION 5.1.1. The number

$$\hat{\rho}(\tilde{B}_0, \tilde{B}_1) = \lim_{r\to\infty} \max_{(d_1,\dots,d_r)\in\{0,1\}^r} \|\tilde{B}_{d_1} \cdots \tilde{B}_{d_r}\|^{1/r}$$

is called the joint spectral radius (in what follows, JSR) of a pair of linear operators $\tilde{B}_0, \tilde{B}_1$. The number

$$\check{\rho}(\tilde{B}_0, \tilde{B}_1) = \lim_{r\to\infty} \min_{(d_1,\dots,d_r)\in\{0,1\}^r} \|\tilde{B}_{d_1} \cdots \tilde{B}_{d_r}\|^{1/r}$$

is called the lower spectral radius (LSR).

(For correctness of this definition and the main properties of JSR and LSR, see Appendix A.6.) Denote by $\rho(\tilde{B})$ the (ordinary) spectral radius of the operator $\tilde{B}$, i.e., the maximal absolute value of its eigenvalues.

LEMMA 5.1.2. *For any operators $\tilde{B}_0, \tilde{B}_1$ and any product $\Pi_k = \tilde{B}_{d_1} \cdots \tilde{B}_{d_k}$, we have*

$$\check{\rho}(\tilde{B}_0, \tilde{B}_1) \le \Big(\rho(\Pi_k)\Big)^{1/k} \le \hat{\rho}(\tilde{B}_0, \tilde{B}_1). \tag{5.1}$$

For the proof see Appendix A.6. In particular, for $k = 1$ we get

$$\check{\rho} \le \min\{\rho(\tilde{B}_0),\, \rho(\tilde{B}_1)\}, \qquad \hat{\rho} \ge \max\{\rho(\tilde{B}_0),\, \rho(\tilde{B}_1)\}.$$

If both operators $\tilde{B}_0, \tilde{B}_1$ are similarity transformations (proportional to orthogonal operators), both inequalities become equalities. If the operators commute or they are self-adjoint or both are defined by upper- (lower)-triangle matrices, then the second equality holds, i.e.,

$$\hat{\rho} = \max\{\rho(\tilde{B}_0)\,,\, \rho(\tilde{B}_1)\}, \tag{5.2}$$

while the first inequality generally remains strict. In the general case, it is very complicated to compute or estimate the joint spectral radius (see §7.3).

In what follows, we shall assume that the collection of operators $\{B_0, B_1\}$ is irreducible, i.e., has no common nontrivial affine subspaces in A_n. Otherwise, we shall consider the restrictions of the operators to this subspace.

PROPOSITION 5.1.3. *Let $\{B_0, B_1\}$ be an irreducible collection of affine operators. If $\hat{\rho}(\tilde{B}_0, \tilde{B}_1) < 1$, this collection has a unique fractal $K \subset A_n$. If $\hat{\rho}(\tilde{B}_0, \tilde{B}_1) > 1$, it has no fractals.*

Proof. If $\hat{\rho}(\tilde{B}_0, \tilde{B}_1) < 1$, then there exists a norm in the space $\tilde{A}_n$ such that $\|\tilde{B}_i\| < 1,\ i = 0, 1$ (see Appendix A.6, Theorem A.6.3). It remains to refer to Hutchinson's theorem about the existence and uniqueness of a fractal (Appendix A.17, Theorem A.17.2), which completes the proof.

Conversely, if K is a fractal of the collection of operators $\{B_0, B_1\}$, then its convex hull conv K has a nonempty interior in A_n. Otherwise, it is contained in some affine plane of a smaller dimension. Hence the affine hull of the set K is a common invariant subspace of the operators B_0, B_1, which contradicts the irreducibility assumption. Thus the convex hull of K has a nonempty interior. Hence for some constant $C > 0$, we have

$$\max_{(d_1,\dots,d_r)\in\{0,1\}^r, r\in\mathbb{N}} \|\tilde{B}_{d_1}\cdots\tilde{B}_{d_r}\| < C.$$

Therefore $\hat{\rho}(\tilde{B}_0, \tilde{B}_1) \le 1$. ◇

In what follows, we shall assume that $\hat{\rho}(\tilde{B}_0, \tilde{B}_1) < 1$. In particular, it follows that $\rho(\tilde{B}_i) < 1,\ i = 0, 1$ (Appendix A.6, Lemma 5.1.2). Hence the operator B_i has a unique fixed point $v_i \in A_n$.

THEOREM 5.1.4. *Suppose an irreducible collection of affine operators satisfies the two conditions*

1) $\hat{\rho}(\tilde{B}_0, \tilde{B}_1) < 1$,

2) $B_0v_1 = B_1v_0$ (v_i is a fixed point of the operator B_i).

Then the fractal K of this collection is a continuous curve in A_n. More precisely, there exists a unique continuous function $v:\ [0,1] \to A_n$ (fractal curve) *such that*

$$v(x) = B_i v(2x - i), \qquad x \in \left[\frac{i}{2}, \frac{i+1}{2}\right], \quad i = 0, 1; \tag{5.3}$$

therefore $v([0,1]) = K$. Values of this function are given explicitly by the formula

$$v(x) = \lim_{k\to\infty} B_{d_1}B_{d_2}\cdots B_{d_k}v_0, \tag{5.4}$$

where $d_1, d_2, \dots$ are digits in the binary expansion of the number x: $x = 0.d_1d_2\dots$. In particular, for a dyadic $x = 0.d_1d_2\dots d_k$, we have

$$v(x) = B_{d_1}B_{d_2}\cdots B_{d_k}v_0. \tag{5.5}$$

The Hölder exponent of smoothness in $C[0,1]$ *for the function* v *is*

$$\alpha_v = -\log_2 \hat{\rho}(\tilde{B}_0, \tilde{B}_1).$$

Moreover, if a collection of linear operators $\tilde{B}_0, \tilde{B}_1$ *is irreducible in* $\tilde{A}_n$*, then*

$$C_1 t^{\alpha_v} \le \omega_1(v,t)_\infty \le C_2 t^{\alpha_v}, \tag{5.6}$$

where C_1, C_2 *are positive constants.*

Conversely, if for an irreducible pair of affine operators $\{B_0, B_1\}$*, equation* (5.3) *has a continuous solution, then each of the operators* B_i *has a unique fixed point* v_i *and conditions* 1), 2) *are satisfied.*

In what follows, we shall use the modulus of continuity of the first order ω_1, omitting the index 1. In addition, for the modulus of continuity on the space $C(\mathbb{R})$, we shall omit the index ∞. Thus $\omega(f,h) = \omega_1(f,h)_\infty$.

Proof. First we prove the existence of the limit in (5.4). At the same time we show that the function v defined by this formula is continuous and $\alpha_v \ge -\log_2 \hat{\rho}$.

Define $v(x)$ at dyadic points x by formula (5.5) and estimate the norm of the difference $v(x) - v(y)$ for arbitrary binary rational x and y. Without loss of generality we can assume that $x < y$ and that the pair x, y is distinct from $\{0,1\}$. Denote by z the binary rational number of the minimal order such that $x \le z \le y$ (it is easy to see that such a number is unique).

Now we estimate the norm $\|v(y) - v(z)\|$ (the norm $\|v(z) - v(x)\|$ is estimated analogously). We have $z = 0.d_1 \dots d_q$ and $y = 0.d_1 \dots d_q d_{q+1} \dots$. Denote by r the minimal index such that $r > q$ and $d_r = 1$. It is clear that $|y - z| \ge 2^{-r}$. Let $P \subset \mathbb{N}$ be the set of all indices $p \in \mathbb{N}$ such that the p-th binary digits (in the binary expansions) of the numbers y and z are different. It is easy to see that r is the minimal number in P. We have

$$\begin{aligned}
\|v(y) - v(z)\| &= \left\| \sum_{p \in P} v(0.d_1 \dots d_{p-1}1) - v(0.d_1 \dots d_{p-1}0) \right\| \\
&\le \left\| \sum_{p \ge r} v(0.d_1 \dots d_{p-1}1) - v(0.d_1 \dots d_{p-1}0) \right\| \\
&= \sum_{p \ge r} \|B_{d_1} \cdots B_{d_{p-1}}(v_1 - v_0)\| \le \sum_{p \ge r} C_\varepsilon (\hat{\rho} + \varepsilon)^{p-1} \\
&= \frac{C_\varepsilon (\hat{\rho}+\varepsilon)^r}{(\hat{\rho}+\varepsilon)(1-\hat{\rho}-\varepsilon)} \le \frac{C_\varepsilon}{(\hat{\rho}+\varepsilon)(1-\hat{\rho}-\varepsilon)} \cdot |y - z|^{-\log_2(\hat{\rho}+\varepsilon)}
\end{aligned} \tag{5.7}$$

for some constant C_ε that depends only on $\hat{\rho}$ and ε. Estimating in the same way the norm $\|v(z) - v(x)\|$, we get

$$\|v(y) - v(x)\| \le \tilde{C}_\varepsilon |y - x|^{-\log_2(\hat{\rho}+\varepsilon)} \tag{5.8}$$

for some constant $\tilde{C}_\varepsilon$ that depends only on $\hat{\rho}$ and ε. This inequality holds for each $\varepsilon > 0$. It follows that the function v is uniformly continuous on the set of dyadic numbers and on this set the Hölder exponent is at least $-\log_2 \hat{\rho}$. Therefore this function can be uniquely and continuously extended to the whole segment $[0,1]$ and $\alpha_v \ge -\log_2 \hat{\rho}$ on the whole segment. Formula (5.4) follows from the continuity of

v. If the collection $\{\tilde{B}_0, \tilde{B}_1\}$ is irreducible, we can use Lemma A.6.4 (Appendix A.6). Reasoning as above for $\varepsilon = 0$, $C_\varepsilon = c_2$, we obtain

$$\|v(y) - v(x)\| \le C_2 |y-x|^{-\log_2 \hat{\rho}}.$$

To prove the inverse inequality $\alpha_v \le -\log_2 \hat{\rho}$, we take a set L that consists of all vectors $u \in \tilde{A}_n$ such that

$$\max_{d_i \in \{0,1\},\, i=1,\dots,k} \|\tilde{B}_{d_1} \cdots \tilde{B}_{d_k} u\| = o(1)\hat{\rho}^k \qquad \text{as } k \to \infty.$$

It is clear that L is a linear subspace in $\tilde{A}_n$, invariant under the operators $\tilde{B}_0$, $\tilde{B}_1$. If $v_1 - v_0 \in L$, then L coincides with the whole of $\tilde{A}_n$; otherwise the operators B_0, B_1 would have a common invariant affine subspace $v_0 + L \in A_n$. Thus $\|\tilde{B}_{d_1} \cdots \tilde{B}_{d_k} u\| = o(1)\hat{\rho}^k$ for all $u \in \tilde{A}_n$. Then choosing an orthonormal basis $\{e_1, \dots, e_n\}$ in $\tilde{A}_n$, for all sufficiently large k we have

$$\|\tilde{B}_{d_1} \cdots \tilde{B}_{d_k} e_r\| \le \frac{1}{2n\sqrt{n}} \hat{\rho}^k, \qquad r = 1, \dots, n$$

(here we use Euclidean norm). Consequently,

$$\|\tilde{B}_{d_1} \cdots \tilde{B}_{d_k}\| \le \sqrt{n} \sum_{r=1}^{n} \|\tilde{B}_{d_1} \cdots \tilde{B}_{d_k} e_r\| \le \sqrt{n} \frac{1}{2n\sqrt{n}} \hat{\rho}^k = \frac{1}{2}\hat{\rho}^k.$$

Then

$$\begin{aligned}
\hat{\rho} = \hat{\rho}\{\tilde{B}_0, \tilde{B}_1\} &= \Big[\hat{\rho}\Big\{B_{d_1} \cdots \tilde{B}_{d_k},\ \{d_1, \dots, d_k\} \subset \{0,1\}^k\Big\}\Big]^{1/k} \\
&\le \Big[\max\Big\{\|B_{d_1} \cdots \tilde{B}_{d_k}\|,\ \{d_1, \dots, d_k\} \subset \{0,1\}^k\Big\}\Big]^{1/k} \\
&\le \left[\frac{1}{2}\hat{\rho}^k\right]^{1/k} = 2^{-1/k}\hat{\rho} < \hat{\rho}.
\end{aligned}$$

This is a contradiction; hence the case $v_1 - v_0 \in L$ is not possible. Thus $v_1 - v_0 \notin L$. This means that for some positive constant C there exist arbitrarily long sequences $d_1, \dots, d_k$ such that

$$\|B_{d_1} \cdots B_{d_k}(v_1 - v_0)\| \ge C\hat{\rho}^k.$$

Then for the points $x = 0.d_1 \dots d_k 0$ and $y = 0.d_1 \dots d_k 1$, we have

$$\|v(y) - v(x)\| \ge C\hat{\rho}^k = C|y-x|^{-\log_2 \hat{\rho}}.$$

Therefore,

$$\alpha_v \le -\log_2 \hat{\rho},$$

which proves the inverse inequality. If the collection of operators $\{\tilde{B}_0, \tilde{B}_1\}$ is irreducible, we use Lemma A.6.4: setting $v = v_1 - v_0$ in inequality (5.6), we get

$$\|v(y) - v(x)\| \ge C_1 |y-x|^{-\log_2 \hat{\rho}},$$

which proves (5.6). Moreover, if $\hat{\rho} \ge 1$, the function v is not uniformly continuous on the set of binary rational numbers; hence it cannot be continuously extended to the whole segment $[0,1]$. Therefore in the case $\hat{\rho} \ge 1$ there exist no continuous functions satisfying equation (5.3). Thus the existence of a continuous solution implies property 2); moreover, it implies property 1) as well. Indeed, it follows from equation (5.3) that $B_i v(i) = v(i)$, $i = 0, 1$. Hence the operators B_0, B_1 have fixed points, namely, $v(0)$ and $v(1)$, respectively. These operators have no other fixed points. For example, if the operator B_0 had another fixed point $a \in A_n$, then we would have $\tilde{B}_0(v(0) - a) = v(0) - a$; hence the operator $\tilde{B}_0$ would have an eigenvalue 1 and $\hat{\rho} \geq 1$, which contradicts condition 2). Thus, $v_i = v(i)$ is the unique fixed point of the operator B_i, $i = 1, 2$.

Finally, it follows from the same equation (5.3) that

$$v(1/2) = B_0 v(1) = B_1 v(0).$$

Thus condition 1) is satisfied. This completes the proof. $\Diamond$

COROLLARY 5.1.5. *If a fractal curve v is a Lipschitz function, then $\hat{\rho}(\tilde{B}_0, \tilde{B}_1) = \frac{1}{2}$. Conversely, if $\hat{\rho}(\tilde{B}_0, \tilde{B}_1) = \frac{1}{2}$ and the collection $\tilde{B}_0, \tilde{B}_1$ is irreducible, then v is a Lipschitz function.*

REMARK 5.1.6. If operators $\tilde{B}_0$ and $\tilde{B}_1$ have a common invariant subspace, the Hölder exponent of the function v may not be sharp. Dubuc's function gives a corresponding example. Its derivative (which is a fractal curve), as was shown in [**113**], is "almost" Lipschitz: its modulus of continuity is $t|\ln t|$. We study this phenomenon in more detail in Example 7.3.14, §7.3 .

REMARK 5.1.7. If each of the affine operators B_0, B_1 has a fixed point (v_0, v_1, respectively), then the cross-condition $B_0 v_1 = B_1 v_0$ always implies that $\hat{\rho}(\tilde{B}_0, \tilde{B}_1) \geq \frac{1}{2}$. Indeed, one can easily check that

$$(\tilde{B}_0 + \tilde{B}_1)[v_1 - v_0] = v_1 - v_0.$$

Thus the operator $\tilde{B}_0 + \tilde{B}_1$ has the eigenvalue 1; hence in any norm $\|\tilde{B}_0 + \tilde{B}_1\| \geq 1$. However, if $\hat{\rho}(\tilde{B}_0, \tilde{B}_1) < \frac{1}{2}$, then in some norm $\|B_i\| < \frac{1}{2}$ for both operators. Therefore $\|\tilde{B}_0 + \tilde{B}_1\| < 1$.

We now represent continuous solutions of refinement equations in the form of fractal curves.

Let a continuous function φ with support $[0, N]$ solve the refinement equation $\varphi(x) = \sum_{k=0}^{N} c_k \varphi(2x - k)$. Then the vector function $v : [0, 1] \to \mathbb{R}^N$ defined by the equality

$$v(x) = \big(\varphi(x), \varphi(x+1), \ldots, \varphi(x+N-1)\big)^T \in \mathbb{R}^N \tag{5.9}$$

solves the equation

$$v(x) = T_i v(2x - i), \quad x \in [i/2, (i+1)/2], \qquad i = 0, 1, \tag{5.10}$$

where the operators T_0, T_1 are defined by $N \times N$-matrices:

$$(T_0)_{ij} = c_{2i-j-1}, \qquad (T_1)_{ij} = c_{2i-j}. \tag{5.11}$$

For $k \in \{0, \ldots, N\}$, c_k is a corresponding coefficient of the refinement equation; for all other k, we set $c_k = 0$. Conversely, if vector function (5.9) satisfies equation (5.10), the function $\varphi(x)$ is a solution of the corresponding refinement equation. The reader can easily check this by direct substitution. Similarly, we show that φ is an L_p-solution if and only if the function v belongs to $L_p[0, 1]$ and satisfies equation (5.10) almost everywhere.

It is easy to see that in the space L_p the functions φ and v have the same smoothness. In the space $C(\mathbb{R})$ problems may arise at the endpoints of the segment. If φ is continuous, $\varphi(0) = \varphi(N) = 0$. Hence its smoothness in the space C on the whole real axis is the same as on the segment $[0, N]$; i.e., it is the same as the smoothness of the function v. Repeating this reasoning for derivatives, we conclude that if φ is differentiable, v is also differentiable and the smoothness of the function φ in $C^1(\mathbb{R})$ is the same as the smoothness of v in $C^1[0,1]$. Thus we have the following assertion.

If a refinable function belongs to $C^m(\mathbb{R})$, $m \geq 0$ (or $W_p^m(\mathbb{R})$, $m \geq 0$, $p \in [1, +\infty]$), then its smoothness in this space coincides with the smoothness of vector function (5.9) *in the space $C^m[0,1]$ (respectively, $W_p^m[0,1]$).*

We can now use the results on fractal curves for refinable functions; however to do this, we have to determine the corresponding affine spaces. First, let a refinable function φ be continuous. Then, for A_n, we take the affine hull of the set $\{v(x),\ x \in [0,1]\}$, i.e., the minimal affine subspace in $\mathbb{R}^N$ that contains this set. From equation (5.10) it follows that $T_i A_n \subset A_n$, $i = 0, 1$. Set $B_i = T_i|_{A_n}$. It follows from the minimality of A_n (see the definition of affine hull) that the collection of operators B_0, B_1 is irreducible. The linear part $\tilde{A}_n$ of this affine space is a common invariant subspace of the operators T_0, T_1. The affine plane A_n can be defined by the operators T_i as well. A_n is the minimal affine plane that contains the vector

$$v_0 = \Big(0, \varphi(1), \dots, \varphi(N-1)\Big)^T$$

(which is an eigenvector of the operator T_0 with the eigenvalue 1) and is invariant under both operators T_0, T_1. Respectively, for these operators, $\tilde{A}_n$ is the minimal common invariant subspace that contains the vector $v_1 - v_0$. Note that A_n is always contained in the hyperplane

$$W = \{(y_1, \dots, y_N) \in \mathbb{R}^N,\ \sum y_k = 1\}.$$

Indeed, each summable refinable function has the partition of unity property (Corollary 5.1.8); hence $\sum_{k=0}^{N} v(x+k) = 1$, so $v(x) \in W$ for each $x \in [0,1]$. Thus A_n does not coincide with the whole space $\mathbb{R}^N$ and its dimension n does not exceed $N-1$. For L_p-refinable functions, the space A_n is defined as the affine hull of the set formed by the vectors

$$\Big\{\Big(2^k \int_x^{x+2^{-k}} \varphi(t)\,dt,\ \dots,\ 2^k \int_x^{x+2^{-k}} \varphi(t+N-1)\,dt\Big)^T\Big\}$$

for all $x \in [0,\ 1-2^{-k}]$, $k \in \mathbb{N} \cup \{0\}$. In terms of the operators T_i, A_n is the minimal affine plane that contains the vector

$$c = \Big(\int_0^1 \varphi(t)\,dt\,,\ \dots,\ \int_{N-1}^{N} \varphi(t)\,dt\Big)^T$$

(which is an eigenvector of the operator $\frac{1}{2}\Big(T_0 + T_1\Big)$ with the eigenvalue 1) and is invariant under both operators T_0, T_1; $\tilde{A}_n$ is the minimal common invariant subspace of these operators that contains the vector $T_0 c - c = c - T_1 c$. From the partition of unity property it follows that in this case $A_n \subset W$. It is easy to show that for a continuous refinable function these two definitions of the space A_n coincide, so we shall denote them by the same symbol.

Thus for the refinable function φ, the function $v(x)$ is a fractal curve for the operators B_0, B_1 and the space A_n defined above. Note that here the cross-condition $B_0 v_1 = B_1 v_0$ holds automatically. Indeed, it is not difficult to see that for any pair of vectors

$$u_0 = (0, y_1, \dots, y_N)^T, \qquad u_1 = (y_1, \dots, y^{N-1}, 0)^T$$

(the vectors v_0 and v_1 form such a pair), we have $B_0 u_1 = B_1 u_0$.

Using Theorem 5.1.4, we obtain the following corollary.

COROLLARY 5.1.8. *If a refinable function φ is continuous, then vector function* (5.9) *is a fractal curve of the affine operators B_0, B_1, where $B_i = T_i|_{A_n}$, $i = 0, 1$, the operators T_0, T_1 are determined in $\mathbb{R}^N$ by their matrices by formulas* (5.11), *and A_n is the minimal affine plane that contains the point $v(0)$ and is invariant under the operators T_0, T_1. Here $\hat{\rho}(\tilde{B}_0, \tilde{B}_1) < 1$ and the Hölder exponent in $C(\mathbb{R})$ of φ is equal to*

$$\alpha_\varphi = -\log_2 \hat{\rho}(\tilde{B}_0, \tilde{B}_1).$$

Moreover, if the collection of linear operators $\tilde{B}_0, \tilde{B}_1$ is irreducible in $\tilde{A}_n$, then

$$C_1 t^{\alpha_\varphi} \le \omega(\varphi, t) \le C_2 t^{\alpha_\varphi}.$$

Conversely, if the operator T_0 has an eigenvector v_0 with the eigenvalue 1, *the first coordinate of T_0 is equal to* 0, *and $\hat{\rho}(\tilde{B}_0, \tilde{B}_1) < 1$ ($\tilde{B}_i$ is the restriction of the operator T_i to the minimal common invariant subspace of the operators T_0, T_1 that consists of the vector $v_1 - v_0$, and v_1 is obtained from v_0 by the left shift of all coordinates by* 1, *setting the N-th coordinate equal to zero), then the corresponding refinement equation has a continuous solution. Values of the refinable function are given by* (5.9), *where*

$$v(x) = \lim_{k\to\infty} T_{d_1} T_{d_2} \cdots T_{d_k} v_0, \tag{5.12}$$

$d_1, d_2, \dots$ are digits in the binary expansion of the number x. In particular, for a dyadic $x = 0.d_1 d_2 \dots d_k$, we have

$$v(x) = T_{d_1} T_{d_2} \cdots T_{d_k} v_0. \tag{5.13}$$

To find a continuous solution of the refinement equation, we need to find an eigenvector of the operator T_0 with the eigenvalue 1 and the first coordinate 0 (such a vector exists due to the existence of a continuous solution) and then find the corresponding affine plane A_n and check that the joint spectral radius of the operators T_0, T_1 is strictly less than 1. We can compute the values of the function φ by formula (5.13) at all binary rational points and then extend this function continuously to the whole segment $[0, N]$.

Generally speaking, the operator T_0 may have several noncollinear eigenvectors with the eigenvalue 1, each generating its own plane A_n, but only for one of them,

$$\hat{\rho}(T_0|_{A_n}, T_1|_{A_n}) < 1.$$

For all the other vectors (and the corresponding planes) this inequality does not hold. Otherwise, in view of Corollary 5.1.8, the refinement equation would have another continuous solution, which is impossible (Theorem 2.4.4). In Chapter 6, the subspace A_n and vector v_0 will be found explicitly for every refinement equation.

5.2. Fractal curves in the space L_p

Any function $v : [0,1] \to A_n$ that belongs to the space $L_p[0,1]$ (i.e., all coordinate functions of the mapping v belong to $L_p[0,1]$, $p \in [1, +\infty]$) is called an L_p-fractal curve of the operators B_0, B_1 if it satisfies equation (5.3) almost everywhere on the segment $[0,1]$.

The existence and smoothness of an L_p-fractal curve for a given pair of operators are determined by the generalized joint spectral radius with coefficient p, or *p-radius*.

DEFINITION 5.2.1. By definition, for a given $p \in [1, +\infty]$, the p-radius of linear operators $\tilde{B}_0, \tilde{B}_1$ is the quantity $\rho_p(\tilde{B}_0, \tilde{B}_1)$, which is equal to

$$\lim_{k\to\infty} \left(2^{-k} \sum_{\{d_1,\dots,d_k\}\subset\{0,1\}^k} \left\| \tilde{B}_{d_1} \cdots \tilde{B}_{d_k} \right\|^p \right)^{1/pk}$$

for finite p and

$$\lim_{k\to\infty} \max_{\{d_1,\dots,d_k\}\subset\{0,1\}^k} \left\| \tilde{B}_{d_1} \cdots \tilde{B}_{d_k} \right\|^{1/k}$$

for $p = +\infty$.

Thus the joint spectral radius $\hat{\rho}$ is equal to the p-radius for $p = +\infty$. The function ρ_p is nondecreasing with respect to p. For any p, we have $\check{\rho} \leq \rho_p \leq \hat{\rho}$.

There are several efficient algorithms for the approximate calculation of ρ_p for a given pair of operators **[115]**, **[116]**. For even integer p, starting from $p = 2$, the value ρ_p is calculated as the (ordinary) spectral radius of some finite-dimensional operator **[115]**, **[117]**, **[118]**.

The proof of the following theorem is analogous to that of Theorem 5.1.4, so we do not give it here.

THEOREM 5.2.2. *Suppose $p \in [1, +\infty)$ and an irreducible collection of affine operators satisfies the condition $\rho_p(\tilde{B}_0, \tilde{B}_1) < 1$; then there exists a unique function $v : [0,1] \to A_n$ that belongs to $L_p[0,1]$ and satisfies equality (5.3) almost everywhere on the segment $[0,1]$. Moreover, there exists a unique point $c \in A_n$ such that $c = \frac{1}{2}B_0 + \frac{1}{2}B_1$. Here*

$$\int_0^1 v(t)\,dt = c$$

and for any dyadic $x = 0.d_1 \dots d_k$, we have

$$2^k \int_x^{x+2^{-n}} v(t)\,dt = B_{d_1} \cdots B_{d_k} c;$$

i.e., the mean value of the function v on the segment $[x, x + 2^{-k}]$ is equal to $B_{d_1} \cdots B_{d_k} c$. If we denote by v^k the piecewise constant function that is identically equal to $B_{d_1} \cdots B_{d_k} c$ on each of the 2^k dyadic segments $[x, x+2^{-k}]$, $x = 0, d_1 \dots d_k$, then v^k tends to v in the metric of $L_p[0,1]$ as $k \to \infty$. If either $\rho_p \geq 2^{-1/p}$ or the operators B_0, B_1 satisfy the cross-condition $B_0 v_1 = B_1 v_0$, then the Hölder exponent of the function v in $L_p[0,1]$ is equal to

$$\alpha_{v,p} = -\log_2 \rho_p(\tilde{B}_0, \tilde{B}_1).$$

Otherwise, $\alpha_{v,p} = \frac{1}{p}$. *Moreover, if the collection of linear operators* $\tilde{B}_0, \tilde{B}_1$ *is irreducible in* $\tilde{A}_n$*, then*

$$C_1 t^{\alpha_{v,p}} \le \omega(v,t)_p \le C_2 t^{\alpha_{v,p}}, \tag{5.14}$$

where C_1, C_2 *are positive constants.*

Conversely, if for an irreducible pair of affine operators $\{B_0, B_1\}$ *equation* (5.3) *has an* L_p*-solution, then* $\rho_p(\tilde{B}_0, \tilde{B}_1) < 1$.

Let us return to refinable functions. Using Theorem 5.2.2, we deduce that if a refinable function φ belongs to the space $L_p(\mathbb{R})$, the vector function (5.9) is an L_p-fractal curve of the affine operators B_0, B_1, where $B_i = T_i|_{A_n}$, $i = 0, 1$, the operators T_0, T_1 are determined in $\mathbb{R}^N$ by their matrices by formulas (5.11), and A_n is the minimal affine plane that contains the point

$$c = \Big(\int_0^1 \varphi(t)\,dt, \dots, \int_{N-1}^{N} \varphi(t)\,dt\Big)^T$$

and is invariant under the operators T_0, T_1. If $p < \infty$, we have

$$\rho_p(\tilde{B}_0, \tilde{B}_1) < 1\,.$$

Since the operators T_0, T_1 always satisfy the cross-condition, the Hölder exponent of the function φ in $L_p[0, N]$ is equal to $-\log_2 \hat{\rho}(\tilde{B}_0, \tilde{B}_1)$.

Conversely, if a vector c satisfies the condition $c = \frac{1}{2}T_0 c + \frac{1}{2}T_1 c$, i.e., it is an eigenvector for the operator $T_0 + T_1$ corresponding to the eigenvalue 2 and $\rho_p(\tilde{B}_0, \tilde{B}_1) < 1$ ($\tilde{B}_i$ is the restriction of the operator T_i to the minimal common invariant space of the operators T_0, T_1 that contains the vector $T_0 c - c$), then the solution φ of the corresponding refinement equation belongs to $L_p(\mathbb{R})$. Moreover, $\int_0^1 v(t)\,dt = c$ and for any dyadic rational $x = 0.d_1 \dots d_k$, we have

$$2^k \int_x^{x+2^{-n}} v(t)\,dt = B_{d_1} \cdots B_{d_k} c; \tag{5.15}$$

i.e., the mean value of the function v on the segment $[x, x + 2^{-k}]$ is equal to $B_{d_1} \cdots B_{d_k} c$.

Thus, in order to construct an L_p-solution of the refinement equation, we need to find an eigenvector of the operator $T_0 + T_1$ with the eigenvalue 2, to find the corresponding affine plane A_n, and check that the p-radius of the operators T_0, T_1 is strictly less than 1 on this plane. Now we can compute the mean value of the function φ on all dyadic intervals. It is clear that in this way the function is determined uniquely. One can proceed in a different way: take a piecewise constant function $c(x)$ that is equal to the k-th coordinate of the vector c on each of the intervals $[k, k+1)$, $k = 0, \dots, N-1$, and apply to it the iterations of the operator T (recall that $Tf(x) = \sum c_k f(2x - k)$). Then by Theorem 5.2.2, $T^m c(\cdot)$ converges to φ in the space $L_p[0, N]$ with the rate of convergence at least $C \cdot (\rho_p)^m$. For c, we can take an arbitrary point of the plane A_n.

The operator $T_0 + T_1$ may also have several eigenvectors with the eigenvalue 2, each of them generating its own plane A_n; however the relation $\rho_p(T_0|_{A_n}, T_1|_{A_n}) < 1$ holds for only one of them.

The main results concerning smoothness of refinable functions in the space L_p will be presented in Chapter 7, where the spaces A_n and the general form of the operators T_0, T_1 on them are expressed in explicit form.

5.3. Smoothness of fractal curves in the spaces W_p^r and C^r

According to Corollary 5.1.5, if a fractal curve is continuously differentiable, $\hat{\rho} = \frac{1}{2}$. In the general case, the converse is not true. Now we shall find sufficient conditions for affine operators to generate a continuously differentiable fractal curve.

PROPOSITION 5.3.1. *A fractal curve corresponding to a pair of affine operators B_0, B_1 is continuously differentiable on the segment $[0, 1]$ if and only if each operator $\tilde{B}_i$, $i = 0, 1$, has an eigenvector $\tilde{v}_i$ corresponding to the eigenvalue $\frac{1}{2}$ and the following conditions hold:*

1) $\tilde{B}_0\tilde{v}_1 = \tilde{B}_1\tilde{v}_0$.
2) $\hat{\rho}\left(\tilde{B}_0|_{\tilde{L}}, \tilde{B}_1|_{\tilde{L}}\right) < \frac{1}{2}$,

where $\tilde{L}$ is the minimal subspace $\tilde{A}_n$ that is invariant under B_0 and B_1 and contains the vector $\tilde{v}_1 - \tilde{v}_0$.

Proof. Suppose a fractal curve $v(x)$ is differentiable. Represent the affine space A_n as an affine plane in $\mathbb{R}^{n+1}$, taking an arbitrary point outside A_n as the origin of coordinates. The operators B_i can be uniquely extended to linear operators in $\mathbb{R}^{n+1}$. Differentiating equation (5.3) with respect to x, we get

$$v'(x) = 2\tilde{B}_i v'(2x - i), \quad x \in \left[\frac{i}{2}, \frac{i+1}{2}\right], \quad i = 0, 1.$$

Thus v' is an affine curve in the space

$$L = \text{aff}\left\{v'(x),\ x \in [0, 1]\right\},$$

corresponding to the operators $2\tilde{B}_0|_L$, $2\tilde{B}_1|_L$. It remains to refer to Theorem 5.1.4.

Conversely, if there exist vectors $\tilde{v}_i$, $i = 0, 1$, such that $\tilde{B}_i\tilde{v}_i = \frac{1}{2}\tilde{v}_i$ and properties 1) and 2) hold, then, by Theorem 5.1.4, there exists a fractal curve $\tilde{v} : [0, 1] \to L$, where $L = \tilde{v}_0 + \tilde{L}$ and L is the minimal subspace $\tilde{A}_n$ that is invariant under B_0 and B_1 and contains the vector $\tilde{v}_1 - \tilde{v}_0$. Then the function

$$v_0 + \int_0^x \tilde{v}(t)\, dt$$

satisfies (5.3) and hence (by Theorem 5.1.4) it coincides with $v(x)$. Therefore v is differentiable and

$$v'(x) = \tilde{v}(x). \qquad \diamondsuit$$

Thus, if a fractal curve is continuously differentiable, its derivative is also a fractal curve. The dimension of the derivative (i.e., the dimension of the affine hull of its image) is 1 less than the dimension of the curve. To verify this, let us return to the proof of Proposition 5.3.1. The function v is a fractal curve in A_n and its derivative is a fractal curve in $L \subset A_n$. These two spaces cannot coincide since the joint spectral radius of the operators $\tilde{B}_0, \tilde{B}_1$ is at least $\frac{1}{2}$ on the whole space $\tilde{A}_n$ (Remark 5.1.7) and at least $\frac{1}{2}$ on the space $\tilde{L}$ (Proposition 5.1.3). Therefore,

$$\dim L \leq \dim A_n - 1.$$

On the other hand, $\dim L \geq \dim A_n - 1$ since the function v is obtained from v' by integration of the function

$$v(x) = v_0 + \int_0^x v'(t)\,dt, \tag{5.16}$$

which either does not increase the dimension of the image (if $v_0 \in L$) or increases it by 1 (if $v_0 \notin L$). Thus,

$$\dim\ \operatorname{Im} v' + 1 \ = \ \dim\ \operatorname{Im} v.$$

The converse is also true; namely, each fractal curve has a primitive. Let v be a fractal curve corresponding to an irreducible pair of operators B_0, B_1 in the space A_n. Represent A_n as an affine hyperplane in the linear space $\mathbb{R}^{n+1}$, $0 \notin A_n$. Then for an arbitrary $v_0 \in \mathbb{R}^{n+1} \setminus \tilde{A}_n$, the function

$$V(x) = V_0 + \int_0^x v(x)\,dx \tag{5.17}$$

is a fractal curve for the affine operators $\frac{1}{2}\mathbf{B_i} + b_i$, $i = 0, 1$, where $\mathbf{B_i}$ is a linear extension of the operator B_i to $\mathbb{R}^{n+1}$, $b_i = \frac{1}{2}\mathbf{B_i}v_i - v_i$, $V_1 = V(1)$. This can be proved by direct substitution into the equation of the fractal curve. Thus we have the following statement.

COROLLARY 5.3.2. *If a fractal curve in A_n is continuously differentiable r times, its r-th derivative is a fractal curve in A_{n-r}. Conversely, any fractal curve in A_{n-r} is the r-th derivative of an appropriate curve in A_n.*

It follows from Proposition 5.3.1 that if a fractal curve v is continuously differentiable, then each of the operators $\tilde{B}_0, \tilde{B}_1$ has the eigenvalue $\frac{1}{2}$. Moreover, in this case these operators have a common invariant subspace of dimension $(n-1)$, where their joint spectral radius is strictly less than $\frac{1}{2}$ (this subspace is $\operatorname{span}\operatorname{Im} v'$). By induction, we conclude that if a fractal curve is differentiable r times, the corresponding operators $\tilde{B}_0, \tilde{B}_1$ have the eigenvalues $\frac{1}{2}, \dots, \frac{1}{2^r}$ and their matrices in an appropriate basis have the form

$$\tilde{B}_i = \begin{pmatrix} \frac{1}{2} & \cdots & 0 & 0 & \cdots & 0 \\ \vdots & \cdots & \vdots & \vdots & \cdots & \vdots \\ * & \cdots & \frac{1}{2^r} & 0 & \cdots & 0 \\ & * & & & \tilde{D}_i & \\ & & & & & \end{pmatrix}. \tag{5.18}$$

Here $\hat{\rho}(\tilde{D}_0, \tilde{D}_1) < \frac{1}{2^r}$ (the operators $2^r\tilde{D}_0, 2^m\tilde{D}_1$ generate the r-th derivative of our curve in the space A_{n-r}). In this basis, the plane of the k-th derivative span $\operatorname{Im} v^{(k)}$ has the equation $x_1 = \cdots = x_k = 0$ (the first k coordinates are zeros) and the eigenvectors of the operator $\tilde{B}_0$ ($\tilde{B}_1$) corresponding to the eigenvalues $\frac{1}{2}, \dots, \frac{1}{2^r}$ coincide with the values of the derivatives, from the first to the r-th, at the point $x = 0$ (respectively, $x = 1$).

COROLLARY 5.3.3. *Let l be the maximal integer such that the fractal curve is continuously differential l times* $(0 \le l \le n-1)$. *Then the smoothness of this curve is* $l - \log_2 \hat{\rho}(\tilde{D}_0, \tilde{D}_1)$, *where the matrices $\tilde{D}_i$ are taken from* (5.18) *for $r = l$.*

In particular, if $l = n$, then the matrix $\tilde{B}_i$ given by (5.18) becomes lower triangular with the elements $\frac{1}{2}, \dots, \frac{1}{2^n}$ on the main diagonal. The "affine space" A_0, where the n-th derivative is now situated, has zero dimension, i.e., consists of a single point, the derivative itself becomes an identical constant. Thus, the initial curve is the n-th primitive of a constant, i.e., a polynomial curve of degree at most n. Recall that a polynomial curve is a curve $v : [0,1] \to A_n$ such that all its coordinate components are algebraic polynomials on $[0,1]$. Therefore, if the smoothness of a fractal curve is strictly greater than n, it is polynomial (since it is differentiable n times). Actually, a stronger assertion holds:

PROPOSITION 5.3.4. *The smoothness of an n-dimensional fractal curve is either strictly less than n or infinite. In the latter case the curve is polynomial of degree at most n. If a curve is nondegenerate (of total dimension n), its degree is n and there exists a basis in which the matrices of the operators B_0 and B_1 are upper triangular with the diagonal elements $\frac{1}{2}, \dots, \frac{1}{2^n}$. The k-th component of the curve in this basis is a polynomial of degree k.*

This assertion needs a proof only in the case when the smoothness is equal to n. If a curve is n times continuously differentiable, the assertion has already been proved. It remains to consider the case where $v^{(n-1)}$ is not continuously differentiable but $\alpha_{v^{(n-1)}} = 1$. Let us show that this case is not possible. Let $v^{(n-1)}$ be a fractal curve on the affine plane A_1, i.e., on a straight line. It is generated by the one-dimensional operators

$$2^{n-1}\tilde{D}_0, \quad 2^{n-1}\tilde{D}_1$$

(formula (5.18) for $r = n-1$), i.e., by the contractions with respect to two points of the straight line. By definition, the maximal contraction coefficient is

$$\hat{\rho}\,(\, 2^{n-1}\tilde{D}_0,\; 2^{n-1}\tilde{D}_1 \,) \;=\; 1/2,$$

since

$$-\log_2 \hat{\rho} = \alpha_{v^{(n-1)}} = 1 \Rightarrow \hat{\rho} = 1/2$$

(Corollary 5.3.3). The sum of these coefficients is equal to 1 (due to the "cross-condition", Remark 5.1.7). Hence each of them is equal to 1/2. Thus $\tilde{D}_0 = \tilde{D}_1 = 2^{-n}$, and the fractal curve v^{n-1} is generated by two contractions, and it is obviously a linear function. Therefore v^{n-1} is a linear function; hence v is a polynomial curve.

In conclusion, let us remark that any polynomial curve is a fractal curve. This fact can be derived from formula (5.17) for the primitive. Indeed, any polynomial curve of degree at most n can be obtained from an identical constant by taking a primitive n times.

It is easy to formulate analogs of Proposition 5.3.1 and Corollaries 5.3.2 and 5.3.3 for L_p-fractal curves. Thus we see that a fractal curve belongs to the Sobolev space $W_p^r[0,1]$ if and only if the functions $v, v', \dots, v^{(r-1)}$ are continuous fractal curves in the spaces $A_n, \dots, A_{n-r+1}$, and the derivative $v^{(r)}$ is an L_p-fractal curve in A_{n-r}. Formulas for smoothness of fractal curves in the space W_p^r are analogous.

COROLLARY 5.3.5. *A refinable function is continuously differentiable on $\mathbb{R}$ (or belongs to W_p^1, $p \in [1, +\infty)$) if and only if each operator T_i, $i = 0, 1$, has an eigenvector v_i' corresponding to the eigenvalue $\frac{1}{2}$ and*

$$\hat{\rho}\left(T_{0|L_1}, T_{1|L_1}\right) < \frac{1}{2},$$

respectively,

$$\rho_p\left(T_{0|L_1}, T_{1|L_1}\right) < \frac{1}{2},$$

where L_1 is the minimal subspace in $\mathbb{R}^N$ that is invariant under the operators T_0 and T_1 and contains the vector $v_1' - v_0'$.

A refinable function belongs to $C^r(\mathbb{R})$, $1 \le r \le N-1$ (respectively, to W_p^r) if and only if there exists a basis such that both operators T_i, $i = 0, 1$, have the form

$$T_i = \begin{pmatrix} 1 & 0 & \cdots & 0 & 0 & \cdots & 0 \\ * & \frac{1}{2} & \cdots & \vdots & 0 & \cdots & 0 \\ \vdots & \cdots & \cdots & 0 & \vdots & \cdots & \vdots \\ * & \cdots & * & \frac{1}{2^{r-1}} & 0 & \cdots & 0 \\ & & & & & & \\ & * & & & & \tilde{D}_i & \end{pmatrix}, \tag{5.19}$$

the matrix $\tilde{D}_i$ has an eigenvector v_i^r with the eigenvalue $\frac{1}{2^r}$, and

$$\hat{\rho}\left(\tilde{D}_0|_{L_r}, \tilde{D}_1|_{L_r}\right) < \frac{1}{2^r},$$

respectively,

$$\rho_p\left(\tilde{D}_0|_{L_r}, \tilde{D}_1|_{L_r}\right) < \frac{1}{2^r},$$

where L_r is the minimal subspace in $\mathbb{R}^{N-r}$ that is invariant under the operators $\tilde{D}_0$ and $\tilde{D}_1$ and contains the vector $v_1^r - v_0^r$.

Let r be the maximal integer such that the refinable function belongs to C^r (respectively, W_p^r). Then its smoothness is equal to

$$-\log_2 \hat{\rho}\left(\tilde{D}_0|_{L_r}, \tilde{D}_1|_{L_r}\right);$$

for $p < \infty$, it is

$$-\log_2 \rho_p\left(\tilde{D}_0|_{L_r}, \tilde{D}_1|_{L_r}\right).$$

For $r = N-1$, we again obtain Corollary 3.2.13: a refinable function with support $[0, N]$ cannot belong to the space C^{N-1}; if it belongs to W_p^{N-1}, it coincides with the cardinal B-spline of order $N-1$.

To prove this statement, we use Proposition 5.3.1: if the fractal curve

$$v(t) = \left(\varphi(t), \dots, \varphi(t+N-1)\right)^T$$

belongs to C^{N-1}, it is a polynomial curve; hence φ is a piecewise constant function of degree at most N with nodes at integer points. Its $(N-1)$-st derivative is piecewise constant. Therefore, if it is continuous on $\mathbb{R}$, it is identically zero. Consequently φ is a polynomial with compact support; i.e., it is also identically zero, which is a contradiction. Hence $\varphi^{(N-1)}$ cannot be continuous. If $\varphi^{(N-1)}$ is piecewise constant (which corresponds to the case $\varphi \in W^{N-1}$), we use Lemma 3.5.2. ◊

5.4. Local smoothness of fractal curves

A characteristic feature of fractal curves is their fractional smoothness. For many fractal curves, their local smoothness is not constant; it varies from one point to another. Recall that the Hölder exponent of a continuous function v at a point $x \in [0,1]$ is defined by the formula

$$\alpha_v(x) = \sup\Big\{\alpha \mid \|v(x) - v(y)\| \le C|x-y|^{\alpha},\ y \in [0,1]\Big\}.$$

It is clear that $\alpha_v \le \inf_x \alpha_v(x)$. In the general case, this inequality is strict. In the case when $\alpha_v = \inf_{x\in[0,1]} \alpha_v(x)$, we shall say that the Hölder exponent is attained on the segment $[0,1]$. The Hölder exponent is attained at a point x if $\alpha_v = \alpha_v(x)$. Note that, unlike the global Hölder exponent, the local smoothness at a point may take any positive value including $+\infty$, even if the function is not constant on any interval. Smoothness on the right at a point x is defined as

$$\alpha_v^+(x) = \sup\{\alpha \mid \|v(x) - v(y)\| \le C|x-y|^{\alpha},\ y \ge x\}.$$

Smoothness $\alpha_v^-(x)$ on the left is defined analogously.

Denote by $(x) = d_1, d_2, \ldots$ a sequence of zeros and ones, $x = 0.d_1d_2\ldots$ being the corresponding dyadic number. We say that a number $x = 0.d_1d_2\ldots$ is *normal* if for any $\varepsilon > 0$, there exists $N(\varepsilon)$ such that for each $m > N(\varepsilon)$, there are two different digits among the digits d_k with numbers $k \in [m, m(1+\varepsilon)]$. Thus normal numbers cannot be approximated "too close" by dyadic rationals. Almost all (in the Lebesgue measure) points of the segment $[0,1]$ are normal. All rational points that are not dyadic are normal.

For a given collection $\tilde{B}_0, \tilde{B}_1 \in \mathcal{L}(\mathbb{R}^n)$ and for a given sequence $(x) = d_1, d_2, \ldots$ of ones and zeros, we set

$$\hat{\rho}_x = \hat{\rho}_x(\tilde{B}_0, \tilde{B}_1) = \limsup_{k\to\infty} \|\tilde{B}_{d_1}\cdots\tilde{B}_{d_k}\|^{1/k}$$

(the joint spectral radius along the sequence (x)).

THEOREM 5.4.1. *For any point $x = 0.d_1d_2\ldots$, we have*

$$\alpha_v(x) \le -\log_2 \hat{\rho}_x.$$

If the point x is normal, then $\alpha_v(x) = -\log_2 \hat{\rho}_x$.

Proof. Set $a = v_1 - v_0$.

1) Prove that $\alpha_v(x) \le -\log_2 \hat{\rho}_x$. If $\hat{\rho}_x = 0$, there is nothing to be proved. Assume that $\hat{\rho}_x > 0$. Let $(x_k)_{m_k} = d_1, \ldots, d_{m_k}$ be finite subsequences of (x) such that

$$\lim_{k\to\infty} \|\Pi_{x_k}\|^{1/m_k} = \hat{\rho}_x,$$

where $\Pi_{x_k} = \tilde{B}_{d_1}\cdots\tilde{B}_{d_{m_k}}$. For any $\beta \in (0, \hat{\rho}_x)$, there exists a constant $C(\beta) > 0$ such that

$$\|\Pi_{x_k}\| \ge C(\beta)\beta^{m_k}$$

for all $k \in \mathbb{N}$. Since the collection $\{B_0, B_1\}$ is irreducible in the space A_n, there exist finite sequences $(y_r)_{l_r}$, $r = 1, \ldots, n$, for which the vectors $\Pi_{y_1}(a), \ldots, \Pi_{y_n}(a)$ are linearly independent. Due to the equivalence of all norms in $\tilde{A}_n$, there exists a constant $\gamma > 0$ such that for any $B \in \mathcal{L}(\tilde{A}_n)$ one can find $l \in \{1, \ldots, n\}$ for which

$$\|B\Pi_{y_l}(a)\| \geq \gamma\|B\|.$$

Then for any $k \in \mathbb{N}$, we have

$$\|\Pi_{x_k}\Pi_{y_l}(a)\| \geq \gamma C(\beta)\beta^{m_k}$$

(l depends on k). Let z_0^k, z_1^k be binary rational numbers corresponding to the sequences

$$(x_k y_l 0)_{m_k+l_k+1}, \qquad (x_k y_l 1)_{m_n+l_k+1},$$

respectively. For any $\varepsilon > 0$ there exists a constant $C_0(\varepsilon)$ such that for any $k \in \mathbb{N}$ we have

$$C_0(\varepsilon)|z_1^k - z_0^k|^{\alpha_v(x)-\varepsilon} \geq \|v(z_1^k) - v(z_0^k)\|.$$

On the other hand,

$$\|v(z_1^k) - v(z_0^k)\| = \|\Pi_{x_k}\Pi_{y_l}(a)\| \geq \gamma C(\beta)\beta^{m_k}.$$

Then

$$\log_2 C_0(\varepsilon) + (\alpha_v(x) - \varepsilon)(-m_k - l_k) \geq \log_2(\gamma C(\beta)) + m_k \log_2 \beta.$$

For $k \to \infty$, we get

$$\alpha_v(x) - \varepsilon \leq -\log_2 \beta.$$

Now for $\varepsilon \to 0$ and $\beta \to \hat{\rho}_x$, we obtain the required inequality.

2) Prove that $\alpha_v(x) \geq -\log_2 \hat{\rho}_x$ at normal points. Choose any $b > \hat{\rho}_x$. Let $x = 0.d_1 d_2 \ldots$. For any k, we have

$$\|\tilde{B}_{d_1} \cdots \tilde{B}_{d_k}\| \leq C(b)b^k.$$

Let us estimate the smoothness of the function v at the point x on the right (the smoothness on the left is estimated analogously). Choose an arbitrary $y > x$ and assume that the first m dyadic digits in the binary expansions of the numbers x and y coincide (it may happen that $m = 0$), while the $(m+1)$-st binary digits are different. Thus $y = 0.d_1 \ldots d_m d'_{m+1} d'_{m+2} \cdots$ and $d'_{m+1} = 1$, $d_{m+1} = 0$ (as $y > x$). Suppose that in the dyadic representation of the rational number x there are l ones in a row after the digit d_{m+1} (it may happen that $l = 0$). Thus $d_{m+l+2} = 0$. Suppose that in the binary expansion of the number y there are r zeros in a row after the digit d'_{m+1} (it is possible that $r = 0$). Without loss of generality we can assume that $l \geq r$ (the case $l \leq r$ is similar). Then

$$|y - x| \geq 2^{-m-r-2}.$$

For a given k, we set $x_k = 0.d_1 \ldots d_k$ and y_k is defined analogously. Under the hypothesis of the theorem, $y_{n+1} = y_{m+r+1}$. Set

$$\Lambda_1 = \Big\{k \geq m + r + 2,\ d'_k = 1\Big\}.$$

Then

$$\|v(y) - v(y_{m+1})\| = \|v(y) - v(y_{m+r+1})\| \le \sum_{k\in\Lambda_1} \|v(y_k) - v(y_{k-1})\|$$
$$= \sum_{k\in\Lambda_1} \|\tilde{B}_{d_1}\cdots\tilde{B}_{d'_{k-1}}a\| \le \|\tilde{B}_{d_1}\cdots\tilde{B}_{d_m}\tilde{B}_1\tilde{B}_0^r\| \cdot \sum_{k\in\Lambda_1} \|\tilde{B}_{d_{m+r+2}}\cdots\tilde{B}_{d'_{k-1}}a\|.$$

Choosing an arbitrary $q \in (\hat{\rho}, 1)$, we get

$$\|\tilde{B}_{d_1}\ldots\tilde{B}_{d_k}\| \le C(q)q^k.$$

Therefore the series

$$\sum_{k\in\Lambda_1} \|\tilde{B}_{d_{m+r+2}}\cdots\tilde{B}_{d'_{k-1}}a\|$$

is dominated by a geometric progression with the ratio q; hence its sum is bounded by some constant C depending only on q. Thus,

$$\|v(y) - v(y_{m+1})\| \le C\|\tilde{B}_{d_1}\cdots\tilde{B}_{d_m}\tilde{B}_1\tilde{B}_0^r\|. \tag{5.20}$$

Now replace the sequence of zeros in the binary expansion of the number y_{m+1} by the sequence of ones: $y_{m+1} = 0, d_1\ldots d_m 011\ldots$. Taking into account that $l \ge r$ and setting

$$\Lambda_0 = \Big\{k \ge m+r+2,\ d_k = 0\Big\},$$

we get

$$\|v(x) - v(y_{m+1})\| \le \sum_{k\in\Lambda_0} \|v(x_k) - v(x_{k-1})\| = \sum_{k\in\Lambda_0} \|\tilde{B}_{d_1}\cdots\tilde{B}_{d_{k-1}}a\|$$
$$\le \|\tilde{B}_{d_1}\cdots\tilde{B}_{d_m}\tilde{B}_0\tilde{B}_1^r\| \cdot \sum_{k\in\Lambda_0} \|\tilde{B}_{d_{m+r+2}}\cdots\tilde{B}_{d_{k-1}}a\| \le C\|\tilde{B}_{d_1}\cdots\tilde{B}_{d_m}\tilde{B}_0\tilde{B}_1^r\|.$$

Thus,

$$\|v(x) - v(y_{m+1})\| \le C\|\tilde{B}_{d_1}\cdots\tilde{B}_{d_m}\tilde{B}_0\tilde{B}_1^r\|. \tag{5.21}$$

By definition, for any $b > \hat{\rho}_x$, there exists a constant $C(b) > 0$ such that

$$\|\tilde{B}_{d_1}\cdots\tilde{B}_{d_k}\| \le C(b)\,b^k$$

for all $k \in \mathbb{N}$. Then the left-hand side of inequality (5.21) has an upper bound $C \cdot C(b)\,b^{n+r+1}$. The left-hand side of inequality (5.20) is estimated by the relation

$$C\|\tilde{B}_{d_1}\cdots\tilde{B}_{d_m}\| \cdot \|\tilde{B}_1\tilde{B}_0^r\| \le C \cdot C(b)\,C(q)\,b^m q^{r+1}.$$

Summing (5.21) and (5.20), we get

$$\|v(x) - v(y)\| \le \|v(x) - v(y_{m+1})\| + \|v(y) - v(y_{m+1})\|$$
$$\le CC(b)b^m(b^{r+1} + C(q)q^{r+1}).$$

Taking into account that $|y - x| \ge 2^{-m-r-2}$, we obtain

$$\frac{\log_2\|v(x) - v(y)\|}{\log_2|y-x|} \ge -\frac{\log_2 C + \log_2 C(b) + n\log_2 b + \log_2(b^{r+1} + C(q)q^{r+1})}{m+r+2}.$$

Since the point x is normal, $r/m \to 0$ for $m \to \infty$; hence,

$$\frac{\log_2(b^{r+1} + C(q)q^{r+1})}{m+r+2} \to 0.$$

Thus for $m \to \infty$, we get

$$\frac{\log_2 \|v(x) - v(y)\|}{|y - x|} \geq - \log_2 b.$$

Directing b to $\hat{\rho}_x$, we obtain the required result. $\Diamond$

If the operators $\tilde{B}_0, \tilde{B}_1$ are nondegenerate and the sequence (x) is periodic with period $(d_1 \dots d_k)$, then $\hat{\rho}_x = \big(\rho(\tilde{B}_{d_1} \cdots \tilde{B}_{d_k})\big)^{1/k}$. Therefore at rational points x local smoothness can be expressed in terms of ordinary spectral radius.

PROPOSITION 5.4.2. *Let operators $\tilde{B}_0, \tilde{B}_0$ be nondegenerate. If x is a rational, but not dyadic, with a period $(d_1 \dots d_k)$ $(k \geq 2)$, then we have*

$$\alpha_v(x) = -\frac{1}{k} \log_2 \rho(\tilde{B}_{d_1} \cdots \tilde{B}_{d_k}).$$

If x is binary rational, then

$$\alpha_v^+(x) = -\log_2 \rho(\tilde{B}_0), \qquad \alpha_v^-(x) = -\log_2 \rho(\tilde{B}_1).$$

Proof. If those numbers are not dyadic, the statement follows from Theorem 5.4.1. Binary rational numbers are not normal; hence in this case we need a special proof. Thus let x be a dyadic number. The inequality

$$\alpha_v^+(x) \leq -\log_2 \rho(\tilde{B}_0)$$

is derived in the same way as in the proof of Theorem 5.4.1. Let us prove the inverse inequality

$$\alpha_v^+(x) \geq -\log_2 \rho(\tilde{B}_0).$$

Choose any $b > \rho(\tilde{B}_0)$. Let $x = 0.d_1 d_2 \dots d_n 00 \dots$. For any k we have

$$\|\tilde{B}_{d_1} \cdots \tilde{B}_{d_k}\| \leq C(b) b^k.$$

For an arbitrary

$$y \in (x,\, x + 2^{-m-1}), \quad y = 0.d_1 d_2 \dots d_m d_{m+1}, \dots$$

set $\Lambda = \{k \geq m+1,\ d_k = 1\}$ and $r = \min \Lambda - m - 1$. Then

$$\|v(y) - v(x)\| = \|v(y_{m+r+1}) - v(x)\| \leq \sum_{k \in \Lambda} \|v(y_k) - v(y_{k-1}\|$$

$$= \sum_{k \in \Lambda} \|\tilde{B}_{d_1} \cdots \tilde{B}_{d_{k-1}} a\| \leq \|\tilde{B}_{d_1} \cdots \tilde{B}_{d_m} \tilde{B}_0^r\| \cdot \sum_{k \in \Lambda} \|\tilde{B}_{d_{m+r+2}} \cdots \tilde{B}_{d_{k-1}} a\|.$$

The series

$$\sum_{k \in \Lambda} \|\tilde{B}_{d_{m+r+2}} \cdots \tilde{B}_{d_{k-1}} a\|$$

is dominated by the series $\sum_l C(q)\, q^l$, where $q \in (\hat{\rho}, 1)$. Thus,

$$\|v(y) - v(x)\| \leq C \|\tilde{B}_{d_1} \cdots \tilde{B}_{d_m} \tilde{B}_0^r\| \leq C_1(b) b^r,$$

where $b > \rho(\tilde{B}_0)$. Taking into account that $|y - x| \geq 2^{-m-r-1}$, we get

$$\frac{\log_2 \|v(x) - v(y)\|}{\log_2 |y - x|} \geq -\frac{\log_2 C_1(b) + n \log_2 b}{m + r + 1}.$$

It follows that as $y \to x + 0$ (i.e., $r \to \infty$) we get

$$\frac{\log \|v(x) - v(y)\|}{|y - x|} \geq -\log_2 b.$$

Directing b to $\hat{\rho}_x$, we obtain the required result.

The equality $\alpha_v^-(x) = -\log_2 \rho(\tilde{B}_1)$ is proved in the same way, by replacing the sequence of zeros at the end of the dyadic x by the sequence of ones. ◊

REMARK 5.4.3. The nondegeneracy assumption for operators $\tilde{B}_0, \tilde{B}_1$ in Proposition 5.4.2 is essential. If at least one of the operators is degenerate, then the local smoothness may be different at different dyadic points.

REMARK 5.4.4. It follows from Theorem 5.4.1 that the inequality

$$\alpha_v(x) \leq -\log_2 \hat{\rho}_x$$

holds at all points of the segment $[0, 1]$, while the inverse inequality generally holds only at normal points. We used the normality of the number x only in inequalities (5.20) and (5.21). More precisely, we used the fact that at normal points the values

$$\|\Pi_m \tilde{B}_1 \tilde{B}_0^r\|^{\frac{1}{m+r+1}}, \quad \|\Pi_m \tilde{B}_0 \tilde{B}_1^r\|^{\frac{1}{m+r+1}}$$

are asymptotically equivalent as $m \to \infty$ for any product $\Pi_m = \tilde{B}_{d_1} \cdots \tilde{B}_{d_m}$. This condition holds if and only if for any product Π_m we have

$$\lim_{r\to\infty} \|\Pi_m \tilde{B}_1 \tilde{B}_0^r\|^{\frac{1}{m+r+1}} = \lim_{r\to\infty} \|\Pi_m \tilde{B}_0 \tilde{B}_1^r\|^{\frac{1}{m+r+1}}. \tag{5.22}$$

The equality $\rho(\tilde{B}_0) = \rho(\tilde{B}_1)$ is a necessary condition for the latter relation. It is not sufficient even for nondegenerate operators. Sufficient (rather complicated) conditions can be formulated in terms of eigenvectors of the operators $\tilde{B}_0, \tilde{B}_1$. In what follows, we shall use the formula for local smoothness only for normal points.

REMARK 5.4.5. It follows from the proof of Theorem 5.4.1 that at normal points local smoothness is the same on both the right and the left, which is not the case at dyadic points.

Theorem 5.4.1 allows a number of conclusions concerning the distribution of points with given local smoothness on the segment $[0, 1]$. Here again we restrict ourselves to nondegenerate operators $\tilde{B}_0, \tilde{B}_1$. Otherwise, sets of points of a given local smoothness may have less regular structure (Remark 5.4.11).

Here and below we denote by $\mathcal{D}$ the set of binary rational numbers on the real axis, i.e., $\mathcal{D} = \{\frac{k}{2^r} \mid k, r \in \mathbb{Z},\ r \geq 0\}$, and by $\mathcal{D}_r$ the set of binary rational numbers of order r. In other words,

$$\mathcal{D}_0 = \mathbb{Z}, \quad \mathcal{D}_r = \left\{\frac{k}{2^r},\ \ k \text{ for},\ r \in \mathbb{N}\right\}.$$

LEMMA 5.4.6. *Suppose the operators $\tilde{B}_0$ and $\tilde{B}_1$ are nondegenerate. Then for any $x, y \in [0, 1]$ such that $2^r x - y \in \mathcal{D}$ for some integer r, the local smoothness of the function v is the same at the points x and y.*

In other words, if dyadic numbers x and y end with the same sequence of zeros and ones, then $\alpha_v(x) = \alpha_v(y)$.

`Proof.` It follows from formula (5.3) that a restriction of the curve $v(x)$ to any segment $[2^{-q}k,\, 2^{-q}(k+1)]$ is affinely similar to the whole curve $v(x)$. Namely,

$$v(2^q x - k) = B_{d_1} \cdots B_{d_q} v(x)$$

for an appropriate product $B_{d_1} \cdots B_{d_q}$, where the operator corresponding to the similarity transformation is nondegenerate. Therefore,

$$\alpha_v(2^q x - k) = \alpha_v(x),$$

which completes the proof. ◊

DEFINITION 5.4.7. A nonempty subset of the segment $[0,1]$ is called a 0-1 set if together with any point x it contains all the points whose binary expansions end with the same sequence of zeros and ones as x.

In other words, a set has the 0-1 property if any point belongs or does not belong to it independently of any finite subset of digits in its binary expansion. Then we say that 0-1 set of a point x (denoted by $\mathcal{M}_x$) is the set of points from the segment $[0,1]$ whose binary expansions end with the same sequence as the binary expansion of x. It is clear that $\mathcal{M}_x$ is the minimal (with respect to inclusion) 0-1 set that contains x. It is easy to see that

$$\mathcal{M}_x = \{2^r x + p,\ r \in \mathbb{Z},\, p \in \mathcal{D}\} \cap [0,1].$$

Any set $\mathcal{M}_x$ is an equivalence class; hence the segment $[0,1]$ is partitioned into disjoint 0-1 sets. Each 0-1 set is everywhere dense on the segment $[0,1]$. Moreover, if it is Lebesgue measurable, its measure is either 0 or 1. This fact is called the "zero-one law" (see, for example, [**120**]). It immediately follows from the following statement: if a set $A \subset [0,1]$ has positive measure and $\big(A+p\big) \cap [0,1] \subset A$ for any binary rational p, then A has measure 1 (the reader can easily prove this fact).

It follows from Lemma 5.4.6 that a set of points where local smoothness of a fractal curve takes a given value is either an empty set or a 0-1 set. Consider the local smoothness $\alpha_v(x)$ as a function of x. This function is measurable. If operators $\tilde{B}_0, \tilde{B}_1$ are nondegenerate, it is bounded almost everywhere (since in this case $\hat{\rho}_x \geq \check{\rho} > 0$). Moreover, all its level sets, i.e., the sets

$$\Big\{x \in [0,1] \;\Big|\; \alpha_v(x) \leq t\Big\},$$

are 0-1 sets. Since measurable functions for which all level sets have measure 0 or 1 are constants, the quantity $\alpha_v(x)$ is the same at almost all points of the segment $[0,1]$. We denote this quantity by α_{av} and call it the *mean smoothness* of v. It follows from Theorem 5.4.1 that the mean smoothness exists if and only if for almost all $x \in [0,1]$ the quantity $\hat{\rho}_x$ is the same. Denote it by $\bar{\rho}$. This value is known in the ergodic theory as the Lyapunov exponent.

DEFINITION 5.4.8. The quantity

$$\bar{\rho} = \lim_{k\to\infty} \left(\prod_{d_1,\dots,d_k} \|\Pi_k\| \right)^{1/2^k k} \tag{5.23}$$

is called the Lyapunov exponent of linear operators $\tilde{B}_0, \tilde{B}_1$.

This limit exists and is independent of the norm. In addition, the following statement holds: for an arbitrary pair of linear operators $\tilde{B}_0, \tilde{B}_1$ the limit $\lim_{k\to\infty} \|\Pi_k\|^{1/k}$ exists and is equal to $\bar{\rho}$ for almost all $x \in [0,1)$, where

$$x = 0.d_1 d_2, \ldots, \quad \Pi_k = \tilde{B}_{d_1} \cdots \tilde{B}_{d_k}$$

(Appendix A.6, Proposition A.6.7). Using Theorem 5.4.1, we get

$$\alpha_{av} = -\log_2 \bar{\rho}. \tag{5.24}$$

What values can the local smoothness take for various x? Consider the formula $\alpha_v(x) = -\log_2 \hat{\rho}_x$. Obviously, $\check{\rho} \leq \hat{\rho}_x \leq \hat{\rho}$ for any x. Indeed, passing to the limit as $k \to \infty$ in the inequality

$$\min_{d_1,\ldots,d_k \in \{0,1\}} \|\Pi_k\|^{1/k} \leq \left\|\tilde{B}_{d_1} \ldots \tilde{B}_{d_k}\right\|^{1/k} \leq \max_{d_1,\ldots,d_k \in \{0,1\}} \|\Pi_k\|^{1/k},$$

we obtain the required inequality. Hence by Theorem 5.4.1 the inequality

$$\alpha_v(x) \leq -\log_2 \check{\rho}$$

holds for all $x \in [0,1]$. Note that $\check{\rho} > 0$ since the operators are nondegenerate. Therefore,

$$-\log_2 \check{\rho} < \infty.$$

On the other hand, the local smoothness at any point is not smaller than the global smoothness; therefore,

$$\alpha_v(x) \geq -\log_2 \hat{\rho}.$$

Thus the values of local smoothness at all points $x \in [0,1]$ belong to the finite segment

$$\left[-\log_2 \hat{\rho},\ -\log_2 \check{\rho}\right].$$

It appears that they fill the whole segment; i.e., for each $\alpha \in [-\log_2 \hat{\rho},\ -\log_2 \check{\rho}]$, there exists $x \in [0,1]$ (hence there exists a 0-1 set of points) such that $\alpha_v(x) = \alpha$.

LEMMA 5.4.9. *Suppose operators $\tilde{B}_0$ and $\tilde{B}_1$ are nondegenerate and $\hat{\rho}$, $\check{\rho}$ are their joint spectral radius and lower spectral radius, respectively. Then for each number $a \in [\check{\rho}, \hat{\rho}]$, there exists a normal sequence (x) of zeros and ones such that the joint spectral radius along this sequence is equal to a.*

Proof. By the definition of lower spectral radius, there exists a sequence of operator products $\check{\Pi}_k$, $k \in \mathbb{N}$ (the length of $\check{\Pi}_k$ is k), such that

$$\check{\beta}_k = \|\check{\Pi}_k\|^{1/k} \to \check{\rho}, \quad k \to \infty.$$

Then

$$\left[\rho\left(\check{\Pi}_k\right)\right]^{1/k} \to \check{\rho}, \quad k \to \infty$$

(recall that by ρ we denote the ordinary spectral radius). Since the operators are nondegenerate, we can assume that for $k \geq 3$, each product $\check{\Pi}_k$ contains both operators $\tilde{B}_0, \tilde{B}_1$. Otherwise, multiply each of them from the right by $\tilde{B}_0 \tilde{B}_1$. Here the property $\|\check{\Pi}_k\|^{1/k} \to \check{\rho}$ remains true. Using Lemma A.6.2, choose a sequence of the products $\widehat{\Pi}_k$, $k \in \mathbb{N}$, such that

$$\widehat{\beta}_k = \left[\rho\left(\widehat{\Pi}_k\right)\right]^{1/k} \to \hat{\rho}, \quad k \to \infty.$$

We can assume that for sufficiently large k, the product Π_k contains both operators $\tilde{B}_0, \tilde{B}_1$. If there exist arbitrarily long products that do not contain $\tilde{B}_1$, then

$$\hat{\rho} = \lim_{k\to\infty} \|\tilde{B}_0^k\|^{1/k} = \rho(\tilde{B}_0).$$

Let us show that there exists $r \in \{1, \dots, n\}$ such that

$$\left[\rho(\tilde{B}_0^{k-r}\tilde{B}_1^r)\right]^{1/k} \to \hat{\rho}$$

as $k \to \infty$. Then we can set

$$\Pi_k = \tilde{B}_1^r \tilde{B}_1^{k-r}, \quad k \geq r+1.$$

Let λ be an eigenvalue of $\tilde{B}_0$ such that $|\lambda| = \hat{\rho}$ and u is the corresponding eigenvector (which may be complex). Denote by $L \subset \mathbb{R}^n$ the linear span of all the root subspaces corresponding to the eigenvalues less than $\hat{\rho}$ in modulus. It is clear that

$$L = \{v \in \mathbb{R}^n, \ \tilde{B}_0^k v = o(\hat{\rho}^k)\}.$$

Since the operator $\tilde{B}_1$ is nondegenerate, one of the vectors $\tilde{B}_1 u, \dots, \tilde{B}_1^n u$ does not belong to L. Let $\tilde{B}_1^r u \notin L$. Then

$$\rho(\tilde{B}_0^{k-r}\tilde{B}_1^r) \geq C|\lambda|^k$$

for some $C > 0$, which was to be proved.

Now we shall construct a required sequence of operators from the products $\check{\Pi}_k, \widehat{\Pi}_k$. To this end, we take two arbitrary sequences of positive numbers $\{\check{\varepsilon}_k\}_{k\in\mathbb{N}}$, $\{\hat{\varepsilon}_k\}_{k\in\mathbb{N}}$ that tend to zero. Choose the product $\check{\Pi}_1$ (it consists of a single operator) and the minimal positive number p_1 such that

$$\left\|\left(\check{\Pi}_1\right)^{p_1}\right\|^{1/p_1} \leq \max\{a, \check{\beta}_1\} + \check{\varepsilon}_1.$$

Such a number exists since

$$\lim_{p\to\infty} \left\|\left(\check{\Pi}_1\right)^{p}\right\|^{1/p} = \check{\beta}_1 \leq \max\{a, \check{\beta}_1\} < \max\{a, \check{\beta}_1\} + \check{\varepsilon}_1.$$

Then our sequence begins with $\check{\Pi}_1 \dots \check{\Pi}_1$ (p_1 factors). Denote the finite sequence of operators corresponding to the product $\check{\Pi}_k$ by the same symbol $\check{\Pi}_k$. Thus we have constructed the first block (containing p_1 operators) of our sequence. Now take the product $\widehat{\Pi}_1$ and the minimal positive integer q_1 such that

$$\left\|\left(\check{\Pi}_1\right)^{p_1}\left(\widehat{\Pi}_1\right)^{q_1}\right\|^{1/(p_1+q_1)} \geq \min\{a, \widehat{\beta}_1\} - \widehat{\varepsilon}_1.$$

Such a number exists since

$$\lim_{q\to\infty} \left\|\left(\check{\Pi}_1\right)^{p_1}\left(\widehat{\Pi}_1\right)^{q}\right\|^{1/(p_1+q)} = \lim_{q\to\infty} \left\|\left(\widehat{\Pi}_1\right)^{q}\right\|^{1/(p_1+q)} = \lim_{q\to\infty} \left\|\left(\widehat{\Pi}_1\right)^{q}\right\|^{1/q}$$
$$= \widehat{\beta}_1 \ \geq \ \min\{a, \widehat{\beta}_1\} \ > \ \max\{a, \check{\beta}_1\} - \widehat{\varepsilon}_1.$$

The first equality in this chain follows from the fact that the operator $\left(\check{\Pi}_1\right)^{p_1}$ is nondegenerate; hence

$$\left\|\left(\check{\Pi}_1\right)^{p_1}\left(\widehat{\Pi}_1\right)^{q}\right\| \geq \left\|C\left(\widehat{\Pi}_1\right)^{q}\right\|$$

for some constant C independent of q. Thus we have constructed the second block of the sequence $\widehat{\Pi}_1 \dots \widehat{\Pi}_1$. It consists of q_1 operators. The third block is constructed similarly to the first one, using $\check{\Pi}_2, \check{\beta}_2, \check{\varepsilon}_2$ instead of $\check{\Pi}_1, \check{\beta}_1, \check{\varepsilon}_1$, respectively. This third block is $\check{\Pi}_2 \dots \check{\Pi}_2$ (p_2 factors), where p_2 is the minimal positive integer such that

$$\left\| \left(\check{\Pi}_1\right)^{p_1} \left(\widehat{\Pi}_1\right)^{q_1} \left(\check{\Pi}_2\right)^{p_2} \right\|^{1/(p_1+q_1+2p_2)} \leq \max\{a, \check{\beta}_2\} + \check{\varepsilon}_2.$$

Such a number exists, since

$$\begin{aligned} &\lim_{p\to\infty} \left\| \left(\check{\Pi}_1\right)^{p_1} \left(\widehat{\Pi}_1\right)^{q_1} \left(\check{\Pi}_2\right)^{p} \right\|^{1/(p_1+q_1+2p)} \\ &\quad \leq \lim_{p\to\infty} \left\| \left(\check{\Pi}_1\right)^{p_1} \left(\widehat{\Pi}_1\right)^{q_1} \right\|^{1/(p_1+q_1+2p)} \left\| \left(\check{\Pi}_2\right)^{p} \right\|^{1/(p_1+q_1+2p)} \\ &\quad = \lim_{p\to\infty} \left\| \left(\check{\Pi}_2\right)^{p} \right\|^{1/(p_1+q_1+2p)} = \lim_{p\to\infty} \left\| \left(\check{\Pi}_2\right)^{p} \right\|^{1/(2p)} = \check{\beta}_2 < \max\{a, \check{\beta}_2\} + \check{\varepsilon}_2. \end{aligned}$$

The fourth block is $\widehat{\Pi}_2 \dots \widehat{\Pi}_2$ (q_2 factors), where q_2 is the minimal positive integer such that

$$\left\| \left(\check{\Pi}_1\right)^{p_1} \left(\widehat{\Pi}_1\right)^{q_1} \left(\check{\Pi}_2\right)^{p_2} \left(\widehat{\Pi}_2\right)^{q_2} \right\|^{1/(p_1+q_1+2p_2+2q_2)} \geq \min\{a, \widehat{\beta}_2\} - \widehat{\varepsilon}_2.$$

The existence of such a number is proved in the same way as for the second block. Continuing our construction, we obtain the sequence

$$(x) = \check{\Pi}_1^{p_1} \dots \check{\Pi}_k^{p_k} \widehat{\Pi}_k^{q_k} \dots$$

(by Π^m we denote the sequence $\Pi \dots \Pi$, m times). This sequence is normal. Let us prove that $\hat{\rho}_x = a$.

1) Show that

$$\limsup_{m\to\infty} \|\tilde{B}_{x_1} \cdots \tilde{B}_{x_m}\|^{1/m} \geq a.$$

To do this, consider the sequence

$$m_k = p_1 + q_1 + \dots + p_k + q_k, \quad k \in \mathbb{N}.$$

Then

$$\|\tilde{B}_{x_1} \cdots \tilde{B}_{x_{m_k}}\|^{1/m_k} = \|\check{\Pi}_1^{p_1} \dots \check{\Pi}_k^{p_k} \widehat{\Pi}_k^{q_k}\|^{1/m_k}.$$

By construction, it is bigger than or equal to $\min\{a, \widehat{\beta}_k\} - \widehat{\varepsilon}_k$. Thus,

$$\begin{aligned} \limsup_{m\to\infty} \|\tilde{B}_{x_1} \cdots \tilde{B}_{x_m}\|^{1/m} &\geq \limsup_{k\to\infty} \|\tilde{B}_{x_1} \cdots \tilde{B}_{x_{m_k}}\|^{1/m_k} \\ &\geq \lim_{k\to\infty} \left[\min\{a, \widehat{\beta}_k\} - \widehat{\varepsilon}_k\right] \end{aligned}$$

(since $\widehat{\varepsilon}_k \to 0$, $\widehat{\beta}_k \to \hat{\rho}$)

$$= \min\{a, \hat{\rho}\} = a.$$

2) Show that

$$\limsup_{m\to\infty} \|\tilde{B}_{x_1} \cdots \tilde{B}_{x_m}\|^{1/m} \leq a.$$

In what follows, we shall denote

$$\Pi_m = \tilde{B}_{x_1} \cdots \tilde{B}_{x_m}.$$

Set $b = \max\{1, \|\tilde{B}_0\|, \|\tilde{B}_1\|\}$. We have to consider two cases.

a) The last operator $\tilde{B}_{x_m}$ belongs to the product $\widehat{\Pi}_k$ for some $k \in \mathbb{N}$. In other words,

$$p_1 + q_1 + \dots + kp_k \; < \; m \; \leq \; p_1 + q_1 + \dots + kp_k + kq_k.$$

Set $n_k = p_1 + q_1 + \cdots + kp_k$. Notice that

$$\|\Pi_{n_k}\|^{1/n_k} \leq \max\{a, \check{\beta}_k\} + \check{\varepsilon}_k.$$

Moreover, $n_k \geq k^2$, since

$$n_k \geq 2 + 2 \cdot 2 + \cdots + 2(k-1) + k = k^2.$$

First suppose that $m \leq n_k + q_k$, i.e., $\tilde{B}_{x_m}$ belongs to the first product of the block

$$\widehat{\Pi}_k \ldots \widehat{\Pi}_k.$$

Then

$$\left\|\tilde{B}_{x_1} \cdots \tilde{B}_{x_m}\right\|^{1/m} = \left\|\Pi_{n_k} \tilde{B}_{x_{n_k+1}} \cdots \tilde{B}_{x_m}\right\|^{1/m} \leq \|\Pi_{n_k}\|^{1/m} \left\|\tilde{B}_{x_{n_k+1}} \cdots \tilde{B}_{x_m}\right\|^{1/m}$$

$$\leq \left[\|\Pi_{n_k}\|^{1/n_k}\right]^{n_k/m} \left(b^{m-n_k}\right)^{1/m} \leq \left[\max\{a, \check{\beta}_k\} + \check{\varepsilon}_k\right]^{n_k/m} b^{(m-n_k)/m}.$$

Note that $m - n_k \leq k$, but on the other hand, $m \geq k^2$; therefore,

$$b^{(m-n_k)/m} \leq b^{1/k} \to 1, \quad k \to \infty.$$

Analogously, $1 - \frac{1}{k} \leq n_k/m \leq 1$, and in addition,

$$\max\{a, \check{\beta}_k\} + \check{\varepsilon}_k \to a, \quad k \in \infty.$$

Hence for an arbitrary $\varepsilon > 0$ we have

$$\left[\max\{a, \check{\beta}_k\} + \check{\varepsilon}_k\right]^{n_k/m} b^{(m-n_k)/m} \leq a + \varepsilon$$

for all sufficiently large k. So for this case we have

$$\left\|\tilde{B}_{x_1} \cdots \tilde{B}_{x_m}\right\|^{1/m} \leq a + \varepsilon$$

for sufficiently large k.

Suppose now that $m > n_k + q_k$. Let s be the maximal integer for which $m > n_k + sq_k$; i.e., $\tilde{B}_{x_m}$ belongs to the $(s+1)$-st product of the block $\widehat{\Pi}_k \ldots \widehat{\Pi}_k$. By the construction of the sequence (x), we have

$$\|\Pi_{n_k+sq_k}\|^{1/(n_k+sq_k)} < \min\{a, \check{\beta}_k\} + \check{\varepsilon}_k$$

(since the inverse inequality would hold only if $\widehat{\Pi}_s$ was the last product of the block, which is not the case). Then

$$\begin{aligned} \|\Pi_m\|^{1/m} &= \left\|\Pi_{n_k+sq_k} \tilde{B}_{x_{n_k+sq_k+1}} \cdots \tilde{B}_{x_m}\right\|^{1/m} \\ &\leq \|\Pi_{n_k+sq_k}\|^{1/m} \left\|\tilde{B}_{x_{n_k+sq_k+1}} \cdots \tilde{B}_{x_m}\right\|^{1/m} \\ &\leq \left[\min\{a, \check{\beta}_k\} + \check{\varepsilon}_k\right]^{(n_k+sq_k)/m} b^{(m-n_k-sq_k)/m}. \end{aligned}$$

Since

$$m \geq k^2 \quad \text{for } m - n_k - sq_k \leq k,$$

we have

$$b^{(m-n_k-sq_k)/m} \leq b^{1/k}.$$

On the other hand,

$$\min\{a, \check{\beta}_k\} + \check{\varepsilon}_k \to a, \quad k \to \infty,$$

and, in addition, $1-\frac{1}{k}\le (n_k+sk)/m\le 1$. Hence,

$$[\min\{a,\rho\beta_k\}+\rho\varepsilon_k]^{(n_k+sk)/m}\, b^{(m-n_k-sk)/m} \;\le\; a+\varepsilon$$

for sufficiently large k. Thus for any m such that the case a) takes place, the upper limit $\|\Pi_m\|^{1/m}$ does not exceed a.

b) The last operator $\tilde{B}_{x_m}$ belongs to the product $\check{\Pi}_{k+1}$ for some k. In other words, $n_k+kq_k<m\le n_{k+1}$. Let s be the maximal number for which $n_k+kq_k+s(k+1)<m$. As was proved in a), we have

$$\|\Pi_{n_k+kq_k}\|^{1/(n_k+kq_k)}\le a+\varepsilon$$

for sufficiently large k. If $s=0$, then

$$\|\Pi_m\|^{1/m}=\left\|\Pi_{n_k+kq_k}\tilde{B}_{x_{n_k+kq_k+1}}\cdots\tilde{B}_{x_m}\right\|^{1/m}\le (a+\varepsilon)^{(n_k+kq_k)/m}\, b^{(m-n_k-kq_k)/m}.$$

Since

$$(m-n_k-kq_k)/m\le k/m<1/k,\quad 1-1/k\le (n_k+kq_k)/m\le 1,$$

it follows that for sufficiently large k,

$$\|\Pi_m\|^{1/m}\le a+2\varepsilon.$$

If $s\ge 1$, we have

$$\begin{aligned}\|\Pi_m\|^{1/m}&=\|\Pi_{n_k+kq_k}\check{\Pi}^s_{k+1}\tilde{B}_{x_{n_k+q_kk+s(k+1)+1}}\cdots\tilde{B}_{x_m}\|^{1/m}\\ &\le\left\|\Pi_{n_k+kq_k}\check{\Pi}^s_{k+1}\right\|^{1/m}\|\tilde{B}_{x_{n_k+q_k+s(k+1)+1}}\cdots\tilde{B}_{x_m}\|^{1/m}.\end{aligned}$$

The second factor can be estimated as before:

$$\left\|\tilde{B}_{x_{n_k+q_kk+s(k+1)+1}}\cdots\tilde{B}_{x_m}\right\|^{1/m}\le b^{(m-n_k+q_kk+s(k+1))/m}\le b^{1/k}.$$

In order to estimate the first factor, we use the following inequality:

$$\|AB\|^{1/(r+p)}\le\max\left\{\|A\|^{1/r},\|B\|^{1/p}\right\}$$

for any operators A,B and any $r,p\ge 1$. To prove this, set $\alpha=\|A\|^{1/r}$, $\beta=\|B\|^{1/p}$; then we get

$$\ln\|AB\|^{1/(r+p)}\le\frac{r}{r+p}\ln\alpha+\frac{p}{r+p}\ln\beta\le\max\{\ln\alpha,\ \ln\beta\}.$$

Thus,

$$\begin{aligned}\left\|\Pi_{n_k+kq_k}\check{\Pi}^s_{k+1}\right\|^{1/m}&=\left[\left\|\Pi_{n_k+kq_k}\check{\Pi}^s_{k+1}\right\|^{1/(n_k+q_kk+s(k+1))}\right]^{(n_k+q_kk+s(k+1))/m}\\ &\le\max\left\{\|\Pi_{n_k+kq_k}\|^{1/(n_k+q_kk)},\ \left\|\check{\Pi}^s_{k+1}\right\|^{1/(s(k+1))}\right\}^{(n_k+q_kk+s(k+1))/m}.\end{aligned}$$

On the right-hand side of the latter inequality we have the maximum of two quantities. As already noted, the first quantity does not exceed $a+\varepsilon$ (for large k). The second one is bounded above as follows:

$$\left\|\check{\Pi}^s_{k+1}\right\|^{1/s(k+1)}\le\left\|\check{\Pi}_{k+1}\right\|^{s\cdot\frac{1}{s(k+1)}}=\left\|\check{\Pi}_{k+1}\right\|^{1/(k+1)}=\check{\beta}_k.$$

Since

$$\lim_{k\to\infty}\check{\beta}_k=\check{\rho}\le a,$$

it follows that for sufficiently large k, this quantity does not exceed $a+\varepsilon$. Thus we see that our maximum does not exceed $a+\varepsilon$ for sufficiently large k. Therefore,

$$\left\|\Pi_{n_k+kq_k}\breve{\Pi}^s_{k+1}\right\|^{1/m} \le (a+\varepsilon)^{(n_k+q_k k+s(k+1))/m}$$

and consequently,

$$\|\Pi_m\|^{1/m} \le (a+\varepsilon)^{(n_k+q_k k+s(k+1))/m} b^{1/k}.$$

It is clear that

$$\lim_{k\to\infty} b^{1/k} = 1.$$

Note that

$$m-(k+1) \ \le\ n_k+q_k k+s(k+1) \ \le\ m;$$

hence we have the following estimate:

$$1-\frac{k+1}{m} \le \Big(n_k+q_k k+s(k+1)\Big)/m \le 1.$$

Since

$$m \ \ge\ n_k \ +\ q_k k\ \ge n_k+k\ \ge\ k^2+k\ =\ k(k+1),$$

we get

$$1-\frac{1}{k} \le \Big(n_k+q_k k+s(k+1)\Big)/m \le 1;$$

hence the exponent tends to 1 as k increases. Hence for all sufficiently large k, we have

$$\|\Pi_m\|^{1/m} \le a+2\varepsilon.$$

Thus for all sufficiently large m for which the case b) takes place, we have

$$\|\Pi_m\|^{1/m} \le a+2\varepsilon.$$

Directing ε to zero, we see that for such m, the upper limit of $\|\Pi_m\|^{1/m}$ does not exceed a. Thus $\hat{\rho}_x = a$. This completes the proof.

The following theorem summarizes the results concerning the distribution of points of local smoothness.

THEOREM 5.4.10. *Suppose operators $\tilde{B}_0$ and $\tilde{B}_1$ are nondegenerate and $\hat{\rho}$, $\check{\rho}$ are their joint spectral radius and lower spectral radius, respectively. Then*

$$\Big\{\alpha_v(x),\ x\in[0,1]\Big\} = [-\log_2\hat{\rho},\ -\log_2\check{\rho}].$$

For each α from this segment, the set of points x for which $\alpha_v(x)=\alpha$ is a nonempty 0-1 set. In particular, it is everywhere dense on the segment $[0,1]$. For $\alpha_{av} = -\log\bar{\rho}$, where $\bar{\rho}$ is the Lyapunov exponent, it is a set of full measure on $[0,1]$, while for all other values of α it has measure zero.

REMARK 5.4.11. Theorems 5.1.4 and 5.4.1, which give the formulas for global and local smoothness of fractal curves, hold under the most general assumptions for operators $\tilde{B}_0, \tilde{B}_1$. Their joint spectral radius must be strictly less than 1 (a necessary condition for continuity) and the sum of the operators $\tilde{B}_0+\tilde{B}_1$ must have an eigenvalue 1 (to satisfy the cross-condition $B_0v_1 = B_1v_0$). These conditions hold whenever a continuous fractal curve exists. On the other hand, the assertions about the distribution of local smoothness (Proposition 5.4.2 and Theorem 5.4.10), in the general case, hold only for nondegenerate operators.

In Chapter 7, all these results are applied to refinable functions and wavelets. However to do this, we must prove that the operators T_0, T_1 (defined by formula (5.11)) are nondegenerate on the space V. Moreover, to estimate the modulus of continuity (Theorem 5.1.4, inequality (5.6)), we need the irreducibility of T_0, T_1 on this space. The above problems will be studied in §6.3. Now we consider three classical examples of fractal curves.

5.5. Examples

EXAMPLE 5.5.1 (The curve of binomial distribution). $A_n = \mathbb{R}$, $B_0 x = \lambda x$ is the dilation with coefficient λ with respect to zero, $B_1 x = (1-\lambda)x + \lambda$ is the dilation with coefficient $(1-\lambda)$ with respect to 1, $\lambda \in \left(0, \frac{1}{2}\right)$. The operators $\tilde{B}_0, \tilde{B}_1$ commute; hence,

$$\hat{\rho} = \max\{\rho(\tilde{B}_0), \rho(\tilde{B}_1)\} = \max\{\lambda,\ 1-\lambda\} = 1-\lambda\ , \qquad \check{\rho} = \min\{\lambda,\ 1-\lambda\} = \lambda.$$

Thus,

$$\alpha_v = -\log_2(1-\lambda).$$

Let

$$r(x) \ = \ \limsup_{k\to\infty} \frac{1}{k}\sum_{j=1}^{k} d_j$$

be the upper density of ones in the sequence (x). Since

$$\hat{\rho}_x = (1-\lambda)^{r(x)} \lambda^{1-r(x)},$$

we have (Theorem 5.2.2)

$$\alpha_v(x) \ = \ -r(x)\log_2(1-\lambda) \ - \ (1-r(x))\log_2 \lambda$$

at each normal point x. Since $\rho(\tilde{B}_0) \neq \rho(\tilde{B}_1)$ (Remark 5.4.4), the formula of local smoothness does not hold for all points $x \in [0,1]$. Smoothness at rational points is determined by Proposition 5.4.2. In particular, at dyadic points the smoothness on the left is $-\log_2(1-\lambda)$, and smoothness on the right is $-\log_2 \lambda$. Values of local smoothness fill the segment

$$[-\log_2 \hat{\rho}\, ,\ -\log_2 \check{\rho}] = [-\log_2(1-\lambda)\, ,\ -\log_2 \lambda].$$

For almost all x, the smoothness is given by the relation

$$\alpha_{av} = -\log_2 \bar{\rho} = -\log_2 \sqrt{\lambda(1-\lambda)}.$$

Note that $\alpha_{av} > 1$. In particular, the function $v(x)$ is purely singular; i.e., $v' = 0$ almost everywhere. Further, we have

$$\rho_p = 2^{-1/p}\Big(\ \lambda^p \ + \ (1-\lambda)^p\ \Big)^{1/p}$$

for any $p \in [1, +\infty)$. So we conclude from Theorem 5.2.2 that

$$\alpha_{v,p} = \frac{1}{p} - \frac{1}{p}\log_2\Big(\lambda^p + (1-\lambda)^p\Big).$$

On the segment $[0,1]$ the function $v(x)$ represents the distribution function of the series $\sum_{k=1}^{\infty} \varepsilon_k 2^{-k}$, where $\{\varepsilon_k\}$ are independent, identically distributed random variables, $P\{\varepsilon_k = 0\} = \lambda$, $P\{\varepsilon_k = 1\} = 1-\lambda$. The generalized derivative of the function v, i.e., the density function, defines so-called binomial measure *on the segment* $[0,1]$ (see, for example, [**123**]).

EXAMPLE 5.5.2 (Koch curve). $A_n = \mathbb{R}^2$, and B_0, B_1 are the similarity transformations:

$$\tilde{B}_0 = \frac{1}{2\cos\gamma}\begin{pmatrix} \cos\gamma & \sin\gamma \\ \sin\gamma & -\cos\gamma \end{pmatrix}, \qquad \tilde{B}_1 = \frac{1}{2\cos\gamma}\begin{pmatrix} \cos\gamma & -\sin\gamma \\ -\sin\gamma & -\cos\gamma \end{pmatrix},$$

where $\gamma \in \left(0, \frac{\pi}{4}\right)$. Here the origin of coordinates is a fixed point of the operator B_0 and the point $(1, 0)^T$ is a fixed point of the operator B_1. Unlike the previous example, the operators $\tilde{B}_0, \tilde{B}_1$ do not commute. However each of them is a product of an orthogonal operator and the contraction with coefficient $\frac{1}{2\cos\gamma}$. Therefore $\hat{\rho}_x = \frac{1}{2\cos\gamma}$ for any $x \in [0,1]$. In particular, $\hat{\rho} = \bar{\rho} = \check{\rho} = \frac{1}{2\cos\gamma}$. Condition (5.22) of Remark 5.4.4 is satisfied; hence the smoothness at any point $x \in [0,1]$ can be computed by the formula $\alpha_v(x) = -\log \hat{\rho}_x$. Thus the smoothness of the Koch curve is the same at all points, namely, equal to $1 + \log_2 \cos\gamma$. If γ decreases from $\frac{\pi}{4}$ to 0, the smoothness of the curve increases from $\frac{1}{2}$ to 1. In particular, for $\gamma = \pi/6$, we have $\alpha_v = \frac{1}{2}\log_2 3$. Since ρ_p is also equal to $\frac{1}{2\cos\gamma}$, the smoothness of the Koch curve is the same in all spaces $L_p(\mathbb{R})$, namely, equal to $1 + \log_2 \cos\gamma$. Since the collection of operators $\tilde{B}_0, \tilde{B}_1$ is irreducible, it follows from Theorems 5.1.4 and 5.2.2 that $\omega(v,t) \asymp \omega(v,t)_p \asymp t^{1+\log_2 \cos\gamma}$.

EXAMPLE 5.5.3 (De Rham curves). $A_n = \mathbb{R}^2$, and the operators B_0, B_1 have the following linear parts:

$$\tilde{B}_0 = \begin{pmatrix} \omega & 0 \\ \omega & 1-2\omega \end{pmatrix}, \qquad \tilde{B}_1 = \begin{pmatrix} 1-2\omega & \omega \\ 0 & \omega \end{pmatrix}, \tag{5.25}$$

where $\omega \in \left(0, \frac{1}{2}\right)$. Here the operator B_0 has a fixed point $(0, -1)^T$, and the operator B_1 has a fixed point $(1,0)^T$. In contrast to the previous examples, the operators $\tilde{B}_0, \tilde{B}_1$ are not products of orthogonal operators and contractions, which essentially complicates the analysis of the curve. Nevertheless, it can be proved (see [**120**]) that in this case we have

$$\hat{\rho} = \max\left\{\rho(\tilde{B}_0), \sqrt{\rho(\tilde{B}_0\tilde{B}_1)}\right\}, \quad \check{\rho} = \min\left\{\rho(\tilde{B}_0), \sqrt{\rho(\tilde{B}_0\tilde{B}_1)}\right\}.$$

Here,

$$\rho(\tilde{B}_0) = \rho(\tilde{B}_1) = \begin{cases} 1-2\omega, & \text{if } \omega \le \frac{1}{3}, \\ \omega, & \text{if } \omega \ge \frac{1}{3}, \end{cases}$$

$$\sqrt{\rho(\tilde{B}_0\tilde{B}_1)} = \frac{\omega + \sqrt{4\omega - 7\omega^2}}{2}.$$

Thus the global smoothness α_v, as well as the bounds for the local smoothness $-\log_2 \hat{\rho}$ and $-\log_2 \check{\rho}$, can be computed explicitly for each ω. As ω increases from 0 to $\frac{1}{4}$, α_v monotonically increases from 0 to 1, then monotonically decreases to the point $\omega = \frac{4+\sqrt{2}}{14} = 0.3867\ldots$, where $\alpha_v = 0.8706\ldots$, then it again monotonically increases, and for $\omega \to 0.5$, we have $\alpha_v \to 1$. Thus only for $\omega = \frac{1}{4}$, the smoothness is equal to 1 (in this case de Rham's curve is a parabola); for the other ω, we have $\alpha_v < 1$. Moreover, it follows from Theorem 5.4.10 that $\alpha_v(x) = \alpha_v$ on an everywhere dense set x; hence de Rham's curve is not piecewise differentiable if $\omega \neq \frac{1}{4}$. For all $\omega \neq \frac{1}{4}$, we have $\check{\rho} \neq \hat{\rho}$; hence the values of local smoothness fill a segment. The local smoothness can be determined by the formula $\alpha_v(x) = -\log_2 \hat{\rho}_x$, which holds for all points x, not only for normal ones (it can be shown that condition (5.22) in this case is satisfied for any ω). In practice, it is not clear how to compute the local smoothness at an arbitrary point; at all rational points it is effectively computed by Proposition 5.4.2. It follows from the formulas for $\check{\rho}$ and $\hat{\rho}$ that for $\omega \leq \frac{1}{3}$, the minimal local smoothness is attained at dyadic points x, and the maximal local smoothness is attained at points of the form $x = \frac{p}{3 \cdot 2^k}$ ($p \neq 0 \pmod 3$), i.e., at rational points with period (01). For $\omega > \frac{1}{3}$, the situation is different: the minimal smoothness is attained at the point of the form $x = \frac{p}{3 \cdot 2^k}$, and the maximal smoothness is at dyadic points. Note that at dyadicl points the smoothness is the same from both sides. The mean smoothness α_{av} satisfies the inequalities

$$-\frac{1}{4}\log_2\bigl(\omega(1-2\omega)\bigr) \;\leq\; \alpha_{av} \;\leq\; -\frac{1}{3}\log_2\bigl(\omega(1-2\omega)\bigr)$$

(see [**120**]). In particular, it follows that as $\omega \to 0$ and as $\omega \to \frac{1}{2}$, the mean smoothness tends to $+\infty$. For the Lyapunov exponent of operators (5.25), we have an integral formula which allows the approximate computation of the mean smoothness α_{av} for any ω [**120**]. In [**124**], the Hausdorff dimension of the set x of points where local smoothness takes given values (so-called multifractal dimension) was studied.

CHAPTER 6

Factorization of Refinement Equations

6.1. Operators corresponding to the clean mask

Let there be given a refinement equation with mask $\mathbf{m}_0(z) = \frac{1}{2}\sum_{k=0}^{N} c_k z^k$, where $\sum_k c_k = 2$, $c_0 c_N \neq 0$. Recall that a pair of numbers $\pm a \in \mathbb{C}$ is called a pair of symmetric roots of $\mathbf{m}_0$ if $\mathbf{m}_0(a) = \mathbf{m}_0(-a) = 0$. A finite set of complex numbers $\mathbf{b} = \{b_1, \dots, b_n\}$ is called a *cycle* of the polynomial $\mathbf{m}_0$ if, firstly, it is cyclic, i.e., $b_{j+1} = b_j^2$, $j = 1, \dots, n-1$, $b_1 = b_n^2$, and, secondly, $\mathbf{m}_0(-\mathbf{b}) = 0$. The cycle $\mathbf{b} = \{1\}$ is called trivial; other cycles are called nontrivial. We say that the mask $\mathbf{m}_0$ is *clean* if it has neither symmetric roots nor cycles (including trivial ones).

THEOREM 6.1.1. *If the mask $\mathbf{m}_0$ is clean, then the linear operators T_0, T_1 are degenerate and have no nontrivial common invariant subspaces.*

The converse is also true, even in a stronger sense. This follows from Propositions 6.1.2 and 6.1.4 proved below. Theorem 6.1.1 follows from these two propositions. Throughout this chapter, by $[z]$ we denote the vector $\left(1, z, \dots, z^{N-1}\right)^T \in \mathbb{R}^N$, where $z \neq 0$ is an arbitrary complex number.

PROPOSITION 6.1.2. *If the mask $\mathbf{m}_0$ has no symmetric roots, the operators T_0, T_1 are nondegenerate. If $\pm a$ are symmetric roots of $\mathbf{m}_0$, both operators are degenerate and $T_0^*[\bar{a}^2] = T_1^*[\bar{a}^2] = 0$.*

`Proof.` Set $p_0(z) = \sum_k \bar{c}_{2k} z^k$ and $p_1(z) = \sum_k \bar{c}_{2k+1} z^k$. If $\pm a$ are symmetric roots of $\mathbf{m}_0$, then $p_0(\bar{a}^2) = p_1(\bar{a}^2) = 0$. Consequently,

$$T_0^*[\bar{a}^2] = \left(p_0(\bar{a}^2), \bar{a}^2 p_1(\bar{a}^2), \bar{a}^2 p_0(\bar{a}^2), \dots\right)^T = 0,$$

and, analogously, $T_1^*[\bar{a}^2] = 0$.

Suppose now that the operator T_0^* is degenerate. Let us show that $\mathbf{m}_0$ has symmetric roots (for the operator T_1^* the reasoning is similar). Let $T_0^* u = 0$ for some nonzero vector $u = (u_1, \dots, u_N)^T$. Multiplying u by the rows of the matrix T_0^* with odd indices, we get

$$\sum_{s=0}^{l} \bar{c}_{2s} u_{s+j} = 0 \quad \text{for } j = 1, \dots, \left[\frac{N+1}{2}\right],$$

where l is the maximal number for which $c_{2l} \neq 0$. Thus the sequence $\{u_k\}$ satisfies the difference equation with the coefficients $\bar{c}_0, \bar{c}_2, \dots, \bar{c}_{2l}$. The general solution to this equation is given by the following formula: if $l = 0$, then $u_k = 0$; if $l > 0$, then

$$u_k = \sum_{\nu, r_\nu} A_{\nu, r_\nu} k^{r_\nu - 1} \nu^k,$$

where ν are roots of the polynomial $p_0(z)$ and the integers r_ν take values from 1 to the multiplicities of corresponding roots and A_{ν,r_ν} are constants. This formula holds for all u_k contained in the difference equation, i.e., for $k = 1, \dots, l + \left[\frac{N+1}{2}\right]$. Now multiplying u by the even rows of the matrix T_0^*, we get

$$\sum_{s=l_1}^{l_2} \bar{c}_{2s+1} u_{s+j+1} = 0$$

for $j = 1, \dots, \left[\frac{N}{2}\right]$, where l_1 and l_2, respectively, are the minimal and maximal numbers, for which $c_{2l_1+1} c_{2l_2+1} \neq 0$ (if $c_{2s+1} = 0$ for all s, then $\mathbf{m}_0(z)$ obviously has symmetric roots). From this difference equation we conclude that $u_k = 0$ if $l_1 = l_2$ and $u_k = \sum\limits_{\mu,\, r_\mu} B_{\mu,r_\mu} k^{r_\mu - 1} \mu^k$ if $l_1 < l_2$, where μ are roots of the polynomial $p_1(z)$. This formula holds for $k = l_1 + 2, \dots, l_2 + \left[\frac{N}{2}\right] + 1$. Thus the first formula holds for $l + \left[\frac{N+1}{2}\right]$ values of k, and the second formula holds for $l_2 + \left[\frac{N}{2}\right] - l_1$ values of k. Since $\left[\frac{N+1}{2}\right] + \left[\frac{N}{2}\right] = N$, it follows that

$$l + \left[\frac{N+1}{2}\right] + \left(l_2 + \left[\frac{N}{2}\right] - l_1\right) - N = l + l_2 - l_1.$$

If this expression is equal to zero, then $l = 0$ and $l_1 = l_2$; hence $u_k = 0$ for all $k = 1, \dots, N$, which contradicts our assumption $u \neq 0$. If $l + l_2 - l_1 > 0$, then at least $l + l_2 - l_1$ successive elements of $\{u_k\}$ satisfy both formulas; i.e.,

$$\sum_{\nu,\, r_\nu} A_{\nu,r_\nu} k^{r_\nu - 1} \nu^k = \sum_{\mu,\, r_\mu} B_{\mu,r_\mu} k^{r_\mu - 1} \mu^k.$$

We have obtained a system consisting of at least $l + l_2 - l_1$ homogeneous linear equations in variables A_{ν,r_ν}, B_{μ,r_μ}. The number of variables is equal to the total number of nonzero roots (taking into account their multiplicities) of the polynomials p_0 and p_1. The polynomial p_0 has l roots, and the polynomial p_1 has $l_2 - l_1$ roots. Thus the number of variables is $l + l_2 - l_1$, which does not exceed the number of equations. However the system has a nontrivial solution; hence it must be degenerate. Therefore (see Appendix A.18, Lemma A.18.1) the set of numbers ν and μ must contain equal numbers. So the polynomials p_0 and p_1 have a common root, which we denote by z. Then $\pm\sqrt{z}$ is a pair of symmetric roots for $\mathbf{m}_0$. $\Diamond$

REMARK 6.1.3. One can prove that the kernels of the operators T_0^*, T_1^* always coincide. Moreover, not only the kernels but all their root subspaces $E_i^k = \operatorname{Ker}\left([T_i^*]^k\right)$, $i = 0, 1$, coincide as well.

Let us now consider common invariant subspaces. If $\mathbf{m}_0$ has symmetric roots $\pm a$, the kernels of the operators T_0^*, T_1^* contain a common vector $[\bar{a}^2]$; hence the operators T_0, T_1 have a common invariant hyperplane $[\bar{a}^2]^\perp$. Moreover, each of them takes the whole space $\mathbb{R}^N$ to this hyperplane: $\operatorname{Im} T_i \subset [\bar{a}^2]^\perp$, $i = 0, 1$. Thus, if the mask has symmetric roots, the collection T_0, T_1 is irreducible. Assume now that there are no symmetric roots.

PROPOSITION 6.1.4. *Suppose the mask* $\mathbf{m}_0$ *has no symmetric roots. Then:*

a) *If* $\mathbf{m}_0$ *has a cycle* $\mathbf{b} = \{b_1, \dots, b_n\}$, *then the operators* T_0^*, T_1^* *have a common invariant subspace* $U_{\mathbf{b}} = \operatorname{span}\left\{[\bar{b}_k],\ k = 1, \dots, n\right\}$ *(hence* T_0, T_1 *have a common* *invariant subspace* $U_{\mathbf{b}}^{\perp}$*). The operators* T_i^*, $i=0,1$, *are cyclic on this subspace:*

$$T_0^*[\bar{b}_{k+1}] = \overline{\mathbf{m}_0(b_k)}[\bar{b}_k], \quad T_1^*[\bar{b}_{k+1}] = \frac{1}{\bar{b}_k}\overline{\mathbf{m}_0(b_k)}[\bar{b}_k], \qquad k=1,\dots,n. \tag{6.1}$$

b) *Each common invariant subspace of the operators* T_0^*, T_1^* *contains a subspace* $U_{\mathbf{b}} = \operatorname{span}\{[\bar{b}_k],\ k=1,\dots,n\}$, *where* $\mathbf{b}$ *is a cycle of the mask* $\mathbf{m}_0$*; on this subspace, the operators* T_i^*, $i=0,1$, *are defined by formula* (6.1). *In particular, if* $\mathbf{m}_0$ *has no cycles, the pair of operators* T_0, T_1 *is irreducible.*

The proof is based on two lemmas.

LEMMA 6.1.5. *For any* $z \in \mathbb{C}\setminus\{0\}$, *we have*

$$\begin{aligned} T_0^*[\bar{z}^2] &= \overline{\mathbf{m}_0(z)}[\bar{z}] + \overline{\mathbf{m}_0(-z)}[-\bar{z}], \\ T_1^*[\bar{z}^2] &= \frac{1}{\bar{z}}\overline{\mathbf{m}_0(z)}[\bar{z}] - \frac{1}{\bar{z}}\overline{\mathbf{m}_0(-z)}[-\bar{z}]. \end{aligned} \tag{6.2}$$

Proof. We shall prove only the first equality since the second one is proved analogously. For a given $k=1,\dots,N$, the k-th coordinate of the vector $T_0^*[\bar{z}^2]$ is equal to

$$\sum_{j=1}^{N} \bar{c}_{2j-k-1}(\bar{z}^2)^{j-1} = \sum_{j=1}^{N} \bar{c}_{2j-k-1}\bar{z}^{2j-2}$$
$$= \bar{z}^{k-1} \sum_{s\equiv(k-1)(\mathrm{mod} 2)} \bar{c}_s \bar{z}^s = \bar{z}^{k-1}\overline{\mathbf{m}_0(z)} + (-\bar{z})^{k-1}\overline{\mathbf{m}_0(-z)},$$

which completes the proof. ◊

LEMMA 6.1.6. *If the polynomial* $\mathbf{m}_0$ *has no symmetric roots, all the eigenvectors of the operator* $T_0^*\,(T_1^*)^{-1}$ *have the form* $[z]$ *(up to normalization) for certain* $z \in \mathbb{C}\setminus\{0\}$.

Proof. According to Proposition 6.1.2, the operators T_0^* and T_1^* are nondegenerate, so the operator $T_0^*\,(T_1^*)^{-1}$ is well defined on $\mathbb{R}^N$ and nondegenerate. For an arbitrary $v \in \mathbb{R}^N$, the i-th coordinate of the vector $T_0^* v$ is equal to the $(i+1)$-st coordinate of the vector $T_1^* v$ $(i=1,\dots,N-1)$. Hence for any $y=(y_1,\dots,y_N)^T \in \mathbb{R}^N$, we have

$$T_0^*\,(T_1^*)^{-1}\, y = (y_2,\dots,y_N, f(y))^T,$$

where $f(y) = \sum_k f_k y_k$ is a linear functional on $\mathbb{R}^N$. To prove this, it suffices to set $v = (T_1^*)^{-1} y$ and to compare the corresponding coordinates of the vectors $T_0^* v$ and $T_1^* v$. Since the operator $T_0\,(T_1^*)^{-1}$ is linear, the functional f is linear as well. Thus we have

$$T_0^*\,(T_1^*)^{-1} = \begin{pmatrix} 0 & 1 & 0 & 0 & \cdots & 0 \\ 0 & 0 & 1 & 0 & \cdots & 0 \\ \vdots & & & & & \vdots \\ 0 & 0 & 0 & 0 & \cdots & 1 \\ f_1 & f_2 & \cdots & & \cdots & f_N \end{pmatrix}.$$

Each eigenvector of such a matrix has the form $[z]$ (up to normalization), where z is the corresponding eigenvalue. All eigenvalues are roots of the polynomial

$$P(z) = z^N - \sum_{k=1}^{N} f_k z^{k-1}.$$ ◊

`Proof of Proposition 6.1.4.` a) If $\mathbf{m}_0$ has a cycle $\mathbf{b} = \{b_1, \dots, b_n\}$, then substituting $z = b_k$ into (6.2), we obtain (6.1).

b) If the operators T_0, T_1 have a nontrivial common invariant subspace, its orthogonal complement (denoted by U) is a common invariant subspace for T_0^*, T_1^*. Since the operators are nondegenerate, the operator $T_0^*(T_1^*)^{-1}$ also has an invariant subspace U containing at least one eigenvector. According to Lemma 6.1.6, we can assume that this eigenvector has the form $[z_0]$ for a certain $z_0 \in \mathbb{C} \setminus \{0\}$. Set

$$A = \Big\{z \in \mathbb{C} \setminus \{0\},\ \mathbf{m}_0(\bar{z}) = 0\Big\}, \quad B = \Big\{z \in \mathbb{C} \setminus \{0\},\ [z] \in U\Big\}.$$

It is clear that both sets are finite and B is nonempty since $z_0 \in B$. Note that for any $b \in B$, both numbers $\pm\sqrt{b}$ belong to $A \cup B$. Indeed, if one of the numbers $\pm\sqrt{b}$ (which we denote by b_0) does not belong to A, then substituting $z = \bar{b}_0$ in (6.2), we get

$$[b_0] = \frac{1}{2\overline{\mathbf{m}_0(\bar{b}_0)}}\Big(T_0^*[b] + z_1 T_1^*[b]\Big);$$

hence $[b_0] \in U$ and $b_0 \in B$. Now by Lemma 3.4.8 we deduce that either $\mathbf{m}_0$ has a pair of symmetric roots (which is impossible) or $\mathbf{m}_0$ has a cycle $\mathbf{b} = \{b_1, \dots, b_n\}$, where $[\bar{b}_j] \in U,\ j = 1, \dots, n$. ◊

6.2. Mask cleaning procedure

As shown above, operators T_0, T_1 corresponding to a clean mask are nondegenerate and have no common invariant subspaces. Now we transform the mask of an arbitrary equation into a clean mask.

In what follows, we confine our consideration to stable refinable functions, since we need only such functions for the construction of wavelets. Then (Corollary 3.4.15) the mask $\mathbf{m}_0$ can have only a trivial cycle and symmetric roots outside the unit circle.

Note that an equation with a clean mask cannot have an L_1-solution, since it has no trivial cycles (Theorem 3.4.16). Hence for equations having smooth solutions, the collection of operators T_0, T_1 is reducible and a preliminary factorization of the equation (transforming the mask into a clean one) is always required.

If a mask $\mathbf{m}_0$ has symmetric roots $\pm a$ (in our case $|a| \neq 1$) or a trivial cycle $\mathbf{b} = \{1\}$, we shall use the notation

$$\mathbf{m}_a(z) = \frac{z - a^2}{z^2 - a^2}\mathbf{m}_0(z), \qquad \mathbf{m}_1(z) = \frac{2}{z+1}\mathbf{m}_0(z),$$

respectively. Since $\mathbf{m}_a(1) = 1$ and $\mathbf{m}_1(1) = 1$, the equations with masks $\mathbf{m}_a$ and $\mathbf{m}_1$ also have compactly supported solutions denoted by φ_a and φ_1, respectively. Denote by $T_i^{(a)}$ $(i = 0, 1)$ the linear operators (in the space $\mathbb{R}^{N-1}$) corresponding to the mask $\mathbf{m}_a$; then their matrices are defined by formula (5.11) with the coefficients of the mask $\mathbf{m}_a$. The operators $T_i^{(1)}$ corresponding to the mask $\mathbf{m}_1$ are defined analogously.

It follows from Proposition 6.1.2 that if a mask has symmetric roots $\pm a$, the operators T_0, T_1 have a common invariant subspace $[\bar{a}^2]^\perp$, where $\operatorname{Im} T_i \subset [\bar{a}^2]^\perp$, $i = 0, 1$. Let us see how the operators act on this subspace.

PROPOSITION 6.2.1. *If a mask $\mathbf{m}_0$ has a pair of symmetric roots $\{\pm a\}$, the restrictions of the operators T_0, T_1 to the subspace $[\bar{a}^2]^\perp$ are isomorphic to the pair* *of operators* $T_0^{(a)}, T_1^{(a)}$. *In the basis*

$\tilde{e}_1 = (a^2, -1, 0, \dots, 0)^T$, $\tilde{e}_2 = (0, a^2, -1, 0, \dots, 0)^T$, ..., $\tilde{e}_{N-1} = (0, \dots, 0, a^2, -1)^T$

the matrices of the operators $T_i|_{[\bar{a}^2]^\perp}$ *have the form* $T_i^{(a)}$ $(i = 0, 1)$; *here*

$$\varphi(x) = a^2\varphi_a(x) - \varphi_a(x-1). \tag{6.3}$$

REMARK 6.2.2. In equality (6.3) it would be more correct to normalize the function φ by the factor $1/(a^2 - 1)$, so that $\int_{\mathbb{R}} \varphi(x)\, dx = 1$. However for simplicity we did not do that.

Proof. The vectors $\tilde{e}_1, \dots, \tilde{e}_{N-1}$ constitute a basis of the space $[\bar{a}^2]^\perp$. Consider the $N \times (N-1)$-matrix C_a composed from the columns $\tilde{e}_1, \dots, \tilde{e}_{N-1}$. Direct calculation gives

$$C_a T_i^{(a)} = T_i C_a, \qquad i = 0, 1$$

(this is equivalent to the equality $(z^2 - a^2)\mathbf{m}_a(z) = (z - a^2)\mathbf{m}_0(z)$); hence the matrices of the operators T_i in the new basis coincide with the matrices $T_i^{(a)}$. Further, we have

$$\left(a^2 - e^{-4\pi i\xi}\right)\widehat{\varphi_a}(2\xi) = \left(a^2 - e^{-4\pi i\xi}\right)m_a(\xi)\widehat{\varphi_a}(\xi) = \left(a^2 - e^{-2\pi i\xi}\right)m_0(\xi)\widehat{\varphi_a}(\xi);$$

therefore the function $a^2\varphi_a(x) - \varphi(x-1)$ satisfies the refinement equation with mask $\mathbf{m}_0$, which completes the proof of (6.3). ◊

If a mask has a trivial cycle, i.e., $\mathbf{m}_0(-1) = 0$, then the operators T_0, T_1 have a common invariant subspace $U_1^\perp$ orthogonal to [1] (Proposition 6.1.2). Thus,

$$U_1^\perp = \Big\{(x_1, \dots, x_N)^T,\ \sum_i x_i = 0\Big\}.$$

PROPOSITION 6.2.3. *If* $\mathbf{m}_0(-1) = 0$, *the restrictions of the operators* T_0, T_1 *to the subspace* $U_1^\perp$ *are isomorphic to the operators* $\frac{1}{2}T_0^{(1)}, \frac{1}{2}T_1^{(1)}$. *In the basis*

$$\tilde{e}_1 = (1, -1, 0, \dots, 0)^T, \ \dots, \tilde{e}_{N-1} = (0, \dots, 0, 1, -1)^T \in \mathbb{R}^N$$

the matrices of the operators $T_i|_{U_1^\perp}$ *coincide with the matrices* $\frac{1}{2}T_i^{(1)}$ $(i = 0, 1)$ *and*

$$\varphi = \varphi_1 * \chi_{[0,1]}. \tag{6.4}$$

Proof. The vectors $\tilde{e}_1, \dots, \tilde{e}_{N-1}$ form a basis of the space $U_1^\perp$. Consider the $N \times (N-1)$-matrix C_1 composed from the columns $\tilde{e}_1, \dots, \tilde{e}_{N-1}$. We have

$$C_1 T_i^{(1)} = 2T_i C_1, \qquad i = 0, 1$$

(this is equivalent to the equality $(z+1)\mathbf{m}_1(z) = 2\mathbf{m}_0(z)$); hence the matrices of the operators T_i in the new basis coincide with the matrices $\frac{1}{2}T_i^{(1)}$. Further, the function $\chi_{[0,1]}$ satisfies the refinement equation with mask $\left(\frac{z+1}{2}\right)$, while the function φ_1 satisfies the refinement equation with mask $\mathbf{m}_1$. Therefore the convolution of these functions satisfies the equation with mask $\left(\frac{z+1}{2}\right)\mathbf{m}_1(z)$. This proves equality (6.4). ◊

REMARK 6.2.4. Propositions 6.1.2 and 6.1.4 allow one to find explicitly all common invariant subspaces of the operators T_0, T_1 for any mask $\mathbf{m}_0$. We again confine our consideration to the case of stable refinable functions, where the mask $\mathbf{m}_0$ can have only a trivial cycle and symmetric roots outside the unit circle. We begin with a simple but the most important case.

PROPOSITION 6.2.5. *Let a mask* $\mathbf{m}_0$ *have neither symmetric roots nor nontrivial cycles, and let the multiplicity of the trivial cycle be equal to* $L+1 \geq 0$, *i.e.,*

$$\mathbf{m}_0(z) = \left(\frac{z+1}{2}\right)^{L+1} q(z), \qquad q(-1) \neq 0.$$

Then the operators T_0, T_1 *have a system of embedded invariant subspaces*

$$G_{L+1} \subset \ldots \subset G_1 \subset G_0 = \mathbb{R}^N.$$

Here $\dim G_k = N - k$ *and*

$$G_k = \Big\{(x_1, \ldots, x_N)^T, \ \sum_k k^j x_j = 0, \ j = 0, \ldots, k-1\Big\}.$$

In addition,

$$\operatorname{spec}(T_i) = \Big\{1, 2^{-1}, \ldots, 2^{-L}, \operatorname{spec}(T_i|_{G_{L+1}})\Big\}.$$

Any common invariant subspace of the operators T_0, T_1 *coincides with one of* G_k.

Proof. The proof is by induction on L. For $L = -1$, the collection of operators T_0, T_1 is irreducible (Theorem 6.1.1), so the statement of the theorem is true (the chain $\{G_k\}$ consists of one element $G_0 = \mathbb{R}^N$). Assume that it is true for $L-1$. By Proposition 6.1.4, the operators T_i have the eigenvalue 1 and the invariant subspace $U_1^{\perp} = G_1$. The operators $T_i|_{G_1}$ are isomorphic to the operators $\frac{1}{2}T_i^{(1)}$ with the transition matrix C_1 (Proposition 6.2.3). For the operators $T_i^{(1)}$, the statement is true by assumption. Let $\{G'_k\}$ be the corresponding chain of embedded subspaces of these operators. Since $G_k C_1 = G'_k$, we conclude that $\{G_k\}$ is a chain of embedded subspaces for T_i. Now let V be a common invariant subspace of the operators T_0, T_1. Let us prove that it coincides with one of G_k. Since $V^{\perp}$ is a common invariant subspace of the conjugate operators T_i^*, it contains some space $U_{\mathbf{b}}$ (Proposition 6.1.4). Since $\mathbf{m}_0$ has only trivial cycles, $\mathbf{b} = \{1\}$; therefore $U_{\mathbf{b}} = [1]$. Thus $[1] \subset V^{\perp}$ and $V \perp [1]$, which implies that $V \subset G_1$. According to Proposition 6.2.3, either $V \subset G_1$ or V is a nontrivial invariant subspace of the operators $T_i|_{G_1}$. By induction, this means that $V = G_k$ for some $k = 2, \ldots, L+1$. ◊

COROLLARY 6.2.6. *Under the conditions of Proposition* 6.2.5 *each common invariant subspace of the operators* T_0, T_1 *contains the subspace* G_{L+1}.

REMARK 6.2.7. Proposition 6.2.5 characterizes all common invariant subspaces of the operators T_0, T_1 if a mask has no symmetric roots. In the general case, it is also possible to classify all common invariant subspaces; however it is more complicated. In this case, all common invariant subspaces form several chains, which constitute several binary trees [**125**].

Using Propositions 6.2.1 and 6.2.3, one can successively eliminate all pairs of symmetric roots $\pm a_1, \ldots, \pm a_r$ and all trivial cycles, i.e., the trivial cycle of the given multiplicity $L+1$. As a result, we obtain a clean mask, which we denote by $\mathbf{m}_c$. It is not hard to show that the mask $\mathbf{m}_c$ obtained in this way is independent of the order in which symmetric roots were eliminated. At each iteration, the degree N of the mask (dimension of the space) decreases by 1 and the matrices of the operators T_0, T_1 are transformed by the formulas given in Propositions 6.2.1 and

6.2.3. Denote by C_{a_k} the corresponding transition matrix at the k-th iteration. It is an $(N-k+1)\times(N-k)$-matrix composed from the columns

$$\left(a_k^2,-1,0,\dots,0\right)^T\ ,\ \dots,\ \left(0,\dots,0,a_k^2,-1\right)^T\ \in\ \mathbb{R}^{N-k+1}.$$

After r iterations the mask will have no symmetric roots. Next, we shall eliminate the trivial cycle of multiplicity $L+1$. At the $(r+j)$-th iteration $(j=1,\dots,L+1)$, we go over from the space $\mathbb{R}^{N-r-j+1}$ to the space $\mathbb{R}^{N-r-j}$ with a new basis

$$\left(1,-1,0,\dots,0\right)^T\ ,\ \dots\ ,\ \left(0,\dots,0,1,-1\right)^T.$$

The transition matrix $C_{1,j}$ is composed from vector columns of the new basis. In this way, after k iterations $(k\le r+L+1)$, we have the operators T_0,T_1 restricted to the $(N-k)$-dimensional subspace G_k by the transition matrix

$$C_k=\begin{cases} C_{a_1}\dots C_{a_k} & \text{if}\ \ k\le r,\\ C_{a_1}\dots C_{a_r}C_{1,1}\dots C_{1,k-r} & \text{if}\ \ k>r.\end{cases}$$

The subspace G_k is thus generated by the columns of the matrix C_k. As a result, after $r+L+1$ iterations, we go over to the subspace $\tilde{G}=G_{r+L+1}$ by the transition matrix

$$\tilde{C}=C_{r+L+1}=C_{a_1}\cdots C_{a_r}C_{1,1}\cdots C_{1,L+1}.$$

Thus we obtain the matrices $2^{-L-1}\tilde{T}_0$, $2^{-L-1}\tilde{T}_1$, where the matrices $\tilde{T}_0,\tilde{T}_1$ correspond to the clean mask $\mathbf{m}_c$. According to Theorem 6.1.1, the operators $2^{-L-1}\tilde{T}_0$, $2^{-L-1}\tilde{T}_1$ are nondegenerate and have no nontrivial common invariant subspaces. The subspace $\tilde{G}$, where the initial operators T_i are isomorphic to the operators $2^{-L-1}\tilde{T}_i$, is given in explicit form by the transition matrix $\tilde{C}$ (this space is spanned by the columns of this matrix). These columns form a required basis in this space.

REMARK 6.2.8. In the literature, the mask cleaning procedure is most often used for nonstable refinable functions. In this case (see Corollary 3.4.15) a mask has nontrivial cycles and symmetric roots on the unit circle. Using Propositions 6.2.1 and 6.2.3, we can clean the mask from nontrivial cycles and symmetric roots on the unit circle by getting a stable refinable function φ_0. Integer shifts of the function φ_0 generate the same space V_0 as the function φ, but φ_0 has a smaller support than φ. In terms of the wavelet theory this means that the functions φ and φ_0 generate the same MRA. Therefore by means of cleaning we can go from any refinable function to a stable one generating the same MRA; i.e., we can find a refinable function with the minimal support for a given MRA. In this sense, nonstable functions are of no use for the construction of wavelets since they do not generate any new MRAs. Due to this fact, we focus on stable refinable functions, though nonstable functions find application, for instance, in the theory of subdivision schemes or probability theory (see [**46**], [**127**], [**128**], [**182**], [**129**], [**130**]). Nevertheless, it always makes sense to continue the procedure of cleaning, applying it to stable refinable functions. In doing so, the MRA will change every time. Thus it follows from relation (6.3) that the space V_0 generated by the function φ is contained in the space generated by φ_a and does not coincide with it (strict embedding). Hence when we eliminate a symmetric root outside the unit circle, the space V_0 expands. By eliminating a trivial cycle, we go over to another MRA; the spaces generated by the functions φ and φ_1 are disjoint (more precisely, their intersection consists of zero) since (see formula (6.4)) the functions φ and φ_1 have different smoothness. Nevertheless, for stable functions, the cleaning procedure allows one to find explicitly common invariant subspaces for the operators T_0, T_1. This procedure is used for computation of smoothness. For a given refinement equation, we shall now see how to find explicitly the space $\tilde{A}_n$, on which a fractal curve $v(x) = \big(\varphi(x), \dots, \varphi(x+N-1)\big)^T$ is located, and find the matrices of the operators T_0, T_1 on this subspace. This will allow the application of all the results concerning the smoothness of fractal curves (Chapter 5) to refinable functions.

6.3. Space $\tilde{A}_n$ and the general form of the operators T_0, T_1 on it

Let there be given a refinement equation with a stable solution; i.e., the mask $\mathbf{m}_0$ has neither trivial cycles nor symmetric roots on the unit circle. The cleaning algorithm described in the previous section allows us to find explicitly the subspace

$$\tilde{A}_n = \operatorname{span}\{v(x) - v(y),\ x, y \in [0,1]\},$$

where $v(x) = \big(\varphi(x), \dots, \varphi(x+N-1)\big)^T$. It appears that $\tilde{A}_n = G_{r+1}$. In other words, the space $\tilde{A}_n$ is obtained by elimination of all symmetric roots and *one* trivial cycle (recall that by Corollary 3.4.17, if $\varphi \in L_1(\mathbb{R})$, the mask has a trivial cycle). It follows that $\tilde{A}_n$ has dimension $N - r - 1$ and is generated by columns of the matrix $C_{r+1} = C_{a_1} \dots C_{a_r} C_{1,1}$. In particular, if the mask has no symmetric roots, then

$$\tilde{A}_n = U_1^{\perp} = \Big\{(x_1, \dots, x_N)^T,\ \sum_k x_k = 0\Big\},$$

and the corresponding affine space A_n coincides with the space

$$W = \Big\{(x_1, \dots, x_N)^T,\ \sum_k x_k = 1\Big\} = v(0) + U_1^{\perp}.$$

First we prove this statement for a continuous function φ. Note that

$$T_i G_k \subset G_{k+1}, \qquad i = 0, 1,$$

for each $k = 0, \dots, r-1$ (we set $G_0 = \mathbb{R}^N$). Indeed, since $\pm a_1$ are symmetric roots of $\mathbf{m}_0$, then $\operatorname{Im} T_i \perp [\bar{a}_1^2]$; hence $T_i G_0 \subset [\bar{a}_1^2]^{\perp} = G_1$. On the subspace G_1 the operators T_i are isomorphic to the operators $T_i^{(a_1)}$ corresponding to the mask $\mathbf{m}_{a_1}$ (Proposition 6.2.1). Since $\pm a_2$ are symmetric roots of $\mathbf{m}_{a_1}$, then, as above, we show that $\operatorname{Im} T_i^{(a_1)} \subset [\bar{a}_2^2]^{\perp}$, which means that

$$T_i G_1 \subset G_2, \qquad i = 0, 1,$$

due to the isomorphism of the operators.

In this way, we successively prove the inclusions $T_i G_k \subset G_{k+1}$ for all $k = 0, \dots, r-1$. It follows that G_r is a common invariant subspace of the operators T_0, T_1, and any eigenvector of T_0 with a nonzero eigenvalue belongs to this subspace. Therefore (as $T_0 v(0) = v(0)$), $v(0) \in G_r$. Since

$$\operatorname{span}\Big\{v(x),\ x \in [0,1]\Big\} = \operatorname{span}\Big\{v(0), \tilde{A}_n\Big\}$$

is the minimal subspace invariant under T_0 and T_1 and containing the vector $v(0)$, we conclude that $\operatorname{span}\{v(0), \tilde{A}_n\} \subset G_r$. Let us show that actually we have the inequality $\operatorname{span}\{v(0), \tilde{A}_n\} = G_r$. Indeed, since $\operatorname{span}\{v(0), \tilde{A}_n\}$ is a common invariant subspace of the operators $T_i, i = 0, 1$, it follows that either $\operatorname{span}\{v(0), \tilde{A}_n\} \subset G_{r+1}$ or $\operatorname{span}\{v(0), \tilde{A}_n\} \subset G_{r+1}$ (Proposition 6.2.5). The latter relation is not possible; otherwise the spectrum of the operator T_0 on the subspace G_{r+1} contains 1 (as

$T_0 v(0) = v(0)$). This spectrum consists of the numbers $2^{-1}, \dots, 2^{-L}$ and of the spectrum of the operator $T_0|_{\tilde{G}}$ (Proposition 6.2.5); hence the spectrum of $T_0|_{\tilde{G}}$ also contains 1. But then $\hat{\rho}(T_0|_{\tilde{G}}, T_1|_{\tilde{G}}) \geq 1$ (here we use Lemma 5.1.2). Therefore,

$$\hat{\rho}(T_0|_{\tilde{A}_n}, T_1|_{\tilde{A}_n}) \geq 1$$

($\tilde{G} \subset A_n$ since by Corollary 6.2.6 each common invariant subspace of the operators T_0 and T_1 on G_r contains $\tilde{G}$). So we have a contradiction: since the solution φ is continuous, it follows that $\hat{\rho}(T_0|_{\tilde{A}_n}, T_1|_{\tilde{A}_n}) < 1$. Thus span $\{v(0), \tilde{A}_n\} = G_r$. Further, $\tilde{A}_n$ is also a common invariant subspace of T_0, T_1, and its dimension is smaller by 1 than the dimension span $\{v(0), \tilde{A}_n\} = G_r$. Hence, due to Proposition 6.2.5, the space $\tilde{A}_n$ coincides with G_{r+1}.

In the case $\varphi \in L_1(\mathbb{R})$ the proof is analogous with the only difference that instead of the vector $v(0)$ we take

$$c = \Big(\int_0^1 \varphi(t)\,dt\ ,\ \dots\ ,\ \int_{N-1}^{N} \varphi(t)\,dt \Big)^T$$

(an eigenvector of the operator $\frac{1}{2}\Big(T_0 + T_1\Big)$ with the eigenvalue 1).

Thus we have proved the following theorem.

THEOREM 6.3.1. *Suppose a solution of a refinement equation with mask* $\mathbf{m}_0(z) = \left(\frac{z+1}{2}\right)^{L+1} q(z)$ *is stable and* T_0, T_1 *are the corresponding linear operators. Then the space*

$$\tilde{A}_n = \mathrm{span}\Big\{ v(x) - v(y),\ x, y \in [0,1] \Big\}$$

coincides with the space G_{r+1} *generated by columns of the matrix*

$$C_{r+1} = C_{a_1} \cdots C_{a_r} C_{1,1}.$$

In particular, if $\mathbf{m}_0$ *has no symmetric roots, then*

$$\tilde{A}_n = \Big\{ (x_1, \dots, x_N)^T,\ \sum x_i = 0 \Big\}.$$

In an appropriate basis the matrices of the operators $T_i|_{\tilde{A}_n}, i = 0, 1$, *have the following form: if* $L = 0$, *then* $T_i|_{\tilde{A}_n} = \frac{1}{2}\tilde{T}_i$, *and if* $L \geq 1$, *then*

$$T_i|_{\tilde{A}_n} = \begin{pmatrix} 2^{-1} & 0 & 0 & \cdots & 0 & 0 & \cdots & 0 \\ * & 2^{-2} & 0 & \cdots & 0 & 0 & \cdots & 0 \\ * & * & 2^{-3} & \cdots & 0 & 0 & \cdots & 0 \\ \vdots & \vdots & & & 0 & 0 & \cdots & 0 \\ * & * & * & \cdots & 2^{-L} & 0 & \cdots & 0 \\ * & * & * & \cdots & * & & & \\ \vdots & & & & \vdots & & 2^{-L-1}\tilde{T}_i & \\ * & * & * & \cdots & * & & & \end{pmatrix}, \qquad i = 0, 1, \tag{6.5}$$

where the matrices $\tilde{T}_i$ *correspond to the clean mask* $\mathbf{m}_c$.

Since the operators $\tilde{T}_i$ are nondegenerate (Theorem 6.1.1), we obtain

COROLLARY 6.3.2. *If a solution of a refinement equation is stable, then the operators T_i on the space $\tilde{A}_n$ are nondegenerate.*

REMARK 6.3.3. Theorem 6.3.1 and Corollary 6.3.2 hold for all refinement equations without the assumption that the solution is stable [**125**].

COROLLARY 6.3.4. *If the mask of a refinement equation has neither cycles nor symmetric zeros, then $\tilde{A}_n = \{(x_1, \dots, x_N),\ \sum x_k = 0\}$.*

REMARK 6.3.5. For any refinement equation, both the space $\operatorname{span}\{v(0), \tilde{A}_n\}$ containing the image of the curve $v(x)$ and the space $\tilde{A}_n$ can be found explicitly: $\tilde{A}_n = G_{r+1}$; it is spanned by the columns of the matrix $C_{r+l} = C_{a_1} \dots C_{a_r} C_{1,1}$. Each matrix $C_a, C_{1,1}$ in this product is expressed by the formulas given in Propositions 6.2.1 and 6.2.3. The same is true for the space $\operatorname{span}\{v(0), V\} = G_r$ spanned by the columns of the matrix C_r. The vector $v(0)$, from which we begin the construction of a continuous refinable function φ, can be defined as the only eigenvector of the operator $T_0|_{G_r}$ with eigenvalue 1. This vector is unique. Otherwise the operator $T_0|_{V_{r+1}} = T_0|_{\tilde{A}_n}$ will also have eigenvalue 1; hence $\hat{\rho}(T_0|_{\tilde{A}_n}, T_1|_{\tilde{A}_n}) \geq 1$, which contradicts the continuity of φ. Moreover, it is not hard to show that 1 is a prime eigenvalue not only for the operator $T_0|_{\tilde{A}_n}$, but for T_0 as well. Indeed, by the reverse transition (from the operator $T_0|_{\tilde{A}_n}$ to the operator T_0) symmetric roots are successively added to the mask. According to Proposition 6.2.1, each addition of a symmetric root increases the multiplicity of the eigenvalue 0 without changing other eigenvalues.

The same is true for an L_p-solution φ. In this case the construction begins with the vector c, which is the only eigenvector of the operator $\frac{1}{2}(T_0+T_1)$ corresponding to the eigenvalue 1. Such a vector is unique since 1 is a prime eigenvalue of the operator $\frac{1}{2}(T_0+T_1)|_{\tilde{A}_n}$ and thus of the operator $\frac{1}{2}(T_0+T_1)$. Otherwise, the operator $\frac{1}{2}(T_0+T_1)|_{\tilde{A}_n}$ also has the eigenvalue 1 and thus $\rho_p(T_0|_{\tilde{A}_n}, T_1|_{\tilde{A}_n}) \geq 1$. The latter inequality is impossible since $\varphi \in L_p(\mathbb{R})$. Therefore we have proved the following statement:

COROLLARY 6.3.6. *If a stable refinable function is continuous, the operator T_i, $i = 0, 1$, has a prime eigenvalue 1 and the corresponding eigenvector is equal to $v(i)$. All other eigenvalues of the operators T_i are less than 1 in modulus. If a stable refinable function is summable, the operator $\frac{1}{2}\Big(T_0 + T_1\Big)$ has a prime eigenvalue 1 and the corresponding eigenvector is equal to c. All other eigenvalues of this operator are less than 1 in modulus.*

EXAMPLE 6.3.7. The equation with mask

$$\mathbf{m}_0(z) = \Big(\frac{z+1}{2}\Big)\Big(\frac{z^2+2}{3}\Big)$$

has a continuous solution φ since all its coefficients are positive [**131**]. The mask has neither nontrivial cycles nor symmetric roots on the unit circle and thus φ is stable. Nevertheless, $\tilde{A}_n \neq \{(x_1, x_2, x_3)^T,\ \sum x_i = 0\}$ since $\mathbf{m}_0$ has a pair of symmetric roots $\pm a = \pm\sqrt{2}i$. Both operators T_i^* are degenerate; their common kernel has dimension 1 and is generated by the vector $[-2] = (1, -2, 4)^T$. After the elimination of the symmetric root and the trivial cycle $\{1\}$, we obtain the clean mask $\mathbf{m}_c(z) = \frac{1}{3}(z+2)$. The operators $\tilde{T}_0 = \frac{4}{3}$, $\tilde{T}_1 = \frac{2}{3}$ are one-dimensional and obviously irreducible. The transition matrices have the form

$$C_a = \Big((-2,-1,0)^T, (0,-2,-1)^T\Big), \qquad C_{1,1} = \Big(-1,1\Big)^T.$$

The space $\tilde{A}_n$ is one-dimensional; it is generated by the vector $C_a C_{1,1} = (2,-1,-1)$.

It remains to formulate the last auxiliary result, which we shall need in Chapter 7 to calculate the smoothness of wavelets. The following lemma is a generalization of Lemma A.6.4.

LEMMA 6.3.8. *Let each of the two $N \times N$-matrices $B_i, i = 0,1,$ be lower triangular with blocks $A_i^1, \dots, A_i^l$ on the diagonal:*

$$B_i = \begin{pmatrix} A_i^1 & 0 & 0 & \cdots & 0 \\ * & A_i^2 & 0 & \cdots & 0 \\ * & * & * & & \vdots \\ \vdots & & & & 0 \\ * & \cdots & \cdots & * & A_i^l \end{pmatrix}, \qquad i = 0,1, \tag{6.6}$$

and let the pairs of operators A_0^j, A_1^j be irreducible in $\mathbb{R}^{r_j}$ for each j ($\sum_{j=1}^{l} r_j = N$). Then for every $p \in [1,+\infty]$, we have

$$\rho_p(B_0,B_1) = \max_{j=1,\dots,l} \rho_p(A_0^j, A_1^j).$$

Moreover, if the maximum is attained at s elements of the given l elements, then

$$\begin{aligned} c_1(\rho_p)^k \le \Big(2^{-k} \sum_{d_1,\dots,d_k} \|B_{d_1}\cdots B_{d_k}\|^p\Big)^{1/p} &\le c_2 k^{s-1}(\rho_p)^k, \quad p<\infty, \\ c_1\hat{\rho}^k \le \max_{d_1,\dots,d_k} \|B_{d_1}\cdots B_{d_k}\| \le c_2 k^{s-1}\hat{\rho}^k, & \qquad\qquad p = \infty, \end{aligned} \tag{6.7}$$

where the positive constants c_1, c_2 are independent of k. In particular, if $s = 1$, then

$$\Big(2^{-k} \sum_{d_1,\dots,d_k} \|B_{d_1}\cdots B_{d_k}\|^p\Big)^{1/p} \asymp (\rho_p)^k, \qquad \max_{d_1,\dots,d_k} \|B_{d_1}\cdots B_{d_k}\| \asymp \hat{\rho}^k.$$

The proof is by induction on the number of blocks l. If $l = 1$, our statement coincides with Lemma A.6.4. Assume that the statement is true for matrices consisting of $l-1$ blocks. Denote by D_i ($i = 0,1$) the $N \times N$-matrix whose first r_1 rows and columns consist of zeros, while the other elements coincide with those of the matrix B_i; by C_i denote the $N \times N$-matrix whose elements at the intersection of the first r_1 rows and last $N - r_1$ columns coincide with the elements of B_i (all other elements are zeros); and denote by A_i the matrix coinciding with B_i (actually, with A_i^1) at the intersection of the first r_1 rows and columns (all other elements are zeros). Thus $B_i = A_i + C_i + D_i$ and

$$\begin{aligned} B_{d_1}\cdots B_{d_k} = {} & D_{d_1}\cdots D_{d_k} + D_{d_1}\cdots D_{d_{k-1}}C_{d_k} \\ & + D_{d_1}\cdots D_{d_{k-2}}C_{d_{k-1}}A_{d_k} + \ldots + D_{d_1}C_{d_2}A_{d_3}\cdots A_{d_k} \\ & + C_{d_1}A_{d_2}\cdots A_{d_k} + A_{d_1}\cdots A_{d_k}. \end{aligned} \tag{6.8}$$

Since

$$\|B_{d_1}\cdots B_{d_k}\| \ge \max\Big\{\|A_{d_1}\cdots A_{d_k}\|,\ \|D_{d_1}\cdots D_{d_k}\|\Big\}$$

and

$$\max_{d_1,\dots,d_k}\|A_{d_1}\cdots A_{d_k}\| \ge c_2\hat{\rho}^k(A_0,A_1)$$

by Lemma A.6.4 and, on the other hand,

$$\max_{d_1,\dots,d_k}\|D_{d_1}\cdots D_{d_k}\| \ge c_2\hat{\rho}^k(D_0,D_1)$$

by the inductive assumption, it follows that

$$\hat{\rho} = \hat{\rho}(B_0,B_1) \ge \max\{\hat{\rho}(A_0,A_1),\ \hat{\rho}(D_0,D_1)\}.$$

Lemma A.6.4 implies that

$$\|A_{d_n}\cdots A_{d_k}\| \le c_2\hat{\rho}^{k-n}(A_0,A_1).$$

By the inductive assumption,

$$\|D_{d_1}\cdots D_{d_{n-1}}\| \le c_2(n-1)^s\hat{\rho}^{n-1}(D_0,D_1),$$

where

$$\hat{\rho}(D_0,D_1) = \max_{j=2,\dots,l}\Big\{\hat{\rho}(A_0^j,A_1^j)\Big\}$$

and s is the number of j for which this maximum is attained. Substituting this into (6.8), we get

$$\|B_{d_1}\cdots B_{d_k}\| \le C\sum_{n=0}^{k}(n-1)^s\hat{\rho}^{n-1}(D_0,D_1)\hat{\rho}^{k-n}(A_0,A_1), \tag{6.9}$$

where $C>0$ does not depend on k nor on $d_1,\dots,d_k$. Therefore,

$$\hat{\rho} \le \max\Big\{\hat{\rho}(A_0,A_1),\hat{\rho}(D_0,D_1)\Big\}.$$

Hence,

$$\hat{\rho} = \max\Big\{\hat{\rho}(A_0,A_1),\hat{\rho}(D_0,D_1)\Big\}.$$

If $\hat{\rho} > \hat{\rho}(D_0,D_1)$, then

$$\|B_{d_1}\cdots B_{d_k}\| \le C\hat{\rho}^k,$$

and if $\hat{\rho} = \hat{\rho}(D_0,D_1)$, then

$$\|B_{d_1}\cdots B_{d_k}\| \le Ck^{s+1}\hat{\rho}^k.$$

This proves (6.7) for $p=\infty$. If $p<\infty$, the lower bound is obtained analogously: the inequality

$$\|B_{d_1}\cdots B_{d_k}\| \ge \max\Big\{\|A_{d_1}\cdots A_{d_k}\|,\|D_{d_1}\cdots D_{d_k}\|\Big\}$$

implies that $\rho_p \ge \max\Big\{\rho_p(A_0,A_1),\rho_p(D_0,D_1)\Big\}$ and

$$\left(2^{-k}\sum_{d_1,\dots,d_k}\|B_{d_1}\cdots B_{d_k}\|^p\right)^{1/p} \ge c_1(\rho_p)^k.$$

To prove the upper bound, we use (6.8):

$$2^{-k} \sum_{d_1,\dots,d_k} \|B_{d_1}\cdots B_{d_k}\|^p$$

$$\le C\sum_{n=0}^{k}\Big(2^{1-n}\sum_{d_1,\dots,d_{n-1}} \|D_{d_1}\cdots D_{d_{n-1}}\|^p\Big)\Big(2^{n-k}\sum_{d_n,\dots,d_k}\|A_{d_n}\cdots A_{d_k}\|^p\Big).$$

It follows by the inductive assumption, that

$$\Big(2^{-k}\sum_{d_1,\dots,d_k}\|B_{d_1}\cdots B_{d_k}\|^p\Big)^{1/p} \le C\sum_{n=0}^{k}(n-1)^s\rho_p^{n-1}(D_0,D_1)\rho_p^{k-n}(A_0,A_1).$$

Further reasoning is the same as in the case $p=\infty$. ◊

6.4. Factorization theorems

THEOREM 6.4.1. *A stable refinable function belongs to $C^l(\mathbb{R})$ (respectively, $W_p^l(\mathbb{R})$), $l \ge 1$, if and only if it can be represented in the form*

$$\varphi = B_{l-1} * \varphi_0, \tag{6.10}$$

where $\varphi_0 \in C(\mathbb{R})$ $(L_p(\mathbb{R}))$ is a refinable function.

Here $\alpha_\varphi = l + \alpha_{\varphi_0}$ (respectively, $\alpha_{\varphi,p} = l + \alpha_{\varphi_0,p}$).

Proof. By Theorem 3.4.16 the mask of the equation at the point $z=-1$ has a zero of order $\ge l+1$, i.e., is represented in the form

$$\mathbf{m}_0(z) = \Big(\frac{z+1}{2}\Big)^{l+1} q(z).$$

Then (Proposition 6.2.3) we can reduce the order by formula (6.4) l times and obtain a refinable function φ_0 satisfying the equation with mask $\big(\frac{z+1}{2}\big)q(z)$. Here φ is the convolution of φ_0 with l functions $\chi_{[0,1]}$, i.e., with the B-spline B_{l-1}. Since the convolution with the function $\chi_{[0,1]}$ increases the smoothness by 1, we get

$$\varphi_0 \in C\ (L_p), \quad \alpha_\varphi = l + \alpha_{\varphi_0}$$

(respectively, $\alpha_{\varphi,p} = l + \alpha_{\varphi_0,p}$). ◊

REMARK 6.4.2. Theorem 6.4.1 reduces the analysis of smooth refinable functions to that of continuous (respectively, $L_p(\mathbb{R})$) refinable functions. In order to increase the smoothness of a refinable function by a certain integer r, it suffices to multiply the mask by the polynomial $\big(\frac{z+1}{2}\big)^r$. Then the function is replaced by its convolution with the B-spline B_{r-1}. The order of the equation and the size of support of the refinable function increase by r. Using Proposition 6.2.3, we derive formulas that define the transform of the operators T_0, T_1. By Theorem 6.4.1 it is the only way to obtain stable smooth refinable functions. Any such function is factorized as a convolution of a continuous (respectively, $L_p(\mathbb{R})$) refinable function with a B-spline.

For nonstable functions, the statement of Theorem 6.4.1, generally speaking, is not true. It is known that the mask of a nonstable refinable function does not have to vanish at the point $z=-1$. In this case, the necessary smoothness condition for a function is replaced by the condition for the mask to have a blocking set of order l (Corollary 3.2.11). For nonstable functions, there is an analog of Theorem 6.4.1:

a C^l (W_p^l)-refinable function is represented as the convolution $S_{l-1} * \varphi_0$, where φ_0 is a continuous (respectively, L_p) refinable function and

$$S_{l-1}(x) = \sum_{k=0}^{r} p_k B_{l-1}(x-k)$$

is a spline of order $l-1$, which is a combination of integer shifts of a B-spline. The factors p_k are coefficients of some polynomial depending on the blocking set of the mask $\mathbf{m}_0$ (see [**110**], [**128**]).

We can continue the process of mask factorization, having eliminated all trivial cycles and symmetric roots, and thus obtain a refinable function φ_c corresponding to the clean mask.

THEOREM 6.4.3. *Any stable refinable function is uniquely represented in the form of convolution*

$$\varphi = S_{L-1} * \varphi_c, \tag{6.11}$$

where φ_c is the refinable function corresponding to the clean mask $\mathbf{m}_c$, $S_L(x) = \sum_{k=0}^{r} p_k B_L(x-k)$ is a spline of order L, which is a combination of integer shifts of the B-spline B_L, L is the order of the Strang-Fix conditions satisfied by φ, in particular, $L \ge \alpha_\varphi - 1$. The factors p_k are the coefficients of the polynomial

$$P(z) = \sum_{k=0}^{r} p_k z^k = \prod_{j=1}^{r} (z - a_j^2),$$

where $a_1, \dots, a_r$ are symmetric roots, which appear in the process of mask cleaning.

Proof. According to Theorem 3.4.16, the mask of an equation has a zero of order $L+1$ at the point $z=-1$. Using Proposition 6.2.3, perform $L+1$ reductions of order by formula (6.4). Thus we express φ as the convolution of the B-spline B_L and some refinable function. Then we successively eliminate all pairs of symmetric roots $\pm a_1, \dots, \pm a_r$. At each step of this process we get a new refinable function by formula (6.3). As a result, we obtain the function φ_c corresponding to the clean mask $\mathbf{m}_c$. ◇

REMARK 6.4.4. Theorem 6.4.3 gives complete factorization of a refinable function. Note that φ_c is always a singular distribution. It cannot be a regular function since the mask $\mathbf{m}_c$ has no trivial cycles; i.e., it does not vanish at the point $z=-1$. One can prove that the polynomial $P(z)$ does not depend on the order of elimination of symmetric roots. This is not quite obvious since after each iteration of the cleaning process new pairs of symmetric roots may appear. For a rigorous proof, see the monograph [**46**, Chapter 8] and the corresponding references in it.

Representation (6.11) will be used to compute smoothness of refinable functions. All methods for computation of smoothness (the matrix method, the method based on the decay rate of the Fourier coefficients) can be employed for clean masks and then extended to all refinable functions by formula (6.11).

CHAPTER 7

Smoothness of Compactly Supported Wavelets

To find the smoothness of scaling functions and of compactly supported wavelets is one of the most complicated problems in the theory of wavelets. At present there are three approaches to this problem. They were developed in 1989–1992 and have been improved repeatedly since then. Each of them has its own advantages and disadvantages. Nevertheless, in the general situation, none of them allows us to find the precise value of smoothness in $C(\mathbb{R})$ or $L_1(\mathbb{R})$. We even cannot always give a comprehensive answer to the "simple" question of whether a solution of a given refinement equation is continuous or summable.

7.1. Matrix method

The matrix method estimates smoothness by using the joint spectral radius of the operators T_0 and T_1. First we use the factorization theorem (Theorem 6.4.1) to reduce the order of the equation. Thus we go over from a smooth refinable function to a continuous one (respectively, to $L_p(\mathbb{R})$). Then the continuous refinable function is represented in the form of the fractal curve corresponding to the operators T_0, T_1 in the space $\tilde{A}_n$. According to the results of Chapter 5, smoothness is computed in terms of the joint spectral radius of the operators T_0 and T_1.

One of the advantages of the matrix method is that it enables (at least, theoretically) exact calculation of not only global but also local smoothness of refinable functions and wavelets at each point. To this end, Theorem 5.4.1 is used to represent local smoothness by means of the spectral radius along sequences. Moreover, the operators T_i are both nondegenerate on the space $\tilde{A}_n$ (Corollary 6.3.2). Therefore one can use Proposition 5.4.2 to calculate smoothness at rational points and Theorem 5.4.10 to characterize sets of fixed local smoothness. Here, due to Theorem 6.3.1, the computation is reduced to the matrices $\tilde{T}_0, \tilde{T}_1$ corresponding to the clean mask $\mathbf{m}_c$.

Let there be given a refinement equation with mask $\mathbf{m}_0$, which has a stable solution satisfying the Strang-Fix conditions of order L. This means that the mask has neither nontrivial cycles nor symmetric roots on the unit circle and has a zero of multiplicity $L+1$ at the point $z = -1$. Suppose $\mathbf{m}_c$ is the corresponding clean mask, and the operators $\tilde{T}_0, \tilde{T}_1$ are defined by their matrices constructed by coefficients of the polynomial $\mathbf{m}_c$ by formula (5.11). Set $\hat{\rho} = \hat{\rho}(\tilde{T}_0, \tilde{T}_1)$, $\rho_p = \rho_p(\tilde{T}_0, \tilde{T}_1)$, $p \in [1, +\infty)$.

THEOREM 7.1.1. a) *A refinable function φ belongs to $C^k(\mathbb{R})$, $k \geq 0$, if and only if*

$$L + 1 - \log_2 \hat{\rho} > k.$$

Here,

$$\alpha_\varphi = L + 1 - \log_2 \hat{\rho}.$$

If α_φ *is not an integer, then*

$$\omega(\varphi^{(l)}, h) \asymp h^{\alpha_\varphi - l},$$

where $l = [\alpha_\varphi]$. *If* $\alpha_\varphi = l + 1$ *is an integer, then*

$$C_1 h \;\le\; \omega(\varphi^{(l)}, h) \;\le\; C_2 h |\ln h|, \qquad h \in (0, 1/2).$$

b) *A refinable function* φ *belongs to* $W_p^k(\mathbb{R})$, $k \ge 0$, *if and only if*

$$L + 1 - \log_2 \rho_p \;>\; k.$$

Here,

$$\alpha_{\varphi,p} = L + 1 - \log_2 \rho_p.$$

If $\alpha_{\varphi,p}$ *is not an integer, then*

$$\omega(\varphi^{(l)}, h)_p \asymp h^{\alpha_\varphi - l},$$

where $l = [\alpha_{\varphi,p}]$. *If* $\alpha_{\varphi,p} = l + 1$ *is an integer, then*

$$C_1 h \le \omega(\varphi^{(l)}, h)_p \le C_2 h |\ln h|, \qquad h \in (0, 1/2).$$

Proof. a) Let $l \ge 0$ be the largest integer for which $\varphi \in C^l$. We can assume $l = 0$; otherwise, invoking Theorem 6.4.1, we decompose φ into the convolution $B_{l-1} * \varphi_0$ and go over to the function φ_0, for which $\alpha_{\varphi_0} = \alpha_\varphi - l$. Thus $l = 0$; i.e., φ is continuous but does not belong to C^1. Then $\alpha_\varphi = -\log_2 \hat\rho(T_0|_{\tilde A_n}, T_1|_{\tilde A_n})$ by Theorem 5.1.4. By Theorem 6.3.1, the matrices of the operators $T_i|_{\tilde A_n}$ have the form (6.5). Applying Proposition 6.2.5 to them, we get

$$\hat\rho(T_0|_{\tilde A_n}, T_1|_{\tilde A_n}) = \max\{2^{-1}, \dots, 2^{-L}, 2^{-L-1}\hat\rho\}.$$

If $2^{-L-1}\hat\rho < 2^{-1}$, then the matrices $T_i|_{\tilde A_n}$ satisfy the assumptions of Proposition 5.3.1. Indeed, by Theorem 6.3.1 we have $\tilde A_n = G_{r+1}$ and

$$\hat\rho(T_0|_{G_{r+2}}, T_1|_{G_{r+2}}) = \max\Big\{2^{-2},\ \dots\ ,\ 2^{-L},\ 2^{-L-1}\hat\rho\Big\} \;<\; \frac{1}{2}.$$

Consequently, each operator $T_i|_{\tilde A_n}$, $i = 0, 1$, has a unique eigenvector $\tilde v_i$ with eigenvalue $\frac{1}{2}$. In addition, Proposition 6.2.5 yields $\tilde L \subset G_{r+2}$; hence $\hat\rho(T_0|_{\tilde L}, T_1|_{\tilde L}) < \frac{1}{2}$. Thus all conditions of Proposition 5.3.1 are satisfied, and so $\varphi \in C^1$, which is impossible. Therefore,

$$2^{-L-1}\hat\rho \ge 2^{-1};$$

hence,

$$\hat\rho(T_0|_{\tilde A_n}, T_1|_{\tilde A_n}) = 2^{-L-1}\hat\rho,$$

which proves the equality

$$\alpha_\varphi = L + 1 - \log_2 \hat\rho.$$

Conversely, if $L + 1 - \log_2 \hat\rho > 0$, then $2^{-L-1}\hat\rho < 1$; hence,

$$\hat\rho(T_0|_{\tilde A_n}, T_1|_{\tilde A_n}) = \max\Big\{2^{-1},\ 2^{-2},\ \dots\ ,\ 2^{-L},\ 2^{-L-1}\hat\rho\Big\} < 1.$$

By virtue of Theorem 5.1.4, we have $\varphi \in C(\mathbb{R})$. Assume now that for some $k \ge 0$ the inequality $L + 1 - \log_2 \hat\rho > k$ holds. Prove that $\varphi \in C^k$. For $k = 0$ this has already been proved. In view of Lemma 5.1.2, $\log_2 \hat\rho \ge 1$; therefore $L > k$. Then the function φ can be represented in the form $\varphi = B_{k-1} * \varphi_0$, where the function φ_0 is continuous (since for this function we have the case $k = 0$). Hence $\varphi \in C^k(\mathbb{R})$.

It remains to check the asymptotics of the modulus of continuity $w_{\varphi,h}$. Without loss of generality we assume again that $l = 0$. If α_φ is not an integer, then $2^{-L-1}\hat{\rho} > 2^{-1}$; if it is an integer, then $2^{-L-1}\hat{\rho} = 2^{-1}$. Hence the maximum of the numbers

$$2^{-1}, \ \dots \ , \ 2^{-L}, \ \ 2^{-L-1}\hat{\rho}$$

is attained on the single element $2^{-L-1}\hat{\rho}$ in the first case and on two elements $2^{-L-1}\hat{\rho}$ and 2^{-1} in the second case. Then, in view of Proposition 6.2.5, in the first case, we have

$$c_1\hat{\rho}^k(T_0|_{\tilde{A}_n}, T_1|_{\tilde{A}_n}) \le \max_{d_1,\dots,d_k} \|T_{d_1}\cdots T_{d_k}|_{\tilde{A}_n}\| \le c_2\hat{\rho}^k(T_0|_{\tilde{A}_n}, T_1|_{\tilde{A}_n}); \tag{7.1}$$

in the second case, we have

$$c_1\hat{\rho}^k(T_0|_{\tilde{A}_n}, T_1|_{\tilde{A}_n}) \le \max_{d_1,\dots,d_k} \|T_{d_1}\cdots T_{d_k}|_{\tilde{A}_n}\| \le c_2 k\hat{\rho}^k(T_0|_{\tilde{A}_n}, T_1|_{\tilde{A}_n}). \tag{7.2}$$

Let us turn now to the proof of Theorem 5.1.4. Since

$$\|T_{d_1}\cdots T_{d_k}(v_1 - v_0)\| \le \|T_{d_1}\cdots T_{d_k}|_{\tilde{A}_n}\| \cdot \|v_1 - v_0\|,$$

it follows that by substituting inequalities (7.1) and (7.2) by turn in inequality (5.7), we get

$$\|v(y) - v(z)\| \le C_2|y - z|^{-\log_2\hat{\rho}} \quad \text{in the first case and}$$

$$\|v(y) - v(z)\| \le C_2|y - z|^{-\log_2\hat{\rho}}\Big|\ln|y - z|\Big| \quad \text{in the second case.}$$

Therefore,

$$w_{\varphi,h} \le C_2 h^{\alpha_\varphi}$$

for a noninteger α_φ, and

$$w_{\varphi,h} \le C_2 h^{\alpha_\varphi}|\ln h|$$

for integer α_φ. To prove the opposite inequalities, note that the vector $v_1 - v_0$ belongs to $\tilde{A}_n$ but it does not belong to any smaller (by inclusion) common invariant subspace of the operators T_0, T_1 (by the definition of $\tilde{A}_n$). It follows that

$$\max_{d_1,\dots,d_k} \|T_{d_1}\cdots T_{d_k}(v_1 - v_0)\| \ge C\hat{\rho}^k(T_0|_{\tilde{A}_n}, T_1|_{\tilde{A}_n})$$

(this is proved in the same way as in the proof of Theorem 5.1.4). Therefore $w_{\varphi,h} \ge C_1 h$, which completes the proof of Theorem 7.1.1 in case a).

The proof of case b) is analogous. ◊

COROLLARY 7.1.2. *If $\varphi \in C^k(\mathbb{R})$ $(k \ge 0)$, then $\alpha_\varphi > k$. If $\varphi \in W_p^k(\mathbb{R})$ $(k \ge 0)$, then $\alpha_{\varphi,p} > k$.*

Thus, to construct a continuous stable refinable function, one needs to do the following:

1) Find the multiplicity of zero of the mask $\mathbf{m}_0$ at the point $z = -1$ and make sure that the mask has neither symmetric zeros on the unit circle nor cycles. Then pass to the clean mask $\mathbf{m}_c$, using the cleaning procedure.

2) Compute the joint spectral radius $\hat{\rho}$ of the operators $\tilde{T}_0, \tilde{T}_1$ corresponding to the coefficients of the clean mask. The function φ is continuous if and only if $\log_2\hat{\rho} < L + 1$. Here, $\alpha_\varphi = L + 1 - \log_2\hat{\rho}$.

3) Find the eigenvector v_0 of the operator T_0 corresponding to the eigenvalue 1 (such a vector is unique by Corollary 6.3.6) and set $v_0 = v(0) = (\varphi(0), \dots,$

$\varphi(N-1))^T$. Then, by formula (5.13), we find the values of φ at dyadic points and extend the function to the whole segment by continuity.

An L_p-solution is constructed similarly; $\varphi \in L_p(\mathbb{R})$ if and only if $\log_2 \rho_p < L+1$. Here,

$$\alpha_{\varphi,p} = L + 1 - \log_2 \rho_p.$$

To construct the function φ, we find the eigenvector c of the operator $\frac{1}{2}(T_0 + T_1)$ corresponding to the eigenvalue 1 and set

$$c = \int_0^1 v(x)\,dx = \left(\int_0^1 \varphi(x)\,dx\;,\;\ldots,\; \int_{N-1}^{N} \varphi(x)\;dx \right)^T.$$

This vector is unique (Corollary 6.3.6). Then by formula (5.15) we find the values of the integrals of the functions φ over dyadic segments $[q2^{-r}, (q+1)2^{-r}]$, $q \in \mathbb{Z}$, $q2^{-r} \in [0, N], r = 1, 2, \ldots$.

Thus, *the construction and calculation of coefficients of smoothness for refinable functions in the spaces C and L_p are reduced to the calculation of the spectral radii of the pair of operators $\tilde{T}_0, \tilde{T}_1$ corresponding to the clean mask.*

7.2. Local smoothness of wavelets

To determine the local smoothness of a function $\varphi(x)$ at various points x, we must go back from the clean mask $\mathbf{m}_c(z)$ to the mask $\mathbf{m}_0(z)$ of the initial equation. We shall determine the local smoothness not for the function φ but for the vector function

$$v(x) = \Big(\varphi(x), \varphi(x+1), \ldots, \varphi(x+N-1)\Big)^T, \quad x \in [0,1].$$

For a given point $x \in [0,1]$, the local smoothness $\alpha_v(x)$ is equal to the minimal value of the local smoothness of the function φ at the points $x, x+1, \ldots, x+N-1$. It follows from formula (6.3) that when we eliminate a symmetric root, the local smoothness of the function $v(x)$ remains the same at all points x; hence without loss of generality we can assume that the mask has no symmetric roots. Then (Corollary 6.3.4) the space $\tilde{A}_n$ coincides with the space

$$\tilde{W} = \Big\{(x_1, \ldots, x_N),\ \sum x_k = 0\Big\},$$

and on this space, the operators $\tilde{T}_i$ have the form (6.5). Using Theorem 5.4.1 about smoothness of fractal curves, we obtain the following theorem. Set

$$\hat{\rho}_x = \hat{\rho}_x(\tilde{T}_0, \tilde{T}_1), \qquad \hat{\rho} = \hat{\rho}(\tilde{T}_0, \tilde{T}_1),$$

where l is the greatest integer for which $\varphi \in C^l$ (in other words, $l = [\alpha_\varphi]$ if α_φ is not integer, and $l = \alpha_\varphi - 1$ if α_φ is integer).

THEOREM 7.2.1. *Suppose that a stable refinable function is continuous. Then:*

a) *If $\hat{\rho} < 2$, then $\alpha_{v^{(l)}}(x) \le 1 - \log_2 \hat{\rho}_x$ at any point $x \in [0,1]$; at the normal points we have $\alpha_{v^{(l)}}(x) = 1 - \log_2 \hat{\rho}_x$.*

b) *If $\hat{\rho} \ge 2$, then*

$$\alpha_{v^{(l)}}(x) \le \min\Big\{L - l + 1 - \log_2 \hat{\rho}_x\,,\ 1\Big\}$$

at any point; at the normal points we have

$$\alpha_{v^{(l)}}(x) = \min\Big\{L - l + 1 - \log_2 \hat{\rho}_x\,,\ 1\Big\}.$$

REMARK 7.2.2. The condition $\hat{\rho} < 2$ is equivalent to the condition $l = L$; in other words, φ has a maximal possible number of continuous derivatives for a given order of the Strang-Fix conditions.

First we prove the following analog of Lemma 6.3.8:

LEMMA 7.2.3. *Let each of the two $N \times N$-matrices $B_i, i = 0, 1$, have the lower triangle block form* (6.6). *Then for any $x \in [0, 1]$, we have*

$$\hat{\rho}_x(B_0, B_1) = \max_{j=1,\dots,l} \hat{\rho}_x(A_0^j, A_1^j).$$

In particular,

$$\bar{\rho}(B_0, B_1) = \max_{j=1,\dots,l} \bar{\rho}(A_0^j, A_1^j).$$

Proof. The proof is similar to that of Lemma 6.3.8. ◊

REMARK 7.2.4. The analogous equality for the lower spectral radius $\check{\rho}(B_0, B_1) = \max_{j=1,\dots l} \check{\rho}(A_0^j, A_1^j)$, in general, is not true (which may seem strange.) Corresponding examples are elementary.

Proof of Theorem 7.2.1. As noted above, it suffices to prove the theorem for a mask without symmetric roots. As for Theorem 7.1.1, we reduce the proof to the case $l = 0$; otherwise, if $l \geq 1$, we use Theorem 6.4.1 and decompose φ into the convolution $B_{l-1} * \varphi_0$ and then go over to the function φ_0, for which $l = 0$. Since this function is a linear combination of integer shifts of the function $\varphi^{(l)}$, we have

$$\alpha_{v^{(l)}}(x) = \alpha_{v_0}(x) - l$$

at each point $x \in [0, 1]$. Thus it suffices to prove the theorem for the case $l = 0$. By virtue of Theorem 5.4.1, we have

$$\alpha_v(x) \leq -\log_2 \hat{\rho}_x(T_0|_{\tilde{A}_n}, T_1|_{\tilde{A}_n})$$

at all points x. At normal points this inequality turns into an equality. Now use Theorem 6.3.1. In the case when $\hat{\rho} < 2$, which is equivalent to the equality $L = 0$ (Remark 7.2.2), we have $T_i|_{\tilde{A}_n} = \frac{1}{2}\tilde{T}_i,\ i = 0, 1$. Therefore,

$$\hat{\rho}_x(T_0|_{\tilde{A}_n}, T_1|_{\tilde{A}_n}) = \frac{1}{2}\,\hat{\rho}_x,$$

which completes the proof. If $\hat{\rho} \geq 2$, i.e., $L \geq 1$, the operators T_i have the block form (6.5). Then by Lemma 7.2.3 we have

$$\hat{\rho}_x(T_0|_{\tilde{A}_n}, T_1|_{\tilde{A}_n}) = \max\{2^{-1}, \dots, 2^{-L}, 2^{-L-1}\hat{\rho}_x\} = \max\{2^{-1}, 2^{-L-1}\hat{\rho}_x\}.$$

It follows that

$$-\log_2 \hat{\rho}_x(T_0|_{\tilde{A}_n}, T_1|_{\tilde{A}_n}) = \min\Big\{L + 1 - \log_2 \hat{\rho}_x\,,\ 1\Big\},$$

which completes the proof of the theorem. ◊

COROLLARY 7.2.5. *If the smoothness of a refinable function does not exceed $L - 1$, then $\alpha_{v^{(l)}}(x) \leq 1$ at all points x.*

In the case $l = L$, the local smoothness of the function v may be arbitrarily large. In particular, it can be greater than 1 at almost all (in the Lebesgue measure) points x.

According to Corollary 6.3.2, both operators T_0, T_1 are nondegenerate on the space $\tilde{A}_n$. This fact enables the extension of all results concerning smoothness of fractal curves from Chapter 5 to refinable functions. First combining Theorem 7.2.1 and Proposition 5.4.2, we obtain a formula for smoothness at rational points:

PROPOSITION 7.2.6. *Suppose that a refinable function is continuous. Then:*
a) *If $\hat{\rho} < 2$, then at each rational point x with period $(d_1 \dots d_k)$ $(k \geq 2)$ we have*

$$\alpha_{v^{(l)}}(x) = -\frac{1}{k} \log_2 \rho(\tilde{T}_{d_1} \cdots \tilde{T}_{d_k});$$

at all dyadic points we have

$$\alpha^{+}_{v^{(l)}}(x) = 1 - \log_2 \rho(\tilde{T}_0), \qquad \alpha^{-}_{v^{(l)}}(x) = 1 - \log_2 \rho(\tilde{T}_1).$$

b) *If $\hat{\rho} \geq 2$, then at each rational point with period $(d_1 \dots d_k)$ $(k \geq 2)$ we have*

$$\alpha_{v^{(l)}}(x) = \min\Big\{1,\ L + 1 - l - \frac{1}{k} \log_2 \rho(\tilde{T}_{d_1} \cdots \tilde{T}_{d_k})\Big\};$$

at all dyadic points we have

$$\alpha^{+}_{v^{(l)}}(x) = \min\Big\{1,\ L+1-l-\log_2 \rho(\tilde{T}_0)\Big\}, \quad \alpha^{-}_{v^{(l)}}(x) = \min\Big\{1,\ L+1-l-\log_2 \rho(\tilde{T}_1)\Big\}.$$

At all rational (but not dyadic) points x the smoothness on the right coincides with the smoothness on the left.

Thus, at all rational points the local smoothness of a refinable function can be found explicitly in terms of the (usual) spectral radius of the corresponding product of operators $\tilde{T}_0, \tilde{T}_1$.

Let us now extend the results concerning the distribution of points of fixed local smoothness to refinable functions. Since the operators T_i on the space $\tilde{A}_n$ are both nondegenerate, we can apply Theorem 5.4.10 to the function φ. According to Lemma 7.2.3 for $\hat{\rho} < 2$, for the Lyapunov exponent of the operators $T_i|_{\tilde{A}_n}$, we have

$$\bar{\rho}(T_0|_{\tilde{A}_n}, T_1|_{\tilde{A}_n}) = \frac{1}{2}\bar{\rho}\ ,$$

and for $\hat{\rho} \geq 2$, we have

$$\bar{\rho}(T_0|_{\tilde{A}_n}, T_1|_{\tilde{A}_n}) = \max\Big\{\frac{1}{2},\ 2^{l-L-1}\bar{\rho}\Big\}.$$

Lemma 7.2.3 does not hold for the lower spectral radius (Remark 7.2.4). Nevertheless, for operators $T_i|_{\tilde{A}_n}$, the following analogous relation holds:

LEMMA 7.2.7. *If $\hat{\rho} < 2$, then $\check{\rho}(T_0|_{\tilde{A}_n}, T_1|_{\tilde{A}_n}) = \frac{1}{2}\check{\rho}$; in the case $\hat{\rho} \geq 2$ we have*

$$\check{\rho}(T_0|_{\tilde{A}_n}, T_1|_{\tilde{A}_n}) = \max\Big\{\frac{1}{2},\ 2^{l-L-1}\check{\rho}\Big\}.$$

Proof. If $\hat{\rho} < 2$, then $l = L$, and by Theorem 6.3.1 the operators $T_i|_{\tilde{A}_n}$ are isomorphic to the operators $\tilde{T}_i$; hence in this case there is nothing to prove. If $\hat{\rho} \geq 2$, then by Lemma 7.2.3 we have

$$\hat{\rho}_x(T_0|_{\tilde{A}_n}, T_1|_{\tilde{A}_n}) = \max\Big\{\frac{1}{2}\ ,\ 2^{l-L-1}\hat{\rho}_x\Big\} \tag{7.3}$$

for any $x \in [0,1]$. According to Lemma 5.4.9, for any pair of nondegenerate operators $\tilde{B}_0, \tilde{B}_1$ we have

$$\check{\rho}(\tilde{B}_0, \tilde{B}_1) = \inf_x \hat{\rho}_x(\tilde{B}_0, \tilde{B}_1).$$

Taking inf of both sides of equality (7.3) over all $x \in [0,1]$, we obtain the required relation. $\Diamond$

Using Theorem 5.4.10, we obtain the following statement about the distribution of the points of a fixed local smoothness for refinable functions:

THEOREM 7.2.8. *Suppose that a stable refinable function φ is continuous and $l \geq 0$ is the largest integer for which $\varphi \in C^l(\mathbb{R})$. Suppose the operators $\tilde{T}_i$, $i = 0, 1$, correspond to the clean mask, $\hat{\rho}$, $\check{\rho}$, $\bar{\rho}$ being their joint spectral radius, the lower spectral radius, and the Lyapunov exponent, respectively. Then*

$$\Bigl\{\alpha_{v^{(l)}}(x),\ x \in [0,1]\Bigr\} = \Bigl[\,L + 1 - l - \log_2 \hat{\rho}\ ,\ b\,\Bigr],$$

where $b = 1 - \log_2 \check{\rho}$ if $l = L$ and otherwise, $b = \min\bigl\{1,\ L + 1 - l - \log_2 \check{\rho}\bigr\}$.

For each α from this segment, the set of points x for which $\alpha_{v^{(l)}}(x) = \alpha$ is a nonempty 0-1 set; in particular, it is everywhere dense on the segment $[0,1]$. For the mean smoothness $\alpha = \alpha_{av}$, where

$$\alpha_{av} = \begin{cases} 1 - \log_2 \bar{\rho}\ , & \text{if } \ l = L, \\ \min\Bigl\{1,\ L + 1 - l - \log_2 \bar{\rho}\Bigr\}, & \text{if } \ l < L, \end{cases}$$

this set has full measure on $[0,1]$; for all other values α, it has zero measure.

As the operators $\tilde{T}_i$ are nondegenerate, $\check{\rho} > 0$, which implies the following

COROLLARY 7.2.9. *For any refinable function, the local smoothness of the function $v(x)$ is uniformly bounded above.*

REMARK 7.2.10. Note that all the obtained results do not deal with the local smoothness of the function φ; they deal with the local smoothness of the vector function

$$v(x) = \Bigl(\varphi(x), \varphi(x+1), \ldots, \varphi(x+N-1)\Bigr)^T, \qquad x \in [0,1];$$

i.e., we define the minimal value of the local smoothness of the function φ at the points x, $x+1$, ..., $x+N-1$, which may seem an essential disadvantage of our method. As for the smoothness of the function φ at a given point x, it can only be estimated from below:

$$\alpha_{\varphi^{(l)}}(x) \geq \alpha_{v^{(l)}}(x).$$

This inequality may not turn into an equality at some points x. There are examples of refinable functions that are equal to identical constants on some segment $[k, k+1]$ inside the support; hence for all x from this segment, the local smoothness of $\varphi(x)$ is equal to $+\infty$, and the local smoothness of $v(x)$ is bounded (Corollary 7.2.9). Thus for all x from this segment the strict inequality

$$\alpha_{\varphi^{(l)}}(x) > \alpha_{v^{(l)}}(x)$$

holds. A more detailed analysis of the asymptotic behavior of the vector $T_{d_1} \cdots T_{d_k}\bigl(v(1) - v(0)\bigr)$ (see Theorem 5.4.1) allows one to determine points x where the equality $\alpha_{\varphi^{(l)}}(x) = \alpha_{v^{(l)}}(x)$ holds and to compute the local smoothness $\alpha_{\varphi^{(l)}}(x)$ in the case of strict inequality. In this book, we do not present results of such analysis, since they are rather cumbersome.

Let us discuss the smoothness of wavelets generated by compactly supported refinable (scaling) functions, in particular, smoothness of compactly supported wavelets. Since every such a wavelet function is a linear combination of the form $\sum_k a_k\varphi(x-k)$, its smoothness coincides with the smoothness of the function φ. Hence the smoothness of wavelets can be defined by Theorem 7.1.1:

THEOREM 7.2.11. *Suppose ψ is a wavelet function generated by a compactly supported refinable function, $\mathbf{m}_0$ is the mask of the refinement equation, $\mathbf{m}_c$ is the corresponding clean mask, $\hat{\rho}$, ρ_p $(p<\infty)$ are the joint spectral radius and p-radius of the operators $\tilde{T}_0, \tilde{T}_1$ corresponding to the clean mask $\mathbf{m}_c$. Then:*

a) $\alpha_\psi = L+1-\log_2\hat{\rho}$. *If α_ψ is not an integer, then*

$$\omega(\psi^{(l)}\,,\,h) \asymp h^{\alpha_\psi - l},$$

where $l=[\alpha_\psi]$. If $\alpha_\varphi = l+1$ is an integer, then

$$C_1 h \;\le\; \omega(\psi^{(l)}\,,\,h) \;\le\; C_2 h|\ln h|\,, \qquad h\in\Bigl(0\,,\,\frac{1}{2}\Bigr).$$

b) $\alpha_{\psi,p} = L+1-\log_2\hat{\rho}_p$. *If $\alpha_{\psi,p}$ is not an integer, then*

$$\omega(\psi^{(l)}, h)_p \;\asymp\; h^{\alpha_\psi - l},$$

where $l=[\alpha_{\psi,p}]$. If $\alpha_{\psi,p}=l+1$ is an integer, then

$$C_1 h \;\le\; \omega(\psi^{(l)}, h)_p \;\le\; C_2 h|\ln h|\,, \qquad h\in\Bigl(0\,,\,\frac{1}{2}\Bigr).$$

COROLLARY 7.2.12. *Let ψ be a wavelet function generated by a compactly supported refinable function. If $\psi\in C^k(\mathbb{R})$ $(k\ge 0)$, then $\alpha_\varphi > k$. If $\psi\in W_p^k(\mathbb{R})$ $(k\ge 0)$, then $\alpha_{\varphi,p}>k$.*

COROLLARY 7.2.13. *Let ψ be a wavelet function generated by a compactly supported refinable function. If its smoothness α_ψ is not an integer, then $\omega(\psi^{(l)},h)\asymp h^{\alpha_\psi - l}$, where $l=[\alpha_\psi]$. An analogous result takes place for smoothness in $L_p(\mathbb{R})$.*

All results concerning local smoothness of refinable functions can be extended to wavelets without any changes. We formulate them for compactly supported wavelets ψ_n in Theorems 7.2.14 and 7.2.15. Recall that $N=2n-1$ and $\operatorname{supp}\psi_n\subset[0,N]$. For $n=1$, the wavelet function ψ_1 coincides with the Haar function; the case $n=2$ will be studied in Example 7.3.7. Therefore we shall formulate Theorems 7.2.14 and 7.2.15 for $n\ge 3$. Note that in this case $l<L$; hence $\hat{\rho}\ge 2$, so we have case b) of Theorem 7.2.1. Set

$$u(x) = \Bigl(\psi_n(x)\,,\;\ldots\;,\,\psi_n(x+N-1)\Bigr)^T.$$

As before, denote by $\mathbf{m}_0$ the mask of the scaling function generating ψ_n, by $\mathbf{m}_c$ the corresponding clean mask, and by $\hat{\rho}_x$, $\hat{\rho}$, $\check{\rho}$, $\bar{\rho}$ the corresponding spectral radii of the operators $\tilde{T}_0, \tilde{T}_1$. Using Theorems 7.2.1, 7.2.8, and Proposition 7.2.6, we get

THEOREM 7.2.14. *Let ψ_n be a compactly supported wavelet function, $n\ge 3$. Then for any point $x\in[0,1]$, we have*

$$\alpha_{u^{(l)}}(x) \le \min\Bigl\{L-l+1-\log_2\hat{\rho}_x\;,\;1\Bigr\};$$

at normal points we have

$$\alpha_{u^{(l)}}(x) \;=\; \min\Big\{L - l + 1 - \log_2 \hat{\rho}_x \,,\, 1\Big\}.$$

In particular, at each rational point with period $(d_1 \dots d_k)$ $(k \geq 2)$, *we have*

$$\alpha_{u^{(l)}}(x) = \min\Big\{1,\; L + 1 - l - \frac{1}{k}\log_2 \rho(\tilde{T}_{d_1} \cdots \tilde{T}_{d_k})\Big\};$$

at all dyadic points, we have

$$\alpha^{+}_{v^{(l)}}(x) = \min\Big\{1,\; L{+}1{-}l{-}\log_2 \rho(\tilde{T}_0)\Big\}, \quad \alpha^{-}_{v^{(l)}}(x) = \min\Big\{1,\; L{+}1{-}l{-}\log_2 \rho(\tilde{T}_1)\Big\}.$$

THEOREM 7.2.15. *Let* ψ_n *be a compactly supported wavelet function,* $n \geq 3$. *Then*

$$\Big\{\alpha_{u^{(l)}}(x)\,,\; x \in [0,1]\Big\} \;=\; \Big[L + 1 - l - \log_2 \hat{\rho}\,,\; \min\Big\{1,\; L + 1 - l - \log_2 \check{\rho}\Big\}\Big].$$

For each α *from this segment, the set of points* x *for which* $\alpha_{u^{(l)}}(x) = \alpha$ *is a nonempty* 0-1 *set; in particular, it is everywhere dense on the segment* $[0,1]$. *For the mean smoothness, we have*

$$\alpha_{av} \;=\; \min\Big\{1,\; L + 1 - l - \log_2 \bar{\rho}\Big\};$$

this set has full measure on $[0,1]$, *and for all other values of* α *it has zero measure.*

7.3. Special cases and examples

As shown above, the matrix method allows the computation of the smoothness of refinable functions in the spaces $C(\mathbb{R})$ and $L_p(\mathbb{R})$ and the local smoothness of the corresponding vector function at all points $x \in [0,1]$. This computation is reduced to finding joint spectral radii $\hat{\rho}$, $\hat{\rho}_p$, and $\hat{\rho}_x$ and the Lyapunov exponent of the matrices $\tilde{T}_0, \tilde{T}_1$ corresponding to the clean mask. In practice, the computation of these quantities presents considerable difficulties. This problem becomes much more complicated as the size of the matrices grows, and in our case, as the degree of the clean mask grows. We start our study with low degrees.

<u>deg $\mathbf{m_c} = \mathbf{0}$</u>. In this case, the clean mask is a constant $\mathbf{m}_c = 1$. It is not hard to show that all stable refinable functions with this clean mask are B-splines. Indeed, if the mask $\mathbf{m}_0$ has no symmetric roots, then

$$\mathbf{m}_0(z) = \Big(\frac{z+1}{2}\Big)^{L+1};$$

therefore $\varphi(x) = B_L(x)$. If the mask has symmetric roots, we start eliminating them using Proposition 6.2.1; let $\pm a$ be the last pair of symmetric roots. After the last iteration, i.e., replacing this pair by the root a^2, we obtain the polynomial $\big(\frac{z+1}{2}\big)^{L+1}$. As a^2 is its root, then $a^2 = -1$, and it follows that $\pm a = \pm i$. This is impossible because the mask cannot have symmetric roots on the unit circle due to the stability of the refinable function.

Thus, this case corresponds to B-splines $B_L(x)$ only. In spite of the fact that the smoothness $\alpha_\varphi = L$ is integer, we have $\omega_{\varphi^{(L-1)}} \asymp h$ (compare with Theorem 7.1.1). The smoothness of φ in the space $L_p(\mathbb{R})$ is equal to $L+1$ for all $p \in [1, +\infty)$.

Note that for nonstable refinable functions, the case deg $\mathbf{m}_c = 0$ corresponds not only to B-splines but also to various refinable splines of the form

$$\varphi = \sum_{k=0}^{r} a_k B_L(x-k), \tag{7.4}$$

where a_k are some coefficients. For a complete classification of refinable splines and for their properties, see [**110**], [**128**].

deg $\mathbf{m_c} = \mathbf{1}$. In this case, the clean mask is a first-degree polynomial $\mathbf{m}_c(z) = \lambda + (1-\lambda)z$, where λ is an arbitrary complex number different from 0 or $\frac{1}{2}$ (otherwise, the degree of the mask can be reduced to zero). Here, $\tilde{T}_i$, $i = 0, 1$, are the operators of multiplication in $\mathbb{R}$, and we have

$$\hat{\rho} = 2\max\{|\lambda|, |1-\lambda|\}, \quad \check{\rho} = 2\min\{|\lambda|, |1-\lambda|\}, \quad \rho_p = 2^{1-\frac{1}{p}}\left(|\lambda|^p + |1-\lambda|^p\right)^{1/p}.$$

By Theorem 7.1.1, we get

$$\alpha_\varphi = L - \log_2\Big(\max\{|\lambda|, |1-\lambda|\}\Big), \quad \alpha_{\varphi,p} = L + \frac{1}{p} - \frac{1}{p}\Big(\log_2\left(|\lambda|^p + |1-\lambda|^p\right)\Big). \tag{7.5}$$

Let us now study the local smoothness. In order to be defined we assume that $|\lambda| \geq |1-\lambda|$, i.e., $\operatorname{Re}\lambda \geq \frac{1}{2}$. For an arbitrary sequence x of zeros and ones, we have $\hat{\rho}_x = 2|\lambda|^{1-\overline{\delta}}|1-\lambda|^{\overline{\delta}}$, where

$$\overline{\delta} \;=\; \overline{\delta}(x) \;=\; \liminf_{k\to\infty} \frac{1}{k}\sum_{j=1}^{k} d_j$$

is the lower density of ones in the sequence x. Using Theorem 7.2.1, we deduce that at any normal point $x \in [0,1]$ inside the disc $|\lambda| < 1$ (where $\hat{\rho} < 2$) the local smoothness of the function $v^{(L)}$ is equal to

$$\alpha_{v^{(L)}}(x) = -(1-\overline{\delta})\log_2|\lambda| - \overline{\delta}\log_2|1-\lambda|.$$

At dyadic points x the smoothness on the right and on the left is equal to

$$\alpha^{+}_{v^{(L)}} = -\log_2|\lambda|, \qquad \alpha^{-}_{v^{(L)}} = -\log_2|1-\lambda|,$$

respectively. Thus, if $\operatorname{Re}\lambda \neq \frac{1}{2}$, then at all dyadic points of the interval $[0,1]$ the smoothness on the left is not equal to the smoothness on the right.

As $\bar{\rho} = 2\sqrt{|\lambda(1-\lambda)|}$, we have

$$\alpha_{av} \;=\; -\;\frac{\log_2|\lambda| + \log_2|1-\lambda|}{2}.$$

In particular, if $|\lambda(1-\lambda)| < \frac{1}{4}$ (i.e., the point λ is located inside the corresponding lemniscate with center at the point $\frac{1}{2}$), then $\alpha_{av} > 1$; hence $\varphi^{L+1}(x) \equiv 0$ almost everywhere. As λ tends to the point 1 (remaining inside the circle $|\lambda| < 1$), the mean smoothness of the function $\varphi^{(L)}$ tends to infinity, though this function is not even differentiable everywhere: according to Theorem 7.2.8, its local smoothness fills the segment $\big[-\log_2|\lambda|, -\log_2|1-\lambda|\big]$; as $\lambda \to 1$, the left endpoint of this segment tends to 0.

Note that if λ belongs to the set

$$\Gamma = \Big\{\lambda \in \mathbb{C}\,,\ |1-\lambda| < \frac{1}{2}\,,\ |\lambda| < 1\Big\}$$

(the intersection of two open discs), the smoothness on the left is greater than 1, and the smoothness on the right is less than 1. Hence at all dyadic points the function $v^{(L)}$ is not differentiable on the right, but it is differentiable on the left, the left-hand derivative being equal to 0.

If $l < L$, i.e., $|\lambda| \geq 1$, we have case b) of Theorem 7.2.1:

$$\alpha_{v^{(l)}}(x) = \min\Big\{L - l - (1-\bar{\delta})\log_2|\lambda| - \bar{\delta}\log_2|1-\lambda|\ ,\ 1\Big\}.$$

In contrast to the previous case, the smoothness cannot exceed 1 at any point. At dyadic points x the smoothness on the right is equal to

$$\alpha^{+}_{v^{(l)}}(x) = \min\Big\{L - l - \log_2|\lambda|,\ 1\Big\},$$

and the smoothness on the left is equal to

$$\alpha^{-}_{v^{(l)}}(x) = \min\Big\{L - l - \log_2|1-\lambda|,\ 1\Big\}.$$

For the mean smoothness, we have

$$\alpha_{av}(x) = \min\left\{L - l - \frac{\log_2|\lambda| + \log_2|1-\lambda|}{2}\ ,\ 1\right\}. \tag{7.6}$$

In particular, if $|\lambda(1-\lambda)| \leq 2^{2(L-l-1)}$, then $\varphi^{(L+1)}(x) = 1$ for almost all x. In the case of the strict inequality

$$|\lambda(1-\lambda)|\ <\ 2^{2(L-l-1)},$$

this fact is easy to explain: here, $\alpha_{\varphi^{(l)},1} > 1$ (formula (7.5)); hence $\varphi^{(l)} \in W_1^1$. Thus the function $\varphi^{(l)}$ is differentiable almost everywhere and its derivative belongs to L_1. By virtue of the Lebesgue theorem (Appendix A.10), such a function is locally Lipschitz almost everywhere; i.e., for almost all x, there exists a constant $C(x)$ such that

$$\Big\|v^{(l)}(x+h) - v^{(l)}\Big\| = \Big\|\int_x^{x+h} v^{(l+1)}(t)\,dt\ \Big\| \leq C(x)|h|.$$

Therefore $\alpha_{v^{(l)}} = 1$.

<u>$\deg \mathbf{m_c} = \mathbf{2}$</u>. In this case, the clean mask is the second-degree polynomial

$$\mathbf{m}_c(z) = \frac{1}{2}\Big(c_2z^2 + c_1z + c_0\Big),$$

where $c_0 + c_1 + c_2 = 2$. For $c_1 = 1$, the mask admits further reduction of degree, and this case can be reduced to one of the previous cases; so we assume that $c_1 \neq 1$. The matrices of the operators $\tilde{T}_i$ have the following form:

$$\tilde{T}_0 = \begin{pmatrix} c_0 & 0 \\ c_2 & c_1 \end{pmatrix}, \qquad \tilde{T}_1 = \begin{pmatrix} c_1 & c_0 \\ 0 & c_2 \end{pmatrix}.$$

The values of $\hat{\rho}$, $\hat{\rho}_p$, $\check{\rho}$, $\bar{\rho}$ are rather hard to calculate. We consider only the case of real coefficients c_0, c_1, c_2 and focus our attention on the computation of the joint spectral radius $\hat{\rho}$, which is responsible for the continuity of the refinable function.

LEMMA 7.3.1. *If the coefficients c_0, c_1, c_2 are real, then the matrices of both operators $\tilde{T}_0, \tilde{T}_1$ are symmetric in an appropriate basis if and only if*

$$c_0c_1c_2(c_1 - 1) \geq 0. \tag{7.7}$$

The reader can easily prove this lemma. ◊

The set of pairs of coefficients (c_0, c_2) satisfying (7.7) (the coefficient c_1 is expressed as $c_1 = 2 - c_0 - c_2$) is easily represented on the coordinate plane. It is not hard to compute the joint spectral radius for each point of this set since for symmetric matrices, we have

$$\hat{\rho} = \max\{\rho(\tilde{T}_0),\, \rho(\tilde{T}_1)\}.$$

Indeed, in view of (5.1) we have $\hat{\rho} \geq \max\{\rho(\tilde{T}_0),\, \rho(\tilde{T}_1)\}$. On the other hand, it is clear that $\hat{\rho} \leq \max\{\|\tilde{T}_0\|,\, \|\tilde{T}_1\|\}$, and for symmetric matrices, $\|\tilde{T}_i\| = \rho(\tilde{T}_i)$ (the norm is Euclidean). Hence,

$$\hat{\rho} = \max\{\rho(\tilde{T}_0),\, \rho(\tilde{T}_1)\}.$$

Thus $\hat{\rho} = \max\{|c_0|,\, |c_1|,\, |c_2|\}$. We obtain

PROPOSITION 7.3.2. *If the coefficients c_0, c_1, c_2 are real and*

$$c_0\, c_1\, c_2\, (c_1 - 1) \;\geq\; 0,$$

then

$$\hat{\rho}(\tilde{T}_0,\, \tilde{T}_1) \;=\; \max\{|c_0|,\, |c_1|,\, |c_2|\}.$$

Note that if both matrices are symmetric in some basis, it is not easy to compute the lower spectral radius $\check{\rho}$ or the p-radius ρ_p. As for the pairs (c_0, c_2) that do not satisfy that inequality, it is not easy to compute the joint spectral radius for them (even by a search of more than two matrix products). The computation made in [**132**], [**133**] shows that for some pairs (c_0, c_2), we cannot obtain the exact value of the joint spectral radius even by an exhaustive search of all products of $k = 20$ matrices. For instance, in the case $(c_0, c_2) = (1.2\,,\, -0.4)$, the value $\big(\max \rho(\Pi_k)\big)^{1/k}$ approximates $\hat{\rho}$ with relative error 10^{-3} for $k \geq 72$. That is, one has to search through more than $6 \cdot 10^{19}$ matrix products! We do not even know whether we can restrict our search to finitely many matrices for each point (c_0, c_2). In other words, we do not know if Lagarias's hypothesis [**134**] is true for matrices $\tilde{T}_0$, $\tilde{T}_1$ with real coefficients.

Only in the case of a symmetric mask ($c_0 = c_2$) both $\hat{\rho}$ and $\check{\rho}$ can be easily found.

PROPOSITION 7.3.3. *If the coefficients c_0, c_1, c_2 are real and $c_0 = c_2$, then in the case $c_1 \in [0, 1]$ we have*

$$\hat{\rho} = \sqrt{\rho(\tilde{T}_0\tilde{T}_1)}, \qquad \check{\rho} = \max\{\rho(\tilde{T}_0),\, \rho(\tilde{T}_1)\}.$$

If $c_1 \notin [0, 1]$, then

$$\hat{\rho} = \max\{\rho(\tilde{T}_0),\, \rho(\tilde{T}_1)\}, \qquad \check{\rho} = \sqrt{\rho(\tilde{T}_0\tilde{T}_1)}.$$

Thus, if $c_1 \leq 0$ or $c_1 \geq 1$, then

$$\hat{\rho} \;=\; \max\{|c_0|,\, |c_1|,\, |c_2|\} = \max\left\{ \Big|c_1\Big|,\, \Big|1 - \frac{c_1}{2}\Big| \right\}$$

(which is consistent with Proposition 7.3.2), and

$$\check{\rho} = \frac{1}{4}\max\left\{ \left|2 - c_1 + \sqrt{4 + 12c_1 - 7c_1^2}\right|,\quad \left|2 - c_1 - \sqrt{4 + 12c_1 - 7c_1^2}\right| \right\}.$$

In the case $0 \leq c_1 \leq 1$, these quantities change places.

To prove Proposition 7.3.3, we need an auxiliary result, which serves as a basis for the computation of the joint spectral radius.

PROPOSITION 7.3.4. *Suppose $\tilde{B}_0, \tilde{B}_1$ are linear operators acting in $\mathbb{R}^n$ and λ is a positive number. Then:*

a) *If there exists a convex body (a compact set with nonempty interior) centrally symmetric with respect to zero $\mathcal{M} \subset \mathbb{R}^n$, for which $\tilde{B}_i\mathcal{M} \subset \lambda\ \mathcal{M}$, then $\hat{\rho}(\tilde{B}_0, \tilde{B}_1) \le \ \lambda$.*

b) *If there exists a closed set $\mathcal{Q} \subset \mathbb{R}^n$ not containing zero, for which $\tilde{B}_i\mathcal{Q} \subset \lambda\mathcal{Q}$, then $\check{\rho}(\tilde{B}_0, \tilde{B}_1) \ge \lambda$.*

`Proof.` a) Denote by $\|\cdot\|_{\mathcal{M}}$ the Minkowski norm corresponding to the convex body $\mathcal{M}$. Since $\|\tilde{B}_i\|_{\mathcal{M}} \le \lambda$, $i = 0, 1$, the norm of any product $\tilde{B}_{d_1} \cdots \tilde{B}_{d_k}$ does not exceed λ^k. Therefore,

$$\max_{d_1,\dots,d_k} \|\tilde{B}_{d_1} \cdots \tilde{B}_{d_k}\|_{\mathcal{M}}^{1/k} \le \lambda$$

for any k; hence $\hat{\rho}(\tilde{B}_0, \tilde{B}_1) \le \lambda$.

b) Let $h = \inf\{\|u\|,\ u \in \mathcal{Q}\}$. It is clear that $h > 0$. Choose an arbitrary point $a \in \mathcal{Q}$. For any $d_1, \dots, d_k$, we have $\tilde{B}_{d_1} \cdots \tilde{B}_{d_k} a \in \lambda^k \mathcal{Q}$; hence $\|\tilde{B}_{d_1} \cdots \tilde{B}_{d_k} a\| \ge h\lambda^k$. Therefore

$$\min_{d_1,\dots,d_k} \|\tilde{B}_{d_1} \cdots \tilde{B}_{d_k}\|^{1/k} \ge \left(\frac{h}{\|a\|}\right)^{1/k} \lambda$$

for any k; hence,

$$\check{\rho}(\tilde{B}_0, \tilde{B}_1) \ge \lambda. \qquad \Diamond$$

The idea of the proof of Proposition 7.3.3 is very simple. Proposition 7.3.4 allows one to obtain upper estimates for $\hat{\rho}$ and lower estimates for $\check{\rho}$. To this end, we only have to find an appropriate convex body $\mathcal{M}$ or a closed set $\mathcal{Q}$, respectively. On the other hand, Lemma 5.1.2 gives opposite estimates. According to (5.1), for any product Π_k, the variable $\Big(\rho(\Pi_k)\Big)^{1/k}$ is a lower estimate for $\hat{\rho}$ and an upper estimate for $\check{\rho}$. If the lower estimate happens to coincide with the upper estimate, then the exact value of $\hat{\rho}$ (or $\check{\rho}$) is found. To do this, we have to find an appropriate product Π_k and a set $\mathcal{M}$ ($\mathcal{Q}$, respectively).

`The proof of Proposition 7.3.3` will be divided into several cases. First we compute $\hat{\rho}$. If $c_1 \le 0$ or $c_1 \ge 1$, the coefficients satisfy (7.7) and by virtue of Proposition 7.3.2 we have $\hat{\rho} = \max\{\rho(\tilde{T}_0),\ \rho(\tilde{T}_1)\}$. Now let $0 < c_1 < 1$. Using inequality (5.1), we get

$$\hat{\rho} \ge \sqrt{\rho(\tilde{T}_0\tilde{T}_1)}.$$

To prove the opposite inequality, we use Proposition 7.3.4. Set

$$y = \left(\sqrt{c_0^2 + 4c_1c_0} - c_0\right)/(2c_1)\ , \quad b = \left(1 + y\Big(\frac{c_1}{c_0} - 1\Big)\right) / \left(2 - \frac{c_1}{c_0}\right).$$

Then the required body $\mathcal{M}$ is the convex hexagon with the vertices

$$(1, y),\ (-1, -y),\ (y, 1),\ (-y, -1),\ (-b, b),\ (b, -b)\,.$$

Direct verification gives

$$\tilde{T}_i\mathcal{M} \subset \lambda\mathcal{M}, \qquad i = 0, 1.$$

Indeed, for $i = 0$, we have

$$\tilde{T}_0(\pm 1, \pm y) = \lambda(\pm y, \pm 1)\,, \quad \tilde{T}_0(\pm y, \pm 1) \in [\lambda(\pm y, \pm 1)\,, \lambda(\pm b, \mp b)]\,,$$
$$\tilde{T}_0(\pm b, \mp b) \in \lambda\mathcal{M}.$$

The case $i = 1$ is analogous. Thus $\hat{\rho} = \lambda$.

Let us now find $\check{\rho}$. If $c_1 \le -\frac{2}{7}$ or $c_1 \ge 2$, then the eigenvalues of the operator $\tilde{T}_0\tilde{T}_1$ are equal in modulus: $|\lambda_1| = |\lambda_2|$. Therefore,

$$\sqrt{\rho(\tilde{T}_0\tilde{T}_1)} = [\det(\tilde{T}_0\tilde{T}_1)]^{1/4} = [\det \tilde{T}_0]^{1/2}$$

(since $\det \tilde{T}_0 = \det \tilde{T}_1$). For any $d_1, \dots, d_k$, we have

$$\left[\rho(\tilde{T}_{d_1} \cdots \tilde{T}_{d_k})\right]^{1/k} \ge \left[\det(\tilde{T}_{d_1} \cdots \tilde{T}_{d_k})\right]^{1/2k} = [\det \tilde{T}_0]^{1/2}.$$

Hence $\check{\rho} \ge [\det \tilde{T}_0]^{1/2}$. On the other hand, according to Lemma 5.1.2, we have

$$\check{\rho} \le \left[\rho(\tilde{T}_0\tilde{T}_1)\right]^{1/2} = [\det \tilde{T}_0]^{1/2}.$$

Thus $\check{\rho} = \left[\rho(\tilde{T}_0\tilde{T}_1)\right]^{1/2}$.

In the case $-\frac{2}{7} < c_1 < 2$, we use Proposition 7.3.4. For $c_1 \in [0, 1]$, it suffices to prove that

$$\check{\rho} \ge \lambda = \max\left\{\rho(\tilde{T}_0),\, \rho(\tilde{T}_1)\right\}.$$

For $c_1 \in \left[0, \frac{2}{3}\right]$, we take $\mathcal{Q} = \{(x_1, x_2) \in \mathbb{R}^2,\ x_1, x_2 \ge 1\}$. If $c_1 \in \left[\frac{2}{3}, 1\right]$, then for $\mathcal{Q}$ we can take the angle with vertex at the point $(1, 1)$ whose sides are parallel to the vectors $(2 - 3c_1,\, 2 - c_1)$ and $(2 - c_1,\, 2 - 3c_1)$.

For $c_1 \in \left[-\frac{2}{7}, 0\right] \cup (1, 2)$, it suffices to prove that $\check{\rho} \ge \lambda = \sqrt{\rho(\tilde{T}_0\tilde{T}_1)}$. If $c_1 \in (1, 2)$, then for $\mathcal{Q}$, we take a figure bounded by the segment $[(y, 1),\, (1, y)]$ (y is defined above) and two rays going from the points $(y, 1)$ and $(1, y)$ in the directions of the vectors $(c_1 - 2,\, 2c_1)$ and $(2c_1,\, c_1 - 2)$, respectively. Finally, for $c_1 \in \left[-\frac{2}{7}, 0\right]$, the set $\mathcal{Q}$ is bounded by the same segment $[(y, 1),\, (1, y)]$ and two rays going from its endpoints in the direction of the vectors

$$\left(2 - c_1 - \sqrt{4 + 12c_1 - 7c_1^2}\;,\; -4c_1\right) \quad \text{and} \quad \left(-4c_1\;,\; 2 - c_1 - \sqrt{4 + 12c_1 - 7c_1^2}\right),$$

respectively. This completes the proof of Proposition 7.3.3. ◊

The authors do not know whether there exist any explicit formulas for the computation of the p-radius ρ_p for finite p or the Lyapunov exponent $\bar{\rho}$ under the conditions of Proposition 7.3.3.

$\deg \mathbf{m_c} \ge \mathbf{3}$. For operators $\tilde{T}_i$ of dimension greater than 3, we do not know whether there exist any ways to compute the joint spectral radii; there are only algorithms for approximate computation. At present, there exist several algorithms of approximate computation for $\hat{\rho}, \check{\rho}, \rho_p, \bar{\rho}$. The computational complexity of these algorithms (the number of arithmetic operations to find a given value with a relative error ε) increases enormously as dimension n increases and ε decreases. In [**135**], it is shown that even for a pair of 2×2-matrices the computation of the joint spectral radius $\hat{\rho}$, in general, cannot be reduced to the computation of the usual spectral radius of any finite product of these matrices. Moreover, in the general case, there exists no algorithm to compute $\hat{\rho}$, which would be polynomial with respect to both the dimension n and $\frac{1}{\varepsilon}$ [**71**], [**136**], [**137**]. Nevertheless, there are algorithms (which are mentioned below) polynomial with respect to $\frac{1}{\varepsilon}$ for fixed dimension n and algorithms polynomial with respect to n for fixed ε.

We do not know whether the specific properties of the operators $\tilde{T}_0, \tilde{T}_1$ can be used to reduce the complexity considerably; we shall formulate all results for the general case for an arbitrary pair of operators $\tilde{B}_1, \tilde{B}_n$ acting in $\mathbb{R}^n$. According to Lemmas 6.3.8 and 7.2.3, without loss of generality it can be assumed that a pair $\tilde{B}_0, \tilde{B}_1$ is irreducible; by virtue of Theorem 6.1.1, the operators $\tilde{T}_0, \tilde{T}_1$ satisfy this condition.

1) *Computation of $\check{\rho}$, $\hat{\rho}$, $\bar{\rho}, \rho_p$ by the exhaustive search of matrix products.* For any $k \geq 1$, we have

$$\max_{d_1,\ldots,d_k} \Big(\rho(\Pi_k)\Big)^{1/k} \leq \hat{\rho} \leq \max_{d_1,\ldots,d_k} \Big\|\Pi_k\Big\|^{1/k}, \tag{7.8}$$

where $\Pi_k = \tilde{B}_{d_1} \cdots \tilde{B}_{d_k}$. The left-hand side of this inequality follows from Lemma 5.1.2; as for the right-hand side, the reader can easily prove it. Both values $\max_{d_1,\ldots,d_k} \Big(\rho(\Pi_k)\Big)^{1/k}$ and $\max_{d_1,\ldots,d_k} \|\Pi_k\|^{1/k}$ tend to $\hat{\rho}$ from opposite sides as $n \to \infty$ (the first one in the sense of upper limit, according to Lemma A.6.2, and the second one by definition). Thus for sufficiently large k, the joint spectral radius will be bounded from both sides with a given relative error ε. In general, the number k will have order at least ε^{-1}; hence the number of matrix products to search is of order $2^{1/\varepsilon}$. The number of operations increases exponentially with respect to $1/\varepsilon$. In [**138, 139, 140**], several modifications of this algorithm are suggested, which allow the reduction of the search. Nevertheless, the complexity remains exponential.

Similarly, the lower spectral radius can be computed approximately with the help of the limit relation

$$\check{\rho} = \lim_{k\to\infty} \min_{d_1,\ldots,d_k} \Big(\rho(\Pi_k)\Big)^{1/k}$$

(Lemma A.6.2). In terms of refinable functions, this means that we compute the maximal and the minimal local smoothness by the exhaustive search of sufficiently many rational points x (since the variable $-\log_2\Big(\rho(\Pi_k)\Big)^{1/k}$ is equal to the local smoothness at each rational point with period $(d_1 \ldots d_k)$, Proposition 7.2.6).

The quantities $\bar{\rho}$ and ρ_p are computed by their definition (formula (5.23) and Definition 5.2.1); the computational complexity is also exponential and exceeds the complexity of computation for $\hat{\rho}$. To determine local smoothness, we also have to compute $\hat{\rho}_x$ by its definition with the exception of rational points x, for which local smoothness can be computed in terms of the usual spectral radius (Proposition 7.2.6).

2) *Computation of ρ_p and estimation of $\hat{\rho}$ with the help of the Kronecker products.* It appears that the p-radius ρ_p can be computed explicitly for even integer values $p = 2r$, in which case the problem is reduced to computing the usual spectral radius for a specific matrix of order n^{2r}. For $r \to \infty$, we obtain an approximation for $\hat{\rho}$. This algorithm was first developed in [**115**] and later in [**117**] and [**118**]. For simplicity we confine our consideration to the case of matrices with real elements.

The Kronecker product of two matrices $A = (a_{ij})$ and $B = (b_{ij})$, where A is a $m \times n$ matrix and B is a $k \times l$ matrix, is the $mk \times nl$ matrix

$$A \otimes B = \begin{pmatrix} a_{11}B & \cdots & a_{1n}B \\ \vdots & * & \vdots \\ a_{m1}B & \cdots & a_{mn}B \end{pmatrix}.$$

Set $B^{\otimes k} = B \otimes \cdots \otimes B$ (k factors). Then

$$\rho_p(\tilde{B}_0, \tilde{B}_1) \;=\; \big(\rho(\mathcal{A}_p)\big)^{1/p}, \tag{7.9}$$

where $\rho(\mathcal{A}_p)$ is the spectral radius of the $n^p \times n^p$ matrix

$$\mathcal{A}_p = \frac{1}{2}\left(\tilde{B}_0^{\otimes 2r} + \tilde{B}_1^{\otimes 2r}\right), \quad p = 2r,$$

is an even integer (for the proofs, see [**115**] or [**117**]). In particular, for $r = 1$, the operator $\mathcal{A}_2$ can be represented as the operator in the space of $n \times n$ matrices acting by the formula

$$\mathcal{A}_2 X = \frac{1}{2}\left(\tilde{B}_0 X \tilde{B}_0^* + \tilde{B}_1 X \tilde{B}_1^*\right). \tag{7.10}$$

The spectral radius of this operator is equal to $(\rho_2)^2$ (Appendix A.6, Theorem A.6.6). One can reduce the dimension a little and consider the restriction of the operator A_2 to the space of symmetric $n \times n$ matrices (the dimension will decrease from n^2 to $\frac{n(n+1)}{2}$); the result remains the same [**115**].

Using Theorem 7.1.1 b), we deduce that *smoothness of refinable functions and wavelets in the space $L_2(\mathbb{R})$, i.e., $\alpha_{\varphi,2}$ is computed in terms of the spectral radius of the $\frac{n(n+1)}{2}$-dimensional operator* (7.10), *while $\alpha_{\varphi,2r}$, for any $r \geq 1$, is computed in terms of the spectral radius of operator* (7.9). The main inconvenience of this approach is the high dimension of the operator $\mathcal{A}_{2r}$ for large r.

This method allows the computation of the joint spectral radius $\hat{\rho}$ within polynomial time (in the dimension n, for fixed ε). Since for any p we have

$$\rho_p \ \leq \ \hat{\rho} \ \leq \ 2^{1/p}\rho_p,$$

it follows that ρ_p approximates the joint spectral radius $\hat{\rho}$ with error $\varepsilon \asymp \frac{1}{p}$. Using (7.9), we see that $\left(\rho(\mathcal{A}_p)\right)^{1/p}$ approximates $\hat{\rho}$ with error $\varepsilon \asymp \frac{1}{p}$. For instance, to approximate the joint spectral radius with accuracy of 70%, it suffices to take $p = 2$ (independently of n) and to compute the joint spectral radius of operator (7.10). For $p = 8$, we obtain the accuracy of 91%. This approach for computation of the joint spectral radii was developed in [**117**] and [**118**]. For a fixed error ε, it has a polynomial complexity with respect to n; the total number of operations is of order $n^{C/\varepsilon}$, where C is an absolute constant. This method is convenient for a rough estimation of $\hat{\rho}$ for high dimensions n. Besides, as shown in [**116**], the function $\rho_{1/x}$ is convex in $x \in (0, 1)$, which allows some reduction of the upper estimate of the complexity of the algorithm.

3) *Geometric method for computation of $\hat{\rho}, \check{\rho}$, and ρ_p.* The joint spectral radius is the exact infimum of the numbers λ, for which there exists a norm $\|\cdot\|$ in $\mathbb{R}^n$ such that $\|\tilde{B}_i\| \leq \lambda$, $i = 0, 1$ (see the proof of Proposition 5.1.3). For any irreducible pair of operators, this infimum is attained, and, in a sense, there exists the minimal norm that has the property $\|\tilde{B}_i\| \leq \lambda$, $i = 0, 1$. We get the following theorem.

THEOREM 7.3.5. *For an arbitrary irreducible pair of linear operators $\tilde{B}_0, \tilde{B}_1$, there exists a convex body M (a convex compact set with nonempty interior) symmetric with respect to zero and a positive number λ such that*

$$\mathrm{Conv}(\tilde{B}_0 M \cup \tilde{B}_1 M) = \lambda M.$$

Moreover, for any convex body with such a property, we have $\lambda = \hat{\rho}$.

`Proof.` For the proof, see [**119**]. ◊

In what follows, a body M will be called invariant. For a given pair of operators, an invariant body is not necessarily unique. The Minkowski norm in $\mathbb{R}^n$ corresponding to an invariant body satisfies the relation $\|\tilde{B}_i\| \leq \lambda$, $i = 0, 1$. A geometric algorithm for the computation of the joint spectral radius based on Theorem 7.3.5 was developed in [**119**]. This algorithm consists of successive approximations of an invariant body by polyhedrons. For a fixed dimension n, the algorithm is polynomial in $\frac{1}{\varepsilon}$. For example, for $n = 2$, its complexity does not exceed $C\varepsilon^{-3/2}$. For small dimensions n, it allows the computation of the joint spectral radius to a high accuracy. For large dimensions, this algorithm is not practically applicable, since its complexity increases exponentially as the dimension increases.

In specific cases it is possible to choose an invariant body M in such a way that the exact value of $\hat{\rho}$ can be found. Sometimes, one can choose the sets $\mathcal{M}$, $\mathcal{Q}$ from Proposition 7.3.4 to obtain estimates for $\hat{\rho}$ and $\check{\rho}$ and even to find their exact values, as was done in Proposition 7.3.3

Theorem 7.3.5 can be extended to p-radii for all $p \in [1, +\infty)$. To this end, instead of the operation Conv of taking a convex hull, we use the Firey addition $\overset{p}{\oplus}$ (which is also called L_p-addition of convex bodies). This is a binary operation on the set of centrally symmetric convex sets. By definition, the equality $M_3 = M_1 \overset{p}{\oplus} M_2$ means that

$$\varphi_{M_3} = \|(\varphi_{M_1}, \varphi_{M_2})\|_p,$$

where $\varphi_M(x) = \sup_{y \in M}(x, y)$ is the function of support of M and $\|\cdot\|_p$ is the L_p-norm in $\mathbb{R}^2$. For example, for $p = \infty$, the Firey addition coincides with taking a convex hull and, for $p = 1$, it is the usual Minkowski addition of convex sets.

THEOREM 7.3.6. *For an arbitrary irreducible pair of linear operators $\tilde{B}_0, \tilde{B}_1$, for any $p \in [1, +\infty)$, there exists a convex body $M \subset \mathbb{R}^n$ symmetric with respect to the origin and a positive number λ such that*

$$\tilde{B}_0 M \overset{p}{\oplus} \tilde{B}_1 M = \lambda M.$$

Moreover, for any convex body possessing this property, we have $\lambda = 2^{1/p}\rho_p$.

Proof. For the proof, see [**115**]. ◊

We shall say that this body M is invariant and shall denote it by M_p. A geometric algorithm for the computation of ρ_p consists of successive approximations of an invariant body M_p by p-zonotopes (the Firey sums of finitely many segments). This algorithm is polynomial in $\frac{1}{\varepsilon}$; its complexity does not exceed $C(n)\varepsilon^{2-2n}$ (see [**115**]).

For even integer values of p, there is an interesting connection between the invariant body M_p and the operator $\mathcal{A}_p$. This operator always has an invariant cone $K \subset \mathbb{R}^{n^p}$ [**115**], [**118**]. Therefore by the Perron-Frobenius theorem (see, for example, [**26**]) it has a real eigenvector $U_p \in K$ with the eigenvalue $\rho(\mathcal{A}_p) = (\rho_p)^p$. The vector U_p is naturally represented as a set of n^{p-1} vectors from $\mathbb{R}^n$. Denote by $M_U \subset \mathbb{R}^n$ the corresponding zonotope, i.e., the $\overset{p}{\oplus}$-sum of all these vectors. Then $M_U = M_p$ is an invariant body of the operators $\tilde{B}_0, \tilde{B}_1$. Thus invariant bodies correspond to the eigenvectors of the operator $\mathcal{A}_p$ with the eigenvalue $\rho(\mathcal{A}_p)$. In particular, if such a vector is unique, the invariant body is also unique. For $p = 2$, the operator $\mathcal{A}_2$ is given by formula (7.10); the set of positive semidefinite symmetric matrices is its invariant cone. Hence the corresponding eigenvector U_2 is also a positive semidefinite (in fact, even positive definite) symmetric matrix. Therefore U_2 defines an ellipsoid in $\mathbb{R}^n$. This ellipsoid is an invariant body M_2 with respect to L_2-addition $\overset{2}{\oplus}$. For the proofs, see [**115**].

EXAMPLE 7.3.7 (The Daubechies wavelet function ψ_2). The Daubechies wavelet function $\psi_2(x)$ (in what follows, we shall omit index 2) is defined by a scaling function φ satisfying the following refinement equation (we omit the index 2):
(7.11)

$$\varphi(x) = \frac{1+\sqrt{3}}{4}\varphi(2x) + \frac{3+\sqrt{3}}{4}\varphi(2x-1) + \frac{3-\sqrt{3}}{4}\varphi(2x-2) + \frac{1-\sqrt{3}}{4}\varphi(2x-3).$$

For this mask, we have $L = 1$, since $\mathbf{m}(-1) = \mathbf{m}'(-1) = 0$. The clean mask is given by the formula

$$\mathbf{m}_c(z) = \frac{1-\sqrt{3}}{2}z + \frac{1+\sqrt{3}}{2}.$$

Thus,

$$\alpha_\psi = \alpha_\varphi = 1 - \log_2\Big(\frac{1+\sqrt{3}}{2}\Big) = 0.5500\ldots.$$

Therefore the function ψ is continuous but does not belong to C^1. Since the smoothness is not integer, it follows that $\omega(\psi, t) \asymp t^{0.5500\ldots}$. The derivative ψ' is summable but does not belong to $L_2(\mathbb{R})$. Indeed,

$$\alpha_{\psi,1} = 1 - \log_2\frac{\sqrt{3}}{2} = 1.2075\ldots,$$

thus the smoothness of ψ' in the space L_1 is equal to $1 - \log_2 \sqrt{3} = 0.2075\ldots$. On the other hand, $\alpha_{\psi,2} = 1$, so that $\psi' \notin L_2(\mathbb{R})$.

Let us compute the local smoothness of the function $v(x) = \big(\psi(x), \ldots, \psi(x+3)\big)^T$. According to (7.6), for almost all $x \in [0,1]$, we have

$$\alpha_v(x) \;=\; \min\Big\{1 - \log_2 \frac{\sqrt{2}}{2}\,,\; 1\Big\} = \min\{\frac{3}{2}, 1\} = 1$$

(which also follows from the fact that ψ belongs to the space W_1^1). The values of the variable $\alpha_v(x)$ for $x \in [0,1]$ fill the segment

$$\Big[1 - \log_2 \frac{1+\sqrt{3}}{2}\;,\; 1\Big] = \big[0.550\ldots\,,\, 1\big].$$

At binary rational points x we have

$$\alpha_v(x-0) = \min\{1 - \log_2 \tfrac{\sqrt{3}-1}{2},\, 1\} = \min\{2.4499\ldots, 1\} = 1,$$
$$\alpha_v(x+0) = \min\{1 - \log_2 \tfrac{1+\sqrt{3}}{2},\, 1\} = \min\{0.5500\ldots,\, 1\} = 0.5500\ldots.$$

Thus, at all binary rational points the right-hand derivative is infinite. It is not hard to show that the left-hand derivative exists and is finite.

EXAMPLE 7.3.8 (The third Daubechies wavelet function ψ_3). The third Daubechies wavelet function is defined by the scaling function satisfying the quintic refinable equation with the coefficients

$$\begin{aligned}
c_0 &= [1 + \sqrt{10} + \sqrt{5 + 2\sqrt{10}}]/16,\\
c_1 &= [5 + \sqrt{10} + 3\sqrt{5 + 2\sqrt{10}}]/16,\\
c_2 &= [10 - 2\sqrt{10} + 2\sqrt{5 + 2\sqrt{10}}]/16,\\
c_3 &= [10 - 2\sqrt{10} - 2\sqrt{5 + 2\sqrt{10}}]/16,\\
c_4 &= [5 + \sqrt{10} - 3\sqrt{5 + 2\sqrt{10}}]/16,\\
c_5 &= [1 + \sqrt{10} - \sqrt{5 + 2\sqrt{10}}]/16.
\end{aligned} \tag{7.12}$$

Here, $L = 2$. The clean mask is (the coefficients are approximate):

$$\mathbf{m}_c(z) = \frac{1}{2}\Big((0.398538\ldots)z^2 - (2.162272\ldots)z + (3.763736\ldots)\Big).$$

The coefficients satisfy (7.7); hence by Lemma 7.3.1 we have

$$\hat{\rho}(\tilde{T}_0, \tilde{T}_1) = \max\{\rho(\tilde{T}_0), \rho(\tilde{T}_1)\} = \max\{0.398538, 2.162272, 3.763736\} = 3.763736.$$

Then

$$\alpha_\psi = L + 1 - \log_2 \hat{\rho} = 3 - \log_2(3.763736) = 1.087833\ldots.$$

Thus $\psi \in C^1(\mathbb{R})$; its derivative is continuous with the Hölder exponent $\alpha_{\psi'} = 0.087833\ldots$. By virtue of Theorem 7.2.11, we have

$$\omega(\psi', t) \;\asymp\; t^{\alpha_\psi - 1} \;=\; t^{0.087833}.$$

The function ψ has the minimal smoothness at dyadic points. Indeed,

$$\alpha_{v'}(x+0) \;=\; 2 - \log_2 \rho(\tilde{T}_0) \;=\; 2 - \log_2(3.763736) \;=\; 0.087833\,.$$

Note that

$$\alpha_{v'}(x-0) \;=\; 2 - \log_2 \rho(\tilde{T}_1) \;=\; 2 - \log_2(2.162272) \;=\; 0.887451\,.$$

It is clear that at dyadic points the smoothness on the right is different from the smoothness on the left. The smoothness on the right is equal to the minimal local smoothness of the function ψ'. It appears that the smoothness on the left is equal to the maximal smoothness. In other words, $\check{\rho}(\tilde{T}_0, \tilde{T}_1) = \rho(\tilde{T}_1) = 2.162272\ldots$. To prove this, we use Proposition 7.3.4 b). For $\mathcal{Q}$, we can take the sum of the two angles: the first one has the vertex at the point $(1,0)$ and the sides parallel to the vectors $(5,1)$ and $(5,-1)$; the second one is symmetric to it with respect to the origin. Thus the local smoothness $v'(x)$ does not exceed $0.887451\ldots$ at each point. In particular, $v'(x)$ is nowhere differentiable.

EXAMPLE 7.3.9 (The Daubechies wavelet function ψ_4). The corresponding refinement equation of order $N = 7$ has the following coefficients:

$$\begin{aligned} c_0 &= 0.325802, & c_4 &= -0.264507,\\ c_1 &= 1.010945, & c_5 &= 0.043616,\\ c_2 &= 0.892200, & c_6 &= -0.046503,\\ c_3 &= -0.039575, & c_7 &= -0.014976. \end{aligned} \tag{7.13}$$

For this equation, $L = 4$; after cleaning, we obtain a cubic equation with the coefficients

$$\tilde{c}_0 = 5.2128\,, \quad \tilde{c}_1 = -4.6762\,, \quad \tilde{c}_2 = 1.7032\,, \quad \tilde{c}_3 = -0.2397.$$

The matrices of the operators $\tilde{T}_0, \tilde{T}_1$ are

$$\tilde{T}_0 = \begin{pmatrix} 5.2128 & 0 & 0 \\ 1.7032 & -4.6762 & 5.2128 \\ 0 & -0.2397 & 1.7032 \end{pmatrix}, \quad \tilde{T}_1 = \begin{pmatrix} -4.6762 & 5.2128 & 0 \\ -0.2397 & 1.7032 & -4.6762 \\ 0 & 0 & -0.2397 \end{pmatrix}.$$

For these matrices, $\hat{\rho} = \rho(\tilde{T}_0) = 5.2128$. Indeed, $\hat{\rho} \geq \rho(\tilde{T}_0)$. To prove the opposite estimate, we make a linear change of variables with the transition matrix

$$B = \begin{pmatrix} 1 & 0 & 0 \\ -0.1447 & 0.3154 & -8.1258 \\ -0.4669 & 2.6559 & -2.2413 \end{pmatrix}.$$

After this change the matrix $B\tilde{T}_0B^{-1}$ becomes diagonal. The Euclidean norm of the matrix $B\tilde{T}_1B^{-1}$ computed as $\|A\| = \sqrt{\rho(AA^T)}$ is equal to $4.7369\ldots$, which is less than $\rho(\tilde{T}_0)$. Thus in Euclidean norm we have

$$\|B\tilde{T}_0B^{-1}\| \;=\; \rho(\tilde{T}_0), \quad \|B\tilde{T}_1B^{-1}\| \;<\; \rho(\tilde{T}_0);$$

therefore $\hat{\rho} = \rho(\tilde{T}_0)$. Thus $\alpha_\psi = 4 - \log_2 \rho(\tilde{T}_0) = 1.6179\ldots$. Hence $\psi_4 \in C^1(\mathbb{R})$, but $\psi_4 \notin C^2(\mathbb{R})$. Since the regularity exponent is not integer, we have

$$\omega(\varphi', t) \asymp t^{0.6179\ldots}.$$

The minimal local smoothness is attained at dyadic points on the right. At the same points, the smoothness on the left is equal to

$$4 - \log_2 \rho(\tilde{T}_1) = 1.8513\ldots.$$

For example, for the lower spectral radius, we have the estimate

$$\check{\rho} \leq \min \|\tilde{T}_{d_1} \cdots \tilde{T}_{d_6}\|^{1/6} = \|\tilde{T}_0^3\tilde{T}_1\tilde{T}_0\tilde{T}_1\|^{1/6} = 3.907\ldots.$$

As $4 - \log_2 3.907 = 2.0339\ldots$, the maximal local smoothness of the function ψ_4' is equal to 1 (Theorem 7.2.14). In particular, ψ_4 is twice differentiable on sets of points dense almost everywhere, for example, at all rational points x with period (000101).

Thus we have found the smoothness of the wavelet functions ψ_r for $r = 2, 3, 4$. Modern methods of the approximate computation of the joint spectral radius allow one to solve this problem for greater sizes of matrices. Below, we give the values of α_{ψ_r} for $r \le 8$:

(7.14)

r	α
2	0.5500
3	1.0878
4	1.6179
5	1.9690
6	2.1891
7	2.4604
8	2.7608

Let us give some other examples of refinable functions.

EXAMPLE 7.3.10. The equation

$$\varphi(x) = \frac{1}{3}\varphi(2x) + \frac{1}{3}\varphi(2x-1) + \frac{2}{3}\varphi(2x-2) + \frac{2}{3}\varphi(2x-3) \tag{7.15}$$

has mask $\mathbf{m}_0(z) = \frac{1}{6}(z+1)(2z^2+1)$. This mask has a pair of symmetric roots $\pm\frac{\sqrt{2}}{2}$. Hence integer shifts of the function φ are linearly dependent. However, the function is stable because the mask has neither symmetric roots on the unit circle nor nontrivial cycles. We have $L = 0$. After cleaning, we obtain the mask $\mathbf{m}_c(z) = \frac{2}{3}z + \frac{1}{3}$. Therefore,

$$\alpha_\varphi = -\log_2(2/3) = 0.5849\ldots.$$

Consequently, φ is continuous but does not belong to $C^1(\mathbb{R})$. Thus $l = L = 0$. The smoothness is not integer; hence $\omega(\varphi, t) \asymp t^{\alpha_\varphi}$. Using (7.5), we conclude that the derivative φ' does not belong to $L_p(\mathbb{R})$ for any $p \ge 1$. For example,

$$\alpha_{\varphi,1} = 1, \quad \alpha_{\varphi,2} \;=\; 1 + \log_2 0.9 \;=\; 0.8479\ldots.$$

According to (7.6), for the mean smoothness, we have

$$\alpha_{av} \;=\; \log_2\sqrt{4.5} \;=\; 1.0849\ldots.$$

Thus $\varphi'(x) = 0$ for almost all x; hence φ is purely singular. It is not hard to show that φ has bounded variation. Further, the values of $\alpha_v(x)$ fill the whole segment

$$[\log_2 3 - 1,\ \log_2 3] \;=\; [0.5849\ldots, 1.5849\ldots].$$

At dyadic points we have

$$\alpha_v(x-0) = \log_2 3 - 1 = 0.5849\ldots, \qquad \alpha_v(x+0) = \log_2 3 = 1.5849\ldots.$$

Thus, at all dyadic points the left-hand derivative is infinite, while the right-hand derivative is equal to zero.

EXAMPLE 7.3.11 (A curve of constant local smoothness). Consider the equation

$$\varphi(x) = \frac{1 + i\tan\beta}{2}\varphi(2x) + \varphi(2x-1) + \frac{1 - i\tan\beta}{2}\varphi(2x-2), \tag{7.16}$$

where β is an arbitrary parameter from the interval $\left(0, \frac{\pi}{3}\right)$. For this equation $L = 0$. After cleaning the mask, we have

$$\mathbf{m}_c(z) = \frac{1 - i\tan\beta}{2}z \;+\; \frac{1 + i\tan\beta}{2}.$$

Note that $|\frac{1-i\tan\beta}{2}| = |\frac{1+i\tan\beta}{2}| = \frac{1}{2\cos\beta}$. The local smoothness of the function $v(x)$ at all points x is the same, being equal to $\log_2(2\cos\beta)$. In particular, v is continuous and $\alpha_v = \log_2(2\cos\beta)$; however it is not differentiable at any point. Here we see that the segment $[1-\log_2\hat{\rho},\ 1-\log_2\check{\rho}]$, which is the range of values of the local smoothness, may turn into a single point.

Using (7.5), we see that the smoothness of φ in the spaces $L_p(\mathbb{R})$ for all p is also equal to $\log_2(2\cos\beta)$.

EXAMPLE 7.3.12 (An almost Lipschitz function). Consider the equation from the previous example and increase its order by 1. We obtain the equation with $L = 1$ and the same clean mask $\mathbf{m}_c$. Put $\beta = \frac{\pi}{3}$ in this equation.

The mask of this equation takes the form

$$\mathbf{m}_0(z) \;=\; \left(\frac{z+1}{2}\right)^2 \left(e^{-\frac{\pi i}{3}}z + e^{\frac{\pi i}{3}}\right), \tag{7.17}$$

and the clean mask takes the form

$$\mathbf{m}_c(z) \;=\; e^{-\frac{\pi i}{3}}z + e^{\frac{\pi i}{3}}.$$

We have $\alpha_\varphi = 1$ and $\alpha_{\varphi,p} = 1$ for all p. The local smoothness $\alpha_v(x)$ is equal to 1 at each point x of the segment $[0,1]$. Nevertheless, the function v_φ is not Lipschitz. At some points of the segment $[0,1]$, it is not even locally Lipschitz. For example, at the point $x = \frac{2}{3} = 0.101010\ldots$ we have

$$\sup_{s\in[0,t]} \|v(x+s) - v(x)\| \;\geq\; C\,t\,|\ln t|\,, \qquad t \in (0, 1/2)\,. \tag{7.18}$$

In particular, this shows that in the case of integer Hölder exponent, the estimate $\omega(\varphi, t) \leq Ct|\ln t|$ cannot be improved even locally. Let $x_k = 0.10101\ldots01$ (k pairs (1, 0) and 1). Let us prove that

$$\|v(x_k) - v(x_{k-1})\| \;\geq\; Ck2^{-2k},$$

where the constant $C > 0$ does not depend on k. Relation (7.18) follows.

Using (5.10), we get

$$v(x_k) - v(x_{k-1}) = \Big(T_1T_0\Big)^k (v(1) - v(0)). \tag{7.19}$$

It is clear that

$$v(1) - v(0) \;\in\; W = \Big\{y \in \mathbb{R}^3, \sum_{j=1}^{3} y_j = 0\Big\}.$$

It is not hard to check that the operator $T_1T_0|_W$ has an eigenvalue $\lambda = 1/2$ of multiplicity 2. This operator has only one (up to normalization) eigenvector. So the Jordan form of the corresponding matrix has one Jordan block of dimension 2. Therefore,

$$\left\|\Big(T_1T_0\Big)^k \Big(v(1) - v(0)\Big)\right\| \geq Ck2^{-k}.$$

EXAMPLE 7.3.13. For the equation

$$\varphi(x) = -\frac{1}{4}\varphi(2x) - \frac{1}{4}\varphi(2x-1) + \frac{3}{4}\varphi(2x-2) + \frac{5}{4}\varphi(2x-3) + \frac{1}{2}\varphi(2x-4), \tag{7.20}$$

we have $L = 2$, and the clean mask is $\mathbf{m}_c(z) = 2z - 1$. Thus,

$$\alpha_\varphi \;=\; 2 - \log_2 2 \;=\; 1.$$

Consequently, $l = 0$. Further, $\alpha_{\varphi,1} = 3 - \log_2 3 = 1.4150\ldots$. Thus $\varphi \in W_1^1(\mathbb{R})$. Moreover, $\varphi \in W_p^1(\mathbb{R})$ for all finite p. Indeed,

$$\alpha_{\varphi,p} \;=\; 2 + \frac{1}{p} - \frac{1}{p}\log_2(2^p+1) \;=\; 1 + \frac{1}{p}\log_2\frac{2\cdot 2^p}{2^p+1} \;>\; 1.$$

It is not hard to show that $\varphi \in W^1_\infty$. Therefore the function φ is Lipschitz with constant $\|\varphi'\|_\infty$.

Thus $\omega(\varphi, t) \asymp t$, in spite of the fact that the Hölder exponent is integer.

EXAMPLE 7.3.14 (Dubuc's interpolation scheme). In Example 3.5.5, we considered an interpolation scheme introduced by S. Dubuc and the corresponding refinable function φ satisfying equation (3.32) with mask (3.33). For this mask, $L = 3$ and $\mathbf{m}_c = -0.5 + 2z - 0.5z^2$. Thus the clean equation has coefficients $c_0 = -1$, $c_1 = 4$, $c_2 = -1$. The mask is symmetric, so we use Proposition 7.3.3 and obtain $\hat{\rho} = 4$, $\check{\rho} = 2$. Thus,

$$\alpha_\varphi \;=\; L + 1 - \log_2\hat{\rho} \;=\; 2;$$

hence $\varphi \in C^1(\mathbb{R})$, but $\varphi \notin C^2(\mathbb{R})$. Moreover, the derivative $\varphi'(x)$ is almost Lipschitz. At each dyadic point $x \in [0,1]$ we have

$$C_1|\ln t| \;\le\; \sup_{|s|\le t}\|v'(x+s) - v'(s)\| \;\le\; C_2\, t\,|\ln t|\,, \qquad t\in(0,1/2)\,. \tag{7.21}$$

The latter relation is proved in the same way as in Example 7.3.12. Both matrices T_0, T_1 have a Jordan block of dimension 2 corresponding to the eigenvalue $\lambda = \frac{1}{4}$. Hence at dyadic points $x \in [0,1)$ we have

$$|\varphi(x + 2^{-k}) - \varphi(x)| \ge C(x)k2^{-k}.$$

Analogously, for $x \in (0,1]$, we get

$$|\varphi(x - 2^{-k}) - \varphi(x)| \ge C(x)k2^{-k}.$$

Relation (7.21) follows. Moreover, at each binary rational point $x \in (0,1)$ the function $v'(x)$ is almost Lipschitz on both sides. However, note that as $|c_0| = |c_1| = 1$, the function $\varphi'(x)$ is Lipschitz at the points 0 and 6; moreover, $\alpha_{\varphi'}(0) = \alpha_{\varphi'}(6) = 3$. For example, $|\varphi'(t)| \le Ct^2$ as $t \to +0$. At the same time, at the other points of the segment $[0,6]$ the function $\varphi'(x)$ is almost Lipschitz. This fact shows that components of a vector function $v(x)$ do not necessarily have the same local smoothness. At some points x, some components may have smoothness greater than $\alpha_{v'}(x)$.

Though $\varphi \notin C^2(\mathbb{R})$, φ is twice differentiable almost everywhere. Moreover, $\varphi'' \in L_p(\mathbb{R})$ for all finite p. To prove this, it suffices to check that $\rho_p(\tilde{T}_0, \tilde{T}_1) < 4$. In L_1-norm ($\|v\|_1 = |v_1| + |v_2|$) we have

$$\|[\tilde{T}_0]^2\|_1 = \|[\tilde{T}_1]^2\|_1 = 16, \quad \|\tilde{T}_0\tilde{T}_1\|_1 = \|\tilde{T}_1\tilde{T}_0\|_1 = 8.$$

Then

$$2^{-2k}\sum_{d_1,\ldots,d_{2k}}\|\tilde{T}_{d_1}\cdots\tilde{T}_{d_{2k}}\|_1^p \le 2^{-2k}\sum_{d_1,\ldots,d_{2k}}\|\tilde{T}_{d_1}\tilde{T}_{d_2}\|_1^p\cdots\|\tilde{T}_{d_{2k-1}}\tilde{T}_{d_{2k}}\|_1^p$$
$$= 2^{-2k}(16^p + 16^p + 8^p + 8^p)^k,$$

which implies that

$$\rho_p(\tilde{T}_0, \tilde{T}_1) \le [(16^p + 8^p)/2]^{1/2p} = 2\left(2^{2p-1} + 2^{p-1}\right)^{\frac{1}{2p}}.$$

Thus,

$$\begin{aligned}\alpha_{\varphi,p} &\ge 4 - \log_2 2\left(2^{2p-1} + 2^{p-1}\right)^{\frac{1}{2p}} \\ &= 3 - \frac{1}{2p}\log_2\left(2^{2p-1} + 2^{p-1}\right) > 2.\end{aligned}$$

Therefore $\varphi \in W_p^2(\mathbb{R})$ for all finite p. At the same time, $\varphi \notin W_\infty^2(\mathbb{R})$, since the function $\varphi'(x)$ is not Lipschitz.

EXAMPLE 7.3.15 (A symmetric cubic equation). For an equation

$$\varphi(x) = \sum_{k=0}^{3} c_k \varphi(2x - k)$$

whose coefficients are real and symmetric ($c_0 = c_3,\ c_1 = c_2$) we can find out if its solution is continuous or not. If the mask has a cycle or a pair of symmetric roots, then the clean mask will have order 1 or 0, and our investigation can be reduced to the cases considered above. So we shall assume that the mask has neither cycles nor symmetric roots. Then for the existence of a regular solution, the necessary condition $c_0 + c_2 = c_1 + c_3 = 1$ must be satisfied. The equation can be written in the following form:

$$\varphi(x) = \omega\varphi(2x) + (1-\omega)\varphi(2x-1) + (1-\omega)\varphi(2x-2) + \omega\varphi(2x-3), \tag{7.22}$$

where $\omega \in \mathbb{R}$ is a parameter. For $\omega \in (0,1)$, this equation defines de Rham's curves (Examples 3.5.4 and 5.5.3).

If we reduce the order by 1, we obtain the clean mask

$$\mathbf{m}_c(z) = \omega z^2 + (1-2\omega)z + \omega.$$

This mask is symmetric, so using Proposition 7.3.3, we can find the smoothness of the solution for any ω. The refinable function φ is continuous if and only if $\omega \in (0,1)$, i.e., all the coefficients of equation (7.22) are positive. Here,

$$\alpha_\varphi = \begin{cases} -\log_2(1-2\omega) & \text{if } 0 < \omega \le \frac{1}{4}, \\ 1 - \log_2\left(\omega + \sqrt{4\omega - 7\omega^2}\right) & \text{if } \frac{1}{4} \le \omega \le \frac{1}{2}, \\ -\log_2 \omega & \text{if } \frac{1}{2} \le \omega < 1. \end{cases} \tag{7.23}$$

For $\omega = \frac{1}{4}$, the function $\varphi(x) = B_2(x)$ is a cardinal B-spline, which belongs to $C^1(\mathbb{R})$. For $\omega = \frac{1}{4}$,

$$\varphi = \frac{1}{2}\chi_{[0,1]} * \chi_{[0,2]}$$

is a piecewise linear function; thus $\varphi \in W_\infty^1(\mathbb{R})$. For all other ω, we have $\alpha_\varphi < 1$; hence the function φ is not Lipschitz. The maximal local smoothness $\alpha_\varphi^{\max}$ is computed analogously. For $\omega > \frac{1}{2}$, we have

$$\alpha_\varphi^{\max} = 1 - \log_2\left(\omega + \sqrt{4\omega - 7\omega^2}\right) < 1;$$

in particular, the function v_φ is nowhere differentiable. On the contrary, for $0 < \omega < \frac{1}{2}$, the function φ is differentiable almost everywhere, and for $\omega \ne \frac{1}{4}$, it is purely singular ($\varphi'(x) \equiv 0$ almost everywhere).

For negative ω, the function φ is no longer continuous but is still a regular function. For example, using formula (7.10) for $p = 2$, we derive that $\rho_2(\tilde{T}_0, \tilde{T}_1)$ is equal to the square root of the (usual) spectral radius of the operator $\mathcal{A}_2$ restricted to a three-dimensional cone of positive definite matrices. It follows that ρ_2 is equal to the spectral radius of the following operator in $\mathbb{R}^2$:

$$A = \begin{pmatrix} 2 - 8\omega + 12\omega^2 & 4\omega^2 \\ 4\omega - 8\omega^2 & 4\omega - 8\omega^2 \end{pmatrix}.$$

It follows that $\varphi \in L_2(\mathbb{R})$ for $\omega > -0.1963\ldots$. To determine for which ω the function $\varphi \in L_1$, we must estimate ρ_1. As in the previous example, using the formula for L_1-norm in $\mathbb{R}^2$ and taking $k = 1$, we deduce that $\varphi \in L_1(\mathbb{R})$ for $\omega \geq (3 - \sqrt{17})/8 = -0.140\ldots$, which is even worse than for $\varphi \in L_2(\mathbb{R})$. However, for $k = 5$, $\varphi \in L_1(\mathbb{R})$ for $\omega \geq -0.22$. On the other hand, for $\omega \leq -\frac{1}{2}$, we have $\varphi \notin L_1(\mathbb{R})$. This is proved in the same way as in the case $\omega > 1$; namely,

$$\check{\rho} = |\omega + \sqrt{4\omega - 7\omega^2}| \geq 2.$$

Thus somewhere on the segment $[-0.5, -0.22]$ there is a point ω_0 such that for all ω on the right of ω_0, functions φ are regular, while for ω on the left of it, φ are singular. It would be interesting to find this boundary value ω_0. For example, the authors do not know whether the function φ is summable for $\omega = -0.25$.

7.4. Method of pointwise estimation of the Fourier transform

As discussed above, the matrix method allows (at least, theoretically) the exact computation of global smoothness for wavelets and gives very sharp information about local smoothness. Its main disadvantage is that it is complicated for large dimensions N. In the latter case, two other methods based on the estimation of the rate of decay of the Fourier transform are more convenient. None of them can find exactly the Hölder exponent, and nothing can be done about that. These methods cannot be applied to estimate local smoothness. Nevertheless, at present they are the only effective methods to estimate smoothness for large N and to obtain asymptotic estimates as $N \to \infty$. The first method is based on the pointwise estimation of the Fourier transform.

For a given function $g(x)$, set

$$\vartheta = \vartheta(g) = \sup\left\{\beta \in \mathbb{R} \,\middle|\, |g(\xi)| \leq C(|\xi| + 1)^{-\beta},\ \xi \in \mathbb{R}\right\}.$$

The following result, which connects the smoothness of a function with the pointwise estimation of its Fourier transform, is well known: for a continuous function $f \in L_1(\mathbb{R})$, the inequality

$$\vartheta - 1 \leq \alpha_f \leq \vartheta \tag{7.24}$$

holds, where $\vartheta = \vartheta(\widehat{f})$. For the proof, see Appendix A.7, Theorem A.7.1.

REMARK 7.4.1. Even for refinable functions both estimates in (7.24) cannot be improved. For example, if $L \geq 1$, for the cardinal spline $\varphi(x) = B_L(x)$ we have $\alpha_\varphi = L$, while $\vartheta(\widehat{\varphi}) = L + 1$, since

$$|\widehat{\varphi}(\xi)| = \left|\frac{\sin \pi\xi}{\pi\xi}\right|^{L+1}.$$

On the other hand, for a given $L \geq 0$, consider a family of masks

$$\mathbf{m}_\gamma(z) = \Big(\frac{z+1}{2}\Big)^{L+1} \mathbf{m}_{c,\gamma}(z)\,,$$

where $\mathbf{m}_{c,\gamma}(z) = \gamma z + 1 - \gamma$ and the parameter γ belongs to the interval $(0,1)$. Applying inequality (3.14) to the corresponding refinable function φ_γ, we get

$$\big|\widehat{\varphi}_\gamma(n2^k + \delta)\big| > C\big|\widehat{\varphi}_\gamma(n + 2^{-k}\delta)\big|,$$

for all k, where n is an arbitrary integer and δ is sufficiently small. Further, we have

$$\big|\widehat{\varphi}_\gamma(n + 2^{-k}\delta)\big| = \left|\frac{\sin\,\pi(n+2^{-k}\delta)}{\pi(n+2^{-k}\delta)}\right|^{L+1} \prod_{j=1}^{\infty} m_{c,\gamma}\big(2^{-j}(n+2^{-k}\delta)\big).$$

If $\gamma \neq \frac{1}{2}$, then the product on the right-hand side does not vanish for any sufficiently small δ; hence,

$$|\widehat{\varphi}_\gamma(n + 2^{-k}\delta)| \geq C(n,\delta)2^{-k(L+1)}.$$

Thus,

$$|\widehat{\varphi}_\gamma(n2^k + \delta)| > C_{n,\delta}C2^{-k(L+1)};$$

hence $\vartheta \leq L+1$. In fact, $\vartheta \geq L+1$, since $|m_{c,\gamma}(\xi)| \leq 1$ for all ξ. It follows that

$$|\widehat{\varphi}_\gamma(\xi)| \;\leq\; |\widehat{B}_L(\xi)|.$$

Thus $\vartheta \leq L+1$ for any $\gamma \neq \frac{1}{2}$. The smoothness of the function φ_γ is given by the relation (formula (7.5))

$$\alpha_\varphi = L - \log_2\big(\max\{\gamma,\ 1-\gamma\}\big).$$

As $\gamma \to \frac{1}{2}$, we have $\alpha_\varphi \to L+1 = \vartheta$. Therefore the second inequality in (7.24) cannot be improved for refinable functions.

In what follows, for an arbitrary refinable function φ, we shall denote the variable $\vartheta(\widehat{\varphi})$ by $\vartheta(m_0)$, where m_0 is the mask of the corresponding equation. Our aim is to express ϑ for an arbitrary mask m_0 in terms of the value $\vartheta(m_c)$ of the corresponding clean mask. First of all note that for a clean mask, the Fourier transform $\widehat{\varphi}_c$ cannot tend to zero at infinity; in particular, $\vartheta(m_c) \leq 0$. Indeed, since the function m_c has neither cycles nor symmetric zeros, the function $\widehat{\varphi}_c(\xi)$ has no periodic zeros. Hence there exists an odd number q for which $|\widehat{\varphi}_c(q/2)| > 0$. Taking into account that $m_c(1/2) \neq 0$, we have

$$\begin{aligned}\Big|\widehat{\varphi}_c(2^{k-1}q)\Big| &= \Big|m_c(2^{k-2}q)\cdots m_c(q)m_c(q/2)\widehat{\varphi}_c(q/2)\Big| \\ &= \Big|m_c^{k-1}(0)m_c(1/2)\widehat{\varphi}_c(q/2)\Big| = \Big|m_c(1/2)\widehat{\varphi}_c(q/2)\Big| > 0\end{aligned}$$

for any $k \geq 1$.

LEMMA 7.4.2. *For any mask m_0, we have*

$$\vartheta(m_0) = L + 1 + \vartheta(m_c),$$

where m_c is the corresponding clean mask.

It follows from this lemma that $\vartheta(m_0) \le L+1$ for any mask m_0. Thus the rate of decay of the Fourier transform for a refinable function does not exceed the order of the Strang-Fix condition, which it satisfies. Though Lemma 7.4.2 may seem simple, its proof is quite technical.

`The proof of Lemma 7.4.2` will be presented only for the case of a stable refinable function, when the mask has neither cycles nor symmetric zeros on the unit circle. If there is a symmetric zero $\{\pm a\}$ outside the unit circle, then we return to the previous level and obtain a new function φ_α, where $e^{-2\pi i\alpha} = a$. According to Proposition 6.2.1,

$$\widehat{\varphi}(\xi) = (1 - e^{-4\pi i(\alpha+\xi)})\, \widehat{\varphi}_\alpha(\xi).$$

Since $\alpha \notin \mathbb{R}$, the factor $1 - e^{-4\pi i(\alpha+\xi)}$ is bounded in modulus above and below by positive constants. Therefore $\vartheta(m_\alpha) = \vartheta(m_0)$. Thus transition to the previous level does not change the rate of decay of the Fourier transform for a refinable function. After necessary transitions, we can assume that the mask has no symmetric roots. Thus,

$$m_0(\xi) = \Big(\frac{e^{-2\pi i\xi}+1}{2}\Big)^{L+1} m_c(\xi) \qquad \text{and} \qquad \widehat{\varphi}(\xi) = \Big(\frac{e^{-2\pi i\xi}-1}{-2\pi i\xi}\Big)^{L+1} \widehat{\varphi}_c(\xi).$$

The inequality $\vartheta(m_0) \ge L+1+\vartheta(m_c)$ is obvious. The proof of the opposite inequality is complicated by the fact that the factor $(e^{-2\pi i\xi}-1)^{L+1}$ has zeros of order $L+1$ at integer points; these zeros may "quench" maximums of the function $(1+|\xi|)^{-L-1}\widehat{\varphi}_c(\xi)$. In what follows, we shall write ϑ instead of $\vartheta(m_c)$. We consider two cases: $\vartheta < 0$ and $\vartheta = 0$. Let $\vartheta < 0$. For any $\beta \in (\vartheta, 0)$, there exist arbitrarily large ξ such that

$$\big|\widehat{\varphi}_c(\xi)\big| \ge C_\beta |\xi|^{-\beta}.$$

Take one such ξ and represent it in the form $\xi = 2^k q + \delta$, where q is odd and $\delta \in \left[-\frac{1}{2}, \frac{1}{2}\right]$ (the number $2^k q$ is the nearest integer for the number ξ). We have

$$\begin{aligned} C_\beta|\xi|^{-\beta} &\le |\widehat{\varphi}_c(\xi)| \\ &= \Big|m_c(2^{k-1}q + 2^{-1}\delta)\cdots m_c(q+2^{-k}\delta) m_c(2^{-1}q+2^{-k-1}\delta)\widehat{\varphi}_c(2^{-1}q+2^{-k-1}\delta)\Big| \\ &= \Big|\prod_{j=1}^{k} m_c(2^{-j}\delta)\Big| \cdot \Big|m_c\big(\frac{1}{2}+2^{-k-1}\delta\big)\Big| \cdot \Big|\widehat{\varphi}_c(2^{-1}q+2^{-k-1}\delta)\Big|. \end{aligned}$$

The products $\Big|\prod_{j=1}^{k} m_c(2^{-j}\delta)\Big| \cdot \Big|m_c(2^{-1}q+2^{-k-1}\delta)\Big|$ are bounded above by a constant C_1 uniformly with respect to all k; hence,

$$\big|\widehat{\varphi}_c(2^{-1}q+2^{-k-1}\delta)\big| \ge \frac{C_\beta}{C_1}|\xi|^{-\beta}.$$

Here, if $\xi \to \infty$, then $q \to \infty$; otherwise (since $\beta < 0$), the continuous function $\widehat{\varphi}_c$ would be unbounded on a segment. In addition, it is obvious that $|\xi| > |q/2|$. Thus there exist arbitrarily large numbers q for which

$$\Big|\widehat{\varphi}_c\Big(2^{-1}q+2^{-k-1}\delta\Big)\Big| \; > \; C|q/2|^{-\beta}.$$

But then

$$\Big|\frac{e^{-2\pi i(2^{-1}q+2^{-k-1}\delta)}-1}{-2\pi i(2^{-1}q+2^{-k-1}\delta)}\Big|^{L+1}\Big|\widehat{\varphi}_c(2^{-1}q+2^{-k-1}\delta)\Big| > C|q/2|^{-\beta-L-1}$$

for some constant C (as $|e^{-2\pi i(2^{-1}q+2^{-k-1}\delta)}-1|\geq\sqrt{2}$ for $\xi\in\left[-\frac{1}{2}\,,\,\frac{1}{2}\right]$). Therefore,

$$\left|\widehat{\varphi}(2^{-1}q+2^{-k-1}\delta)\right|>C|q/2|^{-\beta-L-1};$$

hence $\vartheta(m_0)\ \leq\ \vartheta(m_c)+L+1$.

In the case $\vartheta=0$, take an odd number q for which $\widehat{\varphi}_c(q/2)\neq 0$. Then for any $k\geq 1$, we get

$$\begin{aligned}\widehat{\varphi}_c\Bigl(2^kq+\frac{q}{2}\Bigr)&=\prod_{j=1}^{k+1}m_c(2^{k-j}q+2^{-j-1}q)\widehat{\varphi}_c\Bigl(\frac{q}{2}+2^{-k-2}q\Bigr)\\&=\prod_{j=1}^{k}m_c(2^{-j-1}q)m_c\Bigl(\frac{q}{2}+2^{-k-2}q\Bigr)\,\widehat{\varphi}_c\Bigl(\frac{q}{2}+2^{-k-2}q\Bigr)\\&=\frac{m_c\bigl(\frac{1}{2}+2^{-k-2}q\bigr)\widehat{\varphi}_c(q/2)}{\widehat{\varphi}_c(2^{-k-2}q)}\widehat{\varphi}_c\Bigl(\frac{q}{2}+2^{-k-2}q\Bigr).\end{aligned}$$

As $k\to\infty$, this fraction tends to $m_c\bigl(\frac{1}{2}\bigr)\widehat{\varphi}_c^2(q/2)\neq 0$. Hence for large k, the variable $\left|\widehat{\varphi}_c\bigl(2^kq+\frac{q}{2}\bigr)\right|$ is bounded below by some positive constant. Then

$$\left|\widehat{\varphi}\Bigl(2^kq+\frac{q}{2}\Bigr)\right|=\left|\frac{e^{-2\pi i\left(2^kq+\frac{q}{2}q\right)}-1}{-2\pi i\bigl(2^kq+\frac{q}{2}\bigr)}\right|^{L+1}\left|\widehat{\varphi}_c\bigl(2^kq+\frac{q}{2}\bigr)\right|>C|2^kq|^{-L-1},\tag{7.25}$$

which was to be proved. ◊

COROLLARY 7.4.3. *For any refinable function φ, we have*

$$\alpha_\varphi\in[\vartheta+L\,,\;\vartheta+L+1],$$

where $\vartheta=\vartheta(m_c)$.

Now the problem of computing ϑ for an arbitrary mask m_0 is reduced to that for a clean mask m_c. To do this, we introduce one further notation. For an arbitrary $k\geq 1$, we set

$$\vartheta_k\ =\ -\frac{1}{k}\log_2\Bigl\|m_c(\xi)\cdots m_c(2^{k-1}\xi)\Bigr\|_\infty.$$

PROPOSITION 7.4.4. *For an arbitrary clean mask m_c, we have*

$$\vartheta(m_c)=\lim_{k\to\infty}\vartheta_k=\sup_{k\in\mathbb{N}}\vartheta_k.$$

Moreover, there exist a constant $C>0$ and a sequence $\xi_j\to\infty$ such that

$$|\widehat{\varphi}_c(\xi_j)|\ \geq\ C|\xi_j|^{-\vartheta},\qquad j\in\mathbb{N}.$$

Proof. By Fekete's lemma (Appendix A.6, Lemma A.6.1), any sequence $\{a_k\}$ such that

$$(k+n)a_{k+n}\ \leq\ ka_k+na_n,\qquad k\,,n\in\mathbb{N},$$

converges, and

$$\lim_{k\to\infty}a_k\ =\ \inf_{k\in\mathbb{N}}a_k.$$

It is easy to see that the sequence $a_k\ =\ -\vartheta_k$ has the above property; hence it converges and $\lim\vartheta_k=\sup\vartheta_k$. Denote this limit by b and show that $b=\vartheta$. Prove that $\vartheta\geq\vartheta_k$ for any k; then it will follow that $\vartheta\geq b$. Take sufficiently large ξ (for definiteness, positive) and denote by l an integer such that $2^{k(l-1)} \le \xi \le 2^{kl}$. We have

$$\begin{aligned}|\widehat{\varphi}_c(\xi)| &= \Big|\prod_{s=1}^{l}\Big(\prod_{j=1}^{k} m_c(2^{j-ks-1}\xi)\Big)\Big| \cdot \Big|\widehat{\varphi}_c(2^{-kl}\xi)\Big| \\ &\le 2^{-kl\vartheta_k}\big\|\widehat{\varphi}_c\big\|_\infty \le \xi^{-\frac{l}{l-1}\vartheta_k}\big\|\widehat{\varphi}_c\big\|_\infty.\end{aligned}$$

For $l \to \infty$, we derive that $\vartheta \ge \vartheta_k$, which was required. It remains to prove that there exist a constant $C > 0$ and a sequence $\xi_j \to \infty$ such that $|\widehat{\varphi}_c(\xi_j)| \ge C|\xi_j|^{-b}$. This would imply that $\vartheta \le b$ and $|\widehat{\varphi}_c(\xi_j)| \ge C|\xi_j|^{-\vartheta}$. Note that in the case $b = \vartheta = 0$, the existence of such a sequence has already been proved (7.25). It remains to prove this assertion for $b < 0$. Assume the converse:

$$\xi^b\widehat{\varphi}(\xi) \to 0, \quad \xi \to \infty. \tag{7.26}$$

Let the absolute value of the function $f_k(\xi) = \prod_{j=1}^{k} m_c(2^{j-1}\xi)$ attains its maximal value at the point $\xi_k \in [0,\, 1]$. If necessary, going over to a subsequence, we can assume that ξ_k converges to some $\xi_0 \in [0,\, 1]$. As is known, $|f_k(\xi_k)| = 2^{-k\vartheta_k} \ge 2^{-kb}$. For any k, we have

$$|\widehat{\varphi}_c(2^k\xi_k)| = |f_k(\xi_k)\widehat{\varphi}_c(\xi_k)| \ge 2^{-kb}|\widehat{\varphi}(\xi_k)|. \tag{7.27}$$

Here, $2^k\xi_k \to \infty$; otherwise, it would follow from (7.26) that the function $\widehat{\varphi}_c$ is unbounded on some compact set. Combining (7.26) and (7.27) for $\xi = 2^k\xi_k$, we get $\widehat{\varphi}(\xi_k) \to 0,\ k \to \infty$; therefore $\widehat{\varphi}_c(\xi_0) = 0$. Taking an arbitrary integer n and using the same reasoning for the sequence $\{\xi_k + n\}_{k\in\mathbb{Z}}$, we obtain $\widehat{\varphi}_c(\xi_0 + n) = 0$. Hence the function $\widehat{\varphi}_c$ has a periodic zero, which is impossible (the mask m_c is clean). This completes the proof. ◊

Thus, for each k, ϑ_k gives a lower estimate for ϑ, and this estimate tends to ϑ as k increases. Correspondingly (Corollary 7.4.3), ϑ_k gives a lower estimate for the smoothness of the refinable function:

THEOREM 7.4.5. *For any refinable function* φ, *we have*

$$\alpha_\varphi \ \ge\ \vartheta_k + L$$

for any $k \ge 1$.

As is known, in the general case, this is the estimate for smoothness only up to 1. It does not become sharp even as $k \to \infty$. This is a disadvantage of the method of pointwise estimation which cannot be overcome. Nevertheless, for large degree n of a clean mask, this method is often used in view of its simplicity. Even for $k = 1$ (i.e., $\vartheta_1 = -\log_2 \|m_c\|_\infty$), it may give satisfactory estimates.

COROLLARY 7.4.6.

$$\alpha_\varphi \ \ge\ L - \log_2 \|m_c\|_\infty.$$

We shall study in detail only one example, namely, the family of scaling functions $\{\varphi_r\}$ generating the Daubechies wavelets constructed in Chapter 4 (here we have to replace the index n by r, since by n we now denote the order of a clean mask).

EXAMPLE 7.4.7 (The Daubechies wavelets ψ_r). For a scaling function φ_r and the corresponding mask m_r, we have (formulas (4.10) and (4.11))

$$|m_r(\xi)|^2 = \left(\cos^2 \pi\xi\right)^r P_r\left(\sin^2 \pi\xi\right),$$

where

$$P_r(x) = \sum_{k=0}^{r-1} \binom{r-1+k}{k} x^k.$$

Thus the squared clean mask $m_{c,r}$ is equal to

$$|m_{c,r}(\xi)|^2 = P_r\left(\sin^2 \pi\xi\right). \tag{7.28}$$

Consider some elementary properties of the polynomial P_r.

LEMMA 7.4.8. *The function $x^{-r+1}P_r(x)$ is decreasing in x. In addition, for any $x \in [0,1]$, we have*

$$P_r(x) \le 2^{r-1}\left(\max\{1,\, 2x\}\right)^{r-1}.$$

Proof. If $0 \le x \le y$, then

$$\begin{aligned} x^{-r+1}P_r(x) &= x^{-r+1}\sum_{k=0}^{r-1}\binom{r-1+k}{k}x^k = \sum_{k=0}^{r-1}\binom{r-1+k}{k}x^{-(r-k-1)} \\ &\ge \sum_{k=0}^{r-1}\binom{r-1+k}{k}y^{-(r-k-1)} = y^{-r+1}P_r(y). \end{aligned}$$

Since the polynomial P_r solves the equation

$$(1-x)^r P_r(x) + x^r P_r(1-x) = 0,$$

after the change of variables $x = \frac{1}{2}$ we get $P_r(1/2) = 2^{r-1}$. Since $P_r(x)$ is an increasing function, $P_r(x) \le P_r(1/2) = 2^{r-1}$ for $x \le \frac{1}{2}$. If $x \ge \frac{1}{2}$, then we use the first statement of the lemma:

$$P_r(x) \le x^{r-1}2^{r-1}P_r(1/2) = 2^{r-1}(2x)^{r-1}. \qquad \lozenge$$

Note that

$$\begin{aligned} \|m_{c,r}\|_\infty &= \left[\sum_{k=0}^{r-1}\binom{r-1+k}{k}\right]^{1/2} < \left[2^{r-1}\sum_{k=0}^{r-1}\binom{r-1+k}{k}2^{-k}\right]^{1/2} \\ &= \left[2^{r-1}P_r(1/2)\right]^{1/2} = 2^{r-1}. \end{aligned} \tag{7.29}$$

Corollary 7.4.6 gives $\alpha_{\varphi_r} > 0$. Thus, beginning with $r = 2$, all the Daubechies wavelet functions ψ_r are at least continuous.

For a more detailed analysis, we shall estimate ϑ_2. We have

$$\vartheta_2 = -\frac{1}{2}\log_2\left\|m_{c,r}(\xi)m_{c,r}(2\xi)\right\|_\infty = -\frac{1}{4}\log_2\Big(\sup_{0\le x\le 1} P_r(x)P_r\big(4x(1-x)\big)\Big)$$

(since $\sin^2 2\pi\xi = 4\sin^2(\pi\xi)\big(1-\sin^2(\pi\xi)\big)$). If $x \le \frac{1}{2}$ or $x \ge \frac{1}{2}+\frac{\sqrt{2}}{4}$ (i.e., $4x(1-x) \le \frac{1}{2}$), we derive from Lemma 7.4.8 that

$$P_r(x)P_r\big(4x(1-x)\big) \le 2^{3(r-1)}.$$

On the remaining interval $\frac{1}{2} \le x \le \frac{1}{2} + \frac{\sqrt{2}}{4}$ we have

$$P_r(x)P_r(4x(1-x)) \;\le\; 2^{2(r-1)}[16x^2(1-x)]^{r-1} \;\le\; 2^{6(r-1)}\Big(\frac{4}{27}\Big)^{r-1}$$

(since $x^2(1-x) \le \frac{4}{27}$ on the segment $[0,1]$). Thus,

$$\vartheta_2 \;\ge\; -\frac{1}{4}\log_2\Big(2^{8(r-1)}3^{-3(r-1)}\Big) \;=\; -2r+2+(r-1)\frac{3}{4}\log_2 3. \tag{7.30}$$

Therefore,

$$\vartheta(m_r) \ge 1 + (r-1)\Big(\frac{3}{4}\log_2 3 \;-\; 1\Big),$$

and we have

$$\alpha_{\varphi_r} \;\ge\; (r-1)\Big(\frac{3}{4}\log_2 3 \;-\; 1\Big). \tag{7.31}$$

Asymptotically, for large r, this gives $\alpha_{\varphi_r} \ge \mu r$, where $\mu = 0.1887\ldots$. We now need but a little more to prove the estimate $\alpha_{\varphi_r} \ge 0.2r$ (Theorem 4.1.6). We could continue the process, estimating ϑ_3, etc. However it is more effective to employ the following modification of the method of pointwise estimation, which uses cyclic sets. For the Daubechies wavelets, this method allows computation of the exact value of ϑ.

7.5. Estimation by invariant cycles

A finite set of points on the fundamental period $b = \{\beta_1, \ldots, \beta_k\} \subset \mathbb{T}$ is called a *cycle* with respect to the operation of doubling $\beta \to 2\beta$ if $\beta_{j+1} \equiv 2\beta_j \bmod 1$, $j = 1, \ldots, k$ (here, we set $\beta_{k+1} = \beta_1$). This definition has something in common with the notion of cyclic set introduced in §3.4. If $b = \{\beta_1, \ldots, \beta_k\}$ is a cycle, then the set of the unit circle $\mathbf{b} = \{e^{-2\pi i\beta_1}, \ldots, e^{-2\pi i\beta_k}\}$ is cyclic, and vice versa. The simplest cycle $b = \{0\}$ will be called trivial by analogy with the trivial cyclic set $\mathbf{b} = \{1\}$. By $\mathcal{B}$ we denote the set of all cycles including the trivial one. For an arbitrary mask m_c and cycle $b = \{\beta_1, \ldots, \beta_k\}$, we set

$$\Theta_b = \Theta_b(m_c) = -\frac{1}{k}\log_2\Big|\prod_{j=1}^{k} m_c(\beta_j)\Big|.$$

Thus the value of $-\Theta_b$ is equal to the mean value of the logarithm of the modulus of the clean mask at the points of the cycle b.

THEOREM 7.5.1. *For an arbitrary clean mask m_c, we have*

$$\vartheta(m_c) = \inf_{b\in\mathcal{B}} \Theta_b(m_c).$$

Proof. First we show that $\Theta_b(m_c) \ge \vartheta(m_c)$ for any cycle b (in what follows, we shall omit the symbol m_c). Choose an arbitrary element of the cycle $\beta \in b$. Since the mask m_c is clean, there exists an integer s, for which $\widehat{\varphi}_c(\beta + s) \ne 0$ (otherwise, the function $\widehat{\varphi}_c(\xi)$ would have a periodic zero; see §3.4). If the cycle b is nontrivial, then $\beta + s \ne 0$. If $b = \{0\}$, then there exists a nonzero $s \in \mathbb{Z}$, for which $\widehat{\varphi}_c(s) \ne 0$. Otherwise, the function φ_c would satisfy the Strang-Fix condition, which is impossible since the mask m_c is clean (Theorem 3.2.10 and §3.4). Thus $\beta + s \neq 0$. Then for any p, we have

$$\widehat{\varphi}_c\Big(2^{kp}(\beta+s)\Big) \;=\; \widehat{\varphi}_c(\beta+s)\prod_{r=1}^{kp} m_c(2^{kp-r}(\beta+s)) \;=\; \widehat{\varphi}_c(\beta+s)\left(\prod_{j=1}^{k} m_c(\beta_j)\right)^p .$$

Thus,

$$\left|\widehat{\varphi}_c\Big(2^{kp}(\beta+s)\Big)\right| \geq C2^{-kp\Theta_b}$$

for any k, where $C = |\widehat{\varphi}_c(\beta+s)| > 0$. Therefore $\Theta_b \geq \vartheta$.

Thus, $\inf\limits_{b\in\mathcal{B}} \Theta_b \geq \vartheta$. It remains to prove that there exist cycles b for which the difference $\Theta_b - \vartheta_b$ takes arbitrarily small values. According to Proposition 7.4.4, there exists a sequence $\xi_j \to \infty$ for which $|\widehat{\varphi}_c(\xi_j)| \;\geq\; C|\xi_j|^{-\vartheta}$. First consider the case when the mask $m_0(\xi)$ does not vanish on the real axis, i.e., $\inf\limits_{\xi\in\mathbb{T}} |m_0(\xi)| = a > 0$. Choose sufficiently large ξ_j and denote by n a positive integer for which $2^{n-1} < \xi_j \leq 2^n$. There exists an integer r such that

$$\left|\xi_j - \frac{2^n r}{2^n-1}\right| < 1.$$

Set $\beta = \frac{r}{2^n-1}$. By construction, $\beta \in \left[\frac{1}{3}, \frac{4}{3}\right]$. Since $2^n\beta \equiv \beta \pmod 1$, the set $\{2^s\beta\}_{s=0,\dots,n-1}$ is either a cycle or the sum of p equal cycles $b = \{2^s\beta\}_{s=0}^{k}$, $n = pk$. Set $\beta_i = 2^{i-1}\beta$, $i = 1,\dots,k$. We have

$$\widehat{\varphi}_c(2^{pk}\beta) = \widehat{\varphi}(\beta)\left(\prod_{i=1}^{k} m_c(\beta_i)\right)^p .$$

Thus,

$$|\widehat{\varphi}_c(2^n\beta)| \;=\; |\widehat{\varphi}(\beta)|\,2^{-n\Theta_b} \;\leq\; C_0 2^{-n\Theta_b}, \tag{7.32}$$

where $C_0 = \sup\limits_{1/3\leq\xi\leq 4/3} |\widehat{\varphi}(\xi)|$. Using the inequality

$$|m_c(\xi+h)| \leq |m_c(\xi)| + \|m_c'\|_\infty |h|,$$

we get

$$\begin{aligned}
|\widehat{\varphi}_c(\xi_j)| = \prod_{s=1}^{\infty}\left|m_c(2^{-s}\xi_j)\right| &\leq \prod_{s=1}^{\infty}\left(\left|m_c(2^{n-s}\beta)\right| + 2^{-s}\|m_c'\|_\infty\right) \\
&= \left(\prod_{s=1}^{\infty}\left|m_c(2^{n-s}\beta)\right|\right)\cdot\prod_{s=1}^{\infty}\left|1 + \frac{2^{-s}\|m_c'\|_\infty}{|m_c(2^{n-s}\beta)|}\right| \\
&\leq |\widehat{\varphi}_c(2^n\beta)|\prod_{s=1}^{\infty}\left(1 + \frac{2^{-s}\|m_c'\|_\infty}{a}\right).
\end{aligned}$$

Since $\left(1 + \frac{2^{-s}\|m_c'\|}{a}\right) \leq e^{2^{-s}\|m_c'\|/a}$, we have

$$|\widehat{\varphi}_c(\xi_j)| \;\leq\; e^{\|m_c'\|/a}\,|\widehat{\varphi}_c(2^n\beta)|\,.$$

Combining this inequality with (7.32), we get $|\widehat{\varphi}_c(\xi_j)| \;\leq\; C_1 2^{-n\Theta_b}$, where C_1 is a constant independent of j or n. Thus,

$$2^{-n\Theta_b} \;\geq\; C_2|\xi_j|^{-\vartheta}.$$

Taking the logarithm of both sides of this inequality and recalling that $\log_2 |\xi_j| > n-1$, we get

$$n\Theta_b \leq (n-1)\vartheta - \log_2 C_2,$$

which gives $\Theta_b \leq \vartheta$ as $n \to \infty$. ◊

REMARK 7.5.2. If a mask is not clean, then the statement of Theorem 7.5.1 may not hold. For example, if a mask has cycle b, i.e., $m_0\big(b+\frac{1}{2}\big) = 0$, then Θ_b may be strictly less than ϑ. This is the reason why the subdivision scheme corresponding to a continuous refinable function may be divergent [**127**].

Thus, each cycle gives an upper bound for the rate of decay $\widehat{\varphi}(\xi)$. On the other hand (Proposition 7.4.4), for each k, the value ϑ_k gives a lower bound. The equality

$$\sup_{k\in\mathbb{N}} \vartheta_k = \vartheta = \inf_{b\in\mathcal{B}} \Theta_b \tag{7.33}$$

provides a theoretical possibility of finding ϑ with any accuracy by the usual exhaustive search (of the necessary number of cycles and quantities ϑ_k). In addition, equality (7.33) proves the following useful fact:

COROLLARY 7.5.3. *The variable $\vartheta(m_0)$ is a continuous function of coefficients of the mask at each point $m_0 \sim (c_0, \ldots, c_N)$ such that the mask m_0 is clean.*

Proof. Given $\varepsilon > 0$, we find k and a cycle b such that $\Theta_b(m_0) - \vartheta_k(m_0) < \varepsilon/2$ (it is possible in view of (7.33)). If the mask is clean, both functions Θ_b and ϑ_k are continuous at the point m_0. Hence there exists sufficiently small $\delta > 0$ such that under the condition $|\tilde{c}_i - c_i| < \delta$, $i = 0, \ldots, N$, the mask $\tilde{m}_0$ with the coefficients $\tilde{c}_i$ is clean and

$$|\Theta_b(\tilde{m}_0) - \Theta_b(m_0)| < \varepsilon/2, \qquad |\vartheta_k(\tilde{m}_0) - \vartheta_k(m_0)| < \varepsilon/2.$$

Thus the point $\vartheta(m_0)$ belongs to the segment $[\vartheta_k(m_0)\,,\,\Theta_b(m_0)]$, and the point $\vartheta(\tilde{m}_0)$ belongs to the segment $[\vartheta_k(\tilde{m}_0)\,,\,\Theta_b(\tilde{m}_0)]$, where the distances from the endpoints of the first segment to the corresponding endpoints of the second one are less than $\varepsilon/2$. Therefore $|\vartheta(\tilde{m}_0) - \vartheta(m_0)| < \varepsilon$. ◊

REMARK 7.5.4. If a mask m_0 is not clean, the function $\vartheta(m_0)$ may be discontinuous at this point. The rate of decay of the Fourier transform may have jumps at such points. For example, for the mask $m_0(\xi) = \left(\frac{e^{-2\pi i\xi}+1}{2}\right)^n$ corresponding to the cardinal B-spline B_{n-1}, we have $\vartheta = n$. If we slightly change the coefficients to make the mask clean (for example, setting $m_0 = \left(\frac{e^{-2\pi i\xi}+1+\alpha}{2+\alpha}\right)^n$, $\alpha \to 0$), we get $\vartheta(m_0) \leq 0$.

In many cases, to estimate the rate of decay, the upper bound is taken from equality (7.33); i.e., upper bounds are obtained from cycles. To find the lower bound, instead of the exhaustive search of the variables ϑ_k, the following result is used:

LEMMA 7.5.5. *Suppose $[0,1] = D_1 \cup \cdots \cup D_k$ and there exists $q > 0$ such that*

$$\begin{array}{ll} |m_c(\xi)| \leq q, & \xi \in D_1, \\ |m_c(\xi)m_c(2\xi)| \leq q^2, & \xi \in D_2, \\ \cdots & \cdots \\ |m_c(\xi)\cdots m_c(2^{k-1})| \leq q^k, & \xi \in D_k. \end{array}$$

Then $\vartheta \geq -\log_2 q$.

`Proof.` Take sufficiently large n and estimate the variable $\left|\prod_{s=0}^{n-1} m_c(2^{-s}\xi)\right|$. Since $\eta = 2^{-n+1}\xi \in D_m$ for some $m \in \{1, \dots, k\}$, we have

$$\left|\prod_{s=0}^{n-1} m_c(2^{-s}\xi)\right| = \left|\prod_{l=0}^{n-1} m_c(2^{l}\eta)\right| \le q^m \left|\prod_{l=m}^{n-1} m_c(2^{l}\eta)\right|.$$

Next, we apply the same inequality to $2^l\eta$, etc. As a result, we get

$$\left|\prod_{s=0}^{n-1} m_c(2^{-s}\xi)\right| \le q^{n-r} \left|\prod_{s=0}^{r} m_c(2^{-s}\xi)\right|,$$

where $r \le k-1$. Thus,

$$\left|\prod_{s=0}^{n-1} m_c(2^{-s}\xi)\right| \le q^{j-r}(\|m_c\|_\infty)^{k-1}.$$

Now, for n, take a number such that $2^{n-1} < |\xi| \le 2^n$. We get

$$|\widehat{\varphi}_c(\xi)| \ \le \ q^{j-r}(\|m_c\|_\infty)^{k-1}\|\widehat{\varphi}_c\|_{C[1,2]},$$

which yields the required inequality as $|\xi| \to \infty$. ◊

In practice, to compute the rate of decay for specific refinement equations, it is sometimes possible to find a cycle b and partition $[0,1] = D_1 \cup \dots \cup D_k$ (as a rule, corresponding, in some sense, to this cycle) such that $q = -\log_2 \Theta_b$; hence both variables are equal to $\vartheta(m_c)$. In particular, this method is appropriate for the scaling functions φ_r generating the Daubechies wavelets. In this case, it suffices to take the cycle $b = \{\frac{1}{3}, \frac{2}{3}\}$ and the partition $D_1 = \left[\frac{1}{6}, \frac{5}{6}\right]$, $D_2 = \left[0, \frac{1}{6}\right]$, $\left[\frac{5}{6}, 1\right]$.

EXAMPLE 7.5.6 (The Daubechies wavelets ψ_r). Let $q^2 = \left|m_c\left(\frac{1}{3}\right)m_c\left(\frac{2}{3}\right)\right|$. It is clear that $Q_b(m_c) = -\log_2 q$, where $b = \{\frac{1}{3}, \frac{2}{3}\}$. We prove that for the Daubechies masks the following inequalities hold:

$$\begin{aligned} &|m_c(\xi)| \le q, && \xi \in D_1, \\ &|m_c(\xi)m_c(2\xi)| \le q^2, && \xi \in D_2, \end{aligned} \tag{7.34}$$

where the sets D_1, D_2 are defined above. Then it will follow from Lemma 7.5.5 that $\vartheta(m_c) = -\log_2 q$, and the rate of decay for the Daubechies scaling functions will be computed.

Note first that for the masks $m_{c,r}$, we have $m_{c,r}(\xi) = m_{c,r}(1-\xi)$. So $q = \left|m_{c,r}\left(\frac{1}{3}\right)\right| = \left|m_{c,r}\left(\frac{2}{3}\right)\right|$. First of all we establish several properties of the polynomial $P_r(x)$.

LEMMA 7.5.7. *For all $r \ge 1$, the polynomial $P_r(x) = \sum_{k=0}^{r-1} \binom{r-1+k}{k} x^k$ satisfies the condition*

$$P_r'(x) \ = \ \frac{r}{1-x}\Big[P_r(x) - x^{r-1}P_r(1)\Big]. \tag{7.35}$$

Proof.

$$P_r'(x) = \sum_{k=0}^{r-1} \binom{r-1+k}{k} r x^k = r \sum_{k=0}^{r-2} \binom{r+k}{k} x^k$$
$$= r\left[P_{r+1}(x) - \binom{2r}{r} x^r - \binom{2r-1}{r-1} x^{r-1}\right]. \tag{7.36}$$

Further,

$$(1-x)P_{r+1}(x) = 1 + \sum_{k=1}^{r} \left[\binom{r+k}{k} - \binom{r-1+k}{k-1}\right] x^k - \binom{2r}{r} x^r$$
$$= 1 + \sum_{k=1}^{r} \binom{r-1+k}{k} x^k - \binom{2r}{r} x^r \tag{7.37}$$
$$= P_r(x) + \binom{2r-1}{r} x^r (1-2x).$$

Combining (7.36) and (7.37), we get

$$(1-x)P_r'(x) \;=\; r\left[P_r(x) - \binom{2r-1}{r} x^{r-1}\right].$$

Taking into account that

$$P_r(1) \;=\; \sum_{k=0}^{r-1} \binom{r-1+k}{k} \;=\; \binom{2r-1}{r},$$

we obtain (7.35). ◊

LEMMA 7.5.8. *For all $r \geq 1$, the polynomial $P_r(x) = \sum_{k=0}^{r-1} \binom{r-1+k}{k} x^k$ satisfies the conditions*

$$\begin{aligned} &P_r(x) \leq P_r\left(\tfrac{3}{4}\right), && \text{if } 0 \leq x \leq \tfrac{3}{4},\\ &P_r(x)P_r\big(4x(1-x)\big) \leq \left[P_r\left(\tfrac{3}{4}\right)\right]^2, && \text{if } \tfrac{3}{4} \leq x \leq 1. \end{aligned}$$

Proof. The first inequality follows from the fact that the function $P_r(x)$ increases on the interval $x \in [0,1]$. Let us prove the second inequality. Set $f(x) = P_r(x)P_r\big(4x(1-x)\big)$. It follows from Lemma 7.5.7 that

$$f'(x) \;=\; \frac{r}{(1-x)(2x-1)} g(x),$$

where

$$g(x) = (6x-5)P_r(x)P_r\big(4x(1-x)\big) \;-\; x^{r-1}(2x-1)P_r(1)P_r\big(4x(1-x)\big)$$
$$+\, 4(1-x)\Big[4x(1-x)\Big]^{r-1} P_r(1)P_r(x). \tag{7.38}$$

Note that $4x(1-x) \leq x$ for $x \geq \frac{3}{4}$. Since the function $x^{-r+1}P_r(x)$ decreases in x (Lemma 7.4.8), we have

$$x^{-r+1}P_r(x) \;\leq\; \big[4x(1-x)\big]^{-r+1} P_r\big(4x(1-x)\big),$$

or

$$\big[4(1-x)\big]^{r-1} P_r(x) \;\leq\; P_r\big(4x(1-x)\big).$$

Substituting this into inequality (7.38), we get

$$g(x) \;\le\; (6x-5)P_r\bigl(4x(1-x)\bigr)\Bigl[P_r(x) - x^{r-1}P_r(1)\Bigr].$$

The expression in square brackets is equal to $\frac{1}{r}(1-x)P_r'(x) \ge 0$ for all $x \le 1$; hence $g(x) \le 0$ for $\frac{3}{4} \le y \le \frac{5}{6}$. Thus the function $P_r(x)P_r\bigl(4x(1-x)\bigr)$ decreases on the interval $\bigl[\frac{3}{4}, \frac{5}{6}\bigr]$, which proves the required inequality $P_r(x)P_r\bigl(4x(1-x)\bigr) \le \bigl[P_r(\frac{3}{4})\bigr]^2$ on this interval.

Turn now to the interval $\frac{5}{6} \le x \le 1$. As $P_r(x) \le \bigl(\frac{4x}{3}\bigr)^{r-1}P_r(\frac{3}{4})$, in view of Lemma 7.4.8, it suffices to prove that

$$\left(\frac{4x}{3}\right)^{r-1} P_r\bigl(4x(1-x)\bigr) \;\le\; P_r\Bigl(\frac{3}{4}\Bigr). \tag{7.39}$$

First note that

$$(1-x)^r P_r(x) \;=\; 1 \;-\; x^r P_r(1-x) \;\le\; 1.$$

Therefore,

$$P_r\bigl(4x(1-x)\bigr) \;\le\; \Bigl[1 - 4x(1-x)\Bigr]^{-r} \;=\; (2x-1)^{-2r}.$$

Further, it follows from Lemma 7.4.8 and the estimate

$$P_r(1) \;=\; \binom{2r-1}{r} \;=\; \frac{1}{2}\binom{2r}{r} \;\ge\; \frac{1}{\sqrt{r}}4^{r-1}$$

that

$$P_r\Bigl(\frac{3}{4}\Bigr) \ge \Bigl(\frac{3}{4}\Bigr)^{r-1} P_r(1) \;\ge\; \frac{1}{\sqrt{r}}3^{n-1}. \tag{7.40}$$

Thus, in order to prove inequality (7.39), it suffices to show that

$$\left[\frac{x}{(2x-1)^2}\right]^{r-1} (2x-1)^{-2} \;\le\; \frac{1}{\sqrt{r}}\Bigl(\frac{9}{4}\Bigr)^{r-1}. \tag{7.41}$$

Since the functions $(2x-1)^{-2}$ and $x(2x-1)^{-2}$ decrease on the interval $\bigl[\frac{5}{6}, 1\bigr]$, it is sufficient to check that (7.41) holds at the point $x = \frac{5}{6}$, in other words, that $\bigl(\frac{5}{6}\bigr)^{r-1} \le \frac{4}{9\sqrt{r}}$. This is true for $r \ge 13$. Thus the lemma is proved for $r \ge 13$. It remains to prove it for $r = 1, \ldots, 12$. To this end, we consider two cases.

1) $\frac{5}{6} \le x \le \frac{2+\sqrt{2}}{4}$. In this case we have $4x(1-x) \ge \frac{1}{2}$. By Lemma 7.4.8 it follows that

$$P_r\bigl(4x(1-x)\bigr) \;\le\; \bigl[8x(1-x)\bigr]^{r-1}P_r\Bigl(\frac{1}{2}\Bigr) \;=\; \bigl[16x(1-x)\bigr]^{r-1}.$$

Analogously, $P_r(x) \le \bigl(\frac{6x}{5}\bigr)^{r-1}P_r\bigl(\frac{5}{6}\bigr)$; hence,

$$\begin{aligned} P_r(x)P_r\bigl(4x(1-x)\bigr) \;&\le\; \Bigl(\frac{6}{5}\Bigr)^{r-1} P_r\Bigl(\frac{5}{6}\Bigr)\bigl[16x^2(1-x)\bigr]^{r-1} \\ &\le\; \Bigl(\frac{20}{9}\Bigr)^{r-1} P_r\Bigl(\frac{5}{6}\Bigr) \end{aligned}$$

(here, we used the fact that the function $x^2(1-x)$ decreases on the interval $\frac{5}{6} \le x \le \frac{2+\sqrt{2}}{4}$). It can be numerically checked that $\Bigl(\frac{20}{9}\Bigr)^{r-1} P_r\bigl(\frac{5}{6}\bigr) \le \Bigl[P_r\bigl(\frac{3}{4}\bigr)\Bigr]^2$ for $r \le 12$.

2) $\frac{2+\sqrt{2}}{4} \le x \le 1$. Set $x_0 = \frac{2+\sqrt{2}}{4}$. Using the estimates

$$P_r\big(4x(1-x)\big) \ \le\ (2x-1)^{-2r}, \qquad P_r(x) \ \le\ \Big(\frac{x}{x_0}\Big)^{r-1} P_r(x_0),$$

we get

$$P_r\big(4x(1-x)\big)P_r(x) \ \le\ x_0^{-r+1} P_r(x_0) \left[\frac{x}{(2x-1)^2}\right]^{r-1}.$$

The latter expression does not exceed $2^r P_r(x_0)$ (since both functions $(2x-1)^{-2}$ and $x(2x-1)^{-2}$ decrease on the interval $[x_0,\, 1]$). It can be numerically checked that $2^r P_r(x_0)$ is less than $\left[P_r\big(\frac{3}{4}\big)\right]^2$ for $r = 5, \dots, 12$.

It remains to prove the lemma for $\frac{2+\sqrt{2}}{4} \le x \le 1$ and $r = 1, 2, 3, 4$. For these r, the order of the polynomial $P_r(x)P_r\big(4x(1-x)\big) - \left[P_r\big(\frac{3}{4}\big)\right]^2$ does not exceed 9. Numerically estimating its roots, we deduce that none of them belongs to the interval $\big(\frac{3}{4},\, 1\big)$, which completes the proof of the lemma, since the required inequality holds at the point $x = 1$. Lemma 7.5.8 is thus proved. ◊

Thus, for any r, the mask $m_{c,r}$ satisfies (7.34), hence $\vartheta(m_{c,r}) = -\log_2 q$, where $q = \left|m_{c,r}\big(\frac{1}{3}\big)\right|$. Therefore,

$$\vartheta(m_{c,r}) \ =\ -\frac{1}{2}\log_2 P_r\Big(\frac{3}{4}\Big). \tag{7.42}$$

Estimate $\vartheta(m_{c,r})$ asymptotically for $r \to \infty$. As

$$\frac{1}{\sqrt{r}} 3^{r-1} \le P_r\Big(\frac{3}{4}\Big) \ \le\ 3^{r-1}$$

(the upper bound follows from Lemma 7.4.8, and the lower bound follows from Lemma 7.5.7), we have

$$\frac{1}{2}\log_2 P_r\Big(\frac{3}{4}\Big) \ =\ \Big(\frac{1}{2}\log_2 3\Big)\cdot r\Big(1 - O(r^{-1}\log r)\Big).$$

Thus,

$$\vartheta(m_{c,r}) \ \sim\ (-0.79348\dots)r.$$

It immediately follows that for large r,

$$\alpha_{\varphi_r} \ \ge \mu r,$$

where $\mu = 1 - 0.79348\dots \ =\ 0.2075\dots$. This implies the statement of Theorem 4.1.6 (for large r, the inequality $\alpha_{\varphi_r} \ \ge 0.2r$ follows from our asymptotic estimate, and for small r it is verified directly with the help of the matrix method, Examples

7.3.7–7.3.9). For small r, this method gives the following values of smoothness:

(7.43)

r	α
2	0.339
3	0.636
4	0.913
5	1.177
6	1.432
7	1.682
8	1.927
9	2.168
10	2.406

The precision of this method is not high, as can be seen by comparing the given estimates with real values of smoothness obtained for $r \leq 8$ by the matrix method (see the table in (7.14)). It is not surprising, since the method of pointwise estimation of the decay of the Fourier transform gives the smoothness only up to 1.

CHAPTER 8

Nonstationary Wavelets

8.1. General theory of nonstationary wavelets

In this chapter, the wavelet systems $\{\psi_{jk}(t) := 2^{j/2}\psi(2^j t + k)\}_{j,k\in\mathbb{Z}}$ discussed above will be called stationary. Now we study *nonstationary wavelet systems* $\{\psi_{jk}\}_{j,k\in\mathbb{Z}}$, which, unlike stationary systems, are generated by some sequence of wavelet functions $\{\psi_j\}_{j\in\mathbb{Z}}$. Here, the symbol ψ_{jk} has a new meaning:

$$\psi_{jk}(t) := \psi_j(t + k2^{-j}), \quad j, k \in \mathbb{Z}.$$

If a nonstationary wavelet system constitutes a basis (orthonormal basis) in the space $L_2(\mathbb{R})$, we shall call it a *nonstationary wavelet basis (orthonormal wavelet basis)* and say that this basis is generated by the functions $\{\psi_j\}$, $j \in \mathbb{Z}$.

REMARK 8.1.1. In the stationary case, the role of functions $\{\psi_j\}_{j\in\mathbb{Z}}$ is played by the dilations of a single function $\psi_j(t) = 2^{j/2}\psi(2^j t)$.

Let us present an analog of Theorem 1.3.3 for the nonstationary case.

THEOREM 8.1.2. *Let a sequence of functions $\{\psi_j\}_{j\in\mathbb{Z}}$, $\psi_j \in L_2(\mathbb{R})$, $\|\psi_j\| = 1$, satisfy the condition*

$$\int_K \sum_{j\in\mathbb{Z}} |\widehat{\psi_j}(\xi)|^2 \, d\xi < \infty \tag{8.1}$$

for any compact set $K \subset \mathbb{R}\backslash\{0\}$.

The functions $\{\psi_j\}$, $j \in \mathbb{Z}$, generate a nonstationary orthonormal wavelet basis if and only if

$$\sum_{j\in\mathbb{Z}} 2^j |\widehat{\psi_j}(\xi)|^2 = 1 \ a.e. \tag{8.2}$$

and

$$\sum_{p=0}^{\infty} \widehat{\psi}_{j-p}(\xi)\overline{\widehat{\psi_{j-p}}(\xi + l2^j)} = 0 \ a.e. \tag{8.3}$$

for all odd $l \in \mathbb{Z}$ and all $j \in \mathbb{Z}$.

Proof. Sufficiency. As $\|\psi_{jk}\| = 1$, it is clear that the system $\{\psi_{jk}\}_{j,k\in\mathbb{Z}}$ is an orthonormal basis if and only if for any function $f \in L_2(\mathbb{R})$, the Parseval equality holds:

$$\|f\|^2 = \sum_{j,k\in\mathbb{Z}} |\langle f, \psi_{jk}\rangle|^2. \tag{8.4}$$

Suppose $f \in L_2(\mathbb{R})$ and $\widehat{f}$ is bounded and has compact support contained in $\mathbb{R}\setminus\{0\}$. By Lemma 1.2.11, we have

$$\sum_{k\in\mathbb{Z}} |\langle f, \psi_{jk}\rangle|^2 = 2^j \int_0^{2^j} \Big|\sum_{l\in\mathbb{Z}} \widehat{f}(\xi + l2^j)\overline{\widehat{\psi}_j(\xi + l2^j)}\Big|^2 \, d\xi. \tag{8.5}$$

Continuing the transformations, we get

$$\begin{aligned} \sum_{k\in\mathbb{Z}} |\langle f, \psi_{jk}\rangle|^2 &= 2^j \int_0^{2^j} \sum_{l\in\mathbb{Z}} |\widehat{f}(\xi + l2^j)|^2 |\widehat{\psi}_j(\xi + l2^j)|^2 \, d\xi \\ &\quad + 2^j \int_0^{2^j} \sum_{\substack{l_1,l_2\in\mathbb{Z} \\ l_1\neq l_2}} \widehat{f}(\xi + l_1 2^j)\overline{\widehat{f}(\xi + l_2 2^j)} \\ &\qquad \times \overline{\widehat{\psi}_j(\xi + l_1 2^j)}\widehat{\psi}_j(\xi + l_2 2^j) \, d\xi \\ &= 2^j \int_{\mathbb{R}} |\widehat{f}(\xi)|^2 |\widehat{\psi}_j(\xi)|^2 \, d\xi \\ &\quad + 2^j \int_{\mathbb{R}} \sum_{\substack{l\in\mathbb{Z} \\ l\neq 0}} \widehat{f}(\xi)\overline{\widehat{f}(\xi + l2^j)\widehat{\psi}_j(\xi)}\widehat{\psi}_j(\xi + l2^j) \, d\xi. \end{aligned} \tag{8.6}$$

Thus,

$$\begin{aligned} \sum_{j,k\in\mathbb{Z}} |\langle f, \psi_{jk}\rangle|^2 &= \int_{\mathbb{R}} |\widehat{f}(\xi)|^2 \sum_{j\in\mathbb{Z}} 2^j |\widehat{\psi}_j(\xi)|^2 \, d\xi \\ &\quad + \sum_{j\in\mathbb{Z}} 2^j \int_{\mathbb{R}} \sum_{\substack{l\in\mathbb{Z} \\ l\neq 0}} \widehat{f}(\xi)\overline{\widehat{f}(\xi + l2^j)\widehat{\psi}_j(\xi)}\widehat{\psi}_j(\xi + l2^j) \, d\xi. \end{aligned} \tag{8.7}$$

Any number $l \neq 0$ can be written as $k2^p$, where k is an odd number and p is a nonnegative integer. If we assume that the Fubini theorem can be applied (which will be justified later), we obtain

$$\begin{aligned} &\sum_{j\in\mathbb{Z}} 2^j \int_{\mathbb{R}} \sum_{\substack{l\in\mathbb{Z} \\ l\neq 0}} \widehat{f}(\xi)\overline{\widehat{f}(\xi + l2^j)\widehat{\psi}_j(\xi)}\widehat{\psi}_j(\xi + l2^j) \, d\xi \\ &\quad = \sum_{\substack{k\in\mathbb{Z} \\ k \text{ is odd}}} \int_{\mathbb{R}} \sum_{j\in\mathbb{Z}} \sum_{p=0}^{\infty} 2^j \widehat{f}(\xi)\overline{\widehat{f}(\xi + k2^{j+p})\widehat{\psi}_j(\xi)}\widehat{\psi}_j(\xi + k2^{j+p}) \, d\xi \\ &\quad = \sum_{\substack{k\in\mathbb{Z} \\ k \text{ is odd}}} \int_{\mathbb{R}} \sum_{j\in\mathbb{Z}} 2^j \widehat{f}(\xi)\overline{\widehat{f}(\xi + k2^j)} \sum_{p=0}^{\infty} 2^{-p} \overline{\widehat{\psi}_{j-p}(\xi)}\widehat{\psi_{j-p}}(\xi + k2^j) \, d\xi. \end{aligned} \tag{8.8}$$

This equality together with (8.7) mean that (8.2) and (8.3) imply (8.4).

As the set of bounded functions with compact spectrum contained in $\mathbb{R}\setminus\{0\}$ is everywhere dense in $L_2(\mathbb{R})$, sufficiency will be proved if we justify the possibility of using the Fubini theorem. Apply the Cauchy inequality several times:

$$
\begin{aligned}
&\sum_{\substack{k\in\mathbb{Z}\\ k \text{ is odd}}} \int_{\mathbb{R}} \sum_{j\in\mathbb{Z}} 2^j |\widehat{f}(\xi)||\widehat{f}(\xi+k2^j)| \\
&\qquad \times \sum_{p=0}^{\infty} 2^{-p} |\widehat{\psi}_{j-p}(\xi)||\widehat{\psi}_{j-p}(\xi+k2^j)|\, d\xi \\
&\qquad \le \sum_{\substack{k\in\mathbb{Z}\\ k \text{ is odd}}} \int_{\mathbb{R}} \sum_{j\in\mathbb{Z}} 2^j |\widehat{f}(\xi)||\widehat{f}(\xi+k2^j)| \\
&\qquad \times \left(\sum_{p=0}^{\infty} 2^{-p} |\widehat{\psi}_{j-p}(\xi)|^2\right)^{1/2} \left(\sum_{p=0}^{\infty} 2^{-p} |\widehat{\psi}_{j-p}(\xi+k2^j)|^2\right)^{1/2} d\xi \\
&\qquad \le \sum_{\substack{k\in\mathbb{Z}\\ k \text{ is odd}}} \left(\int_{\mathbb{R}} \sum_{j\in\mathbb{Z}} 2^j |\widehat{f}(\xi)||\widehat{f}(\xi+k2^j)| \sum_{p=0}^{\infty} 2^{-p} |\widehat{\psi}_{j-p}(\xi)|^2\, d\xi\right)^{1/2} \\
&\qquad \times \left(\int_{\mathbb{R}} \sum_{j\in\mathbb{Z}} 2^j |\widehat{f}(\xi)||\widehat{f}(\xi+k2^j)| \sum_{p=0}^{\infty} 2^{-p} |\widehat{\psi}_{j-p}(\xi+k2^j)|^2\, d\xi\right)^{1/2} \\
&\qquad \le \left(\sum_{\substack{k\in\mathbb{Z}\\ k \text{ is odd}}} \int_{\mathbb{R}} \sum_{j\in\mathbb{Z}} 2^j |\widehat{f}(\xi)||\widehat{f}(\xi+k2^j)| \sum_{p=0}^{\infty} 2^{-p} |\widehat{\psi}_{j-p}(\xi)|^2\, d\xi\right)^{1/2} \\
&\qquad \times \left(\sum_{\substack{k\in\mathbb{Z}\\ k \text{ is odd}}} \int_{\mathbb{R}} \sum_{j\in\mathbb{Z}} 2^j |\widehat{f}(\xi-k2^j)||\widehat{f}(\xi)| \sum_{p=0}^{\infty} 2^{-p} |\widehat{\psi}_{j-p}(\xi)|^2\, d\xi\right)^{1/2} \\
&\qquad = \sum_{\substack{k\in\mathbb{Z}\\ k \text{ is odd}}} \int_{\mathbb{R}} \sum_{j\in\mathbb{Z}} 2^j |\widehat{f}(\xi)||\widehat{f}(\xi+k2^j)| \sum_{p=0}^{\infty} 2^{-p} |\widehat{\psi}_{j-p}(\xi)|^2\, d\xi \\
&\qquad = \sum_{\substack{k\in\mathbb{Z}\\ k \text{ is odd}}} \int_{\mathbb{R}} \sum_{j\in\mathbb{Z}} \sum_{p=0}^{\infty} |\widehat{f}(\xi)||\widehat{f}(\xi+k2^{j+p})| 2^j |\widehat{\psi}_j(\xi)|^2\, d\xi.
\end{aligned}
$$

It can be proved in a similar way that for any set of odd numbers I symmetric with respect to 0 and an arbitrary $J \subseteq \mathbb{Z}$,

$$
\begin{aligned}
&\sum_{k\in I} \int_{\mathbb{R}} \sum_{j\in J} 2^j |\widehat{f}(\xi)||\widehat{f}(\xi+k2^j)| \\
&\qquad \times \sum_{p=0}^{\infty} 2^{-p} |\widehat{\psi}_{j-p}(\xi)||\widehat{\psi}_{j-p}(\xi+k2^j)|\, d\xi \\
&\qquad \le \sum_{k\in I} \int_{\mathbb{R}} \sum_{j\in J} 2^j |\widehat{f}(\xi)||\widehat{f}(\xi+k2^j)| \sum_{p=0}^{\infty} 2^{-p} |\widehat{\psi}_{j-p}(\xi)|^2\, d\xi.
\end{aligned}
\tag{8.9}
$$

It is clear that $|\widehat{f}(\xi)||\widehat{f}((\xi+k2^l))|=0$ if $|k2^l|>\operatorname{diam}(\operatorname{supp}(\hat{f}))$. Therefore,

$$\sum_{\substack{k\in\mathbb{Z}\\ k \text{ is odd}}}\sum_{p=0}^{\infty}2^j|\widehat{f}(\xi)||\widehat{f}(\xi+k2^{j+p})|\le\sum_{p=0}^{\infty}\|\widehat{f}\|_\infty^2\operatorname{diam}(\operatorname{supp}(\widehat{f}))2^{-p}$$
$$=2\|\widehat{f}\|_\infty^2\operatorname{diam}(\operatorname{supp}(\widehat{f})).$$

Finally, by (8.1) we get

$$\sum_{\substack{k\in\mathbb{Z}\\ k \text{ is odd}}}\int_{\mathbb{R}}\sum_{j\in\mathbb{Z}}\sum_{p=0}^{\infty}|\widehat{f}(\xi)||\widehat{f}(\xi+k2^{j+p})|2^j|\widehat{\psi}_j(\xi)|^2\,d\xi$$
$$\le C\|\widehat{f}\|_\infty^2\int_{\operatorname{supp}(\widehat{f})}\sum_{j\in\mathbb{Z}}|\widehat{\psi}_j(\xi)|^2\,d\xi<\infty.$$

Necessity. Fix $\xi_0\neq 0$ and define functions $\widehat{f_\delta}=\delta^{-1/2}\chi_{(\xi_0,\xi_0+\delta)}$ in terms of the Fourier transform for $0<\delta<|\xi_0|$. It follows from (8.5) that for any $J\in\mathbb{N}$,

$$\frac{1}{\delta}\int_{\xi_0}^{\xi_0+\delta}\sum_{j=-J}^{J}2^j|\widehat{\psi}_j(\xi)|^2\le 1,$$

which implies that

$$\sum_{j=-\infty}^{\infty}2^j|\widehat{\psi}_j(\xi)|^2\le 1. \tag{8.10}$$

Obviously,

$$|\widehat{f_\delta}(\xi)|\cdot|\widehat{f_\delta}(\xi+k2^{j+p})|\neq 0$$

implies that $|k2^{j+p}|<\delta$. Hence,

$$\sum_{\substack{k\in\mathbb{Z}\\ k \text{ is odd}}}\sum_{p=0}^{\infty}2^j|\widehat{f_\delta}(\xi)||\widehat{f_\delta}(\xi+k2^{j+p})|\le C\chi_{(\xi_0,\xi_0+\delta)}(\xi).$$

Taking into account (8.7) and (8.10), we get

$$\lim_{\delta\downarrow 0}\frac{1}{\delta}\int_{\xi_0}^{\xi_0+\delta}\sum_{j\in\mathbb{Z}}2^j|\widehat{\psi}_j(\xi)|^2=1,$$

which implies (8.2).

It remains to prove (8.3). For arbitrary $\xi_0\neq 0$, $j_0\in\mathbb{Z}$, odd $k_0\in\mathbb{Z}$, and complex α, $|\alpha|=1$, consider the functions

$$\widehat{g_\delta}:=\delta^{-1/2}\chi_{(\xi_0,\xi_0+\delta)}+\alpha\delta^{-1/2}\chi_{(\xi_0+k_02^{j_0},\ \xi_0+k_02^{j_0}+\delta)}.$$

It is easy to see that

$$\sum_{k=\pm k_0} \widehat{g}_\delta(\xi)\overline{\widehat{g}_\delta(\xi+k2^{j_0})}\sum_{p=0}^{\infty} 2^{-p}\overline{\widehat{\psi}_{j_0-p}(\xi)}\widehat{\psi}_{j_0-p}(\xi+k2^{j_0})\,d\xi$$
$$= 2\operatorname{Re}\left(\frac{\alpha}{\delta}\int_{\xi_0}^{\xi_0+\delta}\sum_{p=0}^{\infty}2^{-p}\widehat{\psi}_{j_0-p}(\xi)\overline{\widehat{\psi}_{j_0-p}(\xi+k_0 2^{j_0})}\,d\xi\right).$$

Thus to prove (8.3), it suffices to show that

$$\lim_{\delta\downarrow 0}\sum_{k=\pm k_0}\int_{\mathbb{R}}\sum_{j\in\mathbb{Z}\setminus\{j_0\}} 2^j|\widehat{g}_\delta(\xi)||\widehat{g}_\delta(\xi+k2^j)| \times\sum_{p=0}^{\infty}2^{-p}|\widehat{\psi}_{j-p}(\xi)||\widehat{\psi}_{j-p}(\xi+k2^j)|\,d\xi = 0 \tag{8.11}$$

and

$$\lim_{\delta\downarrow 0}\sum_{\substack{k\in\mathbb{Z}\setminus\{\pm k_0\}\\ k \text{ is odd}}}\int_{\mathbb{R}}\sum_{j\in\mathbb{Z}} 2^j|\widehat{g}_\delta(\xi)||\widehat{g}_\delta(\xi+k2^j)| \times\sum_{p=0}^{\infty}2^{-p}|\widehat{\psi}_{j-p}(\xi)||\widehat{\psi}_{j-p}(\xi+k2^j)|\,d\xi = 0. \tag{8.12}$$

By (8.9), we have

$$\sum_{k=\pm k_0}\int_{\mathbb{R}}\sum_{j\in\mathbb{Z}\setminus\{j_0\}} 2^j|\widehat{g}_\delta(\xi)||\widehat{g}_\delta(\xi+k2^j)|\sum_{p=0}^{\infty}2^{-p}|\widehat{\psi}_{j-p}(\xi)||\widehat{\psi}_{j-p}(\xi+k2^j)|\,d\xi$$
$$\le \sum_{k=\pm k_0}\int_{\mathbb{R}}\sum_{j\in\mathbb{Z}\setminus\{j_0\}} 2^j|\widehat{g}_\delta(\xi)||\widehat{g}_\delta(\xi+k2^j)|\sum_{p=0}^{\infty}2^{-p}|\widehat{\psi}_{j-p}(\xi)|^2\,d\xi$$
$$\le \int_{\mathbb{R}}\left(\sum_{k=\pm k_0}\sum_{p=0}^{\infty}\sum_{j\in\mathbb{Z}\setminus\{j_0-p\}} 2^j|\widehat{g}_\delta(\xi)||\widehat{g}_\delta(\xi+k2^{j+p})|\right)\left(\sum_{l\in\mathbb{Z}}|\widehat{\psi}_l(\xi)|^2\right)d\xi.$$

It is easy to see that

$$\sum_{k=\pm k_0}\sum_{p=0}^{\infty}\sum_{j\in\mathbb{Z}\setminus\{j_0-p\}} 2^j|\widehat{g}_\delta(\xi)||\widehat{g}_\delta(\xi+k2^{j+p})|$$
$$\le C\chi_{(\xi_0,\xi_0+\delta)}+\chi_{(\xi_0+k_0 2^{j_0},\ \xi_0+k_0 2^{j_0}+\delta)};$$

hence (8.1) implies (8.11).

Equality (8.12) can be proved similarly. ◊

REMARK 8.1.3. In the stationary case condition (8.1) holds automatically. Let $0 < a < b < \infty$. Then

$$\int_a^b \sum_{j\in\mathbb{Z}} |\widehat{\psi}_j(\xi)|^2\, d\xi = \int_a^b \sum_{j\in\mathbb{Z}} 2^{-j}|\widehat{\psi}(2^{-j}\xi)|^2\, d\xi$$

$$= \sum_{j\in\mathbb{Z}} \int_{a2^{-j}}^{b2^{-j}} |\widehat{\psi}(\xi)|^2\, d\xi = \int_{\mathbb{R}} \sum_{j\in\mathbb{Z}} \chi_{[a2^{-j}, b2^{-j}]}(\xi)|\widehat{\psi}(\xi)|^2\, d\xi.$$

It can easily be seen that

$$\sum_{j\in\mathbb{Z}} \chi_{[a2^{-j}, b2^{-j}]}(\xi) \le \log_2(b/a).$$

As in the stationary case, the nonstationary analog of multiresolution analysis is the main method for the construction of nonstationary wavelets.

DEFINITION 8.1.4. A collection of closed spaces $V_j \subset L_2(\mathbb{R})$, $j \in \mathbb{Z}$, is called a *nonstationary multiresolution analysis* (in the sequel, NMRA) if the following conditions (axioms) hold:

MR1. $V_j \subset V_{j+1}$ for all $j \in \mathbb{Z}$.

MR2. $\bigcup_{j\in\mathbb{Z}} V_j$ is dense in $L_2(\mathbb{R})$.

MR3. $\bigcap_{j\in\mathbb{Z}} V_j = \{0\}$.

MR4. For any j, there exists a function $\varphi_j \in V_j$ such that the sequence $\{\varphi_j(\cdot + k2^{-j})\}_{k\in\mathbb{Z}}$ constitutes a Riesz basis in V_j.

A sequence $\{\varphi_j\}_{j\in\mathbb{Z}}$ is called a *scaling sequence* for a given NMRA.

If we denote by P_j the orthogonal projector on V_j, then condition **MR2** implies that $\lim_{j\to\infty} P_j f = f$ for any function $f \in L_2(\mathbb{R})$. It follows from **MR4** that for any $f \in V_j$, the function $f(\cdot + 2^{-j}k)$ also belongs to V_j for any $k \in \mathbf{Z}$.

Without loss of generality we can assume that $\{\varphi_j(\cdot + k2^{-j})\}_{k\in\mathbb{Z}}$ constitutes an orthonormal basis in V_j. The orthogonalization is performed similarly to the stationary case (see Theorem 1.1.14) by passing to the functions $\varphi_j^\sharp$ defined in terms of the Fourier transform by the equalities

$$\widehat{\varphi_j^\sharp}(\xi) := \frac{\widehat{\varphi}_j(\xi)}{(2^j \sum_{k\in\mathbb{Z}} |\widehat{\varphi}_j(\xi + 2^j k)|^2)^{1/2}}. \tag{8.13}$$

As before, we shall use the notation $\psi_{jk}(t) := \psi_j(t + k2^{-j})$. The proof of the following theorem is similar to that of Theorem 1.3.1.

THEOREM 8.1.5. *If $\{V_j\}_{j\in\mathbb{Z}}$ is an NMRA, then there exists a nonstationery orthonormal wavelet basis $\{\psi_{jk}\}_{j,k\in\mathbf{Z}}$ such that for any function f from $L_2(\mathbb{R})$,*

$$P_{j+1}f = P_j f + \sum_{k\in\mathbb{Z}} \langle f, \psi_{jk}\rangle \psi_{jk}. \tag{8.14}$$

Proof. Let W_j be an orthogonal complement of V_j to V_{j+1}: $W_j \oplus V_j = V_{j+1}$. By **MR1**, we have

$$W_j \perp W_{j_1} \quad \text{for } j \ne j_1, \tag{8.15}$$

and for any $j_0 < j$,

$$V_j = V_{j_0} \oplus \left(\bigoplus_{l=j_0}^{j-1} W_l \right). \tag{8.16}$$

It follows from **MR2** and **MR4** that

$$L_2(\mathbf{R}) = \bigoplus_{j\in\mathbf{Z}} W_j. \tag{8.17}$$

Formula (8.14) is equivalent to the fact that for fixed j, the sequence $\{\psi_{jk}\}_{k\in\mathbb{Z}}$ forms an orthonormal basis in W_j. By (8.17) the latter means that $\{\psi_{jk}\}_{j,\,k\in\mathbf{Z}}$ is an orthonormal basis in $L^2(\mathbf{R})$. Thus the problem of the construction of a nonstationary wavelet basis with property (8.14) is reduced to finding functions ψ_j such that the sequence $\{\psi_j(\cdot + k2^{-j})\}_{k\in\mathbb{Z}}$ constitutes an orthonormal basis in W_j.

To construct the functions ψ_j, the following properties of φ_j and W_j are used. Since $\varphi_j \in V_j \subset V_{j+1}$ and $\{\varphi_{j+1,k}\}_{k\in\mathbb{Z}}$ is an orthonormal basis in V_{j+1}, it follows that

$$\varphi_j = \sum_{k\in\mathbb{Z}} h_{j+1,k}\, \varphi_{j+1,k}, \tag{8.18}$$

where

$$h_{j+1,k} := \langle \varphi_j,\, \varphi_{j+1,k}\rangle, \quad \sum_{k\in\mathbb{Z}} |h_{j+1,k}|^2 = 1. \tag{8.19}$$

Equalities (8.18) are nonstationary analogs of the refinement equation. Passing to the Fourier transforms, we get

$$\widehat{\varphi_j}(\xi) = m_{j+1}(\xi 2^{-j-1})\widehat{\varphi_{j+1}}(\xi), \tag{8.20}$$

where $m_{j+1}(\xi) := \sum_{k\in\mathbf{Z}} h_{j+1,k} e^{2\pi i k\xi}$. The functions m_j are called nonstationary masks.

By virtue of Proposition 1.1.12,

$$\sum_{l\in\mathbb{Z}} |\widehat{\varphi_j}(\xi + 2^j l)|^2 = 2^{-j} \tag{8.21}$$

for almost all ξ. Substituting (8.20) into (8.21), we obtain

$$\sum_{l\in\mathbf{Z}} |m_{j+1}(\xi 2^{-j-1} + l/2)\widehat{\varphi}_{j+1}(\xi + 2^j l)|^2 = 2^{-j}.$$

Partitioning the sum into two parts (with even and odd l, respectively) and taking into account the 1-periodicity of m_{j+1}, we get

$$|m_{j+1}(\xi)|^2 + |m_{j+1}(\xi + 1/2)|^2 = 2. \tag{8.22}$$

Let us characterize the subspaces W_j in terms of the Fourier transform. Any function f from W_j belongs to V_{j+1} and is orthogonal to V_j. The first property means that

$$f = \sum_{k\in\mathbb{Z}} f_k \varphi_{j+1,k},$$

where $f_k = \langle f,\, \varphi_{j+1,k}\rangle$. In terms of the Fourier transform we have

$$\widehat{f}(\xi) = m_f(\xi 2^{-j-1})\, \widehat{\varphi}_{j+1}(\xi), \tag{8.23}$$

where $m_f(\xi) = \sum_{k\in\mathbf{Z}} f_k\, e^{2\pi ik\xi}$ is a 1-periodic function from $L_2[0,\,1]$. The condition of the orthogonality f to V_j is equivalent to the fact that $f \perp \varphi_{jk}$ for any $k \in \mathbf{Z}$; i.e.,

$$\int_{\mathbb{R}} \widehat{f}(\xi)\overline{\widehat{\varphi_j}(\xi)}\, e^{\pi ik2^{1-j}\xi}\, d\xi = 0.$$

Note that

$$\int_{\mathbb{R}} \widehat{f}(\xi)\overline{\widehat{\varphi_j}(\xi)}\, e^{\pi ik2^{1-j}\xi}\, d\xi = \int_0^{2^j} e^{\pi ik2^{1-j}\xi} \sum_{l\in\mathbf{Z}} \widehat{f}(\xi+2^jl)\overline{\widehat{\varphi_j}}(\xi+2^jl)\, d\xi = 0. \tag{8.24}$$

Since equality (8.24) holds for any $k \in \mathbf{Z}$, it follows that

$$\sum_{l\in\mathbb{Z}} \widehat{f}(\xi+2^jl)\overline{\widehat{\varphi_j}}(\xi+2^jl) = 0. \tag{8.25}$$

The series in (8.25) converges absolutely in $L([0,\,1])$. Substituting (8.20) and (8.23) into (8.25), grouping the sums with even and odd l, and taking into account (8.21), we obtain

$$\begin{aligned} &2^{j+1}\sum_{l\in\mathbb{Z}} \widehat{f}(\xi+2^jl)\overline{\widehat{\varphi}_j}(\xi+2^jl) \\ &\quad = m_f(\xi 2^{-j-1})\overline{m}_{j+1}(\xi 2^{-j-1}) + m_f(\xi 2^{-j-1}+1/2)\overline{m}_{j+1}(\xi 2^{-j-1}+1/2) = 0. \end{aligned}$$

By (8.22), $\overline{m}_{j+1}(\xi)$ and $\overline{m}_{j+1}(\xi+1/2)$ cannot vanish simultaneously; hence there exists a 1-periodic function $\lambda(\xi)$ such that

$$m_f(\xi) = \lambda(\xi)\overline{m}_{j+1}(\xi+1/2) \ \text{ a.e.} \tag{8.26}$$

and $\lambda(\xi)+\lambda(\xi+1/2) = 0$. The latter equality can be rewritten as $\lambda(\xi) = e^{-2\pi i\xi}\nu(2\xi)$, where ν is a 1-periodic function. Thus the Fourier transform of an arbitrary function from W_j has the form

$$\widehat{f}(\xi) = e^{-\pi i\xi 2^{-j}}\overline{m_{j+1}}(\xi 2^{-j-1}+1/2)\,\nu(\xi 2^{-j})\widehat{\varphi}_{j+1}(\xi), \tag{8.27}$$

where ν is a 1-periodic function. In addition,

$$\begin{aligned} \|f\|^2 = \|\widehat{f}\|^2 &= \int_0^{2^j} |\nu(\xi 2^{-j})|^2 \sum_{l\in\mathbb{Z}} |m_{j+1}(\xi 2^{-j-1}+(l+1)2)\widehat{\varphi}_{j+1}(\xi+2^jl)|^2\, d\xi \\ &= \|\nu\|_{L_2([0,\,1])}. \end{aligned}$$

Thus the 1-periodic function ν is square-integrable.

Having description (8.27) of the space W_j, it is not hard to find a function $\psi_j \in W_j$ whose shifts $\{\psi_j(\cdot + k2^{-j})\}_{k\in\mathbf{Z}}$ form an orthonormal basis in W_j. Let ψ_j be the required function. Then we have

$$\widehat{\psi}_j(\xi) = e^{-\pi i\xi 2^{-j}}\overline{m_{j+1}}(\xi 2^{-j-1}+1/2)\nu_{\psi_j}(\xi 2^{-j})\widehat{\varphi}_{j+1}(\xi).$$

Substituting this expression into (8.21) and using (8.22), we get

$$|\nu_{\psi_j}(\xi)|^2 \equiv 1 \text{ a.e.} \tag{8.28}$$

The simplest thing to do is to set $\nu_{\psi_j}(\xi) \equiv 1$. By (8.27) the shifts of the function ψ_j defined by the equality

$$\widehat{\psi}_j(\xi) = e^{-\pi i \xi 2^{-j}} \overline{m_{j+1}}(\xi 2^{-j-1} + 1/2) \widehat{\varphi}_{j+1}(\xi) \tag{8.29}$$

constitute a basis in W_j. Indeed, if

$$\nu(\xi) = \sum_{k \in \mathbf{Z}} \nu_k \, e^{2\pi i k \xi},$$

then

$$\widehat{f}(\xi) = \left(\sum_{k \in \mathbf{Z}} \nu_k \, e^{\pi i k 2^{1-j} \xi} \right) \widehat{\psi}_j(\xi)$$

or

$$f(\cdot) = \sum_{k \in \mathbf{Z}} \nu_k \, \psi_j(\cdot + k 2^{-j}).$$

Note that in view of (8.28) the Fourier transform of any function whose integer shifts constitute an orthonormal basis in W_j differs from the Fourier transform of ψ_j from (8.29) only by a 1-periodic factor equal to 1 in modulus.

Thus, having a nonstationary multiresolution analysis $\{V_j\}_{j \in \mathbf{Z}}$ generated by a scaling sequence $\{\varphi_j\}_{j \in \mathbb{Z}}$, one can construct a nonstationary orthonormal wavelet basis $\{\psi_{jk}\}_{j,\, k \in \mathbf{Z}}$ in $L_2(\mathbf{R})$ possessing property (8.14). ◊

It follows from formula (8.29) that

$$\psi_j(t) = \sum_{k \in \mathbf{Z}} (-1)^{k-1} h_{j+1,-k+1} \cdot \varphi_{j+1,k}(t). \tag{8.30}$$

For the sake of brevity we shall use the formula

$$\psi_j(t) = \sum_{k \in \mathbf{Z}} (-1)^{k} h_{j+1,-k+1} \cdot \varphi_{j+1,k}(t). \tag{8.31}$$

In the nonstationary case the formula for the projector P_j has the following form.

THEOREM 8.1.6. *Let $\{\varphi_j\}_{j \in \mathbb{Z}}$ be the scaling sequence of an NMRA $\{V_j\}_{j \in \mathbb{Z}}$. Then the orthogonal projectors P_j, $j \in \mathbb{Z}$, on V_j satisfy the relations*

$$\widehat{P_j f}(\xi) = \left(\sum_{l \in \mathbb{Z}} \widehat{f}(\xi + l 2^j) \overline{\widehat{\varphi_j}(\xi + l 2^j)} \right) \widehat{\varphi_j}(\xi), \quad f \in L_2(\mathbb{R}). \tag{8.32}$$

`Proof.` It is obvious that

$$\widehat{\varphi_{jk}}(\xi) = e^{\pi i \xi k 2^{1-j}} \widehat{\varphi_j}(\xi).$$

Therefore,

$$\begin{aligned} \langle f, \varphi_{jk} \rangle = \langle \widehat{f}, \widehat{\varphi_{jk}} \rangle &= \int_{\mathbb{R}} \widehat{f}(\xi) e^{-\pi i \xi k 2^{1-j}} \overline{\widehat{\varphi_j}(\xi)} \, d\xi \\ &= \int_0^{2^j} \left(\sum_{l \in \mathbb{Z}} \widehat{f}(\xi + l 2^j) \overline{\widehat{\varphi_j}(\xi + l 2^j)} \right) e^{-\pi i \xi k 2^{1-j}} \, d\xi. \end{aligned}$$

We can interchange summation and integration in the latter equality because $|\widehat{f}|\cdot|\widehat{\varphi_j}|\in L(\mathbb{R})$ by the Cauchy-Bunyakowsky inequality. By the same inequality we have

$$\left|\sum_{l\in\mathbb{Z}}\widehat{f}(\xi+l2^j)\overline{\widehat{\varphi_j}(\xi+l2^j)}\right|^2\le\sum_{l\in\mathbb{Z}}|\widehat{f}(\xi+l2^j)|^2\sum_{l\in\mathbb{Z}}|\widehat{\varphi_j}(\xi+l2^j)|^2.$$

Therefore $\sum_{l\in\mathbb{Z}}\widehat{f}(\cdot+l2^j)\overline{\widehat{\varphi_j}(\cdot+l2^j)}\in L_2([0,2^j])$ and

$$\begin{aligned}\widehat{P_jf}(\xi)&=\sum_{k\in\mathbb{Z}}\langle f,\varphi_{jk}\rangle\widehat{\varphi_{jk}}(\xi)\\&=\left[\sum_{k\in\mathbb{Z}}\left(\int_0^{2^j}(\sum_{l\in\mathbb{Z}}\hat{f}(v+l2^j)\overline{\widehat{\varphi_j}(v+l2^j)})e^{-\pi ivk2^{1-j}}\,dv\right)\cdot e^{\pi i\xi k2^{1-j}}\right]\widehat{\varphi_j}(\xi)\\&=\left(\sum_{l\in\mathbb{Z}}\widehat{f}(\xi+l2^j)\overline{\widehat{\varphi_j}(\xi+l2^j)}\right)\widehat{\varphi_j}(\xi).\end{aligned}$$ ◊

Repeating almost word for word the proof of Theorem 2.3.4, we derive the following result.

THEOREM 8.1.7. *Let a sequence* $\{V_j\}_{j\in\mathbf{Z}}$ *satisfy properties* **MR1** *and* **MR4**. *Then* $\overline{\bigcup_{j\in\mathbf{Z}}V_j}=L_2(\mathbf{R})$ *if and only if* $\Omega_0:=\bigcup_{j\in\mathbf{Z}}\operatorname{supp}\widehat{\varphi}_j=\mathbf{R}$.

8.2. Nonstationary infinitely differentiable orthonormal wavelets with compact support

In this section, we describe a method for the construction of nonstationary orthonormal wavelet bases. This method allows us to obtain nonstationary orthonormal, infinitely differentiable wavelets with compact support. Furthermore, the length of support of the generator of the j-th level in this system is less than $2T(j+2)2^{-j}$, where $\{T(j)\}_{j=1}^{\infty}$ is a sequence of positive integers which satisfy the following conditions:

$$\lim_{j\to\infty}T(j)=\infty,\quad T(j)\le T(j+1)\le T(j)+1,\quad T(1)=1.$$

First of all we note that such a construction is impossible in the stationary case.

THEOREM 8.2.1. *There exists no stationary orthogonal wavelet basis consisting of infinitely differentiable functions with compact support.*

`Proof.` Assume the converse: let ψ be an orthogonal infinitely differentiable wavelet function with compact support. Then by Theorem 1.7.7,

$$\int_{\mathbb{R}}t^l\psi(t)\,dt=0$$

for all $l\in\mathbb{N}$, which is equivalent to the fact that

$$\widehat{\psi}^{(l)}(0)=0,\quad l\in\mathbb{N}.$$

The latter is impossible since by the Paley-Wiener theorem (see Appendix A.7), $\widehat{\psi}$ is an entire function. $\lozenge$

The main result of this section is the following assertion.

THEOREM 8.2.2. *Let* $\{T(N)\}_{N=1}^{\infty}$ *be an arbitrary sequence of positive integers such that*

$$\lim_{N\to\infty} T(N) = \infty, \tag{8.33}$$

$$T(N) \le T(N+1) \le T(N)+1, \quad T(1)=1. \tag{8.34}$$

Then there exists a system of nonstationary wavelets

$$\Psi := \{\varphi_{0k},\ \psi_{jk},\ j,k\in\mathbb{Z},\ j\ge 0\}$$

having the following properties:

i) Ψ *is an orthonormal basis in* $L_2(\mathbb{R})$.

ii) $\varphi_{0k}(t)=\varphi_{00}(t+k)$; $\psi_{jk}(t)=\psi_{j0}(t+k2^{-j})$; $t\in\mathbb{R}$; $j,k\in\mathbb{Z}$, $j\ge 0$.

iii) $\Psi\subset C^{\infty}(\mathbb{R})$.

iv) $\operatorname{supp}\varphi_{00}\subset[-2T(1)-1,0]$; $\operatorname{supp}\psi_{j0}\subset[-(T(j+2)+1)2^{-j},(2T(j+1)-1)2^{-j-1}]$, $j\in\mathbb{Z}$, $j\ge 0$.

This theorem is proved by Lemmas 8.2.7, 8.2.8, 8.2.9, and 8.2.10 given below. First describe the construction of the system Ψ. Let

$$d_N(\xi)=2^{-1/2}\sum_{l=0}^{2N-1} h_N(l)e^{2\pi i l\xi}, \quad h_N(l)\in\mathbb{R}, \quad N\in\mathbb{N},$$

be the Daubechies masks (see (4.17)). It is clear that

$$d_N(\xi)=\left(\frac{1+\exp(2\pi i\xi)}{2}\right)^N Q_N(\exp(2\pi i\xi)),$$

where the trigonometric polynomial

$$Q_N(\xi)=\sum_{l=0}^{N-1} q_n(l)\xi^l, \quad q_n(0)\neq 0, \tag{8.35}$$

satisfies the identity

$$|Q_N(\exp(2\pi i\xi))|^2 = P_N(\sin^2(\pi\xi)) = \sum_{l=0}^{N-1}\binom{N-1+l}{l}\sin^{2l}(\pi\xi),\ \xi\in\mathbb{R}. \tag{8.36}$$

REMARK 8.2.3. The polynomials Q_N are not defined uniquely, but this is of no significance for the construction (see Remark 4.1.5).

As usual, set $h_N(l)=0$ for $l\notin\{0,1,\cdots,2N-1\}$. The identity

$$|d_N(\xi)|^2+|d_N(\xi+1/2)|^2=1 \tag{8.37}$$

is equivalent to the following properties of the coefficients $h_N(l)$:

$$\sum_{l\in\mathbb{N}} h_N(l)=2^{1/2}, \tag{8.38}$$

$$\sum_{l\in\mathbb{N}} h_N(l)h_N(l+2m)=\delta_{0,m}, \quad m\in\mathbb{Z}. \tag{8.39}$$

LEMMA 8.2.4. *The infinite products*

$$G_j(\xi) := \prod_{N=j+1}^{\infty} d_{T(N)}(2^{-N}\xi), \quad j \in \mathbb{Z}_+, \tag{8.40}$$

converge pointwise for all $\xi \in \mathbb{R}$. The convergence is uniform on compact sets.

Proof. By virtue of (8.38) and (8.39) we have

$$\begin{aligned}
|d_N(\xi) - 1| &\leq 2^{-1/2} \sum_{l=0}^{2N-1} |h_N(l)(\exp(2\pi i l\xi) - 1)| \\
&= 2^{1/2} \sum_{l=0}^{2N-1} |h_N(l) \sin(\pi l \xi)| \\
&\leq \pi 2^{1/2} \sum_{l=0}^{2N-1} l\,|h_N(l)\xi| \\
&\leq \pi 2^{1/2} \left(\sum_{l=0}^{2N-1} l^2\right)^{1/2} \left(\sum_{l=0}^{2N-1} (h_N(l))^2\right)^{1/2} |\xi| \\
&\leq \pi 2^{3/2} N^{3/2} |\xi|.
\end{aligned} \tag{8.41}$$

Condition (8.34) means that

$$T(N) \leq N, \quad N \in \mathbb{N}; \tag{8.42}$$

hence inequality (8.41) implies the convergence of $G_j(\xi)$ for any $\xi \in \mathbb{R}$. It is clear that (8.41), (8.42) imply the uniform convergence on compact sets. ◊

Let $I_r := [2^{r-2},\ 2^{r-1}]$, $r \in \mathbb{N}$.

LEMMA 8.2.5. *Let $j \in \mathbb{Z}_+$. If $|\xi| \in I_r$ for $r > j$, $r \in \mathbb{N}$, then*

$$|G_j(\xi)| \leq \alpha_j \exp\left(-\beta \sum_{N=j+1}^{r} T(N)\right), \tag{8.43}$$

where $\beta > 0$ and $\alpha_j > 0$ depends only on j.

Proof. Fix positive integer $r > j$ and set

$$G_{jr}(\xi) := \prod_{N=j+1}^{r} d_{T(N)}(\xi 2^{-N}),$$

$$A_{jr}(\xi) := \prod_{N=j+1}^{r} \left|\frac{1 + \exp(\pi i \xi 2^{1-N})}{2}\right|^{T(N)} = \prod_{N=j+1}^{r} |\cos \pi\xi 2^{-N}|^{T(N)},$$

$$B_{jr}(\xi) := \prod_{N=j+1}^{r} |P_{T(N)}(\sin^2(\pi\xi 2^{-N}))|^{1/2}.$$

By virtue of (8.37) we have

$$|G_j(\xi)| \leq |G_{jr}(\xi)|, \quad \xi \in \mathbb{R},\ j, r \in \mathbb{Z}_+,\ r > j. \tag{8.44}$$

Using the identity $\sin 2\xi = 2\sin\xi\cos\xi$, we get

(8.45)

$$\begin{aligned}A_{jr}(\xi) &= \prod_{N=j+1}^{r}\left|\frac{\sin\pi\xi 2^{1-N}}{2\sin\pi\xi 2^{-N}}\right|^{T(N)} \\ &= 2^{-\sum_{N=j+1}^{r} T(N)} \\ &\quad\times |\sin\pi\xi 2^{-j}|^{T(j+1)}\prod_{N=j+2}^{r}|\sin\pi\xi 2^{1-N}|^{T(N)-T(N-1)}|\sin\pi\xi 2^{-r}|^{-T(r)} \\ &\le 2^{-\sum_{N=j+1}^{r} T(N)}|\sin\pi\xi 2^{-r}|^{-T(r)}.\end{aligned}$$

If $|\xi|\in I_r$, then $|\sin\pi 2^{-r}\xi|\ge 2^{-1/2}$. It follows that

$$A_{jr}(\xi)\le 2^{-\sum_{N=j+1}^{r} T(N)+T(r)/2}, \qquad \xi\in I_r. \tag{8.46}$$

Since $\binom{N-1+l}{l}\le\binom{N+l}{l}$, $l=0,1,\cdots,N-1$, it follows that for any $y\ge 0$,

$$P_N(y)\le P_{N+1}(y), \qquad N\in\mathbb{N}. \tag{8.47}$$

In (7.29) and (7.30) it is proved that

$$\sup_{t\in\mathbb{R}}|P_r(\sin^2\pi t)P_r(\sin^2 2\pi t)|\le 2^{4(r-1)}d^{r-1}, \tag{8.48}$$

$$\sup_{t\in\mathbb{R}}\left|P_r(\sin^2\pi t)\right| < 2^{2(r-1)}, \tag{8.49}$$

where $d := 16/27$. If $r-j=2l$ for some integer l, then by (8.47), (8.48), and (8.34) we have

(8.50)

$$\begin{aligned}B_{jr}(\xi) &= \prod_{N=1}^{l}|P_{T(j+2N-1)}(\sin^2(\pi\xi 2^{-j-2N+1}))P_{T(j+2N)}(\sin^2(\pi\xi 2^{-j-2N}))|^{1/2} \\ &\le \prod_{N=1}^{l}2^{2(T(j+2N)-1)}d^{(T(j+2N)-1)/2}\le\prod_{N=j+1}^{r}2^{T(N)}d^{(T(N)-1)/4}.\end{aligned}$$

If $r-j=2l-1$ for some integer l, then by (8.47), (8.48), (8.49), and (8.34) we obtain

(8.51)

$$\begin{aligned}B_{jr}(\xi) &= |P_{T(j+1)}(\sin^2(\pi\xi 2^{-j-1})) \\ &\quad\times\prod_{N=1}^{l-1}P_{T(j+2N)}(\sin^2(\pi\xi 2^{-j-2N}))P_{T(j+2N+1)}(\sin^2(\pi\xi 2^{-j-2N-1}))|^{1/2} \\ &\le 2^{T(j+1)-1}\prod_{N=1}^{l-1}2^{2(T(j+2N+1)-1)}d^{(T(j+2N+1)-1)/2} \\ &\le d^{(1-T(j+1))/4}\prod_{N=j+1}^{r}2^{T(N)}d^{(T(N)-1)/4}.\end{aligned}$$

Using (8.46), (8.50), and (8.51), we get (8.43). ◊

COROLLARY 8.2.6. *The functions G_j, $j \in \mathbb{Z}_+$, decrease faster at infinity than any negative power.*

`Proof`. Condition (8.33) means that

$$\lim_{r\to\infty} \sum_{N=1}^{r} T(N)/r = \infty.$$

Thus our statement follows from (8.43). ◊

Define a sequence of functions $\{\varphi_{jk}\}_{j,k\in\mathbb{Z},\ j\geq 0}$ by the Fourier transform:

$$\widehat{\varphi}_j(\xi) := 2^{-j/2} G_j(\xi), \quad j \in \mathbb{Z}_+,\ \xi \in \mathbb{R},$$

(8.52)

$$\varphi_{jk}(t) := \varphi_j(t + k2^{-j}), \quad k \in \mathbb{Z},\ t \in \mathbb{R}.$$

In Lemma 8.2.9 it will be proved that the sequence of functions $\{\varphi_j\}_{j\in\mathbb{Z}_+}$ is a nonstationary scaling sequence (see Definition 8.1.4). The nonstationary wavelets ψ_{jk}, $j \in \mathbb{Z}_+$, $k \in \mathbb{Z}$, are defined by the equalities

$$\widehat{\psi}_j(\xi) := 2^{1/2} \exp(-i\xi 2^{-j-1}) d_{T(j+1)}(-\xi 2^{-j-1} - 1/2)\widehat{\varphi}_{j+1}(\xi), \quad \xi \in \mathbb{R},$$

(8.53)

$$\psi_{jk}(t) := \psi_j(t + k2^{-j}), \quad k \in \mathbb{Z},\ t \in \mathbb{R},$$

according to Theorem 8.1.5. Thus the construction of the system Ψ is complete. Set

$$\Psi := \{\varphi_{0k},\ \psi_{jk},\ j \in \mathbb{Z}_+,\ k \in \mathbb{Z}\}.$$

Let us now study the properties of the system Ψ.

LEMMA 8.2.7. *The functions φ_{jk} and ψ_{jk}, $j \in \mathbb{Z}_+$, $k \in \mathbb{Z}$, are infinitely differentiable.*

`Proof`. The infinite differentiability of a function is equivalent to the fact that its Fourier transform decreases at infinity more rapidly than any negative power. So the statement follows from Corollary 8.2.6. ◊

LEMMA 8.2.8. *For any $j \in \mathbb{Z}_+$, k, $k' \in \mathbb{Z}$, we have*

$$\langle \varphi_{jk}, \varphi_{jk'} \rangle = \delta_{k,k'}, \tag{8.54}$$

$$\langle \psi_{jk}, \psi_{jk'} \rangle = \delta_{k,k'}, \quad \langle \varphi_{jk}, \psi_{jk'} \rangle = 0. \tag{8.55}$$

`Proof`. Fix positive integer $j \geq 0$ and integer $r > j$. For $\nu = 1, \dots, r - j$, define the auxiliary functions

$$\eta_{r\nu}^{(j)}(\xi) := 2^{-j/2} \prod_{l=1}^{\nu} d_{T(r-\nu+l)}(\xi 2^{-j-l}) \chi_{[-1/2,1/2]}(2^{-j-\nu}\xi),$$

$$\widehat{\varphi_{r\nu}^{(j)}} := \eta_{r\nu}^{(j)}, \quad \varphi_{r,\nu,k}^{(j)}(t) = \varphi_{r\nu}^{(j)}(t + k2^{-j}).$$

Note that

$$\eta_{r,\nu+1}^{(j)}(\xi) = d_{T(r-\nu)}(\xi 2^{-j-1}) \eta_{r\nu}^{(j)}(\xi/2)$$

or

$$\varphi_{r,\nu+1}^{(j)}(t) = \sqrt{2} \sum_{k=0}^{2T(r-\nu)-1} h_{T(r-\nu)}(k) \varphi_{r\nu}^{(j)}(2t + k2^{-j}). \tag{8.56}$$

By (8.37) for any $k \in \mathbb{Z}$, we have

$$\langle \varphi^{(j)}_{r,1,0}, \varphi^{(j)}_{r,1,k} \rangle = 2^{-j} \int_{\mathbb{R}} \exp(-ik\pi\xi 2^{1-j}) |d_{T(r)}(\xi 2^{-j-1})|^2 \chi_{[-1/2,1/2]}(2^{-j-1}\xi) d\xi$$
$$= \int_0^1 \exp(-ik2\pi\xi)(|d_{T(r)}(\xi)|^2 + |d_{T(r)}(\xi + 1/2)|^2) d\xi = \delta_{k,0}.$$

On the other hand, if

$$\langle \varphi^{(j)}_{r,\nu,0}, \varphi^{(j)}_{r,\nu,k} \rangle = \delta_{k,0},$$

then by (8.56) and (8.39) we get

$$\langle \varphi^{(j)}_{r,\nu+1,0}, \varphi^{(j)}_{r,\nu+1,k} \rangle = 2 \sum_{l,m \in \mathbb{Z}} h_{T(r-\nu)}(l) h_{T(r-\nu)}(m)$$
$$\times \int_{\mathbb{R}} \varphi^{(j)}_{r\nu}(2t + m2^{-j}) \varphi^{(j)}_{r\nu}(2t + l2^{-j} + 2k2^{-j}) dt$$
$$= \sum_{l \in \mathbb{Z}} h_{T(r-\nu)}(l) h_{T(r-\nu)}(l + 2k) = \delta_{k,0}.$$

By induction we obtain

$$\langle \varphi^{(j)}_{r,r-j,0}, \varphi^{(j)}_{r,r-j,k} \rangle = \delta_{k,0}, \quad k \in \mathbb{Z}.$$

Thus (8.54) will be proved if

$$\lim_{r \to \infty} \eta^{(j)}_{r,r-j} = \widehat{\varphi_j}$$

in the sense of convergence in $L_2(\mathbb{R})$. The estimates obtained in Lemma 8.2.5 and (8.39) imply that for any $j \in \mathbb{Z}_+$,

$$\sup_{r>j} |\eta^{(j)}_{r,r-j}| \in L_2(\mathbb{R}).$$

Since $\eta^{(j)}_{r,r-j}$ converges to $\widehat{\varphi_j}$ pointwise by Lemma 8.2.4, it follows that $\eta^{(j)}_{r,r-j}$ converges to $\widehat{\varphi_j}$ in $L_2(\mathbb{R})$ by Lebesgue's dominated convergence theorem. Thus (8.54) is proved.

Let $j \in \mathbb{Z}_+$ and $g_j(l) := (-1)^l h_j(1-l)$, $l \in \mathbb{Z}$. By (8.38) and (8.39) we have

$$\sum_{l \in Z} g_j(l) h_j(l + 2k) = 0,$$
$$\sum_{l \in Z} g_j(l) g_j(l + 2k) = \delta_{k,0}, \quad k \in \mathbb{Z}. \tag{8.57}$$

It follows from the definition of φ_{jk} and ψ_{jk} that

$$\varphi_{jk}(t) = \sum_{l \in \mathbb{Z}} h_{T(j+1)}(l + 2k) \varphi_{j+1,l}(t),$$
$$\psi_{jk}(x) = \sum_{l \in \mathbb{Z}} g_{T(j+1)}(l + 2k) \varphi_{j+1,l}(t), \quad j \in \mathbb{Z}_+, \quad k \in \mathbb{Z}. \tag{8.58}$$

Using (8.54), (8.57), and (8.58), we get (8.55). ◊

Consider the subspaces

$$V_j := \overline{\operatorname{span}\{\varphi_{jk}, k \in \mathbb{Z}\}}, \quad W_j := \overline{\operatorname{span}\{\psi_{jk},\ k \in Z\}}, \quad j \in \mathbb{Z}_+. \tag{8.59}$$

By Lemma 8.2.8 the systems $\{\varphi_{jk}\}_{k\in\mathbb{Z}}$ and $\{\psi_{jk}\}_{k\in Z}$ are orthonormal bases in V_j and W_j, respectively, $j \in \mathbb{Z}_+$.

LEMMA 8.2.9. *For any* $j \in \mathbb{Z}_+$,

$$V_j \subset V_{j+1}, \quad V_j \oplus W_j = V_{j+1}, \tag{8.60}$$

$$\overline{\bigcup_{j=0}^{\infty} V_j} = L_2(\mathbb{R}). \tag{8.61}$$

Proof. The inclusions $V_j \subset V_{j+1}$ and $W_j \subset V_{j+1}$ follow from (8.58). By virtue of (8.39) and (8.57), for any $j \in \mathbb{Z}_+$ and any $m \in \mathbb{Z}$, we have

$$\varphi_{j+1,m} = \sum_{l\in\mathbb{Z}} \left(h_{T(j+1)}(m+2l)\varphi_{jl} + g_{T(j+1)}(m+2l)\psi_{jl}\right).$$

This equality together with (8.55) completes the proof of (8.60). As for (8.61), it follows from Theorem 8.1.7. ◊

LEMMA 8.2.10. *The functions* φ_j *and* ψ_j, $j \in \mathbb{Z}_+$, *have compact support, and*

$$\operatorname{supp}\varphi_j \subseteq [-(2T(j+1)+1)2^{-j},\, 0], \tag{8.62}$$

$$\operatorname{supp}\psi_j \subseteq [-(T(j+2)+1)2^{-j},\ (2T(j+1)-1)2^{-j-1}]. \tag{8.63}$$

Proof. Let us prove that $G_j(\xi)$ is an entire function of exponential type:

$$|G_j(\xi)| \le C_1 \qquad \text{for} \quad \operatorname{Im}\xi \ge 0, \tag{8.64}$$

$$|G_j(\xi)| \le C_2 \exp(-\pi(2T(j+1)+1)2^{1-j}\operatorname{Im}\xi) \quad \text{for} \quad \operatorname{Im}\xi < 0, \tag{8.65}$$

for some constants C_1, C_2. By the Paley-Wiener theorem (see Appendix A.7) these estimates imply (8.62). First of all note the following inequalities based on (8.38). If $\operatorname{Im}\xi \ge 0$, then

$$\begin{aligned} |d_N(\xi)| &\le \frac{1}{\sqrt{2}} \sum_{l=0}^{2N-1} |h_N(l)| \\ &\le \frac{1}{\sqrt{2}} \left(\sum_{l=0}^{2N-1} |h_N(l)|^2\right)^{1/2} (2N-1)^{1/2} \le N^{1/2}, \\ |d_N(\xi) - d_N(\operatorname{Re}\xi)| &\le \frac{1}{\sqrt{2}} \sum_{l=0}^{2N-1} |h_N(l)| \cdot |\exp(-2\pi l \operatorname{Im}\xi) - 1| \\ &\le \sqrt{2}\pi \sum_{l=0}^{2N-1} |h_N(l)| \cdot l \cdot \operatorname{Im}\xi \le 4\pi N^{3/2} \operatorname{Im}\xi. \end{aligned} \tag{8.66}$$

For $\operatorname{Im}\xi \ge 0$, the latter inequality together with (8.37) imply that

$$|d_N(\xi)| \le 1 + 4\pi N^{3/2} \operatorname{Im}\xi. \tag{8.67}$$

If $\operatorname{Im}\xi \geq 1$, then

$$(8.68)\qquad |d_N(\xi)| \leq 2^{-1/2}\left(\sum_{l=0}^{2N-1} |h_N(l)|^2\right)^{1/2}\left(\sum_{l=0}^{2N-1}\exp(-4\pi l)\right)^{1/2} \leq a < 1,$$

where $a := (2(1-\exp(-4\pi)))^{-1/2}$.

Now we can prove (8.64). Fix a positive integer N_1 such that for any $N > N_1$,

$$(8.69)\qquad T(N) \geq 2,$$

$$(8.70)\qquad (N + 2\log_2 N)^{3/2} \leq N^2,$$

$$(8.71)\qquad 2\log_2 N \leq N,$$

$$(8.72)\qquad \log_2 N \ln 2N \leq (1-N)\ln a.$$

If $\operatorname{Im}\xi \in [0, 2^{N_1}]$, then by (8.67) we have

$$(8.73)\qquad |G_0(\xi)| \leq \prod_{N=1}^{\infty}\left(1 + 4\pi(T(N))^{3/2}2^{N_1-N}\right).$$

Let $\operatorname{Im}\xi > 2^{N_1}$. Fix minimal integer N_2 such that $\operatorname{Im}\xi \leq 2^{N_2}$. Then by (8.68) we have

$$(8.74)\qquad \prod_{N=1}^{N_2-1} |d_{T(N)}(\xi 2^{-N})| \leq a^{N_2-1}.$$

Let N_3 be the minimal integer number such that

$$(8.75)\qquad (N_3)^{3/2}2^{-N_3}\operatorname{Im}\xi < 1.$$

By (8.70) we get

$$(8.76)\qquad N_3 \leq N_2 + 2\log_2 N_2.$$

Using (8.66), (8.42), (8.71), and (8.76), we have

$$(8.77)\qquad \begin{aligned}\prod_{N=N_2}^{N_3-1} |d_{T(N)}(\xi 2^{-N})| &\leq \prod_{N=N_2}^{N_3-1} (T(N))^{1/2} \leq N_3^{(N_3-N_2)/2}\\ &\leq \exp(\ln 2N_2 \log_2 N_2) \leq a^{1-N_2}.\end{aligned}$$

It follows from (8.67), (8.75), and (8.69) that

$$(8.78)\qquad \begin{aligned}\prod_{N=N_3}^{\infty} |d_{T(N)}(\xi 2^{-N})| &\leq \prod_{N=N_3}^{\infty} (1 + 4\pi(T(N))^{3/2}2^{-N}\operatorname{Im}\xi)\\ &\leq \prod_{N=N_3}^{\infty}\left(1 + \pi 2^{2+N_3-N}\left(\frac{N}{N_3}\right)^{3/2}\right) \leq \prod_{N=2}^{\infty}\left(1 + \pi 2^{4-N}\left(\frac{N}{2}\right)^{3/2}\right).\end{aligned}$$

Inequalities (8.73), (8.74), (8.77), and (8.78) imply (8.64) for $j = 0$. The proof of (8.64) for $j > 0$ is quite similar.

Note that by (8.34) we have

$$\sum_{N=j+1}^{\infty} (2T(N)-1)2^{-N} \le 2^{-j}(2T(j+1)+1).$$

Thus estimate (8.65) follows from the equality

$$\begin{aligned} G_j(\xi) = &\prod_{N=j+1}^{\infty} 2^{-1/2}\exp(\pi i \xi 2^{1-N}(2T(N)-1)) \\ &\times \sum_{l=0}^{2T(N)-1} h_N(l)\exp(\pi i(l-(2T(N)-1))\xi 2^{1-N}) \\ = &\exp(\pi i \xi \sum_{N=j+1}^{\infty} 2^{2-N}(2T(N)-1)) \\ &\times \prod_{N=j+1}^{\infty} 2^{-1/2} \sum_{l=0}^{2T(N)-1} h_N(l)\exp(\pi i(l-(2T(N)-1))\xi 2^{1-N}). \end{aligned}$$

The inclusion (8.63) is proved similarly. ◊

REMARK 8.2.11. By using the tensor products of elements Ψ, one can construct a nonstationary orthonormal wavelet basis in $L^2(\mathbb{R}^d)$, $d > 1$, consisting of infinitely differentiable functions with compact support.

8.3. Decay rate of the Fourier transforms of elements of a nonstationary scaling sequence

By definition, *the rate of decrease* of a given function $g(\xi)$, $\xi \in \mathbb{R}$ (denoted by $\mathbf{d}(g)$) is the maximal integer l such that $g(\xi) = o(\xi^{-l})$ for $|\xi| \to \infty$; if g decreases more rapidly than any power function, we set $\mathbf{d}(g) = +\infty$; if such integers l do not exist, then we set $\mathbf{d}(g) = -\infty$. For an arbitrary sequence of trigonometric polynomials $\{m_k\}_{k\in\mathbb{N}}$ such that $m_k(0) = 1$, $k \in \mathbb{N}$, define a nonstationary scaling function φ by the formula

$$\widehat{\varphi}(\xi) = \prod_{k=1}^{\infty} m_k(2^{-k}\xi). \tag{8.79}$$

Each polynomial has the form $m_k(\xi) = \sum_{s=0}^{N_k} c_s^{(k)} e^{-2\pi i s}$, where $c_{N_k}^{(k)} \neq 0$ and $N_k = \deg m_k$ is the degree of the polynomial. Under certain rather general conditions (e.g., when the norms and degrees of all the polynomials are uniformly bounded) product (8.79) converges uniformly on each compact set and thus represents an analytic function. How rapidly can this function decrease at infinity? If the sequence $\{m_k\}$ is stationary, i.e., all the polynomials are equal, then the maximal rate of decrease corresponds to the polynomial $m(\xi) = w_{n-1}(\xi) = \left(\frac{e^{-2\pi i}+1}{2}\right)^n$, i.e., the mask of the equation for the B-spline B_{n-1} (in this section we shall denote this mask by w_{n-1}). Recall that in this case $\mathbf{d} = n-1$; moreover, $\widehat{B}_{n-1}(\xi) = O(\xi^{-n})$, $n \to \infty$ (see Corollary 3.2.14 and Example 3.5.1). This brings up the question: how rapidly can function (8.79) decrease in the general case, when the polynomials m_k, $k \in \mathbb{N}$, are different and chosen arbitrarily? Is it possible to obtain a higher rate of growth by choosing a suitable sequence of polynomials of degree at most n? Theorem 8.3.1

proved below gives a negative answer to this question. It appears that the rate of decrease not only cannot be infinite, but as before, it cannot exceed $n-1$. In particular, this means that the smoothness of nonstationary wavelets cannot exceed the maximal length of support of the functions $\varphi_j(2^{-j}t)$, $j \geq 0$. Thus in the nonstationary case the situation is similar to that for the stationary wavelets.

Thus the smoothness of B-splines is maximal for a wide class of polynomial products: products of the form (8.79) for all possible sequences of the polynomials $\{m_k\}$ (with uniformly bounded degrees and norms).

THEOREM 8.3.1. *Let there be given a sequence of trigonometric polynomials $\{m_k\}_{k\in\mathbb{N}}$ and numbers $n \in \mathbb{N}$, $M \geq 1$ such that*

$$m_k(0) = 1, \ \deg m_k \leq n, \ \|m_k(\cdot)\|_\infty \leq M \quad \textit{for all } k \in \mathbb{N}.$$

Then the rate of decrease of the function $\widehat{\varphi}(\xi) = \prod_{k=1}^\infty m_k(2^{-k}\xi)$ does not exceed $n-1$. Moreover, if the rate of decrease is equal to $n-1$, then $m_k \to w_{n-1}$ as $k \to \infty$.

Proof. Assume that the rate of decrease is no less than $n-1$. This means that

$$\xi^{n-1}\widehat{\varphi}(\xi) \to 0 \quad \text{as} \quad |\xi| \to \infty. \tag{8.80}$$

Set $\sigma = \frac{1}{4\pi n M}$. For each $\xi \in [-\sigma, \sigma]$, $j \geq 0$, and $N \in \mathbb{N}$ or $N = +\infty$, we have

$$\left| \prod_{k=1}^{N} m_{j+k}(2^{-k}\xi) \right| > \frac{1}{2}. \tag{8.81}$$

Indeed, by Bernstein's inequality $\|m'_{j+k}\|_\infty \leq 2\pi n M$; hence,

$$\begin{aligned} \left|m_{j+k}(2^{-k}\xi)\right| &\geq |m_{j+k}(0)| - 2^{-k}|\xi| \cdot \|m'_{j+k}\|_\infty \\ &\geq 1 - 2\pi n M 2^{-k}|\xi| \geq 1 - 2\pi n M 2^{-k}\sigma = 1 - 2^{-k-1}. \end{aligned}$$

Therefore $\left| \prod_{k=1}^{N} m_{j+k}(2^{-k}\xi) \right| \geq \prod_{k=1}^{N} \left(1 - 2^{-k-1}\right) > \frac{1}{2}$. Denote by p the least integer such that $2^p \geq 8\pi n M$. For arbitrary $\delta \in (0, \sigma)$ and $l \in \mathbb{N}$, we obtain $\widehat{\varphi}(2^l + \delta) =$

$$\prod_{k=1}^{l} m_k\left(2^{l-k} + 2^{-k}\delta\right) \prod_{k=1}^{p} m_{l+k}\left(2^{-k} + 2^{-l-k}\delta\right) \prod_{k=1}^{\infty} m_{l+p+k}\left(2^{-p-k} + 2^{-l-p-k}\delta\right).$$

Using (8.81) with $\xi = \delta$, $j = 0$, and $N = l$, we get

$$\left| \prod_{k=1}^{l} m_k\left(2^{l-k} + 2^{-k}\delta\right) \right| = \left| \prod_{k=1}^{l} m_k\left(2^{-k}\delta\right) \right| > \frac{1}{2};$$

now using (8.81) with $\xi = 2^{-p} + 2^{-l-p}\delta$, $j = l + p$, and $N = +\infty$, we get

$$\left| \prod_{k=1}^{\infty} m_{l+p+k}\left(2^{-p-k} + 2^{-l-p-k}\delta\right) \right| > \frac{1}{2}.$$

Thus,

$$|\widehat{\varphi}(2^l + \delta)| \geq \frac{1}{4} \left| m_{l+1}\left(\frac{1}{2} + 2^{-l-1}\delta\right) m_{l+2}\left(\frac{1}{4} + 2^{-l-2}\delta\right) \cdots m_{l+p}\left(2^{-p} + 2^{-l-p}\delta\right) \right|. \tag{8.82}$$

By assumption, $2^{l(n-1)} f(2^l+\delta) \to 0$ as $l \to \infty$; therefore,

$$(8.83)\qquad 2^{l(n-1)} m_{l+1}\Big(\frac{1}{2}+2^{-l-1}\delta\Big) m_{l+2}\Big(\frac{1}{4}+2^{-l-2}\delta\Big) \cdots m_{l+p}\Big(2^{-p}+2^{-l-p}\delta\Big) \to 0$$

as $l \to \infty$. The latter relation holds for all $\delta \in (0,\sigma)$. Choose positive R and consider closed circles of radius R with centers at the points $\Big\{e^{-2^{1-k}\pi i}\Big\}_{k=1}^{p}$ and $\Big\{e^{-3\cdot 2^{1-k}\pi i}\Big\}_{k=1}^{p}$. Denote these circles by $\Big\{\beta_k\Big\}_{k=1}^{p}$ and $\Big\{\gamma_k\Big\}_{k=1}^{p}$, respectively. Thus we obtain a family of $2p$ equal circles, where two circles β_1 and γ_1 coincide. Take R so small that all other circles do not intersect.

Take arbitrary $l \geq 0$ and consider the chain of corresponding algebraic polynomials $\mathbf{m}_{l+1}, \dots, \mathbf{m}_{l+p}$. Let the circle β_k contain s_k roots of the polynomial $\mathbf{m}_{l+k}$ (taking into account the multiplicities of the roots). If $s_k > 0$, denote these roots by $z_1^{(k)}, \dots, z_{s_k}^{(k)}$ and all other roots $\mathbf{m}_{l+k}$ by $z_{s_k+1}^{(k)}, \dots, z_{n_k}^{(k)}$, where $n_k = \deg \mathbf{m}_{l+k}$. We have

$$\mathbf{m}_{l+k}(z) = A_k \mathbf{f}_k(z) \mathbf{q}_k(z),$$

where

$$\mathbf{f}_k(z) = \prod_{j=1}^{s_k} (z - z_j^{(k)}), \qquad \mathbf{q}_k(z) = \prod_{\nu=s_k+1}^{n_k} \left(\frac{z-b_k}{z_\nu^{(k)} - b_k} - 1 \right),$$

$b_k = e^{-2^{1-k}\pi i}$ is the center of the circle β_k, and A_k is a constant (if $s_k = 0$, we set $\mathbf{f}_k = 1$; if $s_k = n_k$, we set $\mathbf{q}_k = 1$). Obviously, the quantities s_k, A_k and the polynomials $\mathbf{f}_k, \mathbf{q}_k$ depend not only on k but on l as well. For brevity of notation, we shall not write this. For any point z on the unit circle, we have $|z - z_j| \leq 1 + R$ for $j = 1, \dots, s_k$ and

$$\left| \frac{z-b_k}{z_\nu^{(k)} - b_k} - 1 \right| < \frac{2}{R} + 1$$

for $\nu = s_k + 1, \dots, n_k$. Consequently,

$$\|\mathbf{f}_k\|_\infty \leq (1+R)^{s_k}$$

and

$$\|\mathbf{q}_k\|_\infty \leq \left(\frac{2}{R} + 1\right)^{n_k - s_k}.$$

Here, by $\|\mathbf{p}(\cdot)\|_\infty$ we denote the norm of an algebraic polynomial on the unit circle: $\|\mathbf{p}(\cdot)\|_\infty = \big\{|\mathbf{p}(z)|,\ |z| = 1\big\}$. By using the inequality $\|\mathbf{m}_{l+k}\|_\infty \geq |\mathbf{m}_{l+k}(1)| = 1$, we get

$$(8.84)\qquad \begin{aligned} |A_k| &\geq \frac{\|\mathbf{m}_{l+k}\|_\infty}{\|\mathbf{f}_k\|_\infty \cdot \|\mathbf{q}_k\|_\infty} \geq \frac{1}{(1+R)^{s_k}\left(\frac{2}{R}+1\right)^{n_k - s_k}} \\ &\geq \frac{1}{(1+R)^n\left(\frac{2}{R}+1\right)^n} = \left(\frac{2}{R} + R + 3\right)^{-n}. \end{aligned}$$

Finally, denote by $\bar\beta_k$ the closed circle of radius $R/2$ with center at the point b_k. For each $z \in \bar\beta_k$, $\nu = s_k + 1, \dots, n_k$, we have

$$\left| \frac{z-b_k}{z_\nu^{(k)} - b_k} \right| < \frac{1}{2};$$

hence $|\mathbf{q}_k(z)| > 2^{-n}$. Combining this with (8.84), we obtain

$$|\mathbf{m}_{l+k}(z)| \geq 2^{-n}\Big(\frac{2}{R} + R + 3\Big)^{-n} |f_k(z)| \quad \text{for each } z \in \bar\beta_k.$$

It follows that for sufficiently large l, namely, such that $2\pi \cdot 2^{-l}\sigma \le R$, we have

$$\left|\prod_{k=1}^{p} m_{l+k}\left(2^{-k}+2^{-l-k}\delta\right)\right| \ge \left[2^{-n}\left(\frac{2}{R}+R+3\right)^{-n}\right]^{p}\left|\prod_{k=1}^{p} f_k\left(2^{-k}+2^{-l-k}\delta\right)\right|,$$

where f_k is a trigonometric polynomial corresponding to the polynomial $\mathbf{f}_k$, i.e., $f(\xi)=\mathbf{f}_k(e^{-2\pi i\xi})$. Substituting this into (8.83), we get

$$2^{l(n-1)}\prod_{k=1}^{p} f_k\left(2^{-k}+2^{-l-k}\delta\right)\to 0 \quad \text{as } l\to\infty. \tag{8.85}$$

To complete the proof, consider the circles of radius $r=\frac{\sigma}{2^{l+p-1}S}$, where $S=\sum_{k=1}^{p}s_k$, with centers at the roots of the polynomials $\prod_{k=1}^{p}\mathbf{f}_k(z)$. Each of these circles cuts out an arc of length less than πr on the unit circle. Therefore the total length of the set

$$\Delta_1=\left\{\delta\in(0,\sigma),\quad \exists j\le s_1 \ \ |e^{-2\pi i(\frac{1}{2}+\delta 2^{-l})}-z_j^{(1)}|<r\right\}$$

is less than $2^l r s_1$. Similarly, we can show that for any $k=2,\dots,p$, the total length of the set

$$\Delta_k=\left\{\delta\in(0,\sigma),\quad \exists j\le s_k \ \ |e^{-2\pi i(2^{-k}+\delta 2^{-l-k+1})}-z_j^{(k)}|<r\right\}$$

is less than $2^{l+k-1}rs_k$. Consequently, the total length of all the sets $\Delta_1,\dots,\Delta_p$ is less than

$$2^l r\sum_{k=1}^{p}2^{k-1}s_k\le 2^l r 2^{p-1}\sum_{k=1}^{p}s_k=2^{l+p-1}rS=\sigma.$$

Therefore there exists a point $\delta_0\in(0,\sigma)$ that does not belong to any of these sets. For this point, we have

$$\left|\prod_{k=1}^{p} f_k\left(2^{-k}+2^{1-l-k}\delta_0\right)\right|\ge r^S.$$

Since S does not exceed the total number of roots of the polynomials $\{\mathbf{m}_{l+k}\}_{k=1}^{p}$ (and thus $S\le np$), it follows that

$$\begin{aligned} r^S &= \left(\frac{\sigma}{2^{l+p-1}S}\right)^{S}=\left(2^{l+p+1}Mn\pi S\right)^{-S}\\ &\ge 2^{-S(l+p+1)}\left(Mn\pi np\right)^{-np}\ge 2^{-Sl}\left(2^{p+1}Mn^2\pi p\right)^{-np}. \end{aligned}$$

Thus,

$$\left|\prod_{k=1}^{p} f_k(2^{-k}+\delta_0 2^{1-l-k})\right|\ge 2^{-Sl}\left(2^{p+1}Mn^2\pi p\right)^{-np}.$$

Using (8.85), we get

$$2^{l(n-1)}\cdot 2^{-lS}\to 0 \quad \text{as } l\to+\infty;$$

hence $S\ge n$.

Recall that we have counted the number of roots of the polynomial $\mathbf{m}_{l+k}$ in the circle β_k and calculated the sum S of these numbers for k from 1 to p. We have shown that for each $l\ge 0$, this sum is at least n. In the same way it can be proved that the analogous sum for the circles $\{\gamma_k\}_{k=1}^{p}$ is at least n. Consider now the total number of roots of the polynomials $\mathbf{m}_{l+k}$ in the circles β_k and γ_k, and calculate the sum S_l of such numbers for all k from 1 to p. As shown above, for each $l \geq 0$, only two cases are possible:

1) $S_l = n$, in which case all the corresponding roots are contained in the circle β_0.

2) $S_l \geq n+1$.

Let us prove that for sufficiently large l, case 1) takes place. To do this, take integer N and estimate the total number of roots of the polynomials $\mathbf{m}_{l+1}, \ldots, \mathbf{m}_{l+p+N}$. Obviously, this number is at most $n(p+N)$. On the other hand, the number of roots is at least $\sum_{j=0}^{N} S_j$, since every root in this sum is counted only once. Assume now that for M chains of polynomials $\{\mathbf{m}_{l+j+k}\}_{k=1}^{p}$, we have case 2). Then $\sum_{j=0}^{N} S_l \geq n(N+1)+M$, so $n(p+N) \geq n(N+1)+M$. Thus $M \leq n(p-1)$. Therefore beginning with a number l_0, we have case 1) for all chains. Thus for each $k \geq l_0$, the polynomial $\mathbf{m}_k$ has exactly n roots in the circle β_0. By the conditions of the theorem this means that the polynomial $\deg \mathbf{m}_k$ has degree n and all its n roots are contained in the circle β_0. Thus, for each $R > 0$, there exists a number $l_0(R)$ such that for any $k \geq l_0(R)$, the polynomial $\mathbf{m}_k$ has degree n and all its n roots are contained in the circle $|z+1| \leq R$. Therefore all n roots of the polynomial $\mathbf{m}_k$ tend to the point -1 as $k \to \infty$. But since for all k, $\mathbf{m}_k(1) = 1$, it follows that $\mathbf{m}_k \to (\frac{1+z}{2})^n$ and $m_k \to w_{n-1}$ as $k \to \infty$.

If we assume that the rate of decrease of the function $f(\xi)$ is at least n, then by the same argument we establish that $m_k \to w_n$ as $k \to \infty$. Consequently, $\deg m_k \geq n+1$ for all sufficiently large k. This contradiction completes the proof. $\diamondsuit$

REMARK 8.3.2. In fact, we proved a stronger statement: if the rate of decrease of a function $\widehat{\varphi}$ is at least $n-1$ (i.e., $\widehat{\varphi}(\xi) = o(|\xi|^{1-n})$) either as $\xi \to +\infty$ or $\xi \to -\infty$, then $m_k \to w_{n-1}$.

REMARK 8.3.3. It follows from our construction of nonstationary wavelets that if the degrees of the polynomials m_k are not bounded, then the rate of decrease of the function $\widehat{\varphi}$ can be infinite. There exist sequences of polynomials whose degrees grow arbitrarily slowly and the function $\widehat{\varphi}$ decreases more rapidly than any power function (see Theorem 8.2.2).

REMARK 8.3.4. According to Theorem 8.3.1, for any bounded in norm sequence of trigonometric polynomials of degree n, the function $\widehat{\varphi}$ cannot decrease more rapidly than $O(\xi^{-n})$. This maximal rate of decrease is attained for the stationary sequence $m_k = w_{n-1}, k \in \mathbb{N}$, but not only for it. For example, for arbitrary $a > 0$, the sequence $m_k(\xi) = \left(\frac{e^{-2\pi i\xi}+a^{(1/2)^k}}{1+a^{(1/2)^k}}\right)^n$ has the same property, i.e., its rate of decrease is maximal. In Problem 8.3.12 we ask the reader to prove the criterion of maximal rate of decrease for sequences of polynomials.

REMARK 8.3.5. The natural question arises: how essential is the requirement of the uniform boundedness in norm of polynomials in Theorem 8.3.1? Can it be replaced by the weaker requirement of the uniform convergence of the product $\widehat{\varphi}(\xi) = \prod_{k=1}^{\infty} m_k(2^{-k}\xi)$ on every compact set? The answer is negative, which may seem paradoxical. If for a bounded sequence, the product $\widehat{\varphi}(\xi)$ cannot decrease too rapidly, then for an unbounded sequence, this product must be greater still and thus must decrease more slowly. Nevertheless, in Problem 8.3.9 we shall see that there exists a sequence $\{m_k\}$ such that $\deg m_k = 2$, $m_k(0) = 1$ for all $k \geq 1$, and the product $\widehat{\varphi}(\xi)$ decreases exponentially as $|\xi| \to \infty$.

PROBLEM 8.3.6. Prove the following analog of Theorem 8.3.1 for algebraic polynomials: for an arbitrary sequence of algebraic polynomials $\{\mathbf{p}_k\}$ with uniformly bounded degrees and norms (on the unit circle) and such that $\mathbf{p}_k(0) = 1$, $k \in \mathbb{N}$, the corresponding product $f(x) = \prod_{k=1}^{\infty} p_k(2^{-k}x)$ cannot tend to zero as $x \to \infty$.

However, as in the case of trigonometric polynomials, this analog of Theorem 8.3.1 does not hold unless the norms are not uniformly bounded.

PROBLEM 8.3.7. For an arbitrary $\alpha \in (1, 2)$, consider the family of algebraic polynomials $Q_k(x) = 1 - \frac{2^k}{k^\alpha}x$, $k \in \mathbb{N}$. Prove that the product

$$f_Q(x) = \prod_{k=1}^{\infty} Q_k(2^{-k}x) = \prod_{k=1}^{\infty} \left(1 - \frac{x}{k^\alpha}\right)$$

converges uniformly on every compact set and for any $\varepsilon > 0$, we have

$$|f_Q(x)| \leq C(\varepsilon) e^{x^{1/\alpha}\left(b(\alpha)+\varepsilon\right)}, \tag{8.86}$$

where

$$b(\alpha) = \int_0^{+\infty} \ln|1 - t^{-\alpha}|dt$$

(this quantity is negative for any $\alpha \in (1, 2)$).

PROBLEM 8.3.8. Using the result of Problem 8.3.7, prove that for an arbitrary even number $d \geq 2$ and every $\varepsilon > 0$, there exists a sequence of algebraic polynomials $\{\mathbf{p}_k\}$ such that $\deg \mathbf{p}_k \leq d$, $\mathbf{p}_k(0) = 1$ for all $k \geq 1$, the product

$$f(x) = \prod_{k=1}^{\infty} \mathbf{p}_k(2^{-k}x)$$

converges uniformly on each compact set, and

$$|f(x)| \leq Ce^{-|x|^{d-\varepsilon}}, \quad x \in \mathbb{R}.$$

PROBLEM 8.3.9. For an arbitrary $\beta \in (\frac{1}{2}, 1)$, consider the sequence of polynomials

$$\left\{ m_k(\xi) = e^{-2\pi i\xi} \frac{\cos 2\pi\xi - \cos\frac{k^\beta}{2^k}}{1 - \cos\frac{k^\beta}{2^k}} \right\}_{k\in\mathbb{N}}.$$

It is easy to see that $\deg m_k = 2$, $m_k(0) = 1$ for each $k \geq 1$. Prove that the corresponding product $\widehat{\varphi}(\xi) = \prod_{k=1}^{\infty} m_k(2^{-k}\xi)$ converges on each compact set and $\widehat{\varphi}(\xi) = O\left(e^{-|\xi|^{2-\varepsilon(\beta)}}\right)$.

PROBLEM 8.3.10. Is the function φ constructed in Problem 8.3.9 compactly supported?

Now we formulate the conditions under which a function $\varphi(\xi)$ has maximum possible rate of decrease (Remark 8.3.3). We shall say that a sequence of trigonometric polynomials $\{m_k\}_{k\in\mathbb{N}}$ satisfies condition $(*)$ if there exists a positive constant C_0 such that for all sufficiently large k, we have $\deg m_k = n$ and all n roots of the corresponding algebraic polynomial $\mathbf{m}_k$ are contained in the circle $|z+1| \leq C_0 2^{-k}$.

PROBLEM 8.3.11. Show that if $m_k(0) = 1$, $k \in \mathbb{N}$, then condition $(*)$ implies that $\|m_k - w_{n-1}\|_\infty = O(2^{-k})$, but in general, the converse is not true.

It appears that condition $(*)$ is equivalent to the maximal decrease.

PROBLEM 8.3.12. Prove that under the conditions of Theorem 8.3.1, $\widehat{\varphi}(\xi) = O(\xi^{-n})$ if and only if the sequence $\{m_k\}$ satisfies condition $(*)$.

8.4. Uncertainty constants for Ψ

The system Ψ has an essential disadvantage: the uncertainty constants of nonstationary wavelets ψ_j increase indefinitely as j increases. In this section, we give the corresponding estimates.

Let Φ_j be the autocorrelation function for φ_j. Then

$$\widehat{\Phi}_j(\xi) := 2^{-j} \prod_{N=j+1}^{\infty} B_{T(N)}(\cos(\xi 2^{-N})).$$

For convenience, we shall estimate the uncertainty constants of autocorrelation functions scaled to zero level:

$$\Phi_j^0(t) := 2^j \Phi_j(2^{-j}t).$$

In this section, we again use the description of the Daubechies masks with the help of Bernstein's polynomials (see (4.26)):

$$|d_N(\xi)|^2 = \mathcal{D}_N(\cos 2\pi\xi), \tag{8.87}$$

where

$$\mathcal{D}_N(t) := \sum_{l=N}^{2N-1} \binom{2N-1}{l} \left(\frac{1+t}{2}\right)^l \left(\frac{1-t}{2}\right)^{2N-1-l}.$$

We shall write $\Delta(\Phi)$ instead of Δ_Φ because of the abundance of indices.

THEOREM 8.4.1.

$$\lim_{j\to\infty} \|\widehat{\Phi_j^0} - \chi_{[-1/2,1/2]}\|_{L_p} = 0, \quad 1 \le p < \infty,$$

$$\liminf_{j\to\infty} \Delta(\widehat{\Phi_j^0}) \ge \frac{1}{2\sqrt{3}},$$

$$\lim_{j\to\infty} \Delta(\Phi_j^0) = \infty,$$

and therefore,

$$\lim_{j\to\infty} \Delta(\widehat{\Phi_j^0})\Delta(\Phi_j^0) = \infty.$$

The proof is analogous to the proof of Theorem 4.3.3; it consists of the following lemmas.

LEMMA 8.4.2.

$$\lim_{j\to\infty} \widehat{\Phi_j^0}(\xi) = \chi_{[-1/2,1/2]}(\xi) \quad \textit{for any } \xi \in \mathbb{R}\backslash\{\pm 1/2\}. \tag{8.88}$$

Proof. It is obvious that

$$\widehat{\Phi_j^0}(\xi) := \prod_{N=1}^{\infty} D_{T(j+N)}(\xi 2^{-N})$$

(see (8.52) and (8.87)). By the properties of Bernstein's polynomials we have

$$\lim_{N\to\infty} \mathcal{D}_N(\xi) = \chi_{[0,1]}(\xi) \quad \text{for any } \xi \in [-1,1]\setminus\{0\}. \tag{8.89}$$

Fix ξ and choose $L \in \mathbb{N}$ so that $|\xi| < 2^{L-2}$. Then

$$\prod_{N=L+1}^{\infty} \chi_{[0,1]}(\cos(\pi\xi 2^{1-N})) = 1,$$

$$\prod_{N=1}^{L} \chi_{[0,1]}(\cos(\pi\xi 2^{1-N})) = \chi_{[0,1]}(\cos \pi\xi)\chi_{[-1,1]}(\xi).$$

Hence,

$$\begin{aligned}
&\widehat{\Phi_j^0}(\xi) - \chi_{[-1/2,1/2]}(\xi) \\
&= \left(\left(\prod_{N=1}^{L} D_{T(j+N)}(\xi 2^{-N}) - \prod_{N=1}^{L} \chi_{[0,1]}(\cos(\pi\xi 2^{1-N}))\right) \prod_{N=L+1}^{\infty} D_{T(j+N)}(\xi 2^{-N})\right) \\
&\qquad - \left(1 - \prod_{N=L+1}^{\infty} D_{T(j+N)}(\xi 2^{-N})\right) \chi_{[0,1]}(\cos \pi\xi)\chi_{[-1,1]}(\xi).
\end{aligned}$$

Introduce the notation

$$\begin{aligned}
I_{j,1}(\xi) &:= \left(\prod_{N=1}^{L} D_{T(j+N)}(\xi 2^{-N}) - \prod_{N=1}^{L} \chi_{[0,1]}(\cos(\pi\xi 2^{1-N}))\right) \\
&\qquad \times \prod_{N=L+1}^{\infty} D_{T(j+N)}(\xi 2^{-N}), \\
I_{j,2}(\xi) &:= \left(1 - \prod_{N=L+1}^{\infty} D_{T(j+N)}(\xi 2^{-N})\right) \chi_{[0,1]}(\cos \pi\xi)\chi_{[-1,1]}(\xi).
\end{aligned}$$

It is obvious that

$$0 \le \prod_{N=L+1}^{\infty} D_{T(j+N)}(\xi 2^{-N}) \le 1.$$

Hence it follows from (8.89) that $|I_{j,1}(\xi)| \to 0$ as $j \to \infty$.

To estimate $I_{j,2}$, use inequalities (4.29) and (4.30). We have

$$|I_{j,2}(\xi)| \le \sum_{N=L+1}^{\infty} \frac{1 - D_{T(j+N)}(\xi 2^{-N})}{D_{T(j+N)}(\xi 2^{-N})}.$$

Since $|\xi| < 2^{L-2}i$, it follows that $|\xi 2^{-N-L}| \le 1/4$ for any $N \in \mathbb{N}$. Therefore,

$$|D_{T(j+N)}(\xi 2^{-N-L})| \ge D_{T(j+N)}(1/4) \ge \frac{1}{2}. \tag{8.90}$$

As $\mathcal{D}_N(t) + \mathcal{D}_N(-t) = 1$, by Lemma 2.2.5 we have

$$|1 - \mathcal{D}_{T(j+N)}(\cos(\pi\xi 2^{1-N}))| \le \sin^{2T(j+N)}(\pi\xi 2^{1-N}). \tag{8.91}$$

It follows from (8.90) and (8.91) that $|I_{j,2}(\xi)| \to 0$ as $j \to \infty$. ◊

LEMMA 8.4.3. *For* $1 \le p < \infty$

$$\lim_{j\to\infty} \|\widehat{\Phi_j^0} - \chi_{[-1/2,1/2]}\|_p = 0. \tag{8.92}$$

Proof. By Lebesgue's dominated convergence theorem, (8.88) implies that

$$\lim_{j\to\infty} \int_{-1}^{1} \widehat{\Phi_j^0}(\xi)\,d\xi = \int_{-1}^{1} \chi_{[-1/2,1/2]}(\xi)\,d\xi = 1.$$

Since

$$\int_{\mathbb{R}} \widehat{\Phi_j^0}(\xi)\,d\xi = \Phi_j^0(0) = 1,$$

it follows that

$$\lim_{j\to\infty} \int_{\{|\xi|\ge 1\}} \widehat{\Phi_j^0}(\xi)\,d\xi = 0.$$

Since $|\widehat{\Phi_j^0}(\xi)| \le 1$, it follows that $|\widehat{\Phi_j^0}(\xi)|^p \le |\widehat{\Phi_j^0}(\xi)|$ and

$$\lim_{j\to\infty} \int_{\{|\xi|\ge 1\}} |\widehat{\Phi_j^0}(\xi)|^p\,d\xi = 0$$

for any $p \in [1,\infty)$. ◊

LEMMA 8.4.4.

$$\liminf_{j\to\infty} \Delta_{\widehat{\Phi_j^0}} \ge \frac{1}{2\sqrt{3}}. \tag{8.93}$$

Proof. Using (8.88), (8.92), and Lebesgue's dominated convergence theorem, we get

$$\begin{aligned}\liminf_{j\to\infty} \Delta_{\widehat{\Phi_j^0}} &= \liminf_{j\to\infty} \left\{ \int_{\mathbb{R}} \xi^2 \left|\widehat{\Phi_j^0}(\xi)\right|^2 d\xi \right\}^{1/2} \\ &\ge \liminf_{j\to\infty} \left\{ \int_{-1/2}^{1/2} \xi^2 \left|\widehat{\Phi_j^0}(\xi)\right|^2 d\xi \right\}^{1/2} = \left\{ \int_{-1/2}^{1/2} \xi^2 d\xi \right\}^{1/2} = \frac{1}{2\sqrt{3}}.\end{aligned}$$ ◊

LEMMA 8.4.5.

$$\liminf_{j\to\infty} \Delta_{\Phi_j^0} = \infty. \tag{8.94}$$

Proof. It follows from (8.92) that

$$\int_{\mathbb{R}} \left|\Phi_j^0(t) - \frac{\sin \pi t}{\pi t}\right|^2 dt = \int_{\mathbb{R}} \left|\widehat{\Phi_j^0}(\xi) - \chi_{[-1/2,1/2]}(\xi)\right|^2 d\xi \to 0 \quad \text{as } j \to \infty.$$

Hence by Fatou's lemma we have

$$\liminf_{j\to\infty} \Delta^2_{\Phi_j^0} = \liminf_{j\to\infty} \int_{\mathbb{R}} t^2 \left|\Phi_j^0(t)\right|^2 dt \ge \int_{\mathbb{R}} t^2 \left|\frac{\sin \pi t}{\pi t}\right|^2 dt = \infty.$$ ◊

Using nonstationary analogs of Lemmas 1.5.9 and 1.5.11, we can give an exact (with respect to the order) estimate of increase of the radii of the autocorrelation functions Φ_j^0.

LEMMA 8.4.6. *Let* $\widehat{F}(\xi) := \prod_{l=1}^{\infty} f_l(\xi 2^{-l})$, *where* $f_l(t)$ *are nonnegative* 1*-periodic functions, and suppose that the infinite product converges absolutely and admits termwise differentiation in the sense of almost everywhere convergence. Set*

$$\widehat{F_j^0}(\xi) := \prod_{l=1}^{\infty} f_{j+l}(\xi 2^{-l}). \tag{8.95}$$

Suppose that

$$\sum_{k\in\mathbb{Z}} \left|\widehat{F_j^0}(\xi+k)\right|^2 \le 1. \tag{8.96}$$

Then

$$\left(\int_{\mathbb{R}} t^2 |F_j^0(t)|^2\,dt\right)^{1/2} \le 2\pi \sum_{l=1}^{\infty} 2^{-l/2} \left(\int_{-1/2}^{1/2} |f'_{j+l}(\xi)|^2\,d\xi\right)^{1/2}.$$

Proof. As $\widehat{\cdot F_j^0(\cdot)}(\xi) = -2\pi i \widehat{F_j^0}'(\xi)$, by the Plancherel identity we have

$$\int_{\mathbb{R}} t^2 |F_j^0(t)|^2\,dt = 4\pi^2 \int_{\mathbb{R}} |\widehat{F_j^0}'(\xi)|^2 d\xi.$$

It follows from (8.95) that

$$\widehat{F_j^0}'(\xi) = \left(\sum_{l\in\mathbb{N}} 2^{-l} \frac{f'_{j+l}(2^{-l}\xi)}{f_{j+l}(2^{-l}\xi)}\right) \prod_{p=1}^{\infty} f_{j+p}(2^{-p}\xi).$$

It follows from (8.96) that $f_l(\xi) \le 1$; hence,

$$|\widehat{F_j^0}'(\xi)| \le \sum_{l\in\mathbb{N}} 2^{-l} |f'_{j+l}(2^{-l}\xi)| \left|\prod_{p=l+1}^{\infty} f_{j+p}(2^{-p}\xi)\right|.$$

Using the triangle inequality in $L_2(\mathbb{R})$, we get

$$\begin{aligned}\left(\int_{\mathbb{R}} |\widehat{F_j^0}'(\xi)|^2\,d\xi\right)^{1/2} &\le \sum_{l\in\mathbb{N}} 2^{-l} \left(\int_{\mathbb{R}} |f'_{j+l}(2^{-l}\xi)|^2 \left|\prod_{p=l+1}^{\infty} f_{j+p}(2^{-p}\xi)\right|^2 d\xi\right)^{1/2} \\ &= \sum_{l\in\mathbb{N}} 2^{-l/2} \left(\int_{\mathbb{R}} |f'_{j+l}(\xi)|^2 \left|\prod_{p=1}^{\infty} f_{j+l+p}(2^{-p}\xi)\right|^2 d\xi\right)^{1/2}.\end{aligned}$$

Taking into account the 1-periodicity of the functions f_l and identity (8.96), we obtain

$$\int_{\mathbb{R}} |f'_{j+l}(\xi)|^2 \left|\prod_{p=1}^{\infty} f_{j+l+p}(2^{-p}\xi)\right|^2 d\xi$$

$$= \int_{-1/2}^{1/2} |f'_{j+l}(\xi)|^2 \sum_{l\in\mathbb{Z}} \left|\widehat{F^0_{j+l}}(\xi+l)\right|^2 d\xi \le \int_{-1/2}^{1/2} |f'_{j+l}(\xi)|^2 d\xi. \qquad \Diamond$$

The following lemma is a nonstationary analog of Lemma 1.5.11.

LEMMA 8.4.7. *Let*

$$\widehat{F}(\xi) := \prod_{l=1}^{\infty} f_l(\xi 2^{-l}),$$

where $f_l(t)$ *are nonnegative, even, and 1-periodic functions decreasing on* $[0,1/2]$. *Then*

$$\left(\int_{\mathbb{R}} t^2 |F(t)|^2\, dt\right)^{1/2} \ge \sqrt{2}\pi \prod_{l=2}^{\infty} f_l(\xi 2^{-l}) \left(\int_{-1/2}^{1/2} |f'_1(\xi)|^2\, d\xi\right)^{1/2}.$$

Proof. Define the function F_1 in terms of the Fourier transform

$$\widehat{F_1}(\xi) = \prod_{l=1}^{\infty} f_{1+l}(\xi 2^{-l}).$$

It can easily be seen that

$$\int_{\mathbb{R}} |\widehat{F}'(\xi)|^2 d\xi \ge \int_{-1}^{1} |\widehat{F}'(\xi)|^2 d\xi = \int_{-1}^{1} \left|\frac{1}{2} f'_1(\xi/2)\widehat{F_1}(\xi/2) + \frac{1}{2} f_1(\xi/2)\widehat{F_1}'(\xi/2)\right|^2 d\xi$$

$$= \frac{1}{2}\int_{-1/2}^{1/2} \left|f'_1(\xi)\widehat{F_1}(\xi) + f_1(\xi)\widehat{F_1}'(\xi)\right|^2 d\xi.$$

By the condition of the lemma, $f_l(\xi)$ are even functions and $f'_l(\xi) \le 0$ for $\xi \in [0,1/2]$. Then $\widehat{F_1}'(\xi) \le 0$ for $\xi \in [0,1/2]$ and

$$\int_{-1/2}^{1/2} \left(f'_1(\xi)\widehat{F_1}(\xi) + f_1(\xi)\widehat{F_1}'(\xi)\right)^2 d\xi$$

$$= 2\int_{0}^{1/2} \left(f'_1(\xi)\widehat{F_1}(\xi) + f_1(\xi)\widehat{F_1}'(\xi)\right)^2 d\xi$$

$$\ge 2\int_{0}^{1/2} \left(f'_1(\xi)\widehat{F_1}(\xi)\right)^2 d\xi \ge (\widehat{F_1}(1/2))^2 \int_{-1/2}^{1/2} (f'_1(\xi))^2\, d\xi. \qquad \Diamond$$

The following lemma gives the asymptotics for the radius of the squared modulus of the Daubechies scaling mask.

LEMMA 8.4.8.

$$\|\sin(2\pi\xi)\mathcal{D}'_N(\cos 2\pi\xi)\|_{L_2[-1/2,1/2]} \sim N^{1/4}. \tag{8.97}$$

Proof. It is easy to see that

$$\mathcal{D}'_N(\cos 2\pi\xi) = \frac{1}{2}\frac{(2N+1)!}{4^N(N!)^2}\sin^{2N}(2\pi\xi) = \sqrt{\frac{N}{\pi}}(1+O(N^{-2}))\sin^{2N}(2\pi\xi).$$

It now remains to use the fact that

$$\int_{-1/2}^{1/2} \sin^{2N}(2\pi\xi)d\xi = \frac{\binom{2N}{N}}{4^N} \sim \sqrt{\frac{\pi}{N}}.$$ ◇

COROLLARY 8.4.9.

$$\Delta_{\Phi_j^0} \sim T(j+1)^{1/4}.$$

Proof. It is easy to see that for $t \in [0,1]$,

$$\mathcal{D}_N(t) \geq \frac{1+t}{2}.$$

Hence for any j,

$$\widehat{\Phi_j^0}(1/2) \geq \prod_{N=1}^{\infty} \frac{1+\cos(\pi 2^{-N})}{2} = \prod_{N=1}^{\infty} \cos^2(\pi 2^{-N-1}) = \left(\frac{\sin \pi/2}{\pi/2}\right)^2 = \frac{4}{\pi^2}.$$

Therefore by Lemma 8.4.7 and (8.97),

$$\Delta_{\Phi_j^0} \geq CT(j+1)^{1/4}.$$

The opposite estimate follows from (8.97) and Lemma 8.4.6. ◇

8.5. Nonstationary wavelets with modified Daubechies masks

We can avoid difficulties connected with the unboundedness of the uncertainty constants of nonstationary wavelets based on the Daubechies masks if we use the modified Daubechies masks. In this section we give a description of this construction. Let $\{T(N)\}_{N=1}^{\infty}$ be an arbitrary sequence of positive integers such that

$$\lim_{N\to\infty} T(N) = \infty, \qquad T(N) \leq T(N+1) \leq T(N)+1, \quad T(1)=1, \tag{8.98}$$

and let m_N^a be the modified Daubechies masks defined in (4.42).

The following lemma is proved similarly to Lemma 8.2.4.

LEMMA 8.5.1. *The infinite products*

$$G_j^a(\xi) := \prod_{N=j+1}^{\infty} m_{T(N)}^a(\xi 2^{-N}), \quad j \in N_0, \tag{8.99}$$

converge pointwise for all $\xi \in \mathbb{R}$. The convergence is uniform on compact sets.

To prove an analog of Lemma 8.2.5, we need to improve the notation and estimates of §4.3.2. Let $b_N(a)$ be a root of the equation

$$2^{a+2/N}(1-b)^{(1-a)/2}(1+b)^{(1+a)/2} = 1, \quad b \in (0,1). \tag{8.100}$$

The derivative of the function $(1-t)^{(1-a)/2}(1+t)^{(1+a)/2}$ is equal to $(a-t)(1-t)^{(-1-a)/2}(1+t)^{(-1+a)/2}$; it is nonpositive for $t \in [a,1]$. Since $2^{-a-2/N}$ increases as N increases, it follows that $b_N(a)$ decreases monotonically as $N \to \infty$ and tends to $b(a)$, which is a root of the equation

$$2^a(1-b)^{(1-a)/2}(1+b)^{(1+a)/2} = 1, \quad b \in (0,1).$$

In this new notation estimate (4.60) has the following form:

(8.101)

$$(B_N^a(\cos 2\pi\xi))^{1/N} \le \begin{cases} 1, & |\xi| \le \xi_N(a), \\ 2^{1+a+\frac{2}{N}} \cos^{1-a}(\pi\xi)|\sin(\pi\xi)|^{1+a}, & |\xi| \in (\xi_N(a), 1/2), \end{cases}$$

where $\xi_N(a) \in [1/4, 1/2]$ and $\cos(2\pi\xi_N(a)) = -b_N(a)$. Since $b_N(a)$ decreases monotonically, $\xi_N(a)$ also decreases monotonically as $N \uparrow \infty$ and tends to $\xi(a)$ defined by the conditions

$$\xi(a) \in [1/4, 1/2], \quad \cos(2\pi\xi(a)) = -b(a).$$

Prove that

(8.102)

$$(B_{N+2}^a(\cos 2\pi\xi))^{\frac{1}{N}} \le \begin{cases} 1, & |\xi| \le \xi_N(a), \\ 2^{1+a+\frac{2}{N}} \cos^{1-a}(\pi\xi)|\sin(\pi\xi)|^{1+a}, & |\xi| \in (\xi_N(a), 1/2). \end{cases}$$

This is obvious for $|\xi| \le \xi_N(a)$, because $\xi_{N+2}(a) < \xi_N(a)$; if $|\xi| \in (\xi_N(a), 1/2] \subset (\xi_{N+2}(a), 1/2]$, we have

$$B_{N+2}^a(\cos 2\pi\xi) \le 2^{(N+2)(1+a)+2}(\cos^{1-a}(\pi\xi)|\sin(\pi\xi)|^{1+a})^{N+2}.$$

Therefore,

$$(B_{N+2}^a(\cos 2\pi\xi))^{1/N} \le 2^{1+a+\frac{2}{N}} \cos^{1-a}(\pi\xi)|\sin(\pi\xi)|^{1+a} \times (2^{1+a}\cos^{1-a}(\pi\xi)|\sin(\pi\xi)|^{1+a})^{2/N}.$$

(8.102) follows from the fact that for $|\xi| \in (\xi_N(a), 1/2] \subset (\xi(a), 1/2]$, we have

$$2^{1+a}\cos^{1-a}(\pi\xi)|\sin(\pi\xi)|^{1+a} \le 1.$$

Thus to estimate B_{N+2}^a, we can use the estimate obtained for B_N^a. Hence using the same reasoning as in the proof of (4.66), we get

$$(B_{N+2}^a(\cos 2\pi\xi)B_N^a(\cos 4\pi\xi))^{1/N} \le \alpha 2^{2\beta}|\cos\pi\xi|^\beta |\cos 2\pi\xi|^\beta, \tag{8.103}$$

where $\alpha \in (0,1)$ and $\beta := 1-a$.

Set $I_r := [2^{r-2},\ 2^{r-1}]$, $r \in \mathbb{N}$.

LEMMA 8.5.2. *Let $j \in \mathbb{Z}_+$. If $|\xi| \in I_r$ for $r > j$, $r \in \mathbb{N}$, then*

$$|G_j^a(\xi)| \le \gamma_j \exp\left(-\mu \sum_{N=j+1}^{r} T(N)\right), \tag{8.104}$$

where the constant $\mu > 0$ depends only on a and the constants $\gamma_j > 0$ depend only on j and a.

Proof. Fix positive integer $r > j$ and set

$$G_{jr}^a(\xi) := \prod_{N=j+1}^{r} m_{T(N)}^a(\xi 2^{-N}).$$

By (4.44) we have $|G_j^a(\xi)| \le |G_{jr}^a(\xi)|$. Let $r - j = 2l,\ l \in \mathbb{N}$. Taking into account (8.103), we have

$$
\begin{aligned}
|G_{jr}^a(\xi)|^2 &= \prod_{N=j+1}^{r} |m_{T(N)}^a(\xi 2^{-N})|^2 \\
&= \prod_{N=1}^{l} B^a_{2T(j+2N-1)-1}(\cos\pi\xi 2^{-j-2N+2}) B^a_{2T(j+2N)-1}(\cos\pi\xi 2^{1-j-2N}) \\
&\le \prod_{N=1}^{l} (\alpha 2^{2\beta})^{2T(j+2N-1)-1} |\cos\pi\xi 2^{-j-2N}\cos\pi\xi 2^{-j-2N+1}|^{\beta(2T(j+2N-1)-1)}.
\end{aligned}
$$

Define the sequence $T'(N)$ in the following way:

$$
T'(j+2N-1) := T'(j+2N) := 2T(j+2N-1) - 1.
$$

Then using (8.46), for $|\xi| \in I_r$, we get

$$
\begin{aligned}
&\prod_{N=1}^{l} \left|\cos\pi\xi 2^{-j-2N}\cos\pi\xi 2^{-j-2N+1}\right|^{2T(j+2N-1)-1} = \prod_{N=j+1}^{r} \left|\cos\pi\xi 2^{-N}\right|^{T'(N)} \\
&\le 2^{-\sum_{N=j+1}^{r} T'(N) + T'(r)/2} = 2^{-2\sum_{N=1}^{l}(2T(j+2N-1)-1) + T(r-1) - 1/2}.
\end{aligned}
$$

Finally, we obtain

$$
\begin{aligned}
|G_{jr}^a(\xi)|^2 &\le \alpha^{2\sum_{N=1}^{l}(2T(j+2N-1)-1)} 2^{2\beta\sum_{N=1}^{l}(2T(j+2N-1)-1)} \\
&\times 2^{-2\beta\sum_{N=1}^{l}(2T(j+2N-1)-1)} 2^{\beta(T(r-1)-1/2)} = \alpha^{\sum_{N=1}^{l}(2T(j+2N-1)-1)} 2^{\beta(T(r-1)-1/2)},
\end{aligned}
\tag{8.105}
$$

which implies (8.104).

The case $r - j = 2l - 1,\ l \in \mathbb{N}$, also follows from (8.105) and the inequality

$$
|G_{jr}^a(\xi)| \le |G_{j+1,r}^a(\xi)|, \tag{8.106}
$$

which is obvious by (4.44). ◊

Lemma 8.5.2 implies the following analog of Corollary 8.2.6.

COROLLARY 8.5.3. *The functions G_j^a, $j \in \mathbb{Z}_+$, decrease at infinity more rapidly than any negative power.*

Define the sequence of functions $\{\varphi_{jk}^a\}_{j,k\in\mathbb{Z},\, j\ge 0}$ by the Fourier transform

$$
\begin{aligned}
\widehat{\varphi_j^a}(\xi) &:= 2^{-j/2} G_j^a(\xi), \quad j \in \mathbb{Z}_+, \\
\varphi_{jk}^a(t) &:= \varphi_j^a(t + k2^{-j}), \quad k \in \mathbb{Z}.
\end{aligned}
\tag{8.107}
$$

The functions ψ_{jk}^a, $j \in \mathbb{Z}_+$, $k \in \mathbb{Z}$, are defined by the equalities

$$
\widehat{\psi_j^a}(\xi) := 2^{1/2}\exp(\pi i \xi 2^{-j-1}) m^a_{T(j+1)}(-\xi 2^{-j-1} - 1/2)\widehat{\varphi_{j+1}^a}(\xi), \quad \xi \in \mathbb{R},
$$

$$
\psi_{jk}^a(t) := \psi_j^a(t + k2^{-j}).
$$

The construction of the system Ψ^a is thus complete. Set

$$
\Psi^a := \{\varphi_{0k}^a,\ \psi_{jk}^a,\ j \in \mathbb{Z}_+,\ k \in \mathbb{Z}\}.
$$

The properties of the system Ψ^a are analogous to the properties of Ψ (see p. 284).

LEMMA 8.5.4. *The functions φ^a_{jk} and ψ^a_{jk}, $j \in \mathbb{Z}_+$, $k \in \mathbb{Z}$, are infinitely differentiable.*

Proof. The proof is based on Corollary 8.5.3. ◊

The following assertions are proved in the same way as Lemmas 8.2.8, 8.2.9, and 8.2.10.

LEMMA 8.5.5. *For any $j \in \mathbb{Z}_+$, $k, k' \in \mathbb{Z}$,*

$$\langle \varphi^a_{jk}, \varphi^a_{jk'} \rangle = \delta_{k,k'}, \tag{8.108}$$

$$\langle \psi^a_{jk}, \psi^a_{jk'} \rangle = \delta_{k,k'}, \quad \langle \varphi^a_{jk}, \psi^a_{jk'} \rangle = 0. \tag{8.109}$$

Consider the subspaces

$$V^a_j := \overline{\operatorname{span}\{\varphi^a_{jk},\ k \in \mathbb{Z}\}}, \quad W^a_j := \overline{\operatorname{span}\{\psi^a_{jk},\ k \in Z\}}, \quad j \in \mathbb{Z}_+. \tag{8.110}$$

By Lemma 8.5.5 the systems $\{\varphi^a_{jk}\}_{k\in\mathbb{Z}}$ and $\{\psi^a_{jk}\}_{k\in Z}$ are orthonormal bases in the spaces V^a_j and W^a_j, $j \in \mathbb{Z}_+$, respectively.

LEMMA 8.5.6. *For any $j \in \mathbb{Z}_+$,*

$$V^a_j \subset V^a_{j+1}, \quad V^a_j \oplus W^a_j = V^a_{j+1}, \tag{8.111}$$

$$\overline{\bigcup_{j=0}^{\infty} V^a_j} = L_2(\mathbb{R}). \tag{8.112}$$

LEMMA 8.5.7. *The functions φ^a_j and ψ^a_j, $j \in \mathbb{Z}_+$, have compact support and*

$$\operatorname{supp} \varphi^a_j \subset [-(2T(j+1)+1)2^{-j}, 0], \tag{8.113}$$

$$\operatorname{supp} \psi^a_j \subset [-(T(j+2)+1)2^{-j},\ (2T(j+1)-1)2^{-j-1}]. \tag{8.114}$$

8.6. Uncertainty constants for Ψ^a

Let Φ^a_j be the autocorrelation function for φ^a_j. Then

$$\widehat{\Phi}^a_j(\xi) := 2^{-j} \prod_{N=j+1}^{\infty} B^a_{2T(N)-1}(\cos(\pi\xi 2^{1-N})).$$

As in §8.4, it is convenient to evaluate the uncertainty constants of autocorrelation functions scaled to zero level:

$$\Phi^{a,0}_j(t) := 2^j \Phi^a_j(2^{-j}t).$$

Then

$$\widehat{\Phi^{a,0}_j}(\xi) := \prod_{N=1}^{\infty} B^a_{2T(j+N)-1}(\cos(\pi\xi 2^{1-N})).$$

As in §4.3.2, φ^M denotes the Meyer scaling function defined in (4.74), and Φ^M denotes its autocorrelation function.

THEOREM 8.6.1. *If a function f_a satisfies the conditions*

$$(f_a)''(t) \geq 0, \quad \text{if } t \leq 0, \qquad (f_a)''(t) \leq 0, \quad \text{if } t > 0,$$

$$(f_a)'(t) \leq C\sqrt{f_a(t)} \quad \text{for some constant } C,$$

then

$$\lim_{j\to\infty} \|\widehat{\Phi_j^{a,0}} - \widehat{\Phi}^M\|_p = 0, \quad 1 \leq p < \infty,$$

$$\lim_{j\to\infty} \Delta(\widehat{\Phi_j^{a,0}}) = \Delta(\widehat{\Phi^M}),$$

$$\lim_{j\to\infty} \Delta(\Phi_j^{a,0}) \leq \frac{8\pi^2\sqrt{2}C}{(\sqrt{2}-1)\|\Phi^M\|_2}, \tag{8.115}$$

and consequently,

$$\lim_{j\to\infty} \Delta(\widehat{\Phi_j^{a,0}})\Delta(\Phi_j^{a,0}) \leq \frac{8\pi^2\sqrt{2}C\Delta(\widehat{\Phi^M})}{(\sqrt{2}-1)\|\Phi^M\|_2} < \infty.$$

The proof is divided into the following lemmas.

LEMMA 8.6.2.

$$\lim_{j\to\infty} \widehat{\Phi_j^{a,0}}(\xi) = f_a(\cos(\pi\omega))\chi_{[-1,1]}(\xi). \tag{8.116}$$

Proof. The proof is analogous to the proof of Lemma 8.4.2. By the properties of Bernstein's polynomials we have

$$\lim_{N\to\infty} B_N(\xi) = f_a(\xi) \quad \text{for any } \xi \in [-1,1]. \tag{8.117}$$

Fix ξ and choose $L \in \mathbb{N}$ so that $|\xi| < 2^{L-2}$. Then

$$f_a(\cos(\pi\xi))\chi_{[-1,1]}(\xi) = \prod_{l=1}^{L} f_a(\cos(\pi\omega 2^{1-l}))$$

and therefore,

$$\begin{aligned}
&\widehat{\Phi_j^{a,0}}(\xi) - f_a(\cos(\pi\xi))\chi_{[-1,1]}(\xi) \\
&\quad = \left\{\left(\prod_{N=1}^{L} B^a_{2T(j+N)-1}(\cos(\pi\xi 2^{1-N})) - \prod_{N=1}^{L} f_a(\cos(\pi\omega 2^{1-l}))\right)\right. \\
&\qquad\qquad \left.\times \prod_{N=L+1}^{\infty} B^a_{2T(j+N)-1}(\cos(\pi\xi 2^{1-N}))\right\} \\
&\qquad - \left\{\left(1 - \prod_{N=L+1}^{\infty} B^a_{2T(j+N)-1}(\cos(\pi\xi 2^{1-N}))\right) f_a(\cos(\pi\xi))\chi_{[-1,1]}(\xi)\right\}.
\end{aligned}$$

Set

$$\begin{aligned}
I_{j,1}(\xi) := &\left(\prod_{N=1}^{L} B^a_{2T(j+N)-1}(\cos(\pi\xi 2^{1-N})) - \prod_{N=1}^{L} f_a(\cos(\pi\omega)2^{1-l})\right) \\
&\qquad \times \prod_{N=L+1}^{\infty} B^a_{2T(j+N)-1}(\cos(\pi\xi 2^{1-N})),
\end{aligned}$$

$$I_{j,2}(\xi) := \left(1 - \prod_{N=L+1}^{\infty} B^a_{2T(j+N)-1}(\cos(\pi\xi 2^{1-N}))\right) f_a(\cos(\pi\xi))\chi_{[-1,1]}(\xi).$$

It is obvious that $0 \le \widehat{\Phi^{a,0}_j}(\xi) \le 1$. Hence (8.117) implies that $|I_{j,1}(\xi)| \to 0$ as $j \to \infty$.

To evaluate $I_{j,2}$, we use inequalities (4.29) and (4.30). We have

$$|I_{j,2}(\xi)| \le \sum_{N=L+1}^{\infty} \frac{1 - B^a_{2T(j+N)-1}(\cos(\pi\xi 2^{1-N}))}{B^a_{2T(j+N)-1}(\cos(\pi\xi 2^{1-N}))}.$$

Since $|\xi| < 2^{L-2}$, it follows that $\left|\xi 2^{1-N}\right| \le 1/4$ for any $N > L$. Hence

$$|B^a_{2T(j+N)-1}(\cos(\pi\xi 2^{1-N}))| \ge \frac{1}{2}, \quad N > L. \tag{8.118}$$

Using (4.44) and (4.60), we get

$$\begin{aligned} &|1 - B^a_{2T(j+N)-1}(\cos(\pi\xi 2^{1-N}))| \\ &\quad \le 2^{(2T(j+N)-1)(1+a)+2} \sin^{(2T(j+N)-1)(1+a)}(\pi\xi 2^{-N}). \end{aligned} \tag{8.119}$$

It follows from (8.118) and (8.119) that $|I_{j,2}(\xi)| \to 0$ as $j \to \infty$. ◊

LEMMA 8.6.3. *For any $p \in [1, \infty)$,*

$$\lim_{j\to\infty} \|\widehat{\Phi^{a,0}_j} - \widehat{\Phi^M}\|_p = 0.$$

`Proof.` The proof is analogous to the proof of Lemma 8.4.3. By Lebesgue's dominated convergence theorem, (8.116) implies that

$$\lim_{j\to\infty} \int_{-1}^{1} \widehat{\Phi^{a,0}_j}(\xi)\, d\xi = \int_{-1}^{1} f_a(\cos(\pi\xi)\, d\xi = 1.$$

As

$$\int_{\mathbb{R}} \widehat{\Phi^{a,0}_j}(\xi)\, d\xi = \Phi^{a,0}_j(0) = 1,$$

we have

$$\lim_{j\to\infty} \int_{\{|\xi|\ge 1\}} \widehat{\Phi^{a,0}_j}(\xi)\, d\xi = 0.$$

It follows from $|\widehat{\Phi^{a,0}_j}(\xi)| \le 1$ that $|\widehat{\Phi^{a,0}_j}(\xi)|^p \le |\widehat{\Phi^{a,0}_j}(\xi)|$; hence,

$$\lim_{j\to\infty} \int_{\{|\xi|\ge 1\}} |\widehat{\Phi^{a,0}_j}(\xi)|^p\, d\xi = 0$$

for any $p \in [1, \infty)$. ◊

LEMMA 8.6.4.

$$\lim_{j\to\infty} \Delta_{\widehat{\Phi^{a,0}_j}} = \Delta_{\widehat{\Phi^M}}. \tag{8.120}$$

Proof. By (4.44), for any $L \in \mathbb{N}$, we have

$$|\widehat{\Phi_j^{a,0}}(\xi)| \le \prod_{N=1}^{2L} B^a_{2T(j+N)-1}(\cos(\pi\xi 2^{1-N})). \tag{8.121}$$

By (8.105), for $|\xi| \in [2^{2L-2}, 2^{2L-1}]$, we have

$$|\widehat{\Phi_j^{a,0}}(\xi)| \le \alpha^{\sum\limits_{N=1}^{L}(2T(j+2N-1)-1)} 2^{\beta(T(j+2L-1)-1/2)}. \tag{8.122}$$

This estimate implies that there exists positive integer L_0 such that

$$\lim_{j\to\infty} \int\limits_{\{|\xi|\ge 2^{L_0}\}} \xi^2\, |\Phi_j^{a,0}(\xi)|^2\, d\xi = \lim_{j\to\infty} \int\limits_{\{|\xi|\ge 2^{L_0}\}} |\Phi_j^{a,0}(\xi)|^2\, d\xi = 0. \tag{8.123}$$

Using (8.123), (8.116), and Lebesgue's dominated convergence theorem, we obtain (8.120). ◊

The following lemma implies estimate (8.115).

LEMMA 8.6.5.

$$\left(\int\limits_{\mathbb{R}} |t\Phi_j^{a,0}(t)|^2\, dt\right)^{1/2} \le \frac{8\pi^2\sqrt{2}C}{\sqrt{2}-1}. \tag{8.124}$$

Proof. By Lemma 8.4.6 and (4.89), we have

$$\left(\int\limits_{\mathbb{R}} t^2|\Phi_j^{a,0}(t)|^2\, dt\right)^{1/2}$$

$$\le 4\pi^2 \sum_{l=1}^{\infty} 2^{-l/2} \left(\int\limits_{-1/2}^{1/2} |(B^a_{2T(j+l)-1})'(\cos 2\pi\xi)\sin 2\pi\xi|^2\, d\xi\right)^{1/2} \le \frac{8\pi^2\sqrt{2}C}{\sqrt{2}-1}. \quad ◊$$

8.7. Nonstationary wavelets bases in Sobolev spaces

In this section, we prove that systems of nonstationary infinitely differentiable compactly supported wavelets constructed in §8.2 and §8.5 constitute unconditional bases in all the Sobolev spaces simultaneously.

THEOREM 8.7.1. *The systems*

$$\Psi := \{\varphi_{0k},\ \psi_{jk},\ j,k \in \mathbb{Z},\ j \ge 0\}$$

and

$$\Psi^a := \{\varphi^a_{0k},\ \psi^a_{jk},\ j,k \in \mathbb{Z},\ j \ge 0\}$$

are unconditional bases in the Sobolev space W_2^s for any $s \ge 0$. If the function

$$\begin{aligned} f(x) &= \sum_{k\in\mathbb{Z}} c_k \varphi(x-k) + \sum_{j\ge 0, k\in\mathbb{Z}} d_{jk}\psi_{jk}(x) \\ &= \sum_{k\in\mathbb{Z}} c^a_k \varphi^a(x-k) + \sum_{j\ge 0, k\in\mathbb{Z}} d^a_{jk}\psi^a_{jk}(x) \end{aligned}$$

belongs to W_2^r, *then its wavelet expansions converge in* W_2^s *for all* $s \in [0, r]$, *and the equivalence relation*

$$\|f\|^2_{W_2^r} \sim \sum_k |c_k|^2 + \sum_{j,k} 2^{2rj} |d_{jk}|^2 \sim \sum_k |c_k^a|^2 + \sum_{j,k} 2^{2rj} |d_{jk}^a|^2 \tag{8.125}$$

holds. Moreover, $2^{j(r-s)}\|P_j f - f\|_{W_2^s} \to 0$ *and* $2^{j(r-s)}\|P_j^a f - f\|_{W_2^s} \to 0$ *as* $j \to +\infty$ *for any function* $f \in W_2^r$, *where* P_j *and* P_j^a *are orthogonal projectors of the space* $L_2(\mathbb{R})$ *on* V_j *and* V_j^a, *respectively, of the nonstationary multiresolution analyses defined in* (8.59) *and* (8.110).

Before proving the theorem, we give some auxiliary assertions.

Lemma 8.7.2. *Let*

$$u(\xi) := \prod_{l=1}^{\infty} |q(2^{-l}\xi)|^{T(l)}, \tag{8.126}$$

$$v_j(\xi) := |q(\xi/2 + \pi)|^{T(j+1)} \prod_{l=2}^{\infty} |q(2^{-l}\xi)|^{T(l)}, \quad j \in \mathbb{N}, \tag{8.127}$$

where $q(\xi) := \sqrt{Q(\xi)}$ *and* Q *is the function defined in Lemma 2.2.5. Then*

$$|\widehat{\varphi}_j(\xi)| \le 2^{-j/2} u(2^{-j}\xi) \tag{8.128}$$

for all $j \ge 0$, *and*

$$|\widehat{\psi}_j(\xi)| \le 2^{\frac{1-j}{2}} v_n(2^{-j}\xi) \tag{8.129}$$

for all $j \ge n > 0$.

Proof. By (8.40) and (8.52) we have

$$\widehat{\varphi}_j(\xi) = 2^{-j/2} \prod_{l=j+1}^{\infty} d_{T(l)}(2^{-l}\xi);$$

hence by Lemma 2.2.5,

$$\begin{aligned} |\widehat{\varphi}_j(\xi)| &\le \left| 2^{-j/2} \prod_{l=j+1}^{\infty} (q(2^{-l}\xi))^{T(l)} \right| \\ &\le 2^{-j/2} \prod_{l=j+1}^{\infty} (q(2^{-l+j}2^{-j}\xi))^{T(l-j)} = 2^{-j/2} u(2^{-j}\xi), \end{aligned}$$

where the latter inequality is based on the fact that $q(\xi) \le 1$ and $T(l)$ increases monotonically (see (8.34)).

Similarly, for all $j \ge n > 0$, we have

$$\begin{aligned} |\widehat{\psi}_j(\xi)| &= 2^{\frac{1-j}{2}} \left| d_{T(j+1)}(-\xi 2^{-j-1} - 1/2) \prod_{l=j+2}^{\infty} d_{T(l)}(2^{-l}\xi) \right| \\ &\le 2^{\frac{1-j}{2}} \left| (q(-\xi 2^{-j-1} - 1/2))^{T(j+1)} \prod_{l=j+2}^{\infty} (q(2^{-l+j}2^{-j}\xi))^{T(l-j)} \right| \\ &\le 2^{\frac{1-j}{2}} v_n(2^{-j}\xi). \end{aligned}$$

◊

LEMMA 8.7.3. *The functions u, v_j, $j \in \mathbb{N}$, decrease at infinity more rapidly than any negative power.*

Proof. The identity $\sin 2t = 2\sin t\cos t$ and the monotone increase of $T(l)$ imply that for any $l \in \mathbb{N}$,

$$\prod_{l=1}^{N} |\cos(2^{-l}\pi\xi)|^{T(l)} \leq \prod_{l=1}^{N} \left|\frac{\sin \pi\xi 2^{1-l}}{2\sin \pi\xi 2^{-l}}\right|^{T(l)} \leq 2^{-\sum_{l=1}^{N} T(l)} |\sin \pi\xi 2^{-N}|^{-T(N)}. \tag{8.130}$$

In view of Lemma 2.2.5 we have

$$u(\xi) \leq \prod_{l=1}^{\infty} |\cos(2^{-l}\pi\xi)|^{T(l)} (\widetilde{q}(2^{-l}\xi))^{T(l)}, \tag{8.131}$$

where $\widetilde{q}(\xi) := \sqrt{\widetilde{Q}(\xi)}$ and the function $\widetilde{Q}$ is defined in Lemma 2.2.5. It follows from (8.34) that

$$\frac{T(l) + T(l+1)}{2} \geq T(l).$$

Hence (4.19) implies that

$$\prod_{l=1}^{N} (\widetilde{q}(2^{-l}\xi))^{T(l)} \leq \sup_{\xi} \widetilde{q}(\xi) (\sqrt{\sigma_2})^{\frac{1}{2}\sum_{l=1}^{N} T(l)}. \tag{8.132}$$

Fix $l \in \mathbb{N}$ and let $|\xi| \in [2^{l-2},\ 2^{l-1})$. Then $|\sin 2^{-l}\pi\xi| \geq 2^{-1/2}$ and the inequalities (8.130)–(8.132) mean that

$$\prod_{l=1}^{N} |\cos(2^{-l}\pi\xi)|^{T(l)} (\widetilde{q}(2^{-l}\xi))^{T(l)} \leq \sup_{\xi} \widetilde{q}(\xi) 2^{(\frac{\log_2 \sigma_2}{4} - 1)\sum_{l=1}^{N} T(l) + \frac{T(N)}{2}}. \tag{8.133}$$

It follows from (8.33) and (8.34) that $T(l) \leq l$ and

$$\lim_{N\to\infty} \frac{1}{N} \sum_{l=1}^{N} T(l) = \infty.$$

Moreover, $0 \leq \tilde{q}(\xi) \leq 1$. So (8.133) implies that u decreases at infinity more rapidly than any negative power.

Fast decrease of the functions v_j, $j \in \mathbb{N}$, is proved similarly. ◊

LEMMA 8.7.4 (**Bernstein's inequality**). *For any $r \geq 0$, there exists $K(r)$ such that for all $f \in V_j$, $j \in \mathbb{N}$, the following inequality holds:*

$$\|f\|_{W_2^0} \leq \|f\|_{W_2^r} \leq K(r) 2^{rj} \|f\|_{W_2^0}. \tag{8.134}$$

Proof. Let

$$f = \sum_{k\in\mathbb{Z}} c_k \varphi_{jk}, \qquad \sum_{k\in\mathbb{Z}} |c_k|^2 < \infty.$$

Then using (8.23), we have (see Appendix A.13)

$$\|f\|_{W_2^r}^2 = \int_{\mathbb{R}} \left|\sum_{k\in\mathbb{Z}} c_k e^{2\pi i 2^{-j}k\xi}\right|^2 |\widehat{\varphi}_j(\xi)|^2 (1 + |\xi|^2)^r \, d\xi. \tag{8.135}$$

Using (8.128), we get

$$\begin{aligned}\|f\|^2_{W_2^r} &\le \int_{\mathbb{R}} \left|\sum_{k\in\mathbb{Z}} c_k e^{2\pi i k\xi}\right|^2 |u(\xi)|^2(1+|\xi|^2 2^{2j})^r\, d\xi \\ &\le 2^{2rj}\int_{\mathbb{R}} \left|\sum_{k\in\mathbb{Z}} c_k e^{2\pi i k\xi}\right|^2 |u(\xi)|^2(1+|\xi|^r)^r\, d\xi \\ &\le (K(r))^2 2^{2rj}\int_{-1/2}^{1/2} \left|\sum_{k\in\mathbb{Z}} c_k e^{2\pi i k\xi}\right|^2 d\xi \\ &\le (K(r))^2 2^{2rj}\|f\|^2_{W_2^0},\end{aligned}$$

where

$$(K(r))^2 := \sup_{\xi\in[-1/2,1/2]} \sum_{l\in\mathbb{Z}} (1+|\xi+l|^2)^r u^2(\xi+l) < \infty$$

by virtue of Lemma 8.7.3. ◊

LEMMA 8.7.5. *Let*

$$u^a(\xi) := \prod_{l=1}^{\infty} |q_a(2^{-2l}\xi)|^{2T(2l)-1}, \tag{8.136}$$

$$v_j^a(\xi) := |p_{j+1}(\xi-1/2)|^{T(j+1)} \prod_{l=1}^{\infty} |q_a(2^{-2l-1}\xi)|^{2T(2l-1)-1}, \tag{8.137}$$

where

$$q_a(\xi) := \min\left\{1, \sqrt{\alpha}2^{\beta}(|\cos\pi\xi|\cdot|\cos 2\pi\xi|)^{\beta/2}\right\}, \tag{8.138}$$

$$p_N(\xi) := \begin{cases} 1, & |\xi| \le \xi_N(a), \\ 2^{\frac{1+a}{2}+\frac{1}{N}}|\cos(\pi\xi)|^{\frac{1-a}{2}}|\sin(\pi\xi)|^{\frac{1+a}{2}}, & |\xi| \in (\xi_N(a), 1/2), \end{cases} \tag{8.139}$$

where $\beta := 1-a$, $\alpha \in (0,1)$ *is defined in* (8.103), $\xi_N(a) \in [1/4, 1/2]$, $\cos 2\pi\xi_N(a) = -b_N(a)$, *and* $b_N(a)$ *is a root of equation* (8.100). *Then*

$$|\widehat{\varphi}_j^a(\xi)| \le 2^{-j/2}u^a(2^{-j}\xi) \tag{8.140}$$

for all $j \ge 0$, *and*

$$|\widehat{\psi}_j^a(\xi)| \le 2^{\frac{1-j}{2}} v_n^a(2^{-j}\xi) \tag{8.141}$$

for all $j \ge n > 0$.

`Proof`. In view of (8.99) and (8.107), we have

$$\widehat{\varphi}_j^a(\xi) = 2^{-j/2}\prod_{l=j+1}^{\infty} m^a_{T(l)}(2^{-l}\xi).$$

Taking into account the monotone increase of $T(l)$ (see (8.34)), it follows from estimates (8.103) and $q_a(\xi) \le 1$ that

$$|\widehat{\varphi}_j^a(\xi)| \le \left|2^{-j/2}\prod_{l=1}^{\infty}(q_a(2^{-j-2l}\xi))^{2T(j+2l-1)-1}\right| \le 2^{-j/2}u^a(2^{-j}\xi).$$

Similarly, for all $j \geq n > 0$, using (8.101), we get

$$|\widehat{\psi_j^a}(\xi)| = 2^{\frac{1-j}{2}} \left| m^a_{T(j+1)}(-\xi 2^{-j-1} - 1/2) \prod_{l=j+2}^{\infty} m^a_{T(l)}(2^{-l}\xi) \right|$$
$$\leq 2^{\frac{1-j}{2}} \left| (p^a_{j+1}(-\xi 2^{-j-1} - 1/2))^{T(j+1)} \prod_{l=1}^{\infty} (q_a(2^{-j-2l-1}\xi))^{2T(j+2l)-1} \right|$$
$$\leq 2^{\frac{1-j}{2}} v_n^a(2^{-j}\xi).$$

◊

LEMMA 8.7.6. *The functions u^a, v_j^a, $j \in \mathbb{N}$, rapidly decrease at infinity.*

This statement is proved in the same way as Corollary 8.5.3.

LEMMA 8.7.7 (**Bernstein's inequality**). *For any $r \geq 0$, there exists $K_a(r)$ such that for all $f \in V_j^a$, $j \in \mathbb{N}$, the inequality*

$$\|f\|_{W_2^0} \leq \|f\|_{W_2^r} \leq K_a(r) 2^{rj} \|f\|_{W_2^0} \tag{8.142}$$

holds.

`Proof.` Let

$$f = \sum_{k\in\mathbb{Z}} c_k^a \varphi_{jk}^a, \qquad \sum_{k\in\mathbb{Z}} |c_k^a|^2 < \infty.$$

Then

$$\|f\|^2_{W_2^r} = \int_{\mathbb{R}} \left| \sum_{k\in\mathbb{Z}} c_k^a e^{\pi i 2^{1-j} k\xi} \right|^2 |\widehat{\varphi_j^a}(\xi)|^2 (1 + |\xi|^2)^r \, d\xi. \tag{8.143}$$

Using (8.140), we obtain

$$\|f\|^2_{W_2^r} \leq \int_{\mathbb{R}} \left| \sum_{k\in\mathbb{Z}} c_k^a e^{2\pi i k\xi} \right|^2 |u^a(\xi)|^2 (1 + |\xi|^2 2^{2j})^r \, d\xi$$
$$\leq 2^{2rj} \int_{\mathbb{R}} \left| \sum_{k\in\mathbb{Z}} c_k^a e^{2\pi i k\xi} \right|^2 |u^a(\xi)|^2 (1 + |\xi|^2)^r \, d\xi$$
$$\leq (K_a(r))^2 2^{2rj} \int_{-1/2}^{1/2} \left| \sum_{k\in\mathbb{Z}} c_k^a e^{2\pi i k\xi} \right|^2 d\xi$$
$$\leq (K_a(r))^2 2^{2rj} \|f\|^2_{W_2^0},$$

where

$$(K_a(r))^2 := \sup_{\xi\in[-1/2,1/2]} \sum_{l\in\mathbb{Z}} (1 + |\xi + l|^2)^r \, (u^a)^2(\xi + l) < \infty$$

due to the fact that u^a are rapidly decreasing. ◊

`Proof of Theorem 8.7.1.` Let $f \in W_2^r$; choose $n \in \mathbb{N}$ in such a way that $T(n) < 2r + 2 \leq T(n+1)$. Let R_j be an orthogonal projector from $L_2(\mathbb{R})$ on the space of wavelets $W_j := \overline{\operatorname{span}\{\psi_{jk},\ k \in \mathbb{Z}\}}$. It is clear that

$$\|P_0 f\|^2_{W_2^0} + \sum_{j=0}^{n-1} 2^{2rj} \|R_j f\|^2_{W_2^0} \le C(r) \|f\|^2_{W_2^0}, \tag{8.144}$$

where

$$C(r) := 1 + \sum_{j=0}^{n-1} 2^{2rj}.$$

For $j \ge n$, using Lemma 1.2.11, we have

$$\|R_j f\|^2_{W_2^0} = \sum_{k\in\mathbb{Z}} |\langle f, \psi_{jk}\rangle|^2 = 2^{2j} \int_0^1 \left| \sum_{l\in\mathbb{Z}} \widehat{f}\,(2^j(\xi+l))\, \overline{\widehat{\psi}_j\,(2^j(\xi+l))} \right|^2 d\xi.$$

Using (8.129) and the Cauchy inequality, we get

$$\begin{aligned} \|R_j f\|^2_{W_2^0} &\le 2^{1+j} \int_{-1/2}^{1/2} \left(\sum_{l\in\mathbb{Z}} |\widehat{f}(2^j(\xi+l)) v_n(\xi+l)| \right)^2 d\xi \\ &\le 2^{1+j} A \int_{-1/2}^{1/2} \left(\sum_{l\in\mathbb{Z}} |\widehat{f}(2^j(\xi+l))|^2 v_n(\xi+l) \right) d\xi \\ &\le A \int_{\mathbb{R}} |\hat{f}(\xi)|^2 v_n(2^{-j}\xi) d\xi, \end{aligned}$$

where

$$A := \sup_{\xi} \left[\sum_{l\in\mathbb{Z}} v_n(\xi+l) \right] < +\infty$$

by Lemma 8.7.3.

Since $2r+2 \le T(n+1)$, it follows that

$$B(r) := \sup_{\xi} \left(\sum_{j\in\mathbb{Z}} |2^{-j}\xi|^{-2r} v_n(2^{-j}\xi) \right) < \infty,$$

as v_n has a zero of multiplicity $T(n+1)$ at zero. Summing the previous estimates over $j \ge n$, we obtain

$$\begin{aligned} \sum_{j=n}^{\infty} 2^{2rj} \|R_j f\|^2_{W_2^0} &= 2A \sum_{j=n}^{\infty} 2^{2rj} \int_{\mathbb{R}} |\widehat{f}(\xi)|^2 v_n(2^{-j}\xi) d\xi \\ &= 2A \sum_{j=n}^{\infty} \int_{\mathbb{R}} |\xi|^{2r} |\widehat{f}(\xi)|^2 |2^{-j}\xi|^{-2r} v_n(2^{-j}\xi) d\xi \le 2\,A\,B(r) \|f\|^2_{W_2^r}. \end{aligned} \tag{8.145}$$

Finally, from (8.144) and (8.145) we get

$$\sum_k |c_k|^2 + \sum_{j,k} 2^{2rj} |d_{jk}|^2 \le [2\,A\,B(r) + C(r)] \|f\|^2_{W_2^r}. \tag{8.146}$$

Let us now prove the opposite inequality. Set $f_0 = P_0 f$ and $f_j = R_{j-1} f$ for $j > 0$. It is clear that the convergence of the series

$$\|P_0(f)\|^2_{W_2^0} + \sum_j 2^{2jr}\|R_j(f)\|^2_{W_2^0}$$

implies that in the space W_p^m the decomposition $f = \sum_j f_j$ takes place. According to Lemma 8.7.4,

$$\left\| \left(\frac{\mathrm{d}}{\mathrm{dx}}\right)^m f_j \right\|_{W_2^0} \le \|f_j\|_{W_2^m} \le K(m) 2^{mj}\|f_j\|_{W_2^0} = \varepsilon_j 2^{(m-r)j}, \tag{8.147}$$

where $\varepsilon_j := 2^{rj}\|f_j\|_{W_2^0} K(m)$. It follows (see Appendix A.13) that

$$\|f\|^2_{W_2^r} \le D(r)(\|P_0(f)\|^2_{W_2^0} + \sum_j 2^{2jr}\|R_j(f)\|^2_{W_2^0}), \tag{8.148}$$

where $D(r) = 2^{2r}K^2(m)$, which proves the first equivalence relation in (8.125).

Use (8.148) and (8.146) to estimate $\|P_j f - f\|^2_{W_2^s}$ for $f \in W_2^r$, $r > s$. We have

$$\begin{aligned}
\|P_j f - f\|^2_{W_2^s} &\le D(s) \sum_{l=j}^{\infty} 2^{2ls}\|R_l(f)\|^2_{W_2^0} \\
&\le 2^{2j(s-r)} D(s) \sum_{l=j}^{\infty} 2^{2lr}\|R_l(f)\|^2_{W_2^0} \\
&\le 2^{2j(s-r)} K(r,s)\|f\|^2_{W_2^r}\varepsilon(j,f),
\end{aligned}$$

where $K(r,s) = D(s)(A(r)B(r) + C(r))$, $\varepsilon(j,f) \in [0,1]$ tends to 0 as $j \to \infty$.

We prove the statements of the theorem for the system Ψ^a, using the same scheme with some technical modification. Consider a function $f \in W_2^r$ and choose $n \in \mathbb{N}$ so that

$$\left\lceil \frac{T(n)(1-a)}{2} \right\rceil < 2r + 2 \le \left\lceil \frac{T(n+1)(1-a)}{2} \right\rceil .$$

Let R_j^a be an orthogonal projector from $L_2(\mathbb{R})$ on the space of wavelets $W_j^\alpha :=$ span $\{\psi_{jk}^\alpha,\ k \in \mathbb{Z}\}$. It is clear that

$$\|P_0^a f\|^2_{W_2^0} + \sum_{j=0}^{n-1} 2^{2rj}\|R_j^a f\|^2_{W_2^0} \le C(r)\|f\|^2_{W_2^0}, \tag{8.149}$$

where

$$C(r) := 1 + \sum_{j=0}^{n-1} 2^{2rj}.$$

For $j \ge n$, using Lemma 1.2.11, we get

$$\|R_j^\alpha f\|^2_{W_2^0} = \sum_{k\in\mathbb{Z}} |\langle f, \psi_{jk}^\alpha\rangle|^2 = 2^{2j} \int_0^1 \left| \sum_{l\in\mathbb{Z}} \widehat{f}\left(2^j(\xi + l)\right) \overline{\widehat{\psi_j^\alpha}\left(2^j(\xi + l)\right)} \right|^2 d\xi.$$

Using (8.141) and the Cauchy inequality, we obtain

$$\begin{aligned}\|R_j^a f\|_{W_2^0}^2 &\le 2^{1+j} \int_{-1/2}^{1/2} \left(\sum_{l\in\mathbb{Z}} |\widehat{f}(2^j(\xi+l)) v_n^a(\xi+l)| \right)^2 d\xi \\ &\le 2^{1+j} A_a \int_{-1/2}^{1/2} \left(\sum_{l\in\mathbb{Z}} |\widehat{f}(2^j(\xi+l))|^2 v_n^a(\xi+l) \right) d\xi \\ &\le 2A_a \int_{\mathbb{R}} |\hat{f}(\xi)|^2 v_n^a(2^{-j}\xi) d\xi,\end{aligned}$$

where

$$A_a := \sup_\xi \left[\sum_{l\in\mathbb{Z}} v_n^a(\xi+l) \right] < +\infty$$

by virtue of Lemma 8.7.6. Since

$$2r+2 \le \left[\frac{T(n+1)(1-a)}{2} \right],$$

it follows that

$$B_a(r) := \sup_\xi \left(\sum_{j\in\mathbb{Z}} |2^{-j}\xi|^{-2r} v_n^a(2^{-j}\xi) \right) < \infty,$$

as v_n^a has a zero of multiplicity $\left[\frac{T(n+1)(1-a)}{2}\right]$ at zero. Summing the previous estimates over $j \ge n$, we get

$$(8.150)\qquad \begin{aligned}\sum_{j=n}^{\infty} 2^{2rj} \|R_j^a f\|_{W_2^0}^2 &= 2A_a \sum_{j=n}^{\infty} 2^{2rj} \int_{\mathbb{R}} |\widehat{f}(\xi)|^2 v_n^a(2^{-j}\xi) d\xi \\ &= 2A_a \sum_{j=n}^{\infty} \int_{\mathbb{R}} |\xi|^{2r} |\widehat{f}(\xi)|^2 |2^{-j}\xi|^{-2r} v_n^a(2^{-j}\xi) d\xi \\ &\le 2\, A_a\, B_a(r) \|f\|_{W_2^r}^2.\end{aligned}$$

Finally, from (8.149) and (8.150) we derive

$$(8.151)\qquad \sum_k |c_k|^2 + \sum_{j,k} 2^{2rj} |d_{jk}^a|^2 \le [2\, A_a\, B_a(r) + C(r)] \|f\|_{W_2^r}^2.$$

Set $f_0^a = P_0^a f$ and $f_j^a = R_{j-1}^a f$ for $j > 0$. Clearly, the convergence of the series

$$\|P_0^a(f)\|_{W_2^0}^2 + \sum_j 2^{2jr} \|R_j^a(f)\|_{W_2^0}^2$$

implies that the decomposition $f = \sum_j f_j$ takes place in the space W_p^m. According to Lemma 8.7.7,

$$(8.152)\qquad \left\| \left(\frac{\mathrm{d}}{\mathrm{dx}}\right)^m f_j^a \right\|_{W_2^0} \le \|f_j^a\|_{W_2^m} \le K_a(m) 2^{mj} \|f_j^a\|_{W_2^0} = \varepsilon_j 2^{(m-r)j},$$

where $\varepsilon_j := 2^{rj}\|f_j^a\|_{W_2^0}K_a(m)$. It follows (see Appendix A.13) that

$$\|f\|_{W_2^r}^2 \le D_a(r)(\|P_0^a(f)\|_{W_2^0}^2 + \sum_j 2^{2jr}\|R_j^a(f)\|_{W_2^0}^2), \tag{8.153}$$

where $D_a(r) := 2^{2r}K_a^2(m)$. This completes the proof of the second equivalence relation in (8.125).

The rate of approximation with respect to the system Ψ^a is estimated in the same way as for the system Ψ, using (8.153) and (8.151). For $f \in W_2^r$, $r > s$, we have

$$\begin{aligned}\|P_j^a f - f\|_{W_2^s}^2 &\le D_a(s)\sum_{l=j}^{\infty} 2^{2ls}\|R_l^a(f)\|_{W_2^0}^2 \\ &\le 2^{2j(s-r)}D_a(s)\sum_{l=j}^{\infty} 2^{2lr}\|R_l^a(f)\|_{W_2^0}^2 \\ &\le 2^{2j(s-r)}K_a(r,s)\|f\|_{W_2^0}^2\varepsilon(j,f),\end{aligned}$$

where $K_a(r,s) := D_a(s)(A_a(r)B_a(r) + C(r))$, $\varepsilon(j,f) \in [0,1]$ tends to 0 as $j \to \infty$. $\Diamond$

CHAPTER 9

Periodic Wavelets

9.1. PMRA and scaling sequence

If we periodize an orthonormal wavelet basis in $L_2(\mathbb{R})$ in the natural way, which is possible provided that wavelet functions decrease sufficiently rapidly at infinity, we obtain an orthonormal basis in $L_2(\mathbb{T})$ called a periodic wavelet system. This system consists of shifts of functions ψ_j, but in general ψ_j is not a dilation of the function ψ_0 (except for the Haar system). Analogously, periodizing binary dilations of a scaling function and taking the subspaces spanned by their shifts, we can obtain a structure which is in a sense similar to a multiresolution analysis, though not its complete analog. In a number of books periodic wavelet bases are defined in this way (in the next section, this will be discussed in more detail). However, this approach to the study of periodic objects is not quite natural; moreover, not all bases that are "worth" being called wavelet systems do satisfy this definition. To describe wavelet systems, we shall follow the standard scheme based on MRAs that, unlike the nonperiodic case, will be defined not only in L_2.

We shall present a theory of periodic wavelets for the multidimensional non-separable case. As in Chapter 2, the matrix dilation is an integer $d \times d$ matrix M whose eigenvalues are greater than 1 in modulus. Denote by m the absolute value of its determinant. Recall that for such a matrix, $m \in \mathbb{N}$, $m > 1$, and relations (2.5), (2.6), (2.8) hold.

Denote by X one of the spaces $L_p(\mathbb{T}^d)$, $1 \le p < \infty$, or $C(\mathbb{T}^d)$. Define the shift operator S_p^j, $p \in \mathbb{Z}^d$, $j \in \mathbb{Z}_+$, on X:

$$S_p^j f(x) := f(x + M^{-j}p). \tag{9.1}$$

DEFINITION 9.1.1. Let $V_j \subset X$, $j \in \mathbb{Z}_+$. We say that a collection $\{V_j\}_{j=0}^{\infty}$ is a *periodic multiresolution analysis (PMRA)* in X if the following properties (axioms) hold:

MR1. $V_j \subset V_{j+1}$.

MR2. $\overline{\bigcup_{j=0}^{\infty} V_j} = X$.

MR3. $\dim V_j = m^j$.

MR4. $\dim\{f \in V_j : S_n^j f = \lambda_n f \ \forall n \in \mathbb{Z}^d\} \le 1$ for any set of complex numbers $\{\lambda_n\}_{n\in\mathbb{Z}^d}$.

MR5. $f \in V_j \Leftrightarrow S_n^j f \in V_j$ for all $n \in \mathbb{Z}^d$.

MR6. a) $f \in V_j \Rightarrow f(M\cdot) \in V_{j+1}$.

b) $f \in V_{j+1} \Rightarrow \sum\limits_{s\in D(M)} f(M^{-1} \cdot + M^{-1}s) \in V_j$.

REMARK 9.1.2. Due to the periodicity of f, in conditions **MR4** and **MR5** we can consider only digits of the matrix M^j (instead of all $n \in \mathbb{Z}^d$); i.e., these conditions can be replaced by the following:

MR4$'$. $\dim\{f \in V_j : S_r^j f = \lambda_r f \ \ \forall r \in D(M^j)\} \le 1$ for any set of complex numbers $\{\lambda_r\}_{r\in D(M^j)}$.

MR5$'$. $f \in V_j \Leftrightarrow S_n^j f \in V_j$ for all $n \in D(M^j)$.

Note that in the case $d = 1$, $M = 2$ condition **MR4$'$** takes the form $\dim\{f \in V_j : S_1^j f = \lambda f\} \le 1$ for any complex number λ; i.e., it means that the dimension of any eigenspace of the 2^{-j}-shift operator in V_j does not exceed 1.

DEFINITION 9.1.3. Let $\{V_j\}_{j=0}^{\infty}$ be a PMRA in X. A sequence of functions $\{\varphi_j\}_{j=0}^{\infty}$, $\varphi \in V_j$, is called *scaling* if the functions $S_n^j \varphi_j$, $n \in D(M^j)$, form a basis in the space V_j.

THEOREM 9.1.4. *Functions $\varphi_j \in X$, $j \in \mathbb{Z}_+$, form a scaling sequence for a PMRA in X if and only if:*

Φ1. $\widehat{\varphi}_0(k) = 0$ *for all* $k \ne 0$.

Φ2. *For any $j \in \mathbb{Z}_+$ and any $n \in \mathbb{Z}^d$, there exists $k \equiv n \pmod{M^{*j}}$ such that $\widehat{\varphi}_j(k) \ne 0$.*

Φ3. *For any $k \in \mathbb{Z}^d$, there exists $j \in \mathbb{Z}_+$ such that $\widehat{\varphi}_j(k) \ne 0$.*

Φ4. *For any $j \in \mathbb{N}$, $n \in \mathbb{Z}^d$, there exists a number μ_n^j such that $\widehat{\varphi}_{j-1}(k) = \mu_n^j \widehat{\varphi}_j(k)$ for all $k \equiv n \pmod{M^{*j}}$.*

Φ5. *For any $j \in \mathbb{Z}_+$, $n \in \mathbb{Z}^d$, there exists a number $\gamma_n^j \ne 0$ such that $\gamma_n^j \widehat{\varphi}_j(k) = \widehat{\varphi}_{j+1}(M^* k)$ for all $k \equiv n \pmod{M^{*j}}$.*

First we prove some auxiliary assertions.

LEMMA 9.1.5. *Suppose $V_j \subset X$, $j \in \mathbb{Z}_+$, and axioms* **MR1**, **MR2**, **MR3**, **MR6 b)** *of Definition* 9.1.1 *hold. Then the space V_0 consists of constants.*

Proof. By property **MR3** the space V_0 is one-dimensional. Let $f \in V_0$, $\|f\| \ne 0$. Show that $\widehat{f}(\mathbf{0}) \ne 0$. Assume that $\widehat{f}(\mathbf{0}) = 0$. Introduce the operator $A:\ Af = \sum_{s\in D(M)} f(M^{-1}\cdot + M^{-1}s)$ and set $g = Af$. By Lemma 2.2.5, we have

$$\widehat{g}(\mathbf{0}) = \sum_{s\in D(M)} \int_{\mathbb{T}^d} f(M^{-1}x + M^{-1}s)dx$$

$$= m \sum_{s\in D(M)} \int_{M^{-1}\mathbb{T}^d + M^{-1}s} f(z)dz = m \int_{\mathbb{T}^d} f(z)dz = m\widehat{f}(0).$$

Let $g_0 \in V_j$. Then by **MR6 b)**, $g_1 := Ag_0 \in V_{j-1}, \dots, g_j := Ag_{j-1} \in V_0$. If $\widehat{g}_j(\mathbf{0}) \ne 0$, this contradicts the fact that V_0 is a one-dimensional space, since $\widehat{f}(\mathbf{0}) = 0$. If $\widehat{g}_0(\mathbf{0}) = 0$, then any function from any V_j has zero mean, which contradicts axiom **MR2**.

Assume now that $\widehat{f}(n) \ne 0$ for some $n \ne \mathbf{0}$. Since $f \in V_0$, using **MR1**, we have $f \in V_1$; hence $g \in V_0$ by **MR6 b)**. Since V_0 is a one-dimensional space, it follows that $g = \lambda f$, which implies that $\widehat{g}(n) = \lambda \widehat{f}(n)$, but $\lambda = m$ because $\widehat{g}(0) = m\widehat{f}(0)$.

On the other hand, in view of Lemma 2.2.5 direct calculation gives

$$\begin{aligned}\widehat{g}(n) &= \sum_{s\in D(M)} \int_{\mathbb{T}^d} f(M^{-1}x + M^{-1}s)e^{2\pi i(x,n)}dx \\ &= m \sum_{s\in D(M)} \int_{M^{-1}\mathbb{T}^d+M^{-1}s} f(z)e^{2\pi i(Mz-s,n)}dz \\ &= m \int_{\mathbb{T}^d} f(z)e^{2\pi i(Mz,n)}dz m\widehat{f}(M^*n).\end{aligned}$$

Consequently,

$$\widehat{f}(n) = \widehat{f}(M^*n) = \cdots = \widehat{f}(M^{*l}n) = \cdots ,$$

but since the Fourier coefficients of the function $f \in X$ tend to zero as absolute values of their numbers increase, we get a contradiction due to (2.8). ◊

Let us define the operators ω_n^j, $j \in \mathbb{Z}_+, n \in \mathbb{Z}^d$, on X recursively:

$$\begin{aligned}\omega_n^0 f &:= f, \\ \omega_n^{j+1} f(x) &:= \frac{1}{m} \sum_{s\in D(M)} e^{-2\pi i(M^{-j-1}s,n)}\omega_n^j f(x + M^{-j-1}s).\end{aligned}$$

LEMMA 9.1.6. *Suppose $V_j \subset X$, $j \in \mathbb{Z}_+$, and axiom* **MR5** *of Definition* 9.1.1 *holds. If $f \in V_{j_0}$, then $\omega_n^j f \in V_{j_0}$ for all $j = 0, \ldots, j_0$, $n \in \mathbb{Z}^d$, and*

$$\omega_n^j f \sim \sum_{m\in\mathbb{Z}^d} \widehat{f}(M^{*j}m + n)e^{2\pi i(M^{*j}m+n,\cdot)}; \tag{9.2}$$

i.e., $\widehat{\omega_n^j f}(k) = \widehat{f}(k)$ for $k \equiv n \pmod{M^{*j}}$ and $\widehat{\omega_n^j f}(k) = 0$ for $k \not\equiv n \pmod{M^{*j}}$.

Proof. The proof is by induction on j. For $j = 0$, there is nothing to prove. Suppose $\omega_n^j f \in V_{j_0}$, $0 \le j < j_0$, and (9.2) holds for all $n \in \mathbb{Z}^d$. Then from **MR5** it follows that

$$\omega_n^j f(\cdot + M^{-j-1}s) = \omega_n^j f(\cdot + M^{-j_0}(M^{j_0-j-1}s)) \in V_{j_0},$$

which implies that $\omega_n^{j+1} f \in V_{j_0}$. Further, we have

$$\begin{aligned}\widehat{\omega_n^{j+1} f}(k) &= \frac{1}{m} \int_{\mathbb{T}^d} \sum_{s\in D(M)} e^{-2\pi i(M^{-j-1}s,n)}\omega_n^j f(x + M^{-j-1}s)e^{-2\pi i(x,k)}dx \\ &= \frac{1}{m} \sum_{s\in D(M)} e^{-2\pi i(M^{-j-1}s,n)} \int_{\mathbb{T}^d} \omega_n^j f(t)e^{-2\pi i(t-M^{-j-1}s,k)}dt \\ &= \frac{1}{m} \sum_{s\in D(M)} e^{-2\pi i(M^{-j-1}s,n-k)}\widehat{\omega_n^j f}(k).\end{aligned}$$

If $k \equiv n \pmod{M^{*j+1}}$, then the sum on the right-hand side is obviously equal to m; hence $\widehat{\omega_n^{j+1} f}(k) = \widehat{\omega_n^j f}(k)$. If $k \not\equiv n \pmod{M^{*j+1}}$ and $n \equiv k \pmod{M^{*j}}$, i.e., $n - k = M^{*j}l$, where the vector l is not congruent to zero modulo M^*, then by Lemma 2.2.6, the latter sum is equal to zero. Finally, if $k \not\equiv n \pmod{M^{*j}}$, then by the inductive assumption, $\widehat{\omega_n^j f}(k) = 0$; hence $\widehat{\omega_n^{j+1} f}(k) = 0$. ◊

LEMMA 9.1.7. *Let $\{V_j\}_{j=0}^{\infty}$ be a PMRA in X. Then in each space V_j there exists a basis $\{v_n^j\}_{n\in D(M^{*j})}$ (for the sake of convenience we set $v_l^j := v_n^j$ if $l \equiv n \pmod{M^j}$) such that:*

V1. $\widehat{v}_n^j(k) = 0$ *for all* $k \not\equiv n \pmod{M^{*j}}$.

V2. *If* $\widehat{v}_n^j(k) \neq 0$ *and* $n \equiv k \pmod{M^{*j+1}}$, *then* $\widehat{v}_n^j(\ell) = \widehat{v}_n^{j+1}(\ell)$ *for all* $\ell \equiv k \pmod{M^{*j+1}}$.

V3. $\widehat{v}_n^j(k) = \widehat{v}_{M^*n}^{j+1}(M^*k)$ *for all* $k \in \mathbb{Z}^d$.

Proof. The proof is by induction on j. For $j = 0$ the statement is obvious, since all integer vectors are congruent. Assume that in the spaces V_j, $j = 0, \dots, j_0$, there exist bases satisfying **V1**, **V2**, **V3**. Introduce the spaces

$$V_j^{(n)} := \{f \in V_j : \widehat{f}(k) = 0 \quad \forall k \not\equiv n \pmod{M^{*j}}\}.$$

Let $F \in V_j$. Then

$$F = \sum_{n\in D(M^{*j})} \omega_n^j F = \sum_{n\in D(M^{*j})} F_n, \quad F_n \in V_j^{(n)}.$$

This means that $V_j = \sum_{n\in D(M^{*j})} V_j^{(n)}$. Consequently,

$$m^j = \dim V_j \leq \sum_{n\in D(M^{*j})} \dim V_j^{(n)}. \tag{9.3}$$

Let us find the dimension of the space $V_j^{(n)}$. If $f \in V_j^{(n)}$, then

$$f(x) \sim \sum_{m\in\mathbb{Z}^d} \widehat{f}(M^{*j}m + n)e^{2\pi i(M^{*j}m+n,x)}.$$

Applying the shift operator, we get

$$(S_p^j f)(x) \sim \sum_{m\in\mathbb{Z}^d} \widehat{f}(M^{*j}m + n)e^{2\pi i(M^{*j}m+n,x+M^{-j}p)} \sim e^{2\pi i(n,M^{-j}p)} f(x);$$

i.e., $S_p^j f(x) = e^{2\pi i(n,M^{-j}p)} f(x)$ for all $p \in \mathbb{Z}^d$. Taking into account **MR4** and (9.3), we deduce that $\dim V_j^{(n)} = 1$.

Let us construct a basis $\{v_n^{j_0+1}\}$. If $\widehat{v}_k^{j_0}(k) \neq 0$, we set $v_k^{j_0+1} := \omega_k^{j_0+1} v_k^{j_0}$; then properties **V1**, **V2** are satisfied by Lemma 9.1.6. Check **V3**. Let $k \equiv \mathbf{0} \pmod{M^*}$, $n \equiv k \pmod{M^{*j_0+1}}$. By the inductive assumption, we have

$$\widehat{v}_k^{j_0+1}(n) = \widehat{v}_k^{j_0}(n) = \widehat{v}_{M^{*-1}k}^{j_0-1}(M^{*-1}n) = \widehat{v}_{M^{*-1}k}^{j_0}(M^{*-1}n).$$

Thus we have found the basis functions with the numbers k for which there exists $n \equiv k \pmod{M^{*j_0+1}}$ such that $\widehat{v}_k^{j_0}(n) \neq 0$. Now let $\widehat{v}_k^{j_0}(n) = 0$ for all $n \equiv k \pmod{M^{*j_0+1}}$ and $k \equiv \mathbf{0} \pmod{M^*}$. Set $v_k^{j_0+1}(x) := v_{M^{*-1}k}^{j_0}(Mx)$. Using

Lemma 2.2.5, we get

$$\widehat{v}_k^{j_0+1}(n) = \int_{\mathbb{T}^d} v_k^{j_0+1}(x)e^{-2\pi i(x,n)}dx = \int_{\mathbb{T}^d} v^{j_0}_{M^{*-1}k}(Mx)e^{-2\pi i(x,n)}dx$$

$$= \sum_{s\in D(M)} \int_{M^{-1}\mathbb{T}^d+M^{-1}s} v^{j_0}_{M^{*-1}k}(Mx)e^{-2\pi i(x,n)}dx$$

$$= \frac{1}{m}\sum_{s\in D(M)} \int_{\mathbb{T}^d+s} v^{j_0}_{M^{*-1}k}(t)e^{-2\pi i(t,M^{*-1}n)}dt = \widehat{v}^{j_0}_{M^{*-1}k}(M^{*-1}n).$$

It is clear that property **V3** holds. **V1** holds by the inductive assumption, while **V2** need not be verified. Finally, let $\widehat{v}_k^{j_0}(n) = 0$ for all $n \equiv k \pmod{M^{*j_0+1}}$ and $k \not\equiv \mathbf{0} \pmod{M^*}$. In this case, for $v_k^{j_0+1}$, we take any nonzero element from the space $V^{(k)}_{j_0+1}$. Property **V1** follows from the definition of $V^{(k)}_{j_0+1}$, while **V2** and **V3** have no need of a check. ◊

REMARK 9.1.8. If spaces $V_j \subset X$, $j \in \mathbb{Z}_+$, satisfy only axioms **MR1**, **MR3**, **MR4**, **MR5** of Definition 9.1.1, then in each space V_j there exists a basis $\{v_n^j\}_{n\in D(M^{*j})}$ that satisfies conditions **V1**, **V2**. This can be proved following the same scheme, if for v_0^0, we take any nonzero element of the space V_0. If $\widehat{v}_k^{j_0}(n) = 0$ for all $n \equiv k \pmod{M^{*j_0+1}}$ and $k \equiv \mathbf{0} \pmod{M^*}$, we take any nonzero element of the space $V^{(k)}_{j_0+1}$.

LEMMA 9.1.9. *If in each space* $V_j \subset X$, $j \in \mathbb{Z}_+$, *there exists a basis* $\{v_n^j\}_{n\in D(M^{*j})}$ *that satisfies condition* **V1** *of Lemma* 9.1.7, *then axiom* **MR4** *of Definition* 9.1.1 *holds.*

Proof. Taking into account Remark 9.1.2, it is sufficient to check condition **MR4′**. Suppose $r \in D(M^j)$, f is an eigenvalue of the operator S_r^j, i.e., $S_r^j f = \lambda_r f$, and $f = \sum_{n\in D(M^{*j})} \alpha_n^j v_n^j$. Since **V1** is satisfied, it follows that on v_n^j the operator S_r^j is the operator of multiplication by $e^{2\pi i(M^{*-j}n,r)}$. Applying the operator S_r^j to the function f, we get

$$S_r^j f(x) = \sum_{n\in D(M^{*j})} \alpha_n^j S_n^j v_n^j(x) = \sum_{n\in D(M^{*j})} \alpha_n^j e^{2\pi i(M^{*-j}n,r)} v_n^j(x).$$

Taking into account that f is an eigenvalue of the operator S_r^j, we obtain

$$0 = S_r^j f(x) - \lambda_r f(x) = \sum_{n\in D(M^{*j})} \alpha_n^j [e^{2\pi i(M^{*-j}n,r)} - \lambda_r] v_n^j(x).$$

From the linear independence of v_n^j it follows that $\alpha_n^j[e^{2\pi i(M^{*-j}n,r)} - \lambda_r] = 0$ for all $n \in D(M^{*j})$. Let us assume that $\alpha_{n_0}^j \neq 0$, $\alpha_{n_1}^j \neq 0$ for two different digits n_0, n_1. Then we get

$$(e^{2\pi i(M^{*-j}n_0,r)} - \lambda_r) - (e^{2\pi i(M^{*-j}n_1,r)} - \lambda_r) = 0,$$

or $e^{2\pi i(M^{*-j}(n_0-n_1),r)} = 1$. If f is an eigenvector of all the operators S_r^j, $r \in D(M^j)$, then summing these equalities over all r, we obtain

$$\sum_{r\in D(M^j)} e^{2\pi i(M^{*-j}(n_0-n_1),r)} = m^j. \tag{9.4}$$

On the other hand, since n_0 is not congruent to n_1 modulo M^{*j}, by Lemma 2.2.6, we get

$$\sum_{r\in D(M^{*j})} e^{2\pi i(M^{*-j}(n_0-n_1),r)} = 0, \tag{9.5}$$

which contradicts (9.4). Thus α_n^j can be nonzero only for $n = n_0$; hence f is proportional to $v_{n_0}^j$. In addition, it follows from the aforesaid that

$$\lambda_r = e^{2\pi i(M^{*-j}n_0,r)}, \quad r \in D(M^j). \tag{9.6}$$

Suppose $g = \sum_n \beta_n^j v_n^j$ is another eigenvalue of all the operators S_r^j, $r \in D(M^j)$. For it, too, only one β_n^j is nonzero. Assume that $\beta_{n_1}^j \neq 0$, $n_1 \neq n_0$. Then, analogously to (9.6), we have $\lambda_r = e^{2\pi i(M^{*-j}n_1,r)}$, $r \in D(M^j)$. Combining this with (9.6), we get $e^{2\pi i(M^{*-j}(n_0-n_1),r)} = 1$ for any $r \in D(M^j)$. Summing these equalities over all r, we obtain (9.4), which contradicts (9.5). Consequently, g is also proportional to $v_{n_0}^j$; hence any subspace consisting of all such functions cannot have dimension greater than 1. ◊

PROPOSITION 9.1.10. *Let V_j be a PMRA in X. A sequence $\{\varphi_j\}_{j=0}^{\infty} \subset X$ is a scaling sequence if and only if*

$$\varphi_j = \sum_{n\in D(M^{*j})} \alpha_n^j v_n^j, \quad \alpha_n^j \neq 0 \quad \forall n \in D(M^{*j}), \tag{9.7}$$

where $\{v_n^j\}_{n\in D(M^{*j})}$ is a basis of the space V_j defined in Lemma 9.1.7.

Proof. Necessity. Let $\{\varphi_j\}_{j=0}^{\infty}$ be a scaling sequence and let α_n^j, $n \in D(M^{*j})$, be coefficients of the expansion of φ_j with respect to the basis $\{v_n^j\}_{n\in D(M^{*j})}$. Applying the shift operator S_r^j, $r \in D(M^j)$, to φ_j, we get

$$S_r^j\varphi_j = \sum_{n\in D(M^{*j})} \alpha_n^j e^{2\pi i(M^{*-j}n,r)} v_n^j. \tag{9.8}$$

Assume that $\alpha_{n_0}^j = 0$ for a certain n_0. Then we have

$$V_j = \operatorname{span}\{S_r^j\varphi_j,\ r \in D(M^j)\} = \operatorname{span}\{v_n^j,\ n \in D(M^{*j}),\ n \neq n_0\},$$

which contradicts the minimality of the basis $\{v_n^j\}_{n\in D(M^{*j})}$.

Sufficiency. Let functions φ_j be given by formula (9.7). As above,

$$S_r^j\varphi_j = \sum_{n\in D(M^{*j})} \alpha_n^j e^{2\pi i(M^{*-j}n,r)} v_n^j, \quad r \in D(M^j).$$

Let us consider these equalities as a system of equations for the unknowns $\alpha_n^j v_n^j$. By Corollary 2.2.7, the matrix of this system is unitary (up to a factor). Since the functions $\alpha_n^j v_n^j$ form a basis and any unitary map takes a basis to a basis, it follows that the functions $S_r^j\varphi_j$, $r \in D(M^j)$, form a basis of the space V_j. ◊

COROLLARY 9.1.11. *If $\{\varphi_j\}_{j=0}^{\infty}$ is a scaling sequence, then $\omega_n^j\varphi_j = \alpha_n^j v_n^j$, where $\alpha_n^j \neq 0$. In particular, the functions $\{\omega_n^j\varphi_j\}_{n=0}^{2^j-1}$ form a basis of the space V_j.*

This assertion follows from (9.2) and the fact that $\dim V_j^{(n)} = 1$ ($V_j^{(n)}$ are the spaces defined in Lemma 9.1.7).

COROLLARY 9.1.12. *In any PMRA there exists a scaling sequence.*

To prove this, it suffices to set $\alpha_n^j = 1$ in (9.7).

`Proof of Theorem 9.1.4.` Necessity. Property **Φ1** follows from Lemma 9.1.5. To prove **Φ2**, we use Corollary 9.1.11. The equality $\omega_n^j \varphi_j = \alpha_n^j v_n^j$ implies that

$$\widehat{\varphi}_j(k) = \widehat{\omega_k^j \varphi_j}(k) = \alpha_n^j \widehat{v}_n^j(k)$$

for any $k \equiv n \pmod{M^{*j}}$. The existence of k congruent to n modulo M^{*j} and such that $\widehat{v}_n^j(k) \neq 0$ follows from the fact that $v_n^j \not\equiv 0$ (since v_n^j is a basis vector) and $\widehat{v}_n^j(\ell) = 0$ for all $\ell \not\equiv n \pmod{M^{*j}}$. Taking into account that $\alpha_n^j \neq 0$, we get $\widehat{\varphi}_j(k) \neq 0$. To prove property **Φ3**, assume the converse. Suppose that $\widehat{\varphi}_j(k) = 0$ for all $j \in \mathbb{Z}_+$. This means that none of the functions φ_j contain the term $e^{2\pi i(k,x)}$. Hence none of the shifts $S_k^j \varphi_j$ contain this term, that is, for any function $f \in \bigcup_{j=0}^{\infty} V_j$, the scalar product $\langle f, e^{2\pi i(k,\cdot)} \rangle$ is equal to zero, which contradicts the completeness of the collection of the spaces V_j (axiom **MR4**). To prove **Φ4**, take an arbitrary $n \in \mathbb{Z}^d$. First consider the case where there exists $k \equiv n \pmod{M^{*j}}$ such that $\widehat{\varphi}_{j-1}(k) \neq 0$. By Corollary 9.1.11, $\omega_n^j \varphi_j = \alpha_n^j v_n^j$, $\omega_n^j \varphi_{j-1} = \omega_n^j \omega_n^{j-1} \varphi_{j-1} = \alpha_n^{j-1} \omega_n^j v_n^{j-1}$, $\alpha_n^j, \alpha_n^{j-1} \neq 0$. In view of property **V2** of Lemma 9.1.7, it follows that

$$\widehat{\varphi}_{j-1}(\ell) = \frac{\alpha_n^{j-1}}{\alpha_n^j} \widehat{\varphi}_j(\ell),$$

for all $\ell \equiv n \pmod{M^{*j}}$. It remains to set $\mu_n^j = \alpha_n^{j-1}/\alpha_n^j$. If $\widehat{\varphi}_{j-1}(k) = 0$ for any $k \equiv n \pmod{M^{*j}}$, then we set $\mu_n^j = 0$. To prove **Φ5** , we again use Corollary 9.1.11. For $k \equiv n \pmod{M^{*j}}$, we have

$$\widehat{\varphi}_{j+1}(M^*k) = \widehat{\omega_{M^*n}^{j+1} \varphi_{j+1}}(M^*k) = \alpha_{M^*n}^{j+1} \widehat{v}_{M^*n}^{j+1}(M^*k),$$

where $\alpha_{M^*n}^{j+1} \neq 0$. On the other hand, $\widehat{\varphi}_j(n) = \alpha_k^j \widehat{v}_n^j(k)$, $\alpha_n^j \neq 0$. In view of property **V3** of Lemma 9.1.7, it follows that

$$\widehat{\varphi}_{j+1}(M^*k) = \frac{\alpha_{M^*n}^{j+1}}{\alpha_n^j} \widehat{\varphi}_j(k).$$

It remains to put $\gamma_n^j = \alpha_{M^*n}^{j+1}/\alpha_n^j$.

Sufficiency. Let functions $\varphi_j \in X$ satisfy properties **Φ1**–**Φ5**. Set $V_j = \operatorname{span}\{S_n^j \varphi_j,\ n \in D(M^j)\}$ and show that $\{V_j\}_{j=0}^{\infty}$ is a PMRA in X and $\{\varphi_j\}_{j \in \mathbb{Z}_+}$ is a scaling sequence. First check **MR5**. Let $f \in V_j$. Then $f = \sum_{k \in D(M^j)} \alpha_k S_k^j \varphi_j$. Applying the shift operator S_p^j, $p \in D(M^j)$, to this equality, we get

$$S_p^j f = \sum_{k \in D(M^j)} \alpha_k S_p^j S_k^j \varphi_j. \tag{9.9}$$

The periodicity of the function f implies that $S_p^j S_n^j f = S_r^j f$, where $r \in D(M^j)$, $r \equiv (p+n) \pmod{M^j}$. Substituting these equalities into (9.9), by the definition of V_j we obtain $S_p^j f \in V_j$. If now $S_p^j f \in V_j$, then by the aforesaid $f = S_{-p}^j S_p^j f \in V_j$. Now let us prove **MR3**. Analogously to (9.8), taking into account Lemma 9.1.6, we have

$$S_r^j \varphi_j = \sum_{n \in D(M^{*j})} e^{2\pi i(M^{*-j}n, r)} \omega_n^j \varphi_j. \tag{9.10}$$

It follows that $V_j = \operatorname{span}\{\omega_n^j\varphi_j,\ n\in D(M^{*j})\}$. Let us show that the functions $\omega_n^j\varphi_j$, $n\in D(M^{*j})$, form a basis of V_j. Suppose these functions are linearly independent. Then there exist numbers α_k, $k\in D(M^{*j})$, $\alpha_{k_0}\neq 0$, such that $\sum_{k\in D(M^{*j})}\alpha_k\omega_k^j\varphi_j = 0$. By Lemma 9.1.6, it follows that $\widehat{\omega_n^j\varphi_j}(k) = \widehat{\varphi}_j(k) = 0$ for all $k\equiv k_0 \pmod{M^{*j}}$, which contradicts **Φ2**. To prove **MR3**, it remains to note that the number of the functions $\omega_n^j\varphi_j$ is m^j. Since the number of the functions $S_n^j\varphi_j, n\in D(M^j)$, is also equal to m^j, we deduce that $\{S_n^j\varphi_j\}_{n\in D(M^j)}$ is a basis of the space V_j. To prove **MR1**, we need to verify if $f\in V_j$ implies that $f\in V_{j+1}$. It suffices to restrict our consideration to the basis functions $\omega_n^j\varphi_j, n\in D(M^{*j})$. Lemmas 9.1.6 and 2.2.3 imply that

$$\omega_n^j\varphi_j = \sum_{p\in D(M^*)} \omega_{n+M^{*j}p}^{j+1}\omega_n^j\varphi_j. \tag{9.11}$$

Using Lemma 9.1.6 and **Φ4** and taking into account the M^{*j+1}-periodicity of the sequence μ_n^{j+1} with respect to the subindex, we get

$$\begin{aligned}
\omega_{n+M^{*j}p}^{j+1}\omega_n^j\varphi_j &\sim \sum_{k\in\mathbb{Z}^d}\widehat{\omega_n^j\varphi_j}(M^{*j+1}k+n+M^{*j}p)e^{2\pi i(M^{*j+1}k+n+M^{*j}p,x)}\\
&= \sum_{k\in\mathbb{Z}^d}\widehat{\varphi}_j(M^{*j+1}k+n+M^{*j}p)e^{2\pi i(M^{*j+1}k+n+M^{*j}p,x)}\\
&= \sum_{k\in\mathbb{Z}^d}\mu_{M^{*j+1}k+n+M^{*j}p}^{j+1}\widehat{\varphi}_{j+1}(M^{*j+1}k+n+M^{*j}p)e^{2\pi i(M^{*j+1}k+n+M^{*j}p,x)}\\
&= \mu_{n+M^{*j}p}^{j+1}\sum_{k\in\mathbb{Z}^d}\widehat{\varphi}_{j+1}(M^{*j+1}k+n+M^{*j}p)e^{2\pi i(M^{*j+1}k+n+M^{*j}p,x)}.
\end{aligned}$$

The right-hand side of this equality is the Fourier series of the function

$$\mu_{n+M^{*j}p}^{j+1}\omega_{n+M^{*j}p}^{j+1}\varphi_{j+1}.$$

It follows that

$$\omega_n^j\varphi_j = \sum_{p\in D(M^*)}\mu_{n+M^{*j}p}^{j+1}\omega_{n+M^{*j}p}^{j+1}\varphi_{j+1}. \tag{9.12}$$

It remains to note that $\omega_{n+M^{*j}p}^{j+1}\varphi_{j+1}\in V_{j+1}$ by Lemma 9.1.6.

Before proceeding to the remaining properties, let us show that in V_j there exists a basis satisfying the conditions of Lemma 9.1.7. Define numbers α_n^j, $j\in\mathbb{Z}_+$, $n\in D(M^{*j})$, recursively with respect to j: $\alpha_0^0 := 1$:

If $\mu_n^j\neq 0$, then $\alpha_n^j := \alpha_n^{j-1}/\mu_n^j$.

If $\mu_n^j = 0$, $n\equiv \mathbf{0} \pmod{M^*}$, then $\alpha_n^j := \alpha_{M^{*-1}n}^{j-1}\gamma_{M^{*-1}n}^{j-1}$.

If $\mu_n^j = 0$, $n\not\equiv \mathbf{0} \pmod{M^*}$, then $\alpha_n^j := 1$.

By construction, $\alpha_n^j\neq 0$. Set $v_n^j = \omega_n^j\varphi_j/\alpha_n^j$. As already shown, $\omega_n^j\varphi_j$ form a basis of the space V_j; hence $\{v_n^j\}_{n\in D(M^{*j})}$ is also a basis, and it is not hard to verify that properties **V1**–**V3** of Lemma 9.1.7 hold. Property **MR4** follows from Lemma 9.1.9. To prove **MR6**, it suffices to show that it holds for the functions $\{v_n^j\}_{n\in D(M^{*j})}$.

Using Lemma 9.1.6 and **V3**, we get

$$v_n^j(Mx) \sim \sum_{k\in\mathbb{Z}^d} \widehat{v}_n^j(k)e^{2\pi i(k,Mx)} = \sum_{k\in\mathbb{Z}^d} \widehat{v}_{M^*n}^{j+1}(M^*k)e^{2\pi i(M^*k,x)}$$
$$= \sum_{l\in\mathbb{Z}^d,\ l\equiv\mathbf{0}\ (\mathrm{mod}\, M^*)} \widehat{v}_{M^*n}^{j+1}(l)e^{2\pi i(l,x)} \sim v_{M^*n}^{j+1}(x).$$

To prove **MR6 a**), it remains to note that $v_{M^*n}^{j+1} \in V_{j+1}$. Let us check **MR6 b**). First consider the case $n \equiv \mathbf{0} \pmod{M^*}$. Using **V3**, we obtain

$$\sum_{k\in D(M)} v_n^{j+1}(M^{-1}x + M^{-1}k) = \sum_{k\in D(M)} v_{M^{*-1}n}^j(x+k) \in V_j.$$

Now let $n \not\equiv \mathbf{0} \pmod{M^*}$. Then the function

$$\sum_{k\in D(M)} v_n^{j+1}(M^{-1}x + M^{-1}k)$$

has the Fourier series

$$\sum_{l\in\mathbb{Z}^d}\sum_{k\in D(M)} \widehat{v}_n^{j+1}(M^{*j+1}l+n)e^{2\pi i(M^{*j+1}l+n,M^{-1}x+M^{-1}k)}$$
$$= \sum_{l\in\mathbb{Z}^d} \widehat{v}_n^{j+1}(M^{*j+1}l+n)e^{2\pi i(M^{*j+1}l+n,M^{-1}x)} \sum_{k\in D(M)} e^{2\pi i(M^{*j}l+M^{*-1}n,k)}.$$

The latter sum is equal to zero by Lemma 2.2.6; i.e., **MR6 b**) holds. It remains to prove **MR2**. Since trigonometric polynomials are dense in X, it suffices to prove that any trigonometric polynomial can be approximated by functions from $\bigcup_{j=0}^{\infty} V_j$. It is sufficient to prove this for one harmonic component. Set $f_r(x) = e^{2\pi i(r,x)}$, $r \neq \mathbf{0}$. From **Φ3** it follows that there exists j_0 such that $\widehat{\varphi}_{j_0}(r) \neq 0$, and $\widehat{\varphi}_j(r) \neq 0$ by **Φ4**. Hence $\widehat{v}_r^j(r) \neq 0$ for all $j \geq j_0$. Introduce functions h_j for $j \geq j_0$:

$$h_j(x) := 1 - \frac{v_r^j(x)}{\widehat{v}_r^j(r)}e^{-2\pi i(r,x)};$$

then

$$\widehat{h}_j(n) = \delta_{n0} - \frac{\widehat{v}_r^j(r-n)}{\widehat{v}_r^j(r)}, \quad n \in \mathbb{Z}^d.$$

Thus $\widehat{h}_j(n)$ is different from zero only if $n \neq \mathbf{0}$ and $n \equiv \mathbf{0} \pmod{M^{*j}}$; hence $\widehat{h}_j(n) = 0$ for all integer n from $M^{*j}(-1,1)^d$. Let us select a sequence of embedded parallelepipeds $M^{*j_k}(-1,1)^d$. It follows from (2.5) that there exists n_0 such that $\|M^{*-n_0}\| \leq 1/2$. Setting $j_{k+1} = j_k + n_0$, we get $\|M^{*j_{k+1}}x\| \geq 2\|M^{*j_k}x\|$ for any $x \in \mathbb{R}^d$. For sufficiently large k, we can inscribe into any oblique-angled parallelepiped $M^{*j_k}(-1,1)^d$ a cube K_{j_k} with center at the origin and edges of length a_k parallel to the coordinate axes so that a_k are even integers monotonically and indefinitely increasing as k increases. For convenience, we shall write K_j for our subsequence. As noted above, $\widehat{h}_j(n) = 0$ for all integer n from K_j; hence the partial sums of the Fourier series over K_j for the function h_j are equal to zero. It follows that the corresponding Fejer means $\sigma_{K_j}(h_j)$ over those cubes are also equal to zero. In the sequel we shall need the following equality for $j \geq j_0$:

$$h_j(x) = m^{j_0-j} \sum_{n\in D(M^{j-j_0})} h_{j_0}(x + M^{-j}n). \tag{9.13}$$

To prove it, we apply Lemma 9.1.6 and deduce that the Fourier series of the function

$$m^{j_0-j} \sum_{n\in D(M^{j-j_0})} h_{j_0}(x + M^{-j}n)$$
$$= m^{j_0-j} \sum_{n\in D(M^{j-j_0})} [1 - \frac{v_r^{j_0}(x + M^{-j}n)}{\widehat{v}_r^{j_0}(r)} e^{-2\pi i(r,\, x+M^{-j}n)}]$$

is equal to

$$- \frac{m^{j_0-j}}{\widehat{v}_r^{j_0}(r)} \sum_{n\in D(M^{j-j_0})} e^{-2\pi i(r,\, x+M^{-j}n)} \sum_{l\in\mathbb{Z}^d,\, l\neq\mathbf{0}} \widehat{v}_r^{j_0}(M^{*j_0}l + r)e^{2\pi i(M^{*j_0}l+r,\, x+M^{-j}n)}$$
$$= - \frac{m^{j_0-j}}{\widehat{v}_r^{j_0}(r)} \sum_{l\in\mathbb{Z}^d,\, l\neq\mathbf{0}} \widehat{v}_r^{j_0}(M^{*j_0}l + r)e^{2\pi i(M^{*j_0}l,\, x)} \sum_{n\in D(M^{j-j_0})} e^{-2\pi i(M^{*(j_0-j)}l,\, n)}.$$

By Lemma 2.2.6, the inner sum on the right-hand side is equal to m^{j-j_0} for $l \equiv \mathbf{0}$ (mod M^{j-j_0}); otherwise it is equal to zero. This means that in the outer sum only summands with numbers $l \equiv \mathbf{0}$ (mod M^{j-j_0}) are different from zero; hence it can be rewritten in the form

$$- \frac{m^{j_0-j}}{\widehat{v}_r^{j_0}(r)} \sum_{k\in\mathbb{Z}^d,\, k\neq\mathbf{0}} \widehat{v}_r^{j_0}(M^{*j}k + r)e^{2\pi i(M^{*j}k+r,\, x)} e^{-2\pi i(r,\, x)},$$

which is the Fourier series of the function

$$m^{j-j_0} \left(1 - \frac{v_r^j(x)}{\widehat{v}_r^{j_0}(r)} e^{-2\pi i(r,\, x)}\right).$$

Taking into account that by **V2**, $\widehat{v}_r^{j_0}(r) = \widehat{v}_r^j(r)$ for $j \geq j_0$ and $\widehat{v}_r^{j_0}(r) \neq 0$, we obtain (9.13). This equality and the linearity of the Fejer means imply that

$$\|h_j\| = \|h_j - \sigma_{K_j}(h_j)\| = \|m^{j_0-j} \sum_{n\in D(M^{j-j_0})} [h_{j_0}(x + M^{-j}n) - \sigma_{K_j}(h_{j_0})(x + M^{-j}n)]\|$$
$$\leq m^{j_0-j} \sum_{n\in D(M^{j-j_0})} \|h_{j_0}(x + M^{-j}n) - \sigma_{K_j}(h_{j_0})(x + M^{-j}n)\|$$
$$= m^{j_0-j} \sum_{n\in D(M^{j-j_0})} \|h_{j_0} - \sigma_{K_j}(h_{j_0})\|.$$

The latter expression tends to zero as $j \to \infty$ because the Fejer means of any function from X converge to this function in norm (see Appendix A.16). Thus we establish that

$$\lim_{j\to\infty} \left\| f_r(x) - \frac{v_r^j(x)}{\widehat{v}_r^j(r)} \right\| = \lim_{j\to\infty} \|h_j(x)\| = 0;$$

i.e., f_r is approximated by the functions $\frac{v_r^j(x)}{\widehat{v}_r^j(r)} \in V_j$. $\Diamond$

It follows from this theorem and Remark 9.1.8 that if we exclude axiom **MR6** from the definition of PMRA, then the scaling sequence will be defined by properties **Φ2–Φ4**; i.e., the following assertion holds.

THEOREM 9.1.13. *Let* $\varphi_j \in X,\ j \in \mathbb{Z}_+,$

$$V_j := \operatorname{span}\{S_n^j \varphi_j,\ n \in D(M^j)\}.$$

For the collection of spaces $\{V_j\}_{j=0}^{\infty}$, *axioms* **MR1–MR5** *of Definition* 9.1.1 *are satisfied if and only if the functions* φ_j *satisfy conditions* **Φ2–Φ4** *of Theorem* 9.1.4.

EXAMPLE 9.1.14. Set $\varphi_0 \equiv 1$ and for φ_1, take an arbitrary function from the space X such that $\widehat{\varphi}_1(\mathbf{0}) = 1$, $\widehat{\varphi}_1(k) = 0$ for all $k \equiv \mathbf{0} \pmod{M^*}$, $k \neq \mathbf{0}$, and $\widehat{\varphi}_1(k) \neq 0$ for the rest of $k \in \mathbb{Z}^d$. Put $v_n^1 = \omega_n^j \varphi_1$, $n \in D(M^*)$, and define functions v_n^j, $j > 1$, $n \in D(M^{*j})$, setting their Fourier coefficients by conditions **V1–V3** of Lemma 9.1.7. For $j > 1$, set

$$\varphi_j = \sum_{n \in D(M^{*j})} v_n^j.$$

By construction, it is clear that the sequence $\{\varphi_j\}_{j=0}^{\infty}$ satisfies axioms **Φ1–Φ5** of Theorem 9.1.4; i.e., this sequence is scaling and generates a PMRA. ◊

EXAMPLE 9.1.15. Let $d = 1$, $M = 2$. Let us consider a scaling sequence of Dirichlet kernels with minimum possible symmetric spectrum:

$$\varphi_j(x) = \sum_{k=-2^{j-1}}^{2^{j-1}} e^{2\pi i k x}. \tag{9.14}$$

It is easy to verify that axioms **Φ1–Φ5** of Theorem 9.1.4 are satisfied; i.e., this sequence is scaling and generates a polynomial PMRA, which is a periodic analog of the Kotelnikov-Shannon MRA (the question of whether a PMRA can be generated by a nonperiodic scaling function will be discussed in §9.4). It will follow from the results of §9.2 that if we multiply the first and last summands in sum (9.14) by $\sqrt{2}/2$, the bases $\varphi_j(\cdot + 2^{-j}n)$, $n \in \mathbb{Z}$, will be orthonormal. ◊

9.2. Construction of wavelet functions

In this section, we define wavelet spaces and wavelet functions whose shifts form bases in these spaces. In the orthogonal case, i.e., only for $X = L_2(\mathbb{T}^d)$, the wavelet space is the orthogonal projection of the space V_{j+1} with respect to V_j. In the biorthogonal case we deal with two multiresolution analyses; the space V_{j+1} of one analysis is orthogonally projected with respect to the corresponding component $\widetilde{V}_j$ of the other analysis. Certainly, here the scalar multiplication of functions from V_j by functions from $\widetilde{V}_j$ must make sense.

We shall consider pairs of PMRAs with the first component $\{V_j\}_{j=0}^{\infty}$ in $L_p(\mathbb{T}^d)$, $1 \le p \le \infty$ ($C(\mathbb{T}^d)$ for $p = \infty$) and the second one $\{\widetilde{V}_j\}_{j=0}^{\infty}$ in $L_q(\mathbb{T}^d)$, $1/p + 1/q = 1$ ($C(\mathbb{T}^d)$ for $p = 1$). These pairs will be called (p, q)-pairs.

PROPOSITION 9.2.1. *Let* $\{V_j\}_{j=0}^{\infty}$, $\{\widetilde{V}_j\}_{j=0}^{\infty}$ *form a* (p, q)*-pair,* $\varphi \in V_j$, $\widetilde{\varphi} \in \widetilde{V}_j$. *Systems* $\{S_n^j \varphi\}_{n \in D(M^j)}$, $\{S_k^j \widetilde{\varphi}\}_{k \in D(M^j)}$ *are biorthonormal if and only if the following equalities hold for all* $r \in \mathbb{Z}^d$:

$$\langle \omega_r^j \varphi,\ \omega_r^j \widetilde{\varphi} \rangle = m^{-j}. \tag{9.15}$$

Proof. By Lemma 9.1.6 we have

$$\langle S_n^j\varphi,\ S_k^j\widetilde{\varphi}\rangle = \left\langle \sum_{r\in D(M^{*j})} e^{2\pi i(M^{*-j}r,\ n)}\omega_r^j\varphi,\ \sum_{s\in D(M^{*j})} e^{2\pi i(M^{*-j}s,\ k)}\omega_s^j\widetilde{\varphi}\right\rangle$$
$$= \sum_{r\in D(M^{*j})}\sum_{s\in D(M^{*j})} e^{2\pi i(M^{*-j}r,\ n)}e^{-2\pi i(M^{*-j}s,\ k)}\langle \omega_r^j\varphi,\ \omega_s^j\widetilde{\varphi}\rangle .$$

Since the spectra of the functions $\omega_r^j\varphi$, $\omega_s^j\widetilde{\varphi}$ are disjoint for $r\neq s$, the corresponding summands in the latter sum are equal to zero. Therefore,

$$\langle S_n^j\varphi,\ S_k^j\widetilde{\varphi}\rangle = \sum_{r\in D(M^{*j})} e^{2\pi i(M^{*-j}r,\ n-k)}c_r, \tag{9.16}$$

where $c_r := \langle \omega_r^j\varphi,\ \omega_r^j\widetilde{\varphi}\rangle$. This and Lemma 2.2.6 imply the sufficiency. Let us now assume that $\{S_n^j\varphi_j\}_{n\in D(M^j)}$ and $\{S_k^j\widetilde{\varphi}_j\}_{k\in D(M^j)}$ are biorthonormal systems. Consider equalities (9.16) for some fixed $n\in D(M^j)$ and all $k\in D(M^j)$ as a system of equations for the unknowns c_r. By Corollary 2.2.7, the solution $c_p = m^{-j}$, $p\in D(M^{*j})$, is unique. It remains to note that $c_p = c_{p+M^{*j}}$, i.e., $c_p = m^{-j}$ for all $p\in\mathbb{Z}^d$, which proves the necessity. ◊

Note that in the case $X = L_2(\mathbb{T}^d)$ formula (9.15) gives a simple method for orthogonalization of an arbitrary scaling sequence $\{\varphi_j\}_{j=0}^{\infty}$. It follows from Proposition 9.2.1 that for the functions

$$\varphi_j^{\sharp} := m^{-j/2}\sum_{r\in D(M^{*j})}\frac{\omega_r^j\varphi_j}{\|\omega_r^j\varphi_j\|},$$

which by Proposition 9.1.10 form a scaling sequence, the systems $\{S_n^j\varphi_j^{\sharp}\}_{n\in D(M^j)}$ are orthogonal for all $j\in\mathbb{Z}_+$. Such a scaling sequence is said to be orthogonalized.

COROLLARY 9.2.2. *Let $\{V_j\}_{j=0}^{\infty}$, $\{\widetilde{V}_j\}_{j=0}^{\infty}$ form a (p,q)-pair with scaling sequences $\{\varphi_j\}_{j=0}^{\infty}$, $\{\widetilde{\varphi}_j\}_{j=0}^{\infty}$, and let μ_n^j, $\widetilde{\mu}_n^j$ be the corresponding factors from* **Φ4** *of Theorem 9.1.4. If the systems $\{S_n^j\varphi_j\}_{n\in D(M^j)}$, $\{S_k^j\widetilde{\varphi}_j\}_{k\in D(M^j)}$ are biorthonormal, then*

$$\sum_{k\in D(M^*)}\mu^j_{p+M^{*j-1}k}\overline{\widetilde{\mu}^j_{p+M^{*j-1}k}} = m \tag{9.17}$$

for all $p\in\mathbb{Z}^d$ and all $j\in\mathbb{N}$.

Proof. Similarly to (9.12), we have

$$m^{-j+1} = \langle \omega_p^{j-1}\varphi_{j-1}, \omega_p^{j-1}\widetilde{\varphi}_{j-1}\rangle$$
$$= \left\langle \sum_{k\in D(M^*)}\mu^j_{p+M^{*j-1}k}\omega^j_{p+M^{*j-1}k}\varphi_{j-1},\ \sum_{l\in D(M^*)}\widetilde{\mu}^j_{p+M^{*j-1}l}\omega^j_{p+M^{*j-1}l}\widetilde{\varphi}_{j-1}\right\rangle$$
$$= \sum_{k\in D(M^*)}\mu^j_{p+M^{*j-1}k}\overline{\widetilde{\mu}^j_{p+M^{*j-1}k}}\langle \omega^j_{p+M^{*j-1}k}\varphi_j,\ \omega^j_{p+M^{*j-1}k}\widetilde{\varphi}_j\rangle$$
$$= m^{-j}\sum_{k\in D(M^*)}\mu^j_{p+M^{*j-1}k}\overline{\widetilde{\mu}^j_{p+M^{*j-1}k}}.$$

It remains to multiply both sides of this equality by m^{-j}. $\diamondsuit$

Let us now construct wavelet spaces and bases. First we consider the orthogonal case. Let $\{V_j\}_{j=0}^{\infty}$ be a PMRA in $L_2(\mathbb{T}^d)$ with scaling sequence $\{\varphi_j\}_{j=0}^{\infty}$ generating orthonormal bases of shifts $\{S_n^j\varphi_j\}_{n\in D(M^j)}$ in the spaces V_j. Our aim is to find in the space V_{j+1} functions $\psi^{(\nu)}$, $\nu = 1,\dots,m-1$, such that the corresponding systems $\{S_n^j\psi_j^{(\nu)}\}_{n\in D(M^j)}$ are orthonormal, orthogonal to the space V_j, and in addition, they extend $\{S_n^j\varphi_j\}_{n\in D(M^j)}$ to a basis of the space V_{j+1}. To do this, we need to construct a unitary matrix given its first row. Let complex numbers $a_{00},\dots,a_{0,m-1}$ (the first row of the required unitary matrix A) satisfy the condition

$$\sum_{r=0}^{m-1} |a_{0r}|^2 = 1. \tag{9.18}$$

If $a_{00} = 1$, then $a_{01},\dots,a_{0,m-1} = 0$, and for A, we take the unitary matrix. If $a_{00} \neq 1$, then the remaining elements of A are given by Householder's transform:

$$a_{k0} = \overline{a_{0k}}\frac{1-a_{00}}{1-\overline{a_{00}}}, \qquad a_{kl} = \delta_{kl} - \frac{a_{0l}\overline{a_{0k}}}{1-\overline{a_{00}}}, \qquad k,l = 1,\dots,m-1. \tag{9.19}$$

Fix $j \in \mathbb{Z}_+$ and $n \in D(M^{*j})$. Let s_k, $k = 0,\dots,m-1$, be the renumbered elements of the set $D(M^*)$. Put

$$a_{0k} = \mu^{j+1}_{n+M^{*j}s_k}/\sqrt{m}, \quad k = 0,\dots,m-1,$$

where μ_k^j are the factors from **Φ4** of Theorem 9.1.4. By Corollary 9.2.2, these numbers satisfy relation (9.18); therefore we can extend this row to the unitary matrix A. Set $\alpha^{\nu,j}_{n+M^{*j}s_k} = \sqrt{m}a_{\nu k}$. If n ranges over all $D(M^{*j})$, then, by Lemma 2.2.3, the vectors $n + M^{*j}s_k$, $k = 0,\dots,m-1$, range over all $D(M^{*j+1})$; i.e., we have defined numbers $\alpha_s^{\nu,j}$ for all $s \in D(M^{*j+1})$. Let us extend this sequence (with respect to the subindex) to $\mathbb{Z}^d$, setting $\alpha_l^{\nu,j} = \alpha_s^{\nu,j}$ for all $l \equiv s \pmod{M^{*j+1}}$. Now, for each $\nu = 1,\dots,m-1$, define the wavelet functions $\psi^{(\nu)}$, defining their Fourier coefficients by the formulas

$$\widehat{\psi}_j^{(\nu)}(l) = \alpha_s^{\nu,j}\widehat{\varphi}_{j+1}(l), \quad l \in \mathbb{Z}^d,$$

and define the wavelet spaces by the formulas

$$W_j^{(\nu)} := \operatorname{span}\{S_n^j\psi_j^{(\nu)},\ n \in D(M^j)\}.$$

THEOREM 9.2.3. *Suppose $\{V_j\}$ is a PMRA in $L^2(\mathbb{T}^d)$ with the scaling sequence $\{\varphi_j\}$ and $S_n^j\varphi_j$, $n \in D(M^j)$, is an orthonormal system for any $j \in \mathbb{Z}_+$. Then we have*

$$V_{j+1} = V_j \oplus W_j^{(1)} \oplus \cdots \oplus W_j^{(m-1)}, \qquad j \in \mathbb{Z}_+,$$

and the system $\{S_n^j\psi_j^{(\nu)}\}_{n\in D(M^j)}$ is an orthonormal basis of the space $W_j^{(\nu)}$, $\nu = 1,\dots,m-1$.

The proof of this theorem will be obtained as a consequence of Theorem 9.2.4.

Consider the biorthogonal case. Let there be given a (p,q)-pair satisfying the conditions of Corollary 9.2.2. To construct wavelet functions, we have to extend two appropriate vectors to two mutually inverse matrices. Let there be given numbers

$a_{00}, \dots, a_{0m-1}$ and $\widetilde{a}_{00}, \dots, \widetilde{a}_{0m-1}$ (the first rows of the required matrices A and $\widetilde{A}$, respectively) such that

$$\sum_{r=0}^{m-1} a_{0r}\overline{\widetilde{a}_{0r}} = 1. \tag{9.20}$$

First assume that $a_{00} = \overline{\widetilde{a}_{00}} \neq 1$. Then for $k, l = 1, \dots, m-1$, the remaining elements will be defined by the formulas

$$a_{l0} = \overline{\widetilde{a}_{0l}}, \qquad a_{lk} = \delta_{lk} - \frac{\overline{\widetilde{a}_{0l}}a_{0k}}{1 - a_{00}}, \tag{9.21}$$

$$\widetilde{a}_{l0} = \overline{a_{0l}}, \qquad \widetilde{a}_{lk} = \delta_{lk} - \frac{\overline{a_{0l}}\widetilde{a}_{0k}}{1 - \overline{a_{00}}}. \tag{9.22}$$

It is easy to verify that

$$A\widetilde{A}^* = I_m. \tag{9.23}$$

Assume now that $a_{00}\widetilde{a}_{00} \neq 0$. Consider the numbers $a'_{0k} = Ca_{0k}$, $\widetilde{a}'_{0k} = \widetilde{a}_{0k}/\overline{C}$, $k = 0, \dots, m-1$, choosing C by the condition $a'_{00} = \overline{\widetilde{a}'_{00}} \neq 1$. This can be done by setting $C = \sqrt{\overline{\widetilde{a}_{00}}/a_{00}}$ and taking the complex value of the square root such that $a'_{00} \neq 1$. The new rows satisfy all the requirements of the previous case, so they can be extended to the matrices A', $\widetilde{A}'$ such that $A'\widetilde{A}'^* = I_m$. Replacing the first rows of these matrices by the initial ones, we obtain the required matrices A, $\widetilde{A}$. Finally, assume that $a_{00}\widetilde{a}_{00} = 0$. It follows from (9.20) that there exists r_0 such that $a_{0r_0}\widetilde{a}_{0r_0} \neq 0$. Interchanging a_{0r_0} and a_{00}, and $\widetilde{a}_{0r_0}$ and $\widetilde{a}_{00}$, we reduce our case to the previous one. If we extend the new rows to mutually inverse matrices and interchange the columns with numbers 0 and r_0 in these matrices, we obtain the required matrices A, $\widetilde{A}$.

Fix $j \in \mathbb{Z}_+$ and $n \in D(M^{*j})$. As above, denote by s_k, $k = 0, \dots, m-1$, the renumbered elements of the set $D(M^*)$. Put

$$a_{0k} = \mu^{j+1}_{n+M^{*j}s_k}/\sqrt{m}, \quad \widetilde{a}_{0k} = \widetilde{\mu}^{j+1}_{n+M^{*j}s_k}/\sqrt{m}, \quad k = 0, \dots, m-1.$$

By Corollary 9.2.2, these numbers satisfy (9.20); hence we can extend these rows to matrices A, $\widetilde{A}$ that satisfy (9.23). Set $\alpha^{\nu,j}_{n+M^{*j}s_k} = \sqrt{m}a_{\nu k}$, $\widetilde{\alpha}^{\nu,j}_{n+M^{*j}s_k} = \sqrt{m}\widetilde{a}_{\nu k}$. It follows from (9.23) that

$$\sum_{k=0}^{m-1} \alpha^{\nu,j}_{n+M^{*j}s_k}\overline{\widetilde{\mu}^{j+1}_{n+M^{*j}s_k}} = 0, \quad \sum_{k=0}^{m-1} \widetilde{\alpha}^{\nu,j}_{n+M^{*j}s_k}\overline{\mu^{j+1}_{n+M^{*j}s_k}} = 0, \tag{9.24}$$

for all $\nu = 1, \dots, m-1$,

$$\sum_{k=0}^{m-1} \alpha^{l,j}_{n+M^{*j}s_k}\overline{\widetilde{\alpha}^{p,j}_{n+M^{*j}s_k}} = m\delta_{lp} \tag{9.25}$$

for all $l, p = 1, \dots, m-1$. If n ranges over the set $D(M^{*j})$, then, by Lemma 2.2.3, the vectors $n + M^{*j}s_k$, $k = 0, \dots, m-1$, range over the set $D(M^{*j+1})$; i.e., we have defined the numbers $\alpha^{\nu,j}_s$, $\widetilde{\alpha}^{\nu,j}_s$ for all $s \in D(M^{*j+1})$. Extend these sequences (with respect to the subindex) to $\mathbb{Z}^d$, setting $\alpha^{\nu,j}_l = \alpha^{\nu,j}_s$, $\widetilde{\alpha}^{\nu,j}_l = \widetilde{\alpha}^{\nu,j}_s$, for all $l \equiv s$

(mod M^{*j+1}). For each $\nu = 1, \dots, m-1$, define the wavelet functions $\psi^{(\nu)}$, $\widetilde{\psi}^{(\nu)}$ defining their Fourier coefficients by the formulas

$$\widehat{\psi}_j^{(\nu)}(l) = \alpha_l^{\nu,j}\widehat{\varphi}_{j+1}(l), \quad \widehat{\widetilde{\psi}}_j^{(\nu)}(l) = \widetilde{\alpha}_l^{\nu,j}\widehat{\widetilde{\varphi}}_{j+1}(l), \quad l \in \mathbb{Z}^d.$$

Define wavelet spaces as

$$\begin{aligned} W_j^{(\nu)} &:= \operatorname{span}\{S_n^j\psi_j^{(\nu)},\ n \in D(M^j)\}, \\ \widetilde{W}_j^{(\nu)} &:= \operatorname{span}\{S_n^j\widetilde{\psi}_j^{(\nu)},\ n \in D(M^j)\}. \end{aligned}$$

The above method of constructing matrices A, $\widetilde{A}$ is not unique; therefore the set of wavelet functions and the wavelet spaces are not uniquely defined.

The problem of constructing wavelet functions is far easier for $m = 2$. In this case, to construct a unique pair of wavelet functions $\psi_j = \psi_j^{(1)}$, $\widetilde{\psi}_j = \widetilde{\psi}_j^{(1)}$ for each j, we have to extend the following row to a unitary matrix:

$$\left(\frac{1}{\sqrt{2}}\mu_n^{j+1}, \frac{1}{\sqrt{2}}\mu_{n+M^{*j}s^*}^{j+1}\right), \quad n \in D(M^{*j}),$$

where s^* is a nonzero digit of the matrix M^*. Using Corollary 2.2.7, it is easy to verify that for the second row of such a matrix, we can take

$$\left(\frac{1}{\sqrt{2}}e^{2\pi i(M^{-j-1}s,n)}\overline{\widetilde{\mu}_{n+M^{*j}s^*}^{j+1}}, \frac{1}{\sqrt{2}}e^{2\pi i(M^{-j-1}s,n+M^{*j}s^*)}\overline{\widetilde{\mu}_n^{j+1}}\right),$$

where s is a nonzero digit of the matrix M. Thus the Fourier coefficients of the functions $\psi_j, \widetilde{\psi}_j$ can be found by the formulas

$$\widehat{\psi}_j(k) = e^{2\pi i(k,M^{-j-1}s)}\overline{\widetilde{\mu}_{k+M^{*j}s^*}^{j+1}}\widehat{\varphi}_{j+1}(k),$$

$$\widehat{\widetilde{\psi}}_j(k) = e^{2\pi i(k,M^{-j-1}s)}\overline{\mu_{k+M^{*j}s^*}^{j+1}}\widehat{\widetilde{\varphi}}_{j+1}(k).$$

THEOREM 9.2.4. *Let $\{V_j\}_{j=0}^{\infty}$, $\{\widetilde{V}_j\}_{j=0}^{\infty}$ form a (p,q)-pair with the scaling sequences $\{\varphi_j\}_{j=0}^{\infty}$ and $\{\widetilde{\varphi}_j\}_{j=0}^{\infty}$, respectively, such that $\{S_n^j\varphi_j\}_{n\in D(M^j)}$, $\{S_k^j\widetilde{\varphi}_j\}_{k\in D(M^j)}$ are biorthonormal systems. Then:*

(i) $W_j^{(\nu)} \subset V_{j+1}$ *for all* $\nu = 1, \dots, m-1$.

(ii) *Any* $f \in V_{j+1}$ *can be represented in the form* $f = f_0 + \sum_{\nu=1}^{m-1} f_\nu$, *where* $f_0 \in V_j, f_\nu \in W_j^{(\nu)}$, $\nu = 1, \dots, m-1$.

(iii) $W_j^{(\nu)} \perp \widetilde{V}_j$, $\widetilde{W}_j^{(\nu)} \perp V_j$ *for all* $\nu = 1, \dots, m-1$.

(iv) $W_j^{(\nu)} \perp \widetilde{W}_j^{(\kappa)}$ *for all* $\nu \neq \kappa$, $\nu, \kappa = 1, \dots, m-1$.

(v) $(S_n^j\psi_j^{(\nu)}, S_k^j\widetilde{\psi}_j^{(\nu)}) = \delta_{nk}$ *for all* $\nu = 1, \dots, m-1$, $k \in D(M^j)$.

Proof. For any fixed n and j, from (9.12) we have

$$\omega_n^j\varphi_j(x) = \sum_{l\in D(M^*)} \mu_{n+M^{*j}l}^{j+1}\omega_{n+M^{*j}l}^{j+1}\varphi_{j+1}(x). \tag{9.26}$$

Analogously,

$$\omega_n^j\psi_j^{(\nu)}(x) = \sum_{l\in D(M^*)} \alpha_{n+M^{*j}l}^{\nu,j}\omega_{n+M^{*j}l}^{j+1}\varphi_{j+1}(x), \quad \nu = 1, \dots, m-1. \tag{9.27}$$

Consider equalities (9.26), (9.27) as a system of m equations in m unknowns $\{\omega^{j+1}_{n+M^{*j}l}\varphi_{j+1}\}$, $l \in D(M^*)$. By the construction of the numbers $\{a_{\nu k}\}$, the matrix of this system is invertible; hence its determinant is nonzero. Therefore the variables $\omega^{j+1}_{n+M^{*j}l}\varphi_{j+1}$ can be expressed in terms of $\omega^j_n\varphi_j \in V_j$ and $\omega^j_n\psi^{(\nu)}_j \in W^{(\nu)}_j$, which implies (i). On the other hand, analogously to (9.10), we have

$$S^j_r\psi^{(\nu)}_j = \sum_{n\in D(M^{*j})} e^{2\pi i(M^{*-j}n,r)}\omega^j_n\psi^{(\nu)}_j, \quad \nu = 1,\ldots,m-1.$$

In view of Corollary 2.2.7, it follows that each $\omega^j_n\psi^{(\nu)}_j$ can be expressed in terms of $S^j_r\psi^{(\nu)}_j$, $r \in D(M^j)$. To prove (ii), it remains to note that $\{\omega^{j+1}_{n+M^{*j}l}\varphi_{j+1}\}$ is a basis of the space V_{j+1}. In addition, as the functions $\omega^j_n\psi^{(\nu)}_j$, $n \in D(M^{*j})$ are linearly independent, it follows from the above argument and property **MR3** of Definition 9.1.1 that $\dim W^{(\nu)}_j = m^j$; i.e., both systems $\{\omega^j_n\psi^{(\nu)}_j\}_{n\in D(M^{*j})}$ and $\{S^j_r\psi^{(\nu)}_j\}_{r\in D(M^j)}$ are bases in $W^{(\nu)}_j$. To prove (iii), it suffices to check that the basis functions of the space $\widetilde{V}_j$ are orthogonal to those of the spaces $W^{(\nu)}_j$, $\nu = 1,\ldots,m-1$. Using equality (9.24) and Proposition 9.2.1, we get

$$\begin{aligned}
&\langle\, \omega^j_n\psi^{(\nu)}_j,\ \omega^j_k\widetilde{\varphi}_j\,\rangle \\
&= \left\langle \sum_{l\in D(M^*)} \alpha^{\nu,j}_{n+M^{*j}l}\omega^{j+1}_{n+M^{*j}l}\varphi_{j+1},\ \sum_{k\in D(M^*)} \widetilde{\mu}^{j+1}_{n+M^{*j}k}\omega^{j+1}_{n+M^{*j}k}\widetilde{\varphi}_{j+1}\right\rangle \\
&= \sum_{l\in D(M^*)} \alpha^{\nu,j}_{n+M^{*j}l}\overline{\widetilde{\mu}^{j+1}_{n+M^{*j}l}}\langle\, \omega^{j+1}_{n+M^{*j}l}\varphi_{j+1},\ \omega^{j+1}_{n+M^{*j}l}\widetilde{\varphi}_{j+1}\,\rangle \\
&= m^{-j-1}\sum_{l\in D(M^*)} \alpha^{\nu,j}_{n+M^{*j}l}\overline{\widetilde{\mu}^{j+1}_{n+M^{*j}l}} = 0.
\end{aligned}$$

Analogously, in view of (9.25), we get

$$\langle\, \omega^j_n\psi^{(\nu)}_j,\ \omega^j_n\widetilde{\psi}^{(\kappa)}_j\,\rangle = m^{-j-1}\sum_{l\in D(M^*)} \alpha^{\nu,j}_{n+M^{*j}l}\overline{\widetilde{\alpha}^{\kappa,j}_{n+M^{*j}l}} = m^{-j}\delta_{\nu\kappa}.$$

This implies (iv). Taking into account Proposition 9.2.1, we obtain (v). ◊

9.3. Wavelet packets

The main idea of the construction of orthogonal wavelet systems by using multiresolution analysis is the following. The space V_{j+1} which contains an orthonormal shift basis can be decomposed into the direct sum of the spaces V_j and several wavelet spaces $W^{(\nu)}_j$ which also contain orthonormal shift bases. Using this idea, each wavelet space can be decomposed into the direct sum. As a result, we obtain more wavelet functions whose shifts form an orthonormal basis of the whole space. Such bases may be useful in both theoretical investigations and applications. In particular, they will be used in Chapter 11.

We confine our consideration to the orthogonal case and matrices M such that $m = 2$. Let there be given a PMRA in $L_2(\mathbb{T}^d)$ generated by a scaling sequence $\{\varphi_j\}_{j=0}^{\infty}$. Suppose the system $\{S^j_n\varphi_j\}_{n\in D(M^j)}$ is orthonormal and $\{S^j_n\psi_j\}_{n\in D(M^j)}$ is the corresponding orthonormal wavelet basis in the space W_j.

Set

$$\lambda_r^j = e^{2\pi i(M^{-j}\sigma,r)}\overline{\mu^j_{r+M^{*j-1}\sigma^*}}$$

where σ, σ^* are nonzero digits of the matrices M, M^*, respectively, and μ_k^j are the factors from property **Φ4** of Theorem 9.1.4. By Corollary 9.2.2, the following equalities hold:

$$|\mu_r^j|^2 + |\lambda_r^j|^2 = |\mu_r^j|^2 + |\mu^j_{r+M^{*j-1}\sigma^*}|^2 = |\lambda_r^j|^2 + |\lambda^j_{r+M^{*j-1}\sigma^*}|^2 = 2, \tag{9.28}$$

and obviously,

$$\mu_r^j\overline{\lambda_r^j} + \mu^j_{r+M^{*j-1}\sigma^*}\overline{\lambda^j_{r+M^{*j-1}\sigma^*}} = 0. \tag{9.29}$$

Let $r \in \mathbb{Z}$, $k \in \mathbb{N}$. Denote by Δ_k and Λ_k the multipliers defined on $L_2(\mathbb{T}^d)$ and given by the equalities

$$\widehat{\Delta_k f}(r) = \mu_r^k \widehat{f}(r), \quad \widehat{\Lambda_k f}(r) = \lambda_r^k \widehat{f}(r).$$

Fix a positive integer s. For integer $j \geq s$ and $\nu = (\nu_{j-s+1}, \dots, \nu_j)$, where ν_k takes values 0 and 1, we put

$$\psi_j^{(\nu)} = \psi_j^{(\nu,s)} = \prod_{k=j-s+1}^{j} \Delta_k^{\nu_k}\Lambda_k^{1-\nu_k}\psi_j,$$

$$W_j^{(\nu)} = \operatorname{span}\{S_n^{j-s}\psi_j^{(\nu)},\ n \in D(M^{*j-s})\}.$$

Since both sequences $\{\mu_r^j\}$, $\{\lambda_r^j\}$ are M^j-periodic with respect to the subindex, we have

$$\begin{aligned} \omega_r^j\psi_j^{(\nu)}(x) &= \sum_{l\in\mathbb{Z}^d}\prod_{k=j-s+1}^{j} (\mu^k_{M^jl+r})^{\nu_k}(\lambda^k_{M^jl+r})^{1-\nu_k}\widehat{\psi}_j(M^jl+r)e^{2\pi i(M^jl+r,x)} \\ &= \prod_{k=j-s+1}^{j} (\mu_r^k)^{\nu_k}(\lambda_r^k)^{1-\nu_k}\omega_r^j\psi_j(x). \end{aligned} \tag{9.30}$$

By the same reasoning as in Theorem 9.2.4, we can prove the inclusion $W_j^{(\nu)} \subset W_j$. Let us now show that $W_j^{(\nu)} \perp W_j^{(\nu')}$ for $\nu \neq \nu'$, and for each ν, the system of functions $\{S^n_{j-s}\psi_j^{(\nu)}\}_{n=0}^{2^{j-s}-1}$ is an orthonormal basis of the space $W_j^{(\nu)}$. In view of Proposition 9.2.1, it suffices to prove that

$$\langle \omega_r^{j-s}\psi_j^{(\nu)}, \omega_r^{j-s}\psi_j^{(\nu')}\rangle = 0 \tag{9.31}$$

for $\nu \neq \nu'$ and

$$\|\omega_r^{j-s}\psi_j^{(\nu)}\|^2 = 2^{-j+s}. \tag{9.32}$$

We prove these equalities by induction on s. Let $s = 1$. Using (9.30), we get

$$\begin{aligned} \langle \omega_r^{j-1}\psi_j^{(\nu)}, \omega_r^{j-1}\psi_j^{(\nu')}\rangle &= \langle \omega_r^{j}\psi_j^{(\nu)}, \omega_r^{j}\psi_j^{(\nu')}\rangle + \langle \omega_{r'}^{j}\psi_j^{(\nu)}, \omega_{r'}^{j}\psi_j^{(\nu')}\rangle \\ &= (\mu_r^j)^{\nu_j}(\overline{\mu_r^j})^{\nu'_j}(\lambda_r^j)^{1-\nu_j}(\overline{\lambda_r^j})^{1-\nu'_j}\|\omega_r^j\psi_j\|^2 \\ &\quad + (\mu_{r'}^j)^{\nu_j}(\overline{\mu_{r'}})^{\nu'_j}(\lambda_{r'}^j)^{1-\nu_j}(\overline{\lambda_{r'}})^{1-\nu'_j}\|\omega_{r'}^j\psi_j\|^2, \end{aligned}$$

where $r' = M^{*j-1}\sigma^* + r$. Taking into account that $\|\omega_k^j\psi_j\|^2 = 2^{-j}$ and using equalities (9.28), (9.29), we obtain (9.31), (9.32) for $s = 1$. Assume now that equalities (9.31), (9.32) hold for all $s < s_0$. Let $\nu = (\nu_{j-s_0+1}, \dots, \nu_j)$, $\nu' =$

$(\nu'_{j-s_0+1}, \ldots, \nu'_j)$. Set $\gamma = (\nu_{j-s_0+2}, \ldots, \nu_j)$, $\gamma' = (\nu'_{j-s_0+2}, \ldots, \nu'_j)$, $\kappa = \nu_{j-s_0+1}$, $\kappa' = \nu'_{j-s_0+1}$. Then

$$\begin{aligned}
&\langle \omega_r^{j-s_0}\psi_j^{(\nu)}, \omega_r^{j-s_0}\psi_j^{(\nu')}\rangle \\
&= \langle \omega_r^{j-s_0+1}\psi_j^{(\nu)}, \omega_r^{j-s_0+1}\psi_j^{(\nu')}\rangle + \langle \omega_{r'}^{j-s_0+1}\psi_j^{(\nu)}, \omega_{r'}^{j-s_0+1}\psi_j^{(\nu')}\rangle \\
&= (\mu_r^{j-s_0+1})^{\kappa}(\overline{\mu_r^{j-s_0+1}})^{\kappa'}(\lambda_r^{j-s_0+1})^{1-\kappa}(\overline{\lambda_r^{j-s_0+1}})^{1-\kappa'}\langle \omega_r^{j-s_0+1}\psi_j^{(\gamma)}, \omega_r^{j-s_0+1}\psi_j^{(\gamma')}\rangle \\
&+(\mu_{r'}^{j-s_0+1})^{\kappa}(\overline{\mu_{r'}^{j-s_0+1}})^{\kappa'}(\lambda_{r'}^{j-s_0+1})^{1-\kappa}(\overline{\lambda_{r'}^{j-s_0+1}})^{1-\kappa'}\langle \omega_{r'}^{j-s_0+1}\psi_j^{(\gamma)}, \omega_{r'}^{j-s_0+1}\psi_j^{(\gamma')}\rangle,
\end{aligned}$$

where $r' = M^{*j-s_0}\sigma^* + r$. By the inductive assumption,

$$\langle \omega_k^{j-s_0+1}\psi_j^{(\gamma)}, \omega_k^{j-s_0+1}\psi_j^{(\gamma')}\rangle = \begin{cases} 0, & \text{if } \gamma \neq \gamma', \\ 2^{-j+s_0-1}, & \text{if } \gamma = \gamma'. \end{cases}$$

Using equalities (9.28) and (9.29), we obtain (9.31) and (9.32) for $s = s_0$.

We have thus proved that

$$W_j = \bigoplus_{\nu} W_j^{(\nu)}, \tag{9.33}$$

and the system of functions $\{S_n^{j-s}\psi_j^{(\nu)}\}_{n,\nu}$ is an orthonormal basis of the space W_j. This system is called a *wavelet packet*.

9.4. Generating function

A wide class of PMRA can be constructed by the following scheme: a scaling sequence is obtained by periodization of some function φ defined on $\mathbb{R}^d$ by the formulas

$$\varphi_j(x) = m^j \sum_{k\in\mathbb{Z}^d} \varphi\left(M^j x + M^j k\right), \tag{9.34}$$

where convergence of the series can be understood in various senses. We shall say that this PMRA is *generated by the function* φ. Suppose the following conditions are satisfied for $\varphi \in L_2(\mathbb{R}^d)$:

(i) The function $\widehat{\varphi}$ is continuous at zero and $\widehat{\varphi}(\mathbf{0}) \neq 0$.

(ii) There exist positive constants A, B such that

$$A \leq \sum_{m\in\mathbb{Z}^d} |\widehat{\varphi}(\xi + m)|^2 < B \quad \text{for almost all } \xi \in \mathbb{R}^d.$$

(iii) There exists a function $m_0 \in L_2(\mathbb{T}^d)$ such that

$$\widehat{\varphi}(M^*\xi) = m_0(\xi)\widehat{\varphi}(\xi) \quad \text{for almost all } \xi \in \mathbb{R}^d.$$

It follows from the results of §2.3 that φ is a scaling function for an MRA in $L_2(\mathbb{R}^d)$. If we additionally assume that φ decreases sufficiently rapidly at infinity, for instance,

$$\varphi(x) = O\left(\frac{1}{(1+|x|)^{d+\varepsilon}}\right), \quad \varepsilon > 0,$$

then $\varphi_j \in L_2(\mathbb{T}^d)$ and by the Poisson summation formula we get

$$\varphi_j(x) = \sum_{m\in\mathbb{Z}^d} \widehat{\varphi}(M^{*-j}m)e^{2\pi i(m,x)}. \tag{9.35}$$

It is not hard to check that the functions φ_j satisfy conditions **Φ1–Φ5** of Theorem 9.1.4.

Let us discuss the question of whether a PMRA can be generated by a nonperiodic function in an arbitrary space X. We confine the discussion to the case $d = 1$, $M = 2$. The series on the right-hand side of (9.34) converges in $L(\mathbb{R})$ and its sum φ_j belongs to the space $L(\mathbb{T})$ for any function $\varphi \in L(\mathbb{R})$. However, equality (9.35) does not always hold almost everywhere. It holds almost everywhere if the sum of the series on the right-hand side of (9.35) belongs to the space $L(\mathbb{T})$ (all the more so if this sum belongs to X). Provided this condition is satisfied, the question of whether a PMRA $\{V_j\}_{j=0}^{\infty}$ is generated by a function φ can be rephrased as follows: the functions

$$\varphi_j(t) := \sum_{k\in\mathbb{Z}} \widehat{\varphi}(2^{-j}k)e^{2\pi ikt}, \quad j \in \mathbb{Z}_+, \tag{9.36}$$

form a scaling sequence for $\{V_j\}_{j=0}^{\infty}$. Note that the condition $\varphi_j \in X$ is equivalent to the condition $\omega_n^j\varphi_j \in X$, $n = 0, \dots, 2^j - 1$. Certainly, the fact that a function φ satisfies these conditions does not mean that φ generates a PMRA. Some more conditions of the type (i)–(iii) should be assumed.

PROPOSITION 9.4.1. *Let a function $\varphi \in L(\mathbb{R})$ have the following properties:*

G1. $\widehat{\varphi}(0) \neq 0$.

G2. *For all $j \in \mathbb{Z}_+$ and $n = 0, \dots, 2^j - 1$ the functions*

$$u_n^j(t) := \sum_{m\in\mathbb{Z}} \widehat{\varphi}(m + n2^{-j})e^{2\pi i(2^j m+n)t}$$

belong to the space X and $A \le \|u_n^j\|_X \le B$, where A, B are positive constants.

G3. *There exists a 1-periodic bounded function m_0 such that $\widehat{\varphi}(2x) = m_0(x)\widehat{\varphi}(x)$ for all $x \in \mathbb{R}$.*

Then φ generates a PMRA in X.

`Proof.` Check that the functions φ_j given by equality (9.36) satisfy conditions **Φ1–Φ5** of Theorem 9.1.4. To prove **Φ1**, note that according to **G3** and **G1**, $m_0(n) = m_0(0) = 1$ for all $n \in \mathbb{Z}$. Therefore if $\widehat{\varphi}_0(k) = \widehat{\varphi}(k) \neq 0$ for a certain $k \in \mathbb{Z}$, $k \neq 0$, then $\widehat{\varphi}_0(2^\ell k) = \widehat{\varphi}_0(k) \neq 0$ for all $\ell \in \mathbb{Z}_+$. But this contradicts the inclusion $\varphi_0 \in X$, which follows from **G2**. Properties **Φ2–Φ4** are obvious and $\mu_k^j = m_0(2^{-j}k)$. To prove **Φ5**, it suffices to put $\gamma_k^j = 1$. ◊

REMARK 9.4.2. The scaling sequence $\{\varphi_j\}_{j\in\mathbb{Z}_+}$ constructed in Proposition 9.4.1 has the following properties:

g1. $\gamma_k^j = 1$ for all $j \in \mathbb{Z}_+$, $k \in \mathbb{Z}$, where γ_k^j are the factors from condition **Φ5** of Theorem 9.1.4.

g2. There exist constants $A, B > 0$ such that $A \le \|\omega_n^j\varphi_j\| \le B$ for all $j \in \mathbb{Z}_+$, $n = 0, \dots, 2^j - 1$.

Note that in each PMRA there exist scaling sequences satisfying **g1, g2**. For instance, one can take

$$\varphi_j(x) = \sum_{n=0}^{2^j-1} (L_n^j v_n^j)^{-1} v_n^j,$$

where v_n^j are the functions from Lemma 9.1.7, L_n^j are linear functionals on X such that

$$B^{-1}\|f\| \le |L_n^j f| \le A^{-1}\|f\|$$

for all $f \in X$, $j \in \mathbb{Z}_+$, $n = 0, \dots 2^j - 1$, and $L_{2n}^{j+1}(v_{2n}^{j+1}) = L_n^j(v_n^j)$. In particular, we can set $L_n^j f = \|f\|$. Such functions φ_j will form a scaling sequence in view of Proposition 9.1.10.

THEOREM 9.4.3. *A PMRA in X is generated by a function $\varphi \in L(\mathbb{R})$ that possesses properties* **G1**–**G3** *if and only if there exists a scaling sequence $\{\varphi_j\}_{j\in\mathbb{Z}_+}$ satisfying* **g1, g2** *and such that*

$$\lim_{x\to\infty} \sup_{j\ge\log 2|x|} \int_{2^{-j}|x|\le|t|\le 1/2} |\varphi_j(t)|\, dt = 0. \tag{9.37}$$

Proof. Necessity. Let $\{V_j\}_{j=0}^\infty$ be a PMRA generated by a function $\varphi \in L(\mathbb{R})$ that possesses properties **G1**–**G3**. Then by Proposition 9.4.1 and Remark 9.4.2, the functions φ_j defined by relation (9.36) form a scaling sequence satisfying **g1, g2**. Consider the function $\sum_{\ell\in\mathbb{Z}} \varphi(x + 2^j\ell)$ (here and in the sequel convergence of series is understood as convergence in norm in $L(\mathbb{R})$), which is 2^j-periodic. By the Poisson summation formula its k-th Fourier coefficient is equal to $2^{-j}\widehat{\varphi}(2^{-j}k)$. Hence,

$$\varphi_j(2^{-j}x) = 2^j \sum_{\ell\in\mathbb{Z}} \varphi(x + 2^j\ell)$$

for almost all $x \in \mathbb{R}$. Put

$$g_j(x) = 2^{-j}\varphi_j\left(2^{-j}x\right)\big|_{[-2^{j-1},2^{j-1}]}. \tag{9.38}$$

We have

$$\int_{-\infty}^{\infty} |\varphi - g_j| \le \int_{|t|\ge 2^{j-1}} |\varphi(t)|\, dt + \sum_{\ell\ne 0} \int_{-2^{j-1}}^{2^{j-1}} |\varphi(t + 2^j\ell)|\, dt = 2\int_{|t|\ge 2^{j-1}} |\varphi(t)|\, dt.$$

On the other hand,

$$\begin{aligned}\sup_j \int_{|t|\ge|x|} |g_j(t)|\, dt &= \sup_{j\ge\log_2|2x|} \int_{|t|\ge|x|} |g_j(t)|\, dt \\ &\le \sup_{j\ge\log_2|2x|} \int_{-\infty}^{\infty} |g_j(t) - \varphi(t)|\, dt + \int_{|t|\ge|x|} |\varphi(t)|\, dt,\end{aligned}$$

but for $j \ge \log_2|2x|$,

$$\int_{|t|\ge 2^{j-1}} |\varphi(t)|\, dt \le \int_{|t|\ge|x|} |\varphi(t)|\, dt \underset{x\to\infty}{\longrightarrow} 0.$$

We have thus proved the relation

$$\lim_{x\to\infty} \sup_{j\in\mathbb{Z}_+} \int_{|t|\ge|x|} |g_j(t)|\, dt = 0,$$

which obviously implies (9.37).

Sufficiency. Let us assume that for a PMRA $\{V_j\}_{j=0}^{\infty}$ in X, there exists a scaling sequence $\{\varphi_j\}_{j\in\mathbb{Z}_+}$ that satisfies **g1, g2**, and (9.37). Show that the sequence of functions g_j given by equality (9.38) is a Cauchy sequence in the space $L(\mathbb{R})$. Take $j_1 < j_2$ and consider the 2^{j_1}-periodic function $\sum_{\ell\in\mathbb{Z}} g_{j_2}(x+2^{j_1}\ell)$. It is not hard to see that the k-th Fourier coefficient of this function is equal to $2^{-j_1}\widehat{\varphi}_{j_2}(2^{j_2-j_1}k)$. However by **g1**, $\widehat{\varphi}_{j_2}(2^{j_2-j_1}k) = \widehat{\varphi}_{j_1}(k)$ for all $k \in \mathbb{Z}$. Therefore,

$$g_{j_1}(x) = \sum_{\ell\in\mathbb{Z}} g_{j_2}(x+2^{j_1}\ell)\big|_{[-2^{j_1-1},2^{j_1-1}]}$$

for almost all $x \in \mathbb{R}$. Using these equalities and (9.37), we get

$$\int_{-\infty}^{\infty} |g_{j_1} - g_{j_2}| \le \int_{|t|\ge 2^{j_1-1}} |g_{j_2}(t)|\,dt + \sum_{\ell\ne 0} \int_{-2^{j_1-1}}^{2^{j_1-1}} |g_{j_2}(t+2^{j_1}\ell)|\,dt$$

$$\le 2\sup_j \int_{|t|\ge 2^{j_1-1}} |g_j(t)|\,dt = 2\sup_{j\ge j_1} \int_{2^{j_1-j-1}\le|t|\le 1/2} |\varphi_j(t)|\,dt \underset{j_1\to\infty}{\longrightarrow} 0.$$

Thus $\{g_j\}_{j=0}^{\infty}$ converges in $L(\mathbb{R})$ to a function $\varphi \in L(\mathbb{R})$. Let us show that φ possesses properties **G1**–**G3** and generates the PMRA $\{V_j\}_{j=0}^{\infty}$. Set

$$\phi_j(x) = 2^j \sum_{\ell\in\mathbb{Z}} \varphi(2^j(x+\ell)).$$

Since g_j converges to φ in the space $L(\mathbb{R})$, it follows that

$$\widehat{\phi}_{j_0}(k) = \int_{-\infty}^{\infty} \varphi(t)e^{-2\pi i 2^{-j_0}kt}\,dt = \lim_{j\to\infty}\int_{-\infty}^{\infty} g_j(t)e^{-2\pi i 2^{-j_0}kt}\,dt = \lim_{j\to\infty}\widehat{\varphi}_j(2^{j-j_0}k).$$

Property **g1** implies that

$$\widehat{\varphi}_j(2^{j-j_0}k) = \widehat{\varphi}_{j_0}(k).$$

Therefore the functions φ_j and ϕ_j have the same Fourier coefficients and by the Poisson summation formula,

$$\widehat{\varphi}_j(k) = \widehat{\phi}_j(k) = \widehat{\varphi}(2^{-j}k). \tag{9.39}$$

for all $k \in \mathbb{Z}, j \in \mathbb{Z}_+$. Hence,

$$\varphi_j(x) = \phi_j(x) = \sum_{k\in\mathbb{Z}} \widehat{\varphi}(2^{-j}k)e^{2\pi ikx} \tag{9.40}$$

for almost all $x \in \mathbb{R}$ and all $j \in \mathbb{Z}_+$. This means that $\{V_j\}_{j=0}^{\infty}$ is generated by the function φ. Property **G2** is obviously satisfied in view of **g2**. By (9.39) and **Φ1**, for all $k \in \mathbb{Z}$, $k \ne 0$, we get

$$\widehat{\varphi}(k) = \widehat{\varphi}_0(k) = 0. \tag{9.41}$$

Taking into account **Φ2**, we obtain **G1**. To check **G3**, we have to define the function m_0. First we define and study it on the set $Q = \{x = 2^{-j}r,\ j,r \in \mathbb{Z}, j \ge 0,\ \widehat{\varphi}(x) \ne 0\}$. Set $m_0(x) = \widehat{\varphi}(2x)/\widehat{\varphi}(x)$ for $x \in Q$ and show that m_0 is a 1-periodic bounded function. Let $x = 2^{-j}r \in Q, r = 2^jm+n, n = 0,\dots,2^j-1$. Then by Corollary 9.1.11 and (9.39), we get

$$|m_0(x)| = \frac{|\widehat{v}^j_{n_1}(2^jm_1+n_1)|\|v^j_n\|\|u^j_{n_1}\|}{|\widehat{v}^j_n(2^jm+n)|\|u^j_n\|\|v^j_{n_1}\|}, \tag{9.42}$$

where $m_1 = 2m, n_1 = 2n$ if $n < 2^{j-1}$ and $m_1 = 2m+1, n_1 = 2n - 2^j$ if $n \geq 2^{j-1}$, v_n^j, u_n^j are the functions from Lemma 9.1.7 and condition **G2**, respectively. Thus we have two cases: $\widehat{v}_{n_1}^j(2^j m_1 + n_1) \neq 0$ and by conditions **V2, V3** of Lemma 9.1.7,

$$\widehat{v}_{n_1}^j(2^j m_1 + n_1) = \widehat{v}_{2n}^{j+1}(2^j m_1 + n_1) = \widehat{v}_n^j(2^j m + n),$$

or $\widehat{v}_{n_1}^j(2^j m_1 + n_1) = 0$ and then by **V3**, $\widehat{v}_{n_1}^j(2^{j+1}\ell + 2^j m_1 + n_1) = 0$ for all $\ell \in \mathbb{Z}$. Therefore the right-hand side of (9.42) does not depend on m. Consequently,

$$m_0(x) = m_0(2^{-j}(2^j m + n)) = m_0(2^{-j}(2^j(m+1) + n)) = m_0(x+1), \tag{9.43}$$

whenever $x, x+1 \in Q$; i.e., m_0 is a 1-periodic function on Q. Let us now prove that

$$|m_0(x)| \leq B/A \tag{9.44}$$

for all $x \in Q$. If $\widehat{v}_{n_1}^j(2^j m_1 + n_1) = 0$, then $m_0(x) = 0$. In this case (9.44) is trivial. Assume that $\widehat{v}_{n_1}^j(2^j m_1 + n_1) \neq 0$. Using the above arguments and taking into account **G2**, we get

$$|m_0(x)| = \frac{\|v_n^j\| \|u_{n_1}^j\|}{\|u_n^j\| \|v_{n_1}^j\|} \leq \frac{B\|v_n^j\|}{A\|v_{n_1}^j\|}.$$

From property **V3** of Lemma 9.1.7 it follows that $v_n^j(x) = v_{2n}^{j+1}(x/2)$, which implies that $\|v_n^j\| = \|v_{2n}^{j+1}\|$. Using **V2**, we get

$$\begin{aligned} v_{n_1}^j(x) &= \sum_{\ell \in \mathbb{Z}} \widehat{v}_{n_1}^j(2^j\ell + 2n)e^{2\pi i(2^j\ell + 2n)x} \\ &= \sum_{\ell \in \mathbb{Z}} \widehat{v}_{2n}^{j+1}(2^{j+1}\ell + 2n)e^{2\pi i(2^{j+1}\ell + 2n)x} \\ &\quad + \sum_{\ell \in \mathbb{Z}} \widehat{v}_{n_1}^j(2^j(2\ell+1) + 2n)e^{2\pi i(2^j(2\ell+1)+2n)x} \\ &= v_{2n}^{j+1}(x) + v(x). \end{aligned}$$

Choose $h \in X^*$ such that $\|h\|_{X^*} \leq 1$ and

$$\|v_{2n}^{j+1}\| = \left| \int_0^1 v_{2n}^{j+1} h \right|,$$

and set $h_1 = \omega_{2n}^{j+1} h$. From the definition of the operators ω_n^j it follows that $\|h_1\|_{X^*} \leq \|h\|_{X^*}$; hence $\|h_1\|_{X^*} \leq 1$. The spectra of the functions $h - h_1$ and v_{2n}^{j+1} are disjoint; hence

$$\|v_{2n}^{j+1}\| = \int_0^1 v_{2n}^{j+1} h_1.$$

On the other hand, the spectra of the functions v and h_1 are disjoint; hence we have

$$\|v_{n_1}^j\| \geq \left| \int_0^1 v_{n_1}^j h_1 \right| = \left| \int_0^1 (v_{2n}^{j+1} + v) h_1 \right| = \left| \int_0^1 v_{2n}^{j+1} h_1 \right| = \|v_{2n}^{j+1}\| = \|v_n^j\|,$$

which completes the proof of (9.44).

If $\widehat{\varphi}(x_0) = 0$ and $x_0 \in \overline{Q}$, then $\widehat{\varphi}(2x_0) = 0$ in view of (9.44) and the continuity of $\widehat{\varphi}$. Let $x_0 \in \mathbb{R} \setminus \overline{Q}$; then there exists an interval $I = (x_0 - \delta, x_0 + \delta)$ such that $\widehat{\varphi}(x) = 0$ for all $x \in I$. However, if $\widehat{\varphi}(x) = 0$, $x = 2^{-j}r$, $j, r \in \mathbb{Z}$, $j \geq 0$, then by **Φ4**, **Φ5**, and (9.39) we have $\widehat{\varphi}(2x) = 0$. It follows that $\widehat{\varphi}(x) = 0$ for all $x \in (2x_0 - 2\delta, 2x_0 + 2\delta)$; in particular, $\widehat{\varphi}(2x_0) = 0$. Taking into account that $x_0 \in \overline{Q}$ whenever $\widehat{\varphi}(x_0) \neq 0$ in view of the continuity of $\widehat{\varphi}$, we can define m_0 on $\mathbb{R} \setminus Q$ in the following way: if there exists $k \in \mathbb{Z}$ such that $\widehat{\varphi}(x + k) \neq 0$, then

$$m_0(x) = \lim_{t \in Q,\, t \to x+k} m_0(t);$$

otherwise $m_0(x) = 1$. From the above reasoning it is clear that we have constructed a 1-periodic function, and (9.44) holds for all $x \in \mathbb{R}$. ◊

EXAMPLE 9.4.4. Let the 1-periodic functions φ_j be defined on $[0, 1)$ by the equalities

$$\varphi_j = 2^j \chi_{[0, 2^{-j}]}.$$

These functions form a scaling sequence satisfying **g1, g2**, and for any $x > 1$,

$$\int\limits_{2^{-j}x \leq |t| \leq 1/2} |\varphi_j(t)|\, dt = 0;$$

i.e., condition (9.37) holds. It is not hard to see that the scaling function for the Haar MRA is a generating function.

We have studied the conditions for a PMRA in X to be generated by a summable function. However, a PMRA can be generated by a nonsummable function as well. For example, the function

$$\varphi(x) = \frac{\sin \pi x}{\pi x} \tag{9.45}$$

is a scaling function for the Kotelnikov-Shannon MRA and though $\varphi \notin L(\mathbb{R})$, it can be made periodic since series (9.34) converges in the sense of principal value. As a result, we obtain the PMRA described in Example 9.1.15. Nevertheless, not all PMRAs are generated even in this sense.

EXAMPLE 9.4.5. Let us construct a PMRA where the orthogonalized scaling sequence cannot be obtained by periodization of any function φ in the sense of principal value.

Define a PMRA by setting the bases $\{v_n^j\}_{n=0}^{2^j-1}$, $j \in \mathbb{Z}_+$, from Lemma 9.1.7. Put

$$v_1^1(x) = \sum_{k=0}^{\infty} q^k \left(e^{2\pi i(2k+1)x} + e^{-2\pi i(2k+1)x}\right),$$

where $q \in (0, 1)$. Since all odd Fourier coefficients of v_1^1 are different from zero, all the functions v_n^j are uniquely defined. The orthogonalized scaling sequence that satisfies conditions **g1, g2** has the form

$$\varphi_j := \sum_{n=-2^{j-1}}^{2^{j-1}} \frac{v_n^j}{\|v_n^j\|}.$$

Suppose there exists a function ϕ defined on $\mathbb{R}$ such that for any $x \in \mathbb{R}$, the series

$$\sum_{k=0}^{\infty} (\phi(x + k) + \phi(x - k))$$

converges absolutely and

$$\varphi_j(x) = 2^j \left(\phi(2^j x) + \sum_{k=1}^{\infty} \left(\phi(2^j x + 2^j k) + \phi(2^j x - 2^j k) \right) \right)$$

for all $j \in \mathbb{Z}_+$. Then the functions g_j given by equality (9.38) converge to ϕ on $\mathbb{R}$.

Let n be odd, $0 < n < 2^j$. Direct calculation gives

$$v_n^j(x) = q^{\frac{-n-1}{2}} e^{2\pi i n x} \frac{q^{2^{j-1}} e^{-2\pi i 2^j x}}{1 - q^{2^{j-1}} e^{-2\pi i 2^j x}} + q^{\frac{n-1}{2}} e^{2\pi i n x} \frac{1}{1 - q^{2^{j-1}} e^{2\pi i 2^j x}}.$$

It follows that in view of the equality $v_n^j = v_{n+2^j}^j$, for odd n, $|n| < 2^{j-1}$,

$$v_n^j(x) = q^{\frac{|n|-1}{2}} e^{2\pi i n x} + \alpha_{jn}(x),$$

where the sequence $\alpha_{jn}(x)$ tends to zero as $j \to \infty$ uniformly with respect to all x and n. It is clear that

$$\frac{v_n^j}{\|v_n^j\|} = e^{2\pi i n x} + \beta_{jn}(x),$$

where the sequence $\beta_{jn}(x)$ tends to zero as $j \to \infty$ uniformly with respect to all x and n. In view of property **V3** of Lemma 9.1.7, these relations also hold for even n, but in this case we have

$$\lim_{j\to\infty} g_j(x) = \lim_{j\to\infty} 2^{-j} \sum_{n=-2^{j-1}}^{2^{j-1}-1} \frac{v_n^j(2^{-j}x)}{\|v_n^j\|} = \lim_{j\to\infty} \frac{e^{-\pi i 2^{-j} x} \sin \pi x}{2^j \sin \pi 2^{-j} x} = \frac{\sin \pi x}{\pi x} = \varphi(x).$$

However, as is known, the function φ generates a PMRA different from the PMRA under consideration.

9.5. Kotelnikov-Shannon system

Let us construct an example of a PMRA in $L^2(\mathbb{T}^2)$ with a scaling sequence consisting of trigonometric polynomials with minimum possible symmetric spectra. In the case $d = 1, M = 2$ we can take the PMRA obtained by periodization of the Kotelnikov-Shannon MRA. As noted in §9.4, for this MRA, the Dirichlet kernels form a scaling sequence and (9.45) is a generating function.

Take $M = \begin{pmatrix} 2 & -1 \\ 2 & 1 \end{pmatrix}$ as a matrix dilation. Note that $m = 4$, $M^* = \begin{pmatrix} 2 & 2 \\ -1 & 1 \end{pmatrix}$, $M^{*-1} = \frac{1}{4} \begin{pmatrix} 1 & -2 \\ 1 & 2 \end{pmatrix}$, and fix the set $D(M^*)$ consisting of the vectors

$$s_0 = \begin{pmatrix} 0 \\ 0 \end{pmatrix}, \quad s_1 = \begin{pmatrix} -1 \\ 0 \end{pmatrix}, \quad s_2 = \begin{pmatrix} 0 \\ -1 \end{pmatrix}, \quad s_3 = \begin{pmatrix} -1 \\ -1 \end{pmatrix}.$$

Let Ω_j denote the parallelogram $M^{*j}[-1,1]^2$ without vertices. Set $a_j := M^{*j} \begin{pmatrix} 0 \\ 1 \end{pmatrix}$, $b_j := M^{*j} \begin{pmatrix} 0 \\ -1 \end{pmatrix}$ and define functions φ_j by their Fourier coefficients: $\widehat{\varphi}_0(k) := \delta_{k0}$ for all $k \in \mathbb{Z}^2$, and for $j \in \mathbb{N}$,

$$\widehat{\varphi}_j(k) := \begin{cases} 2^{-j}, & k \in \Omega_{j-1} \setminus \{a_{j-1}, b_{j-1}\}, \\ 2^{-j-1/2}, & k = a_{j-1} \text{ or } k = b_{j-1}, \\ 0, & k \notin \Omega_{j-1}. \end{cases} \tag{9.46}$$

To show that this is a scaling sequence, we need the following lemma.

LEMMA 9.5.1. *The number of integer points in* Ω_j *is equal to* $4^{j+1}+1$. *All integer points from* Ω_j *except* a_j *belong to different cosets of the matrix* M^{*j+1}, *and the point* a_j *is congruent to* b_j *modulo* M^{*j+1}.

Proof. Let us first show that there are no integer points on the midlines of Ω_j except for the boundary points and zero. The midlines of Ω_j are the segments connecting the point $M^{*j}\begin{pmatrix}0\\1\end{pmatrix}$ with $M^{*j}\begin{pmatrix}0\\-1\end{pmatrix}$ and the point $M^{*j}\begin{pmatrix}1\\0\end{pmatrix}$ with $M^{*j}\begin{pmatrix}-1\\0\end{pmatrix}$. These segments pass through the origin of coordinates and are symmetric with respect to it; thereby it suffices to prove our statement for the halves of the segments. Let us do this for one half-segment (for the second half-segment the prove is analogous); i.e., let us show that there are no integer points on the interval $\left(\mathbf{0},\ M^{*j}\begin{pmatrix}1\\0\end{pmatrix}\right)$. Set $\begin{pmatrix}x_j\\y_j\end{pmatrix}:=M^{*j}\begin{pmatrix}1\\0\end{pmatrix}=M^{*}\begin{pmatrix}x_{j-1}\\y_{j-1}\end{pmatrix}$. For all $j>0$, the first coordinate of this vector is even, while the second coordinate is odd; hence x_j and y_j cannot have even common divisors. The formulas

$$x_{j-1}=\frac{x_j-2y_j}{4},\qquad y_{j-1}=\frac{x_j+2y_j}{4}$$

imply that if x_j and y_j had an odd common divisor, then x_{j-1} and y_{j-1} would have the same common divisor. But $x_0=1$ and $y_0=0$ have no common divisors; hence by induction, x_j and y_j have no common divisors for any $j\in\mathbb{Z}_+$. Fix j and represent the segment $\left(\mathbf{0},\ M^{*j}\begin{pmatrix}1\\0\end{pmatrix}\right)$ in the parametric form

$$\begin{cases} x=tx_j,\\ y=ty_j,\end{cases}\qquad t\in(0,1).$$

Assume that there exists an integer point $(x^0,\ y^0)$ contained in this segment, i.e., there exists $t^0\in(0,1)$ such that $x^0=t^0x_j$, $y^0=t^0y_j$, where $x^0,\ y^0,\ x_j,\ y_j\in\mathbb{Z}$. Therefore t^0 cannot be irrational. On the other hand, t^0 cannot be rational, since if we assume $t^0=p/q$, then x_j and y_j are divisible by q and thus have a common divisor, which is not possible.

Note that for $D(M^{*j})$, we can take the set of integer points contained in $M^{*j}\,[0,1)^2$, since for any two elements from $M^{*j}\,[0,1)^2\cap\mathbb{Z}^2$, the difference $M^{*j}r_1-M^{*j}r_2$, where $r_1,\ r_2\in[0,1)^2$, can be congruent to zero modulo M^{*j} only if r_1 and r_2 coincide. However, the number of integer vectors contained in $M^{*j}\,[0,1)^2$ is equal to m^j, i.e., to the number of integer vectors in the set $D(M^{*j})$. Since there are no integer points on the midlines of Ω_j except for the boundary and zero points, the number of integer points in Ω_j is obtained by summing the interior integer points of the set $M^{*j}\,[0,1)^2$ (their number is 4^j-1, since one digit corresponding to the zero coset is excluded from the set $M^{*j}[0,1)^2$) each counted four times, zero, and boundary points. The number of boundary points is equal to 4, since the boundary points of the midlines are integer, the vertices do not belong to Ω_j, and there are no other integer points on the boundary because each edge differs from the corresponding midline by an integer shift. Thus the number of integer points contained in Ω_j is equal to $4(4^j-1)+1+4=4^{j+1}+1$.

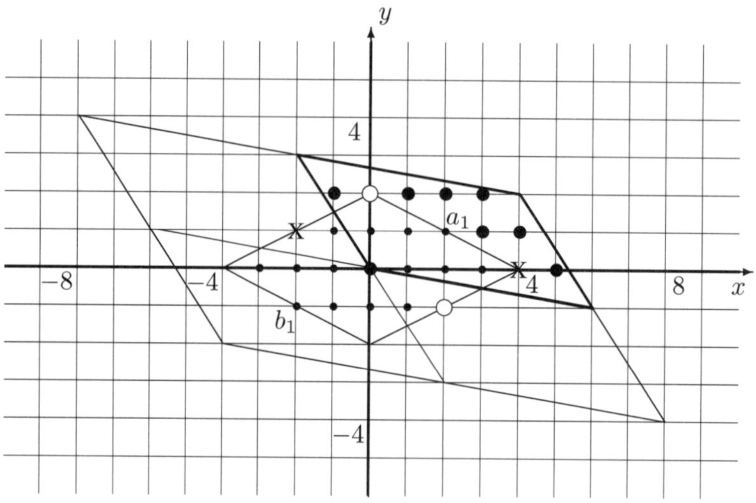

FIGURE 9.1. The interior parallelogram corresponds to Ω_1, the exterior one to Ω_2

Let us show that all the points from Ω_j except a_j belong to different cosets of the matrix M^{*j+1} and $a_j \equiv b_j \pmod{M^{*j+1}}$. By Lemma 2.2.5, the set

$$\bigcup_{k=0}^{3}\left(M^{*-1}\,[0,1)^2\,[0] + M^{*-1}s_k\right)$$

is congruent modulo $\mathbb{Z}^2$ to the set $[0,1)^2$. The relation

$$\bigcup_{k=0}^{3}\left(M^{*-1}\,[0,1)^2 + M^{*-1}s_k\right) = \bigcup_{k=0}^{3}\left(M^{*-1}\left([0,1)^2 + s_k\right)\right) = M^{*-1}[-1,1)^2$$

implies that $M^{*-1}[-1,1)^2$ is congruent to $[0,1)^2$ modulo $\mathbb{Z}^2$. Fix $j \in \mathbb{Z}_+$ and apply the operator M^{*j+1} to each set. We see that the set $M^{*j}[-1,1)^2$ is congruent to $M^{*j+1}\,[0,1)^2$ modulo M^{*j+1}. Since all integer points of the set $M^{*j+1}[0,1)^2$ belong to different cosets of the matrix M^{*j+1}, all integer points of $M^{*j}[-1,1)^2$ belong to different cosets of M^{*j+1} as well.

In Figure 9.1, $M^*[-1,1]^2$ is the interior parallelogram, $M^{*2}[-1,1]^2$ is the exterior parallelogram, and $M^{*2}[0,1]^2$ is the accentuated parallelogram. Integer points of the set Ω_1 are not brightly accentuated. We brightly accentuate the points from the set $M^{*2}[0,1]^2$ which do not belong to Ω_1, i.e., such that there exist points from $M^{*1}[-1,1]^2$ congruent to them modulo M^{*2}.

As shown in the figure, partitioning the large parallelogram into 4 small ones and putting each of these over the right upper parallelogram, we shall see that the coinciding points will be congruent modulo M^{*2}. It is clear that the points a_1 and b_1 are congruent modulo M^{*2}. The points marked by noughts and crosses will also be respectively congruent. The set of integer points contained in $M^{*j}[-1,1)^2$ differs from the set of integer points from Ω_j by the fact that Ω_j does not contain the point $M^{*j}\begin{pmatrix}-1\\-1\end{pmatrix}$ but contains the points $M^{*j}\begin{pmatrix}1\\0\end{pmatrix}$ and $a_j = M^{*j}\begin{pmatrix}0\\1\end{pmatrix}$.

It remains to note that $M^{*j}\begin{pmatrix}-1\\-1\end{pmatrix} \equiv M^{*j}\begin{pmatrix}1\\0\end{pmatrix} \pmod{M^{*j+1}}$ and $a_j \equiv b_j \pmod{M^{*j+1}}$. ◊

Check that the sequence of functions $\{\varphi_j\}_{j=0}^{\infty}$ defined in (9.46) satisfies the conditions of Theorem 9.1.4. Condition **Φ1** follows from the definition, **Φ2** from Lemma 9.5.1, and **Φ3** from relation (2.8), which holds since the eigenvalues of the matrix M are equal to 2 in modulus. To check **Φ4**, find the coefficients μ_k^j from the condition $\widehat{\varphi}_j(k) = \mu_k^{j+1}\widehat{\varphi}_{j+1}(k)$. Note that Ω_j is strictly embedded in Ω_{j+1} for any $j \in \mathbb{Z}_+$. For $j = 0$, the embedding is obvious. Further, applying the operator M^{*j} to $\Omega_0 \subset \Omega_1$, we use the fact that the mapping M^{*j} is nonsingular and thus preserves the character of embedding. Taking for $D(M^{*j+1})$ the set of integer points from Ω_j except for the point a_j, for $k \in D(M^{*j+1})$ we have $\mu_k^{j+1} = 2$ if $k \in \Omega_{j-1}\backslash\{a_{j-1}, b_{j-1}\}$; $\mu_k^{j+1} = \sqrt{2}$, if $k = a_{j-1}$ or $k = b_{j-1}$; $\mu_k^{j+1} = 0$, if $k \notin \Omega_{j-1}$. It is not hard to see that the M^{*j+1}-periodic extension of μ_k^{j+1} with respect to the subindex is a sequence satisfying **Φ4**. For the numbers γ_k^j from **Φ5** of Theorem 9.1.4, we take $\gamma_k^j = 1/2$. Clearly, $\gamma_k^j\widehat{\varphi}_j(n) = \widehat{\varphi}_{j+1}(M^*n)$ for all $j \in \mathbb{Z}_+$, $k \in \mathbb{Z}^2$, and $n \equiv k \pmod{M^{*j}}$. Thus the sequence of functions $\{\varphi_j\}_{j=0}^{\infty}$ satisfies all the conditions of Theorem 9.1.4; hence it is a scaling sequence and the functions φ_j are trigonometric polynomials with minimum possible symmetric spectra. It is not hard to check that

$$\|\omega_r^j\varphi_j\|^2 = \sum_{l\in\mathbb{Z}^d} |\widehat{\varphi}_j(M^{*j}l + r)|^2 = 4^{-j},$$

and hence by Proposition 9.2.1, the systems of functions $\{S_n^j\varphi\}_{n\in D(M^j)}$ are orthonormal.

Let us now construct sequences of wavelet functions. Fix $j \in \mathbb{Z}_+$ and the set $D(M^{*j})$ coinciding with $\Omega_{j-1}\backslash\{a_{j-1}\}$. As above, the set $D(M^*)$ consists of s_0, s_1, s_2, s_3. Let $n \in D(M^{*j})$. If $n \neq b_{j-1}$, then $\mu_{n+M^{*j}s_0}^{j+1} = 2$ and $\mu_{n+M^{*j}s_k}^{j+1} = 0$ for $k = 1, 2, 3$, since the vectors $n+M^{*j}s_k$ are not contained in Ω_{j-1}; moreover, they are not congruent modulo M^{*j+1} to any element from Ω_{j-1}. The corresponding unitary 4×4 matrix is diagonal, its diagonal elements being equal to 2. If $n = b_{j-1}$, then $\mu_{n+M^{*j}s_k}^{j+1} = \sqrt{2}$ for $k = 0$ and $k = 3$ (the vector $b_{j-1} + M^{*j}s_3$ is congruent to a_{j-1} modulo M^{*j+1}, which can be checked directly), and $\mu_{n+M^{*j}s_k}^{j+1} = 0$ for $k = 1$ and $k = 2$, as the vectors $n + M^{*j}s_k$, $k = 1, 2$, are neither contained in Ω_{j-1} nor congruent modulo M^{*j+1} to any element from Ω_{j-1}. The corresponding unitary matrix is

$$\begin{pmatrix} \sqrt{2}/2 & 0 & 0 & \sqrt{2}/2 \\ 0 & 1 & 0 & 0 \\ 0 & 0 & 1 & 0 \\ \sqrt{2}/2 & 0 & 0 & -\sqrt{2}/2 \end{pmatrix}.$$

By Lemma 2.2.3, vectors of the form $n + M^{*j}s_k$, where $s_k \in D(M^*)$ and $n \in D(M^{*j})$, range over the set $D(M^{*j+1})$; therefore it suffices to determine the Fourier coefficients of wavelet functions for all integer vectors l that are congruent modulo

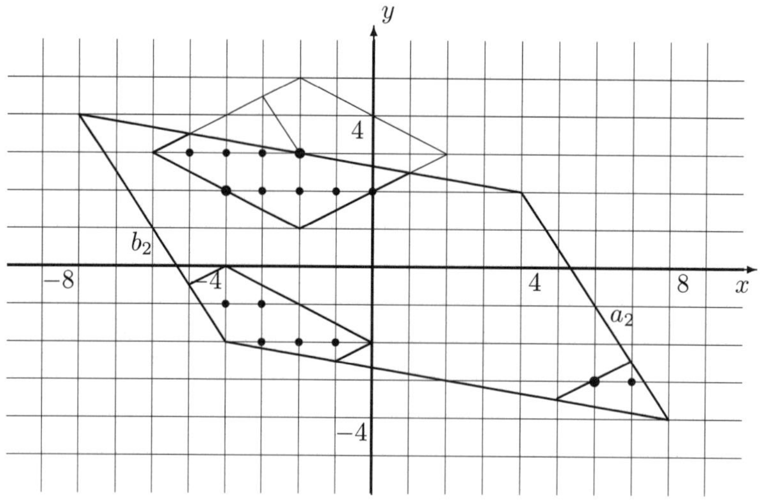

FIGURE 9.2. The spectrum of the function $\psi_2^{(1)}$ is denoted by points; $\hat{\psi}_2^{(1)}$ takes the value 2^{-2} at all accentuated points.

M^{*j+1} to vectors of the form $n + M^{*j} s_k$:

$$\widehat{\psi}_j^{(1)}(l) := \begin{cases} 2^{-j}, & l \equiv n + M^{*j} s_1 \pmod{M^{*j+1}},\ l \in \Omega_j \backslash \{a_j, b_j\}, \\ 2^{-j-1/2}, & l \equiv n + M^{*j} s_1 \pmod{M^{*j+1}},\ l \in \{a_j, b_j\}, \end{cases}$$

if $n \in D(M^{*j})$, and otherwise $\psi_j^{(1)}(l) = 0$;

$$\widehat{\psi}_j^{(2)}(l) := \begin{cases} 2^{-j}, & l \equiv n + M^{*j} s_2 \pmod{M^{*j+1}},\ l \in \Omega_j \backslash \{a_j, b_j\}, \\ 2^{-j-1/2}, & l \equiv n + M^{*j} s_2 \pmod{M^{*j+1}},\ l \in \{a_j, b_j\}, \end{cases}$$

if $n \in D(M^{*j})$, and otherwise $\psi_j^{(1)}(2) = 0$;

$$\widehat{\psi}_j^{(3)}(l) := \begin{cases} 2^{-j}, & l \equiv n + M^{*j} s_3 \pmod{M^{*j+1}},\ l \in \Omega_j \backslash \{a_j, b_j\}, \\ 2^{-j-1/2}, & l \equiv n + M^{*j} s_3 \pmod{M^{*j+1}},\ l \in \{a_j, b_j\}, \end{cases}$$

if $n \in D(M^{*j})$, $n \neq b_{j-1}$;

$$\widehat{\psi}_j^{(3)}(l) := \begin{cases} 2^{-j-1/2}, & l \equiv b_{j-1} \pmod{M^{*j+1}},\ l \in \Omega_j \backslash \{a_j, b_j\}, \\ -2^{-j-1/2}, & l \equiv a_{j-1} \pmod{M^{*j+1}},\ l \in \Omega_j \backslash \{a_j, b_j\}, \end{cases}$$

and otherwise $\psi_j^{(1)}(3) = 0$.

In Figure 9.2, Ω_2 is the large parallelogram, and the domain Ω_1 shifted by $M^* s_1$ is the small one. The whole spectrum of the function $\psi^{(1)}$ is contained in the three accentuated domains in the interior of the large parallelogram. These domains form a set congruent modulo M^{*2} to the small parallelogram. Analogously, Figures 9.3 and 9.4, respectively, present the spectra of the functions $\psi^{(2)}$ and $\psi^{(3)}$.

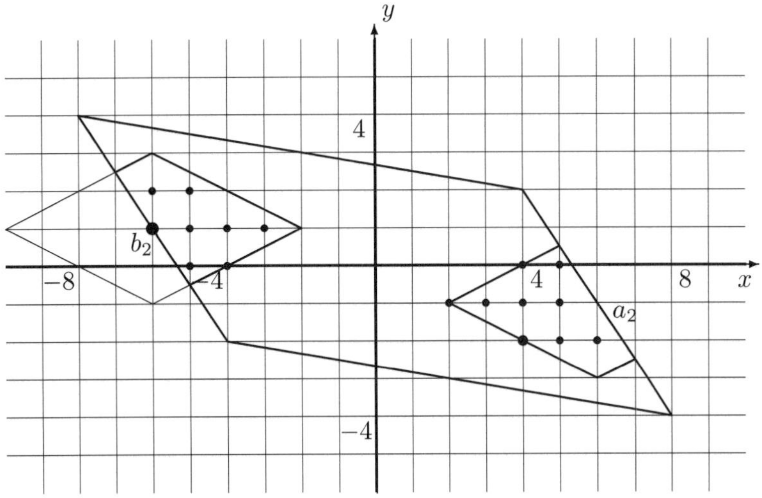

FIGURE 9.3. Points denote the spectrum of the function $\psi_2^{(2)}$; $\hat{\psi}_2^{(2)}$ takes the value 2^{-2} at all accentuated points except for the point $b_2 = (-6, 1)$ and $\hat{\psi}_2^{(2)}(b_2) = 2^{-5/2}$.

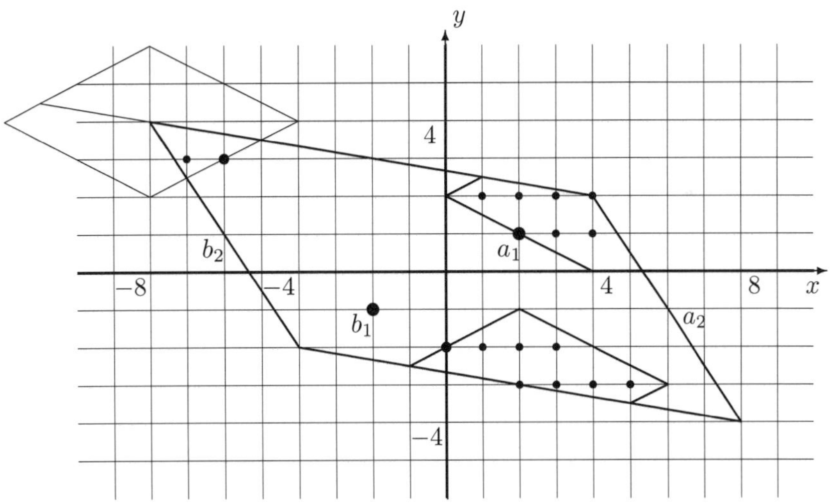

FIGURE 9.4. The points denote the spectrum of the function $\psi_2^{(3)}$; $\hat{\psi}_2^{(3)}$ takes the value 2^{-2} at all accentuated points except the points $a_1 = (2, 1)$ and $b_1 = (-2, -1)$; $\hat{\psi}_2^{(3)}(a_1) = -2^{-5/2}$ and $\hat{\psi}_2^{(3)}(b_1) = 2^{-5/2}$.

CHAPTER 10

Approximation by Periodic Wavelets

10.1. Convergence of wavelet expansions in norm

In this chapter, we study various types of convergence of periodic wavelet expansions.

Fix a (p, q)-pair satisfying the conditions of Theorem 9.2.4. Then the wavelet systems

$$\{S_r^j \psi_j^{(\nu)},\ j \in \mathbb{Z}_+,\ r \in D(M^j),\ \nu = 1, \dots, m-1\},$$

$$\{S_r^j \widetilde{\psi}_j^{(\nu)},\ j \in \mathbb{Z}_+,\ r \in D(M^j),\ \nu = 1, \dots, m-1\}$$

are biorthonormal. For functions $f \in L_p(\mathbb{T}^d)$, we can consider the expansions into the Fourier series

$$\langle f, \widetilde{\varphi}_0 \rangle \varphi_0 + \sum_{j=0}^{\infty} \sum_{\nu=1}^{m-1} \sum_{r \in D(M^j)} \langle f, S_r^j \widetilde{\psi}_j^{(\nu)} \rangle S_r^j \psi_j^{(\nu)}. \tag{10.1}$$

Having renumbered arbitrarily the set of digits of $D(M^j) = \{r_l\}_{l=0}^{m^j-1}$, we denote partial sums of this series by $s_N(f)$ and say that the wavelet expansion converges if the sequence $s_N(f)$ does.

Let $N = (\kappa+1)m^j + n + 1$, $j \in \mathbb{Z}_+$, $\kappa = 0, \dots, m-2$, $n = 0, \dots, m^j - 1$; then the partial sum $s_N(f, x)$ of series (10.1) can be represented in the form

$$\begin{aligned} s_N(f) &= \langle f, \widetilde{\varphi}_0 \rangle \varphi_0 + \sum_{i=0}^{j-1} \sum_{\nu=1}^{m-1} \sum_{r \in D(M^i)} \langle f, S_r^i \widetilde{\psi}_i^{(\nu)} \rangle S_r^i \psi_i^{(\nu)} \\ &\quad + \sum_{\nu=1}^{\kappa} \sum_{r \in D(M^j)} \langle f, S_r^j \widetilde{\psi}_j^{(\nu)} \rangle S_r^j \psi_j^{(\nu)} + \sum_{l=0}^{n} \langle f, S_{r_l}^j \widetilde{\psi}_j^{(\kappa+1)} \rangle S_{r_l}^j \psi_j^{(\kappa+1)} \\ &= s_N^{(0)}(f) + \sum_{\nu=1}^{\kappa} s_N^{(\nu)}(f) + s_N^{(\kappa+1)}(f). \end{aligned} \tag{10.2}$$

Since $s_N^{(0)}$ is a projection operator on the space V_j, the sum $s_N^{(0)}(f)$ on the right-hand side can be reexpanded with respect to the shifts of the function φ_j:

$$s_N^{(0)}(f) = \sum_{r \in D(M^j)} \langle f, S_r^j \widetilde{\varphi}_j \rangle S_r^j \varphi_j. \tag{10.3}$$

LEMMA 10.1.1. *Let η be an even bounded function decreasing on $[0,+\infty)$. If η is summable on $[0,+\infty)$, then for all $u, v \in \mathbb{R}^d$*

$$\sum_{k\in\mathbb{Z}^d}\prod_{l=1}^{d}\eta(u_l+k_l)\eta(v_l+k_l) \le A\eta\left(2^{-3-d/2}|u-v|\right), \tag{10.4}$$

$$\sum_{k\in\mathbb{Z}^d}\prod_{l=1}^{d}\eta(u_l+k_l)\eta(v_l+k_l) \le B\prod_{l=1}^{d}\eta\left(\frac{u_l-v_l}{8}\right), \tag{10.5}$$

where A, B are constants that depend only on η and d. If

$$\int_{\mathbb{R}^d}\eta(|x|)\,dx < \infty, \tag{10.6}$$

then for all $u, v \in \mathbb{R}^d$

$$\sum_{k\in\mathbb{Z}^d}\eta(|u+k|)\eta(|v+k|) \le C\eta\left(2^{-3}|u-v|\right), \tag{10.7}$$

where C depends only on η and d.

Proof. In view of Lemma 1.7.1 we have

$$\sum_{k\in\mathbb{Z}^d}\prod_{l=1}^{d}\eta(u_l+k_l)\eta(v_l+k_l) = \prod_{l=1}^{d}\sum_{k_l\in\mathbb{Z}}\eta(u_l+k_l)\eta(v_l+k_l) \le C^d\prod_{l=1}^{d}\eta\left(\frac{u_l-v_l}{8}\right),$$

which implies (10.5). To prove (10.4), it remains to note that by the monotonicity of the function η we have

$$\prod_{l=1}^{d}\eta\left(\frac{u_l-v_l}{8}\right) \le (\eta(0))^{d-1}\eta\left(\frac{\max\limits_{1\le l\le d}|u_l-v_l|}{8}\right) \le (\eta(0))^{d-1}\eta\left(2^{-3-d/2}|u-v|\right).$$

Inequality (10.7) is proved in the same way as Lemma 1.7.1. ◊

THEOREM 10.1.2. *Let PMRAs $\{V_j\}_{j=0}^{\infty}$, $\{\widetilde{V}_j\}_{j=0}^{\infty}$ form a (p,q)-pair with scaling sequences $\{\varphi_j\}_{j=0}^{\infty}$, $\{\widetilde{\varphi}_j\}_{j=0}^{\infty}$ such that $\{S_n^j\varphi_j\}_{n\in D(M^j)}$ and $\{S_k^j\widetilde{\varphi}_j\}_{n\in D(M^j)}$ are biorthonormal systems, and let $\{\psi_j^{(\nu)}\}_{j=0}^{\infty}$, $\{\widetilde{\psi}_j^{(\nu)}\}_{j=0}^{\infty}$, $\nu = 1,\dots,m-1$, be the corresponding sequences of wavelet functions. If there exists an even, bounded function η decreasing on $[0,\infty)$ that satisfies (10.6) and*

$$\left|m^{-\frac{j}{p}}\varphi_j(x)\right|, \left|m^{-\frac{j}{p}}\psi_j^{(\nu)}(x)\right|, \left|m^{-\frac{j}{q}}\widetilde{\varphi}_j(x)\right|, \left|m^{-\frac{j}{q}}\widetilde{\psi}_j^{(\nu)}(x)\right| \le \eta(|M^jx|) \tag{10.8}$$

for all $x \in [-1/2,1/2]^d$, then for each $f \in L_p(\mathbb{T}^d)$ ($f \in C(\mathbb{T}^d)$ for $p=\infty$), series (10.1) converges to f in L_p-norm (it converges uniformly for $p=\infty$).

Proof. Set

$$g_j(t) = \begin{cases} \varphi_j(M^{-j}t), & t \in M^j\left([-1/2,1/2]^d\right), \\ 0, & t \notin M^j\left([-1/2,1/2]^d\right), \end{cases}$$

and define functions $\widetilde{g}_j$ analogously. It is clear that

$$|m^{-\frac{j}{p}}g_j(x)| \le \eta(|x|), \quad |m^{-\frac{j}{q}}\widetilde{g}_j(x)| \le \eta(|x|) \tag{10.9}$$

and

$$\varphi_j(x) = \sum_{k\in\mathbb{Z}^d} g_j(M^jx + M^jk), \quad \widetilde{\varphi}_j(x) = \sum_{k\in\mathbb{Z}^d} \widetilde{g}_j(M^jx + M^jk) \tag{10.10}$$

for all $x \in \mathbb{R}^d$. Let us estimate the norm of the sum $s_N^{(0)}$ from equality (10.2). Using (10.3), (10.10), (10.9), we get

$$\begin{aligned}
|s_N^{(0)}(f,x)| &= \left| \int_{\mathbb{T}^d} f(t) \sum_{l=0}^{m^j-1} \overline{\widetilde{\varphi}_j(t + M^{-j}r_l)}\varphi_j(x + M^{-j}r_l)\, dt \right| \\
&\le \int_{\mathbb{T}^d} |f(t)| \sum_{l=0}^{m^j-1} \sum_{k'\in\mathbb{Z}^d} \sum_{k\in\mathbb{Z}^d} |g_j(|M^j(t+k+k') + r_l|)\widetilde{g}_j(|M^j(x+k') + r_l|)|\, dt \\
&\le m^j \int_{\mathbb{R}^d} |f(t)| \sum_{l=0}^{m^j-1} \sum_{k'\in\mathbb{Z}^d} \eta(|M^j(t+k') + r_l|)\eta(|M^j(x+k') + r_l|)\, dt \\
&= m^j \int_{\mathbb{R}^d} |f(t)| \sum_{k\in\mathbb{Z}^d} \eta(|M^jt + k|)\eta(|M^jx + k|)\, dt.
\end{aligned}$$

In view of Lemma 10.1.1 it follows that

$$\begin{aligned}
\|s_N^{(0)}(f)\|_p &\le Cm^j \left\| \int_{\mathbb{R}^d} |f(t)| \sum_{k\in\mathbb{Z}^d} \eta(2^{-3-d/2}|M^j(t - \cdot)|)\, dt \right\|_p \\
&= Cm^j \left\| \int_{\mathbb{R}^d} |f(t+\cdot)| \sum_{k\in\mathbb{Z}^d} \eta(2^{-3-d/2}|M^jt|)\, dt \right\|_p \\
&\le C\|f\|_p \int_{\mathbb{R}^d} \eta(2^{-3-d/2}|t|)\, dt.
\end{aligned}$$

It follows from (10.6) that the integral is finite. We have thus established that for $\nu = 0$, the operators $s_N^{(\nu)}$ acting from $L_p(\mathbb{T}^d)$ to $L_p(\mathbb{T}^d)$ (from $C(\mathbb{T}^d)$ to $C(\mathbb{T}^d)$ for $p = \infty$) are uniformly bounded in norm. The uniform boundedness of the operators $s_N^{(\nu)}$ for $\nu = 1, \dots, m-1$ is proved analogously. Consequently,

$$\|s_N(f)\|_p \le C_1, \tag{10.11}$$

where C is an absolute constant.

Choose $\varepsilon > 0$. By property **MR2** of Definition 9.1.1, there exists a function $F \in V_{j_0}$ such that

$$\|f - F\|_p < \varepsilon.$$

According to Theorem 9.2.4, $s_N(F) = F$ for $N \ge m^{j_0}$. Hence by (10.11) we get

$$\|f - s_N(f)\|_p = \|f - F - s_N(f - F)\|_p \le (C_1 + 1)\|f - F\|_p \le (C_1 + 1)\varepsilon. \quad \Diamond$$

REMARK 10.1.3. If PMRAs $\{V_j\}_{j=0}^\infty, \{\widetilde{V}_j\}_{j=0}^\infty$ are generated by nonperiodic functions, then it is natural to give the condition of convergence in terms of these functions. Let $\varphi, \widetilde{\varphi} \in L(\mathbb{R}^d)$ be scaling functions for some MRAs in $L_2(\mathbb{R}^d)$ with biorthonormal integer shifts, and let $\psi^{(\nu)}, \widetilde{\psi}^{(\nu)}$, $\nu = 1, \dots, m$, be the corresponding set of wavelet functions summable on $\mathbb{R}^d$. Assume that the functions

$$\varphi_j(x) = m^{j/p} \sum_{l \in \mathbb{Z}} \varphi(M^j x + M^j l), \tag{10.12}$$

$$\widetilde{\varphi}_j(x) = m^{j/q} \sum_{l, \in \mathbb{Z}} \widetilde{\varphi}(M^j x + M^j l) \tag{10.13}$$

form scaling sequences for a (p, q)-pair, condition (9.15) of Proposition 9.2.1 holds, and the functions

$$\psi_j^{(\nu)}(x) = m^{j/p} \sum_{l \in \mathbb{Z}} \psi^{(\nu)}(M^j x + M^j l), \tag{10.14}$$

$$\widetilde{\psi}_j^{(\nu)}(x) = m^{j/q} \sum_{l \in \mathbb{Z}} \widetilde{\psi}^{(\nu)}(M^j x + M^j l) \tag{10.15}$$

form sequences of wavelet functions in these PMRAs. Repeating the reasoning from the proof of Theorem 10.1.2, replacing the functions $g_j, \widetilde{g}_j$ by φ, $\widetilde{\varphi}$, respectively, in (10.9), (10.10), we deduce that the statement of Theorem 10.1.2 holds if instead of (10.8) we assume that

$$|\varphi(x)|, |\psi^{(\nu)}(x)|, |\widetilde{\varphi}(x)|, |\widetilde{\psi}^{(\nu)}(x)| \le \eta(|x|)$$

for all $x \in \mathbb{R}^d$. It is not hard to see that under this assumption the series on the right-hand sides of equalities (10.12)–(10.15) converge uniformly, so that the functions $\varphi_j, \psi_j^{(\nu)}, \widetilde{\varphi}_j, \widetilde{\psi}_j^{(\nu)}$ are bounded.

In the case $(p, q) = (\infty, 1)$ the requirement for the functions $\widetilde{\varphi}_j, \widetilde{\psi}_j^{(\nu)}$ in the conditions of Theorem 10.1.2 can be weakened.

THEOREM 10.1.4. *Suppose PMRAs $\{V_j\}_{j=0}^{\infty}$, $\{\widetilde{V}_j\}_{j=0}^{\infty}$ form a $(\infty, 1)$-pair with scaling sequences $\{\varphi_j\}_{j=0}^{\infty}$, $\{\widetilde{\varphi}_j\}_{j=0}^{\infty}$ such that $\{S_n^j \varphi_j\}_{n \in D(M^j)}$ and $\{S_k^j \widetilde{\varphi}_j\}_{n \in D(M^j)}$ are biorthonormal systems. Let $\{\psi_j^{(\nu)}\}_{j=0}^{\infty}, \{\widetilde{\psi}_j^{(\nu)}\}_{j=0}^{\infty}$, $\nu = 1, \dots, m-1$, be the corresponding sequences of wavelet functions. If*

$$\sup_j \|\widetilde{\varphi}_j\|_1, \sup_{j,\nu} \|\widetilde{\psi}_j^{(\nu)}\|_1 < \infty \tag{10.16}$$

and there exists an even bounded function η decreasing on $[0, \infty)$ and satisfying condition (10.6) such that

$$|\varphi_j(x)|, \left|\psi_j^{(\nu)}(x)\right| \le \eta(|M^j x|) \tag{10.17}$$

for all $x \in [-1/2, 1/2]^d$, then for each $f \in C(\mathbb{T}^d)$, series (10.1) converges uniformly to f.

Proof. Using (10.3), we get

$$|s_N^{(0)}(f, x)| = \left| \int_{\mathbb{T}^d} f(t) \sum_{l=0}^{m^j - 1} \overline{\widetilde{\varphi}_j(t + M^{-j} r_l)} \varphi_j(x + M^{-j} r_l) \, dt \right|$$

$$\le \|f\|_\infty \sup_j \|\widetilde{\varphi}_j\|_1 \sum_{l=0}^{m^j - 1} |\varphi_j(x + M^{-j} r_l)|.$$

Let g_j have the same meaning as in Theorem 10.1.2. Relations (10.9), (10.10) for the functions g_j, φ_j give

$$\sum_{l=0}^{m^j-1} |\varphi_j(x + M^{-j}r_l)| \le \sum_{l=0}^{m^j-1} \sum_{k\in\mathbb{Z}^d} \eta(|M^j x + M^j k + r_l|) = \sum_{k\in\mathbb{Z}^d} \eta(|M^j x + k|).$$

Since η and (10.6) are monotone functions, it follows that this sum is uniformly bounded. Hence taking into account (10.16), we deduce that for $\nu = 0$, the operators $s_N^{(\nu)}$ acting from $C(\mathbb{T}^d)$ to $C(\mathbb{T}^d)$ are uniformly bounded in norm. The uniform boundedness of the operators $s_N^{(\nu)}$ for $\nu = 1, \dots, m-1$ can be proved analogously. Consequently,

$$\|s_N(f)\|_\infty \le C,$$

where C is an absolute constant. To prove the uniform convergence of series (10.1), it remains to repeat the reasoning from the proof of Theorem 10.1.2. ◊

REMARK 10.1.5. If PMRAs $\{V_j\}_{j=0}^\infty, \{\widetilde{V}_j\}_{j=0}^\infty$ form an $(\infty, 1)$-pair and are generated by nonperiodic functions $\varphi, \widetilde{\varphi}$ and the functions $\varphi_j, \widetilde{\varphi}_j, \psi_j^{(\nu)}, \widetilde{\psi}_j^{(\nu)}$ are defined, respectively, by equalities (10.12), (10.13), (10.14), (10.15), the statements of Theorems 10.1.4 and 10.1.6 will hold if instead of (10.17) we assume that

$$|\varphi(x)|, |\psi^{(\nu)}(x)| \le \eta(|x|)$$

for all $x \in \mathbb{R}^d$. In this case relation (10.16) follows from the condition $\widetilde{\varphi}$, $\widetilde{\psi}^{(\nu)} \in L(\mathbb{R}^d)$.

THEOREM 10.1.6. *Under the conditions of Theorem* 10.1.4, *for any function* $g \in L(\mathbb{T}^d)$, *the series*

$$\langle g, \varphi_0 \rangle \widetilde{\varphi}_0 + \sum_{j=0}^{\infty} \sum_{\nu=1}^{m-1} \sum_{r\in D(M^j)} \langle g, S_r^j \psi_j^{(\nu)} \rangle S_r^j \widetilde{\psi}_j^{(\nu)} \tag{10.18}$$

converges in norm to g *in* $L(\mathbb{T}^d)$.

This theorem is proved following the same scheme as Theorem 10.1.4. In this case, in the estimates of the sums $s_N^{(\nu)}(f)$, the functions φ_j, $\psi_j^{(\nu)}$ and variable x switch roles with the functions $\widetilde{\varphi}_j$, $\widetilde{\psi}_j$ and variable t, respectively.

10.2. Convergence of wavelet expansions almost everywhere

We shall consider wavelet systems generated by a nonperiodic scaling function of a separable MRA in $L_2(\mathbb{R}^d)$.

Suppose $\varphi \in L_2(\mathbb{R})$ is a scaling function whose integer shifts form an orthonormal system and ψ is the corresponding wavelet function. We shall assume that for all $t \in \mathbb{R}$,

$$|\varphi(t)|, |\psi(t)| \le \eta(|t|), \tag{10.19}$$

where η is a bounded decreasing function, summable on $[0, \infty)$.

Introduce the following notation: $E = \{1, \dots, d\}$, $\mathcal{E} = \{e \subset E, e \neq E\}$,

$$g^e(x) = g^e(\varphi, \psi, x) = \prod_{l \in e} \varphi(x_l) \prod_{l \in E \setminus e} \psi(x_l), \quad x \in \mathbb{R}^d,\ e \subset E,$$

$$\Phi_{jn}(x) = 2^{dj/2} g^E(2^j x + n), \quad n \in \mathbb{Z}^d, x \in \mathbb{R}^d,$$

$$\Psi^e_{jn}(x) = 2^{dj/2} g^e(2^j x + n), \quad n \in \mathbb{Z}^d, x \in \mathbb{R}^d, e \in \mathcal{E}.$$

It follows from the results of §2.1 that Φ_{00} is a scaling function and for every fixed j, the functions Φ_{jn}, $n \in \mathbb{Z}^d$, form an orthonormal system; Ψ^e_{00}, $e \in \mathcal{E}$, is the corresponding set of wavelet functions. The functions Φ_{0n}, Ψ^e_{jn}, $j \in \mathbb{Z}_+$, $n \in \mathbb{Z}^d$, $e \in \mathcal{E}$, form an orthonormal basis in $L_2(\mathbb{R}^d)$. As a result of their periodization, which is possible in the standard sense in view of condition (10.19), we obtain a set of functions 1-periodic in each variable:

$$\psi^E_{jn}(x) = \sum_{l \in \mathbb{Z}^d} \Phi_{jn}(x + l),$$

$$\psi^e_{jn}(x) = \sum_{l \in \mathbb{Z}^d} \Psi^e_{jn}(x + l), \quad e \in \mathcal{E}.$$

It is not hard to see that $\{\psi^E_{j0}\}_{j=0}^{\infty}$ is an orthogonalized scaling sequence for a PMRA $\{V_j\}_{j=0}^{\infty}$ in $L_2(\mathbb{T}^d)$ whose matrix dilation is the diagonal matrix M with the diagonal elements equal to 2. For any $j \in \mathbb{Z}_+$ and any $e \subset E$, the functions $\psi^e_{jn}(x)$, $n \in \mathbb{Z}^d \cap [0, 2^j)^d$, form orthonormal bases in the corresponding wavelet spaces W^e_j.

Thus functions ψ^E_{00}, ψ^e_{jn}, $j \in \mathbb{Z}_+$, $n \in \mathbb{Z}^d \cap [0, 2^j)^d$, $e \in \mathcal{E}$, form an orthonormal basis in $L_2(\mathbb{T}^d)$. Moreover, since it follows from (10.19) that these functions are bounded, the Fourier series with respect to such a basis has meaning not only for functions from $L_2(\mathbb{T}^d)$, but for all $f \in L(\mathbb{T}^d)$ as well, and we can discuss its convergence.

Let us define partial sums of the Fourier series of the function $f \in L(\mathbb{T}^d)$ with respect to a wavelet system $\{\psi^e_{jn}\}$. Let $j \in \mathbb{Z}_+$. For any $e \in \mathcal{E}$, we fix an arbitrary subset $\Omega^e = \Omega^e_j$ of the set $[0, 2^j)^d$ and put $\Omega = \Omega_j = \{\Omega^e\}_{e \in \mathcal{E}}$,

$$s_{j\Omega}(f) = \langle f, \psi^E_{00} \rangle \psi^E_{00} + \sum_{e \in \mathcal{E}} \sum_{i=0}^{j-1} \sum_{n \in \mathbb{Z}^d \cap [0, 2^i)^d} \langle f, \psi^e_{in} \rangle \psi^e_{in} + \sum_{e \in \mathcal{E}} \sum_{n \in \mathbb{Z}^d \cap \Omega^e} \langle f, \psi^e_{jn} \rangle \psi^e_{jn}.$$

We say that the series

$$\langle f, \psi^E_{00} \rangle \psi^E_{00} + \sum_{j=0}^{\infty} \sum_{n \in \mathbb{Z}^d \cap [0, 2^j)^d} \sum_{e \in \mathcal{E}} \langle f, \psi^e_{jn} \rangle \psi^e_{in} \tag{10.20}$$

converges if its partial sums $s_{j\Omega_j}(f)$ converge as $j \to \infty$ uniformly over all sequences $\{\Omega_j\}_{j=0}^{\infty}$.

LEMMA 10.2.1. *The identity*

$$s_{j\Omega}(f, x) = 2^{dj} \int_{\mathbb{R}^d} f(t) \sum_{e \subset E} \sum_{r \in \Omega^e} \sum_{k \in \mathbb{Z}^d} \overline{g^e(2^j(t+k) + r)} g^e(2^j(x+k) + r)\, dt$$

takes place, where $\Omega^E = [0, 2^j)^d$.

Proof. First let us show that

$$(10.21)\quad \langle f,\psi^E_{00}\rangle\psi^E_{00}(x)+\sum_{e\in\mathcal{E}}\sum_{i=0}^{j-1}\sum_{r\in[0,2^i)^d}\langle f,\psi^e_{ir}\rangle\psi^e_{ir}(x)=\sum_{r\in[0,2^j)^d}\langle f,\psi^E_{jr}\rangle\psi^E_{jr}(x).$$

If $f\in L_2(\mathbb{T}^d)$, equality (10.21) is obvious since its left and right sides are expansions of the orthogonal projection of the function f on V_j with respect to two different orthonormal bases. Taking into account that by (10.19) the functions ψ^e_{ir} are bounded and approximating $f\in L(\mathbb{T}^d)$ by functions from $L_2(\mathbb{T}^d)$ in the metric of the space L, we deduce that (10.21) holds for any $f\in L(\mathbb{T}^d)$. Thus partial sums can be written in the following form:

$$\begin{aligned}
s_{j\Omega}(f,x)&=\sum_{e\subset E}\sum_{r\in\Omega^e}\langle f,\psi^e_{jr}\rangle\psi^e_{jr}(x)\\
&=2^{dj}\sum_{e\subset E}\sum_{r\in\Omega^e}\int_{\mathbb{T}^d}f(t)\sum_{k\in\mathbb{Z}^d}\overline{g^e(2^j(t+k)+r)}\,dt\sum_{l\in\mathbb{Z}^d}g^e(2^j(x+l)+r)\\
&=2^{dj}\sum_{e\subset E}\sum_{r\in\Omega^e}\sum_{l\in\mathbb{Z}^d}\int_{\mathbb{R}^d}f(t)\overline{g^e(2^jt+r)}g^e(2^j(x+l)+r)\,dt.
\end{aligned}$$

To prove the lemma, it remains to make the change of variables in the integral and interchange summation and integration. ◊

THEOREM 10.2.2. *Let φ and ψ satisfy* (10.19), *where η is an even bounded function decreasing on $[0,\infty)$ and such that*

$$(10.22)\qquad \int_0^\infty \rho^{d-1}\eta(\rho)\,d\rho<\infty.$$

Then for any function $f\in L(\mathbb{T}^d)$, series (10.20) *converges to f at every Lebesgue point. In particular, it converges almost everywhere.*

Proof. Let x be a Lebesgue point of a function $f\in L(\mathbb{T}^d)$. Since the space V_0 consists of constants, $s_{j\Omega}(f)=f$ for $f\equiv\text{const}$ and without loss of generality we can assume that $f(x)=0$. Set

$$\mathcal{N}(u,v)=\sum_{l\in\mathbb{Z}^d}\prod_{s=1}^d\eta(u_s+l_s)\eta(v_s+l_s).$$

By Lemma 10.2.1 and (10.19),

$$\begin{aligned}
|s_{j\Omega}(f,x)|&\le 2^{d(j+1)}\int_{\mathbb{R}^d}|f(x+t)|\mathcal{N}(2^j(x+t),2^jx)\,dt\\
(10.23)\qquad&=2^d\sum_{k\in\mathbb{Z}^d}\int_{\mathbb{T}^d+k}|f(x+2^{-j}t)|\mathcal{N}(2^jx+t,2^jx)\,dt.
\end{aligned}$$

By the definition of a Lebesgue point, for any $\varepsilon>0$ there exists $\delta>0$ such that

$$\frac{1}{h^d}\int_{[-h/2,h/2]^d}|f(x+t)|\,dt<\varepsilon$$

for all $h \in (0, \delta)$. Take j_0 such that $2^{-j_0/2} < \delta$ and

$$\int_{2^{-2+j/2}}^{\infty} \tau^{d-1}\eta(\tau/2^{3+d/2})\,d\tau \le \varepsilon.$$

By Lemma 10.1.1 and (10.23), for all $j \ge j_0$ we have

$$\begin{aligned}|s_{j\Omega}(f,x)| &\le A2^d \Bigg(\eta(0) \int_{[-1/2,1/2]^d} |f(x+2^{-j}t)|\,dt \\ &\quad + \sum_{i=0}^{\infty} \eta(2^{i-4-d/2}) \int_{[-2^j,2^j]^d \setminus [-2^{j-1},2^{j-1}]^d} |f(x+2^{-j}t)|\,dt \Bigg) \\ &\le A2^d \Bigg(\varepsilon\eta(0) + \varepsilon \sum_{0 \le i \le j/2} 2^{id}\eta(2^{i-4-d/2}) + \|f\|_1 \sum_{i > j/2} 2^{id}\eta(2^{i-4-d/2}) \Bigg).\end{aligned}$$

Using the inequality

$$2^{id}\eta(2^{i-4-d/2}) \le 2^{2d} \int_{2^{i-2}}^{2^{i-1}} \tau^{d-1}\eta(2^{-3-d/2}\tau)\,d\tau,$$

we get

$$\begin{aligned}|s_{j\Omega}(f,x)| &\le A2^{3d} \Bigg[\Bigg(\eta(0) + 2^{d(3+d/2)} \int_0^{\infty} \tau^{d-1}\eta(\tau)\,d\tau \Bigg) \varepsilon \\ &\qquad + \|f\|_1 \int_{2^{-2+j/2}}^{\infty} \tau^{d-1}\eta(2^{-3-d/2}\tau)\,d\tau \Bigg] \\ &\le A2^{3d}\varepsilon \Bigg(\eta(0) + 2^{d(3+d/2)} \int_0^{\infty} \tau^{d-1}\eta(\tau)\,d\tau + \|f\|_1 \Bigg).\end{aligned}$$

It follows that for each sequence $\{\Omega_j\}_{j=1}^{\infty}$, the relation $\lim_{j\to\infty} s_{j\Omega_j}(f,x) = 0$ holds uniformly over all $\{\Omega_j\}_{j=1}^{\infty}$. Convergence almost everywhere follows from the Lebesgue theorem (see Appendix A.9). ◊

REMARK 10.2.3. Conditions (10.19), (10.22) imposed on the functions φ, ψ in Theorem 10.2.2 cannot be replaced by $|\varphi(t)|, |\psi(t)| \le Ct^{-d}$. It can be shown that for Meyer's scaling function defined by the equality

$$\widehat{\varphi}(u) = \begin{cases} 1, & \text{if } |u| < 1/3, \\ \left|\sin \frac{3\pi u}{2}\right|, & \text{if } 1/3 < |u| < 2/3, \\ 0, & \text{if } |u| > 2/3, \end{cases}$$

there exists a function $f \in L(\mathbb{T}^2)$ such that series (10.20) (for $d = 2$) diverges at some Lebesgue point, though φ and the corresponding wavelet functions decrease at infinity as u^{-2}.

Therefore the requirements for φ, ψ cannot be weakened essentially without loss of the convergence of series (10.20) for each function $f \in L(\mathbb{T}^d)$ at each of its Lebesgue points. However, the requirements for convergence almost everywhere can be weakened. To do this, we shall investigate convergence on generalized Lebesgue sets (see Appendix A.9).

THEOREM 10.2.4. *Let φ and ψ satisfy* (10.19), *where η is a bounded function decreasing on $[0,\infty]$ and such that*

$$\int_1^\infty (\log \rho)^{d-1+\alpha}\eta(\rho)\,d\rho < \infty, \quad \alpha > 0. \tag{10.24}$$

Then for any function $f \in L(\mathbb{T}^d)$, series (10.20) *converges to f almost everywhere.*

Proof. Let $f \in L(\mathbb{T}^d)$,

$$I_j(f,x,r,h) := h^{-d}\int_{-h}^{h} dt_d \prod_{\ell=1}^{d-1} 2^{j_\ell} \int_{-2^{-j_\ell}h}^{2^{-j_\ell}h} dt_\ell \left|f(x) - f(x+Q_r t)\right|,$$

where Q_r is a nondegenerate linear transformation of the space $\mathbb{R}^d$ that exchanges the roles of the coordinates with numbers r and d. As in Theorem 10.2.2, we can assume that $f(x) = 0$. By Theorem A.9.2 (see Appendix A.9), the relations

$$\lim_{h\to 0} \sup_{l\in\mathbb{Z}_+^{d-1}} \prod_{k=1}^{d-1} (l_k+1)^{-1-\alpha/(d-1)} I_j(f,x,r,h) = 0, \tag{10.25}$$

$$\sup_{h>0} \sup_{l\in\mathbb{Z}_+^{d-1}} \prod_{k=1}^{d-1} (l_k+1)^{-1-\alpha/(d-1)} I_j(f,x,r,h) = C_0 < \infty \tag{10.26}$$

hold for almost all $x \in \mathbb{R}^d$ and all $r = 1,\dots,d$. For every x satisfying (10.25), (10.26) and any $\varepsilon > 0$, there exists $\delta > 0$ such that

$$\sup_{r} \sup_{l\in\mathbb{Z}_+^{d-1}} \prod_{k=1}^{d-1} (l_k+1)^{-1-\alpha/(d-1)} I_l(f,x,r,h) < \varepsilon \tag{10.27}$$

for all $h \in (0,\delta)$. Take positive integer j_0 such that $2^{1-j_0/2} < \delta$ and

$$\int_{2^{-3+j_0/2}}^{\infty} \eta(\tau/8)\log^{d-1+\alpha} 16\tau\,d\tau \le \varepsilon. \tag{10.28}$$

For $i \in \mathbb{Z}_+^d$, set $K(i) = \{k \in \mathbb{Z}^d; [2^{i_s-1}] \le |k_s| < 2^{i_s}, s = 1\dots d\}$, where $[a]$ is the integral part of a. By (10.23) and Lemma 10.1.1 for all $j \ge j_0$ we have

$$\begin{aligned}|S_{j\Omega}(f,x)| &\le B2^d \sum_{i\in\mathbb{Z}_+^d} \sum_{s\in K(i)} \int_{\mathbb{T}^2+s} |f(x+2^{-j}t)| \prod_{k=1}^{d} \eta(|t_k|/8)\,dt \\ &\le B2^d \sum_{i\in\mathbb{Z}_+^d} \prod_{k=1}^{d} \eta([2^{i_k-5}]) \int_{-2^{i_1+1}}^{2^{i_1+1}} dt_1 \dots \int_{-2^{i_d+1}}^{2^{i_d+1}} |f(x+2^{-j}t)|\,dt_d.\end{aligned}$$

In view of symmetry it suffices to estimate the sum

$$\sum_{i_d=0}^{\infty}\sum_{i_1=0}^{i_d}\cdots\sum_{i_{d-1}=0}^{i_d}\prod_{k=1}^{d}\eta([2^{i_k-5}])\int_{-2^{i_1+1}}^{2^{i_1+1}}dt_1\ldots\int_{-2^{i_d+1}}^{2^{i_d+1}}|f(x+2^{-j}t)|\,dt_d$$

$$=\sum_{i_d=0}^{\infty}\sum_{i_1=0}^{i_d}\cdots\sum_{i_{d-1}=0}^{i_d}\prod_{k=1}^{d}\eta([2^{i_k-5}])2^{i_k+1}I_{i^*}(f,x,d,2^{i_d-j+1}),$$

where $i^*=(i_d-i_1,\ldots,i_d-i_{d-1})$. Divide this sum into two parts with respect to i_d: $i_d\le j/2$ and $i_d>j/2$.

Since $2^{i_d-j+1}\le\delta$ for $i_d\le j/2$, by (10.27) the right-hand side does not exceed

$$B2^d\varepsilon\sum_{0\le i_d\le j/2}2^{i_d+1}\eta([2^{i_d-5}])\prod_{k=1}^{d-1}\Bigg[(d-1+i_d)^{1+\alpha}\eta(0)$$

$$+\sum_{i_k=2}^{i_d}(1+i_d-i_k)^{1+\alpha}\int_{2^{i_k-3}}^{2^{i_k-2}}\eta(\tau/8)\,d\tau\Bigg]. \tag{10.29}$$

Since for $i_d\ge 2$,

$$\sum_{i_k=2}^{i_d}(1+i_d-i_k)^{1+\alpha}\int_{2^{i_k-3}}^{2^{i_k-2}}\eta(\tau/8)\,d\tau\le\int_0^{\infty}\eta(\tau/8)\,d\tau(i_d+1)^{\alpha+1},$$

it follows that (10.29) does not exceed

$$B2^d\varepsilon\left(\eta(0)+\int_0^{\infty}\eta(\tau/8)\log^{1+\alpha}16\tau\,d\tau\right)\left(\eta(0)+\int_0^{\infty}\eta(\tau/8)\,d\tau\right)^{d-1}.$$

Similarly, using (10.26) instead of (10.27), we estimate the second part by

$$BC_02^d\left(\int_{2^{-3+j/2}}^{\infty}\eta(\tau/8)\log^{d-1+\alpha}16\tau\,d\tau\right)\left(\eta(0)+\int_0^{\infty}\eta(\tau/8)\,d\tau\right)^{d-1}$$

$$\le C\varepsilon\left(\eta(0)+\int_0^{\infty}\eta(\tau/8)\,d\tau\right)^{d-1}.$$

Taking into account (10.28), it follows that for each sequence $\{\Omega_j\}_{j=1}^{\infty}$, the relation $\lim_{j\to\infty}S_{j\Omega_j}(f,x)=0$ holds and the convergence is uniform over all such sequences.

◊

PROBLEM 10.2.5. Suppose φ satisfies the conditions of Theorem 10.2.4, $\{V_j\}_{j\in\mathbb{Z}}$ is a separable MRA in $L_2(\mathbb{R}^d)$ generated by a scaling function $\Phi:=\varphi\otimes\cdots\otimes\varphi$, and P_j is a projection operator onto V_j. Prove that $P_j(f)\underset{j\to+\infty}{\longrightarrow}f$ almost everywhere for any $f\in L(\mathbb{R}^d)$.

Now we study the convergence of series (10.20) at the strong Lebesgue points of a function f (see Appendix A.9). Unlike the usual Lebesgue points and the points considered in Theorem 10.2.4, which satisfy conditions (10.25), (10.26), the set of strong Lebesgue points does not have full measure for all functions $f \in L(\mathbb{T}^d)$. However, for $f \in L\log L(\mathbb{T}^d)$, and more so for $f \in L_p(\mathbb{T}^d)$, $p > 1$, almost all points are strong Lebesgue points of the function f. Thus having proved the convergence of series (10.20) at every strong Lebesgue point, we shall establish its convergence almost everywhere for a sufficiently wide class of functions.

THEOREM 10.2.6. *If φ and ψ satisfy* (10.19), *where η is an even, bounded, and summable function decreasing on $[0,\infty)$, then for any function $f \in L(\mathbb{T}^d)$, series* (10.20) *converges to f at each strong Lebesgue point.*

`Proof.` Let x be a strong Lebesgue point of a function $f \in L(\mathbb{T}^d)$. As in Theorem 10.2.2, we can assume that $f(x) = 0$. By the definition of strong Lebesgue points, for any given $\varepsilon > 0$ there exists $\delta > 0$ such that

$$\frac{1}{h_1 \dots h_d} \int_{-h_1}^{h_1} \dots \int_{-h_d}^{h_d} |f(x+t)|\, dt < \varepsilon, \text{ for all } |h_i| < \delta,\ i = 1 \dots d,$$

$$\frac{1}{h_1 \dots h_d} \int_{-h_1}^{h_1} \dots \int_{-h_d}^{h_d} |f(x+t)|\, dt \le C_0 \text{ for all } h_i > 0,\ i = 1 \dots d,$$

Choose positive integer j_0 so large that $2^{1-j_0/2} < \delta$ and

$$\int_{2^{-3+j_0/2}}^{\infty} \eta(\tau/8)\, d\tau < \varepsilon.$$

It follows from (10.23) and Lemma 10.1.1 that for all $j \ge j_0$, we have

$$|s_{j\Omega}(f,x)| \le B2^d \sum_{i \in \mathbb{Z}_+^d} \prod_{s=1}^{d} \eta([2^{i_s-5}]) \sum_{k \in K(i)} \int_{\mathbb{T}^d + k} |f(x + 2^{-j}t)|\, dt$$

$$\le B2^d \left(\varepsilon \sum_{i \in \mathbb{Z}_+^d \cap j\mathbb{T}^d} \prod_{s=1}^{d} \eta([2^{i_s-5}]) 2^{i_s+1} + C_0 \sum_{i \in \mathbb{Z}_+^d \setminus j\mathbb{T}^d} \prod_{s=1}^{d} \eta([2^{i_s-5}]) 2^{i_s+1} \right).$$

Here we used the notation of Theorem 10.2.4. For $i_s \ge 3$, using the inequality

$$\eta([2^{i_s-5}]) 2^{i_s+1} \le 16 \int_{2^{i_s-3}}^{2^{i_s-2}} \eta(\tau/8)\, d\tau,$$

we get

$$|s_{j\Omega}(f,x)| \le B2^{5d} \left((\eta(0)+L)^d \varepsilon + C_0 (\eta(0)+L)^{d-1} \int_{2^{-3+j/2}}^{\infty} \eta(\tau/8)\, d\tau \right)$$

$$\le B2^{5d} (\eta(0) + L + C_0)(\eta(0)+L)^{d-1} \varepsilon,$$

where $L = \int_0^\infty \eta(\tau)\, d\tau$. It follows that for each sequence $\{\Omega_j\}_{j=1}^\infty$, the relation $\lim_{j\to\infty} s_{j\Omega_j}(f,x) = 0$ holds, and the convergence is uniform for all $\{\Omega_j\}_{j=1}^\infty$. ◊

Let us consider the case $d = 1$, $M = 2$. Here the assumptions of Theorems 10.2.2 and 10.2.6 are the same; i.e., for the convergence almost everywhere, it suffices to find a majorant η that decreases sufficiently rapidly to provide its summability. Let us extend this statement to arbitrary PMRAs (not necessarily generated by a nonperiodic function).

In the one-dimensional case it is more natural to number the functions $S_k^j\psi_j$, $S_k^j\widetilde{\psi}_j$ in ascending order of indices. Then for some (p,q)-pair, the Fourier series of a function f takes the form

$$\langle f, \widetilde{\varphi}_0\rangle\varphi_0 + \sum_{j=0}^{\infty}\sum_{k=0}^{2^j-1}\langle f, S_k^j\widetilde{\psi}_j\rangle S_k^j\psi_j, \tag{10.30}$$

and its partial sum of order $N = 2^j + n$, $0 \le n < 2^j$, is equal to

$$s_N(f) = \langle f, \widetilde{\varphi}_0\rangle\varphi_0 + \sum_{i=0}^{j-1}\sum_{k=0}^{2^i-1}\langle f, S_k^i\widetilde{\psi}_i\rangle S_k^i\psi_i + \sum_{k=0}^{n}\langle f, S_k^j\widetilde{\psi}_j\rangle S_k^j\psi_j.$$

THEOREM 10.2.7. *Let $d = 1$, $M = 2$, PMRAs $\{V_j\}_{j=0}^{\infty}$, $\{\widetilde{V}_j\}_{j=0}^{\infty}$ form a (p,q)-pair with scaling sequences $\{\varphi_j\}_{j=0}^{\infty}$, $\{\widetilde{\varphi}_j\}_{j=0}^{\infty}$ such that $\{S_n^j\varphi_j\}_{n\in D(M^j)}$ and $\{S_k^j\widetilde{\varphi}_j\}_{n\in D(M^j)}$ are biorthonormal systems, and let $\{\psi_j\}_{j=0}^{\infty}$, $\{\widetilde{\psi}_j\}_{j=0}^{\infty}$ be the corresponding sequences of wavelet functions. If there exists an even, bounded summable function η decreasing on $[0,\infty)$ and such that*

$$|2^{-j/p}\varphi_j(x)|,\ |2^{-j/p}\psi_j(x)|,\ |2^{-j/q}\widetilde{\varphi}_j(x)|,\ |2^{-j/q}\widetilde{\psi}_j(x)| \le \eta(2^jx) \tag{10.31}$$

for all $x \in [-1/2, 1/2]$, then for each function $f \in L(\mathbb{T})$, series (10.30) converges to f at every Lebesgue point. In particular, it converges almost everywhere.

Proof. First of all note that in view of (10.31) the wavelet functions are bounded. Therefore (10.30) makes sense for all $f \in L(\mathbb{T})$. Let x be a Lebesgue point of a function f. Since the space V_0 consists of constants, $s_m(f,x) = f$ for all $f \equiv \text{const}$ and all $m \in \mathbb{Z}_+$. It follows that

$$\begin{aligned} |f(x) - s_N(f,x)| &= \Bigg|\int_0^1 (f(x) - f(x+t)) \\ &\quad\times\left(\sum_{r=0}^{2^j-1}\overline{S_r^j\widetilde{\varphi}_j(t+x)}S_r^j\varphi_j(x) + \sum_{r=0}^{n}\overline{S_r^j\widetilde{\psi}_j(t+x)}S_r^j\psi_j(x)\right)dt\Bigg| \\ &\le \int_0^1 |f(x) - f(x+t)|\sum_{r=0}^{2^j-1}\left|S_r^j\widetilde{\varphi}_j(t+x)S_r^j\varphi_j(x)\right|dt \\ &\quad+ \int_0^1 |f(x) - f(x+t)|\sum_{r=0}^{2^j-1}\left|S_r^j\widetilde{\psi}_j(t+x)S_r^j\psi_j(x)\right|dt = \Sigma_1 + \Sigma_2. \end{aligned} \tag{10.32}$$

Show that

$$\Sigma_1 \underset{j\to\infty}{\longrightarrow} 0. \tag{10.33}$$

Set

$$g_j(t) = \begin{cases} \varphi_j(2^{-j}t), & t \in [-2^{j-1}, 2^{j-1}], \\ 0, & t \notin [-2^{j-1}, 2^{j-1}], \end{cases}$$

and define functions $\widetilde{g}_j$ in a similar way. It is clear that

$$\varphi_j(x) = \sum_{k\in\mathbb{Z}} g_j(2^j x + 2^j k), \quad \widetilde{\varphi}_j(x) = \sum_{k\in\mathbb{Z}} \widetilde{g}_j(2^j x + 2^j k) \tag{10.34}$$

for all $x \in \mathbb{R}$. Hence,

$$\begin{aligned} \Sigma_1 &= \int_0^1 |f(x) - f(x+t)| \\ &\quad \times \sum_{r=0}^{2^j-1} \sum_{k\in\mathbb{Z}} \left|g_j\left(2^j(x+k)+r\right)\right| \sum_{l\in\mathbb{Z}} \left|\widetilde{g}_j\left(2^j(x+t+l+k)+r\right)\right| dt \\ &= \int_{-\infty}^{\infty} |f(x) - f(x+t)| \sum_{r=0}^{2^j-1} \sum_{k\in\mathbb{Z}} \left|g_j\left(2^j(x+k)+r\right)\right| \left|\widetilde{g}_j\left(2^j(x+t+k)+r\right)\right| dt \\ &= \int_{-\infty}^{\infty} |f(x) - f(x+t)| \sum_{k\in\mathbb{Z}} \left|\widetilde{g}_j\left(2^j(x+t)+k\right)\right| \left|g_j\left(2^j x+k\right)\right| dt. \end{aligned}$$

From (10.31) it follows that

$$\left|2^{-j/p} g_j(u)\right|, \left|2^{-j/q}\widetilde{g}_j(u)\right| \le \eta(u)$$

for all $j \in \mathbb{Z}_+$ and $u \in \mathbb{R}$. Hence,

$$\Sigma_1 \le 2^j \int_{-\infty}^{\infty} |f(x) - f(x+t)| \sum_{k\in\mathbb{Z}} \eta\left(2^j(x+t)+k\right) \eta\left(2^j x + k\right) dt.$$

Using Lemma 10.1.1, we get

$$\Sigma_1 \le B 2^j \int_{-\infty}^{\infty} |f(x) - f(x+t)| \eta\left(2^{j-3}t\right) dt. \tag{10.35}$$

Given $\varepsilon > 0$, choose δ such that

$$\frac{1}{h} \int_{-h/2}^{h/2} |f(x+t) - f(x)|\, dt < \varepsilon \tag{10.36}$$

for all $h \in (0, \delta)$. Let $2^{-j} < \delta$, $2^{k_0} \le \delta < 2^{k_0+1}$, $k_0 \in \mathbb{Z}$. Divide the integral on the right-hand side of (10.35) in the following way:

$$\int_{-\infty}^{\infty} = \int_{|t|<2^{-j}} + \int_{2^{-j}<|t|<2^{k_0+1}} + \int_{|t|>2^{k_0+1}} =: I_0 + I_1 + I_2.$$

Using the monotonicity of the function η and (10.36), we get

$$I_0 \le B2^j \int\limits_{|t|<2^{-j}} |f(x+t)-f(x)|\eta(2^{j-3}(t))\,dt \le 2B\eta(0)\varepsilon,$$

$$\begin{aligned} I_1 &\le B2^j \sum_{k=-j}^{k_0} \int\limits_{2^k<|t|<2^{k+1}} |f(x+t)-f(x)|\eta(2^{j-3}(t))\,dt \\ &= B2^j \sum_{k=-j}^{k_0} \int\limits_{2^k<|t|<2^{k+1}} |f(x+t)-f(x)|\eta(2^{j-3}t)\,dt \\ &\le B\sum_{k=-j}^{k_0} 2^{-k-2} \int\limits_{-2^{k+1}}^{2^{k+1}} |f(x+t)-f(x)|\,dt\, 2^{j+k+2}\eta(2^{j+k-3}) \\ &\le 2B\varepsilon \sum_{k=-j}^{k_0} 2^{j+k}\eta(2^{j+k-3}) \le 64B\varepsilon \int\limits_0^\infty \eta(t)\,dt. \end{aligned}$$

Thus I_0 and I_1 are small for sufficiently large j. To estimate I_2, we again use the monotonicity of the function η:

$$\begin{aligned} I_2 &\le B2^j \sum_{k=k_0}^{\infty} \int\limits_{2^k<|t|<2^{k+1}} |f(x+t)-f(x)|\eta(2^{j-3}(t))\,dt \\ &\le B\sum_{k=k_0}^{\infty} 2^{j+k}\eta(2^{j+k_0-3})\,(\|f\|_1+f(x)) \le 32B\,(\|f\|_1+f(x)) \int\limits_{2^{j+k_0-4}}^{\infty} \eta(t)\,dt. \end{aligned}$$

Since η is summable, the integral on the right-hand side tends to zero as j increases, which completes the proof of (10.33). In a similar way, it can be proved that $\Sigma_2 \underset{j\to\infty}{\longrightarrow} 0$. Combining these relations with (10.32), we get

$$\lim_{N\to\infty} s_N(f,x) = f(x). \qquad \diamond$$

10.3. Direct and inverse theorems

Let us now turn to the approximation properties of Fourier series with respect to systems of periodic wavelets in accordance with structural properties of the approximated function. We confine our consideration to the one-dimensional dyadic PMRAs generated by a scaling function φ decreasing sufficiently rapidly and such that its integer shifts form an orthonormal system. Orthogonalized sequences φ_j in such PMRAs have the form

$$\varphi_j(x) = 2^{j/2}\sum_{l\in\mathbb{Z}} \varphi(2^j x + 2^j l).$$

If a corresponding wavelet function ψ decreases sufficiently rapidly, then the functions

$$\psi_j(x) = 2^{j/2}\sum_{l\in\mathbb{Z}} \psi(2^j x + 2^j l)$$

form a sequence of wavelet functions and $\{S_k^j\psi_j\}_{k=0}^{2^j-1}$ is an orthonormal system. It is more convenient to number elements of this system by a single index. Introduce new notation:

$$w_0 = \varphi_0, \quad w_{2^j+n} = S_n^j\psi_j, \quad n = 0, \dots, 2^j - 1.$$

If the functions w_n are bounded, then the Fourier expansions with respect to such a system have meaning for all $f \in L(\mathbb{T})$.

Denote partial sums of the Fourier series of a function f with respect to the wavelet system w_n, $n \in \mathbb{Z}_+$, by

$$s_N(f) = \sum_{n=0}^{N} \langle f, w_n \rangle w_n$$

and introduce *best wavelet approximations* of order N:

$$E_N(f) = E_N(f)_p = \inf \|f - T\|_p,$$

where the lower bound is taken over all functions of the form $T = \sum_{k=0}^{N} \alpha_k w_k$, in particular, for $N = 2^j$, over all elements of the space V_j. Since the spaces V_j, W_j are finite-dimensional, the best approximation is achieved; i.e., for any N there exists a *wavelet polynomial of best approximation* $T_N = \sum_{k=0}^{N} \alpha_k w_k$ of order N for the function f. Note the following obvious properties of best approximations:

B1. $E_n(f) \le \|f\|$.

B2. $E_{n+1}(f) \le E_n(f)$.

B3. If $g = cf$, where $c \in \mathbb{R}$, then $E_n(g) = |c| E_n(f)$.

B4. $E_n(f + g) \le E_n(f) + E_n(g)$.

The behavior of the best approximations characterizes the constructive properties of functions. If a sequence of best approximations tends to zero, this means that the function can be approximated by wavelet polynomials, while the rate of tendency to zero characterizes the error of approximation. Our aim is to investigate the connection between the order of smoothness of a function (a structural property of a function) and its constructive properties. As a structural characteristic of smoothness of a function f, we shall use moduli of continuity $\omega_r(f, h)$, $r \in \mathbb{N}$ (see Appendix A.14).

THEOREM 10.3.1. *Let* $1 \le p \le \infty$, $m \in \mathbb{Z}_+$, $n, r \in \mathbb{N}$, $n \ge m + 1$, $r \le m + 1$, $r < n - 1$, $\varphi \in C(\mathbb{R})$, $\psi \in C^m(\mathbb{R})$, $\varepsilon > 0$,

$$|\varphi(x)| = O\left(\frac{1}{1 + |x|^{1+\varepsilon}}\right),$$

$$|\psi(x)| = O\left(\frac{1}{1 + |x|^{n+\varepsilon}}\right).$$

Then for all $f \in L_p(\mathbb{T})$ *(*$f \in C(\mathbb{T})$ *if* $p = \infty$*) and all positive integers* N *the inequality*

$$E_N(f)_p \le \|f - s_N(f)\|_p \le C\omega_r\left(f, \frac{1}{N}\right)_p \tag{10.37}$$

holds, where C *depends only on* r, p, ε, *and the constants in the symbols* O.

This theorem is an analog of the classical Jackson theorem for trigonometric polynomials (see Appendix A.15). However, unlike the trigonometric case, the modulus of continuity majorizes not only the best approximation but the norm of deviation of a function from a partial sum of its Fourier series as well.

COROLLARY 10.3.2. *Let $1 \le p \le \infty$, and let φ be Meyer scaling function with infinitely differentiable Fourier transform. Then for any $f \in L_p(\mathbb{T})$ ($f \in C(\mathbb{T})$ for $p = \infty$) and for all positive integers N and r, the inequality*

$$\|f - s_N(f)\|_p \le C\omega_r\left(f, \frac{1}{N}\right)_p \tag{10.38}$$

holds, where C depends only on r and p.

Proof. Since $m_0(\xi) = \widehat{\varphi}(2\xi)$ for $\xi \in [-1/2, 1/2]$ and $\widehat{\varphi}(\xi) = 0$ for $|\xi| > (1-\varepsilon)/2$, the infinite differentiability of $\widehat{\varphi}$ implies the infinite differentiability of the mask m_0 and hence of the function $\widehat{\psi}$. Therefore,

$$|\varphi(x)|, |\psi(x)| = O\left(\frac{1}{1+|x|^n}\right)$$

for any $n \in \mathbb{N}$. Since the functions $\widehat{\varphi}$, $\widehat{\psi}$ are compactly supported, $\psi \in C^m(\mathbb{R})$ for any $m \in \mathbb{N}$. Thus the conditions of Theorem 10.3.1 are satisfied for any $r \in \mathbb{N}$. $\Diamond$

Proof of Theorem 10.3.1. The first inequality in (10.37) is trivial. To prove the second inequality, we assume that f is a trigonometric polynomial and show that

$$\|f - s_N(f)\| \le C_1 \frac{\|f^{(r)}\|}{N^r}. \tag{10.39}$$

If $2^j \le N < 2^{j+1}$, then for each $k \ge j$,

$$f - s_N(f) = (f - s_{2^k-1}(f)) + (s_{2^j-1}(f) - s_N(f)) + \sum_{i=j}^{k-1}(s_{2^{i+1}-1}(f) - s_{2^i-1}(f)).$$

Theorem 10.1.4 and Remark 10.1.5 imply the relation

$$\lim_{k\to\infty} \|f - s_{2^k-1}(f)\| = 0;$$

hence to prove (10.39), it suffices to check that

$$\|s_{2^j+L}(f) - s_{2^j-1}(f)\| \le C_2 2^{-jr}\|f^{(r)}\|$$

for all $j \in \mathbb{Z}_+$, $L = 0, \dots, 2^j - 1$. Denote by $\mathbb{Z}(j, L)$ the set of all numbers of the form $2^j l + k$, $l \in \mathbb{Z}$, $k = 0, \dots, L$. Using Lemma 10.2.1, the Taylor formula

$$f(t) = \sum_{k=0}^{r-1} \frac{f^{(k)}(x)}{k!}(t-x)^k + \frac{1}{(r-1)!}\int_x^t f^{(r)}(\tau)(t-\tau)^{r-1}\,d\tau,$$

and Theorem 1.7.7, we get

$$
\begin{aligned}
&s_{2^i+L}(f,x) - s_{2^i-1}(f,x) \\
&= 2^j \int_0^1 f(t) \sum_{k=0}^{L} \sum_{l' \in \mathbb{Z}} \psi(2^j x + 2^j l' + k) \sum_{l \in \mathbb{Z}} \overline{\psi(2^j t + 2l + k)}\, dt \\
&= 2^j \int_{-\infty}^{\infty} f(t) \sum_{\nu \in \mathbb{Z}(j,L)} \psi(2^j x + \nu) \overline{\psi(2^j t + \nu)}\, dt \\
&= \frac{2^j}{(r-1)!} \int_{-\infty}^{\infty} \int_x^t f^{(r)}(\tau)(t-\tau)^{r-1}\, d\tau \sum_{\nu \in \mathbb{Z}(j,L)} \psi(2^j x + \nu) \overline{\psi(2^j t + \nu)}\, dt.
\end{aligned}
$$

Here we interchanged summation and integration, using Lebesgue's dominated convergence theorem. Set $\eta(u) = 1/(1 + |u|^{n+\varepsilon})$. By Lemma 10.1.1,

$$
\left| \sum_{\nu \in \mathbb{Z}(j,L)} \psi(u + \nu)\psi(v + \nu)\, dt \right| \le B\eta\left(\frac{u-v}{8}\right)
$$

for all $u, v \in \mathbb{R}$. Consequently,

$$
\begin{aligned}
&|s_{2^i+L}(f,x) - s_{2^i-1}(f,x)| \\
&\le C_3 2^j \int_{-\infty}^{\infty} \left| \int_x^t f^{(r)}(\tau)(t-\tau)^{r-1}\, d\tau \right| \eta(2^{j-3}(t-x))\, dt \\
&= C_3 2^j \int_{-\infty}^{\infty} \left| \int_x^{x+t} \left| f^{(r)}(\tau) \right| |t + x - \tau|^{r-1}\, d\tau \right| \eta(2^{j-3}t)\, dt.
\end{aligned} \tag{10.40}
$$

Using Jensen's inequality, we get for $p < \infty$,

$$
\begin{aligned}
&\|s_{2^i+L}(f) - s_{2^i}(f)\| \\
&\le C_4 2^j \int_{-\infty}^{\infty} dt \eta(2^{j-3}t) \left(\int_0^1 \left| \int_x^{x+t} \left| f^{(r)}(\tau) \right| |x + t - \tau|^{r-1}\, d\tau \right|^p dx \right)^{1/p} \\
&\le C_4 2^j \int_{-\infty}^{\infty} dt \eta(2^{j-3}t) \left(\int_0^1 |t|^{r(p-1)} \left| \int_x^{x+t} \left| f^{(r)}(\tau) \right|^p |x + t - \tau|^{r-1}\, d\tau \right| dx \right)^{1/p}.
\end{aligned} \tag{10.41}
$$

If $|t| \ge 1$, then taking into account the periodicity of the function $f^{(r)}$, we get

$$
\left| \int_x^{x+t} \left| f^{(r)}(\tau) \right|^p |x + t - \tau|^{r-1}\, d\tau \right| \le |t|^{r-1} \left| \int_x^{x+t} \left| f^{(r)} \right|^p \right| \le 2|t|^r \|f^{(r)}\|^p.
$$

Therefore,

$$
\int_0^1 dx \left| \int_x^{x+t} \left| f^{(r)}(\tau) \right|^p |x + t - \tau|^{r-1}\, d\tau \right| \le C_5 |t|^r \|f^{(r)}\|^p. \tag{10.42}
$$

If $0 \le t < 1$, then

$$\int_0^1 dx \int_x^{x+t} \left|f^{(r)}(\tau)\right|^p |x+t-\tau|^{r-1}\, d\tau$$
$$\le \int_0^{t+1} \left|f^{(r)}(\tau)\right|^p d\tau \int_{\tau-t}^{\tau} (t+x-\tau)^{r-1}\, dx \le \frac{2}{r}\|f^{(r)}\|^p t^r.$$

Thus (10.42) holds in this case. Analogously, if $-1 < t < 0$, then

$$\left|\int_x^{x+t} \left|f^{(r)}(\tau)\right|^p |x+t-\tau|^{r-1}\, d\tau\right|$$
$$\le \int_t^1 \left|f^{(r)}(\tau)\right|^p d\tau \int_{\tau}^{\tau-t} (\tau-x-t)^{r-1}\, dx\, \frac{2}{r}\|f^{(r)}\|^p |t|^r.$$

It follows that (10.42) holds.

Combining (10.42) and (10.41), we get

$$\|s_{2^i+L}(f) - s_{2^i-1}(f)\| \le C_6 \|f^{(r)}\| 2^j \int_{-\infty}^{\infty} |t|^r \eta(2^j t)\, dt$$

for $p < \infty$. For $p = \infty$, the inequality

$$\|s_{2^i+L}(f) - s_{2^i-1}(f)\|_\infty \le C_7 \|f^{(r)}\|_\infty 2^j \int_{-\infty}^{\infty} |t|^r \eta(2^j t)\, dt$$

immediately follows from (10.40). This relation implies (10.39) after the change of variables in the integral if we take into account that the functions $|t|^r \eta(t)$ are summable on $\mathbb{R}$. Thus relation (10.39) holds for all trigonometric polynomials.

Let us now consider an arbitrary function $f \in L_p(\mathbb{T})$ ($f \in C(\mathbb{T})$ for $p = \infty$) and its Valee Poussin mean $v_N(f)$ (see Appendix A.15). Since $v_N(f)$ is a trigonometric polynomial, applying inequality (10.37), we get

$$\|f - s_N(f)\| \le \|f - v_N(f)\| + \|s_N(f - v_N(f))\| + \|v_N(f) - s_N(v_N(f))\|$$
$$\le (\|s_N\| + 1)\|f - v_N(f)\| + C_1 \frac{\|(v_N(f))^{(r)}\|}{N^r}.$$

Relation (A.24) for $h = 1/4N$ and the obvious property of finite differences $\|\Delta_h^r f\| \le 2^r \|f\|$ imply that

$$\frac{\|(v_N(f))^{(r)}\|}{N^r} \le \pi^r \|\Delta_h^r v_N(f)\| \le \pi^r \left(\|\Delta_h^r (f - v_N(f))\| + \|\Delta_h^r f\|\right)$$
$$\le (2\pi)^r \|f - v_N(f)\| + \pi^r \omega_r(f, h).$$

Combining these inequalities, using (A.25), and taking into account that $\|s_N\| \le C_8$ by Theorem 10.1.4, we get

$$\|f - s_N(f)\| \le C_9\, \omega_r\left(f, \frac{1}{N}\right) + \pi^r \omega_r\left(f, \frac{1}{4N}\right) \le C_{10}\, \omega_r\left(f, \frac{1}{N}\right).$$

In conclusion we note that if the condition of the theorem is satisfied for some r, m, n, it is also satisfied for r, $n' = r+2$, $m' = r-1$; hence the constants formally depending on n, m can be assumed to be depending only on r. ◊

THEOREM 10.3.3. *Let $1 \le p \le \infty$, $m \in \mathbb{N}$, $\varepsilon > 0$, $\varphi \in C^m(\mathbb{R})$,*

$$|\varphi(x)|, |\varphi^{(m)}(x)| = O\left(\frac{1}{1+|x|^{1+\varepsilon}}\right).$$

Then for any $k \in \mathbb{Z}_+$, $k \le m$, and all $f \in V_j$, $j \in \mathbb{Z}_+$, the inequality

$$\|f^{(k)}\|_p \le B2^{jk}\|f\|_p \tag{10.43}$$

holds, where B depends only on p, m, ε, and the constant in the symbol O.

Proof. First of all note that the series

$$\sum_{k\in\mathbb{Z}} |\varphi^{(m)}(2^j x + 2^j k)|$$

converges uniformly; hence the elements of the scaling sequence φ_j, and thus all elements of the spaces V_j, have continuous derivatives of order m. It follows from the Landau generalized inequality (see Appendix A.19) that all the derivatives of order less than m have the same order of decrease at infinity as the functions φ, $\varphi^{(m)}$. So it suffices to prove the theorem for $k = m$. Let $f \in V_j$; then $f = s_{2^j-1}(f)$. Therefore,

$$f(x) = 2^j \int_0^1 f(t) \sum_{k=0}^{2^j-1} \sum_{l'\in\mathbb{Z}} \overline{\varphi(2^j t + 2^j l' + k)} \sum_{l\in\mathbb{Z}} \varphi(2^j x + 2^j l + k)\, dt. \tag{10.44}$$

Formal differentiation of the right-hand side m times gives

$$2^{j(m+1)} \int_0^1 f(t) \sum_{k=0}^{2^j-1} \sum_{l'\in\mathbb{Z}} \varphi^{(m)}(2^j x + 2^j l' + k) \sum_{l\in\mathbb{Z}} \overline{\varphi(2^j t + 2^j l + k)}\, dt$$
$$= 2^{jm} \int_{-\infty}^{\infty} f(t) \sum_{\nu\in\mathbb{Z}} \varphi^{(m)}(2^j x + \nu)\overline{\varphi(2^j t + \nu)}\, dt.$$

Set $\eta(u) = 1/(1+|u|^{1+\varepsilon})$. By Lemma 10.1.1,

$$\sum_{\nu\in\mathbb{Z}} |\varphi^{(m)}(u+\nu)\varphi(v+\nu)|\, dt \le C\eta(u-v)$$

for all $u, v \in \mathbb{R}$. This implies the possibility of termwise differentiation and the estimate

$$\|f^{(m)}\| \le C2^{jm} \left\| \int_{-\infty}^{\infty} |f(t)|\eta(x-t)\, dt \right\|$$
$$= C2^{jm} \left\| \int_{-\infty}^{\infty} |f(x-t)|\eta(t)\, dt \right\| \le C2^{jm}\|f\| \int_{-\infty}^{\infty} \eta(t)\, dt = B2^{jm}\|f\|. \qquad \Diamond$$

COROLLARY 10.3.4. *Under the conditions of Theorem* 10.3.3 *for all* $k \in \mathbb{Z}_+$, $k \le m$, *and all* $f \in V_j$, $j \in \mathbb{Z}_+$, *the inequality*

$$\|f^{(k)}\|_p \le \tilde{B} 2^{jk} \omega_k \left(f, 2^{-j}\right)_p \tag{10.45}$$

holds, where $\tilde{B}$ *depends only on* p, m, ε, *and the constant in the symbol* O.

Proof. As in Theorem 10.3.3, it is sufficient to prove this assertion for $k = m$. It is not hard to see that

$$\|f^{(m)}\| \le 2^j \left\| \int_0^{2^{-j}} \Delta_t^m f^{(m)}\, dt \right\| + 2^j \left\| \int_0^{2^{-j}} \sum_{k=1}^{m} (-1)^k C_m^k f^{(m)}(\cdot + kt)\, dt \right\|,$$

where Δ_t^m is a finite difference of order m (see Appendix A.14). From Theorem 10.3.3 it follows that

$$\left\| \int_0^{2^{-j}} |\Delta_t^m f^{(m)}|\, dt \right\| \le \int_0^{2^{-j}} \|\Delta_t^m f^{(m)}\|\, dt$$

$$\le B 2^{jm} \int_0^{2^{-j}} \|\Delta_t^m f\| \le B 2^{j(m-1)} \omega_m \left(f, 2^{-j}\right).$$

Simple calculation gives

$$\int_0^{2^{-j}} \sum_{k=1}^{m} (-1)^k C_m^k f^{(m)}(x + kt)\, dt = \Delta_{2^{-j}}^m f^{(m-1)}(x).$$

Using Theorem 10.3.3 again, we get

$$\left\| \Delta_{2^{-j}}^m f^{(m-1)} \right\| \le B 2^{j(m-1)} \|\Delta_{2^{-j}}^m f\| \le B 2^{j(m-1)} \omega_m \left(f, 2^{-j}\right).$$

Combining these estimates, we get the required inequality. ◊

Inequality (10.43) is an analog of Bernstein's inequality for trigonometric polynomials (see Appendix A.15). The proof of classical inverse theorems is based on Bernstein's inequality. Using the same scheme and Theorem 10.3.3, we prove the following assertions.

THEOREM 10.3.5. *Suppose* $1 \le p \le \infty$ *and a function* φ *satisfies the conditions of Theorem* 10.3.3. *Then for any* $h > 0$ *and all* $f \in L_p(\mathbb{T})$ ($f \in C(\mathbb{T})$ *for* $p = \infty$), *the inequality*

$$\omega_m(f, h)_p \le N h^m \sum_{0 \le l \le h^{-1}} (l+1)^{m-1} E_l(f)_p \tag{10.46}$$

holds, where N *depends only on* p, m, ε, *and the constant in the symbol* O.

Proof. Let T_n be a wavelet polynomial of best approximation of order n for a function f. To prove inequality (10.46) in the case $h > 1$, it suffices to note that $T_0 \equiv \text{const}$ since the space V_0 consists of constants and thus, by property **M3** of the moduli of continuity (see Appendix A.14), we have

$$\omega_m(f, h) = \omega_m(f - T_0, h) \le 2^m \|f - T_0\| = 2^m E_0(f).$$

Now let $h \le 1$. Choose a nonnegative integer k such that

$$2^k \le \frac{1}{h} < 2^{k+1}. \tag{10.47}$$

From properties **M3, M6, M7** of the moduli of continuity we have

$$\begin{aligned}\omega_m(f,h) &\le \omega_m(f - T_{2^k}, h) + \omega_m(T_{2^k}, h)\\ &\le 2^m \|f - T_{2^k}\| + h^m \|T_{2^k}^{(m)}\| = 2^m E_{2^k}(f) + h^m \|T_{2^k}^{(m)}\|.\end{aligned} \tag{10.48}$$

Since $T_0^{(m)} \equiv 0$, the function $T_{2^k}^{(m)}$ can be represented in the form

$$T_{2^k}^{(m)} = T_2^{(m)} - T_0^{(m)} + \sum_{i=2}^{k} \left(T_{2^i}^{(m)} - T_{2^{i-1}}^{(m)} \right). \tag{10.49}$$

Using Theorem 10.3.3 and property **B2** of best approximations, estimate the norm of one of the summands of this sum:

$$\begin{aligned}\|T_{2^i}^{(m)} - T_{2^{i-1}}^{(m)}\| &\le B 2^{im} \|T_{2^i} - T_{2^{i-1}}\|\\ &\le B 2^{im} \left(\|T_{2^i} - f\| + \|T_{2^{i-1}} - f\| \right) = B 2^{im} \left(E_{2^i}(f) + E_{2^{i-1}}(f) \right)\\ &\le B 2^{i(m-1)+3} \sum_{l=2^{i-2}+1}^{2^{i-1}} E_l(f) \le B 2^{2m+1} \sum_{l=2^{i-2}+1}^{2^{i-1}} (l+1)^{m-1} E_l(f).\end{aligned}$$

Analogously,

$$\|T_2^{(m)} - T_0^{(m)}\| \le B 2^m \left(E_2(f) + E_0(f) \right).$$

From these estimates and (10.49) we get

$$\|T_{2^k}^{(m)}\| \le B 2^{2m+2} \sum_{l=0}^{2^k} (l+1)^{m-1} E_l(f). \tag{10.50}$$

Moreover, it follows from (10.47) and **B2** that

$$\begin{aligned}E_{2^k}(f) &\le h^m 2^{(k+1)m} E_{2^k}(f) \le h^m 2^{m+1} 2^{k(m-1)} \sum_{l=2^{k-1}+1}^{2^k} E_l(f)\\ &\le 2^{2m} h^m \sum_{l=2^{k-1}+1}^{2^k} (l+1)^{m-1} E_l(f).\end{aligned} \tag{10.51}$$

Relations (10.47), (10.48), (10.50), and (10.51) imply (10.46). ◊

The conditions of Theorems 10.3.1, 10.3.3, and 10.3.5 are satisfied, e.g., for wavelets generated by a smooth compactly supported scaling function. Using these theorems for such wavelets, we obtain the following assertion.

COROLLARY 10.3.6. *Let $\varphi, \psi \in C^m(\mathbb{R})$ be compactly supported functions, $r \in \mathbb{Z}_+$, $r \le m-1$, $\alpha \in [0,1)$, $\alpha^2 + r^2 > 0$, $1 \le p \le \infty$, $f \in L_p(\mathbb{T})$ ($f \in C(\mathbb{T})$ for $p = \infty$). Then the following relations are equivalent: $E_N(f)_p = O(1/N^{r+\alpha})$, $\|f - s_N(f)\|_p = O(1/N^{r+\alpha})$, $\omega_{r+1}(f,h)_p = O(h^{r+\alpha})$.*

PROBLEM 10.3.7. Let numbers r, α and functions φ, ψ satisfy the conditions of Corollary 10.3.6. Prove that the relations $\omega_{r+1}(f,h)_\infty = O(h^{r+\alpha})$ and $\langle f, w_N \rangle = O(2^{-j(r+\alpha+1/2)})$ are equivalent, where $N = 2^j + n$, $n = 0, \dots, 2^j - 1$.

LEMMA 10.3.8. *Let f_n, $n = 1, 2, \ldots$, be absolutely continuous functions on $[0,1]$ such that $f_n' \in L_p[0,1]$, $1 \le p < \infty$. If the sequence $\{f_n\}$ converges to f in L_1-norm and the sequence $\{f_n'\}$ converges in L_p-norm, then the function f is equivalent to some function F absolutely continuous on $[0,1]$, for which $F' \in L_p[0,1]$.*

Proof. First of all show that $\{f_n(0)\}$ is a Cauchy sequence. Indeed, from the equality

$$f_n(0) = f_n(x) - \int_0^x f_n'$$

we get

$$\begin{aligned} |f_n(0) - f_m(0)| &= |f_n(0) - f_m(0)| \int_0^1 dx \\ &\le \int_0^1 |f_n(x) - f_m(x)|\, dx + \int_0^1 dx \int_0^x |f_n'(t) - f_m'(t)|\, dt \\ &\le \|f_n - f_m\|_1 + \|f_n' - f_m'\|_1 \underset{n,m\to\infty}{\longrightarrow} 0. \end{aligned}$$

Thus the sequence $\{f_n(0)\}$ converges. Denote by a its limit. Let g be the limit of the sequence $\{f_n'\}$ in $L_p[0,1]$. Let us show that the function

$$F(x) := a + \int_a^x g$$

is the required function. By Theorem A.10.2 (see Appendix A.10), F is absolutely continuous and $F'(x) = g(x)$ for almost all x. In addition, we have

$$\begin{aligned} \int_0^1 |f(x) - F(x)|\, dx &\le \int_0^1 \left| f(x) - a - \int_0^x f_n' \right| dx + \int_0^1 dx \int_0^x |f_n' - g| \\ &\le |a - f_n(o)| + \|f - f_n\|_1 + \|g - f_n'\|_1, \end{aligned}$$

which implies the equivalence of the functions f and F. ◊

THEOREM 10.3.9. *Suppose $1 \le p \le \infty$, functions φ and ψ satisfy the conditions of Theorems 10.3.3 and 10.3.1, respectively, $f \in L_p(\mathbb{T})$ ($f \in C(\mathbb{T})$ for $p = \infty$), and*

$$\sum_{l=1}^{\infty} l^{m-1} E_l(f)_p < \infty. \tag{10.52}$$

Then f has the m-th derivative almost everywhere (for $p = \infty$, at each point), $f^{(m)} \in L_p(\mathbb{T})$ (for $p = \infty$, $f^{(m)} \in C(\mathbb{T})$), and for any $k \in \mathbb{Z}_+$, $k \le m$, the inequality

$$\omega_{m-k}\left(f^{(k)}, h\right)_p \le M \left(h^{-k} \omega_m(f,h)_p + \sum_{l=[1/2h]}^{\infty} (l+1)^{k-1} E_l(f)_p \right) \tag{10.53}$$

holds, where M depends only on p, m, ε, and the constants in the symbols O.

Proof. Let T_n be a wavelet polynomial of best approximation of order n for a function f. It follows from Theorem 10.3.1 and properties **M1, M2** of the moduli of continuity (see Appendix A.14) that $\|f - T_n\| \to 0$ for $n \to \infty$; hence for $n > 0$, we have

$$(10.54) \qquad f - T_n = \sum_{i=1}^{\infty} \left(T_{2^i n} - T_{2^{i-1}n}\right),$$

where the series on the right-hand side converges in norm. As noted in the proof of Theorem 10.3.3, all the elements of the spaces V_j, in particular, wavelet polynomials T_n, have continuous m-th derivatives. Using Theorem 10.3.3 and property **B2** of the best approximations, estimate the norm of the m-th derivative of the i-th summand of this sum. For $i > 1$, we have

$$\begin{aligned}\left\|T_{2^i n}^{(m)} - T_{2^{i-1}n}^{(m)}\right\| &\leq B2^{(i+1)m} n^m \left\|T_{2^i n} - T_{2^{i-1}n}\right\| \leq 2B2^{(i+1)m} n^m E_{2^{i-1}n}(f) \\ &\leq 16B2^{(i+1)(m-1)} n^{m-1} \sum_{k=2^{i-2}n+1}^{2^{i-1}n} E_k(f) \leq B2^{3m+1} \sum_{k=2^{i-2}n+1}^{2^{i-1}n} k^{m-1} E_k(f).\end{aligned}$$

Analogously,

$$\begin{aligned}\left\|T_{2n}^{(m)} - T_n^{(m)}\right\| &\leq B2^{2m+1} n^m E_n(f) \\ &\leq B2^{2m+2} n^{m-1} \sum_{k=[n/2]}^{n} E_k(f) \leq B2^{3m+1} \sum_{k=[n/2]}^{n} (k+1)^{m-1} E_k(f).\end{aligned}$$

In view of (10.52) it follows that the series

$$\sum_{i=1}^{\infty} \left\|T_{2^i n}^{(m)} - T_{2^{i-1}n}^{(m)}\right\|$$

converges. Therefore the series obtained by formal differentiating of the right-hand side of (10.54) m times converges in norm. It follows from Lemma 10.3.8 that the left-hand side of (10.54) has the m-th derivative almost everywhere, this derivative belongs to $L_p(\mathbb{T})$ ($C(\mathbb{T})$ for $p = \infty$), and

$$(10.55) \quad \left\|f^{(m)} - T_n^{(m)}\right\| \leq \sum_{i=1}^{\infty} \left\|T_{2^i n}^{(m)} - T_{2^{i-1}n}^{(m)}\right\| \leq B2^{3m+1} \sum_{k=[n/2]}^{\infty} (k+1)^{m-1} E_k(f).$$

Analogously, using the equality

$$f - T_0 = T_1 - T_0 + \sum_{i=1}^{\infty} \left(T_{2^j} - T_{2^{j-1}}\right)$$

instead of (10.54), we get the estimate

$$(10.56) \qquad \left\|f^{(m)}\right\| \leq B2^{3m+1} \sum_{k=0}^{\infty} (k+1)^{m-1} E_k(f).$$

Thus we have proved (10.53) for $k = m$.

Now let $k < m$, $h > 0$, $n_0 = [1/h]$. First we consider the case $n_0 > 0$. By properties **M3, M6, M7** of the moduli of continuity we have

$$(10.57) \qquad \omega_{m-k}\left(f^{(k)}, h\right) \leq 2^{m-k} \|f^{(k)} - T_{n_0}^{(k)}\| + h^{m-k} \|T_{n_0}^{(m)}\|.$$

Apply (10.55) (taking k for m) to the first summand of the right-hand side of (10.57),

$$\|f^{(k)} - T_{n_0}^{(k)}\| \le B2^{3k+1} \sum_{l=[n_0/2]}^{\infty} (l+1)^{k-1} E_l(f) \le B2^{3k+1} \sum_{l=[1/2h]}^{\infty} (l+1)^{k-1} E_l(f).$$

It follows from Theorem 10.3.1, Corollary 10.3.4, and properties **M3, M4, M7** of the moduli of continuity that

$$\begin{aligned} h^{m-k}\|T_{n_0}^{(m)}\| &\le \tilde{B} h^{m-k} (2n_0)^m \omega_m \left(T_{n_0}, \frac{1}{n_0}\right) \\ &\le 4^m \tilde{B} h^{-k} \left(\omega_m \left(T_{n_0} - f, \frac{1}{n_0}\right) + \omega_m \left(f, \frac{1}{n_0}\right)\right) \\ &\le 4^m \tilde{B} h^{-k} \left(2^m E_{n_0}(f) + \omega_m \left(f, \frac{1}{n_0}\right)\right) \\ &\le 8^m \tilde{B}\,(2^m C + 1)\, h^{-k} \omega_m(f, h), \end{aligned}$$

which completes the proof of (10.53) for $n_0 > 0$. If $n_0 = 0$, then by property **M3** of the moduli of continuity, we have

$$\omega_{m-k}\left(f^{(k)}, h\right) \le 2^{m-k}\|f^{(k)}\|.$$

It remains to use (10.56). ◊

The direct and inverse theorems proved above allow one to characterize some important spaces of functions in terms of deviations of functions from partial sums of their wavelet expansions.

Introduce the Lipschitz space $\mathrm{Lip}_p^{r,\alpha}$, where $1 \le p \le \infty$, $0 < \alpha < 1$, $r \in \mathbb{N}$, consisting of 1-periodic functions f that have the absolutely continuous $(r-1)$-st derivative and such that $f^{(r)} \in L_p(\mathbb{T})$ and $\omega\left(f^{(r)}, h\right)_p \le K h^{\alpha}$.

THEOREM 10.3.10. *Let $\varphi, \psi \in C^m(\mathbb{R})$ be compactly supported functions, $r \in \mathbb{N}$, $r \le m-1$, $0 < \alpha < 1$, $1 \le p < \infty$. A function $f \in L_p(\mathbb{T})$ belongs to the space $\mathrm{Lip}_p^{r,\alpha}$ if and only if*

$$\|f - s_n(f)\|_p = O\left(\frac{1}{n^{r+\alpha}}\right). \tag{10.58}$$

`Proof.` If $f \in \mathrm{Lip}_p^{r,\alpha}$, then from Theorem 10.3.1 and property **M6** of the moduli of continuity we have

$$\|f - s_n(f)\| \le C\omega_{r+1}\left(f, \frac{1}{n}\right) \le \frac{C}{n^r}\omega\left(f^{(r)}, \frac{1}{n}\right) \le \frac{CK}{n^{r+\alpha}},$$

which implies necessity. If (10.58), then it follows from Theorems 10.3.5 and 10.3.9 that

$$\begin{aligned} \omega\left(f^{(r)}, h\right) &\le C_1 \left(h \sum_{0 \le l \le h^{-1}} (l+1)^r \|f - s_l(f)\| + \sum_{l=[1/2h]}^{\infty} (l+1)^{r-1} \|f - s_l(f)\|\right) \\ &\le C_2 h^{\alpha}. \end{aligned}$$

◊

Let $\alpha > 0$, $1 \le p, q < \infty$, $r := [\alpha] + 1$. The class of functions $f \in L_p(\mathbb{T})$ such that

$$\int_0^\infty t^{q\alpha} \omega_r^q(f,t)_p \frac{dt}{t} < \infty \tag{10.59}$$

is called the Besov space.

THEOREM 10.3.11. *Let $\varphi, \psi \in C^m(\mathbb{R})$ be compactly supported functions, $r \in \mathbb{N}$, $r \le m$, $1 \le p, q < \infty$, $r - 1 \le \alpha < r$. A necessary and sufficient condition for a function $f \in L_p(\mathbb{T})$ to belong to the space B^α_{qp} is that the series*

$$\sum_{n=1}^\infty n^{q\alpha - 1} \|f - s_n(f)\|^q \tag{10.60}$$

be convergent.

Proof. It is not hard to see that relation (10.59) holds if and only if the series

$$\sum_{n=1}^\infty n^{q\alpha-1} \omega_r^q \left(f, \frac{1}{n}\right)_p \tag{10.61}$$

converges; therefore sufficiency follows from Theorem 10.3.1. Let us prove necessity. First consider the case $r, q > 1$. Set $\varepsilon = \frac{r-\alpha}{2}$. Using Theorem 10.3.5 and Hölder's inequality, we get

$$\begin{aligned}
\sum_{n=1}^\infty n^{q\alpha-1}\omega_r^q\left(f,\frac{1}{n}\right) &\le C \sum_{n=1}^\infty n^{q(\alpha-r)-1} \left(\sum_{k=1}^n k^{r-1}\|f - s_k(f)\|\right)^q \\
&\le C \sum_{n=1}^\infty n^{q(\alpha-r)-1} \left(\sum_{k=1}^n k^{\frac{q\varepsilon}{q-1}-1}\right)^{q-1} \sum_{k=1}^n k^{q(r-\varepsilon)-1}\|f - s_k(f)\|^q \\
&\le C_1 \sum_{n=1}^\infty n^{q(\alpha-r+\varepsilon)-1} \sum_{k=1}^n k^{q(r-\varepsilon)-1}\|f - s_k(f)\|^q \\
&= C_1 \sum_{k=1}^\infty k^{q(r-\varepsilon)-1}\|f - s_k(f)\|^q \sum_{n=k}^\infty n^{q(\alpha-r+\varepsilon)-1} \\
&\le C_2 \sum_{k=1}^\infty k^{q\alpha-1}\|f - s_k(f)\|^q.
\end{aligned}$$

In the case $r = 1$, to estimate sum (10.61), we use the inequality

$$\omega_1^q\left(f, \frac{1}{n}\right) \le \frac{2^{q-1}}{n^q}\|f - s_0(f)\|^q + \frac{2^{q-1}}{n^q}\left(\sum_{k=1}^n \|f - s_k(f)\|\right)^q.$$

For the second summand, we repeat the reasoning used for $r > 1$, while the series corresponding to the first summand obviously converges. For the case $q = 1$, the statement is proved in a similar way, but without using Hölder's inequality. ◊

10.4. Convergence of wavelet expansions at a point

In this section, we again confine our consideration to the case $d = 1$, $M = 2$. Theorem 10.2.7 gives a sufficient condition for the convergence of the Fourier series of a function f with respect to periodic wavelets at each of its Lebesgue points. However, this theorem gives no information about the rate of convergence. We shall now study the approximation of functions whose local smoothness is higher than the smoothness at Lebesgue points and estimate their rate of convergence.

As above, we shall number the functions $S_k^j\psi_j$, $S_k^j\widetilde{\psi}_j$ in ascending order of indices. For a (p,q)-pair, the Fourier series of a function f has the form

$$\langle f, \widetilde{\varphi}_0\rangle \varphi_0 + \sum_{j=0}^{\infty} \sum_{k=0}^{2^j-1} \langle f, S_k^j\widetilde{\psi}_j\rangle S_k^j\psi_j, \tag{10.62}$$

where $s_N(f)$ denotes its partial sum of order N.

DEFINITION 10.4.1. Let $x_0 \in \mathbb{R}$, $\alpha > 0$. A function f belongs to the class $\mathcal{L}_\alpha(x_0)$ if there exists a number s such that

$$\frac{1}{h} \int\limits_{|x-x_0|\le h} |f(x) - s|\, dx = O(h^\alpha), \quad \text{as } h \to 0. \tag{10.63}$$

THEOREM 10.4.2. *Suppose $f \in \mathcal{L}_\alpha(x_0)$, $\alpha > 0$, PMRAs $\{V_j\}_{j=0}^\infty$, $\{\widetilde{V}_j\}_{j=0}^\infty$ form a (p,q)-pair with scaling sequences $\{\varphi_j\}_{j=0}^\infty$, $\{\widetilde{\varphi}_j\}_{j=0}^\infty$ such that $\{S_n^j\varphi_j\}_{n\in D(M^j)}$, $\{S_k^j\widetilde{\varphi}_j\}_{n\in D(M^j)}$ are biorthonormal systems and $\{\psi_j\}_{j=0}^\infty, \{\widetilde{\psi}_j\}_{j=0}^\infty$ are the corresponding sequences of wavelet functions. If*

$$|\varphi_j(x)| \le \frac{C2^{j/2}}{1 + (2^j|x|)^m}, \quad |\widetilde{\varphi}_j(x)| \le \frac{C2^{j/2}}{1 + (2^j|x|)^m}, \quad m > 1, \tag{10.64}$$

for all $x \in [-1/2, 1/2]$, then

$$s_{2^j-1}(f, x_0) - s = O\left(2^{-j\min\{\alpha, m-1\}}\right), \quad j \to \infty, \tag{10.65}$$

for $\alpha \ne m-1$, and

$$s_{2^j-1}(f, x_0) - s = O\left(j 2^{-j\alpha}\right), \quad j \to \infty, \tag{10.66}$$

for $\alpha = m - 1$. If, in addition,

$$|\psi_j(x)| \le \frac{C2^{j/2}}{1 + (2^j|x|)^n}, \quad |\widetilde{\psi}_j(x)| \le \frac{C2^{j/2}}{1 + (2^j|x|)^n}, \quad n > 1, \tag{10.67}$$

for all $x \in [-1/2, 1/2]$, then

$$s_N(f, x_0) - s = O\left(N^{-\min\{\alpha, n-1\}}\right), \quad N \to \infty, \tag{10.68}$$

for $\alpha \ne n - 1$, and

$$s_N(f, x_0) - s = O\left(N^{-\alpha}\log N\right), \quad N \to \infty, \tag{10.69}$$

for $\alpha = n - 1$.

Proof. Consider the case $\alpha \ne m-1$, $\alpha \ne n-1$. Note that it follows from (10.64), (10.67) that the functions $\widetilde{\varphi}_j$, $\widetilde{\psi}_j$ are bounded; hence (10.62) has meaning for all $f \in L(\mathbb{T})$. Repeating the arguments from the proof of Theorem 10.2.7, we get

$$
\begin{aligned}
|s_{2^j-1}(f,x_0)-s| &= \left| \int_0^1 (f(x)-s) \sum_{k=0}^{2^j-1} S_k^j\varphi_j(x_0)\overline{S_k^j\widetilde{\varphi}_j(x)}\,dx \right| \\
&\le \int_{-\infty}^{\infty} |f(x)-s| \sum_{k\in\mathbb{Z}} \left|\widetilde{g}_j\left(2^jx_0+k\right)\right| \left|g_j\left(2^jx+k\right)\right| dx \\
&\le B2^j \int_{-\infty}^{\infty} |f(x)-s|\eta\left(2^{j-3}(x-x_0)\right) dx,
\end{aligned}
$$

where $\eta(u) = C_1(1+|u|^m)^{-1}$. Let j_0 be a positive integer such that

$$
\frac{1}{h} \int_{|x-x_0|\le h} |f(x)-s|\,dx \le C_2 h^{\alpha} \tag{10.70}
$$

for all $h \in (0, 2^{-j_0}]$. By the monotonicity of the function η, for $j \ge j_0$ we have

$$
\begin{aligned}
|s_{2^j-1}(f,x_0)-s| &\le B2^j \Bigg(\eta(0) \int_{|x-x_0|<2^{-j}} |f(x)-s|\,dx \\
&\qquad + \sum_{k=-j}^{\infty} \eta(2^{j+k-3}) \int_{2^k\le|x-x_0|\le 2^{k+1}} |f(x)-s|\,dx \Bigg) \\
&\le B2^j \Bigg(\eta(0) \int_{|x-x_0|<2^{-j}} |f(x)-s|\,dx \\
&\qquad + \sum_{k=-j+1}^{-j_0} 2^{j+1}\eta(2^{j+r}) \int_{|x-x_0|<2^k} |f(x)-s|\,dx \\
&\qquad + \sum_{k=-j_0+1}^{\infty} 2^{j+1}\eta(2^{j+k}) \int_{|x-x_0|<2^k} |f(t)-s|\,dx \Bigg) \\
&= \Sigma_0 + \Sigma_1 + \Sigma_2.
\end{aligned} \tag{10.71}
$$

Using (10.70), we get

$$
\Sigma_0 = O(2^{-j\alpha}) = O(2^{-j\min\{\alpha,m-1\}}), \tag{10.72}
$$

$$
\begin{aligned}
\Sigma_1 &= O(1)2^j \sum_{k=-j}^{-j_0} 2^{k(\alpha+1)}2^{(j+k)m} \\
&= O\left(2^{j(1-m)} \sum_{k=-j+1}^{-j_0} 2^{k(\alpha+1-m)} \right) = O\left(2^{-j\min\{\alpha,m-1\}}\right).
\end{aligned} \tag{10.73}
$$

To estimate Σ_2, note that the function $I(h) = h^{-1} \int\limits_{|t-x_0|<h} |f(t) - s|\, dt$ is bounded on $(0, \infty)$. Therefore,

(10.74)

$$\Sigma_2 = O(1) \sum_{k=-j_0+1}^{\infty} 2^{j+k} \eta(2^{j+k}) = O(1) \sum_{k=-j_0+1}^{\infty} 2^{(j+k)(1-m)}$$

$$= O(1) 2^{j(1-m)} \sum_{k=-j_0+1}^{\infty} 2^{k(1-m)} = O\left(2^{-j(m-1)}\right) = O\left(2^{-j \min\{\alpha, m-1\}}\right).$$

Combining (10.71) with (10.72), (10.73), and (10.74), we obtain (10.65). In particular, (10.65) implies that the sequence $\{s_{2^j-1}(f, x_0)\}_{j=1}^{\infty}$ converges to s. It follows that

(10.75)

$$f(x_0) - s_N(f, x_0) = (s_{2^j-1}(f, x_0) - s_N(f, x_0)) + \sum_{i=j}^{\infty} (s_{2^{i+1}-1}(f, x_0) - s_{2^i-1}(f, x_0)),$$

where $2^j \le N < 2^{j+1}$. Thus to prove (10.68), it suffices to check the relation

$$s_{2^j+L}(f, x_0) - s_{2^j-1}(f, x_0) = O\left(2^{-j \min\{\alpha, m-1\}}\right) \tag{10.76}$$

for all $j = 0, 1, \ldots,$ $L = 0, \ldots, 2^j - 1$. Since the space $\widetilde{W}_j$ is orthogonal to the space V_0, which consists of constants, the left-hand side of this equality can be represented in the form

$$\int_0^1 (f(t) - s) \sum_{k=0}^{L} S_k^j \psi_j(x_0) \overline{S_k^j \widetilde{\psi}_j(t)}\, dt.$$

Estimating this in the same way as $|s_{2^j-1}(f, x_0) - s|$, we prove (10.68). For $\alpha = m-1$ or $\alpha = n - 1$, the proof is analogous. ◊

COROLLARY 10.4.3 (Localization principle). *If under the conditions of Theorem 10.4.2 a function f is identically zero in a neighborhood of the point x_0, then*

$$|s_N(f, x_0)| = O\left(N^{1-n}\right).$$

THEOREM 10.4.4. *Suppose $f \in \mathcal{L}_\alpha(x_0)$, $\alpha > 0$, PMRAs $\{V_j\}_{j=0}^{\infty}$, $\{\widetilde{V}_j\}_{j=0}^{\infty}$ form a (p, q)-pair and are generated, respectively, by scaling functions φ, $\widetilde{\varphi}$ with biorthonormal integer shifts, and ψ, $\widetilde{\psi}$ are the corresponding wavelet functions. If*

$$|\varphi(x)|, |\widetilde{\varphi}(x)| \le \frac{C}{1 + |x|^m}, \quad m > 1,$$

for all $x \in \mathbb{R}$, then

$$|s_{2^j-1}(f, x_0) - s| = O\left(2^{-j \min\{\alpha, m-1\}}\right), \quad j \to \infty, \tag{10.77}$$

for $\alpha \ne m - 1$,

$$|s_{2^j-1}(f, x_0) - s| = O\left(j 2^{-j\alpha}\right), \quad j \to \infty, \tag{10.78}$$

for $\alpha = m - 1$. If, in addition,

$$|\psi(x)|, |\widetilde{\psi}(x)| \le \frac{C}{1 + |x|^n}, \quad n > 1,$$

for all $x \in \mathbb{R}$, *then*

$$|s_N(f,x_0) - s| = O\left(N^{-j\min\{\alpha,n-1\}}\right), \quad N \to \infty, \tag{10.79}$$

for $\alpha \neq n-1$,

$$|s_N(f,x_0) - s| = O\left(N^{-\alpha}\log N\right), \quad N \to \infty, \tag{10.80}$$

for $\alpha = n-1$.

This theorem can be proved analogously to Theorem 10.4.2, using the equalities

$$\varphi_j(x) = 2^{j/2}\sum_{l\in\mathbb{Z}}\varphi(2^jx+2^jl), \qquad \widetilde{\varphi}_j(x) = 2^{j/2}\sum_{l\in\mathbb{Z}}\widetilde{\varphi}(2^jx+2^jl),$$
$$\psi_j(x) = 2^{j/2}\sum_{l\in\mathbb{Z}}\psi(2^jx+2^jl), \qquad \widetilde{\psi}_j(x) = 2^{j/2}\sum_{l\in\mathbb{Z}}\widetilde{\psi}(2^jx+2^jl).$$

Further, we shall show that Theorems 10.4.2 and 10.4.4 cannot be improved in different senses. Consider the case $n-1 > \alpha$, $m-1 > \alpha$. It occurs, for instance, if φ, ψ are compactly supported functions. In this case,

$$|s_N(f,x_0) - s| = O\left(N^{-\alpha}\right). \tag{10.81}$$

The following examples show that estimate (10.81) is sharp with respect to the order.

Consider a Haar system that is a system of periodic wavelets generated by the functions

$$\varphi(x) = \begin{cases} 1, & \text{if } 0 \le x < 1, \\ 0, & \text{otherwise,} \end{cases} \tag{10.82}$$

where the corresponding wavelet function (see §1.4) has the form

$$\psi(x) = \begin{cases} 1, & \text{if } 0 \le x < 1/2, \\ -1, & \text{if } 1/2 \le x < 1, \\ 0, & \text{otherwise.} \end{cases}$$

Choose $f \in L(\mathbb{T})$ such that $f(t) = t^\alpha$ on $[0,\delta]$ and $f(x) = 0$ on $[-\delta, 0]$. It is clear that f satisfies (10.63) for $s=0$, $x_0 = 0$. Consider the (2^j-1)-st partial sum:

$$s_{2^j-1}(f,x) = 2^j\int_0^1 f(t)\sum_{k=0}^{2^j-1}\sum_{l\in\mathbb{Z}}\varphi(2^jx+2^jl+k)\sum_{l'\in\mathbb{Z}}\overline{\varphi(2^jt+2^jl'+k)}\,dt.$$

If $x \in [0, 2^{-j})$, then $\sum_{l\in\mathbb{Z}}\varphi(2^jx+2^jl) = 1$ and $\sum_{l\in\mathbb{Z}}\varphi(2^jx+2^jl+k) = 0$ for $k = 1,\dots,2^j-1$. For sufficiently large j, we have

$$|s_{2^j-1}(f,0)| = 2^j\int_0^{2^{-j}} f(t)\,dt = 2^j\int_0^{2^{-j}} t^\alpha\,dt = \frac{1}{\alpha+1}2^{-j\alpha}.$$

Hence the right-hand side of relation (10.81) cannot be replaced by $o(N^{-\alpha})$.

Now let us show that Theorems 10.4.2 and 10.4.4 cannot be improved if $\alpha = n-1$ or $\alpha = m-1$. Define the function φ as follows:

$$\widehat{\varphi}(u) = \begin{cases} 1, & \text{if } |u| < 1/3, \\ \left|\sin\frac{3\pi u}{2}\right|, & \text{if } 1/3 < |u| < 2/3, \\ 0, & \text{if } |u| > 2/3. \end{cases}$$

This function φ is a scaling function for the Meyer MRA (see §1.4). Simple calculation gives the explicit formula

$$\varphi(u) = \frac{\sin 2\pi u/3}{2\pi u} + \frac{6\cos 4\pi u/3 + 8u \sin 2\pi u/3}{\pi(9 - 16u^2)}.$$

It is clear that $\varphi(u)$ decreases as u^{-2}. The function ψ can be calculated explicitly, using the formula $\widehat{\psi}(u) = e^{i\pi u}(\widehat{\varphi}(u+1)+\widehat{\varphi}(u-1))\widehat{\varphi}(u/2)$, which follows from (1.66) and (1.47). We can show that $\psi(u) = O\left(u^{-2}\right)$. Thus φ generates a PMRA and the functions φ, ψ satisfy the conditions of Theorem 10.4.4 for $n = m = 2$. Show that for any $\gamma(u) = o(u \log u)$, $u \to +0$, there exists a 1-periodic function $f \in L(\mathbb{T})$ that satisfies (10.63) for $x_0 = 0$, $s = 0$, $\alpha = 1$ such that

$$\limsup_{j\to\infty} |\gamma^{-1}(2^{-j}) s_{2^j-1}(f, 0)| = \infty.$$

By virtue of the Banach-Steinhaus theorem, it suffices to find a sequence of positive integers $\{M_j\}$, $M_j \to \infty$, and a sequence of functions $\{f_j\}$, $f_j \in L(\mathbb{T})$, such that

$$\sup_{h>0} \frac{1}{h^2} \int_0^h |f_j| \le C, \tag{10.83}$$

where C is an absolute constant and $|\gamma^{-1}(2^{-j}) s_{2^j-1}(f_j, 0)| \ge M_j$ for some subsequence of positive integers j. Set $N_j = 2^j - 1$, where j is even. By the Poisson summation formula, we have

$$\sum_{l\in\mathbb{Z}} \varphi(2^j(x+l)+k) = \sum_{n\in\mathbb{Z}} \widehat{\varphi}(2^{-j}n) e^{2\pi i n x} e^{2\pi i n k 2^{-j}}.$$

Taking into account that $\sum_{k=0}^{2^j-1} e^{2\pi i k m 2^{-j}} = 0$ for all $m \ne 2^j l$, $l \in \mathbb{Z}$, and $\operatorname{supp} \widehat{\varphi} \cap \operatorname{supp} \widehat{\varphi}(\cdot + k) = 0$ for $k \in \mathbb{Z}, |k| \ge 2$, we get

$$\begin{aligned} s_{N_j}(f, x) = \int_0^1 f(t) \sum_{n\in\mathbb{Z}} \widehat{\varphi}(2^{-j}n) e^{2\pi i n(t-x)} (\widehat{\varphi}(2^{-j}n) \\ + \widehat{\varphi}(2^{-j}n - 1) e^{2\pi i 2^j x} + \widehat{\varphi}(2^{-j}n + 1) e^{-2\pi i 2^j x})\, dt \end{aligned}$$

for all $f \in L(\mathbb{T})$. Consequently,

$$s_{N_j}(f, 0) = \sum_{n\in\mathbb{Z}} \lambda\left(2^{-j}n\right) \widehat{f}(n),$$

where $\lambda(u) = \widehat{\varphi}^2(u) + \widehat{\varphi}(u)(\widehat{\varphi}(u-1) + \widehat{\varphi}(u+1))$. It is not hard to check that $\widehat{\lambda}(u) = \alpha(u) + \beta(u)$, where

$$\alpha(u) = \frac{9(\sin 2\pi u/3 + \sin 4\pi u/3)}{2\pi u(9 - 4u^2)},$$

$$\beta(u) = \frac{3(\cos 2\pi u/3 + \cos 4\pi u/3)}{\pi(9 - 4u^2)}.$$

Set $\sigma_N(f, x) = \sum_{n\in\mathbb{Z}} \lambda(\frac{n}{N}) \widehat{f}(n) e^{2\pi i n x}$. The sequence of means $\sigma_N(f)$ gives a linear method of summation for Fourier series with summable $\widehat{\lambda}$. It is well known that

$\sigma_N(f,x)$ can be represented in the form

$$\sigma_N(f,x) = \int_{-\infty}^{\infty} f\left(x+\frac{t}{N}\right)\widehat{\lambda}(t)\,dt.$$

Since $\sigma_{2^j}(f,0) = s_{N_j}(f,0)$, we have

$$s_{N_j}(f,0) = \int_{-\infty}^{\infty} f(2^{-j}t)\widehat{\lambda}(t)\,dt. \tag{10.84}$$

Set $e_k = [3k, 3k+1/2]$, $k \in \mathbb{Z}$,

$$\omega = \{k_r = 2^{r+j/2}, \dots, 2^{r+j/2} + 2^{2r-1},\ r = 1, \dots, \frac{j}{2} - 2\}, \quad \Omega = \bigcup_{k\in\omega} e_k.$$

It is not hard to see that $\Omega \subset [0, 2^{j-1}]$ and $e_k \cap e_l \neq \emptyset$ for all $k, l \in \omega$, $k \neq l$, for $j \geq 4$. Define even 1-periodic functions f_j on $[0,1/2]$ by the equality

$$f_j(x) = \begin{cases} -1, & \text{if } 2^j x \in \Omega, \\ 0, & \text{if } 2^j x \in [0, 2^{j-1}] \setminus \Omega. \end{cases}$$

Since $f_j \equiv 0$ on $[0, \cdot 2^{1-j/2}]$, relation (10.83) obviously holds for $h < 2^{1-j/2}$. If $2^{l-j/2} \leq h < 2^{l+1-j/2}$, $l \leq \frac{j}{2} - 2$, then

$$\int_0^h |f_j| = \int_{\Omega\cap[0,h]} |f_j| \leq \sum_{r=1}^{l} \sum_{k_r=2^{r+j/2}}^{2^{2r-1}+2^{r+j/2}} 2^{-j-1} \leq 2^{-j-1}\sum_{r=1}^{l} 2^{2r-1}$$
$$= O\left(2^{2l-j}\right) = O(h^2).$$

It is clear that

$$\int_{-\infty}^{\infty} f_j(2^{-j}t)\widehat{\lambda}(t)\,dt = \int_{-\infty}^{\infty} f_j(2^{-j}t)\beta(t)\,dt + O(2^{-j})$$
$$= -\frac{3}{2\pi}\int_0^{\infty} f_j(2^{-j}t)P(t)\frac{dt}{t^2} + O(2^{-j}),$$

where $P(t) = \cos\frac{2\pi t}{3} + \cos\frac{4\pi t}{3}$. Further, we have

$$-\int_0^{\infty} f_j(2^{-j}t)P(t)\frac{dt}{t^2} = \int_{\Omega} P(t)\frac{dt}{t^2} + O(2^{-j}).$$

Since the function $P(t)$ is 3-periodic and nonnegative on Ω, for all even $j > 6$, we have

$$\int_{\Omega} P(t)\frac{dt}{t^2} = \sum_{r=1}^{\frac{j}{2}-2} \sum_{k_r=2^{r+j/2}}^{2^{r+j/2}+2^{2r-1}} \int_{e_{k_r}} P(t)\frac{dt}{t^2} \geq \sum_{r=1}^{\frac{j}{2}-2} \int_{2^{r+j/2}+1}^{2^{r+j/2}+2^{2r-1}+1} \frac{dt}{9t^2} \int_{e_0} P$$

$$\geq \sum_{r=1}^{\frac{j}{2}-2} \frac{2^{2r-1}}{(2^{r+j/2}+1)(2^{2r-1}+2^{r+j/2}+1)} \int_{e_0} P$$

$$\geq 2^{-j-7}\left(\frac{j}{2}-2\right)\int_{e_0} P \geq Mj2^{-j},$$

where M is an absolute constant. Finally, using (10.84), we get

$$\left|\gamma^{-1}(1/N_j)s_{N_j}(f_j,0)\right| \geq M\frac{\log N_j/N_j}{|\gamma(1/N_j)|} = M_j \underset{j\to\infty}{\longrightarrow} \infty,$$

which was to be proved.

The following example shows that the condition $m > 1$ in (10.64) cannot be omitted. Moreover, if we replace it by the assumptions

$$|\varphi_j(x)| \leq \frac{C2^{j/2}}{1+(2^j|x|)^{m_1}},$$

$$|\widetilde{\varphi}_j(x)| \leq \frac{C2^{j/2}}{1+(2^j|x|)^{m_2}},$$

then none of the restrictions $m_1 > 1$, $m_2 > 1$ can be omitted. Consider the following pair of PMRAs $\{V_j\}_{j=0}^{\infty}, \{\widetilde{V}_j\}_{j=0}^{\infty}$. The second component $\{\widetilde{V}_j\}_{j=0}^{\infty}$ is the Haar PMRA generated by the function φ given by relation (10.82). The functions

$$\widetilde{\varphi}_j(x) = 2^{j/2}\sum_{l\in\mathbb{Z}}\varphi(2^jx+2^jl) = 2^{-j/2}\sum_{k\in\mathbb{Z}}\widehat{\varphi}(2^{-j}k)e^{2\pi ikx}, \quad j\in\mathbb{Z}_+,$$

form its scaling sequence. The first component $\{V_j\}_{j=0}^{\infty}$ is determined by the scaling sequence

$$\varphi_j(x) = 2^{-j/2}\sum_{k=-2^{j-1}}^{2^{j-1}}{}' \overline{(\widehat{\varphi}(2^{-j}k))^{-1}}e^{2\pi ikx},$$

where the symbol $\sum'$ means that the first and last summands in the sum are taken with the factor 1/2. Since for each $j \in \mathbb{Z}_+$, the sequences $S_j^k\varphi_j$, $S_j^k\widetilde{\varphi}_j$ satisfy condition (9.15), they are biorthonormal and the corresponding wavelet systems $S_j^k\psi_j$, $S_j^k\widetilde{\psi}_j$ are biorthonormal by Theorem 9.2.4. The functions φ_j, $\widetilde{\varphi}_j$ obviously belong to $L_\infty(\mathbb{T})$. Therefore for any function $f \in L(\mathbb{T})$ we can consider both expansions (10.62) and

$$\langle f,\varphi_0\rangle\widetilde{\varphi}_0 + \sum_{i=0}^{\infty}\sum_{n=0}^{2^j-1}\langle f, S_j^n\psi_j\rangle S_j^n\widetilde{\psi}_j. \tag{10.85}$$

Let $s_N(f), \widetilde{s}_N(f)$ denote the partial sums of order N of series (10.62), (10.85), respectively. Since φ is compactly supported, $\widetilde{\varphi}_j$ satisfies (10.64) for any $m > 1$.

Let us show that φ_j satisfies (10.64) for $m = 1$. Simple calculation gives

$$2^{j/2}\varphi_j(x) = \frac{\pi}{4}\text{ctan}\,\pi x \sin \pi 2^j x + \sum_{k=0}^{2^{j-1}}{}' \gamma_k \cos 2\pi k x, \tag{10.86}$$

where $\gamma_k = \frac{\pi 2^{-j}k}{\sin \pi 2^{-j}k} - \frac{\pi}{2}$. If we apply the Abel transform twice, the latter sum takes the form

(10.87)

$$\frac{1}{2\sin^2 \pi x}\left(\sum_{k=0}^{2^{j-1}-2} \Delta^2 \gamma_k \sin^2 \pi(k+1)x - \Delta\gamma_{2^{j-1}-1}\sin^2 \pi 2^{j-1}x\right) = O\left(\frac{2^{-j}}{x^2}\right).$$

Combining (10.86), (10.87) with the trivial relation $\varphi_j = O(2^{j/2})$, we get

$$|\varphi_j(x)| \le \frac{C\,2^{j/2}}{1+2^j|x|}.$$

Let $\gamma(u) = o(1)$, $u \to \infty$. Show that there exist functions $f_1, f_2 \in L(\mathbb{T})$ with supports on $[1/4, 1/2]$ such that

$$\gamma^{-1}(2^j)s_{2^j-1}(f_1, 0) \underset{j\to\infty}{\longrightarrow} \infty,$$

$$\gamma^{-1}(2^j)\widetilde{s}_{2^j-1}(f_2, 0) \underset{j\to\infty}{\longrightarrow} \infty.$$

By the Banach-Steinhaus theorem it suffices to find a sequence of functions $f_1^{(j)}, f_2^{(j)} \in L(\mathbb{T})$ with supports on $[1/4, 1/2]$ such that $\|f_1^{(j)}\| \le 1$, $\|f_2^{(j)}\| \le 1$,

$$\limsup_{j\to\infty} |\gamma^{-1}(2^j)s_{2^j-1}(f_1^{(j)}, 0)| = \infty, \tag{10.88}$$

$$\limsup_{j\to\infty} |\gamma^{-1}(2^j)\widetilde{s}_{2^j-1}(f_2^{(j)}, 0)| = \infty. \tag{10.89}$$

Define these sequences on the segment $[-1/2, 1/2]$:

$$f_1^{(j)}(x) = \begin{cases} (-1)^l, & \text{if } x \in [2^{-j}l, 2^{-j}(l+1)],\ 2^{j-2} \le l \le \frac{1}{3}2^{j-1}, \\ 0, & \text{otherwise}, \end{cases}$$

$$f_2^{(j)}(x) = \begin{cases} \text{sign}\cos \pi 2^j y, & \text{if } x \in [1/4, 1/3], \\ 0, & \text{otherwise}. \end{cases}$$

By (10.86), (10.87) we have

$$\begin{aligned} s_{2^j-1}(f_1^{(j)}, 0) &= \sum_{k=0}^{2^{j-1}} \langle f_1^{(j)}, S_j^k \widetilde{\varphi}_j\rangle \varphi_j(2^{-j}k) \\ &= 2^{-j}\frac{\pi}{4} \sum_{2^{j-2}\le k\le \frac{1}{3}2^{j-1}} \text{ctan}\,\pi 2^{-j}(k+1/2) + O\left(\sum_{k=2^{j-2}}^{2^{j-1}} \frac{1}{k^2}\right) \\ &\ge 2^{-j}\frac{\pi}{4} \sum_{2^{j-2}\le k\le \frac{1}{3}2^{j-1}} \frac{1/2}{\pi 2^{-j}(k+1/2)} + O\left(2^{-j}\right) \ge \frac{1}{16} + O(2^{-j}). \end{aligned}$$

It follows that (10.88) holds. Taking into account that $S_j^k\widetilde{\varphi}_j(0) = 0$ for all $k = 1, \dots, 2^j - 1$ and using (10.86), (10.87), for $j > 2$, we obtain

$$\begin{aligned}\widetilde{s}_{2^j-1}(f_2^{(j)}, 0) &= \langle f_2^{(j)}, \widetilde{\varphi}_j \rangle \\ &= \frac{\pi}{4}\int_{1/4}^{1/3} \operatorname{ctan}\pi(y + 2^{-j-1})|\cos\pi 2^j y|\,dy + O(2^{-j}) \\ &\geq \frac{\pi}{4}\operatorname{ctan}\frac{11\pi}{24}\int_{1/4}^{1/3} |\cos\pi 2^j y|\,dy + O(2^{-j}) \geq M + O(2^{-j}),\end{aligned}$$

where M is a positive constant.

The classes $\mathcal{L}_\alpha(x_0)$ consist of the functions whose integral means of deviations from some constant tend to zero sufficiently rapidly. Introduce a generalization of these classes by replacing constants by algebraic polynomials.

DEFINITION 10.4.5. Let $x_0 \in \mathbb{R}$, $\alpha > 0$, $r \in \mathbb{Z}_+$, $0 \leq r \leq \alpha$. A function f belongs to the class $\mathcal{L}_{r,\alpha}(x_0)$ if there exists a polynomial P of order r such that

$$\frac{1}{h}\int_{|x-x_0|\leq h} |f(x) - P(x - x_0)|\,dx = O(h^\alpha).$$

THEOREM 10.4.6. *Suppose a PMRA $\{V_j\}_{j=0}^\infty$ is generated by a scaling function φ with integer orthonormal shifts and ψ is the corresponding wavelet function, $\psi \in C^m(\mathbb{R})$,*

$$|\varphi(x)| \leq \frac{C}{1 + |x|^{1+\varepsilon}}, \ \varepsilon > 0, \quad |\psi(x)| \leq \frac{C}{1 + |x|^n}, \ n > m + 1,$$

for all $x \in \mathbb{R}$. If $f \in \mathcal{L}_{r,\alpha}(x_0)$, $\alpha > 0$, $r < m$, then

$$|s_N(f, x_0) - P(0)| = O\left(N^{-\min\{\alpha, n-1\}}\right), \quad N \to \infty, \tag{10.90}$$

for $\alpha \neq n - 1$,

$$|s_N(f, x_0) - P(0)| = O\left(N^{-\alpha}\log N\right), \quad N \to \infty, \tag{10.91}$$

for $\alpha = n - 1$.

Proof. Under the conditions of the theorem, the scaling sequence φ_j corresponding to the function φ can be represented in the form

$$\varphi_j(x) = 2^{j/2}\sum_{l\in\mathbb{Z}} \varphi(2^j x + 2^j l),$$

and the systems $S_k^j\varphi_j$, $k = 0, \dots, 2^j - 1$, are orthonormal. The sequence of corresponding wavelet functions ψ_j can be represented in the form

$$\psi_j(x) = 2^{j/2}\sum_{l\in\mathbb{Z}} \psi(2^j x + 2^j l),$$

and the systems $S_k^j\psi_j$, $k = 0, \dots, 2^j - 1$, are orthonormal as well.

It is clear that x_0 is a Lebesgue point of the function f; hence by Theorem 10.2.2, the sequence $s_N(f, x_0)$ converges to $P(0)$. Therefore,

$$f(x_0) - s_N(f, x_0) = (s_{2^j-1}(f, x_0) - s_N(f, x_0)) + \sum_{i=j}^\infty (s_{2^{i+1}-1}(f, x_0) - s_{2^i-1}(f, x_0)).$$

Consequently, to prove the theorem, it suffices to check that

$$s_{2^j+L}(f,x_0) - s_{2^j-1}(f,x_0) = O\left(2^{-j\min\{\alpha,1-n\}}\right) \tag{10.92}$$

for all sufficiently large $j \in \mathbb{Z}_+$ and $L = 0, 1, \ldots, 2^j - 1$. Using Lemmas 10.2.1 and 10.1.1 and Theorem 1.7.7, we get

$$\begin{aligned}
&|s_{2^j+L}(f,x_0) - s_{2^j-1}(f,x_0)| \\
&\quad \le 2^j \int_{-\infty}^{\infty} |f(x) - P(x-x_0)| \sum_{\nu\in\mathbb{Z}} |\psi(2^j x+\nu)||\psi(2^j x_0+\nu)|\, dx \\
&\quad \le C_1 2^j \int_{-\infty}^{\infty} |f(x) - P(x-x_0)| \eta(2^{j-3}(x-x_0))\, dx,
\end{aligned} \tag{10.93}$$

where $\eta(u) = C(1+|u|^n)^{-1}$. Choose positive integer j_0 such that

$$\frac{1}{h} \int_{|x-x_0|\le h} |f(x) - P(x-x_0)|\, dx \le C\, h^\alpha, \quad 0 < h \le 2^{-j_0}.$$

The monotonicity of η implies that for all $j \ge j_0$ we have

$$\begin{aligned}
&|s_{2^j+L}(f,x_0) - s_{2^j-1}(f,x_0)| \\
&\quad \le A 2^j \Bigg(\eta(0) \int_{|x-x_0|\le 2^{-j}} |f(x) - P(x-x_0)|\, dt \\
&\qquad + \sum_{k=-j+1}^{-j_0} \eta(2^{j+k-3}) \int_{|x-x_0|\le 2^k} |f(x) - P(x-x_0)|\, dt \\
&\qquad + \sum_{k=-j_0+1}^{\infty} \eta(2^{j+k-3}) \int_{|x-x_0|\le 2^k} |f(x) - P(x-x_0)|\, dt \Bigg) \\
&\quad =: \Sigma_0 + \Sigma_1 + \Sigma_2.
\end{aligned} \tag{10.94}$$

The sums Σ_0, Σ_1 can be estimated similarly to the corresponding sums in (10.71),

$$\Sigma_0 + \Sigma_1 = O\left(2^{-j\min\{\alpha,n-1\}}\right). \tag{10.95}$$

Using the obvious relations

$$\int_{|x-x_0|\le h} |f(x) - P(x-x_0)|\, dt = O(h), \quad 0 < h \le 1,$$

$$\int_{|x-x_0|\le h} |f(x) - P(x-x_0)|\, dt = O(h^{r+1}), \quad h \ge 1,$$

we get

$$\begin{aligned}\Sigma_2 &= O(1)\left(\sum_{k=-j_0+1}^{-1} 2^{j+k}\eta(2^{j+k-3}) + \sum_{k=0}^{\infty} 2^{j+k(r+1)}\eta(2^{j+k-3})\right) \\ &\quad + O(1)\Big(2^{j(1-n)} \sum_{k=-j_o+1}^{\infty} 2^{k(1-n)} + 2^{j(1-n)} \sum_{k=0}^{\infty} 2^{k(r-n+1)}\Big) \\ &= O\left(2^{-j\min\{\alpha,n-1\}}\right).\end{aligned}$$

Taking into account (10.93)–(10.95), we obtain (10.92). ◊

CHAPTER 11

Remarkable Properties of Wavelet Bases

11.1. Unconditional wavelet bases

In this section we shall discuss biorthogonal wavelet bases in $L_2(\mathbb{R})$. Theorem 1.1.5 proves that any Riesz basis is unconditional in $L_2(\mathbb{R})$. We now prove that for a wide class of wavelet functions, such bases are unconditional in the spaces $L_p(\mathbb{R})$, $1 < p < \infty$, as well. We shall consider not only wavelet bases generated by some PMRA, but also various bases of the form $\{\psi_{jk}\}$ with the dual system $\{\widetilde{\psi}_{jk}\}$, where ψ, $\widetilde{\psi}$ are functions that belong to the space $L_p(\mathbb{R})$ for any $p \in (1, \infty)$.

THEOREM 11.1.1. *Suppose $\{\psi_{jk}\}$ and $\{\widetilde{\psi}_{jk}\}$ are biorthonormal systems and $\{\psi_{jk}\}_{j,k\in\mathbb{Z}}$ is a Riesz basis in $L_2(\mathbb{R})$. Suppose there exists an even bounded function η decreasing on $[0, \infty)$, satisfying the condition*

$$\int_0^\infty \eta(x) \ln(1 + x)\, dx < \infty, \tag{11.1}$$

and such that

$$|\psi(x)|, |\widetilde{\psi}(x)| \le \eta(x) \tag{11.2}$$

for all $x \in \mathbb{R}$. Then the system $\{\psi_{jk}\}_{j,k\in\mathbb{Z}}$ is an unconditional basis in $L_p(\mathbb{R})$, $1 < p < \infty$.

The main difficulty in the proof of this theorem is to establish the uniform boundedness of the operator norms from $L_p(\mathbb{R})$ to $L_p(\mathbb{R})$ of finite partial sums of the series

$$\sum_{jk} \varepsilon_{jk} \langle f, \widetilde{\psi}_{jk} \rangle \psi_{jk} =: U_\varepsilon f$$

over all sequences of signs $\varepsilon = \{\varepsilon_{jk}\}_{j,k\in\mathbb{Z}}$, $\varepsilon_{jk} = \pm 1$ (see Appendix A.1). We shall prove this assertion by the following scheme. First consider functions $f \in L_1(\mathbb{R}) \cap L_2(\mathbb{R})$. For these functions, the following expansion takes place:

$$f = \sum_{j,k} \langle f, \widetilde{\psi}_{jk} \rangle \psi_{jk}$$

where the series converges in norm in $L_2(\mathbb{R})$. It is clear that the operator U_ε is defined and bounded in norm in $L_2(\mathbb{R})$ (see Corollary 1.8.6 and Theorem 1.1.2). We show that the restriction of the operator U_ε to $L_1(\mathbb{R}) \cap L_2(\mathbb{R})$ is of the weak type $(1, 1)$ and then use the Marcinkiewicz interpolation theorem to prove that this operator is of the strong type (p, p) for all $p \in (1, 2)$ (see Appendix A.4).

Next, it is not hard to check that the operator norms from $L_p(\mathbb{R}) \cap L_2(\mathbb{R})$ to $L_p(\mathbb{R})$ of finite partial sums are uniformly bounded by a constant independent of ε, and this property holds for the extension to the whole space $L_p(\mathbb{T})$, $1 < p < 2$.

It follows from the general properties of bases that it suffices to consider the case $p \in (1, 2)$.

Let us present some auxiliary statements. In what follows, we shall assume that functions ψ, $\widetilde{\psi}$, and η satisfy the conditions of Theorem 11.1.1. We shall denote by $C, C_1, C_2, \dots$ constants depending only on ψ, $\widetilde{\psi}$, and η. Define the operators P_j, Q_j acting from $L_2(\mathbb{R})$ to $L_2(\mathbb{R})$ by the equalities

$$P_j f := \sum_{i<j} \sum_{k\in\mathbb{Z}} \langle f, \widetilde{\psi}_{ik}\rangle \psi_{ik}, \qquad Q_j f := \sum_{i\geq j} \sum_{k\in\mathbb{Z}} \langle f, \widetilde{\psi}_{ik}\rangle \psi_{ik}.$$

It is clear that P_j and Q_j are projection operators onto the spaces

$$\overline{\operatorname{span}\{\psi_{ik}, i<j, k,j\in\mathbb{Z}\}} \quad \text{and} \quad \overline{\operatorname{span}\{\psi_{ik}, i\geq j, k,j\in\mathbb{Z}\}},$$

respectively; they commute with the operator U_ε, and for all $f \in L_2(\mathbb{R})$, the equality

$$Q_j f = f - P_j f \tag{11.3}$$

holds.

LEMMA 11.1.2. *If $f \in L_2(\mathbb{R})$ and $\operatorname{supp} f \subset [0,1]$, then for $|x| > 10$, we have the inequality*

$$|U_\varepsilon Q_0 f(x)| \leq C\|f\|_1 \sum_{j\geq 0} 2^j \eta(2^{j-4}x). \tag{11.4}$$

Proof. First note that the function f is summable on $\mathbb{R}$ since it has compact support. We have

$$U_\varepsilon Q_0 f(x) = \sum_{j\geq 0} \sum_{k\in\mathbb{Z}} \varepsilon_{jk} \langle f, \widetilde{\psi}_{jk}\rangle \psi_{jk}.$$

Divide this sum according to the scheme

$$\sum_{j\geq 0}\sum_{k\in\mathbb{Z}} = \sum_{j\geq 0}\sum_{|k|<5\cdot 2^j} + \sum_{j\geq 0}\sum_{|k|\geq 5\cdot 2^j} =: \Sigma_1(x) + \Sigma_2(x), \tag{11.5}$$

and estimate each sum. Note that

$$\sum_{k\in\mathbb{Z}} |\psi(2^j t + k)| \leq \sum_{k\in\mathbb{Z}} \eta(2^j t + k) \leq 2\eta(0) + \|\eta\|_1, \tag{11.6}$$

for all $j \in \mathbb{Z}$ and $t \in \mathbb{R}$, and

$$|\widetilde{\psi}(2^j x + k)| \leq \eta(2^j x + k) \leq \eta(2^{j-1}x) \tag{11.7}$$

if $|k| < 5\cdot 2^j$, $|x| > 10$. These relations imply that

$$\begin{aligned}
|\Sigma_1(x)| &\leq \sum_{j\geq 0}\sum_{|k|<5\cdot 2^j} \int_{\mathbb{R}} |f(t) 2^{j/2} \widetilde{\psi}(2^j t + k)|\, dt\, 2^{j/2} \eta(2^j x + k) \\
&\leq \sum_{j\geq 0} 2^j \eta(2^{j-1}x) \int_{\mathbb{R}} |f(t)| \sum_{|k|<5\cdot 2^j} |\widetilde{\psi}(2^j t + k)|\, dt \\
&\leq (2\eta(0) + \|\eta\|_1) \sum_{j\geq 0} 2^j \eta(2^{j-1}x) \int_{\mathbb{R}} |f(t)|\, dt \leq C_1 \|f\|_1 \sum_{j\geq 0} 2^j \eta(2^{j-4}x).
\end{aligned}$$

Estimate the sum $\Sigma_2(x)$:

$$
\begin{aligned}
|\Sigma_2(x)| &\le \sum_{j\ge 0}\sum_{|k|\ge 5\cdot 2^j} 2^j \int_0^1 |f(t)||\widetilde{\psi}(2^j t+k)|\,dt\,|\psi(2^j x+k)| \\
&\le \sum_{j\ge 0}\sum_{|k|\ge 5\cdot 2^j} 2^j \|f\|_1 \sup_{t\in[0,1]} \eta(2^j t+k)\,\eta(2^j x+k) \\
&\le \|f\|_1 \sum_{j\ge 0} 2^j \left(\sum_{k\ge 5\cdot 2^j} \eta(k)\eta(2^j x+k) + \sum_{k\le -5\cdot 2^j} \eta(k+2^j)\eta(2^j x+k)\right) \\
&\le \|f\|_1 \sum_{j\ge 0} 2^j \left(\sum_{k\in\mathbb{Z}} \eta(k)\eta(2^j x+k) + \sum_{k\in\mathbb{Z}} \eta(k)\eta(2^j(x-1)+k)\right).
\end{aligned}
$$

Applying Lemma 1.7.1 to the right-hand side of the latter inequality and taking into account that $\eta(2^{j-3}(x-1)) \le \eta(2^{j-4}x)$ for $|x|>10$, we get

$$|\Sigma_2(x)| \le C_2 \|f\|_1 \sum_{j\ge 0} 2^j \eta(2^{j-4}x). \qquad \lozenge$$

LEMMA 11.1.3. *The function* $F(x) := \sum_{j\ge 0} 2^j \eta(2^j x)$ *is even, decreasing on* $(0,\infty)$, *summable and bounded on each interval* (δ,∞), $\delta>0$.

Proof. Let $x>\delta$. It follows from the monotonicity of η that

$$
\begin{aligned}
\sum_{j\ge 0} 2^j \eta(2^j x) &\le \sum_{j\in\mathbb{Z}} 2^j \eta(2^j x) = 2\sum_{j\in\mathbb{Z}} \eta(2^j x) \int_{2^{j-1}}^{2^j} dt \\
&\le 2\sum_{j\in\mathbb{Z}} \int_{2^{j-1}}^{2^j} \eta(tx)\,dt \le 2\int_0^\infty \eta(tx)\,dt \le \frac{1}{x}\|\eta\|_1 \le \frac{1}{\delta}\|\eta\|_1.
\end{aligned}
\tag{11.8}
$$

Since the function $\eta(x)$ is even, nonnegative, and decreasing on $(0,\infty)$, the function F has the same properties. It remains to check that $\int_\delta^\infty F(x)\,dx < \infty$ for any $\delta>0$. Using the monotonicity of η and (11.1) and assuming $\delta<1$, we get

$$
\begin{aligned}
\int_\delta^\infty F(x)\,dx &= \sum_{j\ge 0} \int_\delta^\infty \eta(2^j x) 2^j\,dx = \sum_{j\ge 0} \int_{2^j\delta}^\infty \eta(u)\,du \\
&= \sum_{j\ge 0}\sum_{k\ge j} \int_{2^k\delta}^{2^{k+1}\delta} \eta(u)\,du = \sum_{k\ge 0} (k+1) \int_{2^k\delta}^{2^{k+1}\delta} \eta(u)\,du \\
&\le (1-\log\delta)\|\eta\|_1 + \sum_{k\ge 0} \log 2^k\delta \int_{2^k\delta}^{2^{k+1}\delta} \eta(u)\,du \\
&\le C_1(\delta) + \int_\delta^\infty \log(u+1)\eta(u)\,du =: C_2(\delta). \qquad \lozenge
\end{aligned}
$$

In what follows, we shall need binary intervals. Introduce the following notation:

$$I_{jk} := [-2^{-j}k, 2^{-j}(1-k)], \quad j,k \in \mathbb{Z}.$$

We shall identify any interval I_{jk} with the ordered pair of its numbers (j,k).

COROLLARY 11.1.4. *Let $j,k \in \mathbb{Z}$, $|2^j x + k| > 10$, $f \in L_1(\mathbb{R}) \cap L_2(\mathbb{R})$, $\operatorname{supp} f \subset I_{jk}$. Then*

$$|U_\varepsilon Q_j f(x)| \le C\|f\|_1 2^j \alpha(2^j x + k) \tag{11.9}$$

where the function $\alpha(x)$ is even, summable and decreasing on $(10,\infty)$.

Proof. By an appropriate change of variables in (11.9), we reduce the required statement to the case $j = 0, k = 0$. In this case, the statement of the corollary immediately follows from Lemmas 11.1.2 and 11.1.3. ◊

LEMMA 11.1.5. *Let $f \in L_1(\mathbb{R})$; then*

$$\|U_\varepsilon P_j f\|_\infty \le C \cdot 2^j \cdot \|f\|_1. \tag{11.10}$$

Proof. By an appropriate change of variables in (11.10), we reduce the required statement to the case $j = 0$. Since η is bounded, using (11.6), we have

$$|U_\varepsilon P_0 f(x)| \le \sum_{i<0} 2^i \sum_{k\in\mathbb{Z}} \int_{\mathbb{R}} |f(t)||\widetilde{\psi}(2^i t + k)|\, dt\, |\psi(2^i x + k)|$$

$$\le C_1 \sum_{i<0} 2^i \int_{\mathbb{R}} |f(t)| \sum_{k\in\mathbb{Z}} |\widetilde{\psi}(2^i t + k)|\, dt \le C \sum_{i<0} 2^i \|f\|_1 = C\|f\|_1. \qquad ◊$$

LEMMA 11.1.6. *Let $f \in L_2(\mathbb{R})$, $\operatorname{supp} f \subset I_{jk}$. Then*

$$|P_j f(x)| \le 2^j \|f\|_1 \beta(2^j x + k), \tag{11.11}$$

where β is an even decreasing function that is summable on $[0,\infty)$ and possesses the property

$$\beta(2^l x) \le 2^{4-l}\beta(x) \quad \textit{for } l \in \mathbb{N}, |x| \ge 1. \tag{11.12}$$

Proof. By an appropriate change of variables in (11.11) we reduce the statement to the case $j = 0, k = 0$. Set $\beta(x) = \text{const}$ for $|x| < 10$. We can assume that this constant is greater than $\alpha(10)$, where α is the function from Corollary 11.1.4. For $|x| \ge 10$, it follows from equality (11.3) that $P_0 f(x) = -Q_0 f(x)$. Setting $\beta(x) = \alpha(x)$ for $|x| \ge 10$ and using Corollary 11.1.4, we derive (11.11). Relation (11.12) follows from the obvious inequality $\alpha(2x) \le \frac{1}{2}\alpha(x)$ for $|x| \ge 10$. ◊

In what follows, we shall denote by f^{jk} the restriction of a function f to the binary interval I_{jk}; i.e., $f^{jk} := f\chi_{I_{jk}}$.

LEMMA 11.1.7. *Suppose $f \in L_1(\mathbb{R}) \cap L_2(\mathbb{R})$, $\mathcal{S}$ is a set of mutually disjoint binary intervals I_{jk}, and the inequality $\|f^{jk}\|_1 \le 2^{1-j}\lambda$ holds for any $(j,k) \in \mathcal{S}$. Then*

$$\int_{\mathbb{R}} \Big| \sum_{(j,k)\in\mathcal{S}} P_j f^{jk} \Big|^2 \le C\lambda \|f\|_1. \tag{11.13}$$

Proof. Rewrite the left-hand side of (11.13) in the form

$$\int_{\mathbb{R}} \sum_{(j,k)\in S} \sum_{(j',k')\in S} P_j f^{jk} \overline{P_{j'} f^{j'k'}} =: J.$$

Using Lemma 11.1.6 and making the change of variables in the integral, we get

$$\begin{aligned}
J &\le 2 \sum_{(j',k')\in S} \sum_{\substack{(j,k)\in S\\ j\ge j'}} \int_{\mathbb{R}} \left| P_{j'} f^{j'k'}(x) P_j f^{jk}(x) \right| dx \\
&\le 2 \sum_{(j',k')\in S} 2^{j'} \|f^{j'k'}\|_1 \sum_{\substack{(j,k)\in S\\ j\ge j'}} 2^j \|f^{jk}\|_1 \int_{\mathbb{R}} \beta(2^{j'}x+k')\beta(2^j x+k)\,dx \\
&\le 4\lambda \sum_{(j',k')\in S} 2^{j'} \|f^{j'k'}\|_1 \sum_{\substack{(j,k)\in S\\ j\ge j'}} \int_{\mathbb{R}} \beta(2^{j'}x+k')\beta(2^j x+k)\,dx \\
&= 4\lambda \sum_{(j',k')\in S} \|f^{j'k'}\|_1 \sum_{\substack{(j,k)\in S'\\ j\ge 0}} \int_{\mathbb{R}} \beta(x)\beta(2^j x+k)\,dx,
\end{aligned} \tag{11.14}$$

where the set $S' = S'(j',k')$ consists of the binary intervals $I_{j-j',k-2^{j-j'}k'}$, $(j,k) \in S$. It is clear that these intervals are mutually disjoint.

Prove the uniform boundedness of the inner sum on the right-hand side of (11.14). Denote by S_{nj} the set of binary intervals I_{jk}, $j \ge 0$, $(j,k) \in S'$, contained in the segment $[n, n+1]$ and by κ_{nj} the number of elements in S_{nj}. Then

$$\begin{aligned}
\sum_{\substack{(j,k)\in S'\\ j\ge 0}} \int_{\mathbb{R}} \beta(x)\beta(2^j x+k)\,dx &= \sum_{n\in\mathbb{Z}} \sum_{j=0}^{\infty} \sum_{(j,k)\in S_{nj}} \int_{\mathbb{R}} \beta(x)\beta(2^j x+k)\,dx \\
&= \sum_{n\in\mathbb{Z}} \int_{\mathbb{R}} \sum_{j=0}^{\infty} \sum_{(j,k)\in S_{nj}} \beta(x)\beta(2^j x+k)\,dx.
\end{aligned}$$

Partition the latter integral as

$$\int_{\mathbb{R}} = \int_{n-1}^{n+2} + \int_{-\infty}^{n-1} + \int_{n+2}^{\infty} =: J_{n1} + J_{n2} + J_{n3}.$$

Using the inequality

$$\begin{aligned}
J_{n1} &\le \max_{[n-1,n+2]} \beta \sum_{j=0}^{\infty} \sum_{(j,k)\in S_{nj}} \int_{\mathbb{R}} \beta(2^j x+k)\,dx \\
&= \max_{[n-1,n+2]} \beta \sum_{j=0}^{\infty} 2^{-j}\kappa_{nj} \int_{\mathbb{R}} \beta(x)\,dx \le \max_{[n-1,n+2]} \beta \int_{\mathbb{R}} \beta(x)\,dx
\end{aligned}$$

and taking into account the monotonicity and summability of the function β, we deduce that the sum $\sum_{n\in\mathbb{Z}} J_{n1}$ is uniformly bounded. To estimate the integrals J_{n2}, J_{n3}, notice that the inclusion $I_{jk} \subset [n, n+1]$ implies the inequality $-2^j < 2^j n + k \le 0$. If $x \le n-1$, then $2^j(x-n) \le -2^j$; hence $2^j x + k = 2^j(x-n) + 2^j n + k \le 2^j(x-n) \le 2^{j-1}(x-n)$, which implies that

$$\beta(2^j x+k) \le \beta(2^{j-1}(x-n)). \tag{11.15}$$

If $x \geq n+2$, then $2^j(x-n) \geq 2^{j+1}$; hence $2^j x + k = 2^j(x-n) + 2^j n + k \geq 2^{j-1}(x-n) > 0$, which again implies (11.15). Using (11.15) and (11.12), we get

$$J_{n2} \leq \sum_{j=0}^{\infty} \sum_{(j,k)\in S_{nj}} \int_{-\infty}^{n-1} \beta(x)\beta(2^{j-1}(n-x))\,dx$$

$$= \sum_{j=0}^{\infty} \sum_{(j,k)\in S_{nj}} \int_{1}^{\infty} \beta(n-x)\beta(2^{j-1}x)\,dx$$

$$\leq \sum_{j=0}^{\infty} 2^{5-j}\kappa_{nj} \int_{1}^{\infty} \beta(n-x)\beta(x)\,dx \leq 2^5 \int_{1}^{\infty} \beta(n-x)\beta(x)\,dx.$$

Due to the monotonicity and summability of the function β, it follows that

$$\sum_{n\in\mathbb{Z}} J_{n2} \leq 2^5 \sum_{n\in\mathbb{Z}} \int_{1}^{\infty} \beta(n-x)\beta(x)\,dx \leq C.$$

Analogously, we have

$$\sum_{n\in\mathbb{Z}} J_{n3} \leq \sum_{n\in\mathbb{Z}} \sum_{j=0}^{\infty} \sum_{(j,k)\in S_{nj}} \int_{2}^{\infty} \beta(n+x)\beta(2^{j-1}x)\,dx \leq 2^5 \sum_{n\in\mathbb{Z}} \int_{2}^{\infty} \beta(n+x)\beta(x)\,dx \leq C.$$

Combining the above estimates with (11.14), we obtain the required statement. $\Diamond$

LEMMA 11.1.8 (Calderon-Zygmund decomposition). *For any function $f \in L_1(\mathbb{R})$ and any $\lambda > 0$, there exists a set $\mathcal{S}$ of pairwise disjoint binary intervals I_{jk} such that*

$$\lambda < \frac{1}{\mu I_{jk}} \int_{I_{jk}} |f| \leq 2\lambda \tag{11.16}$$

if $(j,k) \in \mathcal{S}$ and

$$|f(x)| \leq \lambda \tag{11.17}$$

if $(j,k) \notin \mathcal{S}$, for almost all $x \in I_{jk}$.

Proof. Choose $j_0 \in \mathbb{Z}$ such that $2^{j_0}\|f\|_1 \leq \lambda$. Then for any $k \in \mathbb{Z}$, the following inequality holds:

$$\frac{1}{\mu I_{j_0 k}} \int_{I_{j_0 k}} |f| \leq \lambda.$$

Each interval $I_{j_0 k}$ is a union of disjoint intervals $I_{j_0+1,2k}$ and $I_{j_0+1,2k-1}$. Let I be one of these intervals. There are two cases:

(i) $\frac{1}{\mu I}\int_I |f| > \lambda$,

(ii) $\frac{1}{\mu I}\int_I |f| \leq \lambda$.

In case (i), the interval I can be included in the set $\mathcal{S}$ because

$$\lambda < \frac{1}{\mu I}\int_I |f| \leq \frac{2}{\mu I_{j_0 k}} \int_{I_{j_0 k}} |f| \leq 2\lambda.$$

In case (ii), for I, we repeat the same procedure as for I_{j_0k}. Continuing this process, we obtain an infinite or a finite (if at some step we come to case (i)) sequence of embedded intervals which do not belong to $\mathcal{S}$. We include these intervals into the set $\mathcal{S}'$. Repeat the same process for every interval I_{j_0k}, then for every interval $I_{(j_0+1)k}$ that has not been embedded in either $\mathcal{S}$ or $\mathcal{S}'$, etc. Consider the set $F = \mathbb{R} \setminus \bigcup_{(j,k)\in\mathcal{S}} I_{jk}$. For any $x \in F$, there exists a sequence of pairs $(j, k_j) \in \mathcal{S}'$, $j = j_0, j_0+1, \ldots$, such that x belongs to every interval $I_{j,k_j} =: I'_j$. If x is a Lebesgue point of the function $|f|$, then

$$\lim_{j\to\infty} \left| \frac{1}{\mu I'_j} \int\limits_{I'_j} (|f(t)| - |f(x)|)\, dt \right| = 0.$$

Thus for any $\varepsilon > 0$, there exists $j \in \mathbb{N}$ such that

$$|f(x)| \le \frac{1}{\mu I'_j} \int\limits_{I'_j} |f(t)|\, dt + \varepsilon,$$

which implies (11.17) since ε is arbitrary. It remains to note that by the Lebesgue theorem almost all x are Lebesgue points of the summable function $|f|$. ◊

LEMMA 11.1.9. *Let $f \in L_2(\mathbb{R}) \cap L_1(\mathbb{R})$, $\lambda > 0$. Then*

$$\mu\{x \in \mathbb{R} : |U_\varepsilon f(x)| > \lambda\} \le \frac{C}{\lambda}\|f\|_1. \tag{11.18}$$

`Proof.` Let S be the set of binary intervals considered in Lemma 11.1.8. It follows from (11.16) that

$$\mu\left(\bigcup_{(j,k)\in\mathcal{S}} I_{jk}\right) \le \frac{1}{\lambda}\|f\|_1. \tag{11.19}$$

Set $F = \mathbb{R} \setminus \bigcup_{(j,k)\in\mathcal{S}} I_{jk}$. Represent f as the sum of functions

$$g := f \cdot \chi_F + \sum_{(j,k)\in\mathcal{S}} P_j(f^{jk}) \tag{11.20}$$

and

$$h := f - g = \sum_{(j,k)\in\mathcal{S}} f^{jk} - P_j(f^{jk}). \tag{11.21}$$

It is clear that

$$\mu\{x : |U_\varepsilon f(x)| > \lambda\} \le \mu\{x : |U_\varepsilon g(x)| > \lambda/2\} + \mu\{x : |U_\varepsilon h(x)| > \lambda/2\}. \tag{11.22}$$

From (11.17) it follows that

$$\int\limits_F |f(x)|^2\, dx \le \lambda \int\limits_F |f(x)|\, dx \le \lambda\|f\|_1.$$

Now, taking into account Lemma 11.1.7, we get $\|g\|_2 \le C_1\sqrt{\lambda\|f\|_1}$. Using inequality (A.3) (see Appendix A.4), we have

$$\begin{aligned} &\mu\{x : |U_\varepsilon g(x)| > \lambda/2\} = \mu\{x : |U_\varepsilon g(x)|^2 > \lambda^2/4\} \\ &\le \frac{4}{\lambda^2}\|U_\varepsilon g\|_2^2 \le \frac{4}{\lambda^2}\|g\|_2^2 \le \frac{C_2}{\lambda^2}\lambda\|f\|_1 = \frac{C_2}{\lambda}\|f\|_1. \end{aligned} \tag{11.23}$$

Set $A = \bigcup_{(j,k)\in\mathcal{S}} \bigcup_{l=-9}^{10} I_{j,k+l}$. From (11.19) it follows that

$$\mu A \le \frac{20}{\lambda}\|f\|_1. \tag{11.24}$$

On the other hand,

$$\int_{\mathbb{R}\setminus A} |U_\varepsilon h| \le \sum_{(j,k)\in\mathcal{S}} \int_{\mathbb{R}\setminus A} |U_\varepsilon Q_j(f^{jk})| \le \sum_{(j,k)\in\mathcal{S}} \int_{\mathbb{R}\setminus\bigcup_{l=-9}^{10} I_{j,k+l}} |U_\varepsilon Q_j(f^{jk})|.$$

Using Corollary 11.1.4, we obtain

$$\int_{\mathbb{R}\setminus A} |U_\varepsilon h| \le C_3 \sum_{(j,k)\in\mathcal{S}} \|f^{jk}\|_1 \le C_3\|f\|_1. \tag{11.25}$$

Using (A.3) and relations (11.24), (11.25), we get

$$\begin{aligned} \mu\{x : |U_\varepsilon h(x)| > \lambda/2\} &\le \mu A + \mu\{x \in \mathbb{R}\setminus A : |U_\varepsilon h(x)| > \lambda/2\} \\ &\le \frac{20}{\lambda}\|f\|_1 + \frac{2C_3}{\lambda}\|f\|_1 = \frac{C_4}{\lambda}\|f\|_1. \end{aligned} \tag{11.26}$$

To complete the proof, it remains to use (11.22), (11.23), and (11.26). ◊

COROLLARY 11.1.10. *Let $U_{\varepsilon,\Omega} f := \sum_{(j,k)\in\Omega} \varepsilon_{jk}\langle f, \widetilde{\psi}_{jk}\rangle\psi_{jk}$, $p \in (1,2)$. Then there exists a constant M depending only on p, ψ, η and such that*

$$\|U_{\varepsilon,\Omega} f\|_p \le M\|f\|_p \tag{11.27}$$

for all finite sets $\Omega \subset \mathbb{Z}\times\mathbb{Z}$, functions $f \in L_p(\mathbb{R})$, and sequences of signs ε.

Proof. First assume that $f \in L_2(\mathbb{R}) \cap L_p(\mathbb{R})$. In this case, (11.27) follows from Lemma 11.1.9, Theorem A.4.2, and Remark A.4.3 (see Appendix A.4) since the operator $U_{\varepsilon,\Omega}$ can be represented as the half-sum of the operators $U_{\varepsilon'}$, $U_{\varepsilon''}$. An arbitrary function $f \in L_p(\mathbb{R})$ can be approximated by functions from $L_2(\mathbb{R})\cap L_p(\mathbb{R})$. Since Ω is finite, the operator $U_{\varepsilon,\Omega}$ is bounded in $L_p(\mathbb{R})$. Passing to the limit, we obtain (11.27) for f. ◊

Proof of Theorem 11.1.1. By Theorem A.1.8 (see Appendix A.1), it suffices to check that the system $\{\psi_{jk}\}$ is complete and minimal and that the operator norms from $L_p(\mathbb{R})$ to $L_p(\mathbb{R})$ of finite partial sums of the series

$$\sum_{jk} \varepsilon_{jk}\langle f, \widetilde{\psi}_{jk}\rangle\psi_{jk}$$

are uniformly bounded. The latter property was established in Corollary 11.1.10. Minimality follows from the biorthogonality of the systems $\{\psi_{jk}\}$, $\{\widetilde{\psi}_{jk}\}$ (see Appendix A.1, Theorem A.1.3). It remains to prove completeness.

Let $p \in (1,2)$. Set

$$\sigma_j(f,x) = \sum_{|i|<j} \sum_{|k|<2^j} \langle f, \widetilde{\psi}_{ik}\rangle\psi_{ik}$$

and show that for all $f \in L_p(\mathbb{R})$,

$$\|f - \sigma_j(f)\|_p \underset{j\to+\infty}{\longrightarrow} 0. \tag{11.28}$$

Note that for $p = 2$, this relation holds since $\{\psi_{jk}\}$ is an unconditional basis in $L_2(\mathbb{R})$. First assume that the function f is continuous and compactly supported, $\operatorname{supp} f \subset [-R, R]$. Then Hölder's inequality implies that

$$\int_{-2R}^{2R} |f(x) - \sigma_j f(x)|^p \, dx \le (4R)^{1-p/2} \|f - \sigma_j\|_2^p \underset{j\to+\infty}{\longrightarrow} 0 \tag{11.29}$$

and

$$\int_{|x|>2R} |f(x) - \sigma_j f(x)|^p \, dx \le \|f - \sigma_j\|_2^{(p-1)} \left(\int_{|x|>2R} |f(x) - \sigma_j f(x)|^q \, dx \right)^{1/q}, \tag{11.30}$$

where $q = 2/(3-p)$. Prove that the integrals

$$\int_{|x|>2R} |f(x) - \sigma_j f(x)|^q \, dx$$

are uniformly bounded with respect to j. Using Lemme 1.7.1 and taking into account the monotonicity of the function η, we get

$$\begin{aligned}
\int_{|x|>2R} |f(x) - \sigma_j f(x)|^q \, dx &= \int_{|x|>2R} |\sigma_j f(x)|^q \, dx \\
&\le \int_{|x|>2R} dx \left(\int_{-R}^{R} |f(y)| \sum_{i\in\mathbb{Z}} \sum_{k\in\mathbb{Z}} 2^i \eta(2^i x + k)\eta(2^i y + k) \, dy \right)^q \\
&\le C_1^q \int_{|x|>2R} dx \left(\int_{-R}^{R} |f(y)| \sum_{i\in\mathbb{Z}} 2^i \eta\left(2^{i-3}(x-y)\right) \, dy \right)^q \\
&\le (C_1 \|f\|_1)^q \int_{|x|>2R} \left(\sum_{i\in\mathbb{Z}} 2^i \eta\left(2^{i-4} x\right) \right)^q \, dx.
\end{aligned}$$

In view of (11.8) it follows that

$$\int_{|x|>2R} |f(x) - \sigma_j f(x)|^q \, dx \le (C_2 \|f\|_1 \|\eta\|_1)^q \int_{|x|>2R} \frac{dx}{|x|^q}.$$

To prove (11.28) in the general case, it remains to approximate an arbitrary function $f \in L_p(\mathbb{R})$ with respect to the norm in $L_p(\mathbb{R})$ by compactly supported continuous functions and to take into account Corollary 11.1.10.

We have proved the theorem for $p \in (1, 2)$. In view of Corollary 1.8.6 the system $\{\widetilde{\psi}_{jk}\}$ is also a Riesz basis in $L_2(\mathbb{R})$. Hence switching the roles of the systems, we obtain an analogous assertion for $\{\widetilde{\psi}_{jk}\}$. It follows from Theorem A.1.6 (see Appendix A.1) and from the biorthonormality of the systems $\{\psi_{jk}\}$, $\{\widetilde{\psi}_{jk}\}$ that the theorem holds for $2 < p < \infty$ as well. ◊

11.2. Optimal polynomial bases in the space $C(\mathbb{T})$

In this section, we construct orthogonal bases consisting of trigonometric polynomials with minimum possible growth of degrees in the space of continuous periodic functions.

Suppose $0 < \varepsilon \le 1/6$, $\varphi^M = \varphi$ is a scaling function for the Meyer MRA (see §1.4), $\widehat{\varphi} \in C^2(\mathbb{R})$ is its Fourier transform, and ψ is the corresponding wavelet function. Recall that φ has the following properties:

$$\operatorname{supp}\widehat{\varphi} \subset \left[-\frac{1}{2}(1+2\varepsilon), \frac{1}{2}(1+2\varepsilon)\right]; \tag{11.31}$$

$\widehat{\varphi} \equiv 1$ on $[-\frac{1}{2}(1-2\varepsilon), \frac{1}{2}(1-2\varepsilon)]$, $\widehat{\varphi}^2(\xi) + \widehat{\varphi}^2(\xi+1) = 1$ for all $\xi \in [-\frac{3}{2}+\varepsilon, 0]$; the 1-periodic function defined on $[-1/2, 1/2]$ by the equality

$$m_0(\xi) = \widehat{\varphi}(2\xi) \tag{11.32}$$

is its mask; and

$$\widehat{\psi}(\xi) = e^{\pi i \xi} m_0\left(\frac{\xi+1}{2}\right)\widehat{\varphi}\left(\frac{\xi}{2}\right). \tag{11.33}$$

By Proposition 9.4.1, φ generates a PMRA $\{V_j\}_{j=0}^{\infty}$ in $L_2(\mathbb{T})$. The functions

$$\varphi_j(x) := 2^{-j/2}\sum_{l\in\mathbb{Z}} \widehat{\varphi}(2^{-j}l)e^{2\pi i l x}$$

form an orthogonalized scaling sequence, and the functions

$$\psi_j(x) := 2^{-j/2}\sum_{l\in\mathbb{Z}} \widehat{\psi}(2^{-j}l)e^{2\pi i l x}$$

form the corresponding sequence of wavelet functions. In view of relation (11.31), the functions φ_j are trigonometric polynomials and $\deg \varphi_j \le 2^{j-1}(1+2\varepsilon)$.

Let $j, s \in \mathbb{N}$, $s \le j$, ν be a binary vector with coordinates ν_k equal to either 1 or 0, $k = j-s+1, \dots, j$. Define functions $\psi^{j,s,\nu}$ by the equalities

(11.34)

$$\widehat{\psi^{j,s,\nu}}(\xi) = 2^{s/2}\prod_{k=j-s+1}^{j} (m_0(2^{j-k}\xi))^{\nu_k}\left(e^{2\pi i 2^{j-k}\xi} m_0\left(2^{j-k}\xi + \frac{1}{2}\right)\right)^{1-\nu_k}\widehat{\psi}(\xi)$$

and introduce their periodizations

$$\psi_j^{s,\nu}(x) := 2^{-j/2}\sum_{l\in\mathbb{Z}} \widehat{\psi^{j,s,\nu}}(2^{-j}l)e^{2\pi i l x}. \tag{11.35}$$

It follows from the results of §9.3 that the functions $\psi_j^{s,\nu}$ generate a wavelet packet (where the multipliers Δ_k, Λ_k are defined by the equalities

$$\widehat{\Delta_k f}(r) = \sqrt{2}m_0(2^{-k}r)\widehat{f}(r), \qquad \widehat{\Lambda_k f}(r) = \sqrt{2}e^{2\pi i 2^{-k}r}m_0(2^{-k}r + \frac{1}{2})\widehat{f}(r));$$

i.e., the wavelet spaces W_j are decomposed into the direct sum

$$W_j = \bigoplus_{\nu} W_j^{\nu}, \tag{11.36}$$

where

$$W_j^{\nu} = \operatorname{span}\{S_n^{j-s}\psi_j^{s,\nu}, n = 0, \dots, 2^{j-s}-1\}$$

and the system of functions $\{S_n^{j-s}\psi_j^{s,\nu}\}_{n,\nu}$ is an orthonormal basis in W_j.

Let T be a trigonometric polynomial and let $\deg T \le m2^{k-1}(1+2\varepsilon)$, where $m \in \mathbb{N}$, $\varepsilon \le 1/4m$. Since the function m_0 is identically equal to zero on the set

$$\left\{\left[l+\frac{1}{2}-\frac{1}{4}(1-2\varepsilon), l+\frac{1}{2}+\frac{1}{4}(1-2\varepsilon)\right], l \in \mathbb{Z}\right\},$$

it follows that $\deg \Delta_k T \le (m2^{k-1}-2^{k-2})(1+2\varepsilon)$ for odd m and $\deg \Lambda_k T \le (m2^{k-1}-2^{k-2})(1+2\varepsilon)$ for even m. Taking into account that $\deg \psi_j \le 2^j(1+2\varepsilon)$ and assuming that $2\varepsilon \le 2^{-s-2}$, it is not hard to see that the functions $\psi_j^{s,\nu}$ (for fixed j, s) can be renumbered by the single index $l = 1, \dots, 2^s$, so that

$$\deg \psi_j^{s,l} \le (2^{j-1} + l2^{j-s-1})(1+2\varepsilon), \tag{11.37}$$

setting $l = 2^s - (\alpha_1 2^{s-1} + \alpha_2 2^{s-2} + \dots + \alpha_s)$, where

$$\begin{aligned} \alpha_1 &= \begin{cases} 0, & \text{if } \nu_j = 1, \\ 1, & \text{if } \nu_j = 0, \end{cases} \\ \alpha_{k+1} &= \begin{cases} 0, & \text{if } \alpha_k = 1, \nu_{j-k} = 0 \text{ or } \alpha_k = 0, \nu_{j-k} = 1, \\ 1, & \text{if } \alpha_k = 0, \nu_{j-k} = 0 \text{ or } \alpha_k = 1, \nu_{j-k} = 1. \end{cases} \end{aligned}$$

For given $s \in \mathbb{N}$, define the following sequence of trigonometric polynomials:

$$T_1 := \varphi_0; \tag{11.38}$$

for $j < s$, $n = 1, \dots, 2^j$,

$$T_{2^j+n} := \psi_j^{j,n}; \tag{11.39}$$

for $j \ge s$, $l = 1, \dots, 2^s$, $n = 1, \dots, 2^{j-s}$,

$$T_{2^j+(l-1)2^{j-s}+n} := S_{n-1}^{j-s}\psi_j^{s,l}. \tag{11.40}$$

Inequality (11.37) obviously implies that $\deg T_N \le \frac{1}{2}N(1+2\varepsilon)$ for $N \le 2^s$. Representing positive integer $N > 2^s$ as $N = 2^j + m2^{j-s} + n$, where $j \ge s$, $m = 0, \dots, 2^s - 1$, $n = 1, \dots, 2^{j-s}$, and using (11.37), we obtain

$$\begin{aligned} \deg T_N &\le (2^{j-1} + (m+1)2^{j-s-1})(1+2\varepsilon) \\ &= \frac{1}{2}(2^j + m2^{j-s} + 2^{j-s})(1+2\varepsilon) \le \frac{1}{2}N(1+2^{-s})(1+2\varepsilon). \end{aligned}$$

Given $\varepsilon_0 > 0$, we can choose a sufficiently large positive integer s and sufficiently small $\varepsilon > 0$ such that the inequality

$$\deg T_N \le \frac{1}{2}N(1+\varepsilon_0) \tag{11.41}$$

holds for all N. According to the above renumbering of the functions $\psi_j^{s,\nu}$, we replace ν by the single index $l = 1, \dots, 2^s$ in the designations of the spaces W_j^ν and functions $\psi^{j,s,\nu}$.

In this new notation, for $j \ge s$, we have

$$W_j^l = \operatorname{span}\{T_{2^j+(l-1)2^{j-s}+1}, \dots, T_{2^j+l2^{j-s}}\}, \quad l = 1, \dots, 2^s, \tag{11.42}$$

$$W_j = \bigoplus_{l=1}^{2^s} W_j^l = \operatorname{span}\{T_{2^j+1}, \dots, T_{2^{j+1}}\}, \tag{11.43}$$

and for $j < s$, we have

$$W_j^l = \operatorname{span}\{T_{2^j+l}\}, \quad l = 1, \dots, 2^j, \tag{11.44}$$

$$W_j = \bigoplus_{l=1}^{2^j} W_j^l = \operatorname{span}\{T_{2^j+1}, \dots, T_{2^{j+1}}\}. \tag{11.45}$$

By construction, the system of functions $T_{2^j+1}, \dots, T_{2^{j+1}}$ forms a wavelet packet and thus is an orthonormal basis in W_j. In view of Theorem 9.2.3 and property **MR2** of Definition 9.1.1, we have

$$V_{j+1} = V_j \oplus W_j, \quad L_2(\mathbb{T}) = V_0 \oplus W_0 \oplus W_1 \oplus W_2 \oplus \cdots . \tag{11.46}$$

Taking into account the equality $V_0 = \operatorname{span}\{T_1\}$, we deduce that the trigonometric polynomials T_n, $n = 1, 2 \dots$, form an orthonormal basis of the space $L_2(\mathbb{T})$. For any function $f \in L(\mathbb{T})$, the Fourier expansion with respect to this basis

$$\sum_{n=1}^{\infty} \langle f, T_n \rangle T_n \tag{11.47}$$

makes sense. Partial sums of this series will be denoted by $\sigma_N(f)$, $N \in \mathbb{N}$.

Our next aim is to show that the system T_n, $n = 1, 2 \dots$, is a basis in the space $C(\mathbb{T})$.

THEOREM 11.2.1. *For any function $f \in C(\mathbb{T})$, series* (11.47) *converges uniformly and for any $N \in \mathbb{N}$, we have the estimate*

$$\|f - \sigma_N(f)\| \le C \mathcal{E}_{[N/8]}(f)_\infty, \tag{11.48}$$

where C is an absolute constant and $\mathcal{E}_k(f)_\infty$ is the best approximation of the function f by trigonometric polynomials of order k (see Appendix A.15*).*

Proof. Let $N = 2^j + m\,2^{j-s} + r$, $\quad m = 0, \dots, 2^s - 1$, $r = 0, \dots, 2^{j-s} - 1$. Then

$$\begin{aligned} \sigma_N(f) &= \sum_{n=1}^{N} \langle f, T_n \rangle T_n \\ &= \sum_{n=1}^{2^j} \langle f, T_n \rangle T_n + \sum_{k=0}^{m-1} \sum_{n=0}^{2^{j-s}-1} \langle f, S_n^{j-s} \psi_j^{s,k+1} \rangle S_n^{j-s} \psi_j^{s,k+1} \\ &\quad + \sum_{n=0}^{r} \langle f, S_n^{j-s} \psi_j^{s,m+1} \rangle S_n^{j-s} \psi_j^{s,m+1}. \end{aligned} \tag{11.49}$$

It follows from relations (11.43), (11.45), (11.46) that the first sum on the right-hand side is the orthogonal projection of the function f on the space V_j. Its expansion with respect to the orthonormal basis $S_n^j \varphi_j$, $n = 0, \dots, 2^j - 1$, takes the form

$$\sum_{n=0}^{2^j-1} \langle f, S_n^j \varphi_j \rangle S_n^j \varphi_j =: \sigma_N^{(1)}(f).$$

Let $\nu \in \mathbb{Z}$, $|\nu| \le 2^{j-2}$, $f_\nu(x) := e^{2\pi i \nu x}$. By the construction of the functions φ_j, $\psi_j^{s,k}$ and the properties of φ, we deduce that $\widehat{\psi}_j^{s,k}(\nu) = 0$ for $k = 1, \dots, 2^s$,

$\widehat{\varphi}_j(\nu) = 2^{-j/2}$, $\widehat{\varphi}_j(\nu + 2^j n) = 0$ for $n \in \mathbb{Z}_+$. Therefore,

$$\sigma_N(f_\nu) = \sum_{n=0}^{2^j-1} \langle f_\nu, S_n^j \varphi_j \rangle S_n^j \varphi_j$$

$$= \sum_{l\in\mathbb{Z}} \widehat{\varphi}_j(l) \langle f_\nu, f_l \rangle \sum_{k\in\mathbb{Z}} \widehat{\varphi}_j(k) f_k \sum_{n=0}^{2^j-1} e^{2\pi i 2^{-j} n(k-l)}.$$

This equality, relation (2.7), and the orthonormality of the trigonometric system imply that $\sigma_N(f_\nu) = f_\nu$. Hence the operator σ_N does not change trigonometric polynomials of order at most 2^{j-2}; in particular, it does not change the polynomial T of best approximation of order 2^{j-2} of the function f, which implies that

$$\|f - \sigma_N(f)\| \le \|f - T - \sigma_N(f - T)\| \le (1 + \|\sigma_N\|)\mathcal{E}_{2^{j-2}}(f)_\infty,$$

where $\|\sigma_N\|$ is the operator norm for σ_N from $C(\mathbb{T})$ to $C(\mathbb{T})$. To prove the theorem, it remains to check the boundedness of the sequence $\|\sigma_N\|$. It follows from the assumption $\widehat{\varphi} \in C^2(\mathbb{R})$ and relations (11.31), (11.32), (11.33), (11.34) that

$$|\varphi(x)|, |\psi^{j,s,l}(x)| = O\left(\frac{1}{1+x^2}\right). \tag{11.50}$$

This relation, Theorem 10.1.2, and Remark 10.1.3 yield the uniform estimate for $\|\sigma_N^{(1)}\|$. The norms of the operators corresponding to the other sums over n on the right-hand side of (11.49) can be estimated in a similar way. ◊

COROLLARY 11.2.2. *For any $\varepsilon > 0$, in the space $C(\mathbb{T})$ there exists a basis consisting of mutually orthogonal trigonometric polynomials T_n, $n = 1, 2 \dots$, such that* $\deg T_n \le \frac{n}{2}(1 + \varepsilon)$.

This assertion follows from Theorem 11.2.1 (see Appendix A.1) and inequality (11.41).

The order of growth of the polynomials in Corollary 11.2.2 cannot be reduced.

THEOREM 11.2.3. *If the trigonometric polynomials T_n form a basis in $C(\mathbb{T})$, then there exists $\varepsilon > 0$ such that the order of T_n is at least $\frac{n}{2}(1+\varepsilon)$ for $n \ge n_0$.*

The proof of this fact is given in [**242**].

11.3. Optimal polynomial bases in the space $C[-1,1]$

We shall now construct polynomial bases consisting of mutually orthogonal algebraic polynomials with minimum possible growth of degrees in the space of functions continuous on a segment.

Let $\mathcal{L}_n$, $n = 0, 1, \dots$, be the orthonormal Legendre polynomials (see Appendix A.20). Set $X_0 = \mathcal{L}_0$, $X_n = \frac{1}{\sqrt{2}}(\mathcal{L}_{2n} + i\mathcal{L}_{2n-1})$, $X_{-n} = \overline{X_n}$ for $n = 1, 2, \dots$. It is clear that $\{X_n\}_{n\in\mathbb{Z}}$ is an orthonormal basis in $L_2[-1,1]$. Denote by $\deg P$ the degree of an algebraic polynomial P considered as a polynomial with respect to the system X_n; i.e., $\deg P = n$ for polynomials of degree $2n$ and $2n-1$.

Consider the operator $\mathcal{I}$ that maps isomorphically $L_2(\mathbb{T})$ onto $L_2[-1,1]$: $X_n = \mathcal{I} f_n$, $n \in \mathbb{Z}$, where $f_n(x) = e^{2\pi i n x}$. Obviously, the image of any trigonometric polynomial $T = \sum_{n=-N}^{N} \alpha_n f_n$ is an algebraic polynomial of the form $\mathcal{I}T = \sum_{n=-N}^{N} \alpha_n X_n$, $\deg T = \deg \mathcal{I}T$, and the orthogonality of functions $T_1, T_2 \in L_2(\mathbb{T})$ implies that

$\mathcal{I}T_1, \mathcal{I}T_2$ are orthogonal in $L_2[-1,1]$. Let $\varphi_j, \psi_j, \psi_j^{s,l}$ be the functions from §11.2. Set

$$G_j := \mathcal{I}\varphi_j, \quad H_j := \mathcal{I}\psi_j, \quad H_j^{s,l} := \mathcal{I}\psi_j^{s,l}$$

for all values of indices for which the right-hand sides have meaning and define linear operators S_n^j, $j \in \mathbb{Z}_+$, $n = 0, \dots, 2^j - 1$, on $L_2[-1,1]$ by the equalities

$$S_n^j X_k = e^{2\pi i 2^{-j} nk} X_k, \quad k \in \mathbb{Z}.$$

It is clear that the systems $\{S_n^j G_j\}_n$, $\{S_n^j H_j\}_n$, $\{S_n^{j-s} H_j^{s,l}\}_{n,l}$, $s \le j$, are orthonormal for any $j \in \mathbb{Z}_+$ and

$$\begin{aligned}
&\operatorname{span}\{S_n^j G_j,\ n = 0, \dots, 2^j - 1\} = \mathcal{I}(V_j) =: \mathcal{V}_j,\\
&\operatorname{span}\{S_n^j H_j,\ n = 0, \dots, 2^j - 1\} = \mathcal{I}(W_j) =: \mathcal{W}_j,\\
&\operatorname{span}\{S_n^{j-s} H_j^{s,l},\ n = 0, \dots, 2^{j-s} - 1,\ \} = \mathcal{I}(W_j^l) =: \mathcal{W}_j^l.
\end{aligned}$$

It follows from relations (11.45), (11.46) that

$$\mathcal{V}_{j+1} = \mathcal{V}_j \oplus \mathcal{W}_j, \quad \mathcal{W}_j = \mathcal{W}_j^1 \oplus, \dots, \oplus \mathcal{W}_j^{2^s}, \tag{11.51}$$

$$L_2[-1,1] = \mathcal{V}_0 \oplus \mathcal{W}_0 \oplus \mathcal{W}_1 \oplus \mathcal{W}_2 \oplus \cdots. \tag{11.52}$$

Now put $P_n := \mathcal{I}T_n$, where T_n are trigonometric polynomials defined by equalities (11.38), (11.39), (11.40), where the parameters s and ε are chosen in such a way that inequality (11.41) holds for given $\varepsilon_0 > 0$. Then the polynomials P_n, $n \in \mathbb{N}$, form an orthonormal basis in $L_2[-1,1]$,

$$\deg P_n \le \frac{n}{2}(1 + \varepsilon_0); \tag{11.53}$$

for $j \ge s$,

$$\mathcal{W}_j^l = \operatorname{span}\{P_{2^j + (l-1)2^{j-s} + 1}, \dots, P_{2^j + l2^{j-s}}\}, \quad l = 1, \dots, 2^s, \tag{11.54}$$

and for $j < s$,

$$\mathcal{W}_j^l = \operatorname{span}\{P_{2^j + l}\}, \quad l = 1, \dots, 2^j. \tag{11.55}$$

Thus by using the operator $\mathcal{I}$ and starting from a PMRA, we have induced an analogous structure in the space $L_2[-1,1]$, and starting from the trigonometric basis $\{T_n\}_n$, we have induced a basis $\{P_n\}_n$ of algebraic polynomials which inherits some properties of its preimage. Further it will be shown that these polynomials form a basis in the space of continuous functions. However, the proof of an analog of Theorem 11.2.1 appears far more complicated technically. Before turning to this assertion, we shall prove some auxiliary facts.

LEMMA 11.3.1. *Let $\theta \in (0,1)$, $n \in \mathbb{Z}$, $n \ne 0$; then $X_n(\cos \pi\theta)$ is the sum of two summands of the form*

$$\frac{1}{\sqrt{\sin \pi\theta}} \left[A_1(\theta)\xi_1(n)e^{2\pi i n\theta} + A_2(\theta)\xi_2(n)e^{-2\pi i n\theta}\right] + O\left(\frac{1}{|n|(\sin \pi\theta)^{3/2}}\right) \tag{11.56}$$

or

$$\begin{aligned}
&\frac{1}{\sqrt{\sin \pi\theta}} \left[A_1(\theta)\xi_1(n)e^{2\pi i n\theta} + A_2(\theta)\xi_2(n)e^{-2\pi i n\theta}\right]\\
&\quad + \frac{1}{n(\sin \pi\theta)^{3/2}} \left[A_3(\theta)\xi_3(n)e^{2\pi i n\theta} + A_4(\theta)\xi_4(n)e^{-2\pi i n\theta}\right] + O\left(\frac{1}{n^2(\sin \pi\theta)^{5/2}}\right)
\end{aligned} \tag{11.57}$$

or

$$\int_0^\theta \frac{d\tau}{\sqrt{\cos\pi\tau-\cos\pi\theta}}\Big[\sqrt{n}A_5(\tau)\xi_5(n)e^{2\pi in\tau}+\sqrt{n}A_6(\tau)\xi_6(n)e^{-2\pi in\tau} + \frac{1}{\sqrt{n}}A_7(\tau)\xi_7(n)e^{2\pi in\tau}+\frac{1}{\sqrt{n}}A_8(\tau)\xi_8(n)e^{-2\pi in\tau}\Big]+O\left(\frac{1}{|n|^{3/2}}\right) \tag{11.58}$$

or

(11.59)

$$\frac{1}{\sqrt{\sin\pi\theta}}\left[A_1(\theta)\xi_1(n)e^{2\pi in\theta}+A_2(\theta)\xi_2(n)e^{-2\pi in\theta}+O\left(\frac{1}{n^2}\right)\right]$$
$$+\frac{1}{\sqrt{\sin\pi\theta}}\int_\theta^{1/2}\frac{dt}{\sin^{3/2}\pi t}\int_0^t\frac{d\tau}{\sqrt{\cos\pi\tau-\cos\pi t}}\Big[\frac{1}{\sqrt{n}}A_9(t)\xi_9(n)e^{2\pi in(\tau+t-\theta)}$$
$$+\frac{1}{\sqrt{n}}A_{10}(t)\xi_{10}(n)e^{-2\pi in(\tau+t-\theta)}+n^{-3/2}A_{11}(t)\xi_{11}(n)e^{2\pi in(\tau+t-\theta)}$$
$$+n^{-3/2}A_{12}(t)\xi_{12}(n)e^{-2\pi in(\tau+t-\theta)}+O\left(\frac{1}{|n|^{5/2}}\right)\Big],$$

where $x^a = |z|^a e^{ia\arg z}$, $0\le \arg z < 2\pi$, *the functions* ξ_l *are constant on the intervals* $(0,\infty)$ *and* $(-\infty,0)$, $|A_l(\theta)|\le C$.

Representations (11.56), (11.57), (11.58) can readily be derived from relations (A.42), (A.43), (A.39), respectively (see Appendix A.20), by using the Euler formulas and simple trigonometric transformations. Analogously, (11.59) can be obtained by combining (A.40) with (A.39).

LEMMA 11.3.2. *Let* $\gamma\in C^2(\mathbb{R})$, $\operatorname{supp}\gamma\subset[-\Delta,\Delta]$, $\theta\in\mathbb{R}$; *then*

$$\sum_{l\in\mathbb{Z}}\gamma(2^{-j}l)e^{2\pi il\theta}=O\left(\frac{1}{|\sin\pi\theta|}\right), \tag{11.60}$$

$$\sum_{l\in\mathbb{Z}}\gamma(2^{-j}l)e^{2\pi il\theta}=O\left(\frac{1}{2^j|\sin\pi\theta|^2}\right), \tag{11.61}$$

$$\int_0^1\left|\sum_{l\in\mathbb{Z}}\gamma(2^{-j}l)e^{2\pi il\theta}\right|d\theta=O(1). \tag{11.62}$$

Proof. To prove (11.60), apply the Abel transform to the left-hand side and use the explicit representation of the Dirichlet kernel. To prove (11.61), apply the Abel transform twice to the left-hand side, use the explicit representation of the Feyer kernel, the inequality

$$|\Delta_h^2\gamma(x)|\le h^2\|\gamma^{(2)}\|_\infty$$

(see Appendix A.14), and the structural properties of γ. To prove (11.62), note that

$$|\widehat{\gamma}(u)|=O\left(\frac{1}{1+u^2}\right);$$

hence by the Poisson summation formula, we have

$$\int_0^1 \left|\sum_{l\in\mathbb{Z}} \gamma(2^{-j}l)e^{2\pi i l\theta}\right| d\theta \le 2^j \int_0^1 \sum_{k\in\mathbb{Z}} |\widehat{\gamma}(2^j x + 2^j k)|\, dx = \int_{-\infty}^{\infty} |\widehat{\gamma}(x)|\, dx. \qquad \diamondsuit$$

LEMMA 11.3.3. *Under the conditions of Lemma* 11.3.2, *we have*

$$\int_0^1 \left|\sum_{l\in\mathbb{Z}} \gamma(2^{-j}l)e^{2\pi i l\theta}\right| \sqrt{|\sin \pi\theta|}\, d\theta = O(2^{-j/2}), \tag{11.63}$$

$$\int_0^1 \left|\sum_{l\in\mathbb{Z}} \gamma(2^{-j}l)e^{2\pi i l\theta}\right| \frac{d\theta}{\sqrt{|\sin \pi\theta|}} = O(2^{j/2}). \tag{11.64}$$

Proof. To prove (11.63), represent the integral in the form

$$\int_0^1 = \int_{-1/2}^{1/2} = \int_{|\theta|\le 2^{-j}} + \sum_{s=1}^{j-1} \int_{2^{-j+s-1}\le|\theta|\le 2^{-j+s}} = \sum_{s=0}^{j-1} I_s.$$

It follows from (11.62) that $I_0 = O(2^{-j/2})$. For $s > 0$, applying (11.61) to the expression under the integral sign, we get

$$\sum_{s=1}^{j-1} I_s \le C 2^{-j} \sum_{s=1}^{j-1} 2^{\frac{j-s}{2}} = O(2^{-j/2}).$$

Analogously, using (11.60) instead of (11.61) and taking into account the obvious relation $\sum_{l\in\mathbb{Z}} \gamma(2^{-j}l)\, e^{2\pi i l\theta} = O(2^j)$ instead of (11.62), we get (11.64). $\diamondsuit$

Introduce the following notation: $e_\tau := \{t \in \mathbb{R} : |\sin \pi t| \le |\sin \pi\tau|\}$. For fixed $\phi \in [0, 1/2]$, we set

$$h_1(t) = \chi_{e_{2\phi}}(t) \cos \pi t, \quad h_2(t) = \cos \pi t - h_1(t).$$

LEMMA 11.3.4. *Let* $\gamma \in C^2(\mathbb{R})$, $\operatorname{supp}\gamma \subset [-\Delta, \Delta] \setminus [-\delta, \delta]$, $\tau \in \mathbb{R}$,

$$\sigma^{\pm}_{\alpha,\nu}(\theta) = \sum_{l\in\mathbb{Z}} \gamma(2^{-j}l)(2^{-j}l)^{\alpha} \xi_\nu(2^{-j}l) e^{2\pi i l(\tau\pm\theta)},$$

where $\alpha \in \mathbb{R}$, ξ_ν *are the functions considered in Lemma* 11.3.1. *Then*

$$\begin{aligned} &\int_0^1 \left|\sum_{l\in\mathbb{Z}} \gamma(2^{-j}l)e^{2\pi i l\tau} X_l(\cos \pi\theta)\right| dh_1(\theta) \\ &\le C_1\sqrt{\phi} \left\{ \int_{-1/2}^{1/2} \left(|\sigma^{+}_{0,1}(\theta)| + |\sigma^{-}_{0,2}(\theta)|\right) d\theta + O(1) \right\}, \end{aligned} \tag{11.65}$$

$$\int_0^1 \left| \sum_{l\in\mathbb{Z}} \gamma(2^{-j}l) e^{2\pi i l\tau} X_l(\cos\pi\theta) \right| dh_2(\theta)$$

$$\le C_2 \int_{-1/2}^{1/2} \left(|\sigma_{0,1}^{+}(\theta)| + |\sigma_{0,2}^{-}(\theta)| \right) \frac{|h_2'(\theta)|\, d\theta}{\sqrt{|\sin\pi\theta|}} \tag{11.66}$$

$$+ C_3 2^{-j} \int_{-1/2}^{1/2} \left(|\sigma_{-1,3}^{-}(\theta)| + |\sigma_{-1,4}^{-}(\theta)| \right) \frac{|h_2'(\theta)|\, d\theta}{\sqrt{|\sin\pi\theta|^3}} + O(2^{-j/2}).$$

`Proof.` Relation (11.65) can easily be derived from (11.56) if when estimating the remainder term we take into account the structure of support of the function γ and the periodicity of the functions $\sigma_{\alpha,\nu}^{\pm}$. To prove (11.66), represent the integral on the left-hand side in the form

$$\int_{[0,1]\setminus e_{2\phi}} = \int_{(e_{2^{-j}}\setminus e_{2\phi})\cap[0,1]} + \int_{[0,1]\setminus e_{2^{-j}}} = I_1 + I_2.$$

If $\phi \ge 2^{-j-1}$, then $I_1 = 0$. If $\phi \le 2^{-j-1}$, then the same reasoning as in the proof of (11.65) gives

$$I_1 \le C \left\{ \int_{-1/2}^{1/2} \left(|\sigma_{0,1}^{+}(\theta)| + |\sigma_{0,2}^{-}(\theta)| \right) \frac{|h_2'(\theta)|\, d\theta}{\sqrt{|\sin\pi\theta|}} + O(2^{-j/2}) \right\}.$$

Analogously, to estimate I_2, we use (11.57). As before, taking into account the structure of support of γ to estimate the remainder, we obtain (11.66). ◊

Let $\psi \in L(\mathbb{R})$, $\alpha \in \mathbb{R}$, $\nu = 1, \dots, 8$. Set

$$s_{n,\alpha,\nu}^{\pm}(\tau) = s_{n,\alpha,\nu}^{\pm}(\tau,\psi) = \sum_{l\in\mathbb{Z}} e^{2\pi i(n2^{-j}\pm\tau)l} (2^{-j}l)^{\alpha} \xi_\nu(2^{-j}l) \widehat{\psi}(2^{-j}l),$$

where ξ_ν are the functions from Lemma 11.3.1.

LEMMA 11.3.5. *Let $|\tau| \le 1/4$, $|\gamma| \le 1/2$, $\alpha, \beta \in \mathbb{R}$, $\widehat{\psi} \in C^2(\mathbb{R})$, $\operatorname{supp}\widehat{\psi} \subset [-\Delta,\Delta] \setminus [-\delta,\delta]$. Then*

$$I = \int_{-1/2}^{1/2} \sum_{n=0}^{2^j-1} |s_{n,\alpha,\nu}^{\pm}(\tau)| |s_{n,\beta,\mu}^{\pm}(\theta\pm\tau)| |\sin\pi\theta|^{\gamma}\, d\theta = O\left(2^{(1-\gamma)j}\right),$$

where the signs + or − in the relations $s_{n,\beta,\mu}^{\pm}$ and $\theta \pm \tau$ are coordinated in such a way that τ has the same sign in all the exponents of the integrand.

`Proof.` Without loss of generality we can confine our consideration to the signs + instead of each of the three ±. Moreover, we leave only the first half of the summands in the sum under the integral sign. For the second half, the estimation can be made in a similar way. The functions $g_{\alpha,\nu}$, $g_{\beta,\mu}$ defined by the equalities

$\widehat{g}_{\alpha,\nu}(u) = \widehat{\psi}(u)\xi_\nu(u)u^\alpha$, $\widehat{g}_{\beta,\mu}(u) = \widehat{\psi}(u)\xi_\mu(u)u^\beta$ obviously have the same order of decrease as ψ; hence

$$|g_{\alpha,\nu}(u)|, |g_{\beta,\mu}(u)| = O\left(\frac{1}{1+u^2}\right), \tag{11.67}$$

and by the Poisson summation formula,

$$\begin{aligned}&\sum_{n=0}^{2^{j-1}-1} |s^+_{n,\alpha,\nu}(\tau)||s^+_{n,\beta,\mu}(\theta+\tau)| \\ &= 2^{2j}\sum_{n=0}^{2^{j-1}-1}\left|\sum_{l\in\mathbb{Z}} g_{\alpha,\nu}(2^j\tau+2^jl+n)\right|\left|\sum_{k\in\mathbb{Z}} g_{\beta,\mu}(2^j(\theta+\tau)+2^jk+n)\right|.\end{aligned} \tag{11.68}$$

Since in view of (11.67),

$$\sum_{l\neq 0}|g_{\alpha,\nu}(2^j\tau+2^jl+n)| = O(2^{-2j}),$$

$$\sum_{n=0}^{2^{j-1}-1}\sum_{k\in\mathbb{Z}}|g_{\beta,\mu}(2^j(\theta+\tau)+2^jk+n)| = O(1),$$

it follows that the right-hand side of (11.68) does not exceed

$$2^{2j}\sum_{n=0}^{2^{j-1}-1}\sum_{k\in\mathbb{Z}}|g_{\alpha,\nu}(2^j\tau+n)||g_{\beta,\mu}(2^j(\theta+\tau)+2^jk+n)| + O(1).$$

Taking into account Lemma 1.7.1 and (11.68), we obtain

$$\begin{aligned}&\sum_{n=0}^{2^{j-1}-1} |s^+_{n,\alpha,\nu}(\tau)||s^+_{n,\beta,\mu}(\theta+\tau)||\sin\pi\theta|^\gamma \\ &= O\left(1+\sum_{k\in\mathbb{Z}}\frac{2^{2j}|\theta|^\gamma}{1+(2^j(\theta+k))^2}\right) = O\left(1+\frac{2^{2j}|\theta|^\gamma}{1+(2^j(\theta))^2}\right).\end{aligned} \tag{11.69}$$

Representing the integral I in the form

$$\int_{-1/2}^{1/2} = \int_{|\theta|\le 2^{-j}} + \sum_{s=0}^{j-1}\int_{2^{-j+s-1}\le|\theta|\le 2^{-j+s}}$$

and applying (11.69) to the integrand in every summand, we get

$$I = O\left(2^{j(1-\gamma)} + 2^{j(1-\gamma)}\sum_{s=0}^{j-1}2^{-s(1-\gamma)}\right) = O\left(2^{j(1-\gamma)}\right). \qquad \diamond$$

COROLLARY 11.3.6. *Under the conditions of Lemma 11.3.5, the following relation holds:*

$$\int_{-1/2}^{1/2}\sum_{n=0}^{2^j-1}|s^\pm_{n,\alpha,\nu}(\tau)||s^\pm_{n,\beta,\mu}(\theta)|\,d\theta = O\left(2^j\right).$$

The statement of the corollary is a particular case of Lemma 11.3.5 if we make the change of variables $\theta = \theta' \pm \tau$ and take into account the periodicity of the integrand.

LEMMA 11.3.7. *Let the function* ψ *satisfy the conditions of Lemma* 11.3.5, $H_j := 2^{-j/2}\sum_{l\in\mathbb{Z}}\widehat{\psi}(2^{-j}l)X_l$, $j\in\mathbb{Z}_+$, $\alpha\in\mathbb{R}$, $n=0,\dots,2^j-1$. *Then*

(11.70)

$$\int_0^1 \sum_{n=0}^{2^j-1} |s^{\pm}_{n,\alpha,\nu}(\tau)|\left|S^j_n H_j(\cos\pi\theta)\right| dh_1(\theta)$$

$$\le C_1 2^{-j/2}\sqrt{\phi}\int_{-1/2}^{1/2}\sum_{n=0}^{2^j-1}|s^{\pm}_{n,\alpha,\nu}(\tau)|\left(|s^{+}_{n,0,1}(\theta)|+|s^{-}_{n,0,2}(\theta)|\right)d\theta + O\left(2^{j/2}\sqrt{\phi}\right),$$

$$\int_0^1 \sum_{n=0}^{2^j-1} |s^{\pm}_{n,\alpha,\nu}(\tau)|\left|S^j_n H_j(\cos\pi\theta)\right| dh_2(\theta)$$

(11.71)
$$\le C_2 2^{-j/2}\int_{-1/2}^{1/2}\sum_{n=0}^{2^j-1}|s^{\pm}_{n,\alpha,\nu}(\tau)|\left(|s^{+}_{n,0,1}(\theta)|+|s^{-}_{n,0,2}(\theta)|\right)\frac{|h_2'(\theta)|\,d\theta}{\sqrt{|\sin\pi\theta|}}$$
$$+ C_3 2^{-3j/2}\int_{-1/2}^{1/2}\sum_{n=0}^{2^j-1}|s^{\pm}_{n,\alpha,\nu}(\tau)|\left(|s^{+}_{n,-1,3}(\theta)|\right.$$
$$\left.+|s^{-}_{n,-1,4}(\theta)|\right)\frac{|h_2'(\theta)|\,d\theta}{\sqrt{|\sin\pi\theta|^3}} + O(1).$$

Proof. Using (11.56), (11.57) and taking into account that $\widehat{\psi}(2^{-j}l)$ is different from zero only for $2^j\delta\le|l|\le 2^j\Delta$, we get

(11.72)

$$|S^j_n H_j(\cos\pi\theta)| \le C_1\left(\frac{2^{-j}}{\sin\pi\theta}\right)^{1/2}\left(|s^{+}_{n,0,1}(\theta)|+|s^{-}_{n,0,2}(\theta)|\right) + O\left(\frac{2^{-j/2}}{\sin^{3/2}\pi\theta}\right),$$

(11.73)
$$|S^j_n H_j(\cos\pi\theta)| \le C_2\left(\frac{2^{-j}}{\sin\pi\theta}\right)^{1/2}\left(|s^{+}_{n,0,1}(\theta)|+|s^{-}_{n,0,2}(\theta)|\right)$$
$$+ C_3\left(\frac{2^{-j}}{\sin\pi\theta}\right)^{3/2}\left(|s^{+}_{n,-1,3}(\theta)|+|s^{-}_{n,-1,4}(\theta)|\right) + O\left(\frac{2^{-3j/2}}{\sin^{5/2}\pi\theta}\right).$$

Let $g_{\alpha,\nu}$ be the function defined in Lemma 11.3.5. The Poisson summation formula and (11.67) give

$$\sum_{n=0}^{2^j-1}|s^{\pm}_{n,\alpha,\nu}(\tau)| = 2^j\sum_{n=0}^{2^j-1}\left|\sum_{l\in\mathbb{Z}} g_{\alpha,\nu}(\pm 2^j\tau+2^jl+n)\right| = O\left(2^j\right). \tag{11.74}$$

Relation (11.70) follows from relations (11.72) and (11.74) if we take into account the periodicity of the functions $s^{\pm}_{n,\eta,\kappa}(\theta)$. To prove (11.71), represent the integral on the left-hand side in the form

$$\int_{[0,1]\setminus e_{2\phi}} = \int_{(e_{2^{-j}}\setminus e_{2\phi})\cap[0,1]} + \int_{[0,1]\setminus e_{2^{-j}}} = I_1 + I_2.$$

If $|\phi| \geq 2^{-j-1}$, then $I_1 = 0$. If $|\phi| \leq 2^{-j-1}$, then the same reasoning as in the proof of (11.70) gives

$$I_1 \leq C_4 2^{-j/2} \int_{-1/2}^{1/2} \sum_{n=0}^{2^j-1} |s^{\pm}_{n,\alpha,\nu}(\tau)| \left(|s^{+}_{n,0,1}(\theta)| + |s^{-}_{n,0,2}(\theta)|\right) \frac{|h_2'(\theta)|\, d\theta}{\sqrt{|\sin \pi\theta|}} + O(1).$$

Similarly, to estimate I_2, we use (11.73) and (11.74). It remains to note that

$$2^{-j/2} \int_{[0,1]\setminus e_{2^{-j}}} \frac{dh_2(\theta)}{\sin^{5/2} \pi\theta} = O(1). \qquad \Diamond$$

LEMMA 11.3.8. *Let functions ψ, H_j satisfy the conditions of Lemma 11.3.7, $j \in \mathbb{Z}_+$, $n = 0, \ldots, 2^j - 1$. Then*

$$\int_0^1 |S_j^n H_j(\cos \pi\theta)|\, d\cos \pi\theta = O\left(2^{-j/2}\right).$$

Proof. From (11.72) we have

$$\int_0^1 |S_n^j H_j(\cos \pi\theta)|\, d\cos \pi\theta \leq C 2^{-j/2} \int_{-1/2}^{1/2} \left(|s^{+}_{n,0,1}(\theta)| + |s^{-}_{n,0,2}(\theta)|\right) d\theta + O\left(2^{-j/2}\right).$$

Let $g_{0,\nu}$, $\nu = 1, 2$, be the functions defined in Lemma 11.3.5. By the Poisson summation formula we have

$$\int_{-1/2}^{1/2} |s^{\pm}_{n,0,\nu}(\theta)|\, d\theta \leq 2^j \sum_{l\in\mathbb{Z}} \int_{-1/2}^{1/2} |g_{0,\nu}(\pm 2^j\theta + 2^j l + n)|\, d\theta = \int_{-\infty}^{\infty} |g_{0,\nu}(\pm\theta)|\, d\theta.$$

It remains to note that in view of (11.67) the integral on the right-hand side of this equality is finite. $\Diamond$

LEMMA 11.3.9. *For all $N \in \mathbb{N}$, the following relation holds:*

$$\int_{-1}^{1} \left| \frac{1}{N} \sum_{n=0}^{N} \sum_{k=0}^{2n} \mathcal{L}_k(x)\mathcal{L}_k(t) \right| dt = O(1).$$

Proof. Since the Lebesgue functions of the Feyer-Legendre sums are uniformly bounded (see Appendix A.20), in view of the equality

$$\begin{aligned}
\sum_{n=0}^{2N} \sum_{k=0}^{n} \mathcal{L}_k(x)\mathcal{L}_k(t) &= \sum_{n=0}^{N} \sum_{k=0}^{2n} \mathcal{L}_k(x)\mathcal{L}_k(t) + \sum_{n=0}^{N-1} \sum_{k=0}^{2n+1} \mathcal{L}_k(x)\mathcal{L}_k(t) \\
&= 2 \sum_{n=0}^{N-1} \sum_{k=0}^{2n} \mathcal{L}_k(x)\mathcal{L}_k(t) + \sum_{k=0}^{2N} \mathcal{L}_k(x)\mathcal{L}_k(t) + \sum_{n=0}^{N-1} \mathcal{L}_{2n+1}(x)\mathcal{L}_{2n+1}(t) \\
&= 2 \sum_{n=0}^{N} \sum_{k=0}^{2n} \mathcal{L}_k(x)\mathcal{L}_k(t) - \sum_{n=0}^{N} \mathcal{L}_{2n}(x)\mathcal{L}_{2n}(t)
\end{aligned}$$

it suffices to prove the relation

$$I = \frac{1}{N} \int_{-1}^{1} \left| \sum_{n=0}^{N} \mathcal{L}_{2n}(x)\mathcal{L}_{2n}(t) \right| dt = O(1). \tag{11.75}$$

Set $t = \cos \pi\theta$, $x = \cos \pi\phi$, $\phi, \theta \in [0, 1]$. Since $\mathcal{L}_{2n}$ is an even function, we can confine the consideration to the case $\phi \in [0, 1/2]$. Applying (A.42) to $\mathcal{L}_{2n}(\cos \pi\theta)$ and using (A.38) to estimate the remainder term, we get

$$I = \frac{\sqrt{2}}{N\sqrt{\pi}} \int_{0}^{1} \left| \sum_{n=0}^{N} \mathcal{L}_{2n}(\cos \pi\phi) \cos \left[\left(2n + \frac{1}{2} \right) \pi\theta - \frac{\pi}{4} \right] \right| \frac{d \cos \pi\theta}{\sqrt{\sin \pi\theta}} + O(1).$$

Represent the integral on the right-hand side in the form

$$\int_{0}^{1} d \cos \pi\theta = \int_{0}^{1} dh_1(\theta) + \int_{0}^{1} dh_2(\theta) = I_1 + I_2.$$

To estimate I_2 for $\phi \neq 0$, apply the Dirichlet-Mehler formula (A.39) to $\mathcal{L}_{2n}(\cos \pi\phi)$. Rearranging the integrals, we get

(11.76)

$$\begin{aligned} I_2 \leq \frac{1}{2} \int_{0}^{\phi} \frac{d\tau}{\sqrt{\cos \pi\tau - \cos \pi\phi}} \int_{0}^{1} \Bigg\{ & \left| \sum_{n=0}^{N} e^{2\pi i n(\theta+\tau)} \sqrt{n + \frac{1}{4}} \right| \\ & + \left| \sum_{n=0}^{N} e^{2\pi i n(\theta-\tau)} \sqrt{n + \frac{1}{4}} \right| + \left| \sum_{n=0}^{N} e^{-2\pi i n(\theta+\tau)} \sqrt{n + \frac{1}{4}} \right| \\ & + \left| \sum_{n=0}^{N} e^{-2\pi i n(\theta-\tau)} \sqrt{n + \frac{1}{4}} \right| \Bigg\} \frac{dh_2(\theta)}{\sqrt{\sin \pi\theta}}. \end{aligned}$$

Applying the Abel transform to each sum and using the well-known properties of the Dirichlet kernels, we establish that these sums are of order

$$O(\sqrt{N} |\sin \pi(\theta \pm \tau)|^{-1}).$$

Thus the inner integral in (11.76) does not exceed the sum of four summands of the form

$$C_1 \sqrt{N} \int_{0}^{1} \frac{dh_2(\theta)}{|\sin \pi(\theta \pm \tau)| \sqrt{\sin \pi\theta}}. \tag{11.77}$$

Since $\tau \leq \phi$, it follows that $|\sin \pi(\theta \pm \tau)|^{-1} = O\left((\sin \pi\theta)^{-1}\right)$ for $\theta \geq 2\phi$. Therefore,

$$\int_{0}^{1} \frac{dh_2(\theta)}{|\sin \pi(\theta \pm \tau)| \sqrt{\sin \pi\theta}} \leq C_2 \int_{0}^{1} \frac{d\theta}{\sqrt{\sin \pi\theta}} = O(1).$$

Combining this relation with (11.77) and (11.76), we get

$$I_2 = O(\sqrt{N}). \tag{11.78}$$

For $\phi = 0$, the integral I_2 can be estimated in the same way, using equality (A.38) instead of the Dirichlet-Mehler formula. Let us now estimate I_1 in the case $\phi \le 1/N$. Use the Dirichlet-Mehler formula for $\phi \neq 0$:

$$I_1 \le C_3 \int_0^{\phi} \frac{d\tau}{\sqrt{\cos\pi\tau - \cos\pi\phi}} \int_0^1 \sum_{n=0}^{N} \sqrt{n + \frac{1}{4}} \frac{g_1'(\theta)}{\sqrt{\sin\pi\theta}}\, d\theta.$$

Taking into account that $g_1'(\theta)\sin\pi\theta^{-1/2} \le C\sqrt{\phi} = O\left(N^{-1/2}\right)$, we obtain

$$I_1 = O(N). \tag{11.79}$$

Analogously, (11.79) follows from (A.38) for $\phi = 0$. Finally, we consider the case $\phi \ge 1/N$. Combining relations (A.39) and (A.40), we get

$$(\sin\pi\phi)^{1/2}\mathcal{L}_{2n}(\cos\pi\phi) = \sqrt{\frac{2}{\pi}} \cos\left[\left(2n + \frac{1}{2}\right)\pi\phi - \frac{\pi}{4}\right]$$

$$- \frac{\sqrt{2\pi}}{\sqrt{2n + \frac{1}{2}}} \int_{\phi}^{1/2} \frac{\sin\pi\left(2n + \frac{1}{2}\right)(t - \phi)}{4\sin^{3/2}\pi t}\, dt \int_0^t \frac{\cos\pi\left(2n + \frac{1}{2}\right)\tau}{\sqrt{\cos\pi\tau - \cos\pi t}}\, d\tau + O\left(\frac{1}{n^2}\right).$$

Taking into account that $g_1'(\theta)(\sin\pi\theta)^{-1/2} = O\left(\phi^{1/2}\right)$ and $\phi^{-1} \le N$, we obtain

$$I_1 \le C_4 \int_{\phi}^{1/2} \frac{dt}{t^{3/2}} \sum_{n=0}^{N} \frac{1}{\sqrt{n}} + O(1) = O\left(\sqrt{\frac{N}{\phi}} + 1\right) = O(N).$$

Combining this estimate with (11.78) and (11.79), we obtain (11.75) . ◊

Return now to the consideration of the orthonormal basis $\{P_n\}_{n=0}^{\infty}$ constructed at the beginning of this section. For any function f continuous on $[-1, 1]$, the Fourier expansion with respect to this basis has the form

$$\sum_{n=1}^{\infty} \langle f, P_n\rangle P_n. \tag{11.80}$$

Partial sums of this series will be denoted by $\sigma_N(f)$, $N \in \mathbb{N}$.

THEOREM 11.3.10. *For any function $f \in C[-1, 1]$, series (11.80) converges uniformly on $[-1, 1]$, and for any $N \in \mathbb{N}$, the estimate*

$$\|f - \sigma_N(f)\| \le C\mathcal{A}_{[N/4]}(f)_\infty \tag{11.81}$$

holds, where C is an absolute constant, $\mathcal{A}_k(f)_\infty = \inf \|f - P\|_\infty$ with the infimum taken over all possible algebraic polynomials P of order at most k.

`Proof.` From the definitions of the functions $G_j, H_j, H_j^{s,l}$ and $\varphi_j, \psi_j, \psi_j^{s,l}$ we have

$$\begin{aligned} G_j &= 2^{-j/2} \sum_{l \in \mathbb{Z}} \widehat{\varphi}(2^{-j}l) X_l, \\ H_j &= 2^{-j/2} \sum_{l \in \mathbb{Z}} \widehat{\psi}(2^{-j}l) X_l, \\ H_j^{s,l} &= 2^{-j/2} \sum_{l \in \mathbb{Z}} \widehat{\psi}^{s,\nu}(2^{-j}l) X_l. \end{aligned}$$

Using these equalities and relations (11.51), (11.52), (11.54), (11.55) and repeating the reasoning of the proof of Theorem 11.3.10, we obtain the estimate

$$\|f - \sigma_N(f)\| \le (1 + \|\sigma_N\|)\mathcal{A}_{[N/4]}(f)_\infty,$$

where $\|\sigma_N\|$ is the operator norm for σ_N from $C[-1,1]$ to $C[-1,1]$. The representation of the partial sum σ_N with number $N = 2^j + m2^{j-s} + r$, $m = 0, \dots, 2^s - 1$, $r = 0, \dots, 2^{j-s} - 1$, has the form

$$\sigma_N(f) = \sum_{n=0}^{2^j-1} \langle f, S_n^j G_j\rangle S_n^j G_j + \sum_{k=0}^{m-1} \sum_{n=0}^{2^{j-s}-1} \langle f, S_n^{j-s} H_j^{s,k+1}\rangle S_n^{j-s} H_j^{s,k+1} + \sum_{n=0}^{r} \langle f, S_n^{j-s} H_j^{s,m+1}\rangle S_n^{j-s} H_j^{s,m+1}.$$

To prove the theorem, we have to check that the sequence $\|\sigma_N\|$ is bounded. To do this, it suffices to show that the integrals

$$J(x) = \int_{-1}^{1} \left| \sum_{n=0}^{2^j-1} S_n^j G_j(x) \overline{S_n^j G_j(t)} \right| dt,$$

$$J^m(x) = \int_{-1}^{1} \sum_{n=0}^{2^{j-s}-1} \left| S_n^{j-s} H_j^{s,m}(x) S_n^{j-s} H_j^{s,m}(t) \right| dt, \quad m = 1, \dots, 2^s,$$

are uniformly bounded (with respect to $x \in [-1,1]$). Fix m and estimate J^m. Rename the functions $H_j^{s,m}$, $\psi^{s,m}$ by H_j, ψ (in the aforesaid, these designations had different meanings but they will not be used in what follows). Note that the function $\widehat{\psi}$ satisfies the conditions of Lemma 11.3.5. It is clear that

$$J^m(x) \le \int_{-1}^{1} \sum_{n=0}^{2^j-1} |S_n^j H_j(x) S_n^j H_j(t)| \, dt = I.$$

We confine our consideration to the case $x \in [0,1]$. For $x \in [-1,0]$, the proof is similar. Set $x = \cos \pi\phi$, $\phi \in [0,1/2]$. First let $\phi \in [1/4,1/2]$. Applying inequalities (11.72) and (11.74) to $S_n^j H_j(\cos \pi\phi)$, we get

$$I \le C_1 2^{j/2} \max_n \int_{-1}^{1} |S_n^j H_j(t)| \, dt.$$

This inequality and Lemma 11.3.8 imply that $I = O(1)$. Let us now consider the case $\phi \in (0,1/4]$. Making the change of variables $t = \cos \pi\theta$, $\theta \in [0,1]$, in the integral I, we represent it in the form

$$I = \int_0^1 dh_1(\theta) + \int_0^1 dh_2(\theta) = I_1 + I_2.$$

To prove the boundedness of I_1 for $\phi \geq 2^{-j}$, estimate $S_n^j H_j(\cos\phi)$, using (11.59) and taking into account the structure of the support of the function $\widehat{\psi}$:

$$(11.82)\quad \begin{aligned} |S_n^j H_j(\cos\phi)| \leq C_2 \frac{2^{-j/2}}{\sqrt{\phi}} \Bigg[& |s_{n,0,1}^+(\phi)| + |s_{n,0,2}^-(\phi)| \\ & \times 2^{-j/2} \int_\phi^{1/2} \frac{dt}{t^{3/2}} \int_0^t \frac{d\tau}{\sqrt{\cos\pi\tau - \cos\pi t}} \Big(|s_{n,-1/2,9}^+(\tau + t - \phi)| \\ & + |s_{n,-1/2,10}^-(\tau + t - \phi)| \Big) + O(2^{-j/2}) \Bigg]. \end{aligned}$$

Here each summand contains an expression of the form $s_{n,\alpha,\nu}^{\pm}(u)$. The corresponding summand in I_1 can be estimated by the integral

$$\int_0^1 \sum_{n=0}^{2^j-1} |s_{n,\alpha,\nu}^{\pm}(u)||S_n^j H_j(\cos\pi\theta)|\, dh_1(\theta) = I_1^{\alpha,\nu}.$$

Using inequality (11.70) for this integral and applying the corollary of Lemma 11.3.5 to each summand on the right-hand side of (11.70), we prove $I_1^{\alpha,\nu} = O\left(2^{j/2}\sqrt{\phi}\right)$. Using Lemma 11.3.5 to estimate the remainder term, we derive

$$I_1 = O\left(1 + (2^j\phi)^{-1/2}\right) = O(1).$$

To prove the boundedness of I_2 for $\phi \neq 0$, estimate $S_n^j H_j(\cos\phi)$ using (11.58):

$$(11.83)\quad \begin{aligned} |S_n^j H_j(\cos\phi)| \leq C_3 \int_0^\phi \frac{dt}{\sqrt{\cos\pi t - \cos\pi\phi}} \Big(& |s_{n,1/2,5}^+(t)| + |s_{n,1/2,6}^-(t)| \\ & + |s_{n,-1/2,7}^+(t)| + |s_{n,-1/2,8}^-(t)| + O\left(2^{-j/2}\right)\Big). \end{aligned}$$

Each summand of the sum under the integral sign contains an expression of the form $s_{n,\alpha,\nu}^{\pm}(t)$. The corresponding summand in I_2 can be estimated by the integral

$$\int_0^1 \sum_{n=0}^{2^j-1} |s_{n,\alpha,\nu}^{\pm}(t)||S_n^j H_j(\cos\pi\theta)|\, dh_2(\theta) = I_2^{\alpha,\nu}.$$

Apply inequality (11.71) to this integral and make the change of variables $\theta' = \theta \pm t$ in each summand of the right-hand side of (11.71), choosing the sign + or − in such a way that all the exponents of the integrand contain t with the same sign. Taking into account that for $t \leq \phi$,

$$(11.84)\quad |h_2'(\theta)||\sin\pi\theta|^{-1\pm 1/2} \leq C_4 |\sin\pi\theta'|^{\pm 1/2},$$

according to Lemma 11.3.5 for $\gamma = \pm 1/2$, we get $I_2^{\alpha,\nu} = O(1)$. Taking into account Lemma 11.3.8 to estimate the remainder term, we deduce that I_2 is bounded. Estimate now I_1 for $\phi \in (0, 2^{-j})$. Arguing as in the case $\phi \geq 2^{-j}$, we get

$$\int_0^1 \sum_{n=0}^{2^j-1} |s_{n,\alpha,\nu}^{\pm}(u)||S_n^j H_j(\cos\pi\theta)|\, dh_1(\theta) = O\left(2^{j/2}\sqrt{\phi}\right).$$

Combining this relation with (11.83), taking into account Lemma 11.3.8 and the equality $\sqrt{\phi} = O\left(2^{-j/2}\right)$, we again deduce that $I_1 = O(1)$. For $\phi = 0$, the integral I_1 is absent, and I_2 can be estimated as before, using, instead of (11.58), the equality

$$X_l(1) = \sqrt{2il} + \frac{1}{4\sqrt{2il}} + O\left(\frac{1}{|l|^{3/2}}\right), \tag{11.85}$$

where $\sqrt{z} = \sqrt{|z|}e^{i\frac{\arg z}{2}}$, $-\pi \le \arg z \le \pi$. Thus we have proved that J^m is bounded.

To estimate J, transform the expression under the integral sign:

$$\begin{aligned}\sum_{n=0}^{2^j-1} \overline{S_n^j G_j(t)} S_n^j G_j(x) &= 2^{-j} \sum_{l\in\mathbb{Z}} \widehat{\varphi}(2^{-j}l)\overline{X_l(t)} \sum_{k\in\mathbb{Z}} \widehat{\varphi}(2^{-j}k) X_k(x) \sum_{n=0}^{2^j-1} e^{2\pi i 2^{-j} n(k-l)} \\ &= \sum_{m\in\mathbb{Z}} \sum_{l\in\mathbb{Z}} \widehat{\varphi}(2^{-j}l)\overline{X_l(t)}\widehat{\varphi}(2^{-j}l + m) X_{l+2^j m}(x) \\ &= \sum_{m\in\mathbb{Z}} \omega_m(t,x).\end{aligned}$$

Since $\operatorname{supp}\widehat{\varphi} \subset [-1/2 - \varepsilon/2, 1/2 + \varepsilon/2]$, it follows that $\omega_m(t,x) = 0$ for all $m \neq 0, 1, -1$. Show that

$$\int_{-1}^{1} |\omega_m(t,x)|\, dt = O(1) \tag{11.86}$$

for $m = 0, 1, -1$. Rewrite $\omega_0(t,x)$ in the real form:

$$\omega_0(t,x) = 1 + \sum_{n=1}^{\infty} \widehat{\varphi}^2(2^{-j}n)\left(\mathcal{L}_{2n-1}(x)\mathcal{L}_{2n-1}(t) + \mathcal{L}_{2n}(x)\mathcal{L}_{2n}(t)\right).$$

Applying the Abel transform twice to this sum and taking into account Lemma 11.3.9, we obtain (11.86) for $m = 0$.

Let θ, ϕ have the same meaning as above, and confine our consideration to the case $\phi \in [0, 1/2]$. Set $\gamma_\pm(u) = \widehat{\varphi}(u)\widehat{\varphi}(u \pm 1)$; then

$$\int_0^1 |\omega_{\pm 1}(t,x)|\, dt = \int_0^1 \left| \sum_{l\in\mathbb{Z}} \gamma_\pm(2^{-j}l)\overline{X_l(\cos\pi\theta)} X_{l\pm 2^j}(\cos\pi\phi)\right| d\cos\pi\theta.$$

Represent this integral in the form

$$\int_0^1 d\cos\pi\theta = \int_0^1 dh_1(\theta) + \int_0^1 dh_2(\theta) = i_1 + i_1.$$

For $\phi \neq 0$, apply (11.58) to $X_{l\pm 2^j}(\cos\pi\phi)$. Using inequality (A.38) and taking into account the structure of the support of the function $\gamma_\pm$ to estimate the remainder, we get

$$i_2 \le C\,2^{j/2} \int_0^{\phi} \frac{d\tau}{\sqrt{\cos\pi\tau - \cos\pi\phi}}$$

$$\times \int_0^1 \Bigg\{ \Bigg| \sum_{l\in\mathbb{Z}} \gamma_\pm(2^{-j}l)\xi_5(2^{-j}l\pm 1)(2^{-j}l)^{1/2} e^{2\pi i l\tau} X_l(\cos\pi\theta) \Bigg|$$

$$+ \Bigg| \sum_{l\in\mathbb{Z}} \gamma_\pm(2^{-j}l)\xi_6(2^{-j}l\pm 1)(2^{-j}l)^{1/2} e^{-2\pi i l\tau} X_l(\cos\pi\theta) \Bigg|$$

$$+ \Bigg| \sum_{l\in\mathbb{Z}} \gamma_\pm(2^{-j}l)\xi_7(2^{-j}l\pm 1)(2^{-j}l)^{-1/2} e^{2\pi i l\tau} X_l(\cos\pi\theta) \Bigg|$$

$$+ \Bigg| \sum_{l\in\mathbb{Z}} \gamma_\pm(2^{-j}l)\xi_8(2^{-j}l\pm 1)(2^{-j}l)^{-1/2} e^{-2\pi i l\tau} X_l(\cos\pi\theta) \Bigg| \Bigg\}\, dh_2(\theta) + O(1).$$

It is not hard to see that the function $\gamma(u) = \gamma_\pm(u)\xi_\nu(u\pm 1)(u)^\alpha$ satisfies the conditions of Lemma 11.3.4. Apply (11.66) to the inner integral and make the change of variable $\theta' = \tau \pm \theta$ in every summand of the right-hand side of (11.66). Taking into account that for $\tau \le \phi$, (11.84) holds, in view of Lemma 11.3.3 we get $i_2 = O(1)$. For $\phi = 0$, this relation can be proved in a similar way, using equality (11.85) instead of (11.58). Analogously, use (11.58) to estimate i_1 for $\phi \in (0, 2^{-j})$. Using (11.65) and (11.62), we obtain $i_1 = O(1)$. Finally, to prove the boundedness of i_1 for $\phi \in (2^{-j}, 1/2)$, apply relation (11.59) to $X_{l\pm 2^j}(\cos\pi\phi)$. Using inequality (A.38) and taking into account the structure of the support of the function $\gamma_\pm$ to estimate the remainder terms, we have

$$i_1 \le C_5 \frac{1}{\sqrt{\phi}} \int_0^1 \Bigg\{ \Bigg| \sum_{l\in\mathbb{Z}} \gamma_\pm(2^{-j}l)\xi_1(2^{-j}l\pm 1) e^{2\pi i l\phi} X_l(\cos\pi\theta) \Bigg|$$

$$+ \Bigg| \sum_{l\in\mathbb{Z}} \gamma_\pm(2^{-j}l)\xi_2(2^{-j}l\pm 1) e^{-2\pi i l\phi} X_l(\cos\pi\theta) \Bigg| \Bigg\}\, dh_1(\theta) + O\left(\frac{2^{-j/2}}{\sqrt{\phi}}\right)$$

$$+ C_6 \frac{2^{-j/2}}{\sqrt{\phi}} \int_\phi^{1/2} \frac{dt}{t^{3/2}} \int_0^t \frac{d\tau}{\sqrt{\cos\pi\tau - \cos\pi t}}$$

$$\times \int_0^1 \Bigg\{ \Bigg| \sum_{l\in\mathbb{Z}} \gamma_\pm(2^{-j}l)\xi_9(2^{-j}l\pm 1)(2^{-j}l)^{-1/2} e^{2\pi i l(\tau+t-\phi)} X_l(\cos\pi\theta) \Bigg|$$

$$+ \Bigg| \sum_{l\in\mathbb{Z}} \gamma_\pm(2^{-j}l)\xi_{10}(2^{-j}l\pm 1)(2^{-j}l)^{-1/2} e^{-2\pi i l(\tau+t-\phi)} X_l(\cos\pi\theta) \Bigg|$$

$$+ \Bigg| \sum_{l\in\mathbb{Z}} \gamma_\pm(2^{-j}l)\xi_{11}(2^{-j}l\pm 1)(2^{-j}l)^{-3/2} e^{2\pi i l(\tau+t-\phi)} X_l(\cos\pi\theta) \Bigg|$$

$$+\left|\sum_{l\in\mathbb{Z}}\gamma_\pm(2^{-j}l)\xi_{12}(2^{-j}l\pm 1)(2^{-j}l)^{-3/2}e^{-2\pi i l(\tau+t-\phi)}X_l(\cos\pi\theta)\right|\Bigg\}\,dh_1(\theta)$$
$$+O\left(\frac{2^{-j}}{\phi}\right).$$

It remains to apply (11.65) and (11.62) to the inner integrals. ◊

COROLLARY 11.3.11. *For any $\varepsilon > 0$, in the space $C[-1,1]$ there exists a basis consisting of mutually orthogonal algebraic polynomials P_n, $n = 1, 2\dots$, such that the degree of P_n does not exceed $n(1+\varepsilon)$.*

This assertion follows from Theorem 11.3.10 (see Appendix A.1) and inequality (11.53).

The order of growth of polynomials in Corollary 11.3.11 cannot be reduced.

THEOREM 11.3.12. *If algebraic polynomials P_n form a basis in $C[a,b]$, then there exists $\varepsilon > 0$ such that the degree of P_n is at least $n(1+\varepsilon)$ for $n \geq n_0$.*

The proof of this fact is presented in [**242**].

11.4. Wavelet bases in Besov and Lizorkin-Triebel spaces

Wavelet theory is used to describe constructively various functional spaces and to construct unconditional bases in them.

The goal of this section is to construct unconditional bases in anisotropic Besov and Lizorkin-Triebel spaces. As a result, we obtain unconditional bases in anisotropic Sobolev spaces $W_p^{(s)}$, $s = (s_1, \dots, s_n)$, $p \in (1, \infty)$. To construct these bases, we use the Meyer-David wavelets, which are best suited to the decomposition method for spaces of anisotropic smoothness.

11.4.1. Definitions and preliminary results. In this section, we consider functions defined on $\mathbb{R}^n$ and use the following notation:

$a^m := \prod_\lambda a_\lambda^{m_\lambda}$ for $a = (a_1, \dots, a_n)$ where $a_\lambda > 0$ and $m = (m_1, \dots, m_n) \in \mathbb{Z}^n$;

if $\alpha = (\alpha_1, \dots, \alpha_n) \in \mathbb{R}^n$, $\alpha_\lambda \in \mathbb{N}$, then $|\alpha| := \sum_\lambda \alpha_\lambda$ is the order of the multi-index α, $D^\alpha := \dfrac{\partial^{|\alpha|}}{\partial_{x_1}^{\alpha_1}\cdots\partial_{x_n}^{\alpha_n}}$. Recall that F and F^{-1} are the operators that take any function to its direct and inverse Fourier transform, respectively. The relation $A(t) \ll B(t)$ means that $A(t) \leq C\,B(t)$, where the constant C is independent of t. The relation $A(t) \sim B(t)$ is equivalent to $A(t) \ll B(t) \ll A(t)$. In this section, by $|x, X|$ we denote $\|x\|_X$.

Recall the definition of Meyer-David wavelets. Suppose $h \in \mathbb{N}$ is a fixed positive integer and $\vartheta_h(\xi)$ is an odd infinitely differential function that is equal to $\frac{\pi}{4}$ and $-\frac{\pi}{4}$ for $\xi \geq \frac{h}{2(2h+1)}$ and $\xi \leq -\frac{h}{2(2h+1)}$, respectively. For nonnegative ξ, define a function $\theta_h(\xi)$ by the following inequality:

$$\theta_h(\xi) := \begin{cases} \frac{\pi}{4} + \vartheta_h\left(\xi - \frac{h}{2}\right), & \xi \in \left[\frac{h^2}{2h+1}, \frac{h(h+1)}{2h+1}\right), \\ \frac{\pi}{4} - \vartheta_h\left(\frac{h\xi}{h+1} - \frac{h}{2}\right), & \xi \in \left[\frac{h(h+1)}{2h+1}, \frac{(h+1)^2}{2h+1}\right), \\ 0, & \xi \notin \left[\frac{h^2}{2h+1}, \frac{(h+1)^2}{2h+1}\right), \quad \xi \geq 0. \end{cases}$$

The function $\theta_h(\xi)$ is extended to negative ξ as an even function for odd h and as an odd function for even h.

Finally, define the Meyer-David wavelet $\psi_h(t) \in S(\mathbb{R}^n)$ and the corresponding scaling function $\varphi_h(t) \in S(\mathbb{R}^n)$ by their Fourier transforms:

$$F\psi_h(\xi) := e^{-i\pi\xi} \sin \theta_h(\xi),$$

$$F\varphi_h(\xi) := \begin{cases} \frac{1}{\sqrt{h}} \cos \theta_h(\xi), & |\xi| \leq \frac{h(h+1)}{2h+1}, \\ 0, & |\xi| > \frac{h(h+1)}{2h+1}. \end{cases} \tag{11.87}$$

Let $b := \frac{h+1}{h}$; then it is easy to check that the functions

$$b^{j/2}\psi_h(b^j t + k), \quad j \in \mathbb{Z}, \quad k \in \mathbb{Z},$$

form an orthonormal basis in $L_2(\mathbb{R})$.

In the case $n > 1$, to construct an orthonormal basis in $L_2(\mathbb{R}^n)$, the function $\varphi_h(t)$ is also used; the construction is described below.

Recall the definition of anisotropic Besov and Lizorkin-Triebel spaces. Let $(s) := (s_1, \dots, s_n)$, where all s_λ are either simultaneously positive or simultaneously negative, or all of them are equal to zero.

Fix arbitrary numbers $C_\lambda > 1$ and $\gamma > 1$. For $s_\lambda > 0$, introduce the following notation:

$$\gamma_\lambda := \gamma^{1/s_\lambda}, \quad (\gamma) := (\gamma_1, \dots, \gamma_n),$$

$$\Pi_j := \{\xi \in \mathbb{R}^n : |\xi_\lambda| \leq C_\lambda \gamma_\lambda^j\}, \quad j \in \mathbb{N}, \tag{11.88}$$

$$\Gamma_1 := \Pi_2, \quad \Gamma_j := \Pi_{j+1} \backslash \Pi_{j-1}, \quad j = 2, 3, \dots. \tag{11.89}$$

Consider the set Φ of smooth anisotropic partitions of unity, i.e., the set of systems $\Sigma = \{\sigma_j(\xi)\}_{j=1}^\infty$ of nonnegative functions $\sigma_j \in C^\infty(\mathbb{R})$, $j \in \mathbb{N}$, such that

$$\operatorname{supp} \sigma_j \subset \Gamma_j,$$

$$(\gamma)^{j\alpha} |D^\alpha \sigma_j(\xi)| \leq c_\alpha \quad \forall j \in \mathbb{N}, \quad \forall \alpha \in \mathbb{N}_0^n, \quad \forall \xi \in \mathbb{R}^n,$$

$$\sum_{j=1}^\infty \sigma_j(\xi) \equiv 1.$$

Set $V_j(x) := (F^{-1}\sigma_j)(x)$.

DEFINITION 11.4.1. The anisotropic Lizorkin-Triebel space $F_{pq}^{(s)}(\mathbb{R}^n) = F_{pq}^{(s)}$, $0 < p < \infty$, $0 < q \le \infty$, is a Banach space of functions $f \in S'(\mathbb{R}^n)$ such that the norm

$$|f, F_{pq}^{(s)}| = |\{\gamma^j |f * V_j|\}_{j=1}^{\infty}, \ L_p(l_q)| \tag{11.90}$$

is finite. Here, in the case $\min(p,q) < 1$, a Banach space and the norm are replaced by a quasi-Banach space and the quasi-norm, respectively.

It is clear that for $s_\lambda = s_0$, for any λ we obtain the isotropic spaces $F_{pq}^{s_0}$ (see Proposition 11.4.8)

The spaces $F_{\infty q}^{(s)}$ require a different approach.

Let for $j \in \mathbb{Z}$, $k \in \mathbb{Z}^n$,

$$Q_{jk}^{(s)} := \{x \in \mathbb{R}^n : \ x_\lambda \in [k_\lambda \gamma_\lambda^{-j}, \ (k_\lambda + 1)\gamma_\lambda^{-j})\},$$

$$\mathcal{Q}^{(s)} := \{Q_{jk}^{(s)} : \ j \ge 0, \ j \in \mathbb{Z}, \ k \in \mathbb{Z}^n\},$$

$$\dot{\mathcal{Q}}^{(s)} := \{Q_{jk}^{(s)} : \ j \in \mathbb{Z}, \ k \in \mathbb{Z}^n\}.$$

DEFINITION 11.4.2. The anisotropic Lizorkin-Triebel space $F_{\infty q}^{(s)}(\mathbb{R}^n) := F_{\infty q}^{(s)}$, $0 < q \le \infty$, is a (quasi-)Banach space of functions $f \in S'(\mathbb{R}^n)$ such that the (quasi-)norm

$$|f, F_{\infty q}^{(s)}| := \sup_{Q_{jk}^{(s)} \in \mathcal{Q}^{(s)}} \left(|Q_{jk}^{(s)}|^{-1} \int_{Q_{jk}^{(s)}} \sum_{\nu=j}^{\infty} \gamma^{\nu q} |(f * V_\nu)(x)|^q dx \right)^{1/q}$$

is finite, with natural modification for $q = \infty$.

REMARK 11.4.3. If $s_\lambda < 0$, then the weight γ^j in Definitions 11.4.1 and 11.4.2 should be replaced by γ^{-j}, and the numbers γ_λ should be defined as $\gamma_\lambda := \gamma^{-1/s_\lambda}$.

DEFINITION 11.4.4. Replacing the norm in $L_p(l_q)$ by the norm in $l_q(L_p)$ in (11.90) (without stipulation of the case $p = \infty$), one can define Besov spaces $B_{pq}^{(s)}(\mathbb{R}^n) := B_{pq}^{(s)}$, $0 < p, q \le \infty$.

Suppose the system of parallelepipeds (11.88) is considered for $j \in \mathbb{Z}$ and corridors $\dot{\Gamma}_j$ are defined by the equalities

$$\dot{\Gamma}_j := \Pi_{j+1} \backslash \Pi_j, \quad j \in \mathbb{Z}.$$

Denote by $\dot{\Phi}$ the set of systems $\dot{\Sigma} = \{\dot{\sigma}_j(\xi)\}_{-\infty}^{\infty}$ of nonnegative functions $\dot{\sigma}_j \in C^\infty(\mathbb{R}^n)$, $j \in \mathbb{Z}$, such that

$$\operatorname{supp} \dot{\sigma}_j \subset \dot{\Gamma}_j,$$

$$a^{j\alpha} |D^\alpha \dot{\sigma}_j(\xi)| \le c_\alpha \quad \forall j \in \mathbb{Z}, \quad \forall \alpha \in \mathbb{N}_0^n, \quad \forall \xi \in \mathbb{R}^n,$$

$$\sum_{j \in \mathbb{Z}} \dot{\sigma}_j(\xi) \equiv 1, \quad \forall \xi \in \mathbb{R}^n \backslash \{0\}.$$

Introduce a definition of homogeneous anisotropic spaces $\dot{F}_{pq}^{(s)}$ and $\dot{B}_{pq}^{(s)}$.

DEFINITION 11.4.5. $\dot{F}_{pq}^{(s)}(\mathbb{R}^n) := \dot{F}_{pq}^{(s)}$, $0 < p < \infty$, $0 < q \le \infty$, is a (quasi-) Banach space of elements $f \in S'(\mathbb{R}^n)/P(\mathbb{R}^n)$ ($P(\mathbb{R}^n)$ is a set of all polynomials on $\mathbb{R}^n$) for which the (quasi-)norm

$$|f, \dot{F}_{pq}^{(s)}| = |\{\gamma^{\pm j}|f * \dot{V}_j|\}_{j=-\infty}^{\infty}, L_p(l_q)|$$

is finite, where $\dot{V}_j(x) = (F^{-1}\dot{\sigma}_j)(x)$; the weight γ^j corresponds to the case $s_\lambda > 0$ and γ^{-j} to the case $s_\lambda < 0$.

If we replace the norm in $L_p(l_q)$ by the norm in $l_q(L_p)$, we obtain the spaces $\dot{B}_{pq}^{(s)}$ (here, the case $p = \infty$ is not excluded).

DEFINITION 11.4.6. The spaces $\dot{F}_{\infty q}^{(s)}(\mathbb{R}^n) := \dot{F}_{\infty q}^{(s)}$, $0 < q \le \infty$, are defined analogously to Definition 11.4.2:

$$\dot{F}_{\infty q}^{(s)} = \Bigg\{ f \in S'(\mathbb{R}^n)/P(\mathbb{R}^n) :$$

$$|f, \dot{F}_{\infty q}^{(s)}| := \sup_{Q_{jk}^{(s)} \in \dot{\mathcal{Q}}^{(s)}} \left(|Q_{jk}^{(s)}|^{-1} \int_{Q_{jk}^{(s)}} \sum_{\nu=j}^{\infty} \gamma^{\nu q} |(f * \dot{V}_\nu)(x)|^q dx \right)^{1/q} < \infty \Bigg\}.$$

In some specific cases, it is convenient to use the following special choice of parameters C_λ and γ .

For $(s) \neq 0$, let $(a) := (a_1, ..., a_n)$ be a vector with coordinates $a_\lambda := s/s_\lambda$, where s is defined by the relation

$$\frac{1}{s} = \frac{1}{n} \left(\sum_{\lambda=1}^{n} \frac{1}{s_\lambda} \right).$$

For $(s) = 0$, we set the number s to be equal to 0 and all a_λ to be equal to 1.

We say that the number s is the mean smoothness and (a) is the anisotropy vector.

Introduce the following notation:

$\|x\|_{(a)} := \max\limits_\lambda |x_\lambda|^{1/a_\lambda}$ is the anisotropic distance;

$|\alpha|_{(a)} := \sum\limits_\lambda \alpha_\lambda a_\lambda$ is the anisotropic order of multi-index;

$t \circ x := (t^{a_1} x_1, ..., t^{a_n} x_n)$.

Set $\gamma := 2^s$, $C_\lambda := C^{a_\lambda}$, where $C > 1$ is an arbitrary fixed number. Then the sets Π_j and Γ_j defined in (11.88) and (11.89) can be described in the following way:

$$\Pi_j := \{\xi \in \mathbb{R}^n : \ \|\xi\|_{(a)} \le C2^j\}, \quad j \in \mathbb{N}, \tag{11.91}$$

$$\Gamma_1 := \Pi_2, \qquad \Gamma_j := \Pi_{j+1} \backslash \Pi_{j-1}, \quad j \ge 2. \tag{11.92}$$

In this case, the conditions for a smooth anisotropic partition of unity are as follows: $\{\sigma_j(\xi)\}_1^\infty$ is a sequence of nonnegative functions $\sigma_j \in C^\infty(\mathbb{R}^n)$, $j \in \mathbb{N}$, such that

$$\operatorname{supp}\sigma_j \subset \Gamma_j,$$

$$2^{j|\alpha|_{(a)}} \left| D_\xi^\alpha \sigma_j(\xi) \right| \le c_\alpha \text{ for any multi-indices } \alpha,\ j \in \mathbb{N},\ \xi \in \mathbb{R}^n,$$

$$\sum_1^\infty \sigma_j(\xi) \equiv 1.$$

Now we formulate several statements which are proved in the same way as their isotropic analogs.

PROPOSITION 11.4.7. *Suppose* $0 < p < \infty$, $0 < q \le \infty$, $\{f_k(x)\}_{k=0}^\infty$ *is a sequence of entire functions of exponential type* $\mu_k = (\mu_{k1}, \dots, \mu_{kn})$. *Then for* $\kappa > \frac{n}{2} + \frac{n}{\min(p,q)}$, *for any system of functions* $M_k(\xi)$, $\xi \in R^n$, $M_k \in H_2^\kappa$, $k = 0, 1, 2, \dots$, *we have*

$$|\{F^{-1}(M_k F f_k)\}_{k=0}^\infty,\ L_p(l_q)| \ll |\{M_l(\mu_{l1}\xi_1, \dots, \mu_{ln}\xi_n)\}_{l=0}^\infty,\ l_\infty(H_2^\kappa)| \cdot |\{f_k\}_{k=0}^\infty,\ L_p(l_q)|,$$

where H_2^κ *is the space of Bessel potentials* ($:= F_{22}^\kappa$) *and the constant is independent of* f_k *or* M_k.

The following statement follows from Proposition 11.4.7 and its scalar analog.

PROPOSITION 11.4.8. *Up to the equivalence of norms, the spaces* $F_{pq}^{(s)}$, $\dot{F}_{pq}^{(s)}$ *and* $B_{pq}^{(s)}$, $\dot{B}_{pq}^{(s)}$ *do not depend on the choice of systems* $\Sigma \in \Phi$ *and* $\dot{\Sigma} \in \Phi$.

PROPOSITION 11.4.9. *Let* $q \in [1, \infty)$; *then* $(F_{1q}^{(s)})' = F_{\infty q'}^{(-s)}$, *where* $q' := q/(q-1)$.

PROPOSITION 11.4.10. *Suppose* $v \in (0, \infty)$, $d = (d_1, \dots, d_n)$, $d_\lambda > 0$, $f \in S(\mathbb{R}^n)$ *and* $\operatorname{supp} Ff \subseteq \{\xi \in R^n : |\xi_\lambda| \le d_\lambda\}$. *Then for any multi-index* $\alpha = (\alpha_1, \dots, \alpha_n)$,

$$|(D^\alpha f)(x)| \le c^{|\alpha|} d^\alpha \sup_{z \in \mathbb{R}^n} \frac{f(x-z)}{1 + \|(d_1 z_1, \dots, d_n z_n)\|^{n/v}},$$

where the constant c *is independent of* f *or* d.

11.4.2. Construction of wavelet bases. A system of wavelets with compact spectra is an unconditional basis in Besov and Lizorkin-Triebel spaces if the spectra of functions of this system form a set of corridors close to $\{\Gamma_j\}$, which appears in the definition of these spaces. In the isotropic case, this can be easily obtained by using tensor products of Meyer's wavelets. The anisotropic situation would be analogous to the isotropic one if there existed systems of wavelets of single variables obtained by the dilation of the initial function by b^j, $j \in \mathbb{Z}$, for any $b > 1$. In this case, the required unconditional basis would be obtained by tensor multiplication of the systems of wavelets whose coefficient of dilation with respect to the variable x_λ is equal to γ_λ. However, systems of wavelets with compact spectrum have only been constructed for the coefficients of dilation $b = \frac{h+1}{h}$, $h \in \mathbb{N}$; therefore we have to construct an unconditional wavelet basis in the Besov and Lizorkin-Triebel anisotropic spaces by dividing Meyer-David wavelets into groups.

Let h_λ be a positive integer such that

$$\frac{h_\lambda + 1}{h_\lambda} \leq \gamma_\lambda. \tag{11.93}$$

In this section, the choice of h_λ satisfying (11.93) is arbitrary; for more detail, see Remark 11.4.24.

Set

$$b_\lambda := \frac{h_\lambda + 1}{h_\lambda}, \quad (b) = (b_1, ..., b_n), \quad r_\lambda := \frac{h_\lambda^2}{1 + 2h_\lambda}. \tag{11.94}$$

For $j = 2, 3, \dots$, define the numbers $l_j^{(\lambda)}$ as the least integers that satisfy the inequalities

(11.95)

$$\log_{b_\lambda} C_\lambda + (j-1)\log_{b_\lambda} \gamma_\lambda - \log_{b_\lambda} r_\lambda \leq l_j^{(\lambda)} + 1 < \log_{b_\lambda} C_\lambda + j \log_{b_\lambda} \gamma_\lambda - \log_{b_\lambda} r_\lambda.$$

As the difference between the right and left sides in (11.95) is equal to $\log_{b_\lambda} \gamma_\lambda \geq 1$, such numbers exist for any $j \geq 2$. Moreover, it follows from (11.95) that $l_j^{(\lambda)} \uparrow \infty$ for $j \to \infty$ and

$$\log_{b_\lambda} \gamma_\lambda - 1 < l_{j+1}^{(\lambda)} - l_j^{(\lambda)} < \log_{b_\lambda} \gamma_\lambda + 1, \quad j \geq 2. \tag{11.96}$$

Let $\mathcal{N}_j$, $j \geq 2$, be the set of pairs (m, ε) of n-dimensional vectors $m \in \mathbb{Z}^n$, $\varepsilon = (\varepsilon_1, \dots, \varepsilon_n)$, $\varepsilon_\lambda = 0, 1$, $\varepsilon \neq 0$, such that a coordinate m_λ takes the fixed value $m_\lambda = l_j^{(\lambda)} + 1$ if $\varepsilon_\lambda = 0$, ranging over the set $\{l_j^{(\lambda)} + 1, \dots, l_{j+1}^{(\lambda)}\}$ if $\varepsilon_\lambda = 1$.

For example, for $n = 2$, the set $\mathcal{N}_2$ consists of the elements

$$\begin{gathered}
\left((l_2^{(1)} + 1, l_2^{(2)} + 1); (0,1)\right), \quad \left((l_2^{(1)} + 1, l_2^{(2)} + 2); (0,1)\right), \dots, \\
\left((l_2^{(1)} + 1, l_3^{(2)}); (0,1)\right), \\
\left((l_2^{(1)} + 1, l_2^{(2)} + 1); (1,0)\right), \quad \left((l_2^{(1)} + 2, l_2^{(2)} + 1); (1,0)\right), \dots, \\
\left((l_3^{(1)}, l_2^{(2)} + 1); (1,0)\right), \\
\left((l_2^{(1)} + 1, l_2^{(2)} + 1); (1,1)\right), \quad \left((l_2^{(1)} + 2, l_2^{(2)} + 1); (1,1)\right), \\
\left((l_2^{(1)} + 1, l_2^{(2)} + 2); (1,1)\right), \dots, \left((l_3^{(1)}, l_3^{(2)}); (1,1)\right).
\end{gathered}$$

Relation (11.96) obviously implies the following proposition.

PROPOSITION 11.4.11. *The number of elements in any set $\mathcal{N}_j$ does not exceed*

$$N^* := \prod_\lambda ([\log_{b_\lambda} \gamma_\lambda] + 2) < \infty. \tag{11.97}$$

Let

$$\mathcal{N}_1 := \{((l_2^{(1)} + 1, \dots, l_2^{(n)} + 1); (0, \dots, 0))\}, \qquad \mathcal{N} := \bigcup_{j=1}^{\infty} \mathcal{N}_j.$$

Let us now consider the set of triples of n-dimensional vectors:

$$\mathcal{M}_1 := \left\{ \left((l_2^{(1)} + 1, \dots, l_2^{(n)} + 1), k, (0, \dots, 0) \right) : \ k \in \mathbb{Z}^n \right\},$$

$$\mathcal{M}_j := \{(m, k, \varepsilon) : \ (m, \varepsilon) \in \mathcal{N}_j, \ k \in \mathbb{Z}^n\}, \quad j \geq 2,$$

$$\mathcal{M} := \bigcup_{j=1}^{\infty} \mathcal{M}_j.$$

Let $\psi^{(\lambda, \varepsilon_\lambda)}(t)$ be $\psi_{h_\lambda}(t)$ for $\varepsilon_\lambda = 1$ and $\varphi_{h_\lambda}(t)$ for $\varepsilon_\lambda = 0$.

Consider the functions

$$\begin{aligned} \psi_{mk}^{(\varepsilon)}(x) &:= \prod_\lambda b_\lambda^{\frac{m_\lambda}{2}} \psi^{(\lambda, \varepsilon_\lambda)}(b_\lambda^{m_\lambda} x_\lambda - k_\lambda^{(\varepsilon_\lambda)}) \\ &= (b)^{\frac{m}{2}} \prod_\lambda \psi^{(\lambda, \varepsilon_\lambda)}(b_\lambda^{m_\lambda} x_\lambda - k_\lambda^{(\varepsilon_\lambda)}), \quad m, k \in \mathbb{Z}^n, \end{aligned}$$

where

$$k_\lambda^{(\varepsilon_\lambda)} := \begin{cases} k_\lambda, & \varepsilon_\lambda = 1, \\ k_\lambda / h_\lambda, & \varepsilon_\lambda = 0. \end{cases}$$

DEFINITION 11.4.12. For fixed m and ε, we say that the set $\{\psi_{mk}^{(\varepsilon)}\}_{k \in \mathbb{Z}^n}$ is a layer and the set

$$\Psi_j^{(s)} := \{\psi_{mk}^{(\varepsilon)}\}_{(m,k,\varepsilon) \in \mathcal{M}_j}$$

is a j-block.

The functions from the same layer have the same spectrum. The following assertion, which immediately follows from the choice of numbers b_λ and $l_j^{(\lambda)}$, explains why the given functions are united into a block.

PROPOSITION 11.4.13. *For any $j \in \mathbb{N}$, the following inclusion holds:*

$$\bigcup_{\mathcal{M}_j} \operatorname{supp} F\psi_{mk}^{(\varepsilon)} \subset \Gamma_j \cup \Gamma_{j+1}.$$

It is easy to see that

$$\bigcup_{\mathcal{M}_1} \operatorname{supp} F\psi_{mk}^{(\varepsilon)} = \left\{ \xi \in \mathbb{R}^n : \ |\xi_\lambda| < r_\lambda b_\lambda^{l_2 + 2} \right\},$$

$$\bigcup_{\mathcal{M}_j} \operatorname{supp} F\psi_{mk}^{(\varepsilon)} = \left\{ \xi \in \mathbb{R}^n : \ |\xi_\lambda| < r_\lambda b_\lambda^{l_{j+1}+2} \right\} \setminus \left\{ \xi \in \mathbb{R}^n : \ |\xi_\lambda| < r_\lambda b_\lambda^{l_j + 1} \right\},$$

$$j \geq 2.$$

These sets form corridors, which can serve as a basis for the definition of the spaces B and F up to equivalence (see Theorems 11.4.15 and 11.4.16). The spectra of products consisting of the functions ψ_{h_λ} fill the angles of the corridor, and the spectra of the products containing the functions φ_{h_λ} fill the parallelepipeds connecting these angles.

Denote by $\Psi^{(s)}$ the system of functions

$$\Psi^{(s)} := \{\psi_{mk}^{(\varepsilon)}\}_{(m,k,\varepsilon) \in \mathcal{M}}.$$

For arbitrary $j \in \mathbb{Z}$, consider the least integers $\dot{l}_j^{(\lambda)}$ satisfying (11.95); for $j \in \mathbb{Z}$, define the sets $\dot{\mathcal{N}}_j$ and $\dot{\mathcal{M}}_j$ in the same way as the sets $\mathcal{N}_j$ and $\mathcal{M}_j$. Set

$$\dot{\mathcal{M}} := \bigcup_{j\in\mathbb{Z}} \dot{\mathcal{M}}_j, \qquad \dot{\Psi}^{(s)} := \{\psi_{mk}^{(\varepsilon)}\}_{(m,k,\varepsilon)\in\dot{\mathcal{M}}}.$$

The following proposition is established by standard argument.

PROPOSITION 11.4.14. *The systems* $\Psi^{(s)}$ *and* $\dot{\Psi}^{(s)}$ *form orthonormal bases in* $L_2(\mathbb{R}^n)$.

In conclusion, we present a special case of the construction, which is important for the following discussion. It corresponds to the choice of parameters C_λ and γ described above (see page 410).

Let h be chosen in such a way that

$$\frac{h+1}{h} \le \min_\lambda 2^{a_\lambda}. \tag{11.98}$$

Consider the wavelet basis obtained by choosing all h_λ being equal to h.

In this case, the numbers $l_j^{(\lambda)}$, $j \ge 2$, are the least integers that satisfy the inequalities

$$\log_b C_\lambda + (j-1)a_\lambda \log_b 2 - \log_b r \le l_j^{(\lambda)} + 1 < \log_b C_\lambda + j a_\lambda \log_b 2 - \log_b r, \tag{11.99}$$

where $b := \frac{h+1}{h}$, $r := \frac{h^2}{1+2h}$.

The sets $\mathcal{N}_j^{(a)}$, $j \ge 2$, are the sets of pairs (m, ε) of n-dimensional vectors $m \in \mathbb{Z}^n$, $\varepsilon = (\varepsilon_1, ..., \varepsilon_n)$, $\varepsilon_\lambda = 0, 1$, $\varepsilon \ne 0$, such that a coordinate m_λ takes the fixed value $m_\lambda = l_j^{(\lambda)} + 1$ if $\varepsilon_\lambda = 0$ and ranges over the set $\{l_j^{(\lambda)} + 1, \dots, l_{j+1}^{(\lambda)}\}$ if $\varepsilon_\lambda = 1$.

By virtue of Proposition 11.4.11, the number of elements in each set $\mathcal{N}_j^{(a)}$, $j \ge 2$, is finite and does not exceed the number $\prod_\lambda([a_\lambda \log_b 2] + 2)$. Set

$$\mathcal{M}_1^{(a)} := \mathcal{N}_1^{(a)} := \left\{\left((l_2^{(1)} + 1, \dots, l_2^{(n)} + 1), (0, \dots, 0)\right)\right\},$$

$$\mathcal{M}_j^{(a)} := \left\{(m, k, \varepsilon) : (m, \varepsilon) \in \mathcal{N}_j^{(a)},\ k \in \mathbb{Z}^n\right\}, \quad j \in \mathbb{N},\ j \ge 2,$$

$$\mathcal{M}^{(a)} := \bigcup_{j=1}^{\infty} \mathcal{M}_j^{(a)}.$$

Let $\psi^{(\varepsilon_\lambda)}(t)$ be equal to $\psi_h(t)$ for $\varepsilon_\lambda = 1$ and $\varphi_h(t)$ for $\varepsilon_\lambda = 0$, where $\psi_h(t)$ and $\varphi_h(t)$ are the Meyer-David wavelets defined in (11.87) for h from (11.98).

Consider the wavelets

$$\psi_{mk}^{(\varepsilon)}(x) := \prod_\lambda b^{m_\lambda/2} \psi^{(\varepsilon_\lambda)}(b^{m_\lambda} x_\lambda + k_\lambda^{(\varepsilon_\lambda)}),$$

where $m, k \in \mathbb{Z}^n$,

$$k_\lambda^{(\varepsilon_\lambda)} := \begin{cases} k_\lambda, & \varepsilon_\lambda = 1, \\ k_\lambda / h_\lambda, & \varepsilon_\lambda = 0. \end{cases}$$

Denote $\psi_{00}^{(\varepsilon)}(x)$ by $\psi^{(\varepsilon)}(x)$. The set $\Psi_j^{(a)} := \{\psi_{mk}^{(\varepsilon)}\}_{(m,k,\varepsilon)\in\mathcal{M}_j^{(a)}}$ is a j-block, $\Psi^{(a)} := \bigcup_{j=1}^{\infty} \Psi_j^{(a)}$. In view of Proposition 11.4.13,

$$\bigcup_{\psi_{mk}^{(\varepsilon)}\in\Psi_j^{(a)}} \operatorname{supp} F\psi_{mk}^{(\varepsilon)} \subset \Gamma_j \cup \Gamma_{j+1}. \tag{11.100}$$

The system $\Psi^{(a)}$ forms an orthonormal basis in $L_2(\mathbb{R}^n)$ (Proposition 11.4.14).

11.4.3. Equivalent norms in the spaces B and F. For $f \in S'(\mathbb{R}^n)$, consider the functions

$$A_j(x) := A_j(x; f) = \sum_{(m,\varepsilon)\in\mathcal{N}_j} |f * \kappa_m^{(\varepsilon)}|(x), \quad j \geq 1,$$

where

$$\kappa_m^{(\varepsilon)}(x) := (b)^{\frac{m}{2}} \psi_{m,0}^{(\varepsilon)}(-x). \tag{11.101}$$

It is obvious that by virtue of Proposition 11.4.11 the functions $A_j(x)$ are well defined.

In all the theorems stated below, the weight γ^j corresponds to the case $s_\lambda > 0$, and the weight γ^{-j} corresponds to the case $s_\lambda < 0$.

For the proofs of all these theorems, see page 423.

THEOREM 11.4.15. A. *The norm in the space* $F_{pq}^{(s)}$, $0 < p < \infty$, $0 < q \leq \infty$, *is equivalent to the quantity*

$$A(f) := |\{\gamma^{\pm j} A_j(\cdot)\}_{j=1}^{\infty},\ L_p(l_q)|.$$

B. *The norm in the space* $B_{pq}^{(s)}$, $0 < p, q \leq \infty$, *is equivalent to the quantity*

$$|\{\gamma^{\pm j} A_j(\cdot)\}_{j=1}^{\infty},\ l_q(L_p)|.$$

Similarly to A_j, define the functions $\dot{A}_j(\cdot)$ for all $j \in \mathbb{Z}$.

THEOREM 11.4.16. *The norm in the space* $\dot{F}_{pq}^{(s)}$, $0 < p < \infty$, $0 < q \leq \infty$, *is equivalent to the quantity*

$$\dot{A}(f) := |\{\gamma^{\pm j} \dot{A}_j(\cdot)\}_{j\in\mathbb{Z}},\ L_p(l_q)|,$$

and the norm in the space $\dot{B}_{pq}^{(s)}$, $0 < p, q \leq \infty$, *is equivalent to the quantity*

$$|\{\gamma^{\pm j} \dot{A}_j(\cdot)\}_{j\in\mathbb{Z}},\ l_q(L_p)|.$$

11.4.4. Formulations of the results and comments. For $f \in S'(\mathbb{R}^n)$, consider the numbers $c_{mk}^{(\varepsilon)} := \langle f, \psi_{mk}^{(\varepsilon)}\rangle$. It is not hard to check that the Fourier series

$$\sum_{\mathcal{M}} c_{mk}^{(\varepsilon)} \psi_{mk}^{(\varepsilon)}(x) = \sum_{j=1}^{\infty} \sum_{\mathcal{M}_j} c_{mk}^{(\varepsilon)} \psi_{mk}^{(\varepsilon)}(x)$$

converges to f in $S'(\mathbb{R}^n)$ and the Fourier series

$$\sum_{\dot{\mathcal{M}}} c_{mk}^{(\varepsilon)} \psi_{mk}^{(\varepsilon)}(x) = \sum_{j\in\mathbb{Z}} \sum_{\dot{\mathcal{M}}_j} c_{mk}^{(\varepsilon)} \psi_{mk}^{(\varepsilon)}(x)$$

of the element f from the space $S'(\mathbb{R}^n)/P(\mathbb{R}^n)$ (factorization with respect to the set of all polynomials) converges to this element in this space.

Below, in theorems about unconditional bases in the spaces $F^{(s)}_{pq}$, $0 < p < \infty$, $0 < q \le \infty$, and $B^{(s)}_{pq}$, $0 < p, q \le \infty$, in the cases of nonseparable spaces, we consider the closures of the space $S(\mathbb{R}^n)$ in corresponding norms.

For each $m \in \mathbb{Z}^n$, introduce a partition of the space $\mathbb{R}^n$ into parallelepipeds $\{P^{(s)}_{mk}\}_{k\in\mathbb{Z}^n}$ whose edges are parallel to the coordinate axes:

$$P^{(s)}_{mk} = \left\{x \in \mathbb{R}^n : x_\lambda \in \left[b_\lambda^{-m_\lambda} k_\lambda,\ b_\lambda^{-m_\lambda}(k_\lambda + 1)\right)\right\}, \tag{11.102}$$

and denote by $\widetilde{\chi}^{(s)}_{mk}$ the characteristic function, normalized in $L_2(\mathbb{R}^n)$, of the parallelepiped $P^{(s)}_{mk}$.

THEOREM 11.4.17. A. *The orthonormal system $\Psi^{(s)}$ forms an unconditional basis in $F^{(s)}_{pq}$, $0 < p < \infty$, $0 < q \le \infty$; i.e., any function $f \in F^{(s)}_{pq}$ can be expanded into a Fourier series*

$$f(x) = \sum_{j=1}^{\infty} \sum_{(m,k,\varepsilon)\in\mathcal{M}_j} c^{(\varepsilon)}_{mk} \psi^{(\varepsilon)}_{mk}(x),$$

where

$$|f, F^{(s)}_{pq}| \sim \left| \left(\sum_{j=1}^{\infty} \gamma^{\pm jq} \sum_{(m,k,\varepsilon)\in\mathcal{M}_j} |c^{(\varepsilon)}_{mk} \widetilde{\chi}^{(\varepsilon)}_{mk}|^q \right)^{1/q}, L_p \right|$$

with the natural modification for $q = \infty$.

B. *The orthonormal system $\Psi^{(s)}$ forms an unconditional basis in $B^{(s)}_{pq}$, $0 < p < \infty$, $0 < q \le \infty$, or $1 \le q \le \infty$ for $p = \infty$, where the norm of $f \in B^{(s)}_{pq}$ is equivalent to*

$$\left(\sum_{j=1}^{\infty} \gamma^{\pm jq} \sum_{(m,\varepsilon)\in\mathcal{N}_j} b^{m\left(\frac{1}{2}-\frac{1}{p}\right)q} \left(\sum_{k\in\mathbb{Z}^n} |c^{(\varepsilon)}_{mk}|^p \right)^{q/p} \right)^{1/q}$$

with the obvious modification for $q = \infty$.

THEOREM 11.4.18. *The system $\dot{\Psi}^{(s)}$ forms an unconditional basis $\dot{F}^{(s)}_{pq}$, $0 < p < \infty$, $0 < q \le \infty$, where the norm of $f \in F^{(s)}_{pq}$ is equivalent to the quantity*

$$\left| \left(\sum_{j\in\mathbb{Z}} \gamma^{\pm jq} \sum_{(m,k,\varepsilon)\in\dot{\mathcal{M}}_j} |c^{(\varepsilon)}_{mk} \widetilde{\chi}^{(\varepsilon)}_{mk}|^q \right)^{1/q}, L_p \right|.$$

Analogously, for the space $\dot{B}^{(s)}_{pq}$, $p, q \in (0, \infty]$,

$$|f, \dot{B}^{(s)}_{pq}| \sim \left(\sum_{j\in\mathbb{Z}} \gamma^{\pm jq} \sum_{(m,\varepsilon)\in\dot{\mathcal{N}}_j} b^{m\left(\frac{1}{2}-\frac{1}{p}\right)q} \left(\sum_{k\in\mathbb{Z}^n} |c^{(\varepsilon)}_{mk}|^p \right)^{q/p} \right)^{1/q}.$$

Introduce some more notation. For $(m, k) \in \mathcal{N}$ and $j \ge 1$, set

$$\mathcal{A}_{m,k,j} := \left\{ (\nu, l, \delta) \in \mathcal{M}_j : P^{(s)}_{\nu l} \subseteq P^{(s)}_{mk} \right\};$$

for $(m, k) \in \dot{\mathcal{N}}$ and $j \in \mathbb{Z}$, set

$$\dot{\mathcal{A}}_{m,k,j} := \left\{ (\nu, l, \delta) \in \dot{\mathcal{M}}_j : P^{(s)}_{\nu l} \subseteq P^{(s)}_{mk} \right\}.$$

THEOREM 11.4.19. *For any function* $f \in F^{(s)}_{\infty q}$ *(respectively,* $\dot{F}^{(s)}_{\infty q}$*),* $1 < q < \infty$, *the norm of* $|f, F^{(s)}_{\infty q}|$ *(the norm of* $|f, \dot{F}^{(s)}_{\infty q}|$*) is equivalent to the quantity*

$$\sup_{(m,k,\varepsilon)\in\mathcal{M}} \left(\frac{1}{P^{(s)}_{mk}} \int_{P^{(s)}_{mk}} \sum_{j=1}^{\infty} \gamma^{\pm jq} \sum_{\mathcal{A}_{m,k,j}} |c^{(\delta)}_{\nu l} \widetilde{\chi}^{(\delta)}_{\nu l}(x)|^q \, dx \right)^{1/q},$$

where $c^{(\delta)}_{\nu l}$ *are the Fourier coefficients of the function* f *with respect to the orthonormal system* $\Psi^{(s)}$ *(the symbols* $\mathcal{M}$, $\mathcal{A}^{(s)}_{m,k,j}$, $\Psi^{(s)}$ *are replaced by* $\dot{\mathcal{M}}$, $\dot{A}_{m,k,j}$, $\dot{\Psi}^{(s)}$, *respectively).*

For the particular case of anisotropic basis described on page 414, the parallelepipeds $P^{(s)}_{mk}$ used in Theorems 11.4.17 and 11.4.18 to compute the norm of a function by its Fourier coefficients can be replaced by

$$Q^{(a)}_{j\nu} = \left\{x \in \mathbb{R}^n : x_\lambda \in \left[\nu_\lambda 2^{-a_\lambda j}, (\nu_\lambda + 1)2^{-a_\lambda j}\right)\right\}, \quad j \in \mathbb{N}, \quad \nu \in \mathbb{Z}^n. \tag{11.103}$$

The set of parallelepipeds $Q^{(a)}_{j\nu}$ is simpler than that of parallelepipeds $P^{(s)}_{mk}$, because the first subscript of Q is one-dimensional, while the first subscript of P is n-dimensional. Parallelepipeds of the form (11.103) will be called a-cubes.

Introduce the following notation:

$$P^{(a)}_{mk} = \{x \in \mathbb{R}^n : x_\lambda \in [k_\lambda b^{-m_\lambda}, (k_\lambda + 1)b^{-m_\lambda})\},$$

$$\mathcal{L}^{(a)}_{j\nu} := \left\{(m,k,\varepsilon) : \ (m,\varepsilon) \in \mathcal{N}^{(a)}_j, \ k \in \mathbb{Z}^n, \ P^{(a)}_{mk} \cap Q^{(a)}_{j\nu} \neq \emptyset\right\}, \tag{11.104}$$

$$j \in \mathbb{N}, \quad m, k, \nu \in \mathbb{Z}^n;$$

χ_E is the characteristic function of the set E, $\widetilde{\chi}_E := |E|^{-1/2}\chi_E$. If Q is an a-cube of the form (11.103), then $x_Q = (\nu_1 2^{-a_1 j}, ..., \nu_n 2^{-a_n j})$ is the lower left vertex of the a-cube Q . Note that the anisotropic distance $\|x - y\|_{(a)}$ between two neighboring vertices x, y of the a-cube $Q^{(a)}_{j\nu}$ is equal to $l(Q) := 2^{-j}$; $l(Q)$ is an analog of the edge length for an isotropic cube.

Let

$$\Omega_{(a)} := \{Q^{(a)}_{j\nu}, \ j \in \mathbb{N}, \ \nu \in \mathbb{Z}^n\}$$

be a set of a-cubes; if $Q = Q^{(a)}_{j\nu}$, then $\mathcal{L}^{(a)}_Q := \mathcal{L}^{(a)}_{j\nu}$.

DEFINITION 11.4.20. The space $f^{(s)}_{pq}$, $0 < p < \infty$, $0 < q \leq \infty$, is a (quasi-) Banach space of number sequences $c := \{c_Q\}_{Q\in\Omega_{(a)}}$, with the finite (quasi-)norm

$$|c, f^{(s)}_{pq}| := \left| \left\{ 2^{js} \sum_{Q\in\Omega_{(a)}:\ l(Q)=2^{-j}} c_Q \widetilde{\chi}_Q \right\}_{j=1}^{\infty}, \ L_p(l_q) \right|.$$

THEOREM 11.4.21. *The wavelet system* $\Psi^{(a)}$ *is orthonormal in* $L_2(\mathbb{R}^n)$ *and forms an unconditional basis in* $F^{(s)}_{pq}$, $0 < p < \infty, 0 < q < \infty$*; i.e., any function* $f \in F^{(s)}_{pq}$ *can be expanded into the Fourier series*

$$f = \sum_{j\in\mathbb{N}} \sum_{(m,k,\varepsilon)\in\mathcal{M}^{(a)}_j} c^{(\varepsilon)}_{mk} \psi^{(\varepsilon)}_{mk},$$

where

$$|f, F^{(s)}_{pq}| \sim \left\| \left\{ \sum_{(m,k,\varepsilon)\in\mathcal{L}^{(a)}_Q} |c^{(\varepsilon)}_{mk}| \right\}_{Q\in\Omega_{(a)}}, f^{(s)}_{pq} \right\|.$$

If $q = \infty$, the equivalence of norms is preserved but expansion into the Fourier series is possible only for functions from the closure of the space $S(\mathbb{R}^n)$ in $F^{(s)}_{p\infty}$. In what follows, we shall always denote by $F^{(s)}_{p\infty}$ the closure described above.

In a similar way, one can reformulate the assertions about unconditional bases in the spaces $\dot F^{(s)}_{\infty q}$, $F^{(s)}_{\infty q}$, $0 < q < \infty$, $\dot F^{(s)}_{\infty q}$, and in nonhomogeneous and homogeneous Besov spaces.

The essence of Theorem 11.4.21 is that for fixed $j \geq 1$, in the expression of the equivalent norm, the set of cubes $P^{(s)}_{mk}$ corresponding to the j-th block is replaced by the set of cubes $Q^{(a)}_{j\nu}$, $\nu \in \mathbb{Z}^n$, whose size depends only on j (compare with Theorem 11.4.17). This considerably simplifies employment of the norm of the space $f^{(s)}_{pq}$ in the proofs of the theorems in the next section.

REMARK 11.4.22. The basis constructed above depends on the coefficient of smoothness s (more precisely, it depends on the anisotropy vector a).

Let us give an example demonstrating that the isotropic Meyer basis (i.e., the basis consisting of tensor products of wavelets with the same scale) is not a basis in the anisotropic space.

Let $s_1 = 1$, $s_2 = 1/\log_2 \frac{3}{2} = \log_{3/2} 2$. We shall use the choice of parameters C_λ and γ described on page 410. Then $a_1 = 2$, $a_2 := 2^{1/s_2} = 3/2$. Construct a wavelet basis using tensor products of the Meyer-David wavelets with parameters $h_1 = 1$, $h_2 = 2$ (see (11.93)). It has the form

$$\Psi^{(s_1,s_2)} = \left\{\psi^{(\varepsilon_1,\varepsilon_2)}_{m,k}\right\}_{\varepsilon\in E,\, m\in\mathbb{Z}, k\in\mathbb{Z}^2} = \left\{\psi^{2,\varepsilon_1}_{m,k}(x_1)\psi^{3/2,\varepsilon_2}_{m,k}(x_2)\right\}_{\varepsilon\in E,\, m\in\mathbb{Z}, k\in\mathbb{Z}^2},$$

where, for convenience, the superscripts of ψ are taken equal to the dilations (the same is true for the superscripts of the functions θ used to define the Fourier transform for ψ (see (11.87))). Note that $r_1 = \frac{1}{3}$, $r_2 := \frac{4}{5}$ (see (11.94)).

Choose m_0 so that for any $m \geq m_0$,

$$m < [(m-2)s_2], \quad \frac{2^m}{3} \geq \frac{9}{5}\left(\frac{3}{2}\right)^{[(m-2)s_2]}, \tag{11.105}$$

where $[t]$ is the integer part of the number t.

Then for any $\xi \in \operatorname{supp} F\psi_m^{3/2}$, $\cos\theta_2\left(2^{-m}\xi\right) = 1$ and for any $k,\ l \in \mathbb{Z}$, we have

$$\begin{aligned}
&\left\langle \psi^2_{m,k}(x_1)\psi^{3/2}_{m,l3^m}(x_2), \psi^2_{m,k}(x_1)\varphi^2_{m,l2^{2m}}(x_2)\right\rangle \\
&= 3^{-m/2}\int\limits_{\mathbb{R}} \exp\left(i2\pi\xi_2\left(\left(\frac{3}{2}\right)^{-m} l3^m - 2^{-m}l2^{2m}\right)\right) \\
&\quad \times \exp\left(-i\pi\xi_2\left(\frac{3}{2}\right)^{-m}\right)\sin\theta_{3/2}\left(\left(\frac{3}{2}\right)^{-m}\xi_2\right)\cos\theta_2\left(2^{-m}\xi_2\right)d\xi_2 \\
&= 3^{-m/2}\int\limits_{\mathbb{R}} \exp\left(-i\pi\xi_2\left(\frac{3}{2}\right)^{-m}\right)\sin\theta_{3/2}\left(\left(\frac{3}{2}\right)^{-m}\xi_2\right)d\xi_2 \\
&= \left(\frac{3}{4}\right)^{m/2}\int\limits_{\mathbb{R}} \exp\left(-i\pi\xi_2\right)\sin\theta_{3/2}\left(\xi_2\right)d\xi_2 \\
&= -i\left(\frac{3}{4}\right)^{m/2}\int\limits_{\mathbb{R}} \sin\left(\pi\xi_2\right)\sin\theta_{3/2}\left(\xi_2\right)d\xi_2.
\end{aligned}$$

Let

$$A := \int\limits_{\mathbb{R}} \sin\left(\pi\xi\right)\sin\theta_{3/2}\left(\xi\right)d\xi.$$

It is easy to see that $A \neq 0$. Finally, for any $m \geq m_0$, $k \in \mathbb{Z}$, $l \in \mathbb{Z}$, we have

$$\langle\psi^2_{m,k}(x_1)\psi^{3/2}_{m,l3^m}(x_2), \psi^2_{m,k}(x_1)\varphi^2_{m,l2^{2m}}(x_2)\rangle = -i\left(\frac{3}{4}\right)^{m/2} A. \tag{11.106}$$

Let $\mu = m+1, \ldots, [(m-2)s_2]$. It is easy to see that

$$\left(\frac{3}{2}\right)^{\mu+2}\frac{4}{5} \leq 2^{m+1}\frac{1}{3}; \tag{11.107}$$

hence for any $\xi \in \operatorname{supp} F\psi^{3/2}_{\mu,0}$, $\cos\theta_2\left(2^{-m}\xi\right) = 1$. Under these conditions, for any $k \in \mathbb{Z}$, we obtain

$$\begin{aligned}
&\left\langle \varphi^2_{\mu,2^{\mu-m}k}(x_1)\psi^{3/2}_{\mu,3^\mu 2^{[(m-2)s_2]+1-\mu}}(x_2), \psi^2_{m,k}(x_1)\varphi^2_{m,2^{[(m-2)s_2]+1+m}}(x_2)\right\rangle \\
&= 2^{-\frac{\mu+m}{2}}\int\limits_{\mathbb{R}} \exp\left(i2\pi\xi_1\left(2^{-\mu}2^{\mu-m}k - 2^{-m}k\right)\right) \\
&\quad \times \exp\left(i\pi\xi_1 2^{-m}\right)\cos\theta_2\left(2^{-\mu}\xi_1\right)\sin\theta_2\left(2^{-m}\xi_1\right)d\xi_1 \\
&\quad \times \left(\frac{3}{2}\right)^{-\mu/2} 2^{-m/2}\int\limits_{\mathbb{R}} \exp\left(i2\pi\xi_2\left(\left(\frac{3}{2}\right)^{-\mu} 3^\mu 2^{[(m-2)s_2]+1-\mu} - 2^{[(m-2)s_2]+1}\right)\right) \\
&\quad \times \exp\left(-i\pi\xi_2\left(\frac{3}{2}\right)^{-\mu}\right)\sin\theta_{3/2}\left(\left(\frac{3}{2}\right)^{-\mu}\xi_2\right)\cos\theta_2\left(2^{-m}\xi_2\right)d\xi_2
\end{aligned}$$

$$= 2^{\frac{m-\mu}{2}} \int_{\mathbb{R}} \exp(i\pi\xi_1) \sin\theta_2(\xi_1)\, d\xi_1$$

$$\times \left(\frac{3}{2}\right)^{\mu/2} 2^{-m/2} \int_{\mathbb{R}} \exp(-i\pi\xi_2) \sin\theta_{3/2}(\xi_2)\, d\xi_2$$

$$= \left(\frac{3}{4}\right)^{\mu/2} \cdot iA \cdot \int_{\mathbb{R}} \cos\left(\frac{\xi_1}{2}\right) \sin\theta_2(\xi_1)\, d\xi_1.$$

Let

$$B := \int_{\mathbb{R}} \cos\left(\frac{\xi}{2}\right) \sin\theta_2(\xi)\, d\xi.$$

It is clear that $B \neq 0$. Finally, we have

$$\begin{aligned} &\left\langle \varphi^2_{\mu,2^{\mu-m}k}(x_1)\psi^{3/2}_{\mu,3^{\mu}2^{[(m-2)s_2]+1-\mu}}(x_2), \psi^2_{m,k}(x_1)\varphi^2_{m,2^{[(m-2)s_2]+1+m}}(x_2)\right\rangle \\ &\qquad = \left(\tfrac{3}{4}\right)^{\mu/2} \cdot iA \cdot B. \end{aligned} \tag{11.108}$$

Note that in the case under consideration, we have

$$P^{(s)}_{m,k} = Q^{(a)}_{m,k}$$

(see (11.102) and (11.103)).

Using (11.108) and Theorem 11.4.17, we find a lower estimate for

$$\|\psi^2_{m,k}(x_1)\varphi^2_{m,2^{[(m-2)s_2]+1+m}}(x_2)\|_{F^{(1,s_2)}_{pq}}$$

$$\gg \Bigg\| \Bigg(\sum_{\mu=m+1}^{[(m-2)s_2]} 2^{\mu s q} \times \left|\langle \varphi^2_{\mu,2^{\mu-m}k}(x_1)\psi^{3/2}_{\mu,3^{\mu}2^{[(m-2)s_2]+1-\mu}}(x_2), \psi^2_{m,k}(x_1)\varphi^2_{m,2^{[(m-2)s_2]+1+m}}(x_2)\rangle \times \widetilde{\kappa}_{Q^{(a)}_{\mu,2^{\mu-m}k,3^{\mu}2^{[(m-2)s_2]+1-\mu}}}\right|^q \Bigg)^{1/q} \Bigg\|_p .$$

The left lower angles of the rectangles $Q^{(a)}_{\mu,2^{\mu-m}k,3^{\mu}2^{[(m-2)s_2]+1-\mu}}$ coincide; therefore,

$$\Bigg\| \Bigg(\sum_{\mu=m+1}^{[(m-2)s_2]} 2^{\mu s q} \times \left|\langle \varphi^2_{\mu,2^{\mu-m}k}(x_1)\psi^{3/2}_{\mu,3^{\mu}2^{[(m-2)s_2]+1-\mu}}(x_2), \psi^2_{m,k}(x_1)\varphi^2_{m,2^{[(m-2)s_2]+1+m}}(x_2)\rangle \times \widetilde{\kappa}_{Q^{(a)}_{\mu,2^{\mu-m}k,3^{\mu}2^{[(m-2)s_2]+1-\mu}}}\right|^q \Bigg)^{1/q} \Bigg\|_p$$

$$\gg \left(\sum_{j=m+1}^{[(m-2)s_2]} \left(\sum_{\nu=m+1}^{j} 2^{\nu s q} \left(\frac{3}{4}\right)^{\frac{\nu q}{2}} (|A \cdot B|)^q 3^{\frac{\nu q}{2}} \right)^{p/q} (3^{-j} - 3^{-j-1}) \right)^{1/p}$$

$$= |A| \cdot |B| \cdot \left(\sum_{j=m+1}^{[(m-2)s_2]} \left(\sum_{\nu=m+1}^{j} 2^{\nu(s-1)q} 3^{\nu q} \right)^{p/q} 2 \cdot 3^{-j-1} \right)^{1/p}$$

$$\geq |A| \cdot |B| \cdot \left(\sum_{j=m+1}^{[(m-2)s_2]} 2^{j(s-1)p+1} 3^{j(p-1)-1} \right)^{1/p} \gg 2^{ms_2(s-1)} 3^{\frac{(p-1)s_2 m}{p}}.$$

Finally,

$$\|\psi^2_{m,k}(x_1)\varphi^2_{m,2^{[(m-2)s_2]+1+m}}(x_2)\|_{F^{(1,s_2)}_{pq}} \gg 2^{ms_2(s-1)} 3^{\frac{(p-1)s_2 m}{p}}. \tag{11.109}$$

Consider the function

$$f(x_1,x_2) := \sum_{m=m_0}^{\infty} a_m \psi^2_{m,k(m)}(x_1)\psi^{3/2}_{m,2^{[(m-2)s_2]+1-m}3^m}(x_2), \tag{11.110}$$

where $k(m)$ is chosen in such a way that the rectangles $Q^{(a)}_{m,k(m),2^{[(m-2)s_2]+1-m}3^m}$ are mutually disjoint. Then by Theorem 11.4.17, we get

$$\begin{aligned} \|f\|_{F^{(1,s_2)}_{p,q}} &\sim \| \sum_{m=m_0}^{\infty} 2^{ms} a_m \widetilde{\kappa}_{Q^{(a)}_{m,k(m),2^{[(m-2)s_2]+1-m}3^m}} \|_p \\ &= \left(\sum_{m=m_0}^{\infty} \left(2^{ms}3^{m/2}a_m\right)^p 3^{-m} \right)^{1/p} = \left(\sum_{m=m_0}^{\infty} 2^{mps} 3^{m\left(\frac{p}{2}-1\right)} a_m^p \right)^{1/p}. \end{aligned} \tag{11.111}$$

If we expand the function f into a series with respect to

$$\left\{ \psi^{2,\varepsilon_1}_{m,k_1}(x_1), \psi^{2,\varepsilon_2}_{m,k_2}(x_2) \right\}_{m\in\mathbb{Z},\, k\in\mathbb{Z}^2},$$

then this series contains summands of the form

$$a_m \left\langle \psi^2_{m,k(m)}(x_1)\psi^{3/2}_{m,2^{[(m-2)s_2]+1-m}3^m}(x_2), \psi^2_{m,k(m)}(x_1)\varphi^2_{m,2^{[(m-2)s_2]+1+m}}(x_2) \right\rangle \times \psi^2_{m,k(m)}(x_1)\varphi^2_{m,2^{[(m-2)s_2]+1+m}}(x_2).$$

By (11.106) and (11.109) we get

(11.112)

$$\begin{aligned} &\|a_m \langle \psi^2_{m,k(m)}(x_1)\psi^{3/2}_{m,2^{[(m-2)s_2]+1-m}3^m}(x_2), \psi^2_{m,k(m)}(x_1)\varphi^2_{m,2^{[(m-2)s_2]+1+m}}(x_2)\rangle \\ &\quad \times \psi^2_{m,k(m)}(x_1)\varphi^2_{m,2^{[(m-2)s_2]+1+m}}(x_2)\|_{F^{(1,s_2)}_{pq}} \\ &\gg a_m \left(\frac{3}{4}\right)^{m/2} 2^{ms_2(s-1)} 3^{\frac{(p-1)s_2 m}{p}}. \end{aligned}$$

Take $a_m := \left(\frac{3}{4}\right)^{-m/2} 2^{-ms_2(s-1)} 3^{-\frac{(p-1)s_2 m}{p}}$. It follows from (11.112) that the expansion of f with respect to "isotropic" basis does not converge in $F^{(1,s_2)}_{pq}$.

Study the convergence of series (11.110), using (11.111). In (11.111), we have a geometric progression with ratio

$$2^s 3^{\left(\frac{1}{2}-\frac{1}{p}\right)} \left(\frac{3}{4}\right)^{-1/2} 2^{-s_2(s-1)} 3^{-\frac{(p-1)s_2}{p}} = 2^{1+s+s_2-s_2 s} 3^{-\frac{(p-1)s_2+1}{p}}.$$

This ratio is strictly less than 1 if

$$3^{\frac{(p-1)s_2+1}{p}} > 2^{1+s+s_2-s_2 s}$$

or

$$(11.113)\qquad \frac{(p-1)s_2+1}{p} > (1+s+s_2-s_2s)\log_3 2.$$

It is easily seen that $1+s+s_2-s_2s<2$ and $s_2>2\log_3 2$; hence (11.113) holds for sufficiently large p.

Finally, we have a function $f\in F_{pq}^{(1,s_2)}$, whose expansion with respect to an "isotropic system" contains infinitely many terms whose norms are approximately equal to 1.

REMARK 11.4.23. Recall that $F_{p2}^{(l)}=W_p^{(l)}$ for $1<p<\infty$ and $l_\lambda>0$, where by $W_p^{(l)}$ we denote the space of Bessel potentials for fractional l_λ and the Sobolev space for positive integer l_λ. Thus we have constructed a basis of entire functions of exponential type in the anisotropic Sobolev spaces.

REMARK 11.4.24. Naturally, we try to construct a basis such that each of its blocks contains as few layers as possible (see Definition 11.4.12). Let us show that to do this, we should use the Meyer-David wavelets.

It follows from (11.93) and (11.95) that for fixed γ, the number of layers in a block is minimal if h_λ satisfies the condition

$$(11.114)\qquad \gamma_\lambda := \gamma^{1/s_\lambda}\in\left[\frac{h_\lambda+1}{h_\lambda},\ \frac{h_\lambda}{h_\lambda-1}\right)\quad\left(\frac{1}{0}:=\infty\right).$$

Consider the problem of how the number of layers in a block depends on the parameter γ. For simplicity, assume that $n=2$. For definiteness, let $s_1>s_2>0$. If $\gamma=2^{s_1}$, then in view of (11.114), $h_1=h_2=1$; i.e., only the Meyer wavelets are used for the construction of a basis. Here, in view of (11.96) the number of layers in any block is greater than $\log_2 2^{s_1/s_2}=s_1/s_2$. Thus, when s_1 is much greater than s_2, large numbers of layers appear in blocks.

Nevertheless, for any relation between s_1 and s_2, we can obtain $N^*\le 4$, using the Meyer-David wavelets (see (11.97)). To this end, it suffices to set $\gamma=2^{s_2}$, $h_2=1$. Then

$$N^*\le[\log_{(1+h_1^{-1})}2^{s_2/s_1}]+2,$$

where h_1 is such that

$$\frac{h_1+1}{h_1}\le 2^{s_2/s_1}<\frac{h_1}{h_1-1}.$$

It follows that

$$N^*\le\left[\log_{\frac{h_1+1}{h_1}}\left(\frac{h_1}{h_1-1}\right)\right]+2\le 4.$$

It is clear that the latter inequality holds for any $h_1\ge 2$.

For arbitrary $n\in\mathbb{N}$, set:

1) $\gamma=2^{1/s_{\lambda_0}}$, where $s_{\lambda_0}=\min\{s_1,\dots,s_n\}$;

2) for λ, for which $s_\lambda=s_{\lambda_0}$, $h_\lambda=1$;

3) for λ such that $s_\lambda>s_{\lambda_0}$, h_λ is chosen uniquely by condition (11.114).

It is easy to see that here $N^*\le 4^{n-n_1}$, where n_1 is the number of λ for which $s_\lambda=s_{\lambda_0}$.

11.4.5. Isotropic case. In the isotropic case ($s_1 = s_2 = \cdots = s_n = s_0$), the wavelet basis does not depend on the coefficient of smoothness s. Setting $\gamma = 2^{s_0}$, $C_\lambda = \frac{1}{3}$, $h_\lambda = 1$, we obtain the Meyer wavelet basis. Let $\psi^{(1)}(t) = \psi_1(t)$ and $\psi^{(0)}(t) = \varphi_1(t)$, where $\psi_1(t)$ and $\varphi_1(t)$ are the Meyer-David wavelets defined in (11.87) for $h = 1$. Define the wavelets

$$\psi_{jk}^{(\varepsilon)}(x) := \prod_\lambda 2^{j/2}\psi^{(\varepsilon_\lambda)}(2^j x_\lambda + k_\lambda),$$

where $j \in \mathbb{N}, k \in \mathbb{Z}^n$. Let

$$\Psi := \{\psi_{1,k}^{(0,\dots,0)}, \psi_{jk}^{(\varepsilon)}\}_{\{(\varepsilon)\in\{0,1\}^n, \varepsilon\neq 0, j\in\mathbb{N}, k\in\mathbb{Z}^n\}}.$$

THEOREM 11.4.25. *The wavelet system Ψ is orthonormal in $L_2(\mathbb{R}^n)$ and forms an unconditional basis in $F_{pq}^{s_0}$, $0 < p < \infty$, $0 < q < \infty$; i.e., any function $f \in F_{pq}^{s_0}$ can be expanded into the Fourier series*

$$f = \sum_{k\in\mathbb{Z}^n} c_k^{(0,\dots,0)}\psi_{1k}^{(0,\dots,0)} + \sum_{j=1}^\infty \sum_{k\in\mathbb{Z}^n, \varepsilon\in\{0,1\}^n, \varepsilon\neq 0} c_{jk}^{(\varepsilon)}\psi_{jk}^{(\varepsilon)}.$$

Here,

$$|f, F_{pq}^{s_0}| \sim \left|\left(2^{qs_0}\left(\sum_{k\in\mathbb{Z}^n} |c_{1k}^{(0,\dots,0)}|\widetilde{\chi}_{Q_{1,k}}\right)^q + \sum_{j=2}^\infty 2^{jqs_0}\left(\sum_{k\in\mathbb{Z}^n}\left(\sum_{(\varepsilon)\in\{0,1\}^n, \varepsilon\neq 0} |c_{j-1,k}^{(\varepsilon)}|\right)\widetilde{\chi}_{Q_{j,k}}\right)^q\right)^{1/q}, L_p\right|,$$

where $Q_{jk} := \{x \in \mathbb{R}^n : \ x_\lambda \in [k_\lambda 2^{-j}, (k_\lambda + 1)2^{-j}]\}$.

If $q = \infty$, the norms are equivalent, but only functions from the closure of the space $S(\mathbb{R}^n)$ in $F_{p\infty}^{s_0}$ can be expanded into the Fourier series.

The orthonormal system Ψ forms an unconditional basis in $B_{pq}^{s_0}$, $0 < p < \infty$, $0 < q \leq \infty$, or $1 \leq q \leq \infty$ for $p = \infty$, and the norm $f \in B_{pq}^{s_0}$ is equivalent to

$$\left(2^{\left(s_0+\frac{1}{2}-\frac{1}{p}\right)q}\left(\sum_{k\in\mathbb{Z}^n} |c_{1k}^{(0,\dots,0)}|^p\right)^{q/p} + \sum_{j=2}^\infty 2^{j\left(s_0+\frac{1}{2}-\frac{1}{p}\right)q} \sum_{(\varepsilon)\in\{0,1\}^n, \varepsilon\neq 0}\left(\sum_{k\in\mathbb{Z}^n} |c_{j-1,k}^{(\varepsilon)}|^p\right)^{q/p}\right)^{1/q}$$

with the obvious modification for $q = \infty$.

In a similar way, one can reformulate the statements about unconditional bases in the spaces $\dot{F}_{pq}^{s_0}$, $F_{\infty q}^{s_0}$, $0 < q < \infty$, $\dot{F}_{\infty q}^{s_0}$ and the homogeneous Besov spaces.

In this case, the proofs of Theorems 11.4.17, 11.4.18, and 11.4.19 given below are simplified.

11.4.6. Proofs. For simplicity, in what follows, we shall assume that $n = 2$.

`Proof of Theorem 11.4.15.` We shall prove the theorem for the spaces F, since it is easily seen that we come to the same conclusion for the spaces B, using the scalar version of Proposition 11.4.7.

Step 1. Using Proposition 11.4.7, it is not hard to show that the norm in the space $F_{pq}^{(s)}$ can be defined equivalently with the help of the system of functions $\{\sigma_j'(\xi)\}_1^\infty$ satisfying the following conditions:

$$\sigma_j' \geq 0, \quad \sigma_j' \in C^\infty(\mathbb{R}), \quad \operatorname{supp} \sigma_j' \subset \Gamma_j \cup \Gamma_{j+1}, \quad j \in \mathbb{N},$$

$$a^{j\alpha}|D^\alpha \sigma_j'(\xi)| \leq c_\alpha \text{ for any multi-index } \alpha,\ j \in \mathbb{N},\ \xi \in \mathbb{R}^n,$$

there exists a constant $c \in (0, \infty)$ such that $\sum\limits_{j=1}^{\infty} \sigma_j'(\xi) \geq c \quad \forall \xi \in \mathbb{R}^n$.

Step 2. Let $s_\lambda > 0$ (the case $s_\lambda < 0$ is considered analogously). Suppose the functions $\omega_m^{(\varepsilon)}$, $(m, \varepsilon) \in \mathcal{N}_j$, $j \geq 1$, are defined by the condition $F\omega_m^{(\varepsilon)} = |\kappa_m^{(\varepsilon)}|$ (the functions $\kappa_m^{(\varepsilon)}$ are defined in (11.101)). Then the system of functions $\left\{ \sum\limits_{(m,\varepsilon)\in\mathcal{N}_j} F\omega_m^{(\varepsilon)} \right\}_{j=1}^{\infty}$ satisfies the condition of Step 1 (see Proposition 11.4.13). Hence the norm in $F_{pq}^{(s)}$ can be defined equivalently as

$$\mathcal{A}(f) := \left| \left(\sum_{j=1}^{\infty} \gamma^{jq} \left| \sum_{(m,\varepsilon)\in\mathcal{N}_j} f * \omega_m^{(\varepsilon)} \right|^q \right)^{1/q}, L_p \right|.$$

Step 3. Consider the quantities

$$\left| \left(\sum_{j=1}^{\infty} \gamma^{jq} \left| f * \omega_{m_j}^{(\varepsilon_j)} \right|^q \right)^{1/q}, L_p \right|, \tag{11.115}$$

where for any $j \in \mathbb{N}$, (m_j, ε_j) is an arbitrary element from $\mathcal{N}_j$. Let us prove that all these quantities are estimated from above by $\mathcal{A}(f)$ with a constant independent of the choice of (m_j, ε_j) or the function f.

To this end, consider the expression

$$\mathcal{B}(f) := \left| \left(\sum_{j=1}^{\infty} \gamma^{jq} \left| \sum_{\nu=-1}^{1} \sum_{(m,\varepsilon)\in\mathcal{N}_{j+\nu}} f * \omega_m^{(\varepsilon)} \right|^q \right)^{1/q}, L_p \right|,$$

where $\mathcal{N}_0 = \emptyset$.

Note that

$$F\omega_{m_j}^{(\varepsilon_j)} \cdot Ff = \frac{F\omega_{m_j}^{(\varepsilon_j)}}{\sum\limits_{\nu=-1}^{1} \sum\limits_{(m,\varepsilon)\in\mathcal{N}_{j+\nu}} F\omega_m^{(\varepsilon)}} \cdot F\left(\sum_{\nu=-1}^{1} \sum_{(m,\varepsilon)\in\mathcal{N}_{j+\nu}} f * \omega_m^{(\varepsilon)} \right),$$

$$\sum_{\nu=-1}^{1} \sum_{(m,\varepsilon)\in\mathcal{N}_{j+\nu}} (F\omega_m^{(\varepsilon)})(\xi) \geq c > 0 \quad \forall \xi \in \operatorname{supp} F\omega_{m_j}^{(\varepsilon_j)}.$$

In addition,

$$\sum_{\nu=-1}^{1} \sum_{(m,\varepsilon)\in\mathcal{N}_{j+\nu}} (f * \omega_m^{(\varepsilon)})(x)$$

is an entire function of exponential type $(r_1 b_1^{l_{j+2}^{(1)}+2}, r_2 b_2^{l_{j+2}^{(2)}+2})$. Using Proposition 11.4.7, we derive that variables (11.115) are estimated from above by $\mathcal{B}(f)$ with a constant independent of the function f or the choice of (m_j, ε_j). Moreover, it is easy to see that $\mathcal{B}(f) \ll \mathcal{A}(f)$. Thus we have proved the statement of Step 3.

Step 4. Let $g_1^{(\varepsilon)}(\xi) \in C^\infty(\mathbb{R}^2)$, $\xi \in \mathbb{R}^2$, be a compactly supported function that coincides with $\exp(-i\pi(\varepsilon_1\xi_1 + \varepsilon_2\xi_2))$ on $\operatorname{supp} F\psi_{00}^{(\varepsilon)}$. Then it is clear that

$$F\kappa_m^{(\varepsilon)}(\xi) = g_1(b_1^{-m_1}\xi_1, b_2^{-m_2}\xi_2)|F\kappa_m^{(\varepsilon)}(\xi)|.$$

If $g_2^{(\varepsilon)}(\xi) \in C^\infty$, $\xi \in \mathbb{R}^2$, is a compactly supported function that coincides with $\exp(i\pi(\varepsilon_1\xi_1 + \varepsilon_2\xi_2))$ on $\operatorname{supp} F\psi_{00}^{(\varepsilon)}$, then

$$|F\kappa_m^{(\varepsilon)}(\xi)| = g_2(b_1^{-m_1}\xi_1, b_2^{-m_2}\xi_2)F\kappa_m^{(\varepsilon)}(\xi).$$

The latter equalities together with Proposition 11.4.7 lead to the conclusion that we can equivalently replace $f * \omega_m^{(\varepsilon)}$ by $f * \kappa_m^{(\varepsilon)}$ in (11.115).

Step 5. By virtue of Proposition 11.4.11 the variable $\mathcal{A}(f)$ can be estimated by the sum of finitely many ($\leq N^*$) summands of the form (11.115), and in view of Step 4 each of them does not exceed (with some constant) $A(f)$. Thus $\mathcal{A}(f) \ll A(f)$. In turn, by virtue of Proposition 11.4.11 and Step 4, $A(f)$ does not exceed (with some constant) the sum of finitely many ($\leq N^*$) summands of the form (11.115), which by Step 3 can be estimated by $\mathcal{A}(f)$. This completes the proof. $\Diamond$

`Proof of Theorem 11.4.16.` We use the same argument taking into account the assertion corresponding to Proposition 11.4.7 related to the case of two-sided sequences and multipliers in $S'(\mathbb{R}^n)/P(\mathbb{R}^n)$.

`Proof of Theorem 11.4.17.` A. Consider the case $s_\lambda > 0$ (for $s_\lambda < 0$, the argument is analogous). Note that for $(m, k, \varepsilon) \in \mathcal{M}$,

$$c_{mk}^{(\varepsilon)} = \frac{1}{2\pi} \int\limits_{\mathbb{R}^2} Ff(\xi) \cdot \prod_{\lambda=1}^{2} \left(b_\lambda^{-\frac{m_\lambda}{2}} e^{i\xi_\lambda k_\lambda^{(\varepsilon_\lambda)} b_\lambda^{-m_\lambda}} \overline{F\psi^{(\lambda,\varepsilon_\lambda)}}(b_\lambda^{-m_\lambda}\xi_\lambda) \right) d\xi.$$

If we define the functions $(\Delta_m^{(\varepsilon)} f)(x)$ by the equality

$$(F\Delta_m^{(\varepsilon)} f)(\xi) = Ff(\xi)F\kappa_m^{(\varepsilon)}(\xi),$$

then $c_{mk}^{(\varepsilon)} = b^{-\frac{m}{2}} \left(\Delta_m^{(\varepsilon)} f\right) (b_1^{-m_1}k_1^{(\varepsilon_1)}, b_2^{-m_2}k_2^{(\varepsilon_2)})$. Let

$$\mathcal{M}_j' := \{(m,k,\varepsilon) \in \mathcal{M}_j,\ \varepsilon_\lambda \neq 0\}, \quad \mathcal{N}_j' := \{(m,\varepsilon) \in \mathcal{N}_j,\ \varepsilon_\lambda \neq 0\}.$$

Then

$$C(f) := \left| \left(\sum_{j=1}^{\infty} \gamma^{jq} \sum_{(m,k,\varepsilon)\in\mathcal{M}_j'} |c_{mk}^{(\varepsilon)} \widetilde{\chi}_{mk}^{(\varepsilon)}|^q \right)^{1/q}, L_p \right|$$

$$= \left| \left(\sum_{j=1}^{\infty} \gamma^{jq} \sum_{(m,k,\varepsilon)\in\mathcal{M}_j'} |(\Delta_m^{(\varepsilon)} f)(b_1^{-m_1}k_1, b_2^{-m_2}k_2)\chi_{mk}^{(s)}|^q \right)^{1/q}, L_p \right|.$$

Here, $\chi_{mk}^{(s)}$ is the characteristic function of the rectangle $P_{mk}^{(s)}$.

Step 1. Prove that $C(f) \ll A(f)$. It follows from the triangle inequality in $L_p(l_q)$ and Proposition 11.4.11 that

$$C(f) \ll \left| \left(\sum_{j=1}^{\infty} \gamma^{jq} \sum_{\mathcal{M}'_j} |(\Delta_m^{(\varepsilon)} f)(b_1^{-m_1} k_1, b_2^{-m_2} k_2) - (\Delta_m^{(\varepsilon)} f)(x)|^q \chi_{mk}^{(s)}(x) \right)^{1/q}, L_p \right| + A(f).$$

Set

$$E(f) := \left| \left(\sum_{j=1}^{\infty} \gamma^{jq} \sum_{\mathcal{M}'_j} |(\Delta_m^{(\varepsilon)} f)(b_1^{-m_1} k_1, b_2^{-m_2} k_2) - (\Delta_m^{(\varepsilon)} f)(x)|^q \chi_{mk}^{(s)}(x) \right)^{1/q}, L_p \right|.$$

The expansion into the Taylor series gives

$$E(f) = \left| \left(\sum_{j=1}^{\infty} \gamma^{jq} \sum_{(m,k,\varepsilon) \in \mathcal{M}'_j} \left| \sum_{u=1}^{\infty} \frac{1}{u!} \sum_{|\alpha|=u} D^{\alpha}(\Delta_m^{(\varepsilon)} f)(x) \times \prod_{\lambda=1}^{2} (x_\lambda - b_\lambda^{-m_\lambda} k_\lambda)^{\alpha_\lambda} \right|^q \chi_{mk}^{(s)} \right)^{1/q}, L_p \right|.$$

If $x \in P_{mk}^{(s)}$, then $|x_\lambda - k_\lambda b_\lambda^{-m_\lambda}| < b_\lambda^{-m_\lambda}$; hence using the triangle inequality in $L_p(l_q)$ for $p, q \geq 1$, we get the estimate

(11.116)
$$E(f) \leq \sum_{u=1}^{\infty} \frac{1}{u!} \left| \left(\sum_{j=1}^{\infty} \gamma^{jq} \sum_{\mathcal{N}'_j} \left(\sum_{|\alpha|=u} \left| D^{\alpha}(\Delta_m^{(\varepsilon)} f)(x) \prod_{\lambda=1}^{2} b_\lambda^{-m_\lambda \alpha_\lambda} \right| \right)^q \right)^{1/q}, L_p \right|.$$

Let $v \in \Big(0, \min\{p, q\}\Big)$, $\xi_\lambda := r_\lambda b_\lambda^2$. Then, using Proposition 11.4.10 and Proposition 11.4.11, we get

$$E(f) \leq \sum_{u=1}^{\infty} \frac{c^u}{u!} \times \left| \left(\sum_{j=1}^{\infty} \gamma^{jq} \sum_{\mathcal{N}'_j} \left(\sum_{|\alpha|=u} \xi^{\alpha} \sup_z \frac{|(\Delta_m^{(\varepsilon)} f)(x - z)|}{1 + |(\xi_1 z_1 b_1^{m_1},\ \xi_2 z_2 b_2^{m_2})|^{2/v}} \right)^q \right)^{1/q}, L_p \right| \ll A(f).$$

If $\min(p, q) < 1$, then inequality (11.116) is replaced by

$$E(f) \leq \sum_{u=1}^{\infty} \frac{c_{pq}^u}{u!} \left| \left(\sum_{j=1}^{\infty} \gamma^{jq} \sum_{\mathcal{N}'_j} \left(\sum_{|\alpha|=u} \left| D^{\alpha}(\Delta_m^{(\varepsilon)} f)(x) \prod_{\lambda=1}^{2} b_\lambda^{-m_\lambda \alpha_\lambda} \right| \right)^q \right)^{1/q}, L_p \right|,$$

where c_{pq} is a constant in the quasi-triangle inequality for $L_p(l_q)$.

Expressions of the type $C(f)$ for $\varepsilon = (0,0)$, $(0,1)$, and $(1,0)$ are estimated in a similar way; we just note that the characteristic functions of the rectangles

$$P_{mk}^{(s,\varepsilon)} = \left\{x \in \mathbb{R}^n : \; x_\lambda \in \left[b_\lambda^{-m_\lambda} k_\lambda^{(\varepsilon_\lambda)}, \; b_\lambda^{-m_\lambda}(k_\lambda + 1)^{(\varepsilon_\lambda)}\right)\right\}$$

arise. To go over to $\chi_{mk}^{(s)}$, we can use dilation operators, which are continuous in $L_p(l_q)$.

Step 2. In view of the above remark, we obtain the opposite estimates only for the case $\varepsilon_\lambda \neq 0$. For $(m,\varepsilon) \in \mathcal{N} := \bigcup\limits_{j=1}^{\infty} \mathcal{N}_j$, let

$$\mathcal{C}_m^{(\varepsilon)} = \left\{(\nu,\delta) \in \mathcal{N} : \; (\nu,\delta) \neq (m,\varepsilon), \; \operatorname{supp} F\psi_{\nu,0}^{(\delta)} \bigcap \operatorname{supp} F\psi_{m,0}^{(\varepsilon)} \neq \emptyset\right\}.$$

Fix $\tau \in P_{m,0}^{(s)}$ and expand the function $\psi_{m,0}^{(\varepsilon)}(x-\tau)$, $x \in \mathbb{R}^2$, into the Fourier series with respect to the system $\Psi^{(s)}$. It is clear that only the Fourier coefficients of the functions $\psi_{m\mu}^{(\varepsilon)}$ and $\psi_{\nu\mu}^{(\delta)}$, $(\nu,\delta) \in \mathcal{C}_m^{(\varepsilon)}$, $\mu \in \mathbb{Z}^2$, will be different from zero. Thus,

$$\begin{aligned} \psi_{m,0}^{(\varepsilon)}(x-\tau) &= \sum_{\mu \in \mathbb{Z}^2} d(m,\varepsilon,\mu,\tau)\psi_{m,\mu}^{(\varepsilon)}(x) \\ &\quad + \sum_{\mathcal{C}_m^{(\varepsilon)}} \sum_{\mu \in \mathbb{Z}^2} d_1(m,\varepsilon,\mu,\tau,\nu,\delta)\psi_{\nu,\mu}^{(\delta)}(x) \\ &:= \sum_{\mu \in \mathbb{Z}^2} d(m,\varepsilon,\mu,\tau)\psi_{m,\mu}^{(\varepsilon)}(x) + G_{m\tau}^{(\varepsilon)}(x), \end{aligned} \tag{11.117}$$

where $d(m,\varepsilon,\mu,\tau)$ and $d_1(m,\varepsilon,\mu,\tau,\nu,\delta)$ are the corresponding Fourier coefficients.

The function $\psi_{m,0}^{(\varepsilon)}(x-\tau)$, $x \in \mathbb{R}^2$, can be expanded into the Fourier series with respect to $\Psi^{(s)}$ for fixed $\tau \in P_{mk}^{(s)}$, $k \in \mathbb{Z}^2$, using (11.117) and taking into account the simple algebraic structure of the basis $\Psi^{(s)}$. Indeed, if $\tau \in P_{mk}^{(\varepsilon)}$, then

$$\tau_\lambda = k_\lambda b_\lambda^{-m_\lambda} + \{\tau_\lambda b_\lambda^{m_\lambda}\} b_\lambda^{-m_\lambda}.$$

By definition, put

$$\rho_{mk} := \left(k_1 b_1^{-m_1}, \; k_2 b_2^{-m_2}\right), \quad \zeta_m(\tau) := \left(\{\tau_1 b_1^{m_1}\} b_1^{-m_1}, \; \{\tau_2 b_2^{m_2}\} b_2^{-m_2}\right).$$

It follows from (11.117) that for $\tau \in P_{mk}^{(\varepsilon)}$, the following relation holds:

$$\begin{aligned} \psi_{m,0}^{(\varepsilon)}(x-\tau) &= \psi_{m,0}^{(\varepsilon)}(x - \rho_{mk} - \zeta_m(\tau)) \\ &= \sum_{\mu \in \mathbb{Z}^2} d(m,\varepsilon,\mu,\zeta_m(\tau))\psi_{m,\mu}^{(\varepsilon)}(x - \rho_{mk}) + G_{m,\zeta_m(\tau)}^{(\varepsilon)}(x - \rho_{mk}) \\ &= \sum_{\mu \in \mathbb{Z}^2} d(m,\varepsilon,\mu - k,\zeta_m(\tau))\psi_{m,\mu}^{(\varepsilon)}(x) + G_{m,\zeta_m(\tau)}^{(\varepsilon)}(x - \rho_{mk}). \end{aligned}$$

Thus, for an arbitrary $\tau \in \mathbb{R}^2$, we get

$$\begin{aligned}\Delta_m^{(\varepsilon)}(\tau) &= \sum_{k\in\mathbb{Z}^2} \int_{\mathbb{R}^2} f(x)\overline{\psi_{m,0}^{(\varepsilon)}(x-\tau)}\,dx\widetilde{\chi}_{mk}^{(\varepsilon)}(\tau) \\ &= \sum_{k\in\mathbb{Z}^2}\sum_{\mu\in\mathbb{Z}^2} \overline{d(m,\varepsilon,\mu-k,\zeta_m(\tau))}c_{m\mu}^{(\varepsilon)}\widetilde{\chi}_{mk}^{(\varepsilon)}(\tau) \\ &\quad + \sum_{k\in\mathbb{Z}^2}\int_{\mathbb{R}^2} f(x)\overline{G_{m,\zeta_m(\tau)}^{(\varepsilon)}(x-\rho_{mk})}\,dx\cdot\widetilde{\chi}_{mk}^{(\varepsilon)}(\tau) \\ &:= R_1(m,\varepsilon,\tau)+R_2(m,\varepsilon,\tau).\end{aligned}$$

Since the functions $\psi_{h_\lambda}(x_\lambda)$ and $\varphi_{h_\lambda}(x_\lambda)$ decrease at infinity more rapidly than any negative power, it is not hard to show that

$$|d(m,\varepsilon,\mu,\tau)| \ll (1+|\mu|^\alpha)^{-1}$$

for any $\alpha\in\mathbb{R}$ with the constant independent of $\tau\in P_{m,0}^{(s)}$ and $(m,\varepsilon)\in\mathcal{N}$. Let $\beta\in\Big(0,\min(p,q)\Big)$. Then

$$(11.118)\qquad \begin{aligned}R_1(m,\varepsilon,\tau) &\ll \sum_{k\in\mathbb{Z}^2}\sum_{\mu\in\mathbb{Z}^2}\frac{|c_{m\mu}^{(\varepsilon)}|}{1+|\mu-k|^{3+2/\beta}}\,\widetilde{\chi}_{mk}^{(\varepsilon)}(\tau) \\ &\ll \sum_{k\in\mathbb{Z}^2}\sup_{\mu\in\mathbb{Z}^2}\left(\frac{|c_{m\mu}^{(\varepsilon)}|}{1+|\mu-k|^{2/\beta}}\right)\widetilde{\chi}_{mk}^{(\varepsilon)}(\tau).\end{aligned}$$

Set $\eta_m^{(\varepsilon)}f(\tau) := \sum_{k\in\mathbb{Z}^2} c_{mk}^{(\varepsilon)}\,\widetilde{\chi}_{mk}^{(\varepsilon)}(\tau)$. Then

$$b^{\frac{m}{2}}\sup_{\mu\in\mathbb{Z}^2}\frac{|c_{m\mu}^{(\varepsilon)}|}{1+|\mu-k|^{2/\beta}} \ll (M(\eta_m^{(\varepsilon)}f)^\beta)^{1/\beta}(k_1b_1^{-m_1},k_2b_2^{-m_2}),$$

where M is the maximal Hardy-Littlewood operator (over rectangles with sides parallel to the coordinate axes).

Indeed, for any $\mu\in\mathbb{Z}^2$, $\mu\neq k$, we have (for details, see below)

$$(11.119)\qquad \begin{aligned}(M(\eta_m^{(\varepsilon)}f)^\beta)^{1/\beta}(k_1b_1^{-m_1},k_2b_2^{-m_2}) &\geq b^{m/2}|c_{m\mu}^{(\varepsilon)}|\left(4\prod_{\lambda=1}^{2}|k_\lambda-\mu_\lambda|\right)^{-1/\beta} \\ &\gg b^{m/2}|c_{m\mu}^{(\varepsilon)}|(1+|\mu-k|^{2/\beta})^{-1}.\end{aligned}$$

The first inequality is obtained by averaging $(\eta_m^{(\varepsilon)}f)^\beta$ over the rectangle with center at $(k_1b_1^{-m_1},\ k_2b_2^{-m_2})$ and a vertex at $(\mu_1b_1^{-m_1},\ \mu_2b_2^{-m_2})$; the second inequality is obvious. For $\mu=k$, $(\eta_m^{(\varepsilon)}f)^\beta$ should be averaged over $P_{mk}^{(s)}$.

The function $(\eta_m^{(\varepsilon)}f)^\beta$ is a step function over the cubes $P_{mk}^{(s)}$, so it is not hard to show that

$$(11.120)\qquad \sum_{k\in\mathbb{Z}^2}(M(\eta_m^{(\varepsilon)}f)^\beta)^{1/\beta}(k_1b_1^{-m_1},\ k_2b_2^{-m_2})\,\chi_{mk}^{(s)} \ll (M(\eta_m^{(\varepsilon)}f)^\beta)^{1/\beta}.$$

Inequalities (11.118)–(11.120) and Proposition 11.4.11 imply the relation

$$\begin{aligned}
&\left| \left(\sum_{j=1}^{\infty} \gamma^{jq} \sum_{(m,\varepsilon)\in\mathcal{N}_j} |R_1(m,\varepsilon,\tau)|^q \right)^{1/q}, L_p \right| \\
&\ll \left| \left(\sum_{j=1}^{\infty} \gamma^{jq} \sum_{(m,\varepsilon)\in\mathcal{N}_j} |M((\eta_m^{(\varepsilon)} f)^\beta(\tau))^{q/\beta}| \right)^{1/q}, L_p \right| \\
&= \left| \left(\sum_{j=1}^{\infty} \gamma^{jq} \sum_{(m,\varepsilon)\in\mathcal{N}_j} |M((\eta_m^{(\varepsilon)} f)^\beta(\tau))^{q/\beta}| \right)^{\beta/q}, L_{p/\beta} \right|^{1/\beta} \\
&\ll \left| \left(\sum_{j=1}^{\infty} \gamma^{jq} \sum_{(m,\varepsilon)\in\mathcal{N}_j} |(\eta_m^{(\varepsilon)} f)(\tau)|^q \right)^{\beta/q}, L_{p/\beta} \right|^{1/\beta} \\
&= \left| \left(\sum_{j=1}^{\infty} \gamma^{jq} \sum_{(m,\varepsilon)\in\mathcal{N}_j} |(\eta_m^{(\varepsilon)} f)(\tau)|^q \right)^{1/q}, L_p \right| = C(f).
\end{aligned}$$

The summands in the expression for $R_2(m,\varepsilon,\tau)$ can be estimated similarly. ◊

`Proof of Theorem 11.4.17.B.` Consider the case $p < \infty$. Using the same argument as in the proof of Theorem 11.4.17.A for each function $\Delta_m^{(\varepsilon)} f$, $(m,\varepsilon) \in \mathcal{N}$, we obtain the estimates

$$|\eta_m^{(\varepsilon)} f,\ L_p| \ll |\Delta_m^{(\varepsilon)} f,\ L_p| \ll |\eta_m^{(\varepsilon)} f,\ L_p|.$$

These estimates together with Theorem 11.4.15.B imply the statement of Theorem 11.4.17.B. The case $p = \infty$ follows from the duality. ◊

`Proof of Theorem 11.4.18.` This is similar to that of Theorem 11.4.17 (see also the proof of Theorem 11.4.16). ◊

`Proof of Theorem 11.4.19.` The statement of this theorem follows from Theorem 11.4.17.A for $p = 1$, $q \in [1, \infty]$, Proposition 11.4.9, and the duality. ◊

Let

$$Mf(x) = \sup_E |E|^{-1} \int_E |f(x+y)|\, dy,$$

where the *supremum* is taken over all sets of the form $E := \{y \in \mathbb{R}^n : \|y\|_{(a)} \leq t\}$, $t > 0$.

The function Mf is an anisotropic analog of the Hardy-Littlewood maximal function. It is known that for $1 < p, q < \infty$

$$|\{Mf_k\}_1^\infty,\ L_p(l_q)| \ll |\{f_k\}_1^\infty,\ L_p(l_q)|. \tag{11.121}$$

Proof of Theorem 11.4.21. It follows from Theorem 11.4.17 that

$$|f, F_{pq}^{(s)}| \sim \left| \left(\sum_{j\in\mathbb{N}} 2^{jqs} \sum_{(m,k,\varepsilon)\in\mathcal{M}_j^{(a)}} |c_{mk}^{(\varepsilon)} \widetilde{\chi}_{P_{mk}^{(a)}}|^q \right)^{1/q}, L_p \right|. \tag{11.122}$$

Let $\min\{p,q\} > 1$. Show that for $x \in Q_{j\nu}^{(a)} \in \Omega$,

$$M\left(\sum_{(m,k,\varepsilon)\in\mathcal{M}_j^{(a)}} |c_{mk}^{(\varepsilon)}| \widetilde{\chi}_{P_{mk}^{(a)}} \right)(x) \gg \sum_{(m,k,\varepsilon)\in\mathcal{L}_{j\nu}} |c_{mk}^{(\varepsilon)}| 2^{jn/2}. \tag{11.123}$$

Introduce the notation

$$\begin{aligned} d_\lambda &:= \min_{(m,k,\varepsilon)\in\mathcal{L}_{j\nu}^{(a)}} \left\{ \nu_\lambda 2^{-a_\lambda j};\ k_\lambda b^{-m_\lambda} \right\}, \\ e_\lambda &:= \max_{(m,k,\varepsilon)\in\mathcal{L}_{j\nu}^{(a)}} \left\{ (\nu_\lambda + 1) 2^{-a_\lambda j};\ (k_\lambda + 1) b^{-m_\lambda} \right\}, \\ E &:= \{ x \in \mathbb{R}^n :\ x_\lambda \in [d_\lambda, e_\lambda] \}. \end{aligned}$$

It follows from (11.99) that

$$e_\lambda - d_\lambda \le 2^{-a_\lambda j} + 2b^{-l_j^{(\lambda)}-1} \ll 2^{-a_\lambda j}.$$

Moreover, for $(m,k,\varepsilon) \in \mathcal{M}_j^{(a)}$, we have

$$|P_{mk}^{(a)}| \gg \prod_\lambda b^{-l_{j+1}^{(\lambda)}} \gg \prod_\lambda 2^{-a_\lambda j} = 2^{-jn}.$$

Hence,

$$\begin{aligned} &|E|^{-1} \int_E \sum_{(m,k,\varepsilon)\in\mathcal{M}_j^{(a)}} |c_{mk}^{(\varepsilon)}| \widetilde{\chi}_{P_{mk}^{(a)}}(x)\, dx \\ &\ge 2^{jn} \sum_{(m,k,\varepsilon)\in\mathcal{L}_{j\nu}^{(a)}} |c_{mk}^{(\varepsilon)}| \cdot |P_{mk}^{(a)}|^{1/2} \gg 2^{jn/2} \sum_{(m,k,\varepsilon)\in\mathcal{L}_{j\nu}^{(a)}} |c_{mk}^{(\varepsilon)}|. \end{aligned} \tag{11.124}$$

(11.123) follows from (11.124); in turn inequalities (11.121)–(11.123) imply that

$$|f, F_{pq}^{(s)}| \gg \left| \left\{ \sum_{(m,k,\varepsilon)\in\mathcal{L}_Q^{(a)}} |c_{mk}^{(\varepsilon)}| \right\}_{Q\in\Omega}, f_{pq}^{(s)} \right|.$$

Let us prove the opposite inequality. Let $x \in Q_{j\nu}^{(a)}$. Then

$$\begin{aligned} \sum_{(m,k,\varepsilon)\in\mathcal{M}_j^{(a)}} |c_{mk}^{(\varepsilon)}| \widetilde{\chi}_{P_{mk}^{(a)}}(x) &= \sum_{\{P_{mk}^{(a)}:\ x\in P_{mk}^{(a)}\}} |c_{mk}^{(\varepsilon)}| \widetilde{\chi}_{P_{mk}^{(a)}}(x) \\ &\le \sum_{(m,k,\varepsilon)\in\mathcal{L}_{j\nu}^{(a)}} |c_{mk}^{(\varepsilon)}| \widetilde{\chi}_{P_{mk}^{(a)}}(x) \ll 2^{-jn/2} \sum_{(m,k,\varepsilon)\in\mathcal{L}_{j\nu}^{(a)}} |c_{mk}^{(\varepsilon)}|. \end{aligned}$$

Thus we have proved Theorem 11.4.21 for the case $\min\{p,q\} > 1$. If $\min\{p,q\} \le 1$, then the equivalence can be checked similarly, replacing the maximal function Mf by $[M(f^r)]^{1/r}$, $0 < r < \min\{p,q\}$. ◊

11.5. Linear operators in the Lizorkin-Triebel spaces

The aim of this section is to obtain theorems about continuity of anisotropic pseudo-differential operators (PDO) in the scale of anisotropic Lizorkin-Triebel spaces $F_{pq}^{(s)}$ including, for $1 < p < \infty$, $q = 2$, the scale of anisotropic Sobolev spaces. To this end, we use the unconditional basis constructed in $F_{pq}^{(s)}$, which consists of tensor products of appropriate Meyer-David wavelets (see page 407).

11.5.1. Definitions and preliminary information. In this section, we use the same notation as in §11.4. To define the Besov and Lizorkin-Triebel spaces, we use the choice of parameters C_λ and γ described on page 410. In what follows, the anisotropy vector (index (a)) is an arbitrary fixed vector; hence this index will be omitted in the designations $\|x\|_{(a)}$, $\mathcal{N}_j^{(a)}$, $\mathcal{M}_j^{(a)}$, $\Psi^{(a)}$, $\Omega_{(a)}$. This also concerns the term a-cube. The index (a) will be used only for the designation of the anisotropic order of the multi-index to prevent confusion with the usual order.

DEFINITION 11.5.1. For $m \in \mathbb{R}$, $K \in \mathbb{N}_0^n$, $\delta \in [0,1]$, denote by $S_{\delta,a}^{m,K}$ the class of functions (symbols) $\mathcal{P}(x,\xi)$ infinitely differentiable with respect to the first variable and K_λ times continuously differentiable with respect to ξ_λ such that for any multi-indices β, γ, $\beta_\lambda \le K_\lambda$ and any $x, \xi \in \mathbb{R}^n$,

$$|D_\xi^\beta D_x^\gamma \mathcal{P}(x,\xi)| \le c_{\beta,\gamma}(1+\|\xi\|)^{m-|\beta|_a+\delta|\gamma|_a}. \tag{11.125}$$

If $\mathcal{P} \in S_{\delta,a}^{m,K}$, then the pseudo-differential operator

$$(Tf)(x) = \int_{\mathbb{R}^n} \mathcal{P}(x,\xi) Ff(\xi) e^{2\pi i(x,\xi)} d\xi, \quad f \in S(\mathbb{R}^n),$$

is called an operator of the class **APDO** $S_{\delta,a}^{m,K}$ ($T \in$ **APDO** $S_{\delta,a}^{m,K}$).

The proof of the continuity of operators of the class **APDO** in the scale of the spaces $F_{pq}^{(s)}$ is based on the notion of the almost diagonal matrix representation of this operator with respect to the basis Ψ.

DEFINITION 11.5.2. Let $0 < p < \infty$, $0 < q \le \infty$, $\sigma_\lambda := s_\lambda(1-m/s)$, $\delta > 0$, $m \in \mathbb{R}$. An operator A with matrix $\{a_{QR}\}_{Q,R\in\Omega}$ is called almost diagonal from $f_{pq}^{(s)}$ to $f_{pq}^{(\sigma)}$ if

$$\sup_{Q,R\in\Omega} \frac{|a_{QR}|}{\omega_{QR}(\delta)} < \infty, \tag{11.126}$$

where

$$\omega_{QR} := \omega_{QR}(s,m,\delta)$$
$$:= \frac{l(Q)^{s-m}}{l(R)^s \left(1+\frac{\|x_Q-x_R\|}{\max\{l(R),l(Q)\}}\right)^{n+\delta}} \cdot \min\left\{\left(\frac{l(Q)}{l(R)}\right)^{\frac{n+\delta}{2}}, \left(\frac{l(R)}{l(Q)}\right)^{\frac{n+\delta}{2}}\right\}.$$

11.5.2. Formulation of results and comments.

THEOREM 11.5.3. *An almost diagonal operator* $A: f_{pq}^{(s)} \to f_{pq}^{(\sigma)}$ *is bounded.*

The proof of almost diagonality of an operator A with matrix $\{\langle\psi_Q, T\psi_R\rangle\}_{Q,R\in\Omega}$ is based on the following statement, which characterizes the decrease at infinity of images of wavelets under the operator T. By ψ_R, $R \in \Omega$, we denote a function of the form $\psi_{mk}^{(\varepsilon)}$, where $(m,k,\varepsilon) \in \mathcal{L}_R$. Here we use the wavelet basis described on page 414.

THEOREM 11.5.4. *If* $\mathcal{P} \in S_{\delta,a}^{m,K}$ *and* $K := (K_1, ..., K_n)$ *is a multi-index such that*

$$K_\lambda = K_1 \pmod 4, \quad \lambda = 2, 3, ..., n,$$

$$\min_\lambda K_\lambda a_\lambda > n,$$

then for any $R \in \Omega$ *and any multi-index* γ,

$$|D_x^\gamma T\psi_R(x)| \ll |R|^{-\frac{1}{2}-\frac{m+|\gamma|_a}{n}} \left(1 + \frac{\|x - x_R\|}{l(R)}\right)^{-n-\alpha}, \tag{11.127}$$

where $\alpha := \min\limits_\lambda K_\lambda a_\lambda - n$.

REMARK 11.5.5. If $\mathcal{P}(x,\xi) \equiv 1$, then $\mathcal{P} \in S_{1,a}^{0,K}$ for any multi-index K. Hence, in this case, estimate (11.127) holds for any $\alpha > 0$.

The following two assertions are the most important results of this section.

THEOREM 11.5.6. *Let* $T \in$ **APDO** $S_{\delta,a}^{m,K}$ *and* $K_\lambda = K_1 \pmod 4$ *for any* $\lambda = 1, ..., n$. *Then for* $m \le s$ *and* $\min\limits_\lambda K_\lambda a_\lambda > n$, *the operator* T *can be extended to a continuous operator acting from* $F_{pq}^{(s)}$ *to* $F_{pq}^{(\sigma)}$, *where* $1 \le p, q \le \infty$, $\min\{p,q\} < \infty$, *and the coordinates of the vector* (σ) *can be computed by the formulas* $\sigma_\lambda = s_\lambda(1 - m/s)$.

If $m > s$ *and* $\min\limits_\lambda K_\lambda a_\lambda > n + [m - s]$, *then the assertion holds under the assumption that the symbol does not depend on the first variable:* $\mathcal{P}(x,\xi) = \mathcal{P}(\xi)$.

The condition $\mathcal{P}(x,\xi) = \mathcal{P}(\xi)$ in Theorem 11.5.6 (the case $m > s$) essentially restricts the area of its applications. This condition can be replaced by the following: $T^*(x^\gamma) = 0$ for all multi-indices γ with $|\gamma| \le L$, where $L \in \mathbb{N}_0$ is such that $\min_{|\gamma|=L+1} |\gamma|_a > m - s$. Then almost diagonality of the matrix $\{\langle\psi_Q, T\psi_R\rangle\}_{Q,R\in\Omega}$ can be proved by using Lemma 11.5.11. However, even in the simplest case, i.e., $n = 2$, $T = \frac{\partial}{\partial x_1}$, $L = 0$, $a_2 < a_1$, this condition is satisfied only for $a_2 > a_1 - s$. The following theorem is an attempt to remove restrictions of this type.

THEOREM 11.5.7. *Let there be given a differential operator*

$$T := \sum_\lambda c_\lambda(x) \frac{\partial^{\Gamma_\lambda}}{\partial x_\lambda^{\Gamma_\lambda}}, \quad c_\lambda \in L_\infty, \quad \lambda = 1, 2, ..., n.$$

Then T *is a bounded operator from* $F_{pq}^{(s)}$ *to* $F_{pq}^{(\sigma)}$, *where* $\sigma_\lambda := s_\lambda(1 - m/s)$, $m := \max_\lambda \Gamma_\lambda a_\lambda$.

The following relations follow from Theorems 11.5.6 and 11.5.7. For simplicity, set $n = 2$:

$$\frac{\partial}{\partial x_1} : F_{pq}^{(s)} \to F_{pq}^{(s_1-1,\ s_2-\frac{s_2}{s_1})}, \tag{11.128}$$

$$\frac{\partial^2}{\partial x_2^2} : F_{pq}^{(s)} \to F_{pq}^{(s_1-\frac{2s_1}{s_2},\ s_2-2)}, \tag{11.129}$$

$$D_h := \frac{\partial}{\partial x_1} - \frac{\partial^2}{\partial x_2^2} : F_{pq}^{(s)} \to \begin{cases} F_{pq}^{(s_1-1,\ s_2-\frac{s_2}{s_1})}, & s_2 > 2s_1, \\ F_{pq}^{(s_1-\frac{2s_1}{s_2},\ s_2-2)}, & s_2 < 2s_1, \\ F_{pq}^{(s_1-1,\ s_2-2)}, & s_2 = 2s_1. \end{cases} \tag{11.130}$$

THEOREM 11.5.8. *Relations* (11.128)–(11.130) *are precise in the sense that no coordinate of the vector of smoothness of the image space can be increased.*

This theorem characterizes the precision of Theorems 11.5.6 and 11.5.7.

11.5.3. Proofs. In the sequel we shall use the fact that all moments of the function ψ_h,

$$\int_{\mathbb{R}^1} t^k \psi_h(t)\, dt = 0, \quad k \in \mathbb{N} \cup \{0\} := \mathbb{N}_0, \tag{11.131}$$

are equal to zero, which follows from the equality $(F\psi_h)^{(k)}(0) = 0$ (see (11.87)).
To prove Theorem 11.5.3, we shall need the following lemma.

LEMMA 11.5.9. *Let* $0 < d \le g < \infty$ *and* $h > ng/d$. *If* $\mu, \nu \in \mathbb{Z}$ *and* $\mu \le \nu$, *then for any cube* Q *such that* $l(Q) = 2^{-\nu}$, *for any* $x \in Q$, *and any sequence* $\{c_R\}_{R\in\Omega}$, *the inequality*
(11.132)

$$\sum_{\{R:\ l(R)=2^{-\mu}\}} |c_R|^g \left(1 + \frac{\|x_R - x_Q\|}{l(R)}\right)^{-h} \ll \left(M\left(\sum_{\{R:\ l(R)=2^{-\mu}\}} |c_R|^d \chi_R\right)(x)\right)^{g/d}$$

holds, where the constant depends only on n *and* $\left(h - \dfrac{ng}{d}\right)$.

Proof. It suffices to prove (11.132) for $x_Q = 0$. Let

$$A_0 := \{R \in \Omega :\ l(R) = 2^{-\mu},\ \|x_R\| 2^{\mu} \le 1\}.$$

For $k \in \mathbb{N}$, we have

$$A_k := \{R \in \Omega :\ l(R) = 2^{-\mu},\ \|x_R\| 2^{\mu} \in (2^{k-1}, 2^k]\}.$$

Then

$$\begin{aligned} \sum_{R\in A_k} |c_R|^g \left(1 + \frac{\|x_R\|}{l(R)}\right)^{-h} &\ll 2^{-kh} \sum_{R\in A_k} |c_R|^g \ll 2^{-kh} \left(\sum_{R\in A_k} |c_R|^d\right)^{g/d} \\ &\ll 2^{-kh+n\mu g/d} \left(\int_{\mathbb{R}} \sum_{R\in A_k} |c_R|^d \chi_R(x)\, dx\right)^{g/d}, \end{aligned} \tag{11.133}$$

since $|R| = 2^{-n\mu}$. Note that

$$\bigcup_{R\in A_k} R \subset \bigcup_{\{R:\ l(R)=2^{-\mu},\ \|x_R\|\le 2^{k-\mu}\}} R. \tag{11.134}$$

Moreover, if $y \in R$, $l(R) = 2^{-\mu}$, and $\|x_R\| \le 2^{k-\mu}$, then

$$\|y\| \ll 2^{k-\mu}. \tag{11.135}$$

It is easy to see that

$$\left|\{y:\ \|y\| \ll 2^{k-\mu}\}\right| \ll 2^{n(k-\mu)}. \tag{11.136}$$

It follows from (11.134)–(11.136) that the right-hand side of (11.133) does not exceed

$$2^{-kh+n\mu g/d}\left(\int_{\|x\|\ll 2^{k-\mu}} \sum_{R\in A_k} |c_R|^d \chi_R(x)\,dx\right)^{g/d}$$
$$\ll 2^{-k(h-ng/d)}\left(M\left(\sum_{R\in A_k} |c_R|^d\chi_R\right)(x)\right)^{g/d}.$$

Summing these inequalities over k, we get (11.132). ◊

REMARK 11.5.10. Under the conditions of Lemma 11.5.9, the estimate

$$\begin{aligned}&\left(\sum_{\{R:\ l(R)=2^{-\mu}\}} |c_R|^g \left(1 + \frac{\|x_R - x_Q\|}{\max\{l(R), l(Q)\}}\right)^{-h}\right)^{1/g}\\ &\qquad \ll 2^{(\mu-\nu)_+ n/d}\left(M\left(\sum_{\{R:\ l(R)=2^{-\mu}\}} |c_R|^d\chi_R\right)(x)\right)^{1/d}\end{aligned} \tag{11.137}$$

holds without the assumption $\mu \le \nu$.

The proof of inequality (11.137) is obtained in a similar way if we replace A_0 by

$$B_0 := \left\{R \in \Omega:\quad l(R) = 2^{-\mu},\ \frac{\|x_R\|}{l(Q)} \le 1\right\}$$

and A_k by

$$B_k := \left\{R \in \Omega:\quad l(R) = 2^{-\mu},\ \frac{\|x_R\|}{l(Q)} \in (2^{k-1}, 2^k]\right\}.$$ ◊

Proof of Theorem 11.5.3. First of all note that it suffices to prove the theorem for $m = 0$. The fact is that the averaged smoothness σ is equal to $s - m$, and

$$\omega_{QR}(s, m, \delta) := l(Q)^{-m}\omega_{QR}(s, 0, \delta).$$

Hence, if A is an almost diagonal operator from $f_{pq}^{(s)}$ to $f_{pq}^{(\sigma)}$, then the operator $\widetilde{A}$ with matrix $\{l(Q)^m a_{QR}\}_{Q,R\in\Omega}$ is almost diagonal from $f_{pq}^{(s)}$ to $f_{pq}^{(s)}$. In addition, it is easily seen that

$$|\{l(Q)^m c_Q\}_{Q\in\Omega}, f_{pq}^{(s)}| = |\{c_Q\}_{Q\in\Omega}, f_{pq}^{(\sigma)}|.$$

Let $r = \min\{p, q\}$. Consider first the case $r > 1$. Let A be an almost diagonal operator on $f_{pq}^{(s)}$ with matrix $\{a_{QR}\}_{Q,R\in\Omega}$ satisfying (11.126). Let us write $A = A_0 + A_1$, where

$$(A_0 c)_Q = \sum_{\{R:\ l(R)\geq l(Q)\}} a_{QR} c_R,$$

$$(A_1 c)_Q = \sum_{\{R:\ l(R)<l(Q)\}} a_{QR} c_R.$$

Let $l(Q) = 2^{-\nu}$, $\nu \in \mathbb{N}$. Then

$$|(A_0 c)_Q| \ll \sum_{\mu\leq\nu} 2^{(\mu-\nu)(s+\frac{n+\delta}{2})} \sum_{\{R:\ l(R)=2^{-\mu}\}} |c_R| \left(1 + \frac{\|x_R - x_Q\|}{l(R)}\right)^{-n-\delta}.$$

Using Lemma 11.5.9 for $g = 1,\ h = n + \delta,\ d = 1$, we get

$$|(A_0 c)_Q| \ll \sum_{\mu\leq\nu} 2^{(\mu-\nu)(s+\frac{n+\delta}{2})} M\left(\sum_{\{R:\ l(R)=2^{-\mu}\}} |c_R|\chi_R\right)(x), \tag{11.138}$$

where $x \in Q$. As

$$|Q|^{-1/2} = 2^{(\nu-\mu)n/2}|R|^{-1/2}, \tag{11.139}$$

where $l(R) = 2^{-\mu}$, taking into account (11.138) and (11.139), we get

$$\begin{aligned}
|A_0 c, f_{pq}^{(s)}| &= \left|\left(\sum_{Q\in\Omega} |l(Q)^{-s}(A_0 c)_Q \widetilde{\chi}_Q|^q\right)^{1/q}, L_p\right| \\
&\ll \left|\left(\sum_{\nu\in\mathbb{N}}\left(2^{\nu s}\sum_{\mu\leq\nu} 2^{(\mu-\nu)(s+\frac{n+\delta}{2})} M\left(\sum_{l(R)=2^{-\mu}} |c_R|\chi_R\right)|Q|^{-1/2}\chi_Q\right)^q\right)^{1/q}, L_p\right| \\
&= \left|\left(\sum_{k=0}^{\infty} 2^{-\frac{k\delta q}{2}} \sum_{\mu\in\mathbb{N}} M\left(2^{\mu s}\sum_{l(R)=2^{-\mu}} |c_R|\widetilde{\chi}_R\right)^q\right)^{1/q}, L_p\right| \\
&\leq \left|\left(\sum_{\mu\in\mathbb{N}} M\left(2^{\mu s}\sum_{l(R)=2^{-\mu}} |c_R|\widetilde{\chi}_R\right)^q\right)^{1/q}, L_p\right| \ll |c, f_{pq}^{(s)}|.
\end{aligned}$$

The latter inequality is based on (11.121).

The estimate $|A_1 c, f_{pq}^{(s)}|$ is obtained analogously (see Remark 11.5.10).

Thus, for $r > 1$ and $q < \infty$, Theorem 11.5.3 is proved.

Let $r \leq 1$ and $q < \infty$. Take $\widetilde{r} < r$. Then $p/\widetilde{r} > 1$ and $q/\widetilde{r} > 1$. It is not hard to show that the operator $\widetilde{A}$ with matrix

$$\{\widetilde{a}_{QR}\}_{Q,R\in\Omega} := \left\{|a_{QR}|^{\widetilde{r}}\left(\frac{|Q|}{|R|}\right)^{\frac{1-\widetilde{r}}{2}}\left(\frac{l(Q)}{l(R)}\right)^{s(1-\widetilde{r})}\right\}$$

is almost diagonal on $f_{p/\widetilde{r},q/\widetilde{r}}^{(s)}$. Set

$$t_Q := l(Q)^{s(1-\widetilde{r})}|Q|^{\frac{1-\widetilde{r}}{2}}|c_Q|^{\widetilde{r}}.$$

Then

$$|t,\ f^{(s)}_{p/\widetilde{r},q/\widetilde{r}}|^{1/\widetilde{r}} = |c,\ f^{(s)}_{pq}|.$$

Finally, we have

$$\begin{aligned}\left|Ac,\ f^{(s)}_{pq}\right| &= \left|\left\{|(Ac)_Q|^{\widetilde{r}}|Q|^{\frac{1-\widetilde{r}}{2}}l(Q)^{s(1-\widetilde{r})}\right\}_{Q\in\Omega},\ f^{(s)}_{p/\widetilde{r},q/\widetilde{r}}\right| \\ &\le \left|\left\{l(Q)^{s(1-\widetilde{r})}|Q|^{\frac{1-\widetilde{r}}{2}}\sum_R |a_{QR}|^{\widetilde{r}}|c_R|^{\widetilde{r}}\right\}_{Q\in\Omega},\ f^{(s)}_{p/\widetilde{r},q/\widetilde{r}}\right| \\ &\le |\widetilde{A},\ f^{(s)}_{p/\widetilde{r},q/\widetilde{r}}|^{1/\widetilde{r}}\cdot|c,\ f^{(s)}_{pq}|.\end{aligned}$$

The proof for the cases $p > 1$ and $q = \infty$ follows from the duality; the cases $p \le 1$ and $q = \infty$ are proved in a similar way. ◊

`Proof of Theorem 11.5.4.` Suppose $\psi_R = \psi^{(\varepsilon)}_{l,k}$, $R := R_{j,\nu} \in \Omega$, $(l,k,\varepsilon) \in \mathcal{L}_R$, $x_R = (x_{R,1},\dots,x_{R,n})$ is the left lower vertex of the cube R (see page 417). Taking into account the form of the function $F\psi_R$ and the compactness of its support, we get

(11.140)

$$\begin{aligned}|D^\gamma_x T\psi_R(x)| &= \left|D^\gamma_x\int_{\mathbb{R}} e^{2\pi i(x,\xi)}\mathcal{P}(x,\xi)F\psi_R(\xi)\,d\xi\right| \\ &\ll \sum_{\tau\le\gamma}\left|\int_{\mathbb{R}}(D^\tau_x e^{2\pi i(x,\xi)})(D^{\gamma-\tau}_x\mathcal{P}(x,\xi))F\psi_R(\xi)\,d\xi\right| \\ &= \sum_{\tau\le\gamma} b^{-l/2}\left|\int_{\mathbb{R}}(2\pi\xi)^\tau e^{2\pi i((x-x_R),\xi)}D^{\gamma-\tau}_x\mathcal{P}(x,\xi)F\psi^{(\varepsilon)}(b^{-l}\xi)\,d\xi\right|.\end{aligned}$$

For any λ, let $x_\lambda \ge x_{R,\lambda}$. Consider the differential operator

$$\Delta^K_\xi := \sum_\lambda \frac{\partial^{K_\lambda}}{\partial\xi_\lambda^{K_\lambda}},$$

where $K := (K_1,\dots,K_n)$ is the multi-index mentioned in Theorem 11.5.4.

The other quadrants of $\mathbb{R}^n$ are studied similarly by an appropriate arrangement of the signs before the summands of Δ^K_ξ.

Using the first condition for K, we obtain

$$\begin{aligned}|D^\gamma_x T\psi_R(x)| \ll \sum_{\tau\le\gamma} b^{-l/2}&\left(\sum_\lambda |x_\lambda - x_{R,\lambda}|^{K_\lambda}\right)^{-1} \\ &\times\left|\int_{\mathbb{R}}\Delta^K_\xi(e^{2\pi i((x-x_R),\xi)})\xi^\tau(D^{\gamma-\tau}_x\mathcal{P}(x,\xi))F\psi^{(\varepsilon)}(b^{-l}\xi)\,d\xi\right|.\end{aligned}$$

Let $\|x - x_R\| = \max_\lambda |x_\lambda - x_{R,\lambda}|^{1/a_\lambda}$ be attained for some $\lambda = \lambda_0$. Then

$$\sum_\lambda |x_\lambda - x_{R,\lambda}|^{K_\lambda} \geq (|x_{\lambda_0} - x_{R,\lambda_0}|^{1/a_{\lambda_0}})^{K_{\lambda_0} a_{\lambda_0}}$$
$$= \|x - x_R\|^{K_{\lambda_0} a_{\lambda_0}} \geq \min_\lambda\{\|x - x_R\|^{K_\lambda a_\lambda}\}.$$

Therefore,

$$\left(\sum_\lambda |x_\lambda - x_{R,\lambda}|^{K_\lambda}\right)^{-1} \leq \max_\lambda\{\|x - x_R\|^{-K_\lambda a_\lambda}\}.$$

Let

$$C_R(x) := \max_\lambda\{\|x - x_R\|^{-K_\lambda a_\lambda}\}.$$

Thus,

$$|D_x^\gamma T\psi_R(x)| \ll C_R(x)$$
$$\times \sum_{\tau \leq \gamma} b^{-l/2} \left| \int_{\mathbb{R}} \Delta_\xi^K (e^{2\pi i((x-x_R),\xi)})\xi^\tau (D_x^{\gamma-\tau}\mathcal{P}(x,\xi))F\psi^{(\varepsilon)}(b^{-l}\xi)\, d\xi \right|. \tag{11.141}$$

Fix an arbitrary multi-index $\tau \leq \gamma$ and apply the integration by parts formula to the corresponding summand in (11.141). In view of (11.100), we obtain the expression

$$B_R(x) := \left| \int_{\Gamma_j \cup \Gamma_{j+1}} e^{2\pi i((x-x_R),\xi)} \Delta_\xi^K \left\{\xi^\tau (D_x^{\gamma-\tau}\mathcal{P}(x,\xi))F\psi^{(\varepsilon)}(b^{-l}\xi)\right\} d\xi \right|.$$

The function

$$\Delta_\xi^K\{\xi^\tau (D_x^{\gamma-\tau}\mathcal{P}(x,\xi))F\psi^{(\varepsilon)}(b^{-l}\xi)\}$$

is equal to the sum of finitely many summands of the form

$$S(x,\xi,\lambda_0,\mu,\sigma)$$
$$:= \left(\prod_{\lambda\neq\lambda_0} \xi_\lambda^{\tau_\lambda}\right) \xi_{\lambda_0}^{\tau_{\lambda_0}-\mu} \frac{\partial^\sigma}{\partial\xi_{\lambda_0}^\sigma}(D_x^{\gamma-\tau}\mathcal{P}(x,\xi)) \frac{\partial^{K_{\lambda_0}-\mu-\sigma}}{\partial\xi_{\lambda_0}^{K_{\lambda_0}-\mu-\sigma}}(F\psi^{(\varepsilon)}(b^{-l}\xi)),$$

where $\lambda_0 \in \{1,\ldots,n\}$, $0 \leq \mu \leq \tau_{\lambda_0}$, $0 \leq \sigma \leq K_{\lambda_0} - \mu$. The number of summands and theirs coefficients depend only on n and K. By (11.125),

$$|S(x,\xi,\lambda_0,\mu,\sigma)| \ll \left(\prod_{\lambda\neq\lambda_0} |\xi_\lambda^{\tau_\lambda}|\right) |\xi_{\lambda_0}|^{\tau_{\lambda_0}-\mu}(1+\|\xi\|)^{m-\sigma a_{\lambda_0}+|\gamma-\tau|_a}$$
$$\times b^{-l_{\lambda_0}(K_{\lambda_0}-\mu-\sigma)} \left(\frac{\partial^{K_{\lambda_0}-\mu-\sigma}}{\partial\xi_{\lambda_0}^{K_{\lambda_0}-\mu-\sigma}} F\psi^{(\varepsilon)}\right)(b^{-l}\xi)$$
$$\ll \left(\prod_{\lambda\neq\lambda_0} |\xi_\lambda^{\tau_\lambda}|\right) |\xi_{\lambda_0}|^{\tau_{\lambda_0}-\mu}(1+\|\xi\|)^{m-\sigma a_{\lambda_0}+|\gamma-\tau|_a} b^{-l_{\lambda_0}(K_{\lambda_0}-\mu-\sigma)}.$$

For $(l,k,\varepsilon)\in\mathcal{L}_R$, $b^{-l_\lambda}\sim 2^{-ja_\lambda}$. Hence the latter expression does not exceed

$$\left(\prod_{\lambda\neq\lambda_0}|\xi_\lambda^{\tau_\lambda}|\right)|\xi_{\lambda_0}|^{\tau_{\lambda_0}-\mu}(1+\|\xi\|)^{m-\sigma a_{\lambda_0}+|\gamma-\tau|_a}\cdot 2^{-j(K_{\lambda_0}-\mu-\sigma)a_{\lambda_0}}.$$

Since for $\xi\in\Gamma_j\cup\Gamma_{j+1}$,

$$\xi_\lambda\ll 2^{ja_\lambda},\quad (1+\|\xi\|)\sim 2^j,\quad j\in\mathbb{N},$$

it follows that

$$\begin{aligned}\int\limits_{\Gamma_j\cup\Gamma_{j+1}}&|S(x,\xi,\lambda_0,\mu,\sigma)|\,d\xi\\ &\ll 2^{j|\tau|_a-ja_{\lambda_0}\mu+j(m-\sigma a_{\lambda_0}+|\gamma-\tau|_a)-j(K_{\lambda_0}-\mu-\sigma)a_{\lambda_0}}\cdot|\Gamma_j\cup\Gamma_{j+1}|\\ &\ll 2^{nj+j(m+|\gamma|_a)-jK_{\lambda_0}a_{\lambda_0}}.\end{aligned}$$

This inequality means that

$$B_R(x)\ll 2^{nj}\cdot 2^{j(m+|\gamma|_a)}\cdot 2^{-j|K|_a}.$$

Using (11.141) and taking into account that $\left(\prod\limits_\lambda b_\lambda^{-l_\lambda/2}\right)\sim 2^{-nj/2}\sim|R|^{1/2}$, we derive

$$|D_x^\gamma T\psi_R(x)|\ll|R|^{-1/2}C_R(x)\cdot 2^{-j|K|_a}|R|^{-\frac{m+|\gamma|_a}{n}}. \tag{11.142}$$

Note that for $2^j\|x-x_R\|\geq 1$,

$$\begin{aligned}C_R(x)\cdot 2^{-j|K|_a}&=\max_\lambda\{2^{-j|K|_a}\|x-x_R\|^{-K_\lambda a_\lambda}\}\\ &\leq\max_\lambda\{2^{-jK_\lambda a_\lambda}\|x-x_R\|^{-K_\lambda a_\lambda}\}\leq(2^j\|x-x_R\|)^{-\min_\lambda\{K_\lambda a_\lambda\}}.\end{aligned}$$

Thus, if $2^j\|x-x_R\|=\frac{\|x-x_R\|}{l(R)}\geq 1$, then

$$C_R(x)\cdot 2^{-j|K|_a}\leq\left(1+\frac{\|x-x_R\|}{l(R)}\right)^{-\min_\lambda\{K_\lambda a_\lambda\}}=\left(1+\frac{\|x-x_R\|}{l(R)}\right)^{-n-\alpha},$$

where $\alpha:=\min\limits_\lambda\{K_\lambda a_\lambda\}-n>0$. The latter inequality combined with (11.142) gives (11.127) for $l(R)^{-1}\|x-x_R\|\geq 1$. In the case when the opposite inequality holds, estimate (11.127) can be derived immediately from (11.140), passing to modulus under the integral sign. ◊

To prove (using Theorem 11.5.4) that a discrete analog of the pseudo-differential operator is almost diagonal, we need the following lemmas.

LEMMA 11.5.11. *Let* $V>n$, $L\in\mathbb{N}_0$, $\theta:=\min\limits_{|\gamma|=L+1}|\gamma|_a$, $\Theta:=\max\limits_{|\gamma|=L+1}|\gamma|_a$, $j,k\in\mathbb{Z}$, $j\leq k$, $S>\max\{V;\ n+\Theta\}$, *and* $x_0\in\mathbb{R}^n$. *If the functions* $g,h\in L_1(\mathbb{R}^n)$ *are such that*

$$|\partial^\gamma g(x)|\ll 2^{j(n/2+|\gamma|_a)}(1+2^j\|x\|)^{-V},\quad|\gamma|\leq L+1, \tag{11.143}$$

$$|h(x)|\ll 2^{kn/2}(1+2^k\|x-x_0\|)^{-S}, \tag{11.144}$$

$$\int\limits_{\mathbb{R}}x^\gamma h(x)dx=0,\quad|\gamma|\leq L, \tag{11.145}$$

then

$$|g * h(x)| \ll 2^{-(k-j)(\theta+n/2)}(1 + 2^j \|x - x_0\|)^{-V}. \tag{11.146}$$

Proof. Using dilations and shifts, it suffices to prove (11.146) for $j = 0,\ x_0 = 0$. In view of condition (11.145),

$$|g * h(x)| = \left| \int_{\mathbb{R}} g(y)h(x-y)\,dy \right|$$
$$\leq \int_{\mathbb{R}} \left| \left(g(y) - \sum_{|\beta| \leq L} \frac{1}{\beta!} \partial^\beta g(x)(y-x)^\beta \right) h(x-y) \right| dy.$$

Applying the Taylor formula to the integrand, we get

$$G(x,y) := \left| g(y) - \sum_{|\beta| \leq L} \frac{1}{\beta!} \partial^\beta g(x)(y-x)^\beta \right|$$
$$= \left| \sum_{|\beta|=L} \frac{1}{\beta!}(x-y)^\beta (\partial^\beta g(z) - \partial^\beta g(x)) \right| \tag{11.147}$$
$$\ll \sum_{|\beta|=L} |\partial^\beta g(z) - \partial^\beta g(x)| \cdot |x-y|^\beta,$$

where z is a point between x and y, $|x-y| := (|x_1 - y_1|, \dots, |x_n - y_n|)$. Applying the Taylor formula to each difference $\partial^\beta g(z) - \partial^\beta g(x)$ on the right-hand side of (11.147) and taking into account (11.143), we get

$$G(x,y) \ll (1 + \|u\|)^{-V} \sum_{|\gamma|=L+1} \sum_{|\beta|=L,\, \beta\leq\gamma} |x-y|^\beta |z-x|^{\gamma-\beta}$$
$$\ll (1 + \|u\|)^{-V} \sum_{|\gamma|=L+1} |x-y|^\gamma, \tag{11.148}$$

where u is a point between z and x. Since for any λ,

$$|x_\lambda - y_\lambda|^{\gamma_\lambda} = (|x_\lambda - y_\lambda|^{1/a_\lambda})^{\gamma_\lambda a_\lambda} \ll \|x-y\|^{\gamma_\lambda a_\lambda},$$

it follows that

$$G(x,y) \ll (1 + \|u\|)^{-V} \|x-y\|^\theta. \tag{11.149}$$

Set

$$A := \{y \in \mathbb{R}^n : \|x-y\| \leq 1\}, \qquad B := \{y \in \mathbb{R}^n : \|x-y\| > 1, \|y\| \leq \tfrac{\|x\|}{2}\},$$

$$C := \left\{y \in \mathbb{R}^n : \|x-y\| > 1, \|y\| > \tfrac{\|x\|}{2}\right\}.$$

Step 1. Estimate the integral

$$\int_A G(x,y)\,dy.$$

Note that on the set A, $1 + \|u\| \gg 1 + \|x\|$; hence (11.149) implies that

$$G(x,y) \ll (1 + \|x\|)^{-V} \|x-y\|^\theta.$$

By virtue of (11.144),

$$\int\limits_A G(x,y)\,dy \ll 2^{kn/2}(1+\|x\|)^{-V} \int\limits_{\|x-y\|\le 1} \|x-y\|^\theta (1+2^k\|x-y\|)^{-S} dy.$$

Making the change of variables $2^k \circ (x-y) = w$ and passing to polar coordinates, we obtain

$$\begin{aligned}\int\limits_A G(x,y)\,dy &\ll 2^{k(\theta+n/2)}(1+\|x\|)^{-V} \int\limits_{\rho\le 2^k} \rho^{\theta+n-1}(1+\rho)^{-S} d\rho \\ &\ll 2^{-k(\theta+n/2)}(1+\|x\|)^{-V},\end{aligned}$$

as $S > n+\theta$.

Step 2. Estimate the integral

$$\int\limits_B G(x,y)\,dy.$$

In view of (11.143), we have

$$\begin{aligned}G(x,y) &\ll (1+\|y\|)^{-V} + (1+\|x\|)^{-V} \sum_{\beta\le L} |x-y|^\beta \\ &\ll (1+\|y\|)^{-V} + (1+\|x\|)^{-V}\|x-y\|^\Theta.\end{aligned}$$

Taking into account that

$$\frac{3}{2}\|x\| \ge \|x-y\| \ge \frac{1}{2}\|x\|$$

and

$$\|x-y\| \ge \frac{1}{2}(1+\|x-y\|) > \frac{1}{4}(1+\|x\|) \Longrightarrow (1+2^k\|x-y\|)^{-S} \ll 2^{-kS}(1+\|x\|)^{-S}$$

on the set B, we derive that

$$\int\limits_B G(x,y)\,dy \ll 2^{k(n/2-S)}(1+\|x\|)^{-V}\left(\int\limits_{\mathbb{R}} \frac{dy}{(1+\|y\|)^V} + \frac{\|x\|^\Theta}{(1+\|x\|)^S} \int\limits_{\|y\|\le\frac{\|x\|}{2}} dy\right).$$

As $V > n$ and $S > n+\Theta$, we finally get

$$\int\limits_B G(x,y)\,dy \ll 2^{-k(\theta+n/2)}(1+\|x\|)^{-V}.$$

Step 3. It remains to estimate

$$\int\limits_C G(x,y)\,dy.$$

On the set C, the inequality $1+\|y\| \gg 1+\|x\|$ holds; hence we have

$$\int\limits_C G(x,y)\,dy \ll 2^{\frac{kn}{2}}(1+\|x\|)^{-V}\int\limits_C \frac{\|x-y\|^{\Theta}}{(1+2^k\|x-y\|)^S}dy$$

$$\ll 2^{\frac{kn}{2}}(1+\|x\|)^{-V}2^{-kS}\int\limits_1^{\infty}\rho^{-S+\Theta+n-1}d\rho$$

$$\ll 2^{-k(\theta+n/2)}(1+\|x\|)^{-V},$$

as $S > n+\Theta$. ◊

LEMMA 11.5.12. *Let $V > n$, $j,k \in \mathbb{Z}$, $j \le k$, $x_0 \in \mathbb{R}^n$. If the functions g, $h \in L_1(\mathbb{R}^n)$ are such that*

$$|g(x)| \ll 2^{jn/2}(1+2^j\|x\|)^{-V},$$

$$|h(x)| \ll 2^{kn/2}(1+2^k\|x-x_0\|)^{-V},$$

then

$$|g*h(x)| \ll 2^{-(k-j)n/2}(1+2^j\|x-x_0\|)^{-V}.$$

Proof. As in the previous lemma, we can assume that $j=0$, $x_0=0$. For $x \in \mathbb{R}^n$, consider the sets

$$A = \{y:\ \|x-y\| \le 1\}\bigcup\{y:\ \|x-y\|>1,\ \|y\|>\frac{1}{2}\|x\|\},\quad B := \mathbb{R}^n\backslash A.$$

Set

$$I_1 := \int\limits_A |g(y)h(x-y)|\,dy,$$

$$I_2 := \int\limits_B |g(y)h(x-y)|\,dy.$$

It is clear that

$$|g*h(x)| \le I_1 + I_2.$$

On the set A the inequality $1+\|y\| \gg 1+\|x\|$ holds; hence we have

$$I_1 \ll (1+\|x\|)^{-V}2^{kn/2}\int\limits_{\mathbb{R}}(1+2^k\|x-y\|)^{-V}dy.$$

After the change of variables $2^k \circ (x-y) = z$, taking into account that $V>n$, we obtain

$$I_1 \ll 2^{-kn/2}(1+\|x\|)^{-V}.$$

On the set B (see the proof of the previous lemma), we have $2^k\|x-y\| \gg (1+\|x\|)2^k$, $\|y\| \sim \|x\|$; hence,

$$I_2 \ll 2^{k(n/2-V)}(1+\|x\|)^{-V}\int\limits_{\mathbb{R}}(1+\|y\|)^{-V}dy.$$

Since $V>n$, it follows that $I_2 \ll 2^{-kn/2}(1+\|x\|)^{-V}$. ◊

Proof of Theorem 11.5.6. Let $s \geq m$, $l(Q) = 2^{-\nu} \geq 2^{-\mu} = l(R)$, $Q, R \in \Omega$. Taking into account Theorem 11.5.4 and the remark after it, we use Lemma 11.5.12, setting

$$g(x) := \psi_Q(x + x_Q), \quad h(x) := l(R)^m \overline{T\psi_R(-x)}, \quad x_0 := -x_R.$$

Then

$$\begin{aligned} a_{QR} &= \langle \psi_Q, T\psi_R \rangle = l(R)^{-m} g * h(-x_Q) \\ &\ll l(R)^{-m} \left(\frac{l(Q)}{l(R)} \right)^{-n/2} \left(1 + \frac{\|x_Q - x_R\|}{l(Q)} \right)^{-n-\alpha}. \end{aligned} \tag{11.150}$$

Consider the case $l(Q) < l(R)$. Let L be such that $\theta := \min_{|\gamma|=L+1} |\gamma|_a > s - m$. Taking into account (11.131) and Theorem 11.5.4, we use Lemma 11.5.11, setting

$$g(x) := l(R)^m \overline{T\psi_R(x + x_R)}, \quad h(x) := \psi_Q(-x), \quad x_0 := -x_Q.$$

Then

$$\begin{aligned} a_{QR} &= \langle \psi_Q, T\psi_R \rangle = l(R)^{-m} g * h(-x_R) \\ &\ll l(R)^{-m} \left(\frac{l(Q)}{l(R)} \right)^{\theta+n/2} \left(1 + \frac{\|x_Q - x_R\|}{l(R)} \right)^{-n-\alpha}. \end{aligned} \tag{11.151}$$

Estimates (11.150) and (11.151) mean that the matrix $\{a_{QR}\}_{Q,R\in\Omega}$ is almost diagonal from $f_{pq}^{(s)}$ to $f_{pq}^{(\sigma)}$. Combining this fact with Theorems 11.4.21 and 11.5.3, we prove Theorem 11.5.6 for $s \geq m$.

If $s < m$ and $\mathcal{P}(x, \xi) = \mathcal{P}(\xi)$, then the functions $T\psi_R$ satisfy (11.131). Hence in this case the proof of Theorem 11.5.6 is similar to the previous one if we interchange Lemmas 11.5.11 and 11.5.12. ◊

The proof of Theorem 11.5.7 is based on the following lemmas. Below we assume that the sums, where the upper bound of the summation index is negative or equal to zero, are absent.

LEMMA 11.5.13 (Anisotropic version of the Taylor formula). *Let L_λ, M_λ, $T_\lambda \in \mathbb{N} \cup \{0\}$, where $L_1 + \cdots + L_n > 0$ and the function $g(x_1, \ldots, x_n)$ has all the derivatives used below. Then*

$$\begin{aligned} g(y) - g(x) &- \sum_\lambda \sum_{\beta=1}^{L_\lambda} \frac{1}{\beta!} \frac{\partial^\beta g}{\partial y_\lambda^\beta}(x)(y_\lambda - x_\lambda)^\beta \\ &- \sum_{\lambda=1}^{n-1} \sum_{\beta=1}^{L_\lambda - 1} \frac{1}{\beta!} \sum_{0<|\gamma^{(\lambda+1)}| \leq M_\lambda - \beta} \frac{1}{\gamma^{(\lambda+1)}!} D^{\gamma^{(\lambda+1)}} \\ &\times \left(\frac{\partial^\beta g}{\partial y_\lambda^\beta} \right)(x)(y_\lambda - x_\lambda)^\beta (y - x)^{\gamma^{(\lambda+1)}} \\ &- \sum_{\lambda=1}^{n-1} \frac{1}{L_\lambda!} \sum_{0<|\delta^{(\lambda+1)}| \leq T_\lambda} \frac{1}{\delta^{(\lambda+1)}!} D^{\delta^{(\lambda+1)}} \\ &\times \left(\frac{\partial^{L_\lambda} g}{\partial y_\lambda^{L_\lambda}} \right)(x)(y_\lambda - x_\lambda)^{L_\lambda} (y - x)^{\delta^{(\lambda+1)}} \end{aligned}$$

$$= \sum_{\lambda=1}^{n-1} \sum_{\beta=1}^{L_\lambda - 1} \frac{1}{\beta!} \sum_{|\gamma^{(\lambda+1)}| = M_\lambda + 1 - \beta} \frac{1}{\gamma^{(\lambda+1)}!} D^{\gamma^{(\lambda+1)}}$$

$$\times \left(\frac{\partial^\beta g}{\partial y_\lambda^\beta} \right) (x_1, \dots, x_\lambda, u^{(\lambda+1)}) (y_\lambda - x_\lambda)^\beta (y - x)^{\gamma^{(\lambda+1)}}$$

$$+ \sum_{\lambda=1}^{n-1} \frac{1}{L_\lambda!} \sum_{|\delta^{(\lambda+1)}| = T_\lambda + 1} \frac{1}{\delta^{(\lambda+1)}!}$$

$$\times D^{\delta^{(\lambda+1)}} \left(\frac{\partial^{L_\lambda} g}{\partial y_\lambda^{L_\lambda}} \right) (x_1, \dots, x_\lambda, v^{(\lambda+1)}) (y_\lambda - x_\lambda)^{L_\lambda} (y - x)^{\delta^{(\lambda+1)}}$$

$$+ \sum_{\lambda=1}^{n-1} \frac{1}{(L_\lambda + 1)!} \frac{\partial^{L_\lambda + 1} g}{\partial y_\lambda^{L_\lambda + 1}} (w_1, \dots, w_\lambda, y^{(\lambda+1)}) (y_\lambda - x_\lambda)^{L_\lambda} (z_\lambda - x_\lambda)$$

$$+ \frac{1}{(L_n + 1)!} \frac{\partial^{L_n + 1} g}{\partial y_\lambda^{L_n + 1}} (x_1, \dots, x_{n-1}, r) (y_n - x_n)^{L_n + 1}, \tag{11.152}$$

where $y^{(\lambda)} := (y_\lambda, ..., y_n)$, $\gamma^{(\lambda)} := (0, ..., 0, \gamma_\lambda, ..., \gamma_n)$;

$$u^{(\lambda+1)} := u(\beta, \gamma^{(\lambda+1)}, L_\lambda), \qquad v^{(\lambda+1)} := v(\delta^{(\lambda+1)}, T_\lambda, x_1, ..., x_\lambda, y_1, ..., y_\lambda)$$

are points between $x^{(\lambda+1)}$ *and* $y^{(\lambda+1)}$;

$(w_1, ..., w_\lambda)$ *is a point between* $(x_1, ..., x_\lambda)$ *and* $(y_1, \dots, y_\lambda)$, *which depends on the rest of the coordinates* x *and* y;

z_λ *is a number between* x_λ *and* y_λ, $\lambda < n$;

r *is a number between* x_n *and* y_n; z_λ *and* r *depend on* x *and* y.

Proof. For simplicity set $n = 3$. Denote by $F(x, y)$ the left-hand side of equality (11.152). Let

$$F_1(x, y) = g(y) - g(x_1, y_2, y_3) - \sum_{\beta=1}^{L_1} \frac{1}{\beta!} \frac{\partial^\beta g}{\partial y_1^\beta} (x) (y_1 - x_1)^\beta$$

$$- \sum_{\beta=1}^{L_1 - 1} \frac{1}{\beta!} \sum_{0 < |\gamma^{(2)}| \le M_1 - \beta} \frac{1}{\gamma^{(2)}!} D^{\gamma^{(2)}} \left(\frac{\partial^\beta g}{\partial y_1^\beta} \right) (x) (y_1 - x_1)^\beta (y - x)^{\gamma^{(2)}}$$

$$- \frac{1}{L_1!} \sum_{0 < |\delta^{(2)}| \le T_1} \frac{1}{\delta^{(2)}!} D^{\delta^{(2)}} \left(\frac{\partial^{L_1} g}{\partial y_1^{L_1}} \right) (x) (y_1 - x_1)^{L_1} (y - x)^{\delta^{(2)}},$$

$$F_2(x, y) = g(x_1, y_2, y_3) - g(x_1, x_2, y_3) - \sum_{\beta=1}^{L_2} \frac{1}{\beta!} \frac{\partial^\beta g}{\partial y_2^\beta} (x) (y_2 - x_2)^\beta$$

$$- \sum_{\beta=1}^{L_2 - 1} \frac{1}{\beta!} \sum_{0 < \gamma_3 < M_2 - \beta} \frac{1}{\gamma_3!} \frac{\partial^{\beta + \gamma_3} g}{\partial y_3^{\gamma_3} \partial y_2^\beta} (x) (y_2 - x_2)^\beta (y_3 - x_3)^{\gamma_3}$$

$$- \frac{1}{L_2!} \sum_{0 < \delta_3 \le T_2} \frac{1}{\delta_3!} \frac{\partial^{L_2 + \delta_3} g}{\partial y_3^{\delta_3} \partial y_2^{L_2}} (x) (y_2 - x_2)^{L_2} (y_3 - x_3)^{\delta_3},$$

$$F_3(x, y) = g(x_1, x_2, y_3) - g(x) - \sum_{\beta=1}^{L_2} \frac{1}{\beta!} \frac{\partial^\beta g}{\partial y_3^\beta} (x) (y_3 - x_3)^\beta.$$

It is obvious that

$$F(x,y) = F_1(x,y) + F_2(x,y) + F_3(x,y).$$

Transform the expression $F_1(x,y)$, applying the Taylor formula to the difference $g(y) - g(x_1, y_2, y_3)$ with respect to the first variable:

$$\begin{aligned} F_1(x,y) = \Bigg(& \sum_{\beta=1}^{L_1-1} \frac{1}{\beta!} \left(\frac{\partial^\beta g}{\partial y_1^\beta}(x_1,y_2,y_3) - \frac{\partial^\beta g}{\partial y_1^\beta}(x) \right) (y_1 - x_1)^\beta \\ & - \sum_{\beta=1}^{L_1-1} \frac{1}{\beta!} \sum_{0<|\gamma^{(2)}|\le M_1-\beta} \frac{1}{\gamma^{(2)}!} D^{\gamma^{(2)}} \left(\frac{\partial^\beta g}{\partial y_1^\beta} \right) (x)(y_1-x_1)^\beta (y-x)^{\gamma^{(2)}} \Bigg) \\ & + \frac{(y_1-x_1)^{L_1}}{L_1!} \Bigg(\left(\frac{\partial^{L_1} g}{\partial y_1^{L_1}}(z_1,y_2,y_3) - \frac{\partial^{L_1} g}{\partial y_1^{L_1}}(x) \right) \\ & \qquad - \sum_{0<|\delta^{(2)}|\le T_1} \frac{1}{\delta^{(2)}!} D^{\delta^{(2)}} \left(\frac{\partial^{L_1} g}{\partial y_1^{L_1}} \right) (x)(y-x)^{\delta^{(2)}} \Bigg), \end{aligned}$$

where $z_1 := z_1(x_1, y_1, L_1)$ is a point between x_1 and y_1. Let

$$\begin{aligned} F_1^{(1)}(x,y) = & \sum_{\beta=1}^{L_1-1} \frac{1}{\beta!} \left(\frac{\partial^\beta g}{\partial y_1^\beta}(x_1,y_2,y_3) - \frac{\partial^\beta g}{\partial y_1^\beta}(x) \right) (y_1 - x_1)^\beta \\ & - \sum_{\beta=1}^{L_1-1} \frac{1}{\beta!} \sum_{0<|\gamma^{(2)}|\le M_1-\beta} \frac{1}{\gamma^{(2)}!} D^{\gamma^{(2)}} \left(\frac{\partial^\beta g}{\partial y_1^\beta} \right) (x)(y_1-x_1)^\beta (y-x)^{\gamma^{(2)}}, \\ F_1^{(2)}(x,y) = & \left(\frac{\partial^{L_1} g}{\partial y_1^{L_1}}(z_1,y_2,y_3) - \frac{\partial^{L_1} g}{\partial y_1^{L_1}}(x) \right) \\ & - \sum_{0<|\delta^{(2)}|\le T_1} \frac{1}{\delta^{(2)}!} D^{\delta^{(2)}} \left(\frac{\partial^{L_1} g}{\partial y_1^{L_1}} \right) (x)(y-x)^{\delta^{(2)}}, \\ \Sigma^{(1)}(x,y) = & \sum_{\beta=1}^{L_1-1} \frac{1}{\beta!} \left(\frac{\partial^\beta g}{\partial y_1^\beta}(x_1,y_2,y_3) - \frac{\partial^\beta g}{\partial y_1^\beta}(x) \right) (y_1 - x_1)^\beta, \\ \Sigma^{(2)}(x,y) = & \sum_{\beta=1}^{L_1-1} \frac{1}{\beta!} \sum_{0<|\gamma^{(2)}|\le M_1-\beta} \frac{1}{\gamma^{(2)}!} D^{\gamma^{(2)}} \left(\frac{\partial^\beta g}{\partial y_1^\beta} \right) (x)(y_1-x_1)^\beta (y-x)^{\gamma^{(2)}}. \end{aligned}$$

It is clear that

$$\begin{aligned} F_1(x,y) &= F_1^{(1)}(x,y) + \frac{(y_1-x_1)^{L_1}}{L_1!} F_1^{(2)}(x,y), \\ F_1^{(1)}(x,y) &:= \Sigma^{(1)}(x,y) - \Sigma^{(2)}(x,y). \end{aligned}$$

Apply the Taylor formula with respect to the second and third variables to each summand of $\Sigma_1^{(1)}(x,y)$. Then we have

$$\Sigma^{(1)}(x,y) = \sum_{\beta=1}^{L_1-1} \frac{1}{\beta!} \sum_{0<|\gamma^{(2)}|\le M_1-\beta} \frac{1}{\gamma^{(2)}!} D^{\gamma^{(2)}} \left(\frac{\partial^\beta g}{\partial y_1^\beta}\right)(x)(y_1-x_1)^\beta (y-x)^{\gamma^{(2)}}$$
$$+ \sum_{\beta=1}^{L_1-1} \frac{1}{\beta!} \sum_{|\gamma^{(2)}|=M_1+1-\beta} \frac{1}{\gamma^{(2)}!} D^{\gamma^{(2)}} \left(\frac{\partial^\beta g}{\partial y_1^\beta}\right)(x_1,u^{(2)})(y_1-x_1)^\beta (y-x)^{\gamma^{(2)}},$$

where $u^{(2)} := u^{(2)}(\beta,\gamma^{(2)},x,y,)$ is a point between (x_2,x_3) and (y_2,y_3). Therefore,

$$F^{(1)}(x,y) = \sum_{\beta=1}^{L_1-1} \frac{1}{\beta!} \sum_{|\gamma^{(2)}|=M_1+1-\beta} \frac{1}{\gamma^{(2)}!} D^{\gamma^{(2)}} \left(\frac{\partial^\beta g}{\partial y_1^\beta}\right)(x_1,u^{(2)})(y_1-x_1)^\beta (y-x)^{\gamma^{(2)}}.$$

Further, using the Taylor formula again, we obtain

$$F_1^{(2)}(x,y) = \left(\frac{\partial^{L_1} g}{\partial y_1^{L_1}}(z_1,y_2,y_3) - \frac{\partial^{L_1} g}{\partial y_1^{L_1}}(x_1,y_2,y_3)\right)$$
$$+ \left(\frac{\partial^{L_1} g}{\partial y_1^{L_1}}(x_1,y_2,y_3) - \frac{\partial^{L_1} g}{\partial y_1^{L_1}}(x)\right)$$
$$- \sum_{0<|\delta^{(2)}|\le T_1} \frac{1}{\delta^{(2)}!} D^{\delta^{(2)}} \left(\frac{\partial^{L_1} g}{\partial y_1^{L_1}}\right)(x)(y-x)^{\delta^{(2)}}$$
$$= \frac{\partial^{L_1+1} g}{\partial y_1^{L_1+1}}(w_1,y^{(2)})(z_1-x_1) + \sum_{0<|\delta^{(2)}|\le T_1} \frac{1}{\delta^{(2)}!} D^{\delta^{(2)}} \left(\frac{\partial^{L_1} g}{\partial y_1^{L_1}}\right)(x)(y-x)^{\delta^{(2)}}$$
$$+ \sum_{|\delta^{(2)}|=T_1+1} \frac{1}{\delta^{(2)}!} D^{\delta^{(2)}} \left(\frac{\partial^{L_1} g}{\partial y_1^{L_1}}\right)(x_1,v^{(2)})(y-x)^{\delta^{(2)}}$$
$$- \sum_{0<|\delta^{(2)}|\le T_1} \frac{1}{\delta^{(2)}!} D^{\delta^{(2)}} \left(\frac{\partial^{L_1} g}{\partial y_1^{L_1}}\right)(x)(y-x)^{\delta^{(2)}}$$
$$= \frac{\partial^{L_1+1} g}{\partial y_1^{L_1+1}}(w_1,y^{(2)})(z_1-x_1)$$
$$+ \sum_{|\delta^{(2)}|=T_1+1} \frac{1}{\delta^{(2)}!} D^{\delta^{(2)}} \left(\frac{\partial^{L_1} g}{\partial y_1^{L_1}}\right)(x_1,v^{(2)})(y-x)^{\delta^{(2)}}.$$

Thus,

$$F_1(x,y) = \sum_{\beta=1}^{L_1-1} \frac{1}{\beta!} \sum_{|\gamma^{(2)}|=M_1+1-\beta} \frac{1}{\gamma^{(2)}!} D^{\gamma^{(2)}} \frac{\partial^\beta g}{\partial y_1^\beta}(x_1,u^{(2)})(y_1-x_1)^\beta (y-x)^{\gamma^{(2)}}$$
$$+ \frac{1}{L_1!} \frac{\partial^{L_1+1} g}{\partial y_1^{L_1+1}}(w_1,y^{(2)})(z_1-x_1)(y_1-x_1)^{L_1}$$
$$+ \frac{1}{L_1!} \sum_{|\delta^{(2)}|=T_1+1} \frac{1}{\delta^{(2)}!} D^{\delta^{(2)}} \left(\frac{\partial^{L_1} g}{\partial y_1^{L_1}}\right)(x_1,v^{(2)})(y-x)^{\delta^{(2)}}(y_1-x_1)^{L_1}.$$

An analogous expression (with corresponding modification) can be obtained for $F_2(x,y)$.

Applying the Taylor formula with respect to the last variable to $F_3(x,y)$, we obviously get

$$F_3(x,y) = \frac{1}{(L_3+1)!} \frac{\partial^{L_3+1} g}{\partial y_3^{L_3+1}}(x_1, x_2, r)(y_3 - x_3)^{L_3+1}. \qquad \Diamond$$

LEMMA 11.5.14. *Let L_λ, M_λ, T_λ be the same as in the previous lemma. Consider the numbers*

$$\theta := \min\left\{ \min_{\substack{1\le\beta\le L_\lambda-1,\\ |\gamma^{(\lambda+1)}|=M_\lambda+1-\beta,\\ 1\le\lambda<n}} (\beta a_\lambda + |\gamma^{(\lambda+1)}|_a), \right.$$

$$\left. \min_{\substack{|\delta^{(\lambda+1)}|=T_\lambda+1,\\ 1\le\lambda<n}} (a_\lambda L_\lambda + |\delta^{(\lambda+1)}|_a),\ \min_\lambda a_\lambda(L_\lambda+1) \right\},$$

$$\Theta := \max\left\{ \max_{\substack{1\le\beta\le L_\lambda-1,\\ 0<|\gamma^{(\lambda+1)}|=M_\lambda+1-\beta,\\ 1\le\lambda<n}} (\beta a_\lambda + |\gamma^{(\lambda+1)}|_a), \right.$$

$$\left. \max_{\substack{|\delta^{(\lambda+1)}|=T_\lambda+1,\\ 1\le\lambda<n}} (a_\lambda L_\lambda + |\delta^{(\lambda+1)}|_a),\ \max_\lambda a_\lambda L_\lambda \right\}.$$

Let $g, h \in L_1(\mathbb{R}^n)$, $x_0 \in \mathbb{R}^n$, $V > n$, $S > V$ be such that

$$|D^l g(x)| \ll 2^{j(|l|_a+n/2)}(1+2^j\|x\|)^{-V}, \quad l \in \mathbb{N}_0^n, \quad j \in \mathbb{Z},$$
$$|h(x)| \ll 2^{kn/2}(1+2^k\|x-x_0\|)^{-S}, \quad k \in \mathbb{Z};$$

and for any $t_1 = 0,1,2,\dots,L_1$ and any

$$t^{(2)} = (t_2, t_3),\ t_2, t_3 \in \mathbb{N}_0, \quad |t^{(2)}| \le \max_\lambda\{L_\lambda,\ M_\lambda,\ T_\lambda\}$$

the relation

$$\int_{\mathbb{R}} x_1^{t_1}(x^{(2)})^{t^{(2)}} h(x)dx = 0$$

holds, where $x^{(2)} = (x_2, x_3)$. If $S > n + \max\{\theta,\ \Theta\}$, then for $j \le k$, we have

$$|(g*h)(x)| \ll 2^{(j-k)(\theta+n/2)}(1+2^j\|x-x_0\|)^{-V}.$$

`Proof.` Using, if necessary, dilations and shifts, we can assume that $j = 0$, $x_0 = 0$. The condition of vanishing moments implies that

(11.153)

$$(g * h)(x) = \int_{\mathbb{R}} g(y)h(x-y)\,dy$$

$$= \int_{\mathbb{R}} \Bigg\{ g(y) - g(x) - \sum_{\lambda=1}^{n}\sum_{\beta=1}^{L_\lambda} \frac{1}{\beta!}\frac{\partial^\beta g}{\partial y_\lambda^\beta}(x)(y_\lambda - x_\lambda)^\beta$$

$$-\sum_{\lambda=1}^{n-1}\sum_{\beta=1}^{L_\lambda-1} \frac{1}{\beta!} \sum_{0<|\gamma^{(\lambda+1)}|\le M_\lambda-\beta} \frac{1}{\gamma^{(\lambda+1)}!} D^{\gamma^{(\lambda+1)}} \frac{\partial^\beta g}{\partial y_\lambda^\beta}(x)(y_\lambda - x_\lambda)^\beta (y-x)^{\gamma^{(\lambda+1)}}$$

$$-\sum_{\lambda=1}^{n-1} \frac{1}{L_\lambda!} \sum_{0<|\delta^{(\lambda+1)}|\le T_\lambda} \frac{1}{\delta^{(\lambda+1)}!} D^{\delta^{(\lambda+1)}} \left(\frac{\partial^{L_\lambda} g}{\partial y_\lambda^{L_\lambda}}\right)(x)(y_\lambda - x_\lambda)^{L_\lambda}(y-x)^{\delta^{(\lambda+1)}} \Bigg\}$$

$$\times h(x-y)dy$$

$$:= \int_A + \int_B + \int_C,$$

where

$$A := \{y :\ \|x-y\| \le 1\}, \quad B := \{y :\ \|x-y\| > 1,\ \|y\| \le \tfrac{\|x\|}{2}\},$$

$$C := \{y :\ \|x-y\| > 1,\ \|y\| > \tfrac{\|x\|}{2}\}.$$

Let us estimate each integral.

Step 1. To estimate $\int_A$, apply the anisotropic version of the Taylor formula to the expression in braces. Using properties of the derivatives of the function g, we get

$$\left|\int_A\right| \ll \int_A \Bigg\{ \sum_{\lambda=1}^{n-1}\sum_{\beta=1}^{L_\lambda-1} \sum_{|\gamma^{(\lambda+1)}|=M_\lambda+1-\beta} |y_\lambda - x_\lambda|^\beta \prod_{\sigma=\lambda+1}^{n} |y_\sigma - x_\sigma|^{\gamma_\sigma}$$

$$\times (1 + \|(x_1,\dots,x_\lambda,u^{(\lambda+1)})\|)^{-V} + \sum_{\lambda=1}^{n-1} \sum_{|\delta^{(\lambda+1)}|=T_\lambda+1} |y_\lambda - x_\lambda|^{L_\lambda}$$

$$\times \prod_{\sigma=\lambda+1}^{n} |y_\sigma - x_\sigma|^{\delta_\sigma} (1 + \|(x_1,\dots,x_\lambda,v^{(\lambda+1)})\|)^{-V}$$

$$+ \sum_{\lambda=1}^{n-1} |y_\lambda - x_\lambda|^{L_\lambda+1}(1 + \|(w_1,\dots,w_\lambda,y^{(\lambda+1)})\|)^{-V}$$

$$+ |y_n - x_n|^{L_n+1}(1 + \|(x_1,\dots,x_{n-1},r)\|)^{-V} \Bigg\} |h(x) - h(y)|\,dy.$$

As on the set A, $\|y\| \sim \|x\|$; all the expressions $(1 + \|\cdot\|)$ inside the braces are estimated from below by $(1 + \|x\|)$. Hence, taking into account that $\|y - x\| \le 1$

and $S > n + \theta$, we get

$$\left|\int\limits_A\right| \ll (1+\|x\|)^{-V} 2^{kn/2} \int\limits_A \|x-y\|^\theta (1+2^k\|x-y\|)^{-S} dy$$

$$\ll 2^{kn/2}(1+\|x\|)^{-V} \int\limits_0^1 \rho^{\theta+n-1}(1+2^k\rho)^{-S} d\rho$$

$$= 2^{kn/2-k(\theta+n)}(1+\|x\|)^{-V} \int\limits_0^{2^k} \frac{\rho^{\theta+n-1}}{(1+\rho)^S} d\rho$$

$$\ll 2^{-k(\theta+n/2)}(1+\|x\|)^{-V}.$$

Step 2. Estimate $\int_B$. Denote by $H(x,y)$ the expression in braces in equality (11.153); then, using properties of the function g, we obtain

$$|H(x,y)| \ll (1+\|y\|)^{-V} + (1+\|x\|)^{-V}(1+\|x-y\|^\Theta)$$
$$\ll (1+\|y\|)^{-V} + (1+\|x\|)^{-V}\|x-y\|^\Theta.$$

However on the set B, we have $3\|x\|/2 \geq \|x-y\| \geq \|x\|/2$; hence,

$$|H(x,y)| \ll (1+\|y\|)^{-V} + (1+\|x\|)^{-V}\|x\|^\Theta.$$

In addition, $\|x-y\| > (1+\|x\|)/4$. Therefore,

$$(1+2^k\|x-y\|)^{-S} \ll 2^{-kS}(1+\|x\|)^{-S}.$$

Since $S > V$, it follows that

$$\left|\int\limits_B\right| \ll 2^{kn/2-kS}(1+\|x\|)^{-V}\left(\int\limits_B \frac{dy}{(1+\|y\|)^V} + \frac{\|x\|^\Theta}{(1+\|x\|)^S}\int\limits_{\|y\|\leq\|x\|/2} dy\right).$$

As $S > n + \max\{\theta, \Theta\}$, we finally get

$$\left|\int\limits_B\right| \ll 2^{-k(\theta+n/2)}(1+\|x\|)^{-V}.$$

Step 3. It remains to estimate $\int_C$. Taking into account that $1+\|y\| \gg 1+\|x\|$ on the set C and using the argument of the previous step, we derive

$$\left|\int\limits_C\right| \ll 2^{kn/2}(1+\|x\|)^{-V} \int\limits_C \frac{\|x-y\|^\theta}{(1+2^k\|x-y\|)^S} dy$$

$$\ll 2^{kn/2}(1+\|x\|)^{-V}2^{-kS} \int\limits_1^\infty \frac{r^{\theta+n-1}}{r^S} dr \ll 2^{-k(\theta+n/2)}(1+\|x\|)^{-V},$$

as $S > n + \max\{\theta, \Theta\}$. ◊

`Proof of Theorem 11.5.7.` Only the case $m > s$ is of interest (the case $m \leq s$ is included in the previous theorem). Consider the operators

$$T_{\lambda_0} := c_{\lambda_0}(x)\frac{\partial^{r_{\lambda_0}}}{\partial x_{\lambda_0}^{r_{\lambda_0}}},$$

whose symbols $\mathcal{P}_{\lambda_0}(\xi_{\lambda_0}, x)$ satisfy the estimate (11.125) with $m_{\lambda_0} = a_{\lambda_0} r_{\lambda_0}$. If $m_{\lambda_0} \le s$, then

$$T_{\lambda_0} : F_{pq}^{(s)} \to F_{pq}^{(\sigma(\lambda_0))}, \tag{11.154}$$

where

$$\sigma(\lambda_0) = \left(s_1 \left(1 - \frac{m_{\lambda_0}}{s}\right), \dots, s_n \left(1 - \frac{m_{\lambda_0}}{s}\right)\right).$$

If $m_{\lambda_0} > s$, then take into account that

$$T^*_{\lambda_0} \left(x^{\gamma}_{\lambda_0} \bar{x}^{\delta}\right) = 0$$

for $\gamma < r_{\lambda_0}$ and any multi-index

$$\delta = (\delta_1, ..., \delta_{\lambda_0 - 1}, \delta_{\lambda_0 + 1}, \dots, \delta_n),$$

where $\bar{x} = (x_1, ..., x_{\lambda_0-1}, x_{\lambda_0+1}, \dots, x_n)$.

After the change of variables $y_1 = x_{\lambda_0}$, $y_{\lambda_0} = x_1$, $y_i = x_i$, $i \ne 1$, λ_0, the operator T_{λ_0} is the operator of differentiation with respect to the first variable. Let us use Lemma 11.5.14 for $L_1 = r_{\lambda_0} - 1$, taking $L_2, \dots, L_n$, M, T so large that $\theta = (L_1 + 1)a_{\lambda_0} = m_{\lambda_0}$. Then $\theta > m_{\lambda_0} - s$. After the reverse change of variables, using the argument similar to that from the proof of Theorem 11.5.6, we obtain (11.154).

Now take into account that $T = \sum_{\lambda=1}^{n} T_\lambda$; thus this operator acts from $F_{pq}^{(s)}$ to the largest of the spaces $F_{pq}^{(\sigma(\lambda))}$, i.e., to $F_{pq}^{(\sigma)}$. ◊

`Proof of Theorem 11.5.8.` Restrict ourselves to the proof of the precision of relation (11.128), assuming for simplicity that $1 < p < \infty$, $s_1 > 1$. We need to show that for any $\alpha, 0 < \alpha < s_2/s_1$,

$$\frac{\partial}{\partial x_1} : F_{pq}^{(s)} \not\to F_{pq}^{(s_1 - 1, s_2 - \alpha)}.$$

Let us establish a stronger fact; i.e., construct a function $f \in B_{p1}^{(s)}$ such that $\frac{\partial f}{\partial x_1} \notin F_{pq}^{(\sigma)}$, where $\sigma_1 := s_1 - 1$, $\sigma_2 := s_2 - \alpha$ (recall that the inclusion $B_{p1}^{(s)} \subset F_{pq}^{(s)}$ holds). It is not hard to check that for the vector (σ),

$$\sigma = \frac{2(s_1 - 1)(s_2 - \alpha)}{(s_1 + s_2) - (1 + \alpha)}, \quad b_1 := \frac{\sigma}{\sigma_1} = \frac{2(s_2 - \alpha)}{(s_1 + s_2) - (1 + \alpha)},$$

$$b_2 := \frac{\sigma}{\sigma_2} = \frac{2(s_1 - 1)}{(s_1 + s_2) - (1 + \alpha)}, \quad b_1 > a_1, \quad b_2 < a_2.$$

As $1 < p < \infty$, we can define the spaces $B_{p1}^{(s)}$ and $F_{pq}^{(\sigma)}$ by using the dissection method instead of the decomposition method. Consider the set of rectangles

$$\Pi_k^{(s)} = \{\xi \in \mathbb{R}^2 : \ |\xi_\lambda| \le 2^{a_\lambda k}\}, \quad k \in \mathbb{N}_0,$$

$$\Pi_k^{(\sigma)} = \{\xi \in \mathbb{R}^2 : \ |\xi_\lambda| \le 2^{b_\lambda k}\}, \quad k \in \mathbb{N}_0,$$

and a set of corridors

$$\Gamma_k^{(s)} := \Pi_k^{(s)} \backslash \Pi_{k-1}^{(s)} \quad k \in \mathbb{N},$$

$$\Gamma_k^{(\sigma)} := \Pi_k^{(\sigma)} \backslash \Pi_{k-1}^{(\sigma)}, \quad k \in \mathbb{N}.$$

Then it is known that

$$|f, B_{p1}^{(s)}| \sim |F^{-1}(Ff\chi_{\Pi_0^{(s)}}), L_p| + \sum_{k=1}^{\infty} 2^{sk}|V_k^{(s)} * f, L_p|, \tag{11.155}$$

$$V_k^{(s)} := F^{-1}\chi_{\Gamma_k^{(s)}}, \tag{11.156}$$

$$|f, F_{pq}^{(\sigma_1,\sigma_2)}| \sim |F^{-1}(Ff\chi_{\Pi_0^{(\sigma)}}), L_p| + \left|\left(\sum_{k=1}^{\infty}(2^{\sigma k}|V_k^{(\sigma)} * f|)^q\right)^{1/q}, L_p\right|, \tag{11.157}$$

$$V_k^{(\sigma)} := F^{-1}\chi_{\Gamma_k^{(\sigma)}}. \tag{11.158}$$

Consider an entire function $g_j^{(1)}(x_1)$ of exponential type (EFET) $2^{a_1 j}$ such that $|\frac{d}{dx_1}g_j^{(1)}, L_p| \sim 2^{a_1 j}$, $|g_j^{(1)}, L_p| = 1$, and an EFET $g_j^{(2)}(x_1)$, $|g_j^{(2)}, L_p| = 1$, whose spectrum belongs to the segment $[2^{b_2(j-1)}, 2^{b_2 j}]$, $j \in \mathbb{N}$. Then the spectrum of the function

$$g_j(x_1, x_2) := 2^{-sj}j^{-1}g_j^{(1)}(x_1)g_j^{(2)}(x_2)$$

lies in the rectangle $\mathcal{R}_j = [-2^{a_1 j}, 2^{a_1 j}] \times [2^{b_2(j-1)}, 2^{b_2 j}]$; in addition,

$$|g_j, L_p| = 2^{-sj}j^{-1} \tag{11.159}$$

and

$$\left|\frac{\partial}{\partial x_1}g_j,\ L_p\right| \sim 2^{-sj}2^{a_1 j}j^{-1}. \tag{11.160}$$

Estimate $|g_j, B_{p1}^{(s_1,s_2)}|$. The corridor $\Gamma_k^{(s)}$, $k \in \mathbb{N}$, and the rectangle $\mathcal{R}_j$ are disjoint if $\Pi_{k-1}^{(s)} \supset \mathcal{R}_j$, i.e., for $k > j+1$. For any j, the first summand of the norm $|g_j, B_{p1}^{(s_1,s_2)}|$ (see (11.155)) is bounded. The second summand can be estimated in the following way:

$$\sum_{k=1}^{\infty} 2^{sk}|V_k^{(s)} * g_j, L_p| \ll \sum_{k=1}^{j+1} 2^{sk}2^{-sj}j^{-1} \le 2^s \sum_{k=1}^{j+1} 2^{sj}2^{-sj}j^{-1} = 2^s(1 + j^{-1}) \le 2 \cdot 2^s.$$

The first inequality is based on the relation

$$F^{-1}\chi_{\Gamma_k^{(s)}} = F^{-1}\chi_{\Pi_k^{(s)}} - F^{-1}\chi_{\Pi_{k-1}^{(s)}}$$

and the fact that the operators of convolution with the Dirichlet kernels of the rectangles $\Pi_k^{(s)}$ are uniformly bounded with respect to k in L_p, $1 < p < \infty$. For the second inequality, we used (11.159).

Thus,

$$|g_j, F_{pq}^{(s_1,s_2)}| \ll |g_j, B_{p1}^{(s_1,s_2)}| \ll 1, \quad j \in \mathbb{N}. \tag{11.161}$$

Next, since $a_1 < b_1$, it follows that $\mathcal{R}_j \subset \Gamma_j^{(\sigma)}$; therefore,

$$\left|\frac{\partial}{\partial x_1}g_j,\ F_{pq}^{(\sigma_1,\sigma_2)}\right| \ge 2^{\sigma j}\left|\frac{\partial}{\partial x_1}g_j,\ L_p\right| \gg 2^{\sigma j}2^{-sj}2^{a_1 j}j^{-1} = 2^{zj}j^{-1},$$

where $z := \sigma - s + a_1$. The first inequality is based on the fact that the spectrum of the function $\frac{\partial}{\partial x_1}g_j$ coincides with the spectrum of g_j. Moreover, $FV_j^{\sigma} \equiv 1$ in the rectangle Γ_j^σ. For the latter inequality, we used (11.160). It is not hard to check that $z > 0$; therefore,

$$\lim_{j\to\infty} \left| \frac{\partial}{\partial x_1} g_j,\ F_{pq}^{(\sigma_1,\sigma_2)} \right| = \infty.$$

Combining this with (11.161), we prove the required assertion. ◊

APPENDIX A

Auxiliary Facts of the Theory of Functions and Functional Analysis

A.1. Bases

Let X be a Banach space and let X^* be its dual space.

DEFINITION A.1.1. A system of elements $\{x_n\}_{n=1}^{\infty} \subset X$ is called a *complete system* if $\overline{\operatorname{span}\{x_n, n \in \mathbb{N}\}} = X$.

DEFINITION A.1.2. A system of elements $\{x_n\}_{n=1}^{\infty} \subset X$ is said to be *minimal* in X if $x_k \notin \overline{\operatorname{span}\{x_n, n \neq k, n \in \mathbb{N}\}}$ for any $k \in \mathbb{N}$.

THEOREM A.1.3. *A system $\{x_n\}_{n=1}^{\infty}$ is minimal in X if and only if there exists a* dual system, *i.e., a system $\{y_n\}_{n=1}^{\infty} \subset X^*$ such that*

$$\langle x_n, y_k \rangle = \delta_{kn}.$$

DEFINITION A.1.4. A system $\{x_n\}_{n=1}^{\infty} \subset X$ is called a *basis* in X if for any element $x \in X$, there exists a unique representation of the form

$$x = \sum_{n=1}^{\infty} c_n f_n,$$

where c_n are complex numbers and the series converges in norm in X.

THEOREM A.1.5. *A system $\{x_n\}_{n=1}^{\infty}$ is a basis in X if and only if it is complete and minimal in X and for any $x \in X$, the inequality*

$$\left\| \sum_{n=1}^{\infty} \langle x, y_n \rangle x_n \right\| \leq M \|x\|$$

holds, where $\{y_n\}_{n=1}^{\infty}$ is a dual system and M is a positive constant independent of x.

THEOREM A.1.6. *If a system $\{x_n\}_{n=1}^{\infty}$ is a basis in X, then the dual system $\{y_n\}_{n=1}^{\infty}$ is a basis in the space $\overline{\operatorname{span}\{y_n, n \in \mathbb{N}\}}$ (with the norm of the space X^*).*

DEFINITION A.1.7. A basis $\{x_n\}_{n=1}^{\infty}$ in a Banach space X is called *unconditional* if for any rearrangement of the positive integers $\sigma = \{\sigma(n)\}_{n=1}^{\infty}$, $\{x_{\sigma(n)}\}_{n=1}^{\infty}$ is also a basis in X.

THEOREM A.1.8. *A complete minimal system $\{x_n\}_{n=1}^{\infty}$ is an unconditional basis in X if and only if for any sequence $\varepsilon = \{\varepsilon_n\}$, $\varepsilon_n = \pm 1$, the inequality*

$$\left\| \sum_{k=1}^{n} \varepsilon_k \langle x, y_k \rangle x_k \right\| \leq M \|x\|$$

holds for any $n \in \mathbb{N}$ and $x \in X$, where $\{y_n\}_{n=1}^{\infty}$ is a dual system and M is a positive constant independent of x or ε.

The proofs of the above statements can be found, for example, in the book [**1**, Chapter 1].

A.2. Linear functionals in normed spaces

By definition, a linear continuous (bounded) functional on a normed space X is a linear mapping $l : X \to \mathbb{R}$ for which $\|l\| = \sup_{x\in X,\, \|x\|=1} l(x) < \infty$. Sometimes we use the notation $\langle l\,,\, x\rangle = l(x)$.

THEOREM A.2.1 (Hahn-Banach theorem). *Suppose X is a normed space, $X_0 \subset X$ is its subspace, and l_0 is a linear continuous functional on X_0. Then there exists a linear continuous functional l defined on X which coincides with l_0 on X_0, such that $\|l\|_X = \|l_0\|_{X_0}$.*

For the proof, see [**29**].

The set of all linear continuous functionals on X is also a normed linear space. It is called a *dual space* for X and is denoted by X^*. A dual space is always complete; i.e., it is a Banach space. Thus $(L_p)^* = L_q$, $(\ell_p)^* = \ell_q$ for $p \in [1, +\infty)$ $(1/p + 1/q = 1)$. The space $C^*[0, 1]$ is isomorphic to the space $\mathrm{Var}[0, 1]$ of functions of bounded variation or the space of Borelean charges on the segment $[0, 1]$ (Riesz's theorem, see [**29**])

The weak topology in a normed space X is defined by converging sequences: a sequence $\{x_k\}_{k\in\mathbb{N}} \subset X$ weakly converges to $x \in X$ if $\langle y, x_k\rangle \to \langle y, x\rangle$ as $k \to \infty$ for any $y \in X^*$. In the spaces $\ell_p\,, p \in [1, +\infty)$, the weak topology coincides with the topology of coordinatewise convergence: a sequence $\{l_k\} \subset \ell_p$ converges to l if and only if $l_k^j \to l^j$ for all $j \in \mathbb{Z}$.

The *-weak topology in the dual space X^* is defined by converging sequences as follows: a sequence $\{y_k\}_{k\in\mathbb{N}} \subset X^*$ converges to $y \in X^*$ if $\langle y_k, x\rangle \to \langle y, x\rangle$ as $k \to \infty$ for any $x \in X$. If the space X is *reflexive*, i.e., the double dual space $X^{**} = (X^*)^*$ coincides with X, then the *-weak topology is equivalent to the weak topology. This holds, for instance, for all the spaces L_p and l_p, $p \in (1, +\infty)$.

Theorem A.2.2 (Banach, Alaoglu). *The closed unit ball in the dual space X^* is compact in the *-weak topology.*

This means that any sequence of points $\{y_k\}$ from the unit ball B^* has a limit point $y \in B^*$; i.e., for any $x \in X$ there is a subsequence $\{k_i\} \subset \mathbb{N}$ such that $\langle y_{k_i}, x\rangle \to \langle y, x\rangle\,,\ i \to \infty$. If X is separable (possesses a countable everywhere dense subset), then the unit ball B^* of X^* is *sequentially compact*; i.e., any sequence has a *-weakly convergent subsequence. In other words, for any sequence $\{y_k\} \subset X^*$ there is a subsequence $\{k_i\} \subset \mathbb{N}$ such that $\langle y_{k_i}, x\rangle \to \langle y, x\rangle\,,\ i \to \infty$, for any $x \in X$. In particular, for all the spaces $C, L_p, l_p\,,\ p \in [1, +\infty)$, the unit ball of the dual space is sequentially *-weak compact.

The proofs of these facts can be found in [**29**]. For the theory of normed and Banach spaces, see, for example, [**17**].

A.3. Distributions

The foundations of the theory of distributions were developed in the period from the late 1930s until the early 1950s, mainly by S. L. Sobolev and L. Schwartz, though distributions occurred in physical works in the early twentieth century. By definition, *distributions* are linear continuous functionals on *spaces of test functions.* Ordinary locally summable functions are naturally embedded in spaces of distributions. The most frequently used spaces of distributions are $\mathcal{D}'(\mathbb{R}^d)$ and $\mathcal{S}'(\mathbb{R}^d)$.

Consider the case of one variable $d = 1$. The space of test functions $\mathcal{D} = \mathcal{D}(\mathbb{R})$ consists of compactly supported, infinitely differentiable functions on $\mathbb{R}$. Thus,

$$\mathcal{D} \;=\; \Big\{\varphi(\cdot) \in C^{\infty},\ \varphi \ \text{ is a compactly supported function}\Big\}.$$

The topology of the space $\mathcal{D}$ is defined by convergent sequences. A sequence $\{\varphi_k\} \subset \mathcal{D}$ converges to zero if the functions φ_k and all their derivatives tend to zero in the metric of $C(\mathbb{R})$ as $k \to \infty$, and the supports of all functions φ_k are uniformly bounded. Thus for any $l \geq 0$, $\|\varphi_k^{(l)}\|_\infty \to 0$ as $k \to \infty$ and there exist constants $C_1, C_2 \in \mathbb{R}$ such that $\operatorname{supp}\varphi_k \subset [C_1, C_2]$ for any k. This definition can be linearly extended to the convergence of sequences to any element $\varphi \in \mathcal{D}$. The space $\mathcal{D}$ is nonmetrizable and nonnormable. The space of distributions $\mathcal{D}' = \mathcal{D}'(\mathbb{R})$ is the space of linear continuous functionals on $\mathcal{D}$. Thus, $f \in \mathcal{D}'$ if the functional $\varphi \to \langle f, \varphi\rangle$ is linear and $\langle f, \varphi_k\rangle \to 0$ as $\varphi_k \to 0$. Strictly speaking, a distribution is not a function since it has no values at points. Sometimes, by analogy with ordinary functions, the designation $f(x)$ is used for distributions. However, values of distributions at a point are not defined.

A sequence $\{f_j\} \subset \mathcal{D}'$ converges to zero in the space $\mathcal{D}'$ (converges weakly) if $\langle f_j, \varphi\rangle \to 0$ for any element $\varphi \in \mathcal{D}$. The space $\mathcal{D}'$ is separable and metrizable, but not normable. Moreover, $\mathcal{D}'$ has the completeness property (if for any test function φ, the sequence of numbers $\langle f_j, \varphi\rangle$ is a Cauchy sequence, then the sequence f_j converges to an element $f \in \mathcal{D}'$) and the precompactness property: for any bounded sequence $\{f_j\} \subset \mathcal{D}'$ (which means that the sequence $\langle f_j, \varphi\rangle$ is bounded for each element $\varphi \in \mathcal{D}$), there exists a convergent subsequence. Each function from $\mathcal{D}'$ is a limit of an appropriate sequence from $\mathcal{D}$. In this sense, $\mathcal{D}$ is everywhere dense in $\mathcal{D}'$.

On the space of distributions $\mathcal{D}'$, the following operations are defined:

(a) Addition, multiplication by numbers, and passing to the limit.

(b) Multiplication by infinitely differentiable functions. For any $f \in \mathcal{D}'$, $h \in C^\infty(\mathbb{R})$, the distribution hf is defined by the formula $\langle hf, \varphi\rangle \;=\; \langle f, h\varphi\rangle$, $\varphi \in \mathcal{D}$.

(c) Change of variables. For any strictly monotone function $\psi(x) \in C^\infty(\mathbb{R})$, the distribution $f(\psi(x))$ is defined by the formula

$$\langle f(\psi(x)), \varphi(x)\rangle \;=\; \big\langle f(y), [\psi^{-1}(y)]'\varphi(\psi^{-1}y)\big\rangle, \quad \varphi \in \mathcal{D}.$$

For example, for a linear change of variables $\psi(x) = ax + b$, we have

$$\Big\langle f(ax+b), \varphi(x)\Big\rangle \;=\; \Big\langle f(y), \frac{1}{a}\varphi\Big(\frac{y-b}{a}\Big)\Big\rangle, \qquad \varphi \in \mathcal{D}.$$

(d) Differentiation. For any distribution f, the derivative is defined. It is the distribution determined by the formula $\langle f', \varphi\rangle \;=\; -\langle f, \varphi'\rangle$, $\varphi \in \mathcal{D}$.

Thus any distribution is infinitely differentiable. On the other hand, some simple operations (multiplication of two distributions, nonsmooth (or nonmonotone) change of variables) are not defined in the space $\mathcal{D}'$.

The support $\operatorname{supp} f$ of a distribution $f \in \mathcal{D}'$ is a closed set K, minimal with respect to inclusion and such that for any test function φ concentrated outside K (i.e., $\operatorname{supp}\varphi \cap K = \emptyset$), we have $\langle f, \varphi\rangle = 0$. It is easy to see that the limit of a sequence of distributions concentrated on the set K is also concentrated on this set.

An ordinary function summable on each compact set is a distribution. The corresponding linear functional is defined by usual integration:

$$\langle f, \varphi\rangle = \int_{\mathbb{R}} f(x)\varphi(x)\,dx.$$

Two locally summable functions define the same functional (i.e., the same generalized function) if and only if they coincide almost everywhere. For ordinary functions, operations (a)–(d), as well as the notion of support, coincide with the corresponding classical notions. Any function from $C(\mathbb{R})$ or $L_p(\mathbb{R})$ is a distribution.

A distribution that can be defined by an ordinary locally summable function is called *regular*; otherwise it is called *singular*. For example, the regular function $\mathbf{1}(x)$ is defined as $\langle \mathbf{1}, \varphi\rangle = \int_{\mathbb{R}} \varphi(x)\,dx$. The simplest example of a singular function is the Dirac delta function $\delta(x)$, defined as $\langle \delta, \varphi\rangle = \varphi(0)$. The delta function has compact support containing a single point $\{0\}$. Sometimes, even for singular functions, we use the designation $\langle f, \varphi\rangle = \int_{\mathbb{R}} f(x)\varphi(x)\,dx$, without assigning any meaning to the notion of integral. For compactly supported distributions f, we also use quantities $\int_{\mathbb{R}} f(x)\,dx$. By definition, $\int_{\mathbb{R}} f(x)\,dx$ is equal to $\langle f, \varphi\rangle$, where φ is an arbitrary test function identically equal to 1 on the support of f. This quantity does not depend on the choice of function φ.

Each distribution has a primitive. Two arbitrary primitives of one function differ by an identical constant $c\mathbf{1}$. Any compactly supported distribution has singularity of finite order. This means that it is a finite order derivative of a regular function.

A disadvantage of the space of distributions $\mathcal{D}'$ is that the Fourier transform is not defined on it. The reason is that the Fourier transform of a nonzero test function from $\mathcal{D}$ is not compactly supported; hence it does not belong to $\mathcal{D}$. For this reason, to deal with the Fourier transform, we use a narrower space of distributions $\mathcal{S}'$.

The space of test functions $\mathcal{S} = \mathcal{S}(\mathbb{R})$ consists of infinitely differentiable functions on $\mathbb{R}$ that, as well as all their derivatives, decrease at infinity more rapidly than any power function. Thus,

$$\mathcal{S} = \Big\{\varphi(\cdot) \in C^{\infty},\ \Big\|(1+|x|^m)\varphi^{(l)}(x)\Big\|_{\infty} < \infty\Big\}.$$

A sequence $\{\varphi_k\} \subset \mathcal{S}$ converges to zero if for any $m, l \geq 0$ there exists a constant $C(m,l)$ such that

$$\Big\|(1+|x|^m)\varphi_k^l(x)\Big\|_{\infty} < C(m,n), \qquad k \in \mathbb{N},$$

and $\|\varphi_k^{(l)}\|_\infty \to 0$ as $k \to \infty$. The space of distributions $\mathcal{S}' = \mathcal{S}'(\mathbb{R})$ is the space of linear continuous functionals on $\mathcal{S}$. The space of test functions $\mathcal{S}$ is often called *the Schwartz space*; $\mathcal{S}'$ is called *the space of tempered distributions.* Since $\mathcal{D}$ is a subspace of $\mathcal{S}$, it follows that $\mathcal{S}'$ is a subspace of $\mathcal{D}'$. Distributions from $\mathcal{S}'$ have basically the same properties as distributions from $\mathcal{D}'$. Operations (a) and (d) are defined similarly; operation (b) is not defined for each function $h \in C^\infty(\mathbb{R})$ but only for the functions that, together with all their derivatives, increase no faster than any power function as $x \to \infty$. In (d), the conditions for a function $\psi(x)$ are more complicated, so we confine our consideration to linear changes of variables, which are defined by the same formulas as for functions from $\mathcal{D}'$. The notions of support, compactly supported functions, and order of singularity remain the same. Ordinary functions are embedded into $\mathcal{S}'$ by the same scheme. The only difference is that all locally summable functions f do not belong to $\mathcal{S}'$. For example, the function $f(x) = e^x$ does not belong to $\mathcal{S}'$. However this space contains two important classes of functions: all functions from the spaces $L_p(\mathbb{R})$, $p \in [1, +\infty]$, and all functions that grow no faster than any power function at infinity for almost all x. In addition, for such functions, distributional derivatives of all orders also belong to $\mathcal{S}'$. The sets of compactly supported functions of the spaces $\mathcal{S}'$ and $\mathcal{D}'$ coincide.

The main advantage of the space $\mathcal{S}'$ is that the operation of the Fourier transform can be defined on it. The Fourier transform is a one-to-one, bicontinuous linear operator on the space $\mathcal{S}$. Hence the Fourier transform defined for function $f \in \mathcal{S}'$ by the formula

$$\langle \hat{f}, \varphi \rangle = \langle f, \check{\varphi} \rangle$$

is a well-defined linear operator on the space $\mathcal{S}'$ which is one-to-one and bicontinuous. The Fourier transform has the same properties as it has for ordinary functions. Here are some of these properties:

$$\begin{aligned} \langle \hat{f}, \hat{\varphi} \rangle &= \langle f, \varphi \rangle \quad \text{for any } f \in \mathcal{S}',\ \varphi \in \mathcal{S} \qquad \text{(Plancherel theorem)}, \\ \widehat{x^m f}(\xi) &= (-2\pi i)^{-m} \widehat{f}^{(m)}(\xi), \qquad \widehat{f^{(l)}}(\xi) = (2\pi i \xi)^l \widehat{f}(\xi), \\ \widehat{f(ax+b)}(\xi) &= a^{-1} e^{2\pi i b\xi/a} \widehat{f}(a^{-1}\xi), \qquad \widehat{f}(a\xi + b) = a^{-1} \widehat{e^{-2\pi i b x/a} f(a^{-1}x)}. \end{aligned}$$

All the aforementioned operations (the Fourier transform and differentiation) are defined for any $f \in \mathcal{S}'$.

On the multidimensional Euclidean space $\mathbb{R}^d$ the definitions of the spaces of test functions $\mathcal{D}$, $\mathcal{S}$ and distributions $\mathcal{D}'$, $\mathcal{S}'$ are quite similar, as well as the properties of these spaces. As usual, in the definition of differentiation, derivatives $f^{(l)}$ are replaced by partial derivatives $\frac{\partial^l f}{\partial x_1^{l_1} \cdots \partial x_d^{l_d}}$, $\sum_{j=1}^{d} l_j = l$; in the definition of change of variables, instead of the derivative $[\psi^{-1}(y)]'$, the corresponding Jacobian is taken; in the properties of the Fourier transform, x^m is replaced by

$$x_1^{m_1} \cdots x_d^{m_d}, \quad \sum_{j=1}^{d} m_j = m,$$

a is a linear operator in $\mathbb{R}^d$, and the factor a^{-1} before the integral is replaced by $\det a^{-1}$.

The proofs of these facts as well as the theory of distributions can be found in [**15**], [**29**], [**41**].

A.4. Marcinkiewicz interpolation theorem

Let T be an operator from L_p to L_q, $1 \le p, q < \infty$. We say that T is an operator of *strong type* (p, q) if

$$\|Tf\|_q \le A\|f\|_p, \tag{A.1}$$

where A does not depend on f. We say that T is an operator of *weak type* (p, q) if for any $\lambda > 0$

$$\mu\{x : |Tf(x)| > \lambda\} \le \left(B\frac{\|f\|_p}{\lambda}\right)^q, \tag{A.2}$$

where B does not depend on f or λ.

PROPOSITION A.4.1. *Any operator of strong type* (p, q) *is an operator of weak type* (p, q).

The proof follows from the estimate

$$\lambda^q \mu\{x : |Tf(x)| > \lambda\} \le \int_{\mathbb{R}} |Tf|^q = \|Tf\|_q^q \le (A\|f\|_p)^q. \tag{A.3}$$ ◊

We shall need operators defined on several spaces L_p simultaneously. Define $L_{p_1}(\mathbb{R}) + L_{p_2}(\mathbb{R})$ as the set of functions f that can be represented in the form $f = f_1 + f_2$, $f_1 \in L_{p_1}(\mathbb{R})$, $f_2 \in L_{p_2}(\mathbb{R})$. Let $p_1 < p_2$. It is easy to see that $L_p(\mathbb{R}) \subset L_{p_1}(\mathbb{R}) + L_{p_2}(\mathbb{R})$ for all $p \in [p_1, p_2]$.

THEOREM A.4.2 (**Marcinkiewicz interpolation theorem**). *If a linear operator* T *is of weak type* (p_1, p_1) *and simultaneously of weak type* (p_2, p_2), $1 \le p_1 < p_2 < \infty$, *then* T *is of strong type* (p, p) *for any* $p \in (p_1, p_2)$. *The constant* A *in inequality* (A.1) *depends only on* p, p_1, p_2 *and the constants* B_{p_1}, B_{p_2}, *which appear in estimates of weak type.*

The proof of this theorem is given in many books, for example, in [**14**, p. 168] and [**1**, p. 485].

REMARK A.4.3. If in Theorem A.4.2 we assume that inequality (A.2) for $p = p_1$ holds only for functions $f \in L_{p_1}(\mathbb{R}) \cap L_{p_2}(\mathbb{R})$, then inequality (A.1) will hold for all $f \in L_p(\mathbb{R}) \cap L_{p_2}(\mathbb{R})$. One can prove this fact leaving the above reasoning practically unchanged.

A.5. Spectral radius

The spectral radius of an operator A is the number

$$\rho(A) := \lim_{n\to\infty} \|A^n\|^{1/n}.$$

The whole spectrum of any linear bounded operator A is contained inside the circle $|\lambda| \le \rho(A)$, and there is at least one point of the spectrum on the boundary of this circle. The proof of this fact is presented, for example, in the book [**12**] (see pp. 266–267). In a finite-dimensional space a spectrum coincides with the set of eigenvalues. If M is a quadratic $d \times d$ matrix, whose eigenvalues are greater than 1 in modulus, then the matrix M^{-1} has finitely many eigenvalues that are less than 1 in modulus; hence $\rho(M^{-1}) < 1$. It follows that for any $\delta > 0$, the sequence $\|M^{-n}\|^\delta$ is majorized by a geometric progression, which implies relation (2.6).

For the readers unfamiliar with the above-mentioned part of functional analysis, we note that relations (2.5) and (2.6) can be proved directly using the expansions with respect to the basis of eigenvectors and associated vectors of the matrix M.

LEMMA A.5.1. *Suppose B is a linear operator acting in $\mathbb{R}^d$, $\{a_i\}_{i=1}^s$ is a finite subset of $\mathbb{R}^d$. Then if $\rho(B) < 1$, there exists a convex compact set $K \subset \mathbb{R}^d$ centrally symmetric with respect to the origin and having the nonempty interior such that $BK - a_i \subset K$, $i = 1, \dots, s$.*

Proof. Take an arbitrary $q \in (\rho, 1)$ and prove the existence of a compact set Q (centrally symmetric with respect to the origin and having a nonempty interior) such that $BQ \subset qQ$. To this end, we slightly change the operator B in such a way that all its eigenvalues become different, its spectral radius still being less than q. Go over to the basis of eigenvectors $v_1, \dots, v_d$, and denote by $\lambda_1, \dots, \lambda_d$ the corresponding eigenvalues. Next, sort them in the following way: let $\lambda_1, \dots, \lambda_s$ be real eigenvalues, all other eigenvalues being divided into pairs of self-conjugate ones: $(\lambda_{s+1}, \lambda_{s+2}), \dots, (\lambda_{d-1}, \lambda_d)$. Each pair of eigenvectors

$$(v_{s+2j-1}, v_{s+2j}), \quad j = 1, \dots, r = \frac{d-s}{2},$$

generates a two-dimensional space. Denote by L_j its real part. The restriction of the operator B to L_j is the superposition of the rotation by the angle $\arg \lambda_{s+2j}$ and the dilation with coefficient $|\lambda_{s+2j}|$. Denote by γ_j the unit circle in the space L_j with center at zero and by b_k the segment with endpoints $\pm a_k$, $k = 1, \dots, s$. Then

$$Q := \operatorname{conv}(b_1, \dots, b_s, L_1, \dots, L_r)$$

is the required compact set. Indeed, $BQ \subset \rho(B)Q \subset qQ$. Choose sufficiently large positive μ such that $qQ - \frac{1}{\mu} a_i \subset Q$ for any i. Then $q\mu Q - a_i \subset \mu Q$, and as $BQ \subset qQ$, we have $B\mu Q - a_i \subset \mu Q$. Setting $K = \mu Q$, we complete the proof. $\Diamond$

A.6. Joint spectral radius and the Lyapunov exponent

Recall (Definition 5.1.1) that for a given pair of linear operators $\tilde{B}_0, \tilde{B}_1$ acting in $\mathbb{R}^n$, the joint spectral radius and lower spectral radius, respectively, are the quantities

$$\hat{\rho}(\tilde{B}_0, \tilde{B}_1) = \lim_{r\to\infty} \max_{(d_1,\dots,d_r)\in\{0,1\}^r} \|\tilde{B}_{d_1} \cdots \tilde{B}_{d_r}\|^{1/r},$$
(A.4)
$$\check{\rho}(\tilde{B}_0, \tilde{B}_1) = \lim_{r\to\infty} \min_{(d_1,\dots,d_r)\in\{0,1\}^r} \|\tilde{B}_{d_1} \cdots \tilde{B}_{d_r}\|^{1/r}.$$

For any finite number of linear operators the definitions are the same, as well as the properties of joint and lower spectral radii. For simplicity we shall formulate all results for the case of two operators. The joint and lower spectral radii are both well-defined by equalities (A.4). This can be shown by invoking the Fekete lemma:

LEMMA A.6.1. *A sequence of numbers $\{a_k\}$ such that*

$$(k+n)a_{k+n} \leq k a_k + n a_n, \qquad k, n \in \mathbb{N},$$

converges (the limit may be equal to $-\infty$) and

$$\lim_{k\to\infty} a_k = \inf_{k\in\mathbb{N}} a_k.$$

For the proof, see [**32**]. The reader can easily prove this assertion. Setting in the first definition

$$a_k = \ln\left(\max_{(d_1,\dots,d_r)\in\{0,1\}^r} \|\tilde{B}_{d_1}\cdots\tilde{B}_{d_r}\|^{1/r}\right),$$

and in the second one

$$a_k = \ln\left(\min_{(d_1,\dots,d_r)\in\{0,1\}^r} \|\tilde{B}_{d_1}\cdots\tilde{B}_{d_r}\|^{1/r}\right),$$

we prove the existence of both limits in Lemma A.6.1. Note that in view of the equivalence of all norms in a finite-dimensional space, limits (A.4) do not depend on the introduced norm.

`Proof of Lemma 5.1.2.` For any r, we have

$$\left(\min_{d_1,\dots,d_{rk}} \|\Pi_{rk}\|\right)^{\frac{1}{rk}} \le \|(\Pi_k)^r\|^{\frac{1}{rk}} \le \left(\max_{d_1,\dots,d_{rk}} \|\Pi_{rk}\|\right)^{\frac{1}{rk}}.$$

As $r \to \infty$, we obtain the required relations. ◊

The joint spectral radius is not only the limit of norms but the limit of the usual spectral radii as well. The proof of the following fact can be found in [**142**]:

LEMMA A.6.2. *For any operators $\tilde{B}_0, \tilde{B}_1$, we have*

$$\check{\rho} = \limsup_{k\to\infty} \min_{d_1,\dots,d_k} \Big(\rho(\Pi_k)\Big)^{1/k}, \qquad \hat{\rho} = \lim_{k\to\infty} \max_{d_1,\dots,d_k} \Big(\rho(\Pi_k)\Big)^{1/k}.$$

Note that $\|\Pi_k\|^{1/k} \ge \hat{\rho}$ for any product Π_k. On the other hand,

$$(\rho(\Pi_k))^{1/k} \le \hat{\rho}$$

for any product Π_k (Lemma 5.1.2). Hence the quantity $\max\limits_{d_1,\dots,d_k} \|\Pi_k\|^{1/k}$ tends to $\hat{\rho}$ from the right, and $\max\limits_{d_1,\dots,d_k} \Big(\rho(\Pi_k)\Big)^{1/k}$ from the left. For the lower spectral radius $\check{\rho}$, both quantities $\min\limits_{d_1,\dots,d_k} \|\Pi_k\|^{1/k}$ and $\min\limits_{d_1,\dots,d_k} \Big(\rho(\Pi_k)\Big)^{1/k}$ tend to $\hat{\rho}$ from the right (the latter quantity does in the sense of the lower limit).

The following theorem clarifies the geometric sense of the joint spectral radius.

THEOREM A.6.3. $\hat{\rho}(\tilde{B}_0, \tilde{B}_1) < 1$ *if and only if there exists a norm in $\mathbb{R}^n$ in which the operators B_0 and B_1 are both contractions.*

`Proof.` If there exists a norm $\|\cdot\|$ in $\mathbb{R}^n$ in which

$$\|\tilde{B}_i\| < q < 1, \qquad i = 0, 1,$$

then $\|\tilde{B}_{d_1}\cdots\tilde{B}_{d_k}\| < q^k$ for any $d_1,\dots,d_k$. Therefore $\hat{\rho} \le q < 1$. Conversely, let $\hat{\rho} \le q < 1$. Construct a norm in which both operators are contracting. By the definition of joint spectral radius, for any $b \in (q, 1)$ there exists a constant $C_b > 1$ such that

$$\|\tilde{B}_{d_1}\cdots\tilde{B}_{d_k}\| < C_b b^k$$

for any $d_1,\dots,d_k$. Consider the set $M \subset \mathbb{R}^n$ consisting of the points $v \in \mathbb{R}^n$ such that $\|v\| \le 1$ and $\|\tilde{B}_{d_1}\cdots\tilde{B}_{d_k} v\| < 1$ for any k and any $d_1,\dots,d_k$. This set is closed, convex, centrally symmetric with respect to zero, and it contains the ball

$\{v \in \mathbb{R}^n,\ \|v\| \le 1/C_b\}$. In addition, $\tilde{B}_i M \subset bM$ for $i = 0, 1$. Hence in the Minkowski norm corresponding to the set M we have

$$\|v\|_M = \inf\Big\{\lambda > 0,\ \frac{1}{\lambda} v \in M\Big\} \tag{A.5}$$

and $\|\tilde{B}_i\|_M \le b$. This completes the proof. ◊

LEMMA A.6.4. *If the collection of operators $\tilde{B}_0, \tilde{B}_1$ is irreducible, then*
1) $\hat{\rho} > 0$,
2) *there exist constants $c_1, c_2 > 0$ such that for any $k \in \mathbb{N}$ and for any $v \in \mathbb{R}^n$, $\|v\| = 1$, we have*

$$c_1 \hat{\rho}^k \le \max_{d_1,\dots,d_k \in \{0,1\}} \|\tilde{B}_{d_1} \cdots \tilde{B}_{d_k} v\| \le c_2 \hat{\rho}^k. \tag{A.6}$$

Proof. For the proof, see [**119**]. ◊

The direct generalization of the notion of joint spectral radius is that of p-radius (Definition 5.2.1), which was used to estimate the smoothness of refinable functions in L_p (Chapter 5, §5.2). The properties and application of p-radius were discussed in §7.3, Chapter 7. From the practical point of view, the case $p = 2$ is of most interest since it provides the conditions for a refinable function to belong to the space L_2. The following assertion is proved in [**187**] and [**188**]:

THEOREM A.6.5. *A stable refinable function belongs to $L_2(\mathbb{R}^d)$ if and only if $\rho_2(\tilde{T}_0, \dots, \tilde{T}_{m-1}) < 1$, where $\tilde{T}_k = T_k|_W$, $W = \{x \in \mathbb{R}^d,\ \sum_i x_i = 0\}$, and the operators T_k are defined by equality* (5.11).

A special case of this criterion (for $d = 1$) was formulated and used in Chapters 5 and 7 (Theorems 5.2.2 and 7.1.1).

Thus, to determine whether a solution of a given refinement equation belongs to L_2, we need to compute the 2-radius of the operators $\tilde{T}_0, \dots, \tilde{T}_{m-1}$.

The following result reduces the computation of the 2-radius to that of eigenvalues of a special matrix. Let $\tilde{B}_0, \dots, \tilde{B}_{m-1}$ be arbitrary $n \times n$-matrices. Consider the operator $\mathcal{A}_2$ acting in the space of $n \times n$-matrices (i.e., in $\mathbb{R}^{n^2}$) by the formula

$$\mathcal{A}_2 X = \frac{1}{m}\left(\tilde{B}_0 X \tilde{B}_0^* + \cdots + \tilde{B}_{m-1} X \tilde{B}_{m-1}^*\right). \tag{A.7}$$

THEOREM A.6.6. *The 2-radius of the operators $\tilde{B}_0, \dots, \tilde{B}_{m-1}$ is equal to $\sqrt{\rho(\mathcal{A}_2)}$, where $\rho(\mathcal{A}_2)$ is the maximal absolute value of the eigenvalues of the operator $\mathcal{A}_2$.*

For the proof, see [**115**]. One can reduce the dimension from n^2 to $\frac{n(n+1)}{2}$ by considering the restriction of the operator A_2 to the space of symmetric matrices. The result remains the same [**115**]. (See the proof of Theorem 2.5.13.)

Recall (Definition 5.4.8) that the Lyapunov exponent of linear operators $\tilde{B}_0, \tilde{B}_1$ is the quantity

$$\bar{\rho} = \lim_{k\to\infty} \left(\prod_{d_1,\dots,d_k} \|\Pi_k\|\right)^{1/2^k k}.$$

This limit exists and is independent of the norm.

PROPOSITION A.6.7. *For an arbitrary pair of linear operators* $\tilde{B}_0, \tilde{B}_1$, *the limit*

$$\lim_{k\to\infty} \|\Pi_k\|^{1/k} \tag{A.8}$$

exists and is equal to $\bar{\rho}$ *for almost all* $x \in [0,1)$, *where*

$$x = 0.d_1, d_2, \dots, \qquad \Pi_k = \tilde{B}_{d_1} \cdots \tilde{B}_{d_k}.$$

For the proof see, for example, [**121**], [**122**]. This fact can be formulated in terms of probability theory. If independent random variables $d_1, d_2, \dots$ take values 0 and 1 with probability $\frac{1}{2}$, then limit (A.8) almost surely exists and equals $\bar{\rho}$. It follows that the discrete random variables

$$\eta_n = \frac{1}{n} \log_2 \|\Pi_k\|$$

converge in probability to the constant $\log_2 \bar{\rho}$. If $\check{\rho}(B_0, B_1) > 0$, then

$$\log_2 \check{\rho}(B_0, B_1) \ \le \eta_k \ \le \ \log_2 \hat{\rho}(B_0, B_1)$$

for each k with probability 1. Hence the quantities η_k are uniformly bounded and consequently,

$$E\eta_k \ \to \ \log_2 \bar{\rho},$$

where by E we denote the mathematical expectation. Note that

$$E\eta_k = \log_2 \left(\prod_{d_1,\dots,d_k} \|\Pi_k\| \right)^{1/2^k k}.$$

Thus we obtain another definition of the Lyapunov exponent:

$$\bar{\rho} = \lim_{k\to\infty} \left(\prod_{d_1,\dots,d_k} \|\Pi_k\| \right)^{1/2^k k}.$$

This limit exists and is independent of norm. For rigorous proofs and other properties of the Lyapunov exponent, see, for example, [**120, 121, 122**]. Similar to the joint spectral radius, the Lyapunov exponent is very hard to compute and estimate.

At the present time, an extensive literature devoted to the joint spectral radius and its generalizations is available (over 200 titles). The recent brief reviews [**116, 143, 144**] can be recommended to the reader. Properties of the joint spectral radius are described, for example, in [**119, 130, 132, 134, 135, 142**] and [**145**]–[**149**]. Problems of computation of joint spectral radii are discussed in [**135**]–[**137**], [**143**], [**150**]. For algorithms of the computation, see [**115**], [**117**]–[**120**], [**138**]–[**140**]. Applications of the joint spectral radius and its generalizations to the wavelet theory, the theory of approximation, dynamical systems, probability theory, number theory, and mathematical physics are discussed in [**120, 125, 127**], [**130**]–[**133**], [**138**], [**150**]–[**156**].

A.7. Smoothness and the decay rate of the Fourier transform

For a given function $g(x)$, set

$$\vartheta = \vartheta(g) = \inf_{\xi\in\mathbb{R}} \{\beta \in \mathbb{R} \,|\, |g(\xi)| \le C(|\xi|+1)^{-\beta}\}.$$

THEOREM A.7.1. *For a continuous function $f \in L(\mathbb{R})$, we have*

$$\vartheta - 1 \leq \alpha_f \leq \vartheta, \tag{A.9}$$

where $\vartheta = \vartheta(\widehat{f})$.

Proof. Differentiating the function as many times as necessary and taking into account that $\widehat{f^{(k)}}(\xi) = (2\pi i x)^k \widehat{f}(\xi)$, we can assume that $\alpha_f \leq 1$. First show that $\alpha_f \leq \vartheta$. Consider the case of compact support $\operatorname{supp} f \subset [0, N]$ (the reader can easily generalize the proof to the general case $f \in L(\mathbb{R})$). Fix ξ such that $|\xi| \geq 1$ and set $h = \frac{1}{2\pi(|\xi|+1)}$. We have

$$e^{-2\pi i h\xi} \widehat{f}(\xi) = e^{-2\pi i h\xi} \int_{\mathbb{R}} f(x) e^{-2\pi i x\xi}\, dx,$$
$$\widehat{f}(\xi) = e^{-2\pi i h\xi} \int_{\mathbb{R}} f(x+h) e^{-2\pi i x\xi}\, dx.$$

Subtracting the second inequality from the first one, we get

$$(e^{-2\pi i h\xi} - 1)\widehat{f}(\xi) = e^{-2\pi i h\xi} \int_0^N \big(f(x) - f(x+h)\big) e^{-2\pi i x\xi}\, dx.$$

The absolute value of the integrand does not exceed Ch^α, where $\alpha < \alpha_f$, and the length of the segment of integration does not exceed $N + h \leq N + 1$. Therefore the integral does not exceed $C(N+1)h^\alpha$. Since $|2\pi h\xi| \geq \frac{1}{2}$, it follows that $|e^{-2\pi i h\xi} - 1| \geq C_0 > 0$ (an absolute constant). Thus,

$$|\widehat{f}(\xi)| \leq \frac{C}{C_0} C(N+1)h^\alpha.$$

Taking into account that $h = \frac{1}{2\pi(|\xi|+1)}$, we obtain the required inequality.

Show that $\alpha_f + 1 \geq \vartheta$. If $\vartheta \geq 1$, there is nothing to prove, so we assume that $\vartheta > 1$. For an arbitrary $\beta \in (1, \vartheta)$ we need to prove that $\alpha_f \geq \beta - 1$. We have

$$|f(x+h) - f(x)| = \Big| \int_{\mathbb{R}} \widehat{f}(\xi) e^{2\pi i x\xi} (e^{2\pi i h\xi} - 1)\, d\xi \Big| \leq \Big| \int_{|\xi| \leq h^{-1}-1} \Big| + \Big| \int_{|\xi| > h^{-1}-1} \Big|$$

(we assume that $h \in (0, 1)$). For the upper estimate of the first integral, we have

$$\int_{|\xi| \leq h^{-1}-1} C(1+|\xi|)^{-\beta} |h\xi|\, d\xi \leq Ch \int_{|\xi| \leq h^{-1}-1} (1+|\xi|)^{1-\beta}\, d\xi$$
$$= \frac{2C}{2-\beta} h(h^{\beta-2} - 1) \leq \frac{2C}{2-\beta} h^{\beta-1}.$$

Estimate the second integral. Taking into account that $|e^{2\pi i x\xi}(e^{2\pi i h\xi} - 1)| \leq 2$, we get

$$\Big| \int_{|\xi| > h^{-1}-1} \Big| \leq 2 \int_{|\xi| > h^{-1}-1} |\widehat{f}(\xi)|\, d\xi \leq 2 \int_{|\xi| > h^{-1}-1} C(1+|\xi|)^{-\beta}\, d\xi \leq \frac{2C}{1-\beta} h^{\beta-1}.$$

Summing these two estimates, we obtain $|f(x+h) - f(x)| \leq Ch^{\beta-1}$, which was to be proved. ◊

A detailed study of the problem about the connection between the decay rate of the Fourier transform of a function and its smoothness in various spaces is presented, for example, in [**14**].

Let us now describe functions with compactly supported Fourier transform. This description is connected with the theory of entire functions of exponential type, i.e., functions f analytical on the whole complex plane and such that

$$f(z) = O(e^{a|z|}), \quad |z| \longrightarrow \infty,$$

for some positive a. The lower bound σ of all such a is called the type of f.

THEOREM A.7.2 (The Paley-Wiener theorem). *Let $\sigma > 0$. Then*

$$f(x) = \int_{-\sigma}^{\sigma} g(\xi) e^{2\pi i x\xi}\, d\xi$$

for some $g \in L_2(\mathbb{R})$ if and only if f belongs to $L_2(\mathbb{R})$ and can be extended to the complex plane as an entire function of exponential type σ. In particular, a function $f \in L_2(\mathbb{R})$ is an integral function of exponential type if and only if its Fourier transform is compactly supported.

The proof of this theorem is available, for example, in the book [**14**, Chapter 16, §7].

A.8. Wiener theorem for L_2

A space $X \subset L_2(\mathbb{R}^d)$ is called invariant under shifts if $f(\cdot + t) \in X$ for all $f \in X$ and all $t \in \mathbb{R}^d$.

If $X \subset L_2(\mathbb{R}^d)$, then by $\widehat{X}$ we denote the set of the Fourier transforms of all functions from X.

THEOREM A.8.1. *A subset X of the space $L_2(\mathbb{R}^d)$ is invariant under shifts if and only if $\widehat{X} = L_2(\Omega)$ for some measurable set $\Omega \subset \mathbb{R}^d$.*

`Proof.` Let $\widehat{X} = L_2(\Omega)$, $\Omega \subset \mathbb{R}^d$, $f \in X$, $f_t := f(\cdot + t)$; then $\widehat{f_t} = \widehat{f} e_t$, where $e_t(x) := e^{2\pi i (x,t)}$, which implies that $\widehat{f_t} \in L_2(\Omega)$. By the definition of $\widehat{X}$ this means that $f_t \in X$; i.e., the space X is invariant under shifts.

Now let X be invariant under shifts. Then $\widehat{X}$ is invariant under multiplication by the functions e_t, $t \in \mathbb{R}^d$. Denote by P the orthogonal projection on $\widehat{X}$. For any $f, g \in L_2(\mathbb{R}^d)$ and any $t \in \mathbb{R}^d$, we have $f - Pf \perp (Pg)\, e_t$; i.e., the following equality holds:

$$\int_{\mathbb{R}^d} (f(x) - Pf(x)) \overline{Pg(x)} e^{-2\pi i (x,t)}\, dx = 0.$$

Taking into account that the function $F := (f - Pf)\overline{P}g$ is summable on $\mathbb{R}^d$, we get $\widehat{F} \equiv 0$, which implies that $\|F\| = 0$. Thus,

$$(f - Pf)\overline{Pg} = 0 \quad \text{a.e.}$$

It follows that

$$fPg = PfPg \quad \text{a.e.}$$

The right-hand side of this equality is symmetric with respect to the functions f, g; hence switching the roles of f and g, we get

$$fPg = gPf \quad \text{a.e.} \tag{A.10}$$

for any $f, g \in L_2(\mathbb{R}^d)$. For g, take any positive function from $L_2(\mathbb{R}^d)$ and set $\varphi = Pg/g$. From (A.10) we get

$$Pf = \varphi f \quad \text{a.e.} \tag{A.11}$$

for all $f \in L_2(\mathbb{R}^d)$, which implies

$$\varphi^2 f = \varphi P f = P^2 f = Pf = \varphi f.$$

Therefore $\varphi^2 = \varphi$ almost everywhere; i.e., almost all values of the function φ are equal to 0 or 1. Denote by Ω the set on which $\varphi = 1$. If $f \in \widehat{X}$, then $Pf = f$ and from (A.11) we get

$$f = \varphi f \quad \text{a.e.} \tag{A.12}$$

It follows that $f = 0$ almost everywhere on $\mathbb{R}^d \setminus \Omega$; i.e., $f \in L_2(\Omega)$. We have proved the inclusion $\widehat{X} \subset L_2(\Omega)$. If $f \in L_2(\Omega)$, then (A.12) holds, which together with (A.11) implies $Pf = f$, i.e., $f \in \widehat{X}$; hence $\widehat{X} \subset L_2(\Omega) \subset \widehat{X}$. $\Diamond$

The above proof is taken from the book [**13**], where the theorem is formulated for the one-dimensional case.

A.9. Lebesgue sets

The classical Lebesgue theorem states that almost all points are Lebesgue points for a locally summable function f; i.e., the following relation holds:

$$\lim_{h \to 0} \frac{1}{h} \int_{-h}^{h} |f(x+t) - f(x)| \, dt = 0, \quad \text{a.e.}$$

For functions of several variables, a point $x \in \mathbb{R}^d$ is usually called a Lebesgue point if the analogous relation

$$\lim_{h \to 0} \frac{1}{h^d} \int_{[-h,h]^d} |f(x+t) - f(x)| \, dt = 0$$

holds. In this sense the Lebesgue theorem also holds. However there exist other natural ways to generalize the notion of Lebesgue points (Lebesgue sets) to the multidimensional case.

Let $\mathcal{T}$ denote the set of nondegenerate linear transforms of the space $\mathbb{R}^d$.

DEFINITION A.9.1. Let $f \in L(\mathbb{T}^d)$, $h > 0$, $x \in \mathbb{R}^d$, $T \in \mathcal{T}$, $j \in \mathbb{Z}_+^{d-1}$. Set

$$I_j(f, x, T, h) = h^{-d} \int_{-h}^{h} dt_d \prod_{l=1}^{d-1} 2^{j_l} \int_{-2^{-j_l}h}^{2^{-j_l}h} dt_l \, |f(x) - f(x + Tt)| \, .$$

For a sequence of positive numbers $A = \{A_j\}_{j \in \mathbb{Z}_+^{d-1}}$, define the sets of points $A\mathcal{L}(f, T)$ and $A\mathcal{L}^*(f, T)$ (generalized Lebesgue sets) in the following way: a point $x \in \mathbb{T}^d$ belongs to $A\mathcal{L}(f, T)$ if

$$\lim_{h \to 0} \sum_{j \in \mathbb{Z}_+^{d-1}} A_j I_j(f, x, T, h) = 0, \tag{A.13}$$

(A.14) $$\sup_{h>0} \sum_{j\in\mathbb{Z}_+^{d-1}} A_j I_j(f,x,T,h) < \infty;$$

a point $x \in \mathbb{T}^d$ belongs to $A\mathcal{L}^*(f,T)$ if

(A.15) $$\lim_{h\to 0} \sup_{j\in\mathbb{Z}_+^{d-1}} A_j I_j(f,x,T,h) = 0,$$

(A.16) $$\sup_{h>0} \sup_{j\in\mathbb{Z}_+^{d-1}} A_j I_j(f,x,T,h) < \infty.$$

It is clear that $A\mathcal{L}(f,T) \subset A\mathcal{L}^*(f,T)$. If $A_j = \prod_{l=1}^{d-1} 2^{-j_l}$, then for any T, $A\mathcal{L}^*(f)$ is the set of the usual Lebesgue points of the function f. If $A_j \equiv 1$, then $A\mathcal{L}^*(f)$ coincides with the set of so-called strong Lebesgue points of the function f introduced in [**229**], i.e., the points x such that

(A.17) $$\lim_{h_1\to 0,\dots,h_d\to 0} \frac{1}{h_1\dots h_d} \int_{-h_1}^{h_1} \dots \int_{-h_d}^{h_d} |f(x+t) - f(x)|\, dt = 0,$$

(A.18) $$\sup_{h_1>0,\dots,h_d>0} \frac{1}{h_1\dots h_d} \int_{-h_1}^{h_1} \dots \int_{-h_d}^{h_d} |f(x+t) - f(x)|\, dt < \infty.$$

From the results obtained by S. Saks [**231**] (see also [**229**], [**230**]) it follows that if $f \in L\log L(\mathbb{T}^d)$, then almost all point are strong Lebesgue points of the function f; there exist summable functions that have no strong Lebesgue points.

THEOREM A.9.2. *Let $A = \{A_j\}_{j\in\mathbb{Z}_+^{d-1}}$ be a monotone function of positive numbers such that*

$$\sum_{j\in\mathbb{Z}_+^{d-1}} A_j(1 + |\log A_j|) < \infty.$$

If $f \in L(\mathbb{T}^d)$, $T \in \mathcal{T}$, then the set $A\mathcal{L}(f,T)$ has full measure.

The proof of this theorem is based on the following statements.

LEMMA A.9.3. *Let $A = \{A_j\}_{j\in\mathbb{Z}_+^{d-1}}$ be a sequence of positive numbers, $\sum_{j\in\mathbb{Z}_+^{d-1}} A_j = 1$, $\sum_{j\in\mathbb{Z}_+^{d-1}} A_j|\log A_j| = K$. If nonnegative functions g_j, $j \in \mathbb{Z}_+^{d-1}$, defined on $\mathbb{R}^d$ are such that $\mu\{x : g_j(x) \ge \xi\} \le \xi^{-1}$, then*

$$\mu\left\{x : \sum_{j\in\mathbb{Z}_+^{d-1}} A_j g_j(x) \ge \xi\right\} \le \frac{2(K+1)}{\xi}.$$

The proof of this lemma for $d = 2$ is given in [**18**]. It holds for an arbitrary d.

LEMMA A.9.4. *Let W be a d-dimensional parallelepiped, $f \in L(\mathbb{T}^d)$, $\xi > 0$,*

$$E(f,\xi) = \left\{x \in \mathbb{T}^d : \sup_{h>0} \frac{1}{\mu(hW)} \int_{hW} |f(x+t)|\, dt > \xi\right\}.$$

Then $\mu E(f,\xi) \le K\xi^{-1}\|f\|_1$, where K is an absolute constant.

The proof of this assertion (for a more general case) is presented in [**18**].

`Proof of Theorem A.9.2.` Let H, G be sets of points $x \in \mathbb{T}^d$ for which conditions (A.13), (A.14), respectively, do not hold. We need to check that $\mu H \cup G = 0$.

Set

$$\widetilde{I}_j(f, x, T, h) = h^{-d} \int_{-h}^{h} dt_d \prod_{l=1}^{d-1} 2^{j_l} \int_{-2^{-j_l} h}^{2^{-j_l} h} dt_l f(x + Tt),$$

$$G_1 = \left\{ x \in \mathbb{T}^d : \sup_{h>0} \sum_{j \in \mathbb{Z}_+^{d-1}} A_j \widetilde{I}_j(|f|, x, T, h) = \infty \right\},$$

$$G_2 = \{x \in \mathbb{T}^d : \ f(x) = \infty\}.$$

Since for each positive integer n

$$G_1 \subset \{x \in \mathbb{T}^d : \sup_{h>0} \sum_{j \in \mathbb{Z}_+^{d-1}} A_j \widetilde{I}_j(|f|, x, T, h) > n\} = G_{n1},$$

by Lemmas A.9.3 and A.9.4, we have

$$\mu G_1 \leq \mu G_{n1} \leq \frac{C}{n} \|f\|_1.$$

Thus $\mu G_1 = 0$, which implies that $\mu G = 0$ if we take into account that $G \subset G_1 \cup G_2$ and $\mu G_2 = 0$ for each $f \in L(\mathbb{T}^d)$. Let us show that the equality

$$\text{(A.19)} \qquad \lim_{h \to 0} \sum_{j \in \mathbb{Z}_+^{d-1}} A_j \left(\widetilde{I}_j(f, x, T, h) - f(x) \right) = 0$$

holds for almost all $x \in \mathbb{T}^d$. Let E be a set of points $x \in \mathbb{T}^d \setminus G$ for which there exist ε_0 and $h_n \to 0$, $h'_n \to 0$ such that

$$\sum_{j \in \mathbb{Z}_+^{d-1}} A_j \left| \widetilde{I}_j(f, x, T, h_n) - \widetilde{I}_j(f, x, T, h'_n) \right| > \varepsilon_0.$$

Denote by $E_N = E_N(f)$ the set of all $x \in \mathbb{T}^d \setminus G$ for which there exist sequences $h_n \to 0$, $h'_n \to 0$ such that

$$\sum_{j \in \mathbb{Z}_+^{d-1}} A_j \left| \widetilde{I}_j(f, x, T, h_n) - \widetilde{I}_j(f, x, T, h'_n) \right| > 1/N.$$

Show that $\mu E_N(f) = 0$ for any N. Given $\varepsilon > 0$, represent the function f in the form $f = f_1 + f_2$, where f_1 is a trigonometric polynomial, $\|f_2\|_1 < \varepsilon$. It is not hard to see that

$$E_N(f) \subset E_{2N}(f_2)$$

$$\subset \left\{ x \in \mathbb{T}^d : \sup_{h>0} \sum_{j \in \mathbb{Z}_+^{d-1}} A_j \widetilde{I}_j(|f|, x, T, h) > \frac{1}{4N} \right\} = F_N(f_2),$$

but in view of Lemmas A.9.3 and A.9.4, $\mu F_N(f_2) \le CN\|f_2\|_1 \le CN\varepsilon$. Thus $\mu E_N = 0$. Since $E \subset \bigcup\limits_{N=1}^{\infty} E_N$, it follows that $\mu E = 0$. If $x \notin E \cup G$, then for any $\varepsilon > 0$ there exist $n_0 > 0$ and $\delta > 0$ (independent of x) such that

$$\left| \sum_{\substack{j\in\mathbb{Z}_+^{d-1} \\ |j|\ge n_0}} A_j \left(\widetilde{I}_j(f,x,T,h) - f(x)\right)\right| < \varepsilon$$

for all $h \in (0,\delta)$. On the other hand, for any fixed $j \in \mathbb{Z}_+^{d-1}$ and almost all x, the relation

$$\lim_{h\to 0} \widetilde{I}_j(f,x,T,h) = f(x)$$

holds. Relation (A.19) follows.

Suppose r is a rational number and $D(r)$ is the set of points $x \in \mathbb{T}^d$ for which

$$\lim_{h\to 0} \sum_{j\in\mathbb{Z}_+^{d-1}} A_j h^{-d} \int_{-h}^{h} dt_d \prod_{l=1}^{d-1} 2^{j_l} \int_{-2^{-j_l}h}^{2^{-j_l}h} dt_l \left(|f(x+Tt) - r| - |f(x) - r|\right) \neq 0.$$

Applying (A.19) to the function $|f(x) - r|$, we get $\mu D(r) = 0$. Then the set $D = \bigcup\limits_r D(r) \cup G_2$ has zero measure. To prove the theorem, it remains to check that $H \subset D$. Let $x \notin D$, $\varepsilon > 0$. Choose rational j_0 such that $|f(x) - j_0| < \varepsilon$. Since $x \notin D(j_0)$, it follows that for sufficiently small h we have

$$\begin{aligned}\sum_{j\in\mathbb{Z}_+^{d-1}} A_j I_j(f,x,T,h) &\le 2|f(x)-j_0| \sum_{j\in\mathbb{Z}_+^{d-1}} A_j \\ &\quad + \sum_{j\in\mathbb{Z}_+^{d-1}} A_j \cdot h^{-d} \int_{-h}^{h} dt_d \prod_{l=1}^{d-1} 2^{j_l} \int_{-2^{-j_l}h}^{2^{-j_l}h} dt_l \left(|f(x+Tt) - j_0| - |f(x) - j_0|\right) \\ &\le 2|f(x)-j_0| \sum_{j\in\mathbb{Z}_+^{d-1}} A_j + \varepsilon \le \left(2\sum_{j\in\mathbb{Z}_+^{d-1}} A_j + 1\right)\varepsilon,\end{aligned}$$

which implies that $x \notin H$. ◊

THEOREM A.9.5. *Let $A = \{A_j\}_{j\in\mathbb{Z}_+^{d-1}}$ be a monotone sequence of positive numbers, $\sum_{j\in\mathbb{Z}_+^{d-1}} A_j < \infty$. If $f \in L(\mathbb{T}^d)$, $T \in \mathcal{T}$, then the set $A\mathcal{L}^*(f,T)$ has full measure.*

To prove this theorem, it suffices to repeat the reasoning of the proof of Theorem A.9.2, using, instead of Lemma A.9.3, the following trivial statement.

LEMMA A.9.6. *Let $A = \{A_j\}_{j\in\mathbb{Z}_+^{d-1}}$ be a sequence of positive numbers, $\sum_{j\in\mathbb{Z}_+^{d-1}} A_j \le K$. If functions $g_j \ge 0$, $j \in \mathbb{Z}_+^{d-1}$, defined on $\mathbb{R}^d$ are such that $\mu\{x : g_j(x) \ge \xi\} \le \xi^{-1}$, then*

$$\mu\left\{x : \sup_{j\in\mathbb{Z}_+^{d-1}} A_j g_j(x) \ge \xi\right\} \le \frac{K}{\xi}.$$

A.10. Absolutely continuous functions

A function f is called *absolutely continuous* if for any $\varepsilon > 0$ there exists $\delta > 0$ such that for each set of disjoint intervals (a_j, b_j), $j = 1, \dots, n$, satisfying the condition $\sum_{j=1}^{n}(b_j - a_j) < \delta$, the following inequality holds:

$$\sum_{j=1}^{n} |f(b_j) - f(a_j)| < \varepsilon.$$

It is obvious that each absolutely continuous function is continuous and each function having a bounded derivative is absolutely continuous.

THEOREM A.10.1. *Each function that is absolutely continuous on $[a, b]$ is differentiable almost everywhere, its derivative f' is summable on $[a, b]$, and*

$$\int_x^y f' = f(y) - f(x)$$

for all $x, y \in [a, b]$.

THEOREM A.10.2. *A function F defined on $[a, b]$ is absolutely continuous if and only if there exists a function f summable on $[a, b]$ and such that*

$$F(x) = F(a) + \int_a^x f(x)\,dx,$$

for all $x \in [a, b]$.

The proofs of these theorems can be found, for example, in [**20**, Chapter 9].

A.11. Nonnegative trigonometric polynomials, Riesz's lemma

To prove that any nonnegative trigonometric polynomial can be represented in the form of the squared absolute value of a trigonometric polynomial (Theorem A.11.3), we first characterize real-valued polynomials.

LEMMA A.11.1. *A trigonometric polynomial*

$$h(\xi) = \sum_{k=-N}^{N} h_k e^{-2\pi i k \xi} \tag{A.20}$$

is real-valued ($h(\xi) \in \mathbb{R}$ for any $\xi \in \mathbb{R}$) if and only if $h_{-k} = \overline{h_k}$ for all k.

To prove this lemma, it suffices to compare the coefficients of equal powers of $e^{-2\pi\xi}$ for the polynomials $h(\xi)$ and $\overline{h(\xi)}$. ◊

REMARK A.11.2. Without loss of generality Lemma A.11.1 can be used in the above formulation since any trigonometric polynomial can be represented in the form (A.20). One of the coefficients h_N and h_{-N} may be equal to zero.

THEOREM A.11.3 (Riesz's lemma). *For any nonnegative trigonometric polynomial $h(\xi) = \sum_{-N}^{N} h_k e^{-2\pi i k\xi}$ there exists a polynomial $m(\xi) = \sum\limits_{k=0}^{N} c_k e^{-2\pi i k\xi}$ such* *that* $|m_0(\xi)|^2 = h(\xi)$. *In addition, if the coefficients of the polynomial* h *are real, it takes the form*

$$h(\xi) = \sum_{k=0}^{N} a_k \cos 2\pi k\xi, \qquad a_k \in \mathbb{R},$$

and the polynomial m_0 *can be chosen in such a way that its coefficients are also real.*

Proof. Set

$$\mathbf{h}(z) = \sum_{k=-N}^{N} h_k z^k.$$

Thus $h(\xi) = \mathbf{h}(e^{-2\pi i\xi})$. If h is a nonnegative polynomial, then $h_{-k} = \overline{h_k}$ (Lemma A.11.1). Consequently, $\mathbf{h}(1/\bar{z}) = \overline{\mathbf{h}(z)}$. Hence the set of roots of the equation $\mathbf{h}(z) = 0$ is symmetric with respect to inversion (the change of variables $z \to 1/\bar{z}$). If we single out the roots on the unit circle and divide the other roots in pairs, we get

$$\mathbf{h}(z) = az^N \prod_{l=1}^{q}(z - e^{-2\pi i\alpha_l}) \prod_{k=1}^{p}(z - z_k)(z - (\overline{z_k})^{-1}), \tag{A.21}$$

where $q + 2p = 2N$ and α_k are the roots of the polynomial h on the fundamental period. Since $h \geq 0$, every such a root has even multiplicity. Hence $q = 2r$ and $\prod_{l=1}^{q}(z - e^{-2\pi i\alpha_l}) = \prod_{l=1}^{r}(z - e^{-2\pi i\alpha_l})^2$. Equality (A.21) takes the form

$$\mathbf{h}(z) = b\prod_{l=1}^{r}(z - e^{-2\pi i\alpha_l})(z^{-1} - e^{2\pi i\alpha_l}) \prod_{k=1}^{p}(z - z_k)(z^{-1} - \overline{z_k}),$$

where $r + p = N$. Since both products are positive for all $z = e^{-2\pi i\xi}$, it follows that b is a real positive number. It remains to set

$$\mathbf{m}(z) = \sqrt{b}\prod_{l=1}^{r}(z - e^{-2\pi i\alpha_l}) \prod_{k=1}^{p}(z - z_k).$$

Thus the set of roots of $\mathbf{h}$ can be divided into two mutually inverse sets A and $1/\bar{A}$. For every such partition, there is a corresponding polynomial $\mathbf{m}(z) = \sqrt{b} \prod\limits_{z_k \in A} (z - z_k)$.

If all coefficients of h_k are real, the set of roots of $\mathbf{h}(\mathbf{z})$ is symmetric with respect to the real axis. Hence for A, we can also take a set symmetric with respect to the real axis. In this case the coefficients of the polynomial m will be real. ◊

Numerous interesting facts and problems concerning nonnegative trigonometric polynomials can be found in [**14**], [**25**].

A.12. The Enestrem-Kakey theorem about zeros of polynomials

THEOREM A.12.1 ([**24**, p. 114, Problem 23]). *If all coefficients* a_i, $i = 0, \dots, n$, *of a polynomial* $\sum_{i=0}^{n} a_i z^{n-i}$ *are positive, then all its zeros are located in the circular ring* $\alpha \leq |z| \leq \beta$, *where* α *and* β *are the minimal and the maximal values of the numbers* $\frac{a_{i+1}}{a_i}$, $i = 0, \dots, n-1$, *respectively.*

A.13. Sobolev spaces

Denote by W_p^s ($s \in \mathbb{Z}$, $s \geq 0$, $p \in [1, +\infty]$) the Sobolev space consisting of functions defined on $\mathbb{R}$ whose s-th derivatives belong to L_p. For an arbitrary $s > 0$,

$$W_2^s(\mathbb{R}) := \{f : \ \|f\|_{W_2^s} := \|F^{-1}\left((1 + |\cdot|^2)^{s/2} F f\right)\|_{L_2(\mathbb{R})} < \infty\},$$

where F is the Fourier transform. For integer s, these two definitions coincide.

THEOREM A.13.1 ([**40**, p. 106]). *For $r > 0$, the norm of any function f in W_2^r is equivalent to the quantity*

$$\inf_{\{\varepsilon_j\}} \left(\sum_{j=0}^{+\infty} \varepsilon_j^2 \right)^{1/2}, \tag{A.22}$$

where the infimum is taken over all sequences of positive numbers $\{\varepsilon_j\}_{j \geq 0}$ such that for some positive $m > r$, there exists a sequence of functions $\{f_j\}_{j \geq 0}$ in W_2^m that satisfies the conditions

$$f = \sum_j f_j, \quad \left\| \left(\frac{d}{dx}\right)^m f_j \right\|_2 \leq \varepsilon_j 2^{(m-r)j}, \quad \|f_j\|_2 \leq \varepsilon_j 2^{-rj}. \tag{A.23}$$

A.14. Moduli of continuity

Let $X = L_p(\mathbb{T})$, $1 \leq p < \infty$, or $X = C(\mathbb{T})$ (for $p = \infty$). The *modulus of continuity* of a function $f \in X$ of order $r \in \mathbb{N}$ is the function defined on $[0, +\infty)$ by the formula

$$\omega_r(f, h)_p = \omega_r(f, h) = \omega_r(h) := \sup_{|t| \leq h} \|\Delta_t^r f\|_p,$$

where $\Delta_t^r = \sum_{k=0}^{r} (-1)^k C_r^k f(x + (r - 2k)t)$. For convenience, we set

$$\omega_0(f, h)_p := \|f\|_p.$$

Moduli of continuity have the following properties:

M1. The function ω_r monotonically increases on $[0, +\infty)$ and $\omega_r(0) = 0$.

M2. The function ω_r is uniformly continuous on $[0, +\infty)$.

M3. $\omega_{r+s}(f, h) \leq 2^r \omega_s(f, h)$ for all $s, r \in \mathbb{Z}_+$.

M4. $\omega_r(nh) \leq n^r \omega_r(h)$ for any $n \in \mathbb{N}$.

M5. $\omega_r(\lambda h) \leq (\lambda + 1)^r \omega_r(h)$ for any $\lambda > 0$.

M6. If a function f has the $(r-1)$-st absolutely continuous derivative on $\mathbb{R}$ and $f^{(r)} \in X$, then

$$\omega_{r+s}(f, h) \leq h^r \omega_s(f^{(r)}, h)$$

for any $s \in \mathbb{Z}_+$.

M7. $\omega_r(f + g) \leq \omega_r(f) + \omega_r(g)$ for all $f, g \in X$.

The proofs of these properties can be found in any book on approximation theory, for example, in [**19**, 3.3].

A.15. Approximation by trigonometric polynomials

THEOREM A.15.1. *Let $1 \le p \le \infty$, $r \in \mathbb{N}$, $T(\xi) = \sum_{k=-N}^{N} c_k e^{2\pi i k\xi}$ be a trigonometric polynomial of order N. For any $h \in (0, 1/4N)$, the following inequality holds:*

$$\|T^{(r)}\|_p \le \left(\frac{\pi N}{\sin 2\pi N h}\right)^r \|\Delta_h^r T\|_p. \tag{A.24}$$

The proof of this theorem is given in the book [**19**, pp. 228, 230].

Inequality (A.24) generalizes the classical Bernstein inequality

$$\|T^{(r)}\|_p \le (2\pi N)^r \|T\|_p,$$

which immediately follows from (A.24) if we take $h = 1/4N$.

Let $1 \le p \le \infty$; the best (trigonometric) approximation of order N of a function $f \in L_p(\mathbb{T})$ ($f \in C(\mathbb{T})$ for $p = \infty$) is the quantity

$$\mathcal{E}_N(f)_p = \inf \|f - T\|_p,$$

where the lower bound is taken over all trigonometric polynomials of order N.

THEOREM A.15.2 (Jackson's theorem). *Let $1 \le p \le \infty$, $r \in \mathbb{N}$. There exists a constant A depending only on p and r such that for any positive integer N and any function $f \in L_p(\mathbb{T})$ ($f \in C(\mathbb{T})$ for $p = \infty$), the following inequality holds:*

$$\mathcal{E}_N(f)_p \le A\omega_r\left(f, \frac{1}{N}\right)_p.$$

The proof of this theorem can be found in any book on approximation theory, for example, in [**19**, Section 5.1.32]. Statements of this type are called direct theorems of approximation theory. The best approximation is nonlinear. The Vallee Poussin means

$$v_N(f, x) := \sum_{|k|<N} \widehat{f}(k) e^{2\pi i k x} + \sum_{N \le |k| \le 2N} \frac{2N - k}{N} \widehat{f}(k) e^{2\pi i k x}$$

are linear operators which approximate an arbitrary function with the same order as best approximations. The trivial equality $v_N(f) = 2\sigma_{2N}(f) - \sigma_N(f)$, where σ_N are the Fejer means, implies that $\|v_N\| \le 3$ (since $\|\sigma_N\| \le 1$, which follows, for example, from Theorem A.16.1). Combining this fact with Theorem A.15.2 and taking into account that $v_N(T) = T$ for any trigonometric polynomial of order at most N, one can obtain the estimate

$$\|f - v_N(f)\|_p \le 4A\omega_r\left(f, \frac{1}{N}\right)_p. \tag{A.25}$$

THEOREM A.15.3. *Let $1 \le p \le \infty$, $m \in \mathbb{N}$. There exists a constant B depending only on p and m such that for any function $f \in L_p(\mathbb{T})$ ($f \in C(\mathbb{T})$ for $p = \infty$), the inequality*

$$\omega_m(f, h)_p \le B h^m \sum_{0 \le l \le h^{-1}} (l + 1)^{m-1} \mathcal{E}_l(f)_p$$

holds for all $h > 0$ *and* $f \in L_p(\mathbb{T})$ *(*$f \in C(\mathbb{T})$ *for* $p = \infty$*). In addition, if*

$$\sum_{l=1}^{\infty} l^{m-1}\mathcal{E}_l(f)_p < \infty,$$

then the function f *has the* m*-th derivative almost everywhere (at each point for* $p = \infty$*),* $f^{(m)} \in L_p(\mathbb{T})$ *(*$f^{(m)} \in C(\mathbb{T})$ *for* $p = \infty$*), and the inequality*

$$\mathcal{E}_n(f^{(m)})_p \le C \sum_{l=[n/2]}^{\infty} (l+1)^{m-1}\mathcal{E}_l(f)_p$$

holds, where the constant C *depends only on* p *and* m*.*

The proof of this theorem is given in the book [**19**, 6.1.1, 6.1.3].

Statements of the type of Theorem A.15.3 are called inverse theorems of approximation theory.

A.16. Multidimensional Fejer means

The Fejer means can be generalized to the multidimensional case in various ways. We consider the Fejer sums $\sigma_K(f)$, where

$$K = K(n) := \{x \in \mathbb{R}^d : \quad |x_l| \le n, \ l = 1, \dots, d\},$$

which are arithmetic means of partial Fourier sums of a function f over rectangles contained in K with centers at the origin and sides parallel to the coordinate axes. It is not hard to verify that

$$\sigma_K(f, x) = \int\limits_{[-1/2,1/2]^d} f(x+t) \prod_{l=1}^{d} \Phi(t_l)\, dt,$$

where

$$\Phi(\tau) = \sum_{k=-n}^{n} \left(1 - \frac{|k|}{n}\right) e^{2\pi i k \tau} = \frac{\sin^2 \pi n \tau}{n \sin^2 \pi \tau}. \tag{A.26}$$

THEOREM A.16.1. *Let* $X = C(\mathbb{T}^d)$ *or* $X = L_p(\mathbb{T}^d)$*,* $1 \le p < \infty$*,* $f \in X$*; then*

$$\lim_{n\to\infty} \|f - \sigma_{K(n)}(f)\| = 0.$$

Proof. The operators σ_K, which assign to each function $f \in X$ its Fejer sum $\sigma_K(f)$, act from X to X. From (A.26) and the orthogonality of the trigonometric system it follows that

$$\|\sigma_K\|_{X\to X} \le \int\limits_{[-1/2,1/2]^d} \left|\prod_{l=1}^{d} \Phi(t_l)\right| dt$$

$$= \prod_{l=1}^{d} \int\limits_{-1/2}^{1/2} |\Phi(t_l)|\, dt_l = \prod_{l=1}^{d} \int\limits_{-1/2}^{1/2} \Phi(t_l)\, dt_l = 1. \tag{A.27}$$

If f is a trigonometric polynomial, we obviously have

$$\|f - \sigma_{K(n)}(f)\|_X \le \|f - \sigma_{K(n)}(f)\|_\infty \underset{n\to\infty}{\longrightarrow} 0.$$

In view of (A.27) and the Banach-Steinhaus theorem, the required statement follows. ◊

A more detailed study of the means of such type can be found, for example, in [**14**, Chapter 17, §1].

A.17. Self-similar sets

Among the numerous definitions of the notion of fractal, the following definition introduced by Hutchinson in [**141**] is most commonly used.

DEFINITION A.17.1. Let there be given a complete metric space X with metric $r(\cdot\,,\,\cdot)$ and several operators $A_j : X \to X, \quad j = 1, \dots, m$. A compact set $K \subset X$ is called a fractal, or self-similar set, of these operators if

$$K = \bigcup_{j=1}^{m} A_j K. \tag{A.28}$$

A mapping $A : X \to X$ is called a contraction operator if there exists a constant $q < 1$ such that $r(Ax, Ay) \le q\, r(x, y)$ for all $x, y \in X$.

THEOREM A.17.2. *If all operators $A_1, \dots, A_m$ are contraction operators, they have a unique fractal.*

The proof is based on the contraction mapping principle:

LEMMA A.17.3. *A contraction mapping of a complete metric space has a unique fixed point.*

Proof. Let $r(Ax, Ay) \le qr(x, y)$ for all $x, y \in X, \quad q < 1$. For any initial point $x_0 \in X$, consider the sequence $x_k = A^k x_0$. For any positive integers k, l, we have

$$r(x_k, x_{k+l}) \le \sum_{j=0}^{l-1} r(x_{k+j+1}, x_{k+j}) \le \sum_{j=0}^{l-1} q^{k+j} r(x_0, x_1) \le r(x_0, x_1) \frac{q^r}{q-1}.$$

Thus the sequence $\{x_k\}$ is a Cauchy sequence; hence it converges to some $x \in X$. It is easy to check that $Ax = x$. Since A is a contraction mapping, the fixed point is unique. ◊

Proof of Theorem A.17.2. Consider the metric space $\mathcal{X}$ consisting of all compact subsets of the space X. This space is nonempty since it contains all singletons of the space X. The distance between two elements of $Q, R \in \mathcal{X}$ is defined according to Hausdorff:

$$r(Q, R) = \max\Big\{ \sup_{x \in Q} \inf_{y \in R} r(x, y),\ \sup_{y \in R} \inf_{x \in Q} r(x, y) \Big\}. \tag{A.29}$$

Define the following mapping $\mathcal{A} : \mathcal{X} \to \mathcal{X}$:

$$\mathcal{A} Q = \bigcup_{j=1}^{m} A_j Q. \tag{A.30}$$

Since A_j are contraction mappings, the mapping $\mathcal{A}$ is well defined on $\mathcal{X}$. It is a contraction mapping with the same constant q (the proof is left to the reader). By Lemma A.17.3, this mapping has a unique fixed point, which was to be proved. ◊

Owing to its generality, Definition A.17.1 includes a lot of fractals and self-similar objects and is used in various areas of mathematics. A disadvantage of this definition is, in a sense, its excessive generality. For example, according to this definition, any segment is a fractal. Triangles and squares are fractals as well. The reader can easily find corresponding sets of contraction operators for each of these figures.

A compact subset K of an affine space is called an affine fractal if there exists a set of contraction linear operators $A_1, \dots, A_m$ satisfying (A.28).

PROBLEM A.17.4. Prove that any convex polyhedron in $\mathbb{R}^d$ is an affine fractal. Is this true for nonconvex polyhedra?

PROBLEM A.17.5. Give an example of a compact subset that belongs to the real line but is not a fractal.

PROBLEM A.17.6. Is a circle on the Euclidean plane a fractal? Is it an affine fractal?

PROBLEM A.17.7. Let $K \subset X$ be a fractal corresponding to contraction operators $A_1, \dots, A_m$. Prove that for any compact set $K_0 \subset X$,

$$r(\mathcal{A}^n K_0\,,\, K) \to 0, \quad n \to \infty,$$

where the mapping $\mathcal{A}$ is defined in (A.30) and the limit is taken in the Hausdorff metric (A.29).

The literature on the theory of fractals and self-similar sets is quite extensive. We recommend to the reader, for example, the books [**30**], [**31**].

A.18. Difference equations

A difference equation is an equation of the form

$$\sum_{k=0}^{n} a_k x_{k+j} = b_j, \qquad j \in \mathbb{Z}, \tag{A.31}$$

where $a_0, \dots, a_n$ are given complex numbers, $a_0 a_n \neq 0$, $\{b_k\}_{k\in\mathbb{Z}}$ is a given sequence, and $x = \{x_k\}_{k\in\mathbb{Z}}$ is an unknown sequence. To each difference equation there corresponds a homogeneous equation:

$$\sum_{k=0}^{n} a_k x_{k+j} = 0, \qquad j \in \mathbb{Z}. \tag{A.32}$$

The difference of two arbitrary solutions to equation (A.31) satisfies homogeneous equation (A.32). Hence all solutions of difference equation (A.31) have the form

$$x = x^{\mathrm{par}} + x^{\mathrm{hom}}, \tag{A.33}$$

where x^{par} is a particular solution of the equation and x^{hom} is an arbitrary solution of the corresponding homogeneous equation. It is clear that the set of solutions of a homogeneous equation form a linear space of dimension n. Indeed, each solution of the equation (A.32) is uniquely defined by its n coefficients $x_0, \dots, x_{n-1}$ (by applying (A.32) successively, we find all x_k for $k \geq n$ and $k \leq -1$), where the coefficients can be chosen arbitrarily. To describe the space of solutions, we need the following auxiliary assertion.

LEMMA A.18.1. *Suppose $\lambda_1, \dots, \lambda_q$ are given complex numbers and $r_1, \dots, r_q$ are given positive integers, $\sum_{j=1}^{q} r_j = n+1$. Consider $n+1$ vectors $v_{j,s}$, $j = 1, \dots, q$, $s = 0, \dots, r_j - 1$, defined by the following relations:*

$$v_{j,s}^k = k^s \lambda_j^k, \tag{A.34}$$

where $k = 0, \dots, n$ (here we set $0^0 = 1$). These vectors are linearly dependent if and only if some of the numbers $\lambda_1, \dots, \lambda_q$ are equal.

Proof. If $\lambda_i = \lambda_j$, then $v_{i,0} = v_{j,0}$; i.e., if some of the vectors $v_{j,s}$ coincide, then they are linearly dependent. Conversely, if the vectors are linearly dependent, then there exists a nonzero vector $p \in \mathbb{R}^{n+1}$ orthogonal to all vectors $v_{j,s}$. Thus,

$$\sum_{k=0}^{n} p_k k^s \lambda_j^k = 0$$

for all $j = 1, \dots, q$, $s = 0, \dots, r_j - 1$. Therefore the polynomial $P(z) = \sum_{k=0}^{n} p_k z^k$ has zeros of multiplicity at least r_j at the points λ_j. The latter is impossible as $\sum_j r_j = n+1$, while the degree of the polynomial is equal to n. ◇

REMARK A.18.2. If all r_j are equal to 0, i.e., $q = n+1$, then the vectors $v_1, \dots, v_{n+1}$ (where $v_j = v_{j,0}$) constitute an $(n+1) \times (n+1)$ matrix $A_{\lambda_1, \dots, \lambda_{n+1}}$ whose determinant is equal to the classical Vandermonde determinant

$$\det A_{\lambda_1, \dots, \lambda_{n+1}} = \prod_{1 \le i < j \le n+1} (\lambda_j - \lambda_i).$$

It is clear that this determinant is equal to zero only if some of the numbers λ_j are equal. In the general case, the determinant of the matrix composed from the vectors $v_{j,s}$ is called the generalized Vandermonde determinant.

Let us return to difference equations. The polynomial

$$\chi_a(z) = \sum_{k=0}^{n} a_k z^k$$

is called the characteristic polynomial of equation (A.31) or equation (A.32).

PROPOSITION A.18.3. *Let the characteristic polynomial of an equation have roots $\lambda_1, \dots, \lambda_q$ of multiplicities $r_1, \dots, r_q$, respectively. Then $n+1$ sequences $v_{j,s} = \{v_{j,s}^k\}_{k \in \mathbb{Z}}$ defined by formula (A.34) form a basis of the space of solutions to the corresponding homogeneous equation.*

Proof. Substituting the sequences $v_{j,s}$ into the homogeneous equation (A.32), we see that they satisfy this equation. According to Lemma A.18.1, these sequences are linearly independent; hence the dimension of their linear span is equal to $n+1$, which implies that they form a basis of the space of solutions. ◇

A particular solution of a nonhomogeneous equation can be found by successively applying formula (A.31): first we choose the coefficients $x_0, \dots, x_{n-1}$ arbitrarily and then find the other x_i. If the right-hand side of the equation has the form

$$b_j = j^d t^j, \quad t \in \mathbb{C},\ r \in \mathbb{N},$$

then a particular solution is found in the form

(A.35)
$$x_k^{\text{par}} = \begin{cases} p(k)\,t^k\,, & \chi_a(t) \neq 0, \\ k^r p(k)\,t^k\,, & t \text{ is a root of the polynomial } \chi_a(z) \text{ of multiplicity } r, \end{cases}$$

where p is a polynomial of degree at most d. In conclusion, we prove another useful assertion which characterizes bounded solutions of homogeneous difference equations.

PROPOSITION A.18.4. *A linear combination $g_k = \sum_{j,s} f_{j,s} v_{j,s}$ of several sequences of the form (A.34) belongs to ℓ_∞ if and only if all λ_j (such that at least one coefficient $f_{j,s}$ is nonzero) are equal to 1 in modulus and all the exponents s are equal to 0.*

Proof. Denote by M the maximal of $|\lambda_j|$. Suppose $M \geq 1$. Without loss of generality we can assume that

$$|\lambda_j| = M \quad \text{if } j \leq m, \qquad |\lambda_j| < M \quad \text{if } j \geq m+1$$

for some $l \geq 1$. Set $\lambda_j = Me^{2\pi i \alpha_j}$. Denote by S the maximal of exponents s of λ_j^k, $j \leq m$; let it be attained for $j \leq l$, where $l \leq m$. Then

$$g_k = k^S M^k \sum_{j \leq l} f_{j,S}\; e^{2\pi i k \alpha_j} + o(k^S M^k), \quad k \to +\infty. \tag{A.36}$$

According to Lemma A.18.1, for some $r \leq l$, the sum $\sum_{j \leq l} f_{j,S}\; e^{2\pi i k \alpha_j}$ does not vanish. Thus,

$$\left| \sum_{j \leq l} f_{j,S}\; e^{2\pi i r \alpha_j} \right| > \varepsilon \tag{A.37}$$

for some $\varepsilon > 0$. Set $\sum_{j \leq l} |f_{j,S}| = A$. There exist arbitrarily large $t \in \mathbb{N}$ such that

$$|e^{2\pi i t \alpha_j} - 1| < \varepsilon/2A, \quad j = 1, \ldots, l.$$

The proof is by the Dirichlet principle: among $l^n + 1$ points

$$(e^{2\pi i k \alpha_1}, \ldots, e^{2\pi i k \alpha_l}) \in \mathbb{T}^l, \quad k \leq l^n + 1,$$

we can find two points (let them correspond to the parameters $k = k_n$ and $k = k_n'$) such that

$$|e^{2\pi i k_n' \alpha_j} - e^{2\pi i k_n \alpha_j}| < 2\pi/n, \quad j = 1, \ldots, l.$$

Taking any n such that $\frac{2\pi}{n} < \frac{\varepsilon}{2A}$ and setting $t_n = k_n' - k_n$, we obtain the required inequality. Substituting this into (A.37), we get

$$\left| \sum_{j \leq l} f_{j,S}\; e^{2\pi i (r + t_n) \alpha_j} \right| > \varepsilon/2.$$

Consequently, for such k, we have

$$|g_k| \geq k^S M^k \varepsilon/3.$$

Hence, if the sequence g_k is bounded, then $M = 1$ and $S = 0$. Thus,

$$\max_{j=1,\ldots,q} |\lambda_j| \leq 1,$$

and for all j such that $|\lambda_j| = 1$ we have $s = 0$. Similarly, considering the sequence g_k as $k \to \infty$, we show that $\min_{j=1,\dots,q} |\lambda_j| \geq 1$, where for all j such that $|\lambda_j| = 1$, we have $s = 0$. This completes the proof. $\Diamond$

In the same way we can prove

COROLLARY A.18.5. *If all λ_j are greater than of equal to 1 in modulus, then no linear combination of sequences of the form* (A.34) *can tend to zero as $k \to +\infty$.*

For the theory of difference equations, see, for example, the books [**27**], [**28**].

A.19. Landau-Kolmogorov inequality

For functions $f \in C^m[0, \infty)$, $m \in \mathbb{N}$, the following analog of the well-known Kolmogorov inequality holds:

$$\|f^{(k)}\|_\infty \leq C\|f\|_\infty^{(1-k/m)}\|f^{(m)}\|_\infty^{k/m},$$

where $k = 1, \dots, m$ and C is a constant depending only of m and k (see [**21**, p. 393]). For $m = 2$, $k = 1$, we have the classical Landau inequality, where $C = 2$ is the sharp constant.

A.20. Legendre polynomials

Algebraic polynomials $\mathcal{L}_n$, $n = 0, 1, \dots$, where each polynomial $\mathcal{L}_n$ is of degree n and has a positive leading coefficient, are called the Legendre polynomials if they are pairwise orthogonal and normalized on $[-1, 1]$. The Legendre polynomials form a complete system in the space $L_2[-1, 1]$ (see, for example, [**22**, Chapter 1]); hence they form an orthonormal basis in this space. Expansions into the Fourier series with respect to this basis are called the Fourier-Legendre series.

For the Legendre polynomials, the following relations hold (see, for example, [**22**, Chapter 4]).

For all $x \in [-1, 1]$ the following inequality holds:

$$|\mathcal{L}_n(x)| \leq \mathcal{L}_n(1) = \sqrt{n + \frac{1}{2}}. \tag{A.38}$$

The Dirichlet-Mehler formula is

$$\mathcal{L}_n(\cos \pi\theta) = \sqrt{2\left(n + \frac{1}{2}\right)} \int_0^\theta \frac{\cos\left(n + \frac{1}{2}\right)\pi\tau \, d\tau}{\sqrt{\cos \pi\tau - \cos \pi\theta}}, \qquad \theta \in (0, 1). \tag{A.39}$$

The Volterra equation of the second kind is

$$(\sin \pi\theta)^{1/2}\mathcal{L}_n(\cos \pi\theta) = \lambda_n \cos\left(\left(n + \frac{1}{2}\right)\pi\theta - \frac{\pi}{4}\right)$$

$$- \frac{\pi}{\left(n + \frac{1}{2}\right)} \int_\theta^{1/2} \frac{\sin\left(n + \frac{1}{2}\right)\pi(t - \theta)}{4\sin^2 \pi t} (\sin \pi t)^{1/2}\mathcal{L}_n(\cos \pi t) \, dt, \tag{A.40}$$

where $\theta \in (0, 1)$,

$$\lambda_n = \sqrt{\frac{2}{\pi}} + O\left(\frac{1}{n^2}\right). \tag{A.41}$$

For the same values of θ and λ_n, the application of the Liouville-Steklov method (see [**23**, §8.61]) gives the asymptotic formulas:

$$(\sin\pi\theta)^{1/2}\mathcal{L}_n(\cos\pi\theta) = \lambda_n \cos\left(\left(n+\frac{1}{2}\right)\pi\theta - \frac{\pi}{4}\right) + O\left(\frac{1}{n\sin\pi\theta}\right), \tag{A.42}$$

$$\begin{aligned}(\sin\pi\theta)^{1/2}\mathcal{L}_n(\cos\pi\theta) &= \lambda_n \cos\left(\left(n+\frac{1}{2}\right)\pi\theta - \frac{\pi}{4}\right)\\ &+ \frac{\lambda_n}{8\left(n+\frac{1}{2}\right)}\sin\left(\left(n+\frac{1}{2}\right)\pi\theta - \frac{\pi}{4}\right)\operatorname{ctan}\pi\theta + O\left(\frac{1}{n^2\sin^2\pi\theta}\right).\end{aligned} \tag{A.43}$$

It follows from the results of [**248**] that for any continuous function, its Fourier-Legendre series is Cesaro summable, which means that the corresponding Lebesgue functions of the Fejer-Legendre sums

$$\int_{-1}^{1}\left|\frac{1}{N+1}\sum_{n=0}^{N}\sum_{k=0}^{n}\mathcal{L}_k(x)\mathcal{L}_k(t)\right| dt$$

are uniformly bounded.

APPENDIX B

Historical Comments

Wavelet theory made its appearance in the 1980–1990s. For 15 years of its development thousands of papers on the themes discussed in this book have been written. We do not give a complete list of these papers, citing only the authors of the most essential results presented in the book and their predecessors. The list of the cited books is not complete either. We focus on the mathematical papers devoted, fully or partially, to wavelet theory and do not refer to numerous books of a more applied character, where wavelet theory is considered from the viewpoint of signal processing and other engineering applications.

Chapter 1

The notion of basis in Banach spaces was introduced by J. Schauder (for more detailed history of the problems concerning bases and unconditional bases, see [**17**]). The notion of the Riesz system and Theorem 1.1.2 belong to N. K. Bari [**47**]. The ortogonalization procedure described in Theorem 1.1.14 was proposed by H. Schweinler and E. Wigner [**48**].

The notion of MRA was introduced by Y. Meyer [**49**] and S. Mallat [**50**] and further developed in the works [**52**], [**53**], [**54**]. When constructing an MRA generated by a scaling function, we follow the book by I. Daubechies [**2**, Chapter 5]. Theorem 1.2.14 is presented without the assumption $\widehat{\varphi}(0) \neq 0$; such a possibility was established by R. Lorents, W. Madych, A. Sahakian [**56**]. Theorem 1.2.10 was proved by C. de Boor, R. DeVore, A. Ron [**55**].

The scheme of the construction of an orthogonal wavelet basis by its generating function described in §1.3 was first realized by Y. Meyer [**49**] for a specific situation. The general construction was described by S. Mallat [**50**]. Biorthogonal wavelet bases were first studied by P. Ouscher [**57**] and C. Chui, J. Wang [**58**]; they constructed two dual wavelet systems on the basis of one multiresolution analysis with orthogonal wavelet spaces. We present another approach proposed by I. Daubechies, A. Cohen, J. Feauveau [**59**] and briefly described in [**2**, Chapter 8], where a pair of biorthogonal wavelet systems is generated by a pair of MRAs. Interpolating MRAs were first considered by C. Chui, J. Wang [**58**].

The Haar system, which is the first orthogonal wavelet basis, was constructed by A. Haar [**61**] in 1910. A first spline wavelet basis (which is the first system of continuous wavelets as well) was obtained by J. Strömberg [**62**] before the theory of MRAs and the corresponding terminology were developed. This system proved to be very close to the orthogonal spline wavelets obtained independently by G. Battle [**63**] and P.-G. Lemarié [**64**] in the framework of the theory of MRAs, with using B-splines as generating scaling functions. As noted above, Y. Meyer's wavelet system was constructed in [**49**]. Equality (1.69) is often ascribed to V.

A. Kotelnikov [**66**] or C. Shannon [**65**]. In fact, the interpolating formula (1.69) was known in mathematics before these works appeared (H. Black [**42**] asserts that, in particular, it was known to Cauchy); however in [**66**] it was first applied for signal processing, and C. Shannon also used it for signal processing.

Schemes of recalculation of coefficients similar to the algorithms of wavelet expansions and recovery described in §1.6 have long been used by engineers for signal and image processing. S. Mallat [**51**], [**52**] developed such an algorithm in the framework of the theory of MRAs.

Y. Meyer [**4**] first began investigating the convergence of wavelet expansions; in particular, he established the convergence in L_p, $1 \le p \le \infty$, for so-called r-regular MRAs, i.e., under the assumption that the decay rates of a scaling function and its r derivatives are more than that of any polynomial at infinity. G. Walter [**68**], [**69**] proved the pointwise convergence for continuous functions under the assumption of regularity. S. Kelly, M. Kon, L. Raphael [**70**] extended these results to a wider class of MRAs and proved the convergence almost everywhere; their requirements for a scaling function appear in Theorems 1.7.4–1.7.6. Theorem 1.7.7 is presented in the work by G. Battle [**71**]; however, in I. Daubechies' opinion [**2**], this fact has already been known by many mathematicians and can be considered as "popular wisdom".

The notion of frame was introduced by R. Daffin and A. Schaeffer [**73**]; they established fundamental facts of the theory of frames, in particular, Proposition 1.8.2 and Corollaries 1.8.5 and 1.8.6. Theorem 1.8.4 was proved by B. Kashin, T. Kulikova [**74**]. Theorem 1.8.9 was actually proved by A. Ron, Z. Shen in [**104**], though they assumed that integer shifts of a scaling function constitute a Bessel system, which actually follows from other assumptions. A more precise formulation of this theorem and Corollary 1.8.10 belong to A. Petukhov [**76**], who proposed a new proof of this theorem as well. Similar formulations for polynomial masks were given independently by C. Chui and W. He [**75**]. A particular case of Theorem 1.8.9—Theorem 1.8.11—was previously proved by W. Lawton [**77**] (for polynomial masks) and M. Bownik [**78**] (for arbitrary masks).

Chapter 2

Actually, the construction of multidimensional wavelets by means of tensor multiplication of one-dimensional MRAs was used by J. Strömberg [**62**] for the Haar system. The general scheme of the construction of separable MRAs is rather natural. It is discussed in many papers and books; see [**2**], [**4**], [**8**]. Nonseparable MRAs with binary dilations (the dilation is a diagonal matrix with the elements equal to 2 on the diagonal) are described in [**4**]. For an arbitrary integer matrix (with natural restrictions), nonseparable MRAs were first considered in [**79**], [**80**], [**81**]. Theorem 2.3.4 (for a binary dilation) was proved by C. de Boor, R. DeVore, A. Ron [**55**].

Theorem 2.5.6 belongs to A. Cohen [**82**]; he considered the one-dimensional binary orthogonal case and assumed that the mask is a trigonometric polynomial. This result was transferred to the biorthogonal case by I. Daubechies, A. Cohen, J. Feauveau [**59**]. A multidimensional analog of this theorem with matrix dilation for masks of a special form and the orthogonal case was proved in [**8**]; for arbitrary polynomial masks, it was obtained by I. E. Maksimenko [**83**]. Theorem 2.5.11 for the one-dimensional orthogonal binary case and for polynomial masks was proved by W. Lawton [**84**]; it was extended to the multidimensional orthogonal case with the matrix dilation by W. Lawton, S. Lee, Z. Shen [**86**] and to the biorthogonal case by I. Maksimenko [**83**].

Wavelet functions in the orthogonal binary case were first constructed by K. Gröchenig [**88**] for regular MRAs. R. Jia, C. Micchelli [**89**], [**90**] proved the existence of a set of wavelet functions under weaker restrictions; however their proof was not constructive. C. de Boor, R. DeVore, A. Ron [**55**] obtained a constructive proof but only under the assumption that the function $\widehat{\varphi}$ is different from zero almost everywhere. A general scheme of the construction of orthogonal wavelets with binary dilation was described by R. Jia, Z. Shen in [**91**] and for an arbitrary coefficient of dilation by P. Wojtaszczyk [**8**, Chapter 5]. In the one-dimensional case methods of the construction of orthogonal and biorthogonal compactly supported wavelets for an arbitrary integer dilation were discussed by W. Lawton, S. Lee, Z. Shen [**93**], [**94**]. In particular, they obtained an algorithm for the extension of a row of Laurent polynomials to a unitary matrix. A method of the construction of a unimodular matrix by its first row using the Euclidean algorithm has long been well known. A method of the construction of multidimensional biorthogonal compactly supported wavelets based on the solution of Serre's problem was proposed by H. Ji, S. Riemenschneider, Z. Shen [**95**], [**96**], [**97**].

Problems related to bases and frames of wavelet systems were investigated by A. Ron, Z. Shen in [**102**]–[**105**]. In particular, they developed techniques for obtaining frame lower bounds (Lemma 2.7.4). The most general sufficient condition for the upper bound was derived by B. Han [**106**]; a particular case of this condition (suitable for the construction of wavelet functions) was previously obtained by G. Weis [**9**] (Lemma 2.7.3). However, both conceptual aspects and the techniques of the proof of these conditions were developed by A. Cohen, I. Daubechies [**108**], who investigated one-dimensional compactly supported wavelet systems with binary dilation. Lemma 2.7.7 was proved by R. Jia [**109**] for a more general case.

Multidimensional Haar MRAs and wavelets with matrix dilation were constructed and studied by K. Gröchenig, W. Madych [**79**].

Chapter 3

The intensive study of refinement equations with a finite number of terms began in the late 1980s. Such equations arose in approximations theory, computer design (the construction of subdivision algorithms for approximation and interpolation of functions), and wavelet theory (the construction of compactly supported wavelets) almost simultaneously and independently. The works by de Rham [**158**], [**159**], [**160**], where limiting curves of the cutting angle algorithm were studied, can be considered as the origin of the theory of refinement equations. The first subdivision algorithm (related to de Rham's algorithm) was investigated by G. Chaikin in [**111**]. In this sense, Chaikin's function (Example 3.5.3, Chapter 3) can be considered as the first refinable function. Previously, functions satisfying difference equations with dilation of the argument appeared in works by P. Erdos for the investigation of the smoothness of Bernoulli convolutions [**161**], [**162**], V. L Rvachev and V. A. Rvachev (Rvachev's functions [**163**]), G. A. Derfel [**164**] (in the latter work the existence and uniqueness of generalized solutions to equations with nonnegative coefficients and arbitrary shifts and dilations was proved).

Refinement equations and related problems were first systematically investigated in the pioneer works by S. Dubuc and G. Deslauriers [**113**], [**114**], N. Din and D. Levin [**112**], [**165**], C. Micchelli and H. Prautzsch [**131**]. In 1991 A. Cavaretta, W. Dahmen, and C. Micchelli published the detailed book [**46**] devoted to refinement equations, which (in spite of numerous works in this area) is still the only book on the theory of refinement equations. In I. Daubechies' book [**2**] refinement equations were first introduced in wavelet theory; later they were thoroughly studied in the works by I. Daubechies and J. Lagarias [**138**], [**166**].

Now the bibliography on the theory of refinement equations consists of hundreds of works, so most of them cannot be mentioned here. Some works will be mentioned below in connection with particular problems of the theory.

The existence and uniqueness of a compactly supported solution of a refinement equation in the space of distributions was proved by various methods in many works, in particular, in [**46**] (where the multidimensional case was considered). In [**138**], it was shown that if a solution belongs to L_1, then it has compact support. In [**114**], it was proved that the support is concentrated on the segment $[0, N]$. In the present book, we use a new scheme of the proof based on the iterations of a transition operator and establish that the iterations weakly converge to the solution.

The Strang-Fix conditions were first formulated in [**168**] and then investigated in more detail in the book [**33**]. In Theorem 3.2.1, we generalize several statements concerning the equivalence of the Strang-Fix conditions to the various properties of compactly supported functions proved in [**46**], [**133**], [**167**]. A criterion of the Strang-Fix conditions for refinable functions (Theorem 3.2.10) was established by V. Protasov in [**168**], where he developed a general approach to the investigation of refinable functions by means of binary trees. The necessity of the Strang-Fix conditions for the smoothness of refinable functions (Theorem 3.2.3) was proved in [**46**] (for stable functions) and in [**168**] for the general case. The maximal smoothness of B-splines among all the refinable functions with a given length of support was proved in [**166**].

The notion of order of approximation by shifts of a compactly supported function and the connection of the approximating properties of compactly supported functions with the Strang-Fix conditions were studied by R. Jia [**169**] and thoroughly investigated by C. de Boor, R. DeVore and A. Ron [**170, 171, 172**]. In all these works the order of approximation was studied in L_2. Section 3.3 consists mainly of results of these works generalized to linear combinations with arbitrary sequences of coefficients (not necessarily belonging to L_2); some of the results are strengthened. For instance, we obtain the exact connection of the order of approximation with the modulus of continuity of the function being approximated (to this end, we had to use a new technique of the proof).

The conditions of linear independence and the Riesz basis property (l_2-stability) for integer shifts of compactly supported functions have been formulated in various terms by many authors. For l_2-stability, the absence of real periodic zeros of the Fourier transform was proved in [**6**], [**46**], [**173**], and other works. R. Jia and J. Wang [**174**] showed that for compactly supported functions, the ℓ_p-stability for all finite p is equivalent to the linear independence of integer shifts over ℓ_∞. In [**6**], the space of bounded sequences, whose convolution with a given nonstable refinable function is equal to zero, was described. To the above statements we added a criterion of linear independence over the space ℓ of sequences (the absence of complex periodic zeros, Theorem 3.4.2) and the explicit form of the kernel of the convolution operator in the space of sequences (formula (3.23)). To do this, we used a new method based on difference equations. Various criteria of stability for refinable functions in terms of masks of equations were obtained by W. Lawton [**175**] and A. Cohen [**176**]. A criterion in terms of symmetric zeros and cycles of masks (only for linear independence over bounded sequences) was first obtained by A. Cohen [**177**], then independently by L. Villemoes [**173**], though some similar results can be found in [**46**], [**178**]. Necessary conditions for the smoothness of a stable refinable function (the mask has a zero of a corresponding multiplicity at the point $\xi = \frac{1}{2}$) were proved in [**168**]. Criteria of the orthonormality for integer shifts of refinable functions were derived in [**175**], [**177**].

Chapter 4

In 1988 I. Daubechies constructed the orthogonal compactly supported wavelets (see the translation of the book [**2**]) using the results obtained by A. Cohen [**176**] (the criterion of the linear independence for integer shifts) and S. Mallat [**179**] (sufficiency of the condition $|m_0(\xi)|^2 + |m_0(\xi + 1/2)|^2 \equiv 1$ for a refinable function to belong to the space L_2). Note that the latter result, though in a different form, was obtained by S. Dubuc and G. Deslauriers [**114**].

§4.3.1. Lemma 4.3.2 was proved in [**198**]. Theorem 4.3.3 was obtained in [**200**].

§4.3.2. The construction of the modified Daubechies wavelets belongs to I. Novikov [**201**]. Theorem 4.3.12 was proved in [**202**].

§4.4. The asymptotic behavior of the Bernstein polynomials corresponding to the classical Daubechies wavelets was studied in [**257**], [**203**], [**204**], [**205**].

Theorem 4.4.5 and Corollary 4.4.6 are generalizations of Theorem 1 and equality (14) from [**203**]. The proof of Lemma 4.4.8 is based on the ideas of [**206**]

Chapter 5

The classical and currently most used definition of fractal (in terms of affine contraction operators) was introduced by Hutchinson in [**141**], where the theorem about the existence and uniqueness of a fractal for a given family of contraction operators is proved. The notion of fractal curve of a pair of affine operators and the proof of the fact that a fractal is a continuous curve if the cross-condition $B_0v_1 = B_1v_0$ holds belongs to M. Barnsley [**30**], though many authors ascribe this proof to de Rham [**158, 159, 160**]. De Rham studied a particular case (the curves obtained by passing to the limit in the angle cutting algorithm) and first introduced the canonical parametrization for fractal curves. The fractal curves obtained by passing to the limit in subdivision algorithms (which can be considered as a generalization of de Rham's construction) were studied in detail in [**131**], [**154**]. In the same works sufficient conditions for the existence of such curves and the lower bounds for their smoothness were derived. The theory of fractal curves was applied to the investigation of refinement equations by I. Daubechies and J. Lagarias in [**138**], [**166**], where they first introduced a formula representing the smoothness of refinable functions in terms of the joint spectral radius, which was later made more precise by D. Collela and C. Heil [**132, 133**]. A generalization of this formula to arbitrary fractal curves and a criterion of the existence of a continuous fractal curve for a given pair of operators were first proved in [**84**]. The results concerning the moduli of continuity of fractal curves in C and L_p were obtained by V. Protasov in [**125**]. Fractal curves were first applied to the investigation of wavelet smoothness by I. Daubechies and J. Lagarias [**138**], [**166**]. In the same works, local smoothness of refinable functions was studied, and explicit formulas for the local smoothness at a given point were derived in simple cases (for one-dimensional operators). For arbitrary fractal curves and refinable functions, a formula for the local smoothness was derived by V. Protasov in [**126**], where sets of points of local smoothness were studied as well. All the other results concerning local smoothness presented in this book (mean smoothness, minimal and maximal smoothness, 0-1 level sets, etc.) are new and have not been published so far. The detailed investigation of the local smoothness of the de Rham curves can be found in [**120**].

Chapter 6

The cleaning procedure for refinement equations with linearly dependent integer shifts was developed by C. Chui [**6**], L. Berg and G. Plonka [**181**], [**182**], L. Villemoes [**173**], V. Protasov [**168**], [**127**], [**183**]. The notion of clean mask and a proof of the theorem about nondegeneracy and irreducibility of corresponding linear operators T_0, T_1 is presented in [**54**]. In the same work, the structure of invariant subspaces of these operators for arbitrary masks is described. By using these results, the explicit form of the operators T_0, T_1 on their common eigenspace W was found and the moduli of continuity of wavelets were calculated. Theorems about the factorization of smooth refinable functions with orthogonal integer shifts occur in many works, e.g., in [**2**], or [**138**], [**166**]. In [**128**], these theorems were proved for arbitrary refinable functions.

Chapter 7

The matrix method for the investigation of the smoothness of refinable functions has been developed by many mathematicians in more than a hundred works. Among them we mention I. Daubegies and J. Lagarias [**138**], [**166**], D. Collela and C. Heil [**132, 133**], N. Dyn and D. Levin [**112**], [**165**], C. Micchelli and H. Prautzsch [**131**], O. Rioul [**184**]. We added some new results concerning the moduli of continuity of wavelets and their local smoothness. The case $\deg m_c = 1$ (the degree of the clean mask) is studied in detail in [**166**]. For the case $\deg m_c = 2$, the smoothness α_φ was first computed by L. Villemoes in [**173**]. All the other results presented in this book for the latter case are new.

The joint spectral radius first appeared in the work by G. Rota and G. Strang [**185**] and then was completely forgotten for almost 30 years. It was not before the early 1990s, when its crucial role in the study of wavelets was revealed, since it is the joint spectral radius that gives an exact Hölder exponent. For the last 15 years more than 200 works devoted to the investigation of this object and methods of its calculation have been written. The brief review by G. Strang [**144**] contains one and a half pages, while the bibliography contains 5 pages! The properties of the joint spectral radius are studied in [**119**] [**130**], [**132**], [**134**], [**135**], [**142**], [**145**]–[**149**].

The computation of the joint spectral radius for various matrices is described in [**133**], [**138**], [**139**], and other works. The geometric approach to this problem was developed by V. Protasov in [**115**], [**119**]. The geometric algorithm of approximate computation of the joint spectral radius is of polynomial complexity with respect to the quantity $1/\varepsilon$, where ε is the accuracy of the approximation. A polynomial (with respect to dimension) algorithm was constructed independently by V. Protasov [**115**], D. Zhow [**117**], V. Blondel and Yu. Nesterov [**118**]. Recent investigations show that it is very unlikely to construct a better algorithm. There is no algorithm for approximate computation of the joint spectral radius, which would be polynomial with respect to both $1/\varepsilon$ and dimension. In particular, this means that the joint spectral radius cannot be represented in the form of algebraic functions of matrix coefficients [**136**], [**137**]. V. Blondel, J. Theys and A. Vladimirov [**135**] constructed an example of two 2×2-matrices whose joint spectral radius is not equal to the usual spectral radius of any finite product of these matrices. For applications of the joint spectral radius and its generalizations in wavelet theory, approximation theory, dynamic systems, probability theory, number theory, and mathematical physics, see [**120**], [**125**], [**127**], [**130**]–[**133**], [**138**], [**150**]–[**156**].

A method of pointwise estimation of decay was developed by I. Daubechies [**2**] and A. Cohen [**176**]. A method of estimation by means of invariant cycles was obtained by A. Cohen [**177**]. We add a statement about the equality $\vartheta = \inf_{b \in \mathcal{B}}$ (Cohen considered only the inequality $\leq$) and proved the continuity of ϑ as a function of the coefficients of the clean mask.

Chapter 8

The notion of nonstationary wavelets was introduced independently in [**207**] (in this paper the term almost-wavelet was proposed), [**208**], and [**55**]. Applications of nonstationary filters for signal processing were considered in [**217**].

The nonstationary refinable sequences studied in this chapter can be considered as a modification of *up*-functions introduced by V. L. Rvachev and V. A. Rvachev. The functions up_j defined by the Fourier transform:

$$\widehat{up_j}(t) = \left(\frac{\sin \pi\xi 2^{-j}}{\pi\xi 2^{-j}}\right)^{T(j+1)} \prod_{N=j+2}^{\infty} \left(\frac{\sin \pi\xi 2^{1-N}}{\pi\xi 2^{1-N}}\right)^{T(N)-T(N-1)}$$

for $T(N) = N$ are studied in [**212**], [**213**]. The corresponding nonstationary multiresolution analysis and nonstationary wavelets are investigated in [**214**]. The general theory of refinable functions satisfying the equation $\varphi(t) = \int_{\mathbb{R}} \varphi(2t - s)\, d\mu(s)$ for some Borel measure μ is considered, e.g., in [**215**]. Theorem 8.7.1 for the system Ψ with $T(N) = N$ was proved in [**210**].

Chapter 9

In the books [**2**], [**4**] one-dimensional periodic wavelets are defined as periodized wavelets in $L_2(\mathbb{R})$ under the assumption that both the wavelet function and generating refinable function decrease sufficiently rapidly at infinity. C. Chui, H. Mhaskar [**218**] constructed the periodic wavelets generated by the sequence of Dirichlet kernels (in our terminology, Kotelnikov-Shannon wavelets). C. Chui, J. Wang [**219**] gave an asymptotic definition of PMRA where the existence of a scaling sequence was postulated. V. A. Zheludev [**220**] studied PMRAs of spline-functions, for which he described scaling sequences and wavelet functions in terms of the Fourier coefficients. A. Petukhov [**221**] developed his ideas and defined PMRA (not only in $L_2(\mathbb{R})$, but in any Banach space from a sufficiently wide class) in terms of shift and dilation operators and gave a method of its construction and of finding wavelet systems. However, his definition did not include polynomial PMRAs, in particular, the periodized Meyer MRAs. The most general definition of PMRA in the spaces L_p, $1 \le p < \infty$, and C was proposed by M. A. Skopina [**222**]; in the framework of this definition scaling sequences were characterized and a method of their construction was described. This theory was extended to the multidimensional case with matrix dilation by I. Maksimenko and M. A. Skopina [**223**]. Wavelet packets in the non-periodic case were introduced by R. Coifman, Y. Meyer, V. Wickerhauser [**225**]. Periodic packets of one-dimensional Meyer wavelets were studied and used for the construction of optimal bases by R. Lorentz, A. Sahakian [**244**] (see commentary to Chapter 11). Criteria of the existence of a nonperiodic function generating a PMRA were investigated in [**222**]. An example of PMRA where the orthogonalized scaling sequence cannot be obtained by periodization in the sense of principal value (§9.4) was given by A. Petukhov [**224**].

Chapter 10

Theorems 10.2.2, 10.2.4, 10.2.6 were proved by M. A. Skopina [**226**]. In the nonperiodic case a criterion for the convergence of separable wavelet expansions at the Lebesgue points (analogous to Theorem 10.2.2) was previously obtained by S. Kelly, M. Kon, L. Raphael [**70**]. Strong Lebesgue points were introduced by E. Belinskii [**229**]. He developed a technique for finding criteria of summability at these points, which provide the convergence of approximation almost everywhere for a wide class of functions. The generalized Lebesgue sets are introduced by M. A. Skopina [**230**].

The direct approximation theorem of Jackson's type for the periodic Kotelnikov-Shannon wavelets was presented by C. Chui, H. Mhaskar [**218**]. I. Novikov [**227**] proved that for the Meyer wavelets, Jackson's inequality with the modulus of continuity of arbitrary order takes place (Corollary 10.3.2 from Theorem 10.3.1). In the general case Theorem 10.3.1 as well as inverse Theorems 10.3.5 and 10.3.9 was proved by M. A. Skopina [**228**]. Similar investigation of approximation properties of wavelet expansions in the nonperiodic case was carried out by R. Jia, J. Lei, E. Cheny [**232**], [**233**], [**234**].

The results of §10.4 were obtained by M. A. Skopina [**228**], [**236**]. The classes $\mathcal{L}_{r,\alpha}(x_0)$ used in this book were introduced by A. Calderón, A. Zygmund [**235**].

Chapter 11

Y. Meyer [**4**] proved that wavelet systems are unconditional bases in the spaces L_p, $1 < p < \infty$, under the assumption that the wavelet function has some smoothness (see also [**3**]). Theorem 11.1.1, where only a minimal decay rate of a wavelet function is required, was proved by P. Wojtaszczyk [**237**], [**238**].

The problem of the existence of polynomial bases in the space of continuous functions with minimum possible growth of the degrees of polynomials (algebraic or trigonometric) has been studied intensively by many mathematicians for forty years (see [**239**]). G. Faber [**240**] proved that no sequence of algebraic polynomials P_n of degree n can constitute a basis in $C[a,b]$. A. Privalov [**241**] obtained a stronger result; namely, he showed that the degrees of the polynomials P_n that constitute a basis in $C[a,b]$ grow at most as $n(1+\varepsilon)$, $\varepsilon > 0$; in [**242**] he constructed an example of a polynomial basis with the optimal growth of degrees. Similar statements were proved in the trigonometric case as well. However the bases constructed by Privalov were not orthogonal. D. Offin, K. Oskolkov [**243**] noticed that the periodic Meyer wavelets constitute an orthogonal basis, where the degree of the polynomials exceeds the minimal degree only twice. By replacing the usual wavelet expansion by a wavelet packet, R. Lorentz and A. Sahakian [**244**] solved the problem for trigonometric polynomials completely. Optimal orthogonal bases of algebraic polynomials were constructed by M. A. Skopina [**245**], [**246**]; another construction was independently obtained by K. Woźniakowski [**247**].

§11.4. In [**249**], it is proved that by using the Meyer wavelets, one can construct an unconditional basis in $L_p(\mathbb{R})$ for any $p \in (1, \infty)$ and in the spaces $B^s_{pq}(\mathbb{R}^n)$, $s \in \mathbb{R}$, $1 \leq p, q < \infty$, and $VMO(\mathbb{R}^n)$ as well. In [**250**], [**251**], [**257**], [**252**], the wavelet systems constructed by using the Dirichlet and Fejer kernels in various spaces of periodic functions were studied. Some remarks concerning periodic analogs of the Meyer wavelets occur in [**253**]. M. Friezier and B. Jawerth investigated properties of the Besov and Lizorkin-Triebel spaces by using nonorthogonal systems of functions, which are similar to wavelet systems [**254**] (see also [**255**], [**256**]).

It is known (see [**40**, Remarks 2.3.1/4] or [**254**]) that the spaces $F^{\alpha}_{\infty q}$ require a different approach. A corresponding definition for $q \in (1, \infty]$ is given in [**40**, Definition 2.3.4]; a definition for $q \in (0, \infty]$ is given in [**254**].

The Meyer-David wavelets are described in [**253**]. For $h = 1$, the construction belongs to Y. Meyer (see page 27); for $h > 1$, it belongs to G. David [**253**].

The anisotropic Besov and Lizorkin-Triebel spaces are considered, e.g., in [**40**, Chapter 10], [**258**].

For the case where some $s_\lambda = 0$ or s_λ have different signs, definitions of anisotropic spaces are given in [**260**]; however these spaces are not described in the framework of the decomposition method presented in this book.

A definition of isotropic spaces $F^{s_0}_{pq}$ can be found in [**40**, Definitions 2.3.1/2]. It is known (see [**40**, Remarks 2.3.1/4] or [**254**]) that the definitions of the spaces $F^{\alpha}_{\infty q}$ require a different approach. A corresponding definition for $q \in (1, \infty]$ is given in [**40**, Definitions 2.3.4], and for $q \in (0, \infty]$ in [**254**]. The latter work is used to introduce the anisotropic spaces $F^{(s)}_{\infty q}$.

Below we give references for isotropic analogs of the auxiliary results used in the present section: Proposition 11.4.7—[**40**, Theorem 1.6.3]; Proposition 11.4.8—[**40**, Proposition 2.3.2/1], [**254**]; Proposition 11.4.9—[**254**]; Proposition 11.4.10—[**40**, Proposition 1.3.2 and Remarks 1.3.2/1].

Proposition 11.4.14 immediately follows from the results of [**253**].

The proofs of Theorems 11.4.17, 11.4.18, and 11.4.19 are simpler for the isotropic case; the statements for the isotropic spaces B coincide with the corresponding results of the work [**249**]; the statements concerning the basis property of Meyer wavelet systems in the isotropic spaces F have not been formulated so far, though they can be proved using the reasoning of the work [**254**].

The assertion related to bilateral sequences and multipliers in $S'(\mathbb{R}^n)/P(\mathbb{R}^n)$ corresponding to Proposition 11.4.7 used in the proof of Theorem 11.4.16 can be found in [**40**, Chapter 5].

The duality used in the proof of Theorem 11.4.19 is proved in Theorem 5.9 of [**254**] for the isotropic case. The anisotropic version can be proved in a similar way.

The function Mf is an anisotropic analog of the Hardy-Littlewood maximal function. Inequality (11.121) is proved in [**262**], [**263**].

§11.5. The part of the theory of pseudo-differential operators (PDO) which was intensively developed in the 1960–1980s mainly considered elliptic problems. These PDO were mainly studied on the scale of Sobolev spaces W_p^s, $1 < p < \infty$, $s \in \mathbb{R}$ (see, e.g., [**34**], [**35**]). To investigate parabolic problems (not in the framework of the theory of PDO), some authors used certain modifications of the spaces W_p^s under the assumption that functions have sufficiently many mixed derivatives, so that the heat conduction operator $D_h = \frac{\partial}{\partial x_1} - a^2 \frac{\partial^2}{\partial x_1^2}$ can be considered as a sum of operators with "good" resolvent sets [**36**]. In the book [**37**], a more natural space was used for the operator D_h, namely, the anisotropic Sobolev space $W_2^{(1,2)}$ of functions whose first generalized derivatives (with respect to the first variable) and second generalized derivatives (with respect to the second variable) belong to L_2. However, anisotropic differential operators, as well as anisotropic PDO, in the spaces $W_p^{(s_1,\dots,s_n)} := W_p^{(s)}$, $1 < p < \infty$, have not been systematically studied. We mention only the books [**38**], [**39**], where the results were obtained, which allow one to derive the conditions for continuous action of the operators of differentiation and their sums in the spaces $B_{pq}^{(s)}$, $F_{pq}^{(s)}$.

In recent years there have been developed two approaches to the study of the isotropic Calderon-Zygmund operators and isotropic PDO in the scale of spaces of differentiable functions. One of them is based on the ideas of atomic and molecular decomposition proposed by R. Coifman and C. Fefferman. An atom is a compactly supported function which is estimated together with its derivatives by the measure of its support; a molecule is a function which decreases sufficiently rapidly at infinity together with some of its derivatives. Each function from the corresponding space can be decomposed into a sum of atoms depending on this function nonlinearly (the atomic decomposition theorem). Then it is proved that the image of an atom under the operator being investigated is a molecule, from which the continuity of the operator is concluded, by using the molecular decomposition theorem. This scheme is described in detail in [**256**] (see also [**254**]); it should be noted that in this scheme the atomic decomposition is a weakened analog of the basis decomposition.

The second approach to the study of isotropic PDO (see [**264**], [**265**]) is based on the existence of unconditional wavelet bases in L_p and the Besov spaces. The matrix $\langle \psi_R, T\psi_Q \rangle_{R,Q}$, where R, Q are diadic cubes, ψ_R, ψ_Q are the corresponding wavelets, T is a PDO, appears almost diagonal, which provides the continuity of T. It is this idea that is investigated for the anisotropic case in the present book. The difference from [**264**] is that we use the fact that the Fourier transforms of the basis elements are compactly supported, and we investigate the images of wavelets under the operator T, proceeding directly from the definition of T, without traditional passage to the integral operator with the Schwartz kernel. This allows us not only to simplify the proofs but to weaken the conditions for the symbol even in the isotropic case.

Anisotropic PDO in the scales of the spaces $F_{pq}^{(s)}$ were studied in [**266**]; the classes of symbols considered there are not so wide as in the present book, which is due to the multiplicative technique we use.

For $m = 0$ and $a_\lambda = 1$ (the isotropic case), Theorem 11.5.3 is proved in [**254**]. In general, the proof is analogous to that given in [**254**]; the most essential difference is that we use the maximal function over parallelepipeds.

Lemmas 11.5.11 and 11.5.12 are anisotropic analogs of Lemmas B.1 and B.2 from [**254**].

The inclusion $B_{p1}^{(s)} \subset F_{pq}^{(s)}$ used in the proof of Theorem 11.5.8 is proved, e.g., in [**259**]. The dissection method is described in [**267**]. The boundedness of the convolution operators with the Dirichlet kernels in L_p, $1 < p < \infty$, is proved in [**38**, c. 308]).

Bibliography

BOOKS

[1] Kashin, B. S. and Sahakian, A. A., Ortogonal′nye ryady (in Russian) [Orthogonal series]. 2nd ed. Izdatelstvo Nauchno-Issledovatelskogo Aktuarno-Finansovogo Tsentra (AFTs), Moscow, 1999; English transl., Translations of Mathematical Monographs, Vol. 75, Amer. Math. Soc., Providence, RI, 1989.

[2] Daubechies, I., Ten lectures on wavelets, CBMS-NSR Series in Appl. Math., SIAM, 1992.

[3] Daubechies, I., Ten lectures on wavelets, Izhevsk: NIZ "Regulyarnaya and khaoticheskaya dinamika", 2001 (Russian translation of [**2**]).

[4] Meyer, Y., Ondelettes. Herman, Paris, 1990.

[5] Meyer, Y., Wavelets and operators, Cambridge University Press, Cambridge, 1992 (English translation of [**4**]).

[6] Chui, C.K., An introduction to wavelets, Academic Press, New York, 1992.

[7] Chui, C., An introduction to wavelets, M.: Mir, 2001 (Russian translation of the book [**6**]).

[8] Wojtaszczyk, P., A mathematical introduction to wavelets, London Math. Soc. Student Texts 37, 1997.

[9] Hernandes, E. and Weis, G. A., A first course of wavelets, CRC Press, Boca Raton, FL, 1996.

[10] Petukhov, A. P., Vvedenie v teoriyu bazisov vspleskov (in Russian). [Introduction to the theory of wavelet bases], St. Petersburg Univ., St. Petersburg, 1999.

[11] Berdyshev, V. I. and Petrak, L.V., Approksimatsiya funktsiĭ, szhatie chislennoi informatsii, prilozheniya (in Russian). [Approximation of functions, compression of numerical information, applications], Inst. Mat. Mekh., Ekaterinburg, 1999.

[12] Sadovnichii, V. A., Teoriya operatorov (in Russian). [Theory of operators], M.: Vysshaya Shkola, 1999.

[13] Rudin, W., Real and complex analysis, McGraw-Hill, New York, 1974.

[14] Zygmund, A., Trigonometric series. 2nd ed., Vols. I and II, Cambridge University Press, New York, 1959.

[15] Stein, E. and Weiss, G., Introduction to Fourier analysis on Euclidean spaces. Vol. 32 of Princeton Mathematical Series, Volume 1 of Monographs in Harmonic Analysis, Princeton University Press, 1971.

[16] Stein, E., Singular Integrals and Differentiability Properties of Functions, Princeton, 1970.

[17] Singer, I., Bases in Banach spaces. I, Springer-Verlag, Berlin, 1970.

[18] De Guzman, M., Differentiation of integrals in R^n, with appendices by Antonio Cordoba and Robert Fefferman and two by Roberto Moriyon. Lecture Notes in Mathematics, Vol. 481, Springer-Verlag, Berlin-New York, 1975.

[19] Timan, A. F., Teoriya priblizheniĭ funktsiĭ deĭstvitel′nogo peremennogo (in Russian). [Theory of approximation of functions of a real variable], M.: Fizmatgiz, 1960; English transl., Macmillan, New York, 1963.

[20] Natanson, I. P., Theory of functions of a real variable, New York, 1964.
[21] Hardy, H., Littlewood, J. E., and Polya, G., Inequalities, 2nd ed, Cambridge Univ. Press, Cambridge, 1952.
[22] Suetin, P. K., Klassicheskie ortogonal′nye mnogochleny (in Russian). [Classical orthogonal polynomials], Nauka, Moscow, 1976.
[23] Szegő, G., Orthogonal Polynomials, American Mathematical Society Colloquium Publications, Vol. XXIII, Providence, R.I., 1975.
[24] Pólya, G. and Szegő, G., Problems and theorems in analysis, Vol. I: Series, integral calculus, theory of functions, Springer-Verlag, Berlin, 1998.
[25] Pólya, G. and Szegő, G., Problems and theorems in analysis. Vol. I: Theory of functions, zeros, polynomials, determinants, number theory, geometry, Springer-Verlag, Berlin, 1998.
[26] Schaefer, H. H., Topological vector spaces, Macmillan, New York; Collier-Macmillan, London, 1966.
[27] Mirolyubov, A. A. and Soldatov, M. A., Lineĭnye neodnorodnye raznostnye uravneniya (in Russian). [Linear inhomogeneous difference equations], Nauka, Moscow, 1986.
[28] Godunov, S. K. and Ryaben′kiĭ, V. S., Raznostnye skhemy. Vvedenie v teoriyu (in Russian). [Difference schemes. An introduction to the underlying theory], Nauka, Moscow, 1977; English transl., Studies in Mathematics and its Applications, Vol. 19, North-Holland, Amsterdam, 1987.
[29] Kirillov, A. A. and Gvishiani, A. D., Teoremy i zadachi funktsional′nogo analiza (in Russian). [Theorems and problems of functional analysis], 2nd ed. Nauka, Moscow, 1988; French transl., Mir, Moscow, 1982
[30] Barnsley, M., Fractals everywhere. Academic Press, Boston, 1988.
[31] Mandelbrot, B. B., Fractals and multifractals. Selecta, Vol. 1, Springer, New York, 1991.
[32] Gelfand, I., Raikov, D., and Shilov, G., Kommutativnye normirovannye kol′tsa (in Russian). [Commutative normed rings], Fizmatgiz, Moscow, 1960; English transl., Chelsea, New York, 1964.
[33] Strang, G. and Fix, G. J., An analysis of the finite element method, Prentice-Hall Series in Automatic Computation, XIV, Prentice-Hall, Englewood Cliffs, New Jersey, 1973.
[34] Cordes, H. O., Elliptic pseudo-differential operators: An abstract theory, Springer-Verlag, Berlin, 1979.
[35] Taylor, M., Pseudodifferential operators, Princeto University Press, Princeton, 1981.
[36] Lions, J.-L. and Magenes, E., Nonhomogeneous boundary value problems and applications. Vol. I, Die Grundlehren der mathematischen Wissenschaften, Vol. 181, Springer-Verlag, New York-Heidelberg, 1972.
[37] Ladyzhenskaya, O. A., Solonnikov, V. A., and Uraltseva, N. N., Lineĭnye i kvazilineĭnye uravneniya parabolicheskogo tipa (in Russian). [Linear and quasilinear equations of parabolic type], Nauka, Moscow, 1967; English transl., Translations of Mathematical Monographs, Vol. 23, American Mathematical Society, Providence, R.I., 1967.
[38] Nikolskii, S. M., Priblizhenie funktsiĭmnogikh peremennykh i teoremy vlozheniya (in Russian). [Approximation of functions of several variables and imbedding theorems], Nauka, Moscow, 1977; English transl., Die Grundlehren der Mathematischen Wissenschaften, Vol. 205, Springer-Verlag, New York-Heidelberg. 1975.
[39] Besov, O. V., Il′in, V. P., and Nikolskii, S. M., Integral′nye predstavleniya funktsiĭ i teoremy vlozheniya (in Russian). [Integral representations of functions, and embedding theorems], 2nd ed., Nauka, Moscow, 1975; English transl. of Chapters IV–VI, Scripta Series in Mathematics. Edited by Mitchell H. Taibleson, V. H. Winston and

Sons, Washington, D.C.; Halsted Press [John Wiley and Sons], New York-Toronto-London, 1979.
[40] Tribel, Kh., Theory of function spaces. II, Monographs in Mathematics, Vol. 84, Birkha"user Verlag, Basel, 1992.
[41] Gelfand, I. M. and Shilov, G. E., Generalized functions. Vol. 2. Spaces of fundamental and generalized functions. Academic Press [Harcourt Brace Jovanovich, Publishers], New York-London, 1968 [1977].
[42] Black, H. C., Modulation Theory. New York, 1953.

SURVEYS

[43] Novikov, I. Ya. and Stechkin, S. B., Basic constructions of wavelets (in Russian). Fundam. Prikl. Mat. 3 (1997), no. 4, 999–1028.
[44] Novikov, I. Ya. and Stechkin, S. B., Fundamentals of wavelet theory (Russian). Uspekhi Mat. Nauk 53 (1998), no. 6(324), 53–128; translation in Russian Math. Surveys 53 (1998), no. 6, 1159–1231.
[45] Astaf'eva, N. M., Wavelet analysis, foundations of the theory, and application examples. Usp. Fiz. Nauk., Vol. 166, No. 11, 1998.
[46] Cavaretta, D., Dahmen, W., and Micchelli, C., Stationary subdivision. Mem. Amer. Math. Soc., Vol. 93, 1991, pp. 1–186.

ARTICLES

[47] Bari, N. K., Biorthogonal systems and bases in Hilbert space (in Russian). Moskov. Gos. Univ. Ucenye Zapiski Matematika 148(4), (1951), 69–107.
[48] Schweinler, H. C. and Wigner, E. P., Orthogonalization methods. J. Math. Phys. 1970, Vol. 11., pp. 1693–1694.
[49] Meyer, Y., Ondelettes and fonctions splines. Seminaire EDP. Paris, December 1986.
[50] Mallat, S., Multiresolution representation and wavelets. Ph. D. Thesis, University of Pennsylvania, Philadelphia, PA, 1988.
[51] Mallat, S., An efficient image representation for multiscale analysis. In Proc. of Machine Vision Conference, Lake Taho, 1987.
[52] Mallat, S., A theory of multiresolution signal decomposition: the wavelets representation. IEEE Trans. Pattern Anal. Machine Intell. 1989. Vol. 11, pp. 674–693.
[53] Mallat, S., Multiresolution approximation and wavelets. Trans. Amer. Math. Soc. 1989, Vol. 315, pp. 69–88.
[54] Lemarié, P. G. and Meyer, Y., Ondelettes et bases Hilbertiennes. Rev. Math. Iber. 1987, Vol. 2, N. 1/2, pp. 1–18.
[55] De Boor, C., DeVore, R., and Ron, A., On construction of multivariate (pre) wavelets. Constr. Approx. 1993, Vol. 9, pp. 123–166.
[56] Lorents, R. A., Madych, W. R., and Sahakian, A. A., Translation and dilation invariant subspaces of $L_2(\mathbb{R})$ and multiresolution analysis, Appl. Comput. Harmon. Anal. 1998, Vol. 5, No. 4, pp. 75–388.
[57] Ouscher, P., Ondelettes fractales et applications. Ph. D. Thesis, Université Paris Dauphine, Paris, France, 1989.
[58] Chui, C. K. and Wang, J. Z., A cardinal spline approach to wavelets. Proc. Amer. Math. Soc. 1991.
[59] Cohen, A., Daubechies, I., and Feauveau, J. C., Biorthogonal Bases of Compactly Supported Wavelets. Communications on Pure and Applied Mathematics. 1992, Vol. XLV, pp. 2485–560.
[60] Gripenberg, G., A necessary and sufficient condition for the existence of father wavelet. Studia Mathematica, 1995, 114, 3, pp. 207–226.
[61] Haar, A., Sur Theorie de orthogonalen Funktionensysteme. Math. Ann. 1910, Vol. 69, pp. 331–371.

[62] Strömberg, J.-O, A modified Franklin system and higher-order spline on $\mathbb{R}^n$ as unconditional basis for Hardy spaces. In "Conference in Harmonic Analysis in Honor of A. Zygmund", Chicago, 1981, Vol. 2, 1983, pp. 475–494.
[63] Battle, G., A block spin construction of ondelettes. Part 1: Lemarié functions. Comm. Math Phys. 1987, Vol. 110, pp. 601–615.
[64] Lemarié, P.-G., Ondelettes á localization exponentialle. J. Math, Pures Appl. (9) 1988. Vol. 67, No. 3, pp. 227–236.
[65] Shannon, C. E., Communication in the presence of noise. Proc. of the IRE. 1949, Vol. 37, pp. 10–21.
[66] Kotelnikov, V. A., On the transmission capacity of the 'ether' and of cables in electrical communications (in Russian). Proceedings of the first All-Union Conference, Upravlenie svyazi RKK, 1933.
[67] Meyer, Y., Principle d'incertitude, bases hilbertiennes et algebres d'operateurs. Seminaire Bourbaki. 1985–1986, Vol. 38, No. 662.
[68] Walter, G., Approximation of the delta-function by wavelets. Local convergence for wavelet expansions. Preprint, 1992.
[69] Walter, G., Pointwise convergence of wavelet expansions. Local convergence for wavelet expansions. Preprint, 1992.
[70] Kelly, S. E., Kon, M. A., and Raphael, L. A., Pointwise convergence of wavelet expansions. Bull. Amer. Math. Soc. 30, 1994, pp. 87–94.
[71] Battle, G., Phase space localization theorem for ondelettes. J. Math Phys. 1989, Vol. 30, pp. 2195–2196.
[72] Battle, G., Heisenberg inequalities for wavelet states. ACHA. 1997, Vol. 4, pp. 119–146.
[73] Duffin, R. J. and Schaeffer, A. S., A class of nonharmonic Fourier series. Trans. Amer. Math. Soc. 1952, Vol. 72. pp. 341–366.
[74] Kashin, B. S. and Kulikova, T. Y., Springer Mathematical Notes, vol. 72, no. 2, 2002, pp. 281–284.
[75] Chui, C. K. and He, W., Compactly supported tight frames associated with refinable functions. Appl. and Comp. Harm. Anal. 2000, Vol. 8, pp. 293–319.
[76] Petukhov, A., Explicit construction of framelets. Appl. and Comp. Harm. Anal. 2001, Vol. 11, pp. 313–327.
[77] Lawton, W., Tight frames of compactly supported affine wavelets. J. Math. Phys. 1990, Vol. 31, pp. 1898–1901.
[78] Bownik, M., Tight frames of multidimensional wavelets. J. Fourier Anal. Appl. 1997, Vol. 3, pp. 525–542.
[79] Gröchenig, K and Madych, W. R., Maltiresolution analysis, Haar bases and self-similar tillings of R^n. IEEE Trans. Inform. Theory. 1992, Vol. 38, No. 2, pp, 556–568.
[80] Madych, W. R., Some elementary properties of multiresolution analysis – a tutiroial in theory and applications. C. K. Chui, ed. Academic Press. 1992, pp. 259–294.
[81] Cohen, A. and Daubechies, I., Nonseparable bi-dimensional wavelet bases. Revista Mat. Iberoamericana (I). 1993, pp. 51–137.
[82] Cohen, A., Ondelettes, analyses multirésolutions et filtres miroir en quadrature. Ann Inst. H. Poincaré, Anal. Non linéaire. 1990, Vol. 7, pp. 439–459.
[83] Maksimenko, I. E., Biorthogonality of multivariable refinable functions (in Russsian). "Voprosy sovremennoi teorii approksimazii", Izd. SPbGU. 2004, pp. 132–145.
[84] Lawton, W., Necessary and sufficient conditions for constructing orthonormal wavelet bases. J. Math. Phys. 1991, Vol. 32, pp. 57–81.
[85] Strang, G., Eigenvalues of $(\downarrow 2)H$ and convergence of the cascade algorithm. IEEE. Trans. SP 1996, Vol. 44.
[86] Lawton, W., Lee, S. N., and Shen, Z., Stability and orthonormality of multivariate refinable functions. SIAM J. Math. Anal. 1997, Vol. 28, No. 4, pp. 999–1114.

[87] Lawton, W., Lee, S. N., and Shen, Z., Convergence of multidimensional cascade algorithm. Numerische Mathematik, 1998, Vol. 78, No. 3. pp. 427–438.

[88] Gröchenig, K, Analyse multi-échelles et bases d'ondelettes. C. R. Acad. Sci. Paris, Sér. I Math. 1987, Vol. 305, No. 1, pp. 13–15; 1992, Vol. 38, No. 2, pp. 556–568.

[89] Jia, R. Q and Micchelli, C. A., Using the refinement equations for the construction of pre-wavelets II: Powers of two. Curves and Surfaces (P. J. Laurent, A. Le Méhauté, and L. L. Schumaker, eds.), Academic Press, New York, 1991, pp. 209–246.

[90] Jia, R. Q and Micchelli, C. A., Using the refinement equations for the construction of pre-wavelets V: extensibility of trigonometric polynomials. Computing. 1992, Vol. 48, pp. 61–72.

[91] Jia, R. Q. and Shen, Z., Multiresolution and wavelets, Proceedings of the Edinburgh Mathematical Society. 1994, Vol. 37, pp. 271–300.

[92] James, I. M., The topology of Stiefel manifolds. LMS Lecture Note Series 24. Cambridge University Press, Cambridge, 1976.

[93] Lawton, W., Lee, S. L., and Shen, Z., An algorithm for matrix extension and wavelet construction. Math. Comp. 1996, Vol. 37, pp. 271–300.

[94] Shen, Z., Extension of matrices with Laurent polynomial entries. Proceedings of the 15th IMACS World Congress on Scientific Computation Modeling and Applied Mathematics, Ashim Syclow, eds. 1997, pp. 57–61.

[95] Riemenschneider, S. D. and Shen, Z. W., Construction of compactly supported biorthogonal wavelets in $L_2(\mathbb{R}^s)$. I. Physics and modern topics in mechanical and electrical engineering. Scientific and Engineering Society Press. 1999, pp. 201–206.

[96] Ji, H., Riemenschneider, S. D., and Shen, Z., Multivariate compactly supported fundamental refinable functions, dual and biorthogonal wavelets. Studies of Applied Mathematics. 1999, Vol. 102, pp. 173–204.

[97] Riemenschneider, S. D. and Shen, Z. W., Construction of compactly supported biorthogonal wavelets in $L_2(\mathbb{R}^s)$. II. Wavelet applications signal and image processing VII. Proceedings of SPIE. 1999, Vol. 3813, pp. 264–272.

[98] Lam, T. Y., Serre's conjecture. Lecture Notes in Mathematics, 635, Springer-Verlag, New York, 1978.

[99] Suslin, A. A., The structure of the special linear group over rings of polynomials (in Russian). Izv. Akad. Nauk SSSR Ser. Mat. 41 (1977), no. 2, 235–252, 477.

[100] Park, H. and Woodburn, C., An algorithmic proof of Suslin's stability theorem for polynomial rings. Journal of Algebra. 1995, Vol. 178, pp. 277–298.

[101] Packer, J. A. and Rieffel, M. A., Wavelet filter functions, matrix completion problem, and projective modules over $C(\mathbb{T}^n)$. J. Fourier. Anal. Appl. 2003, Vol. 9, No. 3, pp. 101–116.

[102] Ron, A. and Shen, Z., Gramian analysis of affine bases and affine frames. In Approximation Theory VIII, Vol. 2: Wavelets (C. K. Chui and L. Schumaker, eds). World Scientific Publishing Co. Inc, Singapore, 1995, pp. 375–382.

[103] Ron, A. and Shen, Z., Frame and stable bases for shift-invariant subspaces of $L_2(R^d)$. Canad. J. Math. 1995, Vol. 47, No. 5, pp. 1051–1094.

[104] Ron, A. and Shen, Z., Affine systems in $L_2(R^d)$: the analysis of the analysis operator. J. Func. Anal. 1997, Vol. 148, pp. 408–447.

[105] Ron, A. and Shen, Z., Affine systems in $L_2(R^d)$: dual systems. J. Fourier. Anal. Appl. 1997, Vol. 3, pp. 617–637.

[106] Han, B., On dual wavelet tight frames. ACHA. 1997, Vol. 4, pp. 380–413.

[107] Han, B., Compactly supported tight wavelet frames and orthonormal wavelets of exponential decay with a general dilation matrix. J. Comput. Appl. Math. 2003, Vol. 155, pp. 43–67.

[108] Cohen, A. and Daubechies, I., A stability criterion for biorthogonal wavelet bases and their related subband coding schemes. Duke Math. J. 1992, Vol. 68, pp. 313–335.

[109] Jia, R. Q., Approximation properties of multivariate wavelets. Math. Comp. 1998, Vol. 67, pp. 647–655.

[110] Lawton, W., Lee, S. L., and Shen, Z., Characterization of compactly supported refinable splines. Adv. Comput. Math. 1995, Vol. 3, No. 1-2, pp. 137–145.

[111] Chaikin, G. M., An algorithm for high speed curve generation. Computer Graphics and Image Processing, 1974, Vol. 3, pp. 346–349.

[112] Dyn, N., Gregory, J. A., and Levin, D., Analysis of linear binary subdivision schemes for curve design. Constr. Approx. 1991, Vol. 7, pp. 127–147.

[113] Dubuc, S., Interpolation through an iterative scheme. J. Math. Anal. Appl. 1986, Vol. 114, pp. 185–204.

[114] Deslauriers, G. and Dubuc, S., Symmetric iterative interpolation processes. Constr. Approx. 1989, Vol. 5, pp. 49–68.

[115] Protasov, V. Yu., A generalized joint spectral radius. A geometric approach (in Russian). Izv. Ross. Akad. Nauk Ser. Mat. 61 (1997), no. 5, 99–136; translation in Izv. Math. 61 (1997), no. 5, 995–1030.

[116] Zhow, D. X., The p-norm joint spectral radius and its applications in wavelet analysis. International conference in wavelet analysis and its applications, AMS/IP Studies in Advanced Mathematics, 2002, Vol. 25, pp. 305–326.

[117] Zhow, D. X., The p-norm joint spectral radius for even integers. Methods Appl. Anal. 1998, Vol. 5, pp. 39–54.

[118] Blondel, V. D. and Nesterov, Yu. P., Computationally efficient approximations of the joint spectral radius. Preprint, 2004.

[119] Protasov, V. Yu., The joint spectral radius and invariant sets of linear operators (in Russian). Fundam. Prikl. Mat. 2 (1996), no. 1, 205–231.

[120] Protasov, V. Yu., On the smoothness of de Rham curves (in Russian). Izv. Ross. Akad. Nauk Ser. Mat. 68 (2004), no. 3, 139–180; translation in Izv. Math. 68 (2004), no. 3, 567–606.

[121] Oseledets, V. I., A multiplicative ergodic theorem. Lyapunov characteristic numbers for dynamical systems. Trans. Moscow. Math. Soc. 1968, Vol. 19, pp. 197-231.

[122] Ravishankar, K., Power law scaling of the top Lyapunov exponent of a product of random matrices. J. Stat. Phys. 1989, Vol. 54, No.1/2, pp. 531–537.

[123] Mandelbrot, B. B., Fisher, A. J., and Calvet, L. E., A Multifractal Model of Asset Returns. Cowles Foundation Discussion Paper, 1997, No. 1164, http://ssrn.com/abstract=78588.

[124] Nikitin, P., The Hausdorff dimension of the harmonic measure on a de Rham curve (in Russian). Zap. Nauchn. Sem. S.-Peterburg. Otdel. Mat. Inst. Steklov.

[125] Protasov, V., Refinement equations and corresponding linear operators. International Journal of Wavelets, Multiresolution and Information Processing 4. 2006, No. 3, pp. 461-474.

[126] Protasov, V., Fractal curves and wavelets, Izv. Math. 70 (2006), no. 5, 975-1013.

[127] Protasov, V., The stability of subdivision operator at its fixed point. SIAM J. Math. Analysis, 2001, Vol. 33, No. 2, pp. 448–460.

[128] Protasov, V. Yu., Piecewise-smooth refinable functions (in Russian). Algebra i Analiz 16 (2004), no. 5, 101–123; translation in St. Petersburg Math. J. 16 (2005), no. 5, 821–835.

[129] Derfel, G. A., Dyn, N., and Levin, D., Generalized refinement equations and subdivision processes. Journal of Approx. Theory, 1995, Vol. 80, pp. 272–297.

[130] Protasov, V., Refinement equations with nonnegative coefficients. J. Fourier Anal. Appl., 2000, Vol.6 , No. 6, pp. 55–77.

[131] Micchelli, C. A. and Prautzsch, H., Uniform refinement of curves. Linear Alg. Appl. 1989, Vol. 114/115, pp. 841–870.

[132] Collela, D. and Heil, C., Characterization of scaling functions. I. Continuous solutions. SIAM J. Matrix Anal. Appl. 1994, Vol. 15, pp. 496–518.

[133] Collela, D. and Heil, C., Dilation equations and the smoothness of compactly supported wavelets. In Wavelets: Mathematics and applications, J. Benedetto and M. Frazier, eds. CRC Press, Boca Raton, FL, 1993, pp. 161–200.

[134] Lagarias, J. C. and Wang, Y., The finiteness conjecture for the generalized spectral radius. Linear Alg. Appl. 1995, Vol. 214, pp. 17–42.

[135] Blondel, V. D., Theys, J., and Vladimirov, A. A., An elementary counterexample to the finiteness conjecture. SIAM J. Matrix Anal. 2003, Vol 24, No. 4, pp. 963–970.

[136] Blondel, V. D. and Tsitsiklis, J. N., Approximating the spectral radius of sets of matrices in the max-algebra is NP-hard. IEEE Trans. Autom. Control. 2000, Vol. 45, No. 9, pp. 1762–1765.

[137] Blondel, V. D. and Tsitsiklis, J. N., The boundedness of all products of a pair of matrices is undecidable. Syst. Control Lett. 2000, Vol. 41, No. 2, pp. 135–140.

[138] Daubechies, I. and Lagarias, J. Two-scale difference equations. II. Local regularity, infinite products of matrices and fractals. SIAM. J. Math. Anal. 1992, Vol. 23, pp. 1031–1079.

[139] Gripenberg, G., Computing the joint spectral radius. Lin. Alg. Appl. 1996, Vol. 234, pp. 43–60.

[140] Maesumi, M., An efficient lower bound for the generalized spectral radius. Linear Alg. Appl. 1996, Vol. 240, pp. 1-7.

[141] Hutchinson, J. E., Fractals and self-similarity. Indiana Univ. Math. J. 1981, Vol. 30, No. 5, pp. 713–747.

[142] Berger, M. A. and Wang Y., Bounded semi-groups of matrices. Linear Alg. Appl. 1992, Vol. 166, pp. 21–27.

[143] Daubechies, I. and Lagarias, J., Corrigendum/addendum to: Sets of matrices all infinite products of which converge. Linear Alg. Appl. 2001, Vol. 327, pp. 69–83.

[144] Strang, G., The joint spectral radius. Commentary by Gilbert Strang on paper number 5 in "Collected works of Gian-Carlo Rota", 2001.

[145] Ando, T. and Shih, M., Simultaneous contractibility. SIAM J. Math. Anal. 1998, Vol. 19, No. 2, pp. 487–498.

[146] Barabanov, N. E., On the Lyapunov exponent of discrete inclusions. III (in Russian). Avtomat. i Telemekh. 1988, no. 5, 17–24; translation in Automat. Remote Control 49 (1988), no. 5, part 1, 558–565.

[147] Gurvits, L., Stability of discrete linear inclusions. Linear Alg. Appl. 1995, Vol. 231, pp. 47–85.

[148] Vladimirov, A. A., Elsner, L., and Beyn, W.-J., Stability and paracontractivity of discrete linear inclusions. Linear Alg. Appl. 2000, Vol. 312, pp. 125–134.

[149] Wirth, F., The generalized spectral radius and extremal norms. Linear Alg. Appl. 2002, Vol. 342, pp. 17–40.

[150] Kozyakin, V. S., Algebraic unsolvability of a problem on the absolute stability of desynchronized systems (in Russian). Avtomat. i Telemekh. 1990, no. 6, 41–47; translation in Automat. Remote Control 51 (1990), no. 6, part 1, 754–759.

[151] Tsitsiklis, J. N., The stability of the products of finite set of matrices. Open problems in communication and computation, T. M. Cover and B. Copinath, eds., 1987, Springer-Verlag, New York, pp. 161–163.

[152] Guglielmi, N. and Zennaro, M., On the zero-stability of variable stepsize multistep methods: the spectral radius approach. Numer. Math. 2001, Vol. 88, pp. 445–458.

[153] Moision, B. E., Orlitsky, A., and Siegel, P. N., On codes that avoid specified differences. IEEE Trans. Inform. Theory, 2001, Vol. 47, No. 1, pp. 433–442.

[154] Dyn, N., Gregory, J. A., and Levin, D., Analysis of linear binary subdivision schemes for curve design. Constr. Approx. 1991, Vol. 7, pp. 127–147.

[155] Yu, T., Han, B., and Overton, M., Design of Hermite subdivision schemes aided by spectral radius optimization. Preprint, 2004.

[156] Protasov, V. Yu., Asymptotics of the partition function (in Russian). Mat. Sb. 191 (2000), no. 3, 65–98; translation in Sb. Math. 191 (2000), no. 3-4, 381–414.

[157] Protasov, V. Yu., On the problem of the asymptotics of the partition function (in Russian). Mat. Zametki 76 (2004), no. 1, 151–156; translation in Math. Notes 76 (2004), no. 1-2, 144–149.

[158] de Rham, G., Une peu de mathematique a propos d'une courbe plane (in French). Revue de Mathematiques Elementaires, 1947, II, Nos. 4, 5, pp. 678–689.

[159] de Rham, G., Sur une courbe plane (in French). J. Math. Pur. Appl. 1956, 35, pp. 25–42.

[160] de Rham, G., Sur les courbes limit de polygones obtenus par trisection (in French). Enseign. Math. 1959, II, 5, pp. 29–43.

[161] Erdös, P., On a family of symmetric Bernuolli convolutions. Amer. J. Math. 1939, Vol. 61 , pp. 974–975.

[162] Erdös, P., On the smoothness properties of Bernuolli convolutions. Amer. J. Math. 1940, Vol. 62, pp. 180–186.

[163] Rvachev, V. L. and Rvachev, V. A., On a function with compact support (in Russian). Dokl. Akad. Nauk. USSR, 1971.

[164] Derfel, G. A., A probabilistic method for studying a class of functional-differential equations (in Russian). Ukrain. Mat. Zh. 41 (1989), no. 10, 1322–1327, 1436; translation in Ukrainian Math. J. 41 (1989), no. 10, 1137–1141 (1990).

[165] Dyn, N. and Levin, D., Interpolatory subdivision schemes for the generation of curves and surfaces. Multivariate approximation and interpolation, 1990 (Duisburg 1989), pp. 91–106, Birkhäuser, Basel.

[166] Daubechies, I. and Lagarias, J., Two-scale difference equations. I. Global regularity of solutions. SIAM. J. Math. Anal. 1991, Vol. 22, pp. 1388–1410.

[167] Fix, G. and Strang, G., Fourier analysis of the finite element method in Ritz-Galerkin theory. Stud. Appl. Math. 1969, Vol. 48, pp. 265–273.

[168] Protasov, V., A complete solution characterizing smooth refinable functions. SIAM Journal of Math. Analysis, 2000, Vol. 31, No. 6, pp. 1332–1350.

[169] Jia, R.-Q., Refinable shift-invariant spaces: From splines to wavelets. C. K. Chui et al., eds., Approximation theory VIII. Vol. 2. Wavelets and multilevel approximation. Papers from the 8th Texas international conference, College Station, TX, USA, January 8–12, 1995. Singapore: World Scientific. Ser. Approx. Decompos. 1995, Vol. 6, pp. 179–208.

[170] De Boor, C., DeVore, R., and Ron, A., The structure of finitely generated shift-invariant spaces in $L_2(\mathbb{R}^d)$. J. Funct. Anal. 1994, Vol. 119, No. 1, pp. 37–78.

[171] De Boor, C., DeVore, R., and Ron, A., Approximation from shift-invariant subspaces of $L_2(\mathbb{R}^d)$. Trans. Amer. Math. Soc. 1994, Vol. 341, No. 2, pp. 787–806.

[172] De Boor, C., DeVore, R., and Ron, A., Approximation orders of FSI spaces in $L_2(\mathbb{R}^d)$. Constructive Approximation, 1998, Vol. 14, No. 3, pp. 411–427.

[173] Villemoes, L., Wavelet analysis of refinement equations. SIAM J. Math. Anal. 1992, Vol. 25, No. 5, pp. 1433–1460.

[174] Jia, R. Q. and Wang, J., Stability and linear independence associated with wavelet decomposition. Proc. Amer. Math. Soc. 1993, Vol. 117, pp. 1115–1124.

[175] Lawton, W. M., Necessary and sufficient conditions for constructing orthonormal wavelets. J. Math. Phys. 1991, Vol. 32, pp. 57–61.

[176] Cohen, A., Ondelettes, analyses multirésolutions et filtres miroir en quadrature. Ann. Inst. H. Poincaré, Anal. linéaire, 1990, Vol. 7, pp. 439–459.

[177] Cohen, A., Ondelettes, analyses multirésolutions et traitement numérique du signal. Ph.D. Thesis, 1990, Université Paris, Dauphine.

[178] Zhou, D.-X., Stability of refinable functions, multiresolution analysis, and Haar bases. SIAM J. Math. Anal. 1996, Vol. 27, No. 3, pp. 891–904.

[179] Mallat, S., Multiresolution approximation and wavelets. Trans. Amer. Math. Soc. 1989, Vol. 315, pp. 69–88.

[180] Protasov, V., Fractal curves and their applications to wavelets. Proceeding of the International Workshop on Self-similar Systems, July 30–August 7, 1998, Dubna, Russia, 1999, pp. 120–125.

[181] Berg, L. and Plonka, G., Spectral properties of two-slanted matrices. Results Math. 1999, Vol. 35, No. 3-4, pp. 201–215.

[182] Berg, L. and Plonka, G., Some notes on two-scale difference equations. Functional equations and inequalities, Math. Appl. 2000, No. 518, pp. 7–29.

[183] Protasov, V., The correlation between the convergence of subdivision processes and solvability of refinement equations. "Algorithms for Approximation IV, Proceedings of the 2001 International Symposium", Huddersfield, England, July 15–20, 2001, pp. 394–401, J. Levesley, I. J. Anderson, and J. C. Mason, eds., Huddersfield University, ISBN No. 186-218-04-07.

[184] Rioul, O., Simple regularity criteria for subdivision schemes. SIAM J. Math. Anal. 1992, Vol. 23, pp. 1544–1576.

[185] Rota, G. C. and Strang, G., A note on the joint spectral radius. Kon. Nederl. Acad. Wet. Proc. 1960, Vol. 63, pp. 379–381.

[186] Cohen, A. and Conze, J. P., Régularité des bases d'ondelettes et mesures ergodiques. Rew. Math. Iberoamer. 1992, Vol. 8, pp. 351–366.

[187] Lau, K. S., Ma, M. F., and Wang, J., Of some sharp regularity estimations of L_2 scaling functions. SIAM J. Math. Anal. 1996, Vol. 27, pp. 835–864.

[188] Lau, K. S. and Wang, J., Characterization of L_p solutions for two-scale dilation equations. SIAM J. Math. Anal. 1995, Vol. 26, pp. 1018–1046.

[189] Battle, G., Heisenberg Inequalities for Wavelets States. Appl. Comp. Harm. Analysis 1997, 4, pp. 119–146.

[190] Novikov, I. Ya, Wavelets (brief survey of the foundations of the theory) (in Russian). Materialy 12 Sibirskoi Shkoly, Novosibirsk, 18–23 July, 1998. Novosibirsk: Izd-vo instituta matematiki im. S. L. Soboleva, 1999. pp. 92–111.

[191] Lemarie-Rieusset, P. G., Existence de "fonction-pere" pour le ondelettes a support compact. C. R. Acad. Sci. Paris I. 1992, Vol. 314, pp. 17–19.

[192] Deslauriers, G. and Dubuc, S., Interpolation dyadique. Fractals, dimensions non entiéres et applications. G. Cherbit, ed., Masson, Paris, 1987, pp. 44–55.

[193] Daubechies, I., Orthonormal basis of compactly supported wavelets. Comm. Pure Appl. Math. 1988, 46, pp. 909–996.

[194] Taswell, C., The Systematized Collection of Wavelet Filters Computable by Spectral Factorization of the Daubechies Polynomial. Technical Report CT-1998-08.

[195] Lemarie-Rieusset, P. G. and Zahrouni, E., More regular wavelets. Applied and Computational Harmonic Analysis. 1998, Vol. 5, pp. 92–105.

[196] Villemoes, L., Energy moments in time and frequency for two-scale difference equation solutions and wavelets. SIAM J. Math. Anal. 1992. Vol. 23, 6, pp. 1519–1543.

[197] Volkmer, H., Asymptotic regularity of compactly supported wavelets. SIAM J. Math. Anal. 1995, Vol. 26, 4, pp. 1075–1087.

[198] Volkmer, H., On the regularity of wavelets. IEEE Trans. Inf. Theory. 1992, Vol. 38, 2, pp. 872–876.

[199] Ojanen, H., Orthonormal compactly supported wavelets with optimal Sobolev regularity. Technical Report math.CA/9807089, 1998.

[200] Chui, C. K. and Wang, J., High-Order Orthonormal Scaling Functions and Wavelets Give Poor Time-Frequency Localization. CAT Report #322. 1994, pp. 1–24.

[201] Novikov, I. Ya., Modified Daubechies wavelets preserving localization with growth of smoothness. East J. Approximation. 1995, Vol. 1, No. 3, pp. 341–348.

[202] Novikov, I. Ya., Uncertainty constants for modified Daubechies wavelets (in Russian). Proceedings of the International Conference "Approximation Theory and Harmonic Analysis" (in Russian) (Tula, 1998). Izv. Tul. Gos. Univ. Ser. Mat. Mekh. Inform. 4 (1998), no. 1, 107–111, 165.

[203] Shen, J. and Strang, G., The zeros of the Daubechies polynomials. Proc. Amer. Math. Soc. 1995, Vol. 124, No. 12, pp. 3819–3833.

[204] Temme, N. M., Asymptotics and numerics of zeros of polynomials that are related to Daubechies wavelets. Appl. Comp. Harmonic Anal. 1997, Vol. 4, No. 4, pp. 414–428.

[205] Temme, N. M., Asymptotic invertion of the incomplete beta function. J. Comp. Appl. Math. 1992, Vol. 41, pp. 145–157.

[206] Kantorovich, L. V., On the convergence of a sequence of Bernstein polynomials outside the main interval (in Russian). Izv. Akad. Nauk SSSR, 1931.

[207] Berkolaiko, M. Z. and Novikov, I. Ya., Infinitely smooth almost-wavelets with compact support (in Russian). Dokl. Akad. Nauk 326 (1992), no. 6, 935–938; translation in Russian Acad. Sci. Dokl. Math. 46 (1993), no. 2, 378–382.

[208] Berkolaiko, M. Z. and Novikov, I. Ya., Infinitely smooth compactly supported near-wavelets (in Russian). Mat. Zametki 56 (1994), no. 3, 3–12, 157; translation in Math. Notes 56 (1994), no. 3-4, 877–883 (1995).

[209] Novikov, I. Ya., On the construction of nonstationary orthonormal infinitely differentiable compactly supported wavelets. Functional Differential Equations. 1994, Vol. 2, pp. 145–156.

[210] Cohen, A. and Dyn, N., Nonstationary subdivision schemes and multiresolution analysis. SIAM J. Math. Anal. 1996, Vol. 27, pp. 1745–1769.

[211] Novikov, I. Ya., Nonstationary orthonormal infinitely differentiable compactly supported wavelets with uniformly bounded uncertainty constants. Proceedings of the International Workshop (July 30–August 7, 1998, Dubna, Russia). (JINR, E5-99-38, Dubna, 1999), pp. 110–116.

[212] Rvachev, V. L. and Rvachev, V. A., Nonclassical methods of approximation theory in boundary value problems. Naukova Dumka, Kiev, 1979.

[213] Rvachev, V. A., Compactly-supported solutions of functional-differential equations and their applications (in Russian). Uspekhi Mat. Nauk 45 (1990), no. 1(271), 77–103, 222; translation in Russian Math. Surveys 45 (1990), no. 1, 87–120.

[214] Dyn, N. and Ron, A., Multiresolution analysis by infinitely differentiable compactly supported functions. Applied and Computational Harmonic Analysis. 1995, 2, pp. 15–20.

[215] Chui, C. K. and Shi, X., Continuous two-scale equations and dyadic wavelets. Advances in Comp. Math. 1994, 2, pp. 185–213.

[216] Schoenberg, I. J., Contributions to the problem of approximation of equidistant data by analytic functions. Quart. Appl. Math. 1946, 4, pp. 45–99, 112–141.

[217] Soman, A. K. and Vaidyanathan, P. P., Orthonormal wavelets and paraunitary filter banks. IEEE Trans. on Sign. Proc. 1993, Vol. 41. No. 3. pp. 1170–1183.

[218] Chui, C. K. and Mhaskar, H. N., On trigonometric wavelets. Constr. Approx. 1993, Vol. 9, pp. 167–190.

[219] Chui, C. K. and Wang, J., A general framework of compact supported splines and wavelets. J. Approx. Th. 1992, Vol. 71, pp. 263–304.

[220] Zheludev, V. A., Periodic splines and wavelets. Proc. of the Conference "Math. Analysis and Signal Processing", Cairo, January 1994, pp. 2–9.

[221] Petukhov, A. P., Periodic wavelets (in Russian). Mat. Sb. 188 (1997), no. 10, 69–94; translation in Sb. Math. 188 (1997), no. 10, 1481–1506.

[222] Skopina, M., Multiresolution analysis of periodic functions. East J. Approximation. 1997, Vol. 3, No. 2, pp. 203–224.

[223] Maksimenko, I. E. and Skopina, M. A., Multidimensional periodic wavelets (in Russian). Algebra i Analiz 15 (2003), no. 2, 1–39; translation in St. Petersburg Math. J. 15 (2004), no. 2, 165–190.

[224] Petukhov, A., Trigonometric rational wavelet bases. Proceedings of the International Workshop (July 30–August 7, 1998, Dubna, Russia) (JINR, E5-99-38, Dubna, 1999), pp. 116–119.

[225] Coifman, R. R., Meyer, Y., and Wickerhauser, V. M., Size properties of wavelet packets. In "Wavelets and their applications", Beylkin et al., eds., Jones and Bartlett, 1992, pp. 453–470.

[226] Skopina, M., Local convergence of Fourier series with respect to periodized wavelets. J. Approx. Theory. 1998, Vol. 94, pp. 191–202.

[227] Novikov, I. Ya., Wavelets of Y. Meyer—an optimal basis in $C(0,1)$ (in Russian). Mat. Zametki 52 (1992), no. 5, 88–92, 143; translation in Math. Notes 52 (1992), no. 5-6, 1137–1140 (1993).

[228] Skopina, M., Wavelet approximation of periodic functions. J. Approx. Theory, 2000, Vol. 104, pp. 302–329.

[229] Belinskii, E. S., Summability of multiple Fourier series at Lebesgue points (in Russian). Teor. Funkcii Funkcional. Anal. i Prilozen. No. 23 (1975), 3–12, 169.

[230] Skopina, M. A., The generalized Lebesgue sets of functions of two variables. Proceedings. Colloquia Math. Societ. Janos Bolyai. 58, Conference on Approximation Theory. Hungary. Kecskemet. August 6–11, 1990, pp. 615–625.

[231] Saks, S., On the strong derivatives of functions of intervals. Fund. Mat. 1935, Vol. 25, pp. 245–252.

[232] Jia, R. Q., A Bernshtein-type inequality associated with wavelet decomposition. Constr. Approx. 1993, Vol. 9, pp. 299–318.

[233] Jia, R. Q. and Lei, J., On approximation by multi-integer translates of functions having global support. J. Approx. Th. 1993, Vol. 72, pp. 2–23.

[234] Lei, J., Jia, R. Q., and Cheny, E. W., Approximation from shift invariant spaces by integral operators. SIAM Journal of Approx. Th. 1997, Vol. 28, pp. 481–498.

[235] Calderón, A. P. and Zygmund, A., Local properties of solutions of elliptic partial differential equation. Studia Math. 1961, Vol. 20, pp. 171–227.

[236] Skopina, M., Localization principle for wavelet expansions. Proceedings of the International Workshop (July 30–August 7, 1998, Dubna, Russia) (JINR, E5-99-38, Dubna, 1999), pp. 125–130.

[237] Wojtaszczyk, P., Wavelets as unconditional bases in $L_p(R)$. J. Fourier Anal. Appl. 1999, Vol. 5, No. 1, pp. 73–85.

[238] Wojtaszczyk, P. and Figiel, T., Special bases in function spaces. Chapter 14 in Handbook of the Geometry of Banach Spaces, Vol. I, Elsevier Science, 2001.

[239] Ul′yanov, P. L., On some results and problems from the theory of bases (in Russian). Zapiski nauchnyh seminarov LOMI. Vol. 170, 1989.

[240] Faber, G., Über die interpolatorische Darstelung stetiger Functionen. Jahresber Deutsch. Math.-Verein, 1914, Vol. 23, pp. 192–210.

[241] Privalov, Al. A., The growth of the powers of polynomials, and the approximation of trigonometric projectors (in Russian). Mat. Zametki 42 (1987), no. 2, 207–214, 343. (Reviewer: L. Gogoladze)

[242] Privalov, Al., A. Growth of powers of polynomial bases (in Russian). Mat. Zametki 48 (1990), no. 4, 69–78, 159; translation in Math. Notes 48 (1990), no. 3-4, 1017–1024 (1991).

[243] Offin, D. and Oskolkov, K., A note on orthonormal polynomial bases and wavelets. Constr. Appr. 1993, Vol. 9, No. 1, pp. 319–325.

[244] Lorentz, R. A. and Sahakian, A. A., Orthogonal trigonometric Shauder bases of optimal degree for $C(0,2\pi)$. J. Fourier Anal. Appl. 1994, Vol. 1, No. 1, pp. 103–112.

[245] Skopina, M. A., On polynomial bases in the space $C[-1,1]$ (in Russian). Zap. Nauchn. Sem. S.-Peterburg. Otdel. Mat. Inst. Steklov. (POMI) 262 (1999), Issled. po Linein. Oper. i Teor. Funkts. 27, 223–226, 236; translation in J. Math. Sci. (New York) 110 (2002), no. 5, 3027–3028.

[246] Skopina, M. A., Orthogonal polynomial Schauder bases in $C[-1,1]$ with optimal growth of degrees (in Russian). Mat. Sb. 192 (2001), no. 3, 115–136; translation in Sb. Math. 192 (2001), no. 3-4, 433–454.

[247] Woźniakowski, K., On an orthonormal polynomial bases $C[-1,1]$. Studia Math. 2001, Vol. 144, No. 2, pp. 181–196.

[248] Chanillo, S. and Muckenhoupt, B., Weak type estimates for Cesaro sums of Jacobi polynomial series. Mem. Amer. Math. Soc. 1993, Vol. 102, No. 487, pp. 1–90.

[249] Lemarie, P. G. and Meyer, Y., Ondelettes et bases hilbertiennes. Revista Matematica Iberoamericana, 1986, 2, 1-2, pp. 1–18.

[250] Lizorkin, P. I., Bases and multipliers in the spaces $B^r_{p,\theta}(\Pi)$ (in Russian). Analytic number theory, mathematical analysis and their applications (dedicated to I. M. Vinogradov on his 85th birthday). Trudy Mat. Inst. Steklov. 143 (1977), 88–104, 209.

[251] Orlovskii, D. G., On multipliers in the spaces $B^r_{p,\theta}$ (Russian summary). Anal. Math. 5 (1979), no. 3, 207–218.

[252] Bochkarev, S. V., Bases in function spaces and the Franklin system (Russian). Translated in Proc. Steklov Inst. Math. 1992, no. 1, 19–37. Theory of functions (in Russian) (Amberd, 1987). Trudy Mat. Inst. Steklov. 190 (1989), 22–39.

[253] Meyer, Y., Principe d'incertitude, bases hilbertiennes et algebres d'operateurs. Asterisque, 1987, 145–146, pp. 206–223.

[254] Friezier, M. and Jawerth, B., A discrete transform and decomposition of distribution spaces. J. Funct. Anal. 1990, 93, No. 1, pp. 34–170.

[255] Friezier, M., Jawerth, B., and Weiss, G., Littlewood-Paley theory and the study of function spaces. CBMS-AMS Regional Conf. Ser., No. 79, 1991.

[256] Torres, R. H., Boundedness results for operators with singular kernels on distribution spaces. Mem. Amer. Math. Soc. 1991, 90, 442, pp. 1–172.

[257] Kateb, D., Lemarie-Rieusset P. G. Asymptotic behavior of the Daubechies filters. Appl. Comp. Harmonic Anal. 1995, Vol. 2, No. 4, pp. 398–399.

[258] Berkolaiko, M. Z., Traces of functions in generalized Sobolev spaces with a mixed norm on an arbitrary coordinate subspace. II (in Russian). Trudy Inst. Mat. (Novosibirsk) 9 (1987), Issled. Geom. "v tselom" i Mat. Anal., 34–41, 206.

[259] Berkolaiko, M. Z., Convexity and concavity of Banach ideal spaces, and embedding theorems (in Russian). Sibirsk. Mat. Zh. 31 (1990), no. 3, 11–18, 216; translation in Siberian Math. J. 31 (1990), no. 3, 373–379 (1991).

[260] Lizorkin, P. I., Fourier transformation in Besov spaces. The zero scale of ${B_{p,\theta}}^0$ (in Russian). Dokl. Akad. Nauk SSSR 1965, 163, pp. 1318–1321.

[261] Strömberg, J.-O., A modified Franklin system and higher order spline systems on $\mathbb{R}^n$ as unconditional basis of Hardy spaces. Wadsworth Math. Ser. 1982, 2, pp. 475–493.

[262] Seeger, A., A note on Triebel-Lizorkin spaces. Approx. and Func. Spaces. Banach Center Publ., 1989, 22, pp. 391–400.

[263] Riviere, N. M., Singular integrals and multiplier operators. Ark. Mat. 1971, 9, pp. 243–278.

[264] Beylkin, G., Coifman, R., and Rokhlin, V., Fast wavelet transforms and numerical algorithms. I. Res. Rep. JALEU/DCS/RR-696, 1989, pp. 1–46.

[265] Meyer, J., Wavelets and operators. Lect. Notes Math. 1989, Vol. 137.

[266] Marshall, J., Weighted parabolic Triebel spaces of product type. Forum Math. 1991, 3, pp. 479–511.

[267] Lizorkin, P. I., Properties of functions in the spaces $\Lambda^r_{p,\theta}$. (in Russian). Trudy Mat. Inst. Steklov. 131 (1974), 158–181.

Index